Chemistry

Palgrave Foundations

A series of introductory texts across a wide range of subject areas to meet the needs of today's lecturers and students

Foundations texts provide complete yet concise coverage of core topics and skills based on detailed research of course requirements suitable for both independent study and class use – *the firm foundations for future study.*

Published

A History of English Literature
Biology
Chemistry
Contemporary Europe
Economics
Economics for Business
Modern British History
Nineteenth-Century Britain
Physics
Politics

Forthcoming

British Politics
Maths for Science and Engineering
Modern European History
Sociology

Chemistry

Second edition

ROB LEWIS AND WYNNE EVANS

First edition 1997
Reprinted twice
Second edition 2001
Published by
PALGRAVE MACMILLAN
Houndmills, Basingstoke, Hampshire RG21 6XS and
175 Fifth Avenue, New York, N.Y. 10010
Companies and representatives throughout the world

PALGRAVE MACMILLAN is the new global academic imprint of St. Martin's Press LLC Scholarly and Reference Division and Palgrave Publishers Ltd (formerly Macmillan Press Ltd).

ISBN 0–333–96257–5

This book is printed on paper suitable for recycling and made from fully managed and sustained forest sources.

A catalogue record for this book is available from the British Library.

10 9 8 7 6 5
10 09 08 07 06 05 04

Typeset by Footnote Graphics Limited, Warminster, Wilts
Printed in China

Contents

Preface

THIS book is designed to provide a course in 'chemical literacy' for those studying access and foundation courses at university or college and for non-specialists on courses such as Engineering Science, Engineering or Biology, for whom chemistry is a subsidiary subject in their first year at university. Much of the material will be suitable also for other chemistry students, particularly those without strong mathematical or chemistry backgrounds.

This book starts 'from scratch', assuming only a minimal experience of mathematics and science. A key feature of each unit are the exercises which consolidate topics before the student proceeds further. The questions at the end of units are designed to extend and reinforce learning. To help students working mainly alone, the answers to all exercises and problems are provided at the back of the book.

The Second Edition has been revised and reformatted, and contains an expanded section on NMR. The website is a new feature of the second edition and may be found at **http://www.palgrave.com/foundations/lewis**. The website contains Appendices containing additional examples and also optional material (such as reaction mechanisms and free energy) for more advanced students who have already mastered the basic material contained in units of the book. We also include a selection of short case studies on our website, which may be given to individual students, or groups of students. Each case study starts with a list of prerequisites (back referenced to the corresponding units in the book) which outlines the background required to cope with that case study.

Although this is a book for students, we hope that it will help and not hinder the teacher's task of coping with large groups, often with widely ranging ability and with differing subject interests. Accordingly, we welcome suggestions for improvements to the book and website from students and teachers alike.

E.W.E.
Rh.L.

To the Student: How to use this Book

- This book is designed so that you can work with the minimum of help.
- It has a glossary of key definitions.
- Use the index to cross reference other key terms and ideas.
- Obtain advice on the topics (and within units, the sections) which are relevant to your studies at this stage. If you possess a sound knowledge of GCSE chemistry, you will find that much of Units 2, 4 and 6 is already known to you.
- Do not move on to a new topic until you are happy with the previous one.
- There are exercises embedded in the text. Make a serious effort to answer these before carrying on. Check your answers with those at the end of the book.
- The Boxes provide additional material (including some applications) but are not essential to the main body of the text and may be returned to after you have mastered the rest of the chapter.
- A website is provided at **http://www.palgrave.com/foundations/lewis**. The appendices included in the website contain extension material which you may require after mastering the basics: consult with your tutor as to *which material* you will need and *when* you should study it. Similar remarks apply to the Case Studies. Concentrate on the basics in the book first before worrying about the Appendices!

Acknowledgements

The authors and publishers wish to thank the following for permission to use copyright material:

CRC Press, LLC for Table 11.6 from *Environmental Chemistry*, Fifth Edition, by S. E. Manahan, Table 5.1, p. 94; Hutchinson Education for Figure 11.3 from *Surface Phenomena* (1972) by S. R. Rao, Figure 27, p. 111; and McGraw Hill Publishers, New York, for Figure 14.9 from *Chemistry for Environmental Engineering* Fourth Edition by C. Sawyer, P. McCarty and G. Parkin, Figure 6.2, p. 294.

Photographs:
Associated Press, p. 173; J. Allen Cash Ltd, pp. 78, 217, 250; Colorsport, p. 232; Chip Clark, p. 18; Frances Arnold, p. 414; Philip Harris, pp. 5, 286, 301; Hulton Getty, p. 165; Laboratory of the Government Chemist, pp. 9, 147; Angharad and Catrin Lewis, p. 36; Lion Laboratories Ltd, p. 106; Popperfoto, p. 405; Steve Redwood, p. 185; Rex Features, pp. 27, 92, 213, 346, 353; Gordon Roberts Photography, pp. 88, 243, 379; Science Museum/Science and Society Picture Library, p. 27; Dr Jeremy Burgess/Science Photo Library, p. 50; University of California, p. 61; UKAEA, p. 30; Valence (Dover Publications Ltd), p. 61; Varian (Australia Pty) Ltd, p. 373; Wolf Blass Wines International, p. 149.

We thank the numerous reviewers and correspondents who have helped us prepare the second edition. Special thanks are due to Dave Rich (Neath Port Talbot College), Drs P. G. Hall, A. J. Berry and P. S. McIntyre (University of Glamorgan) and Dr J. Carnduff (University of Glasgow).

Every effort has been made to contact all the copyright-holders but if any have been inadvertently overlooked the publishers will be pleased to make the necessary arrangements at the first opportunity.

Numbers, Units and Measurement

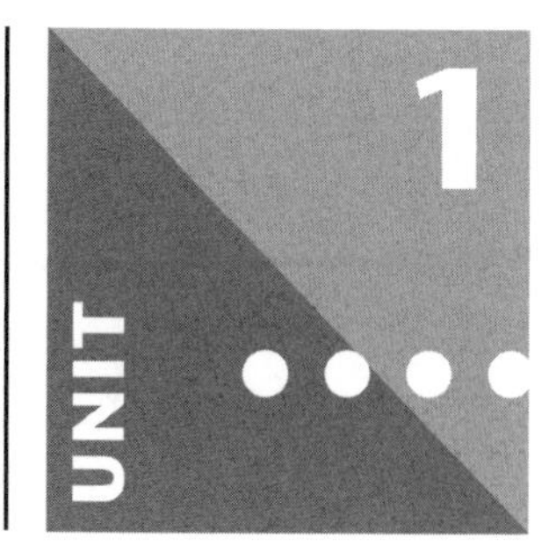

Objectives

- ▶ Explains standard notation
- ▶ Tests you on the use of your calculator
- ▶ Describes how to work out the units of a quantity
- ▶ Defines accuracy and precision
- ▶ Looks at errors and the use of significant figures

Contents

1.1 Very small and very big numbers

Science often involves very large and very small numbers. Such numbers may be cumbersome to write down, and an abbreviated notation (known as 'standard' or 'scientific' notation) is often used. This relies upon the following mathematical symbolism:

$10^{-6} = 0.000\,001$

$10^{-5} = 0.000\,01$

$10^{-4} = 0.000\,1$

$10^{-3} = 0.001$

$10^{-2} = 0.01$

$10^{-1} = 0.1$

$10^{0} = 1$

$10^{1} = 10$

$10^{2} = 100$

$10^{3} = 1\,000$

$10^{4} = 10\,000$

$10^{5} = 100\,000$

$10^{6} = 1\,000\,000$

Now let us look at an example of standard notation. Think of the number 100. This is the same as 1×100. In standard notation we write this as 1×10^2. Similarly,

2300 becomes 2.3×1000 or 2.3×10^3

6 749 008 becomes 6.749008×10^6

0.001 245 0 becomes 1.2450×10^{-3}

Exercise 1A

Standard notation

Express the following in standard notation

(i) 0.000 0345

(ii) 300 000 000

(iii) 0.082 057 5

(iv) 3.5

(v) 602 200 000 000 000 000 000 000

(vi) 17.

1.2 Logarithms

Logarithms to the base 10

The logarithm (or 'log') of a number to the base ten is the power that the number 10 has to be raised to in order to equal that number. For example, $100 = 10^2$. Therefore, the log of 100 is 2. Similarly, since $0.0001 = 1 \times 10^{-4}$, the log of 0.0001 is -4.

What is the log of 150? The log of 150 is the value of x in the expression

$$150 = 10^x$$

We carry out this operation on a calculator. In many types of calculator, this is done by entering the *number*, and then pressing the *log* button. The log of 150 is 2.176. We write this as

$$\log(150) = 2.176$$

What if you are provided with the log of x and asked to find x? Using the above example, how do we get back to 150 from 2.176? To do this we would need to evaluate $10^{2.176}$. (To carry out this operation on a scientific calculator we use the 10^x key, a common sequence of operations being *number*, *shift*, and *10^x*.) We then write

$$10^{+2.176} = 150$$

Similarly,

$$10^{-0.9104} = 0.1229$$

Logarithms to the base e (natural logs)

The symbol 'e' is a mathematical constant (like π) where

$$e = 2.718\ldots$$

The logarithm of a number to the base e is the power that the number e has to be raised to in order to equal that number. For example, $e^{3.912} = 50$ so that the natural log of 50 is 3.912. In this book we symbolize natural logs as 'ln':

$$\ln(50) = 3.912$$

It follows from the definition of natural logs that $\ln(e^x) = x$.

Manipulating logarithms

The following general formulae are useful and apply to logs of any base:

$$\log(ab) = \log a + \log b$$

$$\log\left(\frac{a}{b}\right) = \log a - \log b$$

For example,

$$\log\left(\frac{yz}{km}\right) = \log y + \log z - \log k - \log m$$

Using your calculator

We are now in a position to summarize the type of calculations you need to be able to do on your calculator in preparation for later chapters. You will need to be able to

1. enter numbers in standard notation form;
2. add, subtract, divide and multiply numbers;
3. square numbers and find their square roots;
4. use the calculator memory;
5. calculate $\log x$, $\ln x$, e^x and 10^x.

The way you carry out such calculations varies slightly according to the make of your calculator. Refer to the calculator instructions for further information – or ask a knowledgeable friend! Now try Exercise 1B.

Exercise 1B

Quick test on calculator use

Use your calculator to evaluate the following:

(i) $\left(\frac{45.6}{2.34}\right)^2$

(ii) $\sqrt{300.7}$

(iii) $\log(1.2 \times 10^{-2})$

(iv) $10^{-4.56}$

(v) $\ln(0.178 \times 8.456)$

(vi) $e^{-5.20}$

(vii) $e^{-E/RT}$, where $E = 30\,000$, $R = 8.3145$ and $T = 298$.

1.3 Units

International system of units

The international system of units (usually known as SI units, from the French *Système International*) consists of several *base units* from which all other units (such as those of volume or energy) are derived. Some of the base units are shown in Table 1.1.

Because the base units are sometimes too large or too small for use, SI prefixes (Table 1.2) are used to produce smaller or bigger units. For example, the milligram (= 0.001 g, and symbolized mg) is used if we are reporting small masses.

The cubic metre (written m^3) is too large for most purposes in chemistry, and the cubic decimetre, dm^3 (or litre) is commonly used. There are $1000\,dm^3$ in $1\,m^3$. Also, there are 1000 cubic centimetres (cm^3) in a cubic decimetre (see Fig. 1.1). Summarizing,

$$1\,m^3 = 1000\,dm^3 = 1\,000\,000\,cm^3$$

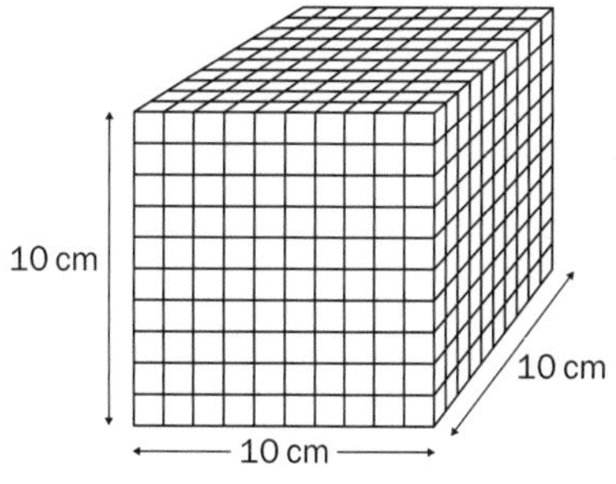

Fig. 1.1 There are $1000\,cm^3$ in $1\,dm^3$.

Table 1.1 Base units

Property	Base unit	Symbol for unit
Mass	kilogram	kg
Length	metre	m
Time	second	s
Temperature	kelvin	K
Amount of substance	mole	mol

Table 1.2 SI prefixes

Value	Prefix	Symbol	Value	Prefix	Symbol
10^9, billion	giga	G	10^{-3}	milli-	m
10^6, million	mega	M	10^{-6}	micro-	μ
10^3, thousand	kilo-	k	10^{-9}	nano-	n
10^{-1}	deci-	d	10^{-12}	pico-	p
10^{-2}	centi-	c	10^{-15}	femto-	f

Exercise 1C

Working with different units

(i) The radius of the hydrogen atom is approximately 40 pm (picometres). Convert this to nm (nanometres).

(ii) 1 cm^3 of a solution contains 0.01 g of salt. What is the mass of salt contained in 1 dm^3 of the same solution?

(iii) The wavelength of orange-yellow light is roughly 600 nm. Express the wavelength in metres.

You will also meet units raised to negative powers, such as m^{-3} (read as 'per metre cubed'). Remember,

$$m^{-3} = \frac{1}{m^3}$$

Now try Exercise 1C.

Amount of substance

One of the most important physical quantities in chemistry is the *amount of substance*, which has the unit of **mole** (symbol mol). The greater the number of particles (atoms, ions or molecules) in a piece of material, the greater is the amount of substance. **Concentration** is another physical quantity which is particularly important to chemists. Concentration is a measure of the packing of particles per unit volume, and is commonly expressed in the units of mol per dm^3, usually written as mol dm^{-3}. Moles and concentration are dealt with in more detail in Chapters 8 and 9.

Temperature

The hotness or coldness of a material is called its temperature. The units of temperature used in science are degrees **Celsius** (°C) or **kelvin** (K). For more details, see page 154.

Force and energy

The quantities of force and energy occur throughout chemistry and we will need to be familiar with their units.

The SI unit of *force* is the **newton** (N). 1 N is the force needed to give a mass of 1 kg an acceleration of 1 m s^{-2}. (If the 1 kg mass were stationary before applying the force, this would mean that the mass would have a velocity of 1 m s^{-1} after 1 second, 2 m s^{-1} after 2 seconds and so on.) The formal definition of the newton is therefore

$$1\,N = 1\,kg\,m\,s^{-2}$$

The SI unit of *energy* is the **joule** (J). 1 joule is the energy used up in pushing against a force of 1 newton over a distance of 1 metre. This means that we can write

$$1\,J = 1\,N\,m$$

Units of physical quantities

A physical quantity consists of a number and a unit. For example, suppose we measure the volume of a block and find it to be $4.5\,\text{cm}^3$:

$$\underset{\text{the number}}{\underset{\uparrow}{4.5}} \qquad \underset{\text{the unit}}{\underset{\uparrow}{\text{cm}^3}}$$

Mathematically, the physical quantity consists of a number multiplied by a unit:

$$\text{physical quantity} = \text{number} \times \text{unit} \qquad \textbf{(1.1)}$$

In our example,

$$\text{physical quantity (i.e. volume)} = 4.5 \times \text{cm}^3$$

For convenience, the physical quantity is usually written without the multiplication sign – here as $4.5\,\text{cm}^3$. This may be compared with the algebraic expression 4.5y (i.e. $4.5 \times y$).

A laboratory balance. This balance measures masses as low as 0.0001 g.

Labelling axes on graphs

Suppose that we are plotting the volume of a gas (V, in dm^3) against the temperature of the gas (T, in kelvin, K). First, consider the y-axis. We might be tempted to label this axis as 'V (dm^3)'. However, we are not actually plotting the quantity *volume* but simply the *number* part of the quantity. Re-arrangement of equation (1.1) shows that

$$\text{number} = \frac{\textit{physical quantity}}{\text{unit}}$$

This shows that in plotting the number part we are really plotting $\frac{\textit{physical quantity}}{\text{unit}}$.

Hence the y-axis is labelled 'Volume/dm^3. This is usually written as V/dm^3. Similar reasoning leads to the x-axis being labelled Temperature/K or T/K.

Deriving the units of a quantity

To illustrate the derivation of the units of a quantity, consider the following question: with mass in kg and the volume in m^3, what are the units of density?

We start with the definition of **density**:

$$\text{density} = \frac{\text{mass}}{\text{volume}}$$

We find the units of density by substituting the units of mass and volume into the equation defining density:

$$\text{units of density} = \frac{\text{units of mass}}{\text{units of volume}} = \frac{\text{kg}}{\text{m}^3} = \text{kg}\,\text{m}^{-3}$$

The *units of density* are therefore kilograms per cubic metre (see Example 1.1).

Example 1.1

The ratio $\frac{E}{RT}$

occurs frequently in chemistry. *E* is an energy per mole of substance (in units of joules per mole, symbolized $J\,mol^{-1}$), *T* is the temperature (in K) and *R* is a universal constant with the units of $J\,mol^{-1}K^{-1}$ (read as joule per mole per kelvin). What are the units of *E/RT*?

▶ Answer

$$\text{units of } \frac{E}{RT} = \frac{J\,mol^{-1}}{J\,mol^{-1}K^{-1}K}$$

Cancelling like units,

$$\frac{\cancel{J\,mol^{-1}}}{\cancel{J\,mol^{-1}}\cancel{K^{-1}}\cancel{K}}$$

so that E/RT is unitless, i.e. it is simply a number.

▶ Comment

The most commonly encountered form of this expression is $e^{-E/RT}$ (e = 2.718). Note that as E increases, $e^{-E/RT}$ *decreases*. Now try Exercise 1D.

Exercise 1D

Deriving units

(i) The energy (q joules) needed to raise the temperature of a material (of mass m grams) by ΔT kelvin may be calculated by the equation

$$q = m \times C \times \Delta T$$

where C is the specific heat capacity of the material. What are the units of C?

(ii) The mass m (in grams) of an amount of substance n (in moles) is related to the molar mass M of that substance by the expression

$$m = n \times M$$

What are the units of M?

1.4 Errors in experiments

Types of experimental error

Suppose we measure the temperature of a liquid. The difference between a *single* measurement of temperature and the *true* temperature is the *error* of the measurement. Generalizing:

Error = experimental value − true value

It is always good practice to repeat measurements. For example, if we are measuring the concentration of pesticide in a lake, we might fill a large bottle with lake water and later (back in the laboratory) withdraw 50 cm^3 portions of the lake water from the bottle and analyse each portion for pesticide using the same analytical technique. The *mean* pesticide concentration is then obtained by averaging the concentrations found in each 50 cm^3 portion. In this case, the error in the pesticide measurements is the difference between the *mean* measurement and the true concentration.

There are two main kinds of error that we need to consider in experimetal measurements:

1. The first type are called **random errors**, random because they cause repeat measurements on the same sample to go up and down.

   Random errors will cause successive measurements to be scattered, although averaging a large number of such measurements will produce a mean measurement which will not be greatly affected by random error. However, repeating a measurement many times may not be practical (for example, there may be too little sample available, or the measurements might be too time consuming or expensive to carry out). It is for this reason that precise measurements are highly desirable.

2. The second type are called **systematic errors**. A systematic error affects all measurements, and makes all measurements either higher or lower than the 'true value'. Systematic errors do not average out, no matter how many repeat measurements are made.

Examples of random errors

Random errors are introduced whenever there is a *subjective* part to the experiment (such as estimating when a solution has reached the mark in a pipette, or recognizing the onset of a colour change during a titration), *or* where the experimental measurements are influenced by rapidly fluctuating conditions (e.g. air draughts).

Examples of systematic errors

A simple example of a systematic error is provided by a balance. Balances are often set to read a mass of zero before being used to weigh a sample. Suppose a speck of dust falls upon the pan of a balance *after* zeroing. This will cause the indicated mass of any object to be greater than the true mass. For example, if the speck has a mass of 0.0001 g, all objects will have an apparent mass which is 0.0001 g too high.

Another example of systematic error involves the analysis of chromium in blood. If the blood samples are stored in stainless steel vessels prior to analysis, then some chromium may dissolve out of the steel into the sample. This introduces a systematic error which causes the measurement (here the chromium concentration) to be overestimated.

Systematic errors are often difficult to recognize, particularly in measurements of the concentration of substances (**quantitative analysis**) in which the concentrations of materials is being found in the presence of substances which *interfere* with the measurement (see Box 1.1).

Exercise 1E

Systematic errors

In order to compare the alcohol content of several wines, a student poured samples of each wine into open test-tubes and the next day analysed each for alcohol using a standard analytical technique. Comparison of the student's results with those obtained by other laboratories showed that her alcohol concentrations were consistently low. Suggest one reason for the systematic error. (**Hint** What happens to wine when it is left in the open air?)

Accuracy and precision

Repeat measurements on the same sample which are close together are said to be *precise*:

Precise measurements have a small random error.

A measurement which is close to the true value is said to be *accurate*:

Accurate measurements have a small systematic error.

The 'rifle shooting analogy' helps us to distinguish between accuracy and precision (Fig. 1.3). In a rifle competition, the aim is to hit the bullseye. Competitor A is a precise shot (the shots are close together) but inaccurate (no bullseye); B is a precise and accurate shot (three bullseyes); C is neither precise or accurate.

We have already noted that measurements are usually repeated several times and that the random errors will be nearly completely cancelled out in the mean measurement provided that enough repeat measures are made. This is why the precision of measurements is important: *the greater the precision, the fewer the number of repeat measurements that need to be made in order for the random errors to be nearly*

BOX 1.1

Example of systematic errors – the analysis of aluminium ions (Al^{3+}) in tea

The level of Al^{3+} consumption by humans has been linked to Alzheimer's disease. One way of determining the concentration of Al^{3+} in tea is to add a **complexing agent** (usually a complicated organic compound) to the tea (Fig. 1.2). This combines with the Al^{3+} ion to produce a red coloured substance (a coloured **complex**):

$$Al^{3+} + \text{complexing agent} \rightarrow \underset{\text{(red colour)}}{Al^{3+} \text{ complex}}$$

The greater the concentration of Al^{3+} in the tea, the stronger will be the intensity of the red colour.

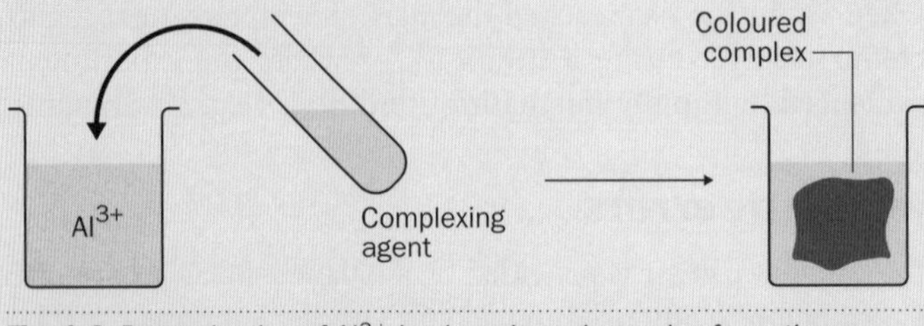

Fig. 1.2 Determination of Al^{3+} ion by coloured complex formation.

However, if the tea contains traces of heavy metal ions (such as copper), it happens that these ions will also form red coloured compounds with the complexing agent. If we presumed that all the red colour is due to the Al^{3+} complex, this would produce a systematic error in which the Al^{3+} concentration in the tea is *overestimated.*

If the tea contains traces of fluoride ions (F^-), these ions react directly with the Al^{3+} producing stable aluminium fluoride complexes and so preventing the Al^{3+} ions from reacting with the complexing agent. This leads to a systematic error in which the Al^{3+} concentration in the tea is *underestimated.*

In an ideal measurement, we would separately measure the concentrations of ions (F^-, Cu^{2+} etc.) which interfere with the measurements of the Al^{3+} ion, and correct the measured Al^{3+} concentration accordingly.

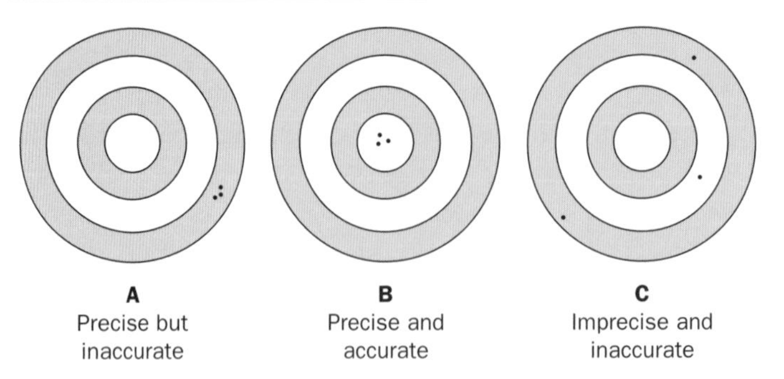

Fig. 1.3 Accuracy and precision – the rifle shooting analogy.

eliminated. The fewer the repeat measurements that are required, the quicker and cheaper are the measurements.

One way of showing the precision of a set of repeat measurements is to quote the **standard deviation** of those results. The lower the standard deviation of a set of results, the better is the precision of those results. Appendix 1 in the website shows how to calculate and use standard deviations.

What do we mean by the true value of a measurement?

If a measurement is accurate, it must give a true value. But how do we know the true value? If we are using a new analytical instrument which detects lead, we might test the accuracy of the instrument by analysing solutions whose lead concentrations [Pb] are known (i.e. **standard solutions**). If the analysis is accurate, the difference

$$[\mathrm{Pb}]^{\mathrm{in}}_{\mathrm{standard}} - [\mathrm{Pb}]^{\mathrm{found\ by}}_{\mathrm{instrument}}$$

should be close to zero. If we are analysing mixtures where the true concentrations are unknown, the absence of systematic errors is essential if we are to have faith in the final result. In such cases, accuracy may be estimated by comparing the results of different analytical methods for the same sample.

Professional organizations (such as the British Standards Institute) often publish the most reliable analytical methods in the form of **standard methods**, in which the likely sources of experimental error are highlighted.

Quality of analytical measurements in laboratories

In the UK, the department of trade and industry (DTI) estimates that the cost of chemical analysis is over £4 billion per year, and involves over 50 000 staff and 1000 laboratories. Poor quality analysis is a barrier to international trade, technological advancement and to the development of government policies such as health and safety. Many laboratories are currently approved under the British Standards accreditation schemes. The DTI's initiative on Valid Analytical Measurement (VAM) seeks to improve the quality of measurements in all analytical chemistry laboratories, including those in the agriculture, building, materials processing and pharmaceutical industries. One of the most important features of the initiative is the encouragement given to companies to compare their analytical methods and procedures. A useful pack on the VAM Initiative may be obtained by writing to the Laboratory of the Government Chemist, Queens Road, Teddington, Middlesex TW11 0LY.

Quality control is very important in analytical laboratories, so that customers know that they can rely upon the results of any analysis.

1.5 Reporting measurements

Significant figures

If we asked someone to measure the length of a piece of wire with a standard ruler and they reported its length as 19.843 cm, we would have every right to be sceptical. 19.843 contains *five significant figures*, a number of figures which cannot be justified when we are using a ruler.

We might estimate the **uncertainty** in the length measurement as ±0.2 cm. This means that the measurement is at worst 0.2 cm too high or 0.2 cm too low. It follows that we are justified only in including the first decimal place of the measurement and we then report the length as 19.8 ± 0.2 cm. Alternatively, we might report the measurement as simply 19.8 cm, a number which contains *three* significant figures. Neglecting the ±0.2 cm is less informative, but because of an agreement between scientists about the meaning of significant figures, even simply writing 19.8 cm carries with it some information about the minimum uncertainties involved in the measurement.

To explain this further, suppose that you report the length of the wire to a friend as 19.8 cm but provide no further information. What could your friend say about the likely uncertainties in the experiment? By general agreement, it is assumed that the uncertainty in the measurement is *equal to at least one digit in the last significant figure*. In our example, reporting the length as 19.8 cm implies that the total uncertainty in the measurement is equal to at least one digit in the first decimal place. In other words, the *minimum* uncertainty is ±0.1 cm. As we have seen, the actual uncertainty is estimated to be greater, as ±0.2 cm.

In order to report the correct number of significant figures in a measurement, an estimate of the uncertainties is obviously required. Sometimes this will be nothing more than an informed guess of the likely effect of random errors. In more sophisticated measurements, further experiments may need to be carried out in order to assess the importance of both random and systematic errors.

Example 1.2

Suppose the mass of a lump of metal is displayed on a balance as 10.0078 g. There are six numbers in this display, so the measurement involves four decimal places and a total of *six significant figures*. In the case of many commercially available four-decimal-point balances, the uncertainty in the measured mass is typically ±0.0002 g. How many significant figures are we justified using when reporting the mass of metal?

▶ Answer

The uncertainty shows that four decimal places can be justified in the measurement. This means that we are justified in reporting the mass of the metal to six significant figures, i.e. as 10.0078 g. This implies that the minimum uncertainty in the measurement is ±0.0001 g.

BOX 1.2

Recognizing the number of significant figures

The easiest way to recognize the number of significant figures in a number is to express the number in standard notation and count the number of digits (including zeros) in the number that multiplies the 10^x part. For example, 0.002 33 becomes 2.33×10^{-3} in standard notation. Since there are three digits in 2.33 the number of significant figures is three. Other examples are as follows:

Number	Standard notation	Number of significant figures
0.002 330	2.330×10^{-3}	4
235.5	2.355×10^{2}	4
0.000 056 767 6	$5.676 76 \times 10^{-5}$	6
14	1.4×10^{1}	2
1302	1.302×10^{3}	4
150	1.50×10^{2} or 1.5×10^{2}	3 or 2

The number of significant figures in the number 150 is ambiguous. If we mean 1.50×10^2, then there are three significant figures. If we mean 1.5×10^2 then there are only two significant figures.

Number of significant figures in a quantity calculated by multiplication or division

Suppose we carry out an experiment to find the density of a lump of metal. We require two measurements, namely: (i) the mass of the metal and (ii) its volume. Suppose the mass of the metal was reported as 10.0078 g whereas the volume of the metal (which is more difficult to determine accurately) was reported as 2.8 cm^3. The density is now calculated as

$$\text{density} = \frac{\text{mass}}{\text{volume}} = \frac{10.0078}{2.8} = 3.574\,214\,29\,\text{g cm}^{-3}$$

where the number 3.574 214 29 is the one that might be displayed on a calculator. It is absurd to report the density as 3.574 214 29 g cm^{-3}, since this would suggest an uncertainty of about $\pm 0.000\,000\,01$ g cm^{-3}! Again, there is a rule to guide us: the number of *significant figures* in the final calculated figure is *set equal to the number of significant figures in the most uncertain contributing measurement.* The density calculation depends upon measurements of volume and mass, but the volume measurement was only reported to two significant figures and is therefore the most uncertain of the two measurements. Accordingly, the density should also be reported to two significant figures:

$$\text{density} = 3.\mathbf{6}\,\text{g cm}^{-3}$$

(where the **5** of the 3.**5**7421429 is rounded up to **6** as explained in Box 1.3).

Number of significant figures in a quantity calculated by addition or subtraction

The rule here is that the number of *decimal places* in the final calculated figure is *set equal to the smallest number of decimal places in the contributing measurements.*

As an example, consider two different samples of water whose volumes were determined by two different methods as 41.66 cm^3 and 2.1 cm^3, respectively. Following the rule, and rounding up, the total volume is reported as 43.8 cm^3.

BOX 1.3

Rounding up

Suppose the mass of a coin is incorrectly reported as 5.6489 g (i.e. five significant figures), and that the uncertainties involved only justify the use of four significant figures. This means that we must *round up* to the fourth significant figure.

The rules we use are:

1. In considering the rounding up of the *n*th significant figure, we consider the next (i.e. the $(n + 1)$th) significant figure *only*.
2. The *n*th significant figure is only rounded up if the $(n + 1)$th figure is *equal to or greater than* 5.

So, rounding 5.6489 g (five significant figures) to four significant figures gives 5.649. However, if we wanted to report the mass of 5.6489 g to three significant figures our mass becomes 5.65, with the fourth significant figure (5.64**8**9) causing the 4 to round up to 5 and the fifth significant figure having no part to play. If we wanted to report the original mass to two significant figures, we go from 5.6489 to 5.6 because 4 is not equal or greater than 5 and the other original figures (the 8 and 9 in 5.64**89**) are irrelevant. Finally, the original mass becomes 6 when expressed to one significant figure with the 6 in 5.**6**489 causing rounding up. In summary:

Coin mass	Number of significant figures
5.6489	5
5.649	4
5.65	3
5.6	2
6	1

Exercise 1F

Significant figures and rounding up

(i) How many significant figures are present in the following numbers: (a) 0.02, (b) 20.02, (c) 890, (d) 0.00765?

(ii) The atomic mass of the oxygen-16 atom is 15.9949 atomic mass units, but a student uses an approximate value of 16.10. Is this justified?

(iii) Round up 0.03467 to (a) three significant figures, (b) two significant figures and (c) one significant figure.

Number of significant figures in a logarithmic quantity

The log of 1.97×10^3 is 3.294 466, but how many decimal places should the answer contain? The rule here is that the number of *decimal places* in the answer is equal to the number of *significant figures* in the initial number. 1.97×10^3 contains three significant figures, so log (**1.97** $\times 10^3$) = 3.**294**.

The reverse also applies. If we wish to find the number whose log is 0.8234, we calculate $10^{0.8234}$ which is 6.658 86. The number of significant figures in the answer equals the number of decimal places in the initial number, so that $10^{\mathbf{0.8234}} = \mathbf{6.659}$.

Quantities are often expressed in logarithmic form so as to compress them and make very big or very small numbers more manageable. An example is found in the quantity known as pH which is calculated from the hydrogen ion concentration (symbolized $[H^+(aq)]$) in a solution by the equation:

$$\text{pH} = -\log [H^+(aq)]$$

This equation is read as 'pH equals the negative of the log of the hydrogen ion concentration'. The reverse of this equation is

$$[H^+(aq)] = 1 \times 10^{-\text{pH}}$$

read as 'the hydrogen ion concentration equals 10 to the power of the negative of the pH value'.

Applying the above rules to significant figures, if the hydrogen ion concentration is 4.403×10^{-3} mol dm^{-3} then

$$\text{pH} = -\log (\mathbf{4.403} \times 10^{-3}) = -(-2.356\,25) = 2.356\,25$$
$$= 2.\mathbf{3563}$$

i.e. four decimal places and rounded up.

If the pH of a solution was expressed as 6.81 then

$$[H^+(aq)] \text{ (in mol dm}^{-3}) = 1 \times 10^{-\text{pH}} = 1 \times 10^{-6.\mathbf{81}} = 1.5488 \times 10^{-7}$$
$$= \mathbf{1.5} \times 10^{-7}$$

i.e. two significant figures and rounded down.

Further examples of pH calculations are found in Unit 9 (page 150) and in Unit 16 (page 286).

Exercise 1G

Examples in the use of significant figures

(i) Significant figures in multiplication

The concentration of hydronium ions in a solution was calculated using the equation

$$[H^+(aq)] = \sqrt{2.04 \times 10^{-8} \times c} \quad \text{mol dm}^{-3}$$

Experiments show that c has a mean value of 0.0108. Only two significant figures are justified in c. Report $[H^+(aq)]$ to the correct number of significant figures.

(ii) Significant figures in addition

The mass of metals in a sample of waste water was determined by analysis to be as follows: Cu^{2+} 0.132 mg, Pb^{2+} 0.3 mg, Zn^{2+} 10.00 mg. What is the total mass of metal present?

(iii) Significant figures in logarithmic quantities

The pK_a of an acid is calculated by the equation:

$$pK_a = -\log K_a$$

where K_a is the *ionization* constant of the acid. At 25°C, the K_a value of ethanoic acid is 8.4×10^{-4} mol dm^{-3}. Calculate pK_a.

Revision questions

1.1. The radii of several atoms and ions (in different units) are as follows: Cr^{3+} 0.069 nm, F^- 1.36×10^{-6} cm, O 1.40×10^{-5} mm. Express the radii in metres and arrange the particles in order of increasing size.

1.2. The temperature of a water bath was reported as 27.1 °C. What does the number of significant figures tell you about the minimum uncertainty involved in this measurement?

1.3. Report the following measurements to four significant figures:

(i) 0.123 47 V,

(ii) 12.45 m,

(iii) 0.003 557 57 cm,

(iv) 1200.5 K.

1.4. Calculate $e^{-E/RT}$ (with $E = 20\,000\ \text{J mol}^{-1}$ and $R = 8.3145\ \text{J mol}^{-1}\text{K}^{-1}$) at

(i) $T = 300$ K and

(ii) $T = 3000$ K.

In each case, express the result to three significant figures.

1.5. A pain killing tablet contains 154 mg caffeine, 101 mg aspirin and 0.23 g filler. Express the total mass of the tablet (in grams) to the correct number of significant figures.

1.6. The **ionic product** of an aqueous solution (a solution with water as the solvent), symbolized K_w, is defined by the equation

$$K_w = [H^+(aq)] \times [OH^-]$$

where the brackets denote the concentrations (in mol dm^{-3}) of $H^+(aq)$ and OH^- ions at that temperature. What are the units of K_w? Calculate K_w for a solution at 25 °C if $[H^+(aq)] = 5 \times 10^{-10}$ mol dm^{-3} and $[OH^-] = 2.0 \times 10^{-5}$ mol dm^{-3}. Express your answer to the correct number of significant figures.

1.7. In spectroscopy, the **absorbance** (A) of a solution of a substance is defined by the expression

$$A = \varepsilon \times c \times b$$

where ε (pronounced epsilon) is the molar absorption coefficient of the substance, c the concentration of the substance (in moles per metre cubed) and b the thickness of the sample (in metres); A is unitless. What are the units of ε?

1.8. Label the following as random or systematic errors:

(i) The variation of the mass recorded by a balance because of air draughts in the laboratory.

(ii) A thermostat in a water bath registers 30 °C when the real temperature of the water is 25 °C.

(iii) The leaking of a gas cell which is being used to hold a sample of gas so that its pressure can be measured.

1.9.

(i) The hydrogen ion concentration of a solution is 8.987×10^{-6} mol dm^{-3}. Calculate the pH of the solution.

(ii) The pH of a solution was 11.344. Calculate the hydrogen ion concentration of the solution.

Extension material to support this unit is available on our website at **http://www.palgrave.com/foundations/lewis.**

Elements, Compounds and Reactions

Objectives

- ▶ Names the various forms of matter
- ▶ Describes the composition of matter
- ▶ Distinguishes between chemical and physical changes
- ▶ Shows you how to construct formulae and equations

Contents

2.1 Matter and energy

The universe is composed of matter and energy. Matter is anything that occupies **space** and has **mass** – rocks, oceans, the air that we breathe and we, ourselves, are all composed of matter. Energy has no shape or form – it is defined as the ability to do **work**.

Mass and weight

Suppose you have a solid rubber ball, which contains a definite amount of matter. If you compress it, its volume reduces but the quantity of matter it contains remains the same. If you took the ball to the Moon, it would still contain the same amount of matter but its **weight** would be different. The weight of a body depends upon the attractive force exerted upon it by gravity and the force of gravity on the Moon is much less than on Earth. The mass of a body, however, depends on the amount of matter it contains and *does not vary with location*.

Although mass and weight are often used interchangeably, this is sloppy because they are not the same. Chemists are concerned with the measurement of masses, and here we shall use the term 'mass' in its proper context.

Work

If we try to move a mass, it resists our efforts. **Work** is done when a force acts on a mass and moves it and this requires energy. **Energy** shows itself in many different forms:

- **Heat energy** is released when fuels burn.
- **Electrical energy** drives videos, freezers and computers.

- **Light energy** is emitted by the sun.
- **Sound energy** is produced by a person speaking.
- **Nuclear energy** generates electrical power.
- **Chemical energy** is stored in a battery.

Forms of matter

Matter is generally observed in three physical states: **solids**, **liquids** and **gases**. A fourth state of matter, **plasma**, is formed at very high temperatures and may be found in the atmospheres of stars.

Matter is composed of **elements**, **compounds** and **mixtures**. The composition of matter is summarized in Fig. 2.1.

1. Elements

Elements are basic forms of matter and cannot be broken down into simpler substances by chemical reactions.

Commonly known elements include copper, gold, oxygen, mercury and sulfur. At present, there are 109 named elements (see the Periodic Table, shown on the inside front cover, which is an arrangement of all known elements) of which 90 have been found naturally; the others have been made by scientists.

2. Compounds

Compounds consist of two or more elements joined together in fixed proportions to form a new substance.

For example, sodium and chlorine combine together to form sodium chloride, or salt. Sugar is formed from a combination of the elements carbon, hydrogen and oxygen. Chemical reactions, or chemical changes occur when **new substances** are formed: elements react together to form compounds, or compounds can be broken down into their constituent elements by chemical changes.

Both elements and compounds are **substances**; the composition of a substance is always the same, regardless of its source. Notice that the scientific meaning of the word 'substance' is very precise – a substance is a single pure form of matter, and not a mixture of different kinds of matter. Substances are relatively rare; however, most naturally occurring materials on Earth are mixtures.

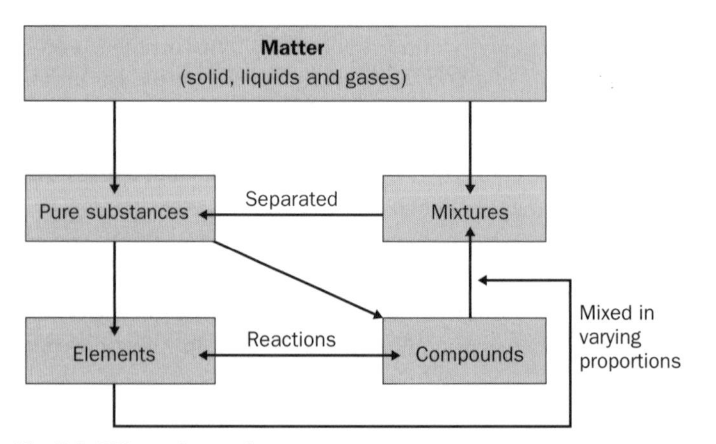

Fig. 2.1 Different forms of matter.

BOX 2.1

Purity of substances

Purity is very important in chemistry. Pure elements and compounds, for example a pure sample of the element copper or pure salt (the compound sodium chloride), have *reproducible* properties. In other words, different samples of salt behave in exactly the same way under identical conditions: they have the same melting points, undergo the same chemical reactions and so on. Mixtures, because they can have varying compositions, often do not behave in the same way under identical conditions. Different samples of soil, for example, may have very different properties although they are all referred to as 'soil'.

Guaranteed pure reagents were first advertised by the German chemical and pharmaceutical firm of Merck in 1888. During the First World War, when the British found themselves unable to obtain these pure chemicals, British chemical manufacturers were encouraged to produce chemicals to new standards of purity. These chemicals were classified AR (or Analytical Reagent). Today, very pure chemicals conform to a British standard and are called 'AnalaR'. Look at reagent bottles in your laboratory and see if you can spot AnalaR chemicals!

3. Mixtures

Mixtures contain more than one substance (elements and/or compounds). The composition of a mixture can be varied.

Air is a mixture of the elements oxygen and nitrogen, the compounds water vapour and carbon dioxide, together with trace amounts of other gases. However, the composition of a sample of air taken in an industrial, polluted, area would be different from that of a sample taken on Mt Everest.

Elements: metals and non-metals

Elements can be roughly divided into metals and non-metals.

1. Metals

Metals usually have the following properties:

- They have a lustre (a 'shiny' appearance).
- They conduct heat and electricity.
- They can be drawn into wire (are **ductile**).
- They can be hammered into thin sheets (are **malleable**).
- They make a 'clanging' sound when you hit them (are **sonorous**).
- In chemical reactions metals react with acids, form **basic oxides** and form **positively charged ions** (you may not understand these terms at this point, but will be able to refer back to these properties later on).

Examples of metals are gold, silver, copper and iron. Mercury is a liquid metal at room temperature.

2. Non-metals

Non-metals usually have the following physical properties:

- They are poor conductors of heat and electricity.
- They cannot be hammered into sheets or drawn into wire, because they are brittle.

- They are not lustrous and not sonorous.
- In chemical reactions non-metals do not react with acids, they form **acidic oxides** and form either **negatively charged ions** or else they form **covalent compounds**.

Examples of non-metals are iodine, oxygen, nitrogen and carbon.

As you will see later, some elements have intermediate properties between those of metals and non-metals and are called **metalloids** or **semimetals**. Examples include arsenic, silicon and germanium.

Atoms and molecules

The 'building bricks' of elements are called **atoms**. They are the smallest part of an element that chemically reacts like a bulk sample of the element. If it were possible to take a piece of copper and divide it into smaller and smaller pieces, eventually we would have one atom of copper. Atoms are incredibly small – 35 000 000 copper atoms, laid end to end in a line, would cover a distance of about 1 cm. The existence of atoms has been proposed since the time of the early Greeks and it explains the behaviour of substances. The **scanning tunnelling microscope**, developed in 1981, enables us to see atoms (see Box 2.2).

Two or more atoms can link together to form new particles called **molecules**. The chemical behaviour of the new molecules is different from that of the atoms which make them up. The molecules may be combinations of atoms of the same element. For example, under normal conditions oxygen gas contains **diatomic** oxygen molecules (two atoms joined together). For this reason, the oxygen molecule is sometimes called 'dioxygen'. If the atoms joined are of different elements, the molecules form part of a compound. The compound known as water consists of molecules in which two hydrogen atoms are joined to one oxygen atom (see Box 2.3).

BOX 2.2

Scanning tunnelling microscope

We cannot look at atoms using a microscope that uses visible light because particles with diameters less than the wavelength of visible light (about 400 nm) cannot be distinguished.

Electron microscopes use electrons (small particles found inside the atom) to produce images of tiny objects. The scanning tunelling microscope (STM) is the most sophisticated of this type of instrument and it can produce images of the surfaces of elements which show the individual atoms. The atoms show up as blurred spheres. The STM has applications in physics, chemistry and biology.

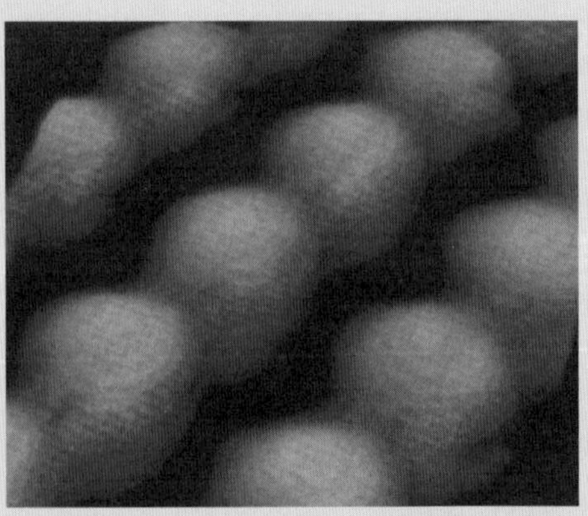

An 'STM' image of the surface of gallium arsenide.

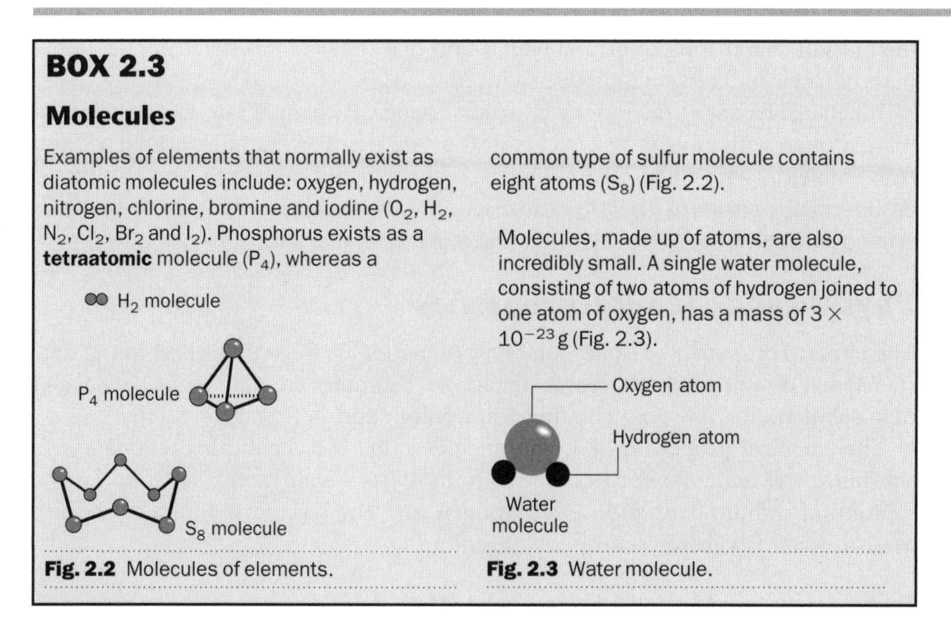

BOX 2.3

Molecules

Examples of elements that normally exist as diatomic molecules include: oxygen, hydrogen, nitrogen, chlorine, bromine and iodine (O_2, H_2, N_2, Cl_2, Br_2 and I_2). Phosphorus exists as a **tetraatomic** molecule (P_4), whereas a common type of sulfur molecule contains eight atoms (S_8) (Fig. 2.2).

Fig. 2.2 Molecules of elements.

Molecules, made up of atoms, are also incredibly small. A single water molecule, consisting of two atoms of hydrogen joined to one atom of oxygen, has a mass of 3×10^{-23} g (Fig. 2.3).

Fig. 2.3 Water molecule.

2.2 Physical and chemical changes

Physical changes

The boiling of water to produce steam is an example of a physical change. The water changes state, from a liquid to a gas, but it does not change into a different type of matter. When salt is dissolved in water, it appears to disappear – but we know it is still there because the water tastes 'salty'. The salt has broken down into small particles which are thoroughly mixed with the water molecules, but the chemical identity of the salt has not changed. **Dissolving** is another example of a physical change. The substance which has dissolved (in this case, salt) is called the **solute**; the substance in which it has dissolved (in this case, water) is called the **solvent**.

Exercise 2A

Physical and chemical changes

Which of the following changes do you think are physical or chemical?

(i) the burning of coal
(ii) the freezing of water
(iii) the melting of solder
(iv) the dissolving of sugar in tea
(v) the magnetization of iron.

Chemical changes (chemical reactions)

A **chemical reaction** involves the conversion of one type of matter into another. New substances are formed. After a chemical reaction there **is no change in the total mass from that present initially**. This is **the law of conservation of mass**. Rust is a compound formed from the chemical combination of iron, oxygen and water. If a piece of iron is allowed to rust completely:

total mass of rust =
mass of iron + mass of combined water + mass of combined oxygen

For example, the amount of money in your bank account may get smaller between pay cheques, but the money hasn't been destroyed – it has simply gone into other accounts! The same principle applies to chemical reactions; atoms are not destroyed but they rearrange in different ways. Charcoal, burning on your barbeque, may appear to do a vanishing act:

charcoal + oxygen (from air) $\rightarrow$ carbon dioxide

However, if you added the mass of oxygen, from the air, used up in the reaction and

Exercise 2B

Physical and chemical properties

Classify the following as physical or chemical properties:

(i) the **hardness** of diamond
(ii) the low **density** of aluminium
(iii) wood **burns** in air
(iv) arsenic is **poisonous**
(v) milk goes **sour**.

Exercise 2C

Extensive and intensive properties

Classify the following properties as extensive or intensive:

(i) length (ii) colour
(iii) mass (iv) density.

the mass of charcoal used up, you would find that the sum was exactly equal to the mass of (invisible) carbon dioxide produced.

When a substance undergoes a chemical change, it changes into another type of matter, often with quite different chemical properties. For example, heated sodium reacts violently with chlorine. Sodium is a highly reactive metal, whereas chlorine is an extremely poisonous gas. After the reaction sodium chloride (salt) is formed; this is not a particularly reactive substance and it is essential to life!

Physical and chemical properties

The physical properties of a particular type of matter are those properties of the substance that do not involve a chemical reaction. Examples are state (**solid**, **liquid**, **gas** or **in solution**), melting point, boiling point, colour and electrical conductivity.

The chemical properties of a substance describe the chemical reactions that a substance will undergo. For example sodium reacts violently with water to form a solution of sodium hydroxide and hydrogen gas. The reactivity shown by sodium towards water is a chemical property of sodium.

Extensive and intensive properties

Any property of matter that can be measured can be classified as **extensive** or **intensive**.

Extensive properties depend on the amount of substance present; intensive properties do not.

Volume is an extensive property. If you have 100 cm^3 of water and add 50 cm^3 more, the total volume is 150 cm^3 – the volumes can be added to give a total.

Temperature is an intensive property. If you have water at a temperature of 50 °C and add more water at 50 °C the temperature of the whole is still 50 °C and *not* 100 °C.

2.3 Chemical formulae

Symbols

Chemists use a kind of shorthand to describe elements and compounts. All elements have characteristic *symbols*. These symbols sometimes consist of the first letter(s) of the name of the element; for example, hydrogen is given the symbol H, whereas the chemical symbol for bromine is Br. Some symbols, however, are not derived from the modern or English name of the element: gold has the chemical symbol Au (from the Latin *Aurum*), whereas tungsten is denoted by W (from *Wolfram*, a German word). Note that *the first letter only of a chemical symbol is written as a capital letter*. Box 2.4 gives the symbols of some elements.

Exercise 2D

Using symbols and formulae

1. **Use Information Box 2.4 to name the following elements:**
(i) Ba (ii) Si (iii) Ar (iv) F (v) Li (vi) Hg (vii) Sn (viii) Ag (ix) B (x) Ti.

2. **Name the elements in the following compounds and state how many atoms of each element are present:**
(i) CsBr (ii) $CuCl_2$ (iii) Fe_3O_4 (iv) CH_4 (v) $PbSO_4$ (vi) $K_2Cr_2O_7$ (vii) Na_3PO_4 (viii) MnO_2.

BOX 2.4

Some symbols for elements (you are strongly advised to learn these!)

Element	Symbol	Element	Symbol	Element	Symbol	Element	Symbol
Aluminium	Al	Cobalt	Co	Manganese	Mn	Sodium	Na
Argon	Ar	Copper	Cu	Mercury	Hg	Sulfur	S
Arsenic	As	Fluorine	F	Neon	Ne	Tin	Sn
Barium	Ba	Gold	Au	Nitrogen	N	Titanium	Ti
Boron	B	Helium	He	Oxygen	O	Tungsten	W
Bromine	Br	Hydrogen	H	Phosphorus	P	Uranium	U
Caesium	Cs	Iodine	I	Platinum	Pt	Vanadium	V
Calcium	Ca	Iron	Fe	Potassium	K	Xenon	Xe
Carbon	C	Lead	Pb	Radon	Rn	Zinc	Zn
Chlorine	Cl	Lithium	Li	Silicon	Si		
Chromium	Cr	Magnesium	Mg	Silver	Ag		

Exercise 2E

Main elements in the human body

Can you identify the main elements in the human body from Fig. 2.4?

Fig. 2.4 Main elements in the human body (diagram from F. C. Hess, *Chemistry Made Simple*, W. H. Allen).

Exercise 2F

Symbols of elements

1. Using library books at your disposal, find out from what names the symbols of the following elements were derived:

(i) sodium (Na) **(ii)** iron (Fe) **(iii)** lead (Pb) **(iv)** tin (Sn) **(v)** potassium (K) **(vi)** silver (Ag).

2. Some of the latest elements discovered derive their names from famous scientists. See if you can find out the main contributions to our knowledge made by the following scientists:

(i) meitnerium (Mt) named after Lise Meitner;
(ii) seaborgium (Sg) named after Glenn Seaborg;
(iii) bohrium (Bh) named after Niels Bohr;
(iv) rutherfordium (Rf) named after Ernest Rutherford.

Formulae

The **formulae** for elements and compounds are constructed using these symbols. The symbol for the element oxygen is written as O. Since oxygen is normally found as oxygen gas, in which the smallest particles are diatomic molecules, the chemical formula for oxygen gas, or dioxygen, is written as O_2. Compounds are written using a collection of different symbols. For example, water is H_2O (indicating that two hydrogen atoms have joined with one oxygen atom to form a molecule of water), whereas sodium chloride is written as NaCl (this compound is formed when sodium reacts with chlorine in a ratio of 1 atom to 1 atom). Chemical formulae therefore give us two important pieces of information:

1. the elements that are present in a substance;

2. the relative number of atoms that make up that substance.

Note that '$2H_2O$' refers to *two separate molecules* of water.

Writing chemical formulae

Although it is necessary for a chemist to learn the symbols of the common elements, it is possible to work out the chemical formulae of common substances without having to learn them parrot fashion. Different elements have atoms with different combining powers. The combining power of an atom is known as its **valency**, and these valencies can be used to find out how many atoms of one element will combine with another. A table of the valencies of common elements and groups is shown on page 431. Again, these need to be *learned*.

To obtain a clearer picture of how a chemical formula is constructed, imagine that the valencies of atoms are 'arms', which link up with the arms of other elements' atoms.

Some elements have more than one valency, but the name of the compound should help you work out the formula. For example, in the compound iron(III) chloride, the iron atoms have a valency of three and its formula is $FeCl_3$, whereas in iron(II) chloride, the iron combines with a valency of two and the formula is $FeCl_2$.

Example 2.1

Construct the formula of magnesium oxide (a compound of magnesium and oxygen).

▶ Answer

The symbol for magnesium is Mg and it has a valency of two. Imagine it as:

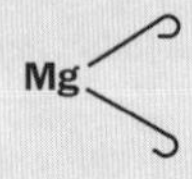

Similarly, oxygen could be thought of as:

O

When magnesium combines with oxygen, the arms link up with one another and *there are no free arms left over*. Therefore, one atom of magnesium combines with one atom of oxygen to form magnesium oxide:

Mg O

The formula is written as MgO. (Leave out the arms, they are only used to help you to construct the formula!)

BOX 2.5

Naming compounds

1. When a compound consists of a metal and a non-metal, the metal is written and named first. The Periodic Table will help you to distinguish whether an element is a metal or non-metal. Metals are shown to the left of the table, whereas non-metals are found to the right. The border between the two is shown as a zig-zag line. The non-metal part is named by putting 'ide' at the end, thus chlorine becomes chloride and oxygen becomes oxide.

Examples:
MgO magnesium oxide
Na_2O sodium oxide
$AlCl_3$ aluminium chloride
CaS calcium sulfide

2. If a compound is composed of two non-metals, the non-metal that is the closest to being a metal (closer to the metal–non-metal border in the table) is written first. If more than one compound of the same two non-metals exist then the prefixes mono(1), di(2), tri(3), tetra(4), penta(5), hexa(6) and so on are put in front of the non-metal to indicate the number of its atoms present in the compound.

Some examples:
CO carbon monoxide
CO_2 carbon dioxide
N_2O_3 dinitrogen trioxide
N_2O dinitrogen monoxide
P_4O_6 tetraphosphorus hexaoxide*
P_4O_{10} tetraphosphorus decaoxide*

* Often written hexoxide and decoxide (for ease of pronunciation).

Example 2.2

What is the formula of aluminium chloride?

▶ Answer

Aluminium chloride is a compound of aluminium and chlorine. The symbols and valencies for atoms of both these elements are

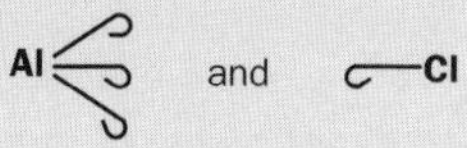

Al (three 'arms') and (one 'arm') Cl

When they combine together,

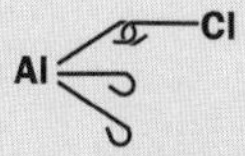

Al—Cl (Al with two 'arms' left over)

aluminium has two 'arms' left over, so another two chlorine atoms, with one 'arm' each, must be added:

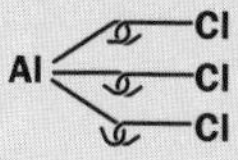

Al—Cl, Al—Cl, Al—Cl

The formula of aluminium chloride is therefore $AlCl_3$. The subscript after Cl shows that *three* chlorine atoms combine with one atom of aluminium. There is *no subscript* after Al, meaning that only *one* atom of aluminium is involved in the combination.

Exercise 2G

Simple formulae

Write formulae for the following compounds, using the valency table on page 411.

(i) calcium oxide
(ii) hydrogen sulfide
(iii) magnesium fluoride
(iv) magnesium chloride
(v) aluminium fluoride
(vi) aluminium oxide
(vii) hydrogen chloride
(viii) lithium bromide
(ix) magnesium sulfide
(x) magnesium nitride.

Exercise 2H

More simple formulae

Write formulae for the following compounds (in (i)–(x), the valencies of the metals atoms are shown in brackets after the name of the metal; you should be able to work out the formulae of (xi)–(xx) just by looking at their names!):

(i) copper(I) oxide
(ii) copper(II) oxide
(iii) lead(IV) oxide
(iv) chromium(III) sulfide
(v) manganese(IV) oxide
(vi) chromium(VI) fluoride
(vii) cobalt(III) sulfide
(viii) chromium(III) oxide
(ix) vanadium(V) oxide
(x) titanium(III) fluoride
(xi) nitrogen monoxide
(xii) arsenic trichloride
(xiii) sulfur hexafluoride
(xiv) dinitrogen
(xv) disulfur dichloride
(xvi) dinitrogen pentoxide
(xvii) dioxygen difluoride
(xviii) tellurium trioxide
(xix) dichlorine hexaoxide (hexoxide)
(xx) silicon tetrachloride.

Groups of atoms

For the purposes of writing formulae, sometimes groups of atoms act as if they have a valency of their own. Their valencies are also shown on page 431.

Example 2.3

Write the formula of ammonium sulfate

▶ Answer

The ammonium group is a combination of one nitrogen atom and four hydrogen atoms, but together they act as if the group has a valency of one:

NH_4—

Similarly, sulfate has a valency of two:

>SO_4

Two ammonium groups therefore combine with one sulfate group:

NH_4—
>SO_4
NH_4—

The formula is written as $(NH_4)_2SO_4$.

▶ Comment

Note that *if more than one group of atoms* are to be represented, the group is enclosed in brackets and the number of groups present is written as a subscript outside the brackets.

Exercise 2I

Formulae involving groups of atoms

Write formulae for the following compounds:

(i) copper(II) sulfate
(ii) copper(II) nitrate
(iii) ammonium chloride
(iv) sodium phosphate
(v) calcium phosphate
(vi) sulfuric acid (dihydrogen sulfate)
(vii) nitric acid (hydrogen nitrate)
(viii) aluminium nitrate
(ix) lithium carbonate
(x) ammonium carbonate
(xi) calcium hydroxide
(xii) potassium hydrogencarbonate
(xiii) calcium hydrogencarbonate
(xiv) sodium hydrogensulfate
(xv) iron(III) hydroxide.

2.4 Writing and balancing equations

Equations for chemical changes can be constructed by setting out the formulae of the reactants and products of a reaction. The same numbers of atoms of each element should appear on each side of the equation – the equation is then **balanced**.

Chemical equations are best written in three stages:

1. Write the names of the reactants and products on each side of the equation.
2. Replace the names with formulae.
3. Balance the equation.

Example 2.4

Hydrogen gas (dihydrogen) reacts with oxygen gas (dioxygen) to form water. Write a balanced equation for the reaction.

▶ Answer

1. hydrogen gas + oxygen gas → water

2. $H_2 + O_2 \rightarrow H_2O$

3. Count the number of atoms of each element on each side of the equation:

left-hand side – two atoms of H and two atoms of O
right-hand side – two atoms of H and one atom of O

We need to increase the number of atoms of O on the right-hand side. Numbers *in front* of the formulae must be added, thus $2H_2O$ means two molecules of water (containing four atoms of H and two atoms of O), so we can increase the number of O atoms on the right-hand side:

$$H_2 + O_2 \rightarrow 2H_2O$$

Now there are more hydrogens on the right-hand side (four) than on the left-hand side (two), but writing 2 in front of H_2 remedies the situation:

$$2H_2 + O_2 \rightarrow 2H_2O$$

The equation is now balanced – there are the same number of atoms of hydrogen and oxygen on each side. This equation tells us that two molecules of H_2 react with one molecule of O_2 to form two molecules of H_2O.

▶ Comment

Note that although chemists often use formulae instead of names (a bottle of water might be labelled H_2O) the formula H_2O, strictly speaking, refers to *one molecule* of water. The balanced equation is represented by Fig. 2.5.

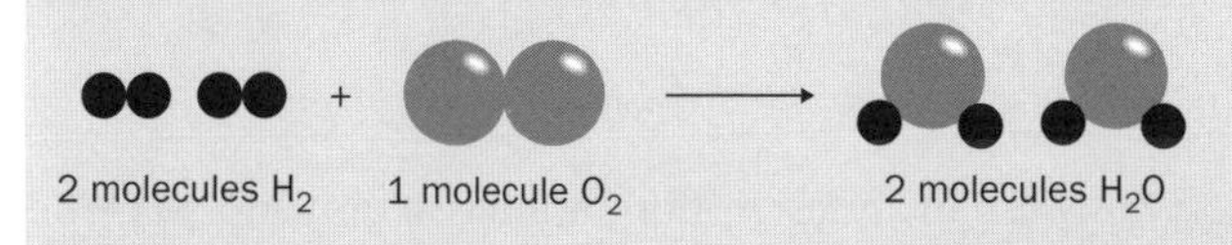

Fig. 2.5 Balanced equation for the reaction of hydrogen with oxygen.

▶ Additional comment

Before you try to balance an equation it is very important that you write correct formulae for the products and reactants. For example, you will come across a chemical with the formula H_2O_2. This compound is commonly called hydrogen peroxide and it is *very different* from water in its chemical behaviour. The extra oxygen in the formula makes a big difference to the nature of the substance!

Exercise 2J

Balancing equations

Balance the following equations:

(i) $N_2 + H_2 \rightarrow NH_3$

(ii) $Na + Cl_2 \rightarrow NaCl$

(iii) $Al + O_2 \rightarrow Al_2O_3$

(iv) $Al + H_3PO_4 \rightarrow AlPO_4 + H_2$

(v) $Na + O_2 \rightarrow Na_2O$

(vi) $Mg_3N_2 + H_2O \rightarrow Mg(OH)_2 + NH_3$

(vii) $Cr_2O_3 + Si \rightarrow Cr + SiO_2$

(viii) $CS_2 + O_2 \rightarrow CO_2 + SO_2$

(ix) $Fe_2O_3 + CO \rightarrow Fe + CO_2$

(x) $FeS_2 + O_2 \rightarrow Fe_2O_3 + SO_2$.

Exercise 2K

Writing equations

Write balanced equations for the following reactions:

(i) The reaction of potassium with oxygen gas (dioxygen) to form potassium oxide.

(ii) The reaction of iron with chlorine gas (dichlorine) to form iron(III) chloride.

(iii) The reaction of lithium with water to form lithium hydroxide and hydrogen gas (dihydrogen).

(iv) The reaction of magnesium with nitrogen gas (dinitrogen) to form magnesium nitride.

(v) The reaction of calcium hydroxide with carbon dioxide to form calcium carbonate and water.

(vi) The reaction of sodium hydroxide with sulfuric acid to form sodium sulfate and water.

(vii) The decomposition of gold(III) oxide into its elements (gold atoms and dioxygen) when heated.

(viii) The reaction of copper with nitric acid to produce copper(II) nitrate, nitrogen dioxide and water.

Example 2.5

Sodium reacts with water to form sodium hydroxide and hydrogen gas. Write a balanced equation for the reaction.

▶ Answer

1. sodium + water → sodium hydroxide + hydrogen gas

2. $Na + H_2O \rightarrow NaOH + H_2$

3. $2Na + 2H_2O \rightarrow 2NaOH + H_2$

The equation in 3. is the balanced equation.

▶ Comment

Sometimes a bit of juggling is required, but when you are balancing equations *never add subscripts to the formulae because you will change the formulae.* Formulae are fixed – *only numbers in front of the formulae* can be changed in order to balance an equation. Although equations are usually balanced using the simplest ratio of numbers, an equation is *not* incorrect if higher numbers are used, as long as the correct ratio is maintained. For example

$$2Na + 2H_2O \rightarrow 2NaOH + H_2$$

means the same as

$$4Na + 4H_2O \rightarrow 4NaOH + 2H_2$$

or

$$Na + H_2O \rightarrow NaOH + \tfrac{1}{2}H_2$$

because in each case the ratio of the species in the equation is 2:2:2:1.

Note that although it is acceptable to use $\tfrac{1}{2}$ in an equation do not be tempted to use smaller fractions – *apart from this one exception, stick to whole numbers.*

State symbols

Substances in balanced equations often have **state symbols** written in brackets after them. The state symbols are:

(s) solid
(l) liquid
(g) gas
(aq) aqueous, i.e. dissolved in water

These are included to give a more detailed description of the reaction and so to aid the reader's understanding.

A historical perspective on this chapter is given by Box 2.6.

Example 2.6

The equation

$$AgNO_3(aq) + NaCl(aq) \rightarrow AgCl(s) + NaNO_3(aq)$$

shows that silver nitrate and sodium chloride solutions react together to form solid silver chloride and sodium nitrate solution. It is important to include the state symbols in this equation, because solid silver nitrate and solid sodium chloride *do not react* – they must first be dissolved in water or no reaction takes place.

Exercise 2L

Equations using state symbols

Write balanced equations (including state symbols) for the following reactions:

(i) Solid calcium carbonate decomposes on heating into solid calcium oxide and carbon dioxide gas.

(ii) Solutions of lead(II) nitrate and sodium iodide react together to form solid lead(II) iodide and sodium nitrate solution.

(iii) When chlorine gas is passed over heated aluminium, solid aluminium chloride is formed.

(iv) North Sea gas chiefly consists of a compound called **methane** (CH_4). Methane burns in oxygen gas to form carbon dioxide and steam.

BOX 2.6

John Dalton

John Dalton (1766–1844) was the son of a poor Quaker couple. He spent most of his life in Manchester teaching and researching. Although the Greeks, as early as 400 BC, believed that matter consisted of small particles which they called atoms, the theory was revived and elaborated on by Dalton. In 1808 he presented his **atomic theory** in which he stated:

- Elements are made up of particles called atoms, which are indestructible.
- The atoms of an element are all the same, but different from those of other elements. (We now know this is not quite true because of the existence of isotopes (Chapter 3).)
- 'Compound atoms' (molecules as we know them) are formed by small, whole-number combinations of atoms.
- A chemical change involves a new arrangement of atoms.

The theory, with a few important modifications, still holds true today. To help him in his work, Dalton used balls with different colours for atoms. For example he used one black ball (carbon) and one white ball (oxygen) to represent CO, or carbon monoxide. Although Dalton proposed symbols for the elements (see Fig. 2.6), the symbols he proposed were found to be difficult to use.

It was left to Berzelius of Sweden to propose that each element be given a letter symbol and that this should represent an *atom* of the element. He proposed that all elements be given a symbol corresponding to the first letter of their names. Any repetitions were avoided by giving some symbols a second letter from the name. This is the system we use today.

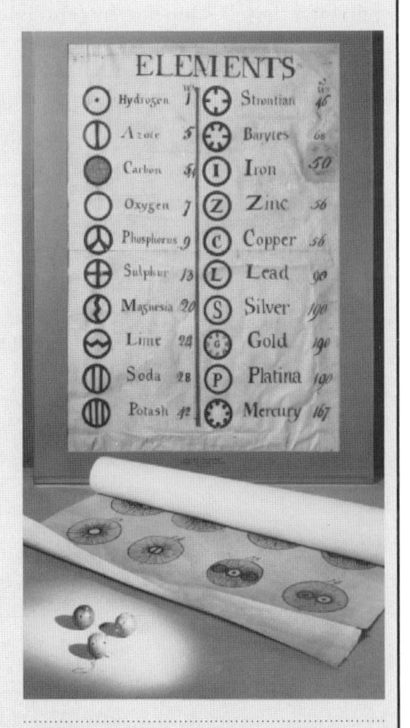

Fig. 2.6 Early symbols for the elements as proposed by Dalton.

Revision questions

2.1. Classify each of the following as an element, compound or mixture:

(i) milk
(ii) white sand, or silicon dioxide
(iii) rubidium
(iv) soil
(v) carbon dioxide
(vi) brass
(vii) nickel
(viii) ink
(ix) barium
(x) chalk.

2.2. The element lithium has a melting point of 181 °C and a boiling point of 1342 °C. It is a soft metal that corrodes rapidly in air. Lithium also conducts electricity and reacts violently with water. Classify all of these properties of lithium as physical or chemical.

2.3. Ethanol (common alcohol) is a compound with a melting point of −114°C and a boiling point of 78 °C. What is its physical state at room temperature (room temperature is about 20 °C)?

2.4. Butane is the chemical name for the compound used in lighter fuel. It melts at −138 °C and boils at 0 °C. What is its physical state at room temperature?

2.5. Use your library to look up the words from which the symbols for the following elements are derived:

(i) copper **(ii)** antimony **(iii)** mercury.

2.6. Write the correct symbols or formulae for each of the following:

(i) one atom of oxygen
(ii) two atoms of oxygen
(iii) two molecules of oxygen gas
(iv) three molecules of sulfur trioxide
(v) four atoms of neon.

2.7. How many atoms of each element are present in the following?

(i) Li_2CO_3 **(ii)** $Ca(NO_3)_2$ **(iii)** $(NH_4)_3PO_4$.

2.8. Write formulae for the following compounds:

(i) barium fluoride
(ii) tin(II) bromide
(iii) tin(IV) chloride
(iv) sodium ethanoate
(v) aluminium sulfide
(vi) potassium hydroxide
(vii) sodium bromate
(viii) potassium dichromate
(ix) ammonium phosphate
(x) lead(II) cyanide.

2.9. Find out a common use for hydrogen peroxide (H_2O_2).

2.10. Liquid hydrogen peroxide slowly decomposes into water and oxygen gas. Write a balanced equation for the reaction.

2.11. Balance the following equations:

(i) $KClO_4 \rightarrow KCl + O_2$
(ii) $S_8 + AsF_5 \rightarrow S_{16}(AsF_6)_2 + AsF_3$
(iii) $Hg + NH_4I \rightarrow HgI_2 + H_2 + NH_3$
(iv) $Al_4C_3 + H_2O \rightarrow Al(OH)_3 + CH_4$
(v) $Zn + H_3PO_4 \rightarrow Zn_3(PO_4)_2 + H_2$
(vi) $Fe_2O_3 + Na_2CO_3 \rightarrow NaFeO_2 + CO_2$.

2.12. Construct balanced equations (including state symbols) to describe the following reactions:

(i) Zinc metal burns in oxygen gas (dioxygen) to form solid zinc oxide.
(ii) Potassium metal reacts with water to form potassium hydroxide solution and hydrogen gas (dihydrogen).
(iii) Carbon dioxide gas reacts with magnesium metal to give solid magnesium oxide and carbon.

Inside the Atom

Objectives

- ▶Describes atomic structure and the evidence for quantization
- ▶Defines isotopic mass and the atomic mass of elements
- ▶Looks at the construction and use of mass spectrometers
- ▶Discusses the electronic configuration of atoms
- ▶Introduces the idea of an electron acting as a wave

Contents

In Chapter 2 we saw that atoms are not created or destroyed in chemical reactions, but that they are simply reshuffled to produce new molecules. Experiments show that the law of conservation of mass is found to apply to all chemical reactions, even to those carried out at high temperature. Do these observations mean that atoms are completely indestructible?

This chapter shows how atoms have been forced to reveal their inner secrets. Atoms are not indestructible – they do indeed have an inner structure, and although they often behave as if they were solid, they are mostly empty space!

3.1 Atomic structure

Pieces of history

In 1897, Sir J. J. Thomson (1856–1940) investigated the way that gases at low pressure conduct electricity. When high voltages were used, streams of minute particles were detected on a fluorescent screen. These particles possessed a negative charge and were even lighter than hydrogen atoms. The particles were called **electrons.**

Thomson realized that the electrons must have been squeezed out of the atoms of gas by the high voltages used in his experiments. (This is one piece of evidence that atoms are not indestructible. Another piece of evidence is provided by radioactivity, discussed in Chapter 21.) In other words, electrons are **subatomic particles**. Thomson realized that since atoms are usually observed to be electrically neutral, there must be positive charges in the atom too. He proposed that atoms consisted of a positively charged 'cement' in which rings of electrons were embedded. Thomson's

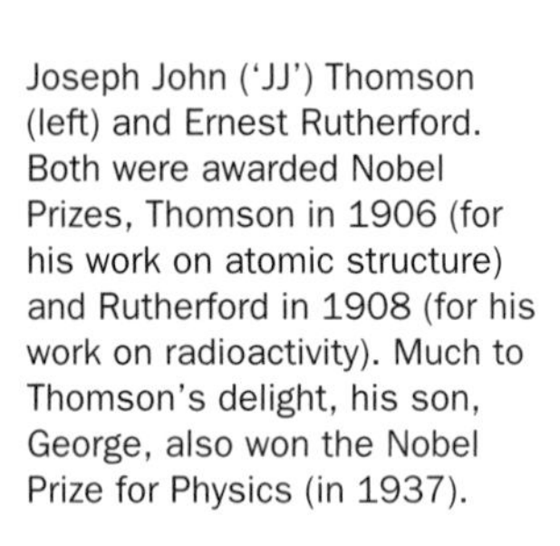

Joseph John ('JJ') Thomson (left) and Ernest Rutherford. Both were awarded Nobel Prizes, Thomson in 1906 (for his work on atomic structure) and Rutherford in 1908 (for his work on radioactivity). Much to Thomson's delight, his son, George, also won the Nobel Prize for Physics (in 1937).

model has been compared with currants in a bun, with the electrons as the currants and the dough as the positive charge.

In 1909 a group led by Lord Rutherford (1871–1937) carried out an experiment in which alpha particles from a radioactive material were fired at a very thin layer of gold metal in an evacuated bulb. (Alpha particles are much smaller than gold atoms.)

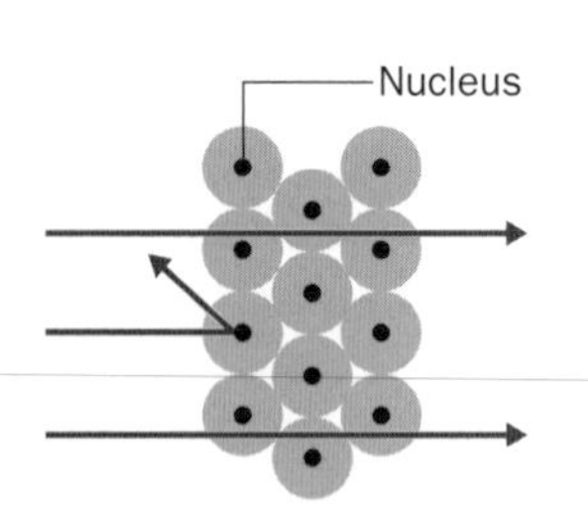

Fig. 3.1 Explanation of the Rutherford experiment. The path of the alpha particles is shown by arrows. Only if an alpha particle makes a direct hit on a nucleus does it suffer a large deflection.

Rutherford's team repeated the experiment many times. They found that although most of the alpha particles passed through the foil undisturbed, very occasionally an alpha particle underwent a massive deflection – like a table tennis ball bouncing off a wall. Only about 1 in every 20 000 alpha particles bounced back in this way, and the deflections were so rare that many experimenters might have simply ignored them. But Rutherford and his workers realized that such deflections were very significant indeed. Rutherford proposed that the mass of an atom was concentrated in its centre (the **nucleus**). The chances of an alpha particle striking the nucleus were very small but, when such a collision did occur, the considerable mass of the nucleus ensured that the alpha particle was greatly deflected (Fig. 3.1).

The 'insides' of atoms: a summary

The Rutherford model may be summarized as follows:

1. Atoms consist of electrons circling the nucleus of the atom. Electrons are negatively charged. By agreement, electrons are assigned a charge of −1. The circular paths of the electrons are called **shells** or **orbits**. The radius of an atom is equal to the radius of the outermost electron shell.

2. The nucleus consists of two kinds of particles, protons (which possess a relative charge of +1) and **neutrons** (with a mass approximately equal to that of the proton but with no electrical charge). Protons and neutrons are collectively known as **nucleons.**

3. Protons and neutrons are very much heavier than electrons. This means that most of the mass of an atom lies in its nucleus. The number of protons added to the number of neutrons is termed the **mass number** (symbolized A) of the atom.

4. Most of the atom is empty space. The picture we should keep in our mind is of minute 'specks' (the electrons) circling huge distances from a small (but relatively heavy) nucleus.

5. Atoms are electrically neutral. This is because the number of protons and electrons are equal. The number of protons in an atom is called its **atomic number** (symbolized Z).

6. The atoms of each element are different because they have different numbers of protons (i.e. different atomic numbers). The helium atom consists of a nucleus of two protons and two neutrons surrounded by two electrons, so we can write

$Z = 2 \qquad A = 4$

Such an atom is shown diagrammatically in Fig. 3.2. The position of the electrons is indicated by 'e^-'. This diagram is not to scale – if it were, the electrons would be several hundred metres from the symbols we have used to represent the nucleus.

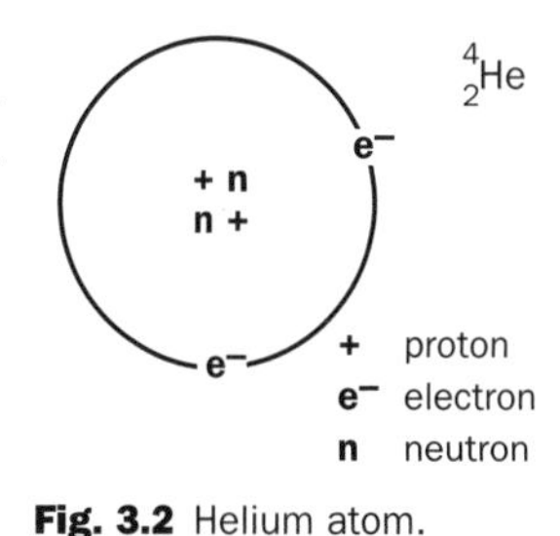

Fig. 3.2 Helium atom.

The helium atom is represented as $^4_2$He. This symbol is built up as follows:

- He is the chemical symbol for helium.
- The superscript '4' is the mass number.
- The subscript '2' is the atomic number.

Generalizing:

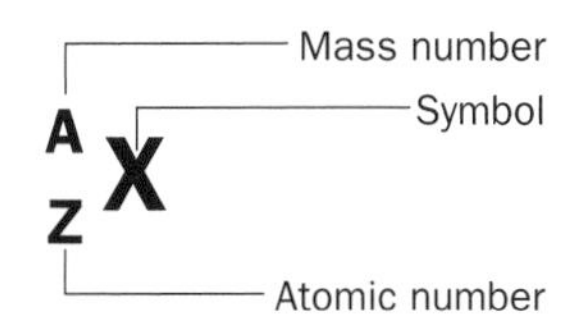

The number of neutrons N is the number of protons Z subtracted from the number of protons and neutrons A:

$N = A - Z$

Structure of the atom

Atoms of carbon are represented as $^{12}_6$C. Sketch the atom, following the pattern of Fig. 3.2.

Atomic mass scale

Table 3.1 gives more information about subatomic particles. In the fifth column, the mass of the particles is given in 'atomic mass units', symbolised as u, where

$1\,u = 1.660\,54 \times 10^{-24}\,g$

Table 3.1 The mass and charge of subatomic particles

Particle	Symbol	Charge*	Mass/g	Mass/u	Approx. mass/u
Proton	p	+1	$1.672\,62 \times 10^{-24}$	1.007 27	1
Electron	e^-	−1	$9.109\,39 \times 10^{-28}$	0.000 54	0
Neutron	n	0	$1.674\,93 \times 10^{-24}$	1.008 67	1
Alpha	α	+2	$6.644\,66 \times 10^{-24}$	4.001 51	4

* Relative to the electron, which has an absolute charge of 1.602×10^{-19} coulombs.

Exercise 3B

Masses of particles

Answer these questions using Table 3.1:

(i) How much heavier is the proton than the electron?

(ii) How much heavier is the alpha particle than the proton?

This tiny mass is *exactly equal to one-twelfth of the mass of one atom of carbon (isotope mass number 12)*. This means that the mass of one atom of $^{12}_{6}C$ is exactly 12 on the atomic mass scale. This is written as $m(^{12}_{6}C) = 12\,u$.

The alpha particle is also included in Table 3.1. This particle is a helium nucleus and may also be symbolized as $^{4}_{2}He^{2+}$. Now try Exercise 3B.

3.2 Isotopes

Introduction

Atoms with the *same number of protons* but with *different numbers of neutrons* are called **isotopes**. In other words, isotopes of an element possess the same atomic number but different mass numbers. For example, consider two isotopes of carbon, $^{12}_{6}C$ and $^{13}_{6}C$. Carrying out a 'particle count' for atoms of both isotopes, we have

Count	$^{13}_{6}C$	$^{12}_{6}C$
number of electrons	6	6
number of protons	6	6
number of neutrons	7	6

In summary, both isotopes of carbon contain six protons and six electrons, but atoms of carbon-13 are heavier because they each contain an extra neutron. (Try Exercise 3C.)

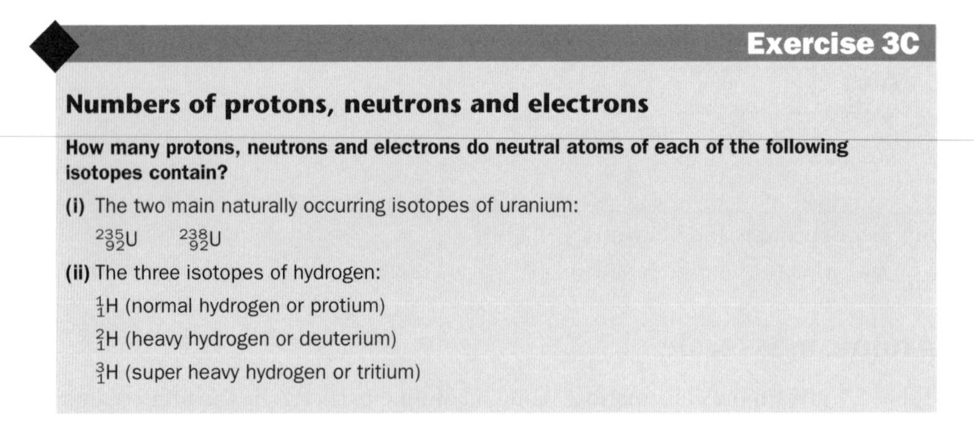

Exercise 3C

Numbers of protons, neutrons and electrons

How many protons, neutrons and electrons do neutral atoms of each of the following isotopes contain?

(i) The two main naturally occurring isotopes of uranium:

$^{235}_{92}U$ $^{238}_{92}U$

(ii) The three isotopes of hydrogen:

$^{1}_{1}H$ (normal hydrogen or protium)

$^{2}_{1}H$ (heavy hydrogen or deuterium)

$^{3}_{1}H$ (super heavy hydrogen or tritium)

Key points about isotopes

1. Isotopes have identical chemical reactions, although their compounds may have slightly differing physical properties (such as boiling point, or density). For example, hydrogen gas composed of normal hydrogen atoms ($^{1}_{1}H_2$) and hydrogen gas composed of deuterium atoms ($^{2}_{1}H_2$) both display identical chemical reactions and both burn in oxygen making normal water ($^{1}_{1}H_2O$) and heavy water ($^{2}_{1}H_2O$), respectively. However, the density and boiling point of $^{2}_{1}H_2$ is greater than that of $^{1}_{1}H_2$, and that of $^{2}_{1}H_2O$ is greater than that of $^{1}_{1}H_2O$.

2. Some isotopes are **radioactive** (discussed further in Chapter 21).

3. Atoms in a naturally occurring sample of the pure element (or of its compounds) may consist of one or more isotopes. The abundance of an isotope is defined as the percentage of atoms of that isotope in a sample of the element. The abundance of isotopes in nature (their **natural abundance**) is fixed. For example, about 75% of all chlorine atoms in nature are atoms of $^{35}_{17}Cl$ (Table 3.2). This percentage does not change appreciably whether the chlorine occurs as chlorine gas, as sodium chloride or as hydrochloric acid, or (in the case of chlorine compounds) whether they were mined in Norway, Australia or Canada (see Fig. 3.3).

Exercise 3D

Isotopes

(i) Use Table 3.2 to answer the following:

(a) Which isotope of uranium is the most abundant in nature?

(b) Which element exists only as one isotope in nature?

(c) Which isotope of hydrogen is radioactive?

(d) Which isotope of hydrogen does not occur in nature?

(ii) Bromine consists of two naturally occurring isotopes, $^{81}_{35}Br$ and $^{79}_{35}Br$, and 50.69% of all bromine atoms are of mass number 79. What percentage of bromine atoms has a mass number of 81?

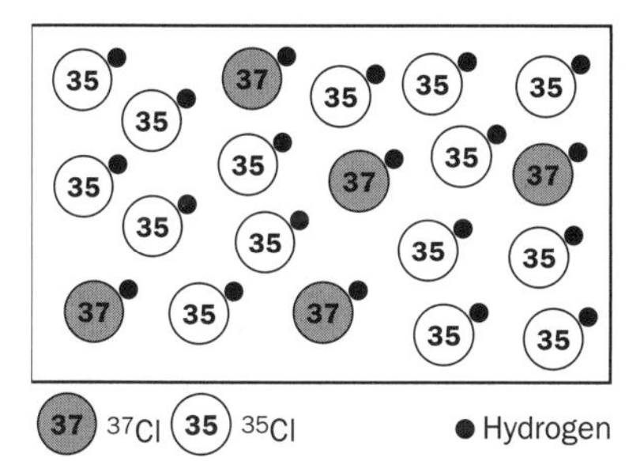

Fig. 3.3 A collection of hydrogen chloride molecules (HCl) as found in nature. About 75% of the HCl molecules contain the ^{35}Cl isotope.

Table 3.2 Isotopic and atomic data for selected elements

Element	Isotope	Natural abundance/%	Isotopic mass/u	Atomic mass of element/u
Hydrogen	$^{1}_{1}H$	99.985	1.0078	1.008
	$^{2}_{1}H$	0.01492	2.0140	
	$^{3}_{1}H$*	0	3.01605	
Chlorine	$^{35}_{17}Cl$	75.770	34.9689	35.45
	$^{37}_{17}Cl$	24.229	36.9659	
Uranium	$^{235}_{92}U$*	0.7205	235.0439	238.0
	$^{238}_{92}U$*	99.274	238.0508	
Carbon	$^{12}_{6}C$	98.893	12.0000	12.01
	$^{13}_{6}C$	1.107	13.0034	
Oxygen	$^{16}_{8}O$	99.759	15.9949	16.00
	$^{17}_{8}O$	0.0374	16.9991	
	$^{18}_{8}O$	0.2039	17.9992	
Fluorine	$^{19}_{9}F$	100	18.9984	18.9984

With the exception of fluorine, the atomic masses of the elements are expressed to four significant figures. Abundances and isotopic masses are taken from the *American Handbook of Physics*.
* shows that the isotope is radioactive.

Atomic mass of elements and of isotopes

The mass of one atom of an isotope on the atomic mass scale is called its **isotopic mass** m. Some isotopic masses are listed in Table 3.2. Remember that the mass of atoms is defined with reference to a 'standard atom', i.e. one atom of carbon-12 for which $m(^{12}_{6}\mathrm{C}) = 12.0000$ u.

Notice that isotopic masses (in atomic mass units) are very nearly equal in size to the mass number of an isotope. This is because the mass of the electrons, protons and neutrons are approximately 0, 1 and 1, respectively, on the atomic mass scale. Note also that the units of m and A are not the same – A is unitless.

If more than one isotope of an element exists in nature, the **average mass** of one atom of that element will be weighted towards the mass of the most abundant isotope. The average mass of one atom of an element in nature is called the **atomic mass** of that element, symbolized ***m*(element)**. The atomic mass of an element is worked out using the equation

$$\text{atomic mass} = \frac{(m_1P_1) + (m_2P_2) + (m_3P_3) + \ldots}{100} \qquad \textbf{(3.1)}$$

where m_1, m_2 and m_3 are the isotopic masses of isotopes 1, 2 and 3, and P_1, P_2 and P_3 are the percentage natural abundances of those isotopes.

An example: chlorine

Chlorine atoms in nature consist of those with mass number 35 and those with mass number 37. We start by taking the isotopic masses as equal to the mass numbers:

$$m(^{35}_{17}\mathrm{Cl}) = 35\,\mathrm{u}$$
$$m(^{37}_{17}\mathrm{Cl}) = 37\,\mathrm{u}$$

If there were equal numbers of both isotopes in nature, the atomic mass of chlorine would be the average of 35 u and 37 u, i.e. 36 u. In doing this averaging, we are really using equation (3.1) with $P_1 = 50\%$ and $P_2 = 50\%$:

$$\text{atomic mass of Cl} = \frac{(35 \times 50) + (37 \times 50)}{100} = 36\,\mathrm{u}$$

The actual abundances (Table 3.2) are 75.770% and 24.229%. Substituting these values in equation (3.1) gives

$$m(\mathrm{Cl}) = \frac{(35 \times 75.770) + (37 \times 24.229)}{100} = 35.48\,\mathrm{u}$$

Use of exact isotopic masses (Table 3.2) gives the accepted value for the atomic mass of chlorine:

$$m(\mathrm{Cl}) = \frac{(34.9689 \times 75.770) + (36.9659 \times 24.22)}{100} = 35.45\,\mathrm{u}$$

Exercise 3E

Working out atomic mass of oxygen

Using the isotopic masses and abundances shown in Table 3.2, confirm that the atomic mass of oxygen is 16.00 u.

For many purposes, the errors involved in using the mass numbers instead of the isotopic masses may be ignored.

Table 3.2 shows that there is only one isotope of fluorine that occurs naturally. Accordingly, for this element:

$$m(^{19}_{17}\text{F}) = m(\text{F})$$

Masses of molecules

The masses of molecules may also be expressed in atomic mass units. For example, use of the isotopic masses in Table 3.2 allows us to calculate the exact molecular mass of $^1H^{35}Cl$ as follows:

$$m(^1\text{H}^{35}\text{Cl}) = (1.0078\,\text{u}) + (34.9689\,\text{u}) = 35.9767\,\text{u}$$

Commercially available compounds usually contain atoms of isotopes of elements in the percentages of their natural abundances, i.e. such compounds are *not isotopically pure*. Therefore, in calculations of the yield of a chemical reaction, it is necessary to use the average atomic masses of each element; we cannot use the isotopic masses. For example, for HCl we would use $m(\text{H}) = 1.008\,\text{u}$ and $m(\text{Cl}) = 35.45\,\text{u}$. The average mass of the HCl molecule in nature is then

$$m(\text{HCl}) = (1.008\,\text{u}) + (35.45\,\text{u}) = 36.46\,\text{u} \approx 36.5\,\text{u}$$

3.3 Mass spectrometer

How a mass spectrometer works and what it is used for

The isotopic masses and abundance of isotopes in a sample of an element are found using an instrument called a **mass spectrometer** (Fig. 3.4).

A mass spectrometer causes atoms to lose electrons, a process called **ionization**. Loss of an electron means that the positive charge of the protons in the nucleus is not balanced by the remaining electrons, and the atom is now positively charged

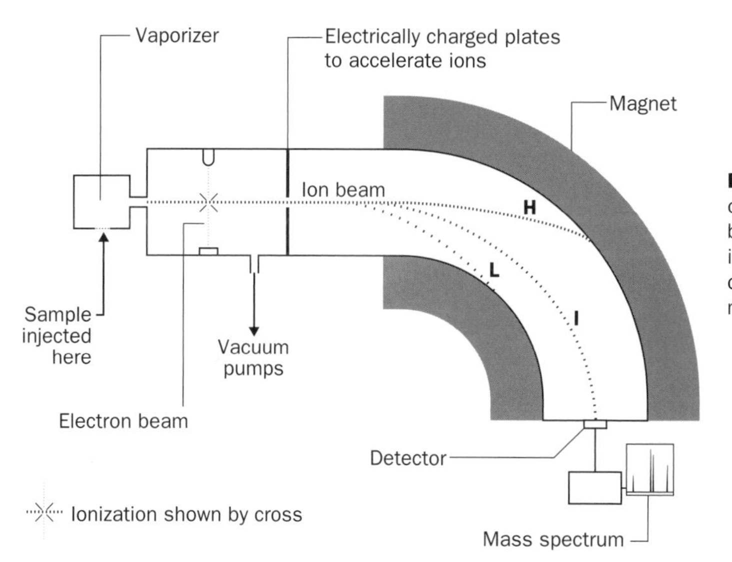

Fig. 3.4 Diagram of a mass spectrometer. This type of spectrometer is said to be 'magnetic scanning' because it uses a varying magnetic field to sort the ions according to their charge and mass. The paths of ions of light (L), intermediate (I) and heavy (H) mass are shown.

A magnetic scanning mass spectrometer. The spectrometer chamber is wedged between the poles of a powerful electromagnet, which bends the beam of ions, and whose magnetic strength may be varied. Because the curvature of the chamber is fixed, only ions of one m/e value reach the detector for a particular magnetic field strength.

(a **positive ion**). The positive ions are separated out in a mass spectrometer according to the ratio of the mass to charge m/e of the ion, where e is the relative charge of the ion (e.g. $e = +1$ for O^+ and $e = +2$ for Cl^{2+}) and m is the mass of the ion in atomic mass units. In the following discussion we will use the mass number of ions instead of the exact ion masses. For example,

$$\frac{m}{e}(^{35}Cl^+) = \frac{35}{1} = 35$$

$$\frac{m}{e}(^{37}Cl^{2+}) = \frac{37}{2} = 18.5$$

Stages involved in a mass spectrometer

As an example, suppose that neon gas (consisting of atoms of the isotopes neon-22, neon-21 and neon-20) has been injected into the mass spectrometer. The purpose of the main parts of the mass spectrometer are as follows:

1. The vaporizer is an oven which ensures that any sample is turned into a gas. (This is unnecessary with neon, which is a gas at room temperature.)
2. Electrons are fired at the gaseous sample. If one of these electrons collides with an atom, it may push out an electron from the atom, making a positive ion:

   atom + e^- (from gun) $\rightarrow$ ion + e^- (from gun) + e^- (from atom)

   See Fig. 3.5. For neon:

   $^{22}Ne(g) + e^- \rightarrow {}^{22}Ne^+(g) + e^- + e^-$

   $^{21}Ne(g) + e^- \rightarrow {}^{21}Ne^+(g) + e^- + e^-$

   $^{20}Ne(g) + e^- \rightarrow {}^{20}Ne^+(g) + e^- + e^-$

   (Some doubly charged ions, such as $^{20}Ne^{2+}$, will also be produced but these will be less common than singly charged ions because more energy is required to make them.)
3. The ions are accelerated using charged plates. A narrow beam of ions is allowed to enter into the rest of the spectrometer 'chamber'. The chamber is evacuated using

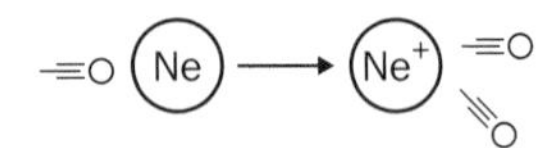

Fig. 3.5 Ionization of a neon atom by an electron (○).

powerful pumps which remove unionized atoms, and which ensure that there are very few molecules of air present which might collide with the ion beam.

4. The beam of ions would continue to travel in a straight line but, because a powerful electromagnet surrounds the chamber, the ions are bent downwards (Fig. 3.6). The degree of bend experienced by an ion depends upon its *m/e* value and upon the strength of the magnetic field. For ions of a particular *m/e* ratio, there will be only *one* value of magnetic field strength which will bend those ions so that they strike the detector.

5. By slowly adjusting the strength of a magnetic field (a procedure called **scanning**), ions of a particular *m/e* ratio are brought down to the ion detector one by one. The size of the electronic signal produced by the detector increases with the number of ions arriving per second. Since the atoms of all the neon isotopes are equally easy to ionize, the signal at each *m/e* value is proportional to the number of atoms of each isotope present in the original injected sample. A plot of signal intensity against *m/e* (called a **mass spectrum**), shows the relative abundance of isotopes for that element (Fig. 3.7).

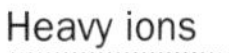

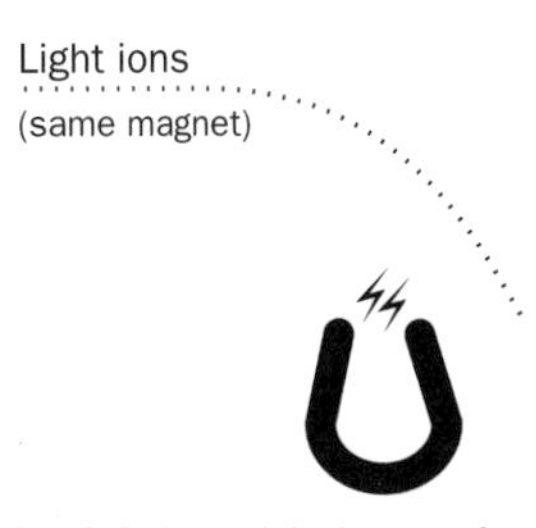

Fig. 3.6 A straight beam of ions (····) is deflected by a magnet. For the same magnetic field strength, lighter ions (small *m*) bend from their path more than heavier ones.

Application of mass spectrometry to molecules

If molecules are injected into a mass spectrometer, the molecules become ionized. For hydrogen chloride ($^{1}H^{35}Cl$):

$$^{1}H^{35}Cl + e^- \rightarrow (^{1}H^{35}Cl)^+ + e^- + e^-$$
$$m/e = 36$$

The ion obtained by ionizing the starting (i.e. parent) molecule is called the **parent ion**. Here, the parent ion is $(^{1}H^{35}Cl)^+$.

However, it often happens that the molecular ions produced in mass spectrometers have so much energy that they break up (i.e. **fragment**) as well as ionize. For example,

$$(^{1}H^{35}Cl)^+ \rightarrow {}^{1}H + {}^{35}Cl^+$$
$$m/e = 36 \qquad m/e = 35$$

As HCl gas consists of both $^{1}H^{35}Cl$ and $^{1}H^{37}Cl$ molecules its mass spectrum will include peaks at $m/e = 36$ (due to $(^{1}H^{35}Cl)^+$), at 38 ($(^{1}H^{37}Cl)^+$), at 37 ($^{37}Cl^+$) and 35 ($^{35}Cl^+$). Peaks derived from species containing ^{35}Cl, the most abundant chlorine

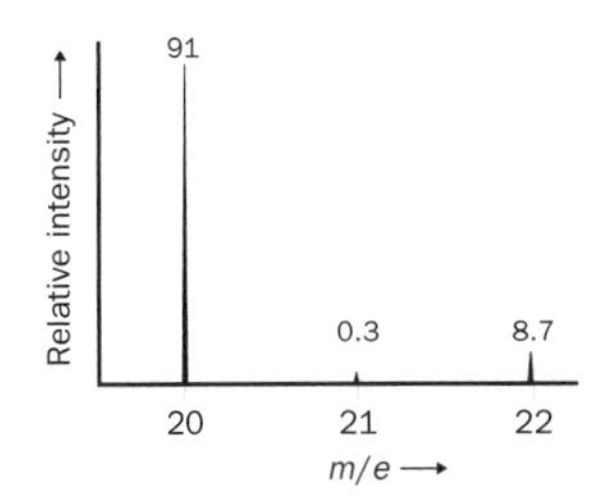

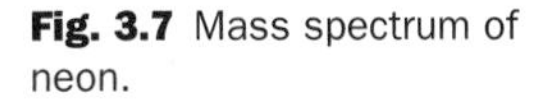

Fig. 3.7 Mass spectrum of neon.

Exercise 3F

Calculation of the atomic mass of neon from its mass spectrum

Use Fig. 3.7 to estimate the atomic mass of the element neon.

Exercise 3G

Mass spectra

(i) Calculate the *m/e* values of the following ions: $(^{12}C^{16}O_2)^+$, $(^{12}C^{1}H_3)^+$ and $(^{14}N_2)^{2+}$.

(ii) The mass spectrum of HF contains peaks at $m/e = 20$ and 19; explain this.

(iii) The mass spectrum of the organic compounds benzene (C_6H_6), nitrobenzene ($C_6H_5NO_2$) and phenol (C_6H_5OH) all contain an intense peak at $m/e = 77$ – why?

isotope, will be the most intense. Generalizing, the intensity and *m*/*e* value of peaks in a mass spectrum are very useful in helping chemists to decide what molecules were present in the original sample. For more information, see Appendix 2 on the website.

3.4 Electronic structure of atoms

Bohr model of the atom

In 1913 Niels Bohr (1885–1962) suggested that the electrons circling the nucleus could only possess certain energies. In other words, the energy of the electron is **quantized**. The key features of the Bohr model (Fig. 3.8) are:

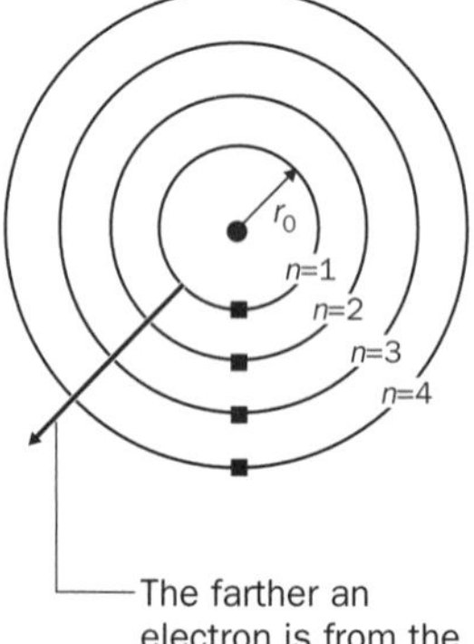

Fig. 3.8 The Bohr model of the atom. The electrons are shown as ■. The nucleus is represented as a dot.

1. Each shell of electrons lies at a definite radius (called the **Bohr radius**) from the nucleus. The electrons in each shell are at the *same energy*.
2. Only some values of Bohr radius are allowed. This means that only certain energy levels are allowed.
3. Each shell is labelled with a **principal quantum number** *n*. For example, the shell closest to the nucleus is labelled $n = 1$ and in the hydrogen atom its Bohr radius, r_0, is 52.9 pm from the nucleus. The higher the value of *n*, the higher the energy of the electron.
4. In order to change from one energy level E_1 to another E_2, the electron must gain (or lose) an amount of energy which is *exactly equal* to the difference in energy (symbolized ΔE) between the two levels:

   $$\Delta E = E_2 - E_1$$
5. Each shell can only hold a fixed number of electrons. For example, the first shell can hold up to two electrons and the second up to eight electrons. (The total number of electrons allowed in a shell is $2n^2$.)
6. Electrons will occupy as low an energy level as possible. This means that if electrons are fed into the shells of an atom the *lowest energy shells are filled first*.

BOX 3.1

Quantum theory

Most people have heard of the 'quantum theory' and the mere mention of it is guaranteed to stop most conversations in their tracks! Yet, many of the ideas behind the quantum theory are beautifully simple.

A quantity is quantized when it is only allowed to take on certain values. Time can take any value (2.5 seconds, 123.6787 7 seconds, 0.0143 seconds etc.) and is not quantized. On the other hand, the position of a painter on a ladder is quantized since he can only stand where there are rungs (Fig. 3.9).

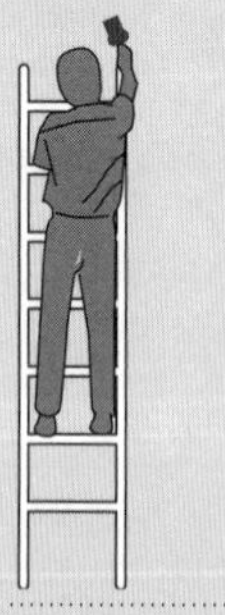

Fig. 3.9 Whether he knows it or not, the position of the painter on the ladder is **quantized**.

The idea that the energy of electrons is quantised was first put forward by Max Planck in 1900. One feature of the mathematics of the quantum theory is that it produces 'magic numbers' which can only take on certain values: such numbers are known as **quantum numbers**. For example, in the case of the hydrogen atom, the energy *E* of an electron is proportional to the **principal quantum number** *n* of the electron's shell. This quantum number can only take the values, 1, 2, 3, 4 and so on. The fact that the value of *n* is restricted to certain values, means that the value of *E* is also limited to certain values. Similar relationships apply to all atoms and in every case **the quantization of the energy of an electron is a natural consequence of the existence of quantum numbers.**

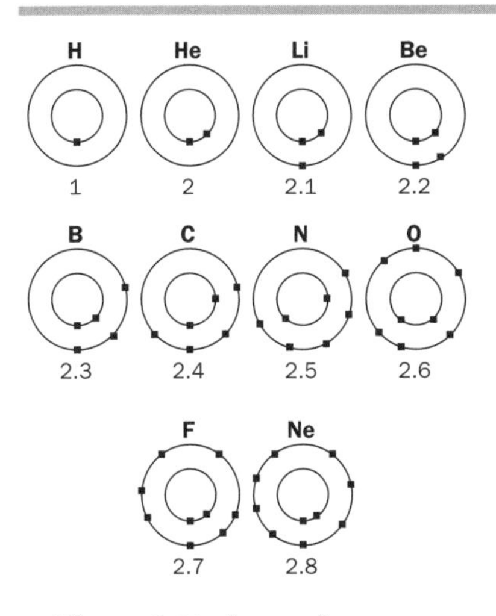

Fig. 3.10 The electronic structure of atoms for elements with atomic numbers 1–10.

Figure 3.10 shows the arrangement of electrons in shells (the so-called **electronic structure** or **electronic configuration**) for atoms of elements with atomic numbers from 1 to 10 in their lowest energy levels. Also shown is the shorthand for such electronic structures; for example, 2.6 (read as 'two dot six') means that there are two electrons in the first shell (for which $n = 1$) and six electrons in the second shell (for which $n = 2$). The configurations of atoms containing the same shells of electrons as in helium, neon (or other 'noble gases' such as argon) are sometimes abbreviated, as in the following examples:

Li 2.1 becomes [He].1

Mg 2.8.2 becomes [Ne].2

Cl 2.8.7 becomes [Ne].7

Table 3.3 gives electronic structures for elements with atomic numbers 1–20.

Table 3.3 The electronic structure for elements with atomic numbers 1–20. The last column will be filled in later on in this unit

Element and symbol	Atomic number	Bohr structure	s,p,d,f structure
Hydrogen H	1	1.	
Helium He	2	2. or [He]	
Lithium Li	3	2.1 or [He].1	
Beryllium Be	4	2.2 or [He].2	
Boron B	5	2.3 or [He].3	
Carbon C	6	2.4 or [He].4	
Nitrogen N	7	2.5 or [He].5	
Oxygen O	8	2.6 or [He].6	
Fluorine F	9	2.7 or [He].7	
Neon Ne	10	2.8 or [Ne]	
Sodium Na	11	2.8.1 or [Ne].1	
Magnesium Mg	12	2.8.2 or [Ne].2	
Aluminium Al	13	2.8.3 or [Ne].3	
Silicon Si	14	2.8.4 or [Ne].4	
Phosphorus P	15	2.8.5 or [Ne].5	
Sulfur S	16	2.8.6 or [Ne].6	
Chlorine Cl	17	2.8.7 or [Ne].7	
Argon Ar	18	2.8.8 or [Ar]	
Potassium K	19	2.8.8.1 or [Ar].1	
Calcium Ca	20	2.8.8.2 or [Ar].2	

3.5 Evidence for the existence of energy levels in atoms

You may have wondered what experimental evidence is available to support the idea that atoms contain electrons which lie in shells at definite energy levels. The truth is that the quantum theory of atoms was not arrived at quickly or easily, and a short account of such evidence will inevitably appear sketchy and incomplete. That said, there are two pieces of evidence which are difficult to explain without quantization, and we shall briefly review them as they introduce important ideas which will be useful later. They are **successive ionization energies** and **emission spectra**.

Successive ionization energies

Evidence for the existence of energy levels is provided by studies of ionization energies (ionization enthalpies) which may be measured electrically or spectroscopically. The **first ionization energy** (symbolized I_1 or $\Delta H^{\ominus}_{ion(1)}$) of an atom in the gaseous state is the minimum energy needed to completely remove the most easily removed electron from the attraction of the atomic nucleus. The outermost electrons possess more energy than the electrons in inner shells, and as a consequence are least tightly held by the nucleus and require less energy to completely break away from the attraction of the nucleus. It is for this reason that if energy is continually added to an atom, the *outer electrons are ejected first.*

Exercise 3H

Ionization energy

Write down an equation representing the fourth ionization energy of beryllium (Be). Is there a fifth ionization energy?

Taking the potassium atom as an example, the **first ionization energy** is the energy needed for the change:

$$\underset{2.8.8.1}{K(g)} \rightarrow \underset{2.8.8}{K^+(g)} + e^-$$

The **second ionization energy** (symbolized I_2 or $\Delta H^{\ominus}_{ion(2)}$) is the energy required for the change:

$$\underset{2.8.8}{K^+(g)} \rightarrow \underset{2.8.7}{K^{2+}(g)} + e^-$$

The third and successive ionization energies are defined in a similar way.

Figure 3.11 shows how the ionization energies for potassium change as we

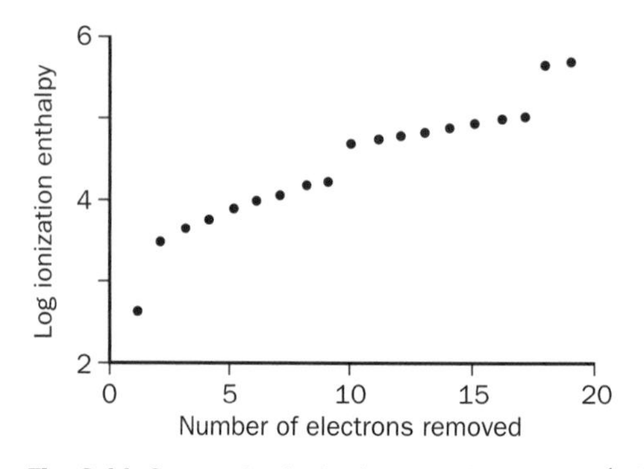

Fig. 3.11 Successive ionization energies (kJ mol^{-1}) for the potassium atom (2.8.8.1).

increase the number of electrons removed. (The log of the ionization energies has been plotted in order to compress the ionization energies so as to make the *y*-axis a reasonable size.)

Each point in Fig. 3.11 represents one electron. The obvious feature is that the points are clumped together into four groups. *This provides direct evidence that the 19 electrons in the potassium atom belong to four groups. These groups are the electron shells.* Moving from left to right in Fig. 3.11, the groups contain one, eight, eight and two electrons, respectively. This is the reverse of the familiar 2.8.8.1 configuration for potassium (*reverse* because in ionization the outer electrons are removed *first*). Although we have only plotted the ionization energies for potassium, similar patterns are seen for atoms of the other elements.

Emission spectra of atoms

Evidence for the quantization of electrons in atoms also comes from **emission spectra**. If a sample of atoms is heated, or exposed to light of a suitable wavelength, the outer electrons of some atoms may gain energy, and the atom is then said to be in an **excited state**. One way in which this energy may be lost is by **emission**, in which the electron loses energy as **light**. (The yellow colour seen when common salt is sprinkled into a flame is an example of such an emission.)

The frequency (ν, pronounced 'nu') of the emitted light depends upon the amount of energy lost by the electron, ΔE. The exact relationship between the two quantities is described by the **Planck equation**:

$$\Delta E = h\nu$$

where h is a universal constant known as the **Planck constant**.

If the emission of light from an element is studied using an instrument which allows the measurement of the frequency of light (a **spectrometer**), it is noticed that several frequencies of light are usually emitted: these frequencies make up the emission spectrum of the element. Just as importantly, however, there are some frequencies which are not emitted. This means that only some values of energy change ΔE are allowed. *The simplest way of explaining this is to assume that the energy of the electron in the atom may only possess certain values.* In other words, the energy of electrons in atoms is quantized. (More details about light emission, and calculations using the Planck equation are given in Chapter 20.)

3.6 More advanced ideas about electronic structure[a]

Sublevels of energy

According to the Bohr picture of the atom, all electrons in the same shell possess the same energy. More detailed experiments (also involving ionization energies and spectra) show that with the exception of electrons in the first shell this is not the

[a] These ideas are not essential for an understanding of Unit 4.

case, and the main energy levels of atoms possess **sublevels** of energy. The sublevels include the four types **s**, **p**, **d** and **f**:

s sublevels contain up to **2** electrons; the electrons are said to be 's electrons'.
p sublevels contain up to **6** electrons; the electrons are said to be 'p electrons'.
d sublevels contain up to **10** electrons; the electrons are said to be 'd electrons'.
f sublevels contain up to **14** electrons; the electrons are said to be 'f electrons'.

A full description of a sublevel also requires the principal quantum number n. For example, the third shell of electrons ($n = 3$) contains s, p and d sublevels of energy, and these are labelled 3s, 3p and 3d. The symbol $3s^1$ would show that there is one electron in the 3s sublevel, and $3p^6$ would show that the 3p sublevel is full.

The order in which the sublevels are filled is:

1s 2s 2p 3s 3p 4s 3d . . .
Low energy → High energy

Note that (rather oddly) the 4s sublevel is filled before the 3d. Using superscripts to show the maximum number of electrons in these levels, the sublevels may be written as follows:

$1s^2\ 2s^2\ 2p^6\ 3s^2\ 3p^6\ 3d^{10}\ 4s^2$ . . .

You are strongly advised to learn this before carrying on!

Note that despite the order of energies of the sublevels, when *writing* the electronic structure of an atom, sublevels with the same value of n are grouped together (i.e. we *write* 3d before 4s).

Think of the aluminium atom, with an electronic configuration of 2.8.3. A more detailed description of its energy levels is obtained by feeding all 13 electrons into the sublevels, giving

$1s^2\ 2s^2\ 2p^6\ 3s^2\ 3p^1$

or

$[Ne]\ 3s^2\ 3p^1$

Exercise 3I

Electron configuration of atoms of the elements: s,p,d notation

Complete the last column of Table 3.3 by entering the detailed electronic structures of the first 20 elements using the s,p,d notation.

Electronic structure of the 'first transition series' of elements

The elements titanium to copper are known as the **first transition series**. The electronic structures of these atoms (and of zinc and scandium) in their lowest energy states are shown in Table 3.4.

Table 3.4 Electronic structure of the first transition series of metals

Element and symbol	Atomic number	Bohr structure	s,p,d structure
Scandium	21	2.8.9.2 or [Ar].3	$1s^2\ 2s^2\ 2p^6\ 3s^2\ 3p^6\ 3d^1\ 4s^2$ or $[Ar]\ 3d^1\ 4s^2$
Titanium	22	[Ar].4	$[Ar]\ 3d^2\ 4s^2$
Vanadium	23	[Ar].5	$[Ar]\ 3d^3\ 4s^2$
Chromium	24	[Ar].6	$[Ar]\ 3d^5\ 4s^1$
Manganese	25	[Ar].7	$[Ar]\ 3d^5\ 4s^2$
Iron	26	[Ar].8	$[Ar]\ 3d^6\ 4s^2$
Cobalt	27	[Ar].9	$[Ar]\ 3d^7\ 4s^2$
Nickel	28	[Ar].10	$[Ar]\ 3d^8\ 4s^2$
Copper	29	[Ar].11	$[Ar]\ 3d^{10}\ 4s^1$
Zinc	30	[Ar].12	$[Ar]\ 3d^{10}\ 4s^2$

Notice that the 4s and 3d levels do not fill up in a completely regular fashion. This is because the $3d^5$ and $3d^{10}$ configurations are particularly stable, and chromium and copper would 'prefer' to possess these structures, even if the 4s sublevel remains unfilled.

The electronic structure of all the known elements is given in the back of this book.

Wave nature of electrons

Experiments have shown that electrons sometimes behave as if they were waves. The idea that the electron can behave as a particle or as a wave is called **wave–particle duality**.

Think of a water wave produced by dropping a stone in a pond. Freeze this picture in your mind and ask yourself the question: where is the water wave? The answer, of course, is that the water wave consists of many ripples which are spread out upon the surface of the water. Some ripples are more pronounced than others, but the wave does not belong to a single spot.

This idea of an 'uncertainty' in the position of a wave is carried over into the atom. If electrons behave like waves we cannot pinpoint their exact position. All we can do is to consider the **probability** of finding the electron at a point or in a region of space.

Orbitals

Suppose we were able to take a snapshot photograph of the 1s electron in the hydrogen atom at an instant. A fraction of a second later we take another snapshot. By taking thousands of such snapshots, and superimposing all of them on a single frame we would end up with a single picture showing thousands of points where the electron had been located. The greater the number of points in a region, the greater is the probability of finding the electron there in the future.

Such snapshots would show that the **volume** in which it is 90% probable that the s electron would be found is spherical. The volume is given the name **atomic orbital**, usually abbreviated to **orbital**. The orbital of a 1s electron is called a 1s-orbital, and it possesses a radius of about 100 pm (100×10^{-12} m). However, the *single* radius at which the 1s electron is most likely to be found is at 52.9 pm (the Bohr radius of the $n = 1$ shell). But – and this is the crucial difference between the wave model and the Bohr model – the wave model states there is always a chance that the electron will be somewhere outside this radius (Fig. 3.12).

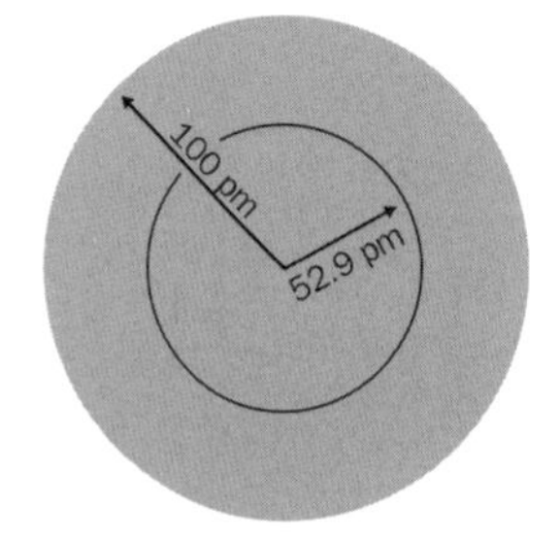

Fig. 3.12 The orbital for a 1s electron in the hydrogen atom. The radius of the sphere in which it is 90% likely that the electron will be found, is about 100 pm. The single radius at which the 1s electron is *most likely* to be found is a distance 52.9 pm from the nucleus. This may be compared with the Bohr theory of the atom, where it was assumed that the electron was *certain* to be found at a radius of 52.9 pm.

All s orbitals are spherical. 2s orbitals have larger radii than 1s orbitals.

Similarly, orbitals containing p electrons are termed **p orbitals**. There are three types of p orbital (labelled p_x, p_y and p_z), which are normally of equal energy but which have different directions in space. The shape of a p orbital is often described as a 'dumb-bell' (Fig. 3.13). Orbitals containing d electrons are termed **d orbitals**. There are five types of d orbitals, and each is normally of equal energy. The shape of d orbitals is complicated and will not concern us here.

Each orbital (whether $2p_z$, 1s or 2s etc.) in an atom holds **a maximum of two electrons**. If there are two electrons present, the electrons must be spinning opposite ways (one anticlockwise, one clockwise) and we say that the electrons are **paired**. This is shown using a box for the orbital and arrows for the electrons. Up and down arrows confirm that their spins are opposite, as follows:

| ↑↓ |

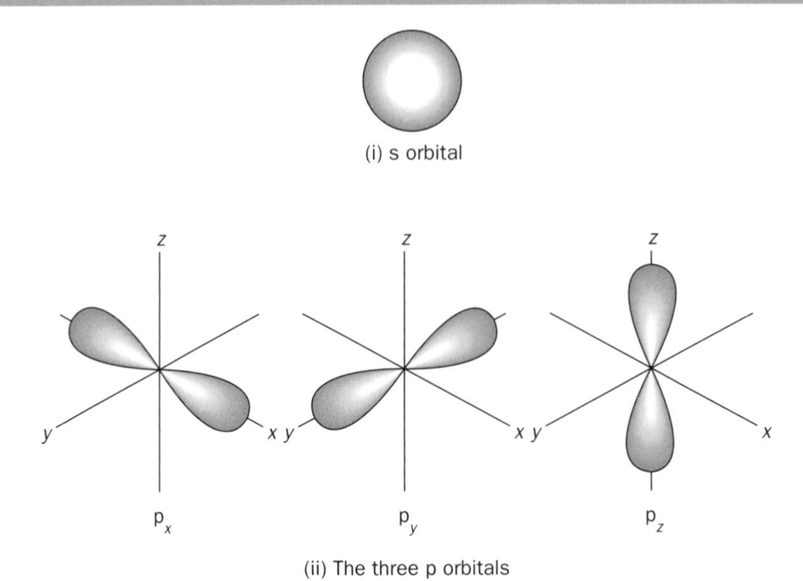

Fig. 3.13 The shape of an s orbital and of the three types of p orbitals.

Orbitals and sublevels

We now return to the order in which the sublevels of electrons in atoms are filled,

$1s^2 \quad 2s^2 \quad 2p^6 \quad 3s^2 \quad 3p^6 \quad 4s^2 \quad 3d^{10} \ldots$

where the superscripts show the maximum number of electrons in each sublevel. Electrons in a 1s orbital occupy the 1s energy sublevel, electrons in a 2s orbital occupy the 2s sublevel and electrons in a 3s orbital occupy the 3s sublevel. Electrons in any of the three p orbitals are normally equal in energy and occupy the 2p sublevel. There are five d orbitals: electrons in these orbitals will occupy the 3d sublevel.

The use of 'boxes' may help make this clearer. In the aluminium atom (electronic configuration $1s^2\,2s^2\,2p^6\,3s^2\,3p^1$) the 13 electrons occupy the following orbitals:

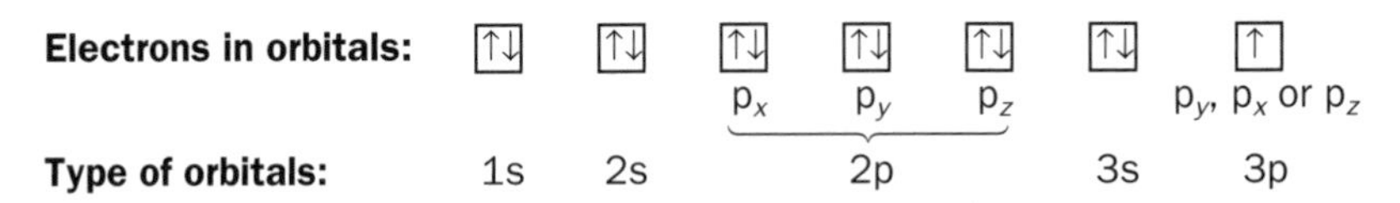

where the $3p^1$ electron occupies either the $3p_y$, $3p_x$ or $3p_z$ orbital, all of which are of equal energy.

The 1s, 2s, 2p and 3s orbitals in the phosphorus atom ($1s^2\,2s^2\,2p^6\,3s^2\,3p^3$) are filled in the same way as for aluminium, with the complication that it is unclear whether the 3p electrons feed into two of the 3p orbitals:

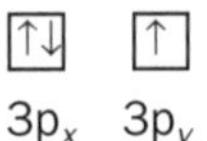

or whether they enter all three p orbitals:

↑ ↑ ↑

$3p_x$ $3p_y$ $3p_z$

It turns out that the second distribution is the correct one – to minimize repulsion, electrons 'prefer' to occupy orbitals singly. This is known as **Hund's rule**.

Exercise 3J

Detailed electron configuration of atoms of the elements

By following the pattern for Al, write out (i) the electronic 'box structures' for atoms with atomic numbers 1–10 and (ii) for the transition metals Mn, Fe and Cu.

Revision questions

3.1. Which of the following is false?

(i) The atoms of an element all have the same atomic number.

(ii) 1 u = one-twelfth of the mass of one atom of $^{13}_{6}C$.

(iii) mass number = atomic number + number of neutrons.

(iv) The atoms of an element are identical, but different from atoms of other elements.

(v) A mass spectrometer detects both ions and neutral atoms.

3.2. The following table gives information on the four naturally occurring isotopes of sulfur:

Isotope	Number of neutrons	Number of electrons	Number of protons	Natural abundance /%
$^{32}_{16}S$				95.0
$^{33}_{16}S$				0.76
$^{34}_{16}S$				4.2
$^{36}_{16}S$				0.021

(i) Fill in the second, third and fourth columns of the table.

(ii) Estimate the atomic mass of sulfur to two significant figures. Why is your answer an estimate?

3.3. Write equations showing the formation of Mg^{+} and Mg^{2+} in the chamber of a mass spectrometer. What are **(i)** the m/e values and **(ii)** the 2.8.1-type electronic configurations, of these ions?

3.4. Explain the following, writing equations for the formation of any ions in the mass spectrometer chamber:

(i) The mass spectrum of fluorine gas (F_2) contains two peaks, at m/e = 19 and 38.

(ii) The mass spectrum of hydrogen cyanide (HCN) contains peaks at m/e = 27, 26 and 13.5.

(iii) The mass spectrum of chlorobenzene (C_6H_5Cl) contains peaks at m/e = 77, 112 and 114.

(iv) A strong line at m/e = 28 is observed when an attempt was made to run the mass spectrum of a compound. (*Hint* it is suspected that the spectrometer is leaking.)

(v) A mixture of methane (CH_4) and chlorine gas (Cl_2) yielded a mass spectrum with peaks at 70, 72, 74, 16, 52 and 50.

3.5. Write equations for the burning of normal hydrogen gas ($^{1}_{1}H_2$) and deuterium gas ($^{2}_{1}H_2$) in oxygen. Would you expect the density of normal and heavy water to be the same?

3.6. Use Table 3.2 to calculate the exact mass in **(i)** atomic mass units, **(ii)** grams, of **(a)** a molecule of heavy water ($^{2}_{1}H_2{}^{16}_{8}O$) and **(b)** an average water molecule in nature (work to four significant figures).

3.7. Briefly explain why atomic emission spectra provide experimental evidence for the existence of electron energy levels (shells) in atoms.

3.8. The successive ionization energies (in kJ mol^{-1}) for argon are 1521, 2666, 3931, 5771, 7328, 8781, 11 996, 13 842, 40 761, 46 188, 52 003, 59 654, 66 201, 72 920, 82 570, 88 620, 397 710 and 426 910. Plot the log of the ionization energy against the number of electrons removed (as in Fig. 3.11). Why does your graph support the electronic structure of argon as being 2.8.8? What is the total amount of energy (in kJ mol^{-1}) required to remove all 18 electrons from the argon atom?

3.9. What is an orbital? Comment upon the size of an orbital in which it is *certain* that an electron will be found.

3.10. Write down the electronic configuration of **(i)** Cl, **(ii)** Cl^{-} (a chlorine atom which has gained an electron), **(iii)** Cl^{+}, **(iv)** He, **(v)** Ne and **(vi)** Cu^{2+}, using the 2.8.1 notation. What is special about the electronic configurations **(ii)**, **(iv)** and **(v)**?

3.11. Repeat question 3.10 using the s,p,d,f notation. Write out the structure for **(iii)** using the electrons in boxes notation.

Extension material to support this unit is available on our website at **http://www.palgrave.com/foundations/lewis.**

Bonding Between Atoms

Contents

Objectives

- ▶Explains why elements react
- ▶Describes the extreme types of bonding between atoms
- ▶Links the properties of compounds to the bonding within those compounds
- ▶Extends the description of bonding in compounds to include 'intermediate' bonding types

4.1 Why atoms combine

You have already seen that atoms of different elements react together to form compounds. The forces that hold these atoms together in compounds are called **chemical bonds**. In order to understand the nature of these bonds, perhaps we should first ask the question: Why do atoms bother to combine in the first place?

Noble gases

Clues to the answer to this question may be found by considering a group of elements known as the **noble** (or **inert**) gases: helium (He), neon (Ne), argon (Ar), krypton (Kr), xenon (Xe) and radon (Rn). Inert means 'inactive' or 'idle', and describes the chemical reactivity of these elements. They are unreactive and for this reason are said to be **chemically stable**. He, Ne and Ar do not react with any other elements, whereas Kr and Xe react only with highly reactive fluorine. What is it that makes them so unwilling to react?

The electronic structures of the atoms of these elements are listed in Table 4.1. With the exception of helium, the atoms have eight (or ns^2 np^6, where n is a whole number greater than one) electrons in the outer shells of their atoms. The arrangement is called a **stable octet** of electrons. There seems to be a stability associated with this arrangement, and with the unique electron configuration of helium, which makes the gases unreactive.

Lewis symbols

In 1916, Kossel and Lewis proposed that atoms of elements react with each other in order to achieve the stable electron arrangements of the noble gases. In order to

Table 4.1 Electronic structures of the noble gases*

Inert gas	Electronic structure (Bohr model)	Electronic structure (s,p,d,f notation)
He	2	$1s^2$
Ne	2.8	[He] $2s^2\ 2p^6$
Ar	2.8.8	[Ne] $3s^2\ 3p^6$
Kr	2.8.18.8	[Ar] $3d^{10}\ 4s^2\ 4p^6$
Xe	2.8.18.18.8	[Kr] $4d^{10}\ 5s^2\ 5p^6$
Rn	2.8.18.32.18.8	[Xe] $4f^{14}\ 5d^{10}\ 6s^2\ 6p^6$

* To save space, the convention [noble gas] is used and refers to the electronic configuration of that gas. So, for example, [He] replaces $1s^2$.

obtain these stable arrangements, atoms may gain, lose or share electrons. These processes are generally represented by using **Lewis symbols** – an atom of an element is represented by the symbol for that element and the electrons in *the outer shell* of the atom are shown as dots, small circles or crosses. For example, the Lewis symbol for the sodium atom is

Na°

The symbol for sodium is Na and the electronic structure of the sodium atom is 2.8.1, so there is **one** electron in the last electron shell which we represent by one dot (or cross – the choice is yours!).

4.2 Ionic bonding

When an atom of sodium reacts with an atom of chlorine, sodium chloride is formed. The Lewis symbols for sodium and chlorine are

Na° ×× × **Cl** ×× ××

2.8.1 2.8.7

Although the complete electron arrangements of each atom are written underneath each Lewis symbol, *only the outer electrons* are shown by dots and crosses because they are the only ones that take part in bonding.

When a sodium atom reacts with a chlorine atom, the sodium atom loses one electron to chlorine (to obtain the stable electron arrangement of neon), while the chlorine atom, in gaining an extra electron, attains the stable electron arrangement of argon. This can be represented in an equation:

Na° + ×**Cl**×× (×× ××) ⟶ **Na⁺** + [×°**Cl**××]⁻ (×× ××)

2.8.1 2.8.7 2.8 2.8.8

Before the reaction the neutral sodium atom had 11 electron and 11 protons. After giving an electron to chlorine, the sodium atom has 10 electrons and 11 protons, so it is not electrically neutral – it carries an overall positive charge equal to the charge of one proton (+1). The new species is written Na^+ and is called a sodium **ion**

Exercise 4A

Lewis symbols

Write Lewis symbols (using dots or crosses) for the following neutral atoms. The number of electrons in each atom is shown in brackets:

(i) F (9)
(ii) K (19)
(iii) Be (4)
(iv) S (16)
(v) P (15)
(vi) Ca (20)
(vii) C (6)
(viii) Al (13)
(ix) Ar (18)
(x) H (1).

BOX 4.1

Ions

Ions are atoms (or groups of atoms) that carry electrical charges. Examples of ions are K^+, F^-, CO_3^{2-}, N^{3-}, Ca^{2+} and Al^{3+}.

Positively charged ions (e.g. K^+, Ca^{2+} and Al^{3+}) are called **cations**, whereas negatively charged ions (e.g. F^-, CO_3^{2-} and N^{3-}) are called **anions**.

(see Box 4.1). The chlorine atom that has accepted an electron now has 17 protons and 18 electrons, so it has an overall charge of -1, is written as Cl^- and is called a chloride ion.

Note that once a sodium atom has given an electron to a chlorine atom, that electron is indistinguishable from the other electrons present in the outer shell of the chloride ion. All eight electrons now 'belong' to the chloride ion. Crosses and dots are used to represent the electrons, simply to show which atoms they originated from. The use of a cross or a dot, does not imply that the electrons are different.

These changes are shown below:

Before

Na atom	Cl atom
electrons = $11 \times (-1)$	electrons = $17 \times (-1)$
protons = $11 \times (+1)$	protons = $17 \times (+1)$
Neutral	Neutral

After

Na^+ ion	Cl^- ion
electrons = $10 \times (-1)$	electrons = $18 \times (-1)$
protons = $11 \times (+1)$	protons = $17 \times (+1)$
Charge = +1	Charge = -1

Sodium chloride is therefore composed of Na^+ and Cl^- ions in a 1:1 ratio. The compound is called an **ionic** compound (because it is composed of ions), and the bonding in the compound is termed **ionic**. Ionic compounds are formed whenever one atom (or group of atoms) transfers electrons to another. Note that, when sodium and chlorine react, they both end up with the electron configuration of a noble gas *but they do not change into noble gases*; they are still the same elements. This is because the identity of an atom depends on the number of protons in its nucleus and, as you can see, these remain the same for sodium and chlorine both before and after the reaction.

The 'ordinary' chemical equation for this reaction tells us that the product of the reaction is sodium chloride, but gives no information about the bonding in the compound formed:

$$2Na(s) + Cl_2(g) \rightarrow 2NaCl(s)$$

For this reason, it is better written as

$$2Na(s) + Cl_2(g) \rightarrow 2Na^+, 2Cl^-(s)$$

Example 4.1

Magnesium reacts with fluorine to form magnesium fluoride. Write an equation to describe the bonding using Lewis symbols.

▶Answer

The bonding in the compound can be shown by writing an equation using Lewis symbols, as follows:

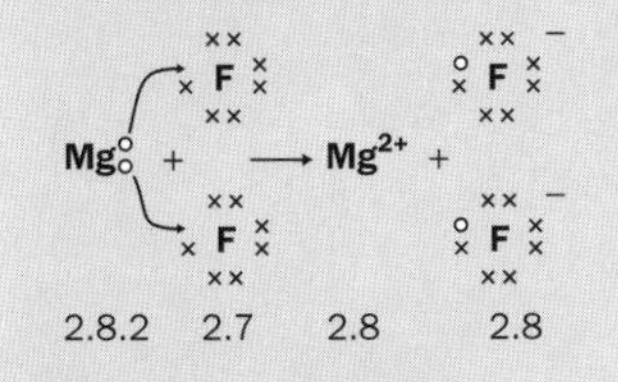

▶Comment

Magnesium needs to lose two electrons to attain the stable electronic configuration of neon, so two fluorine atoms accept one electron each to also gain the same electron arrangement as neon. Magnesium ions (10 electrons, 12 protons) with a charge of 2+ are formed, while fluorine forms F^- ions. Magnesium fluoride is an ionic compound in which the Mg^{2+} ions and F^- ions are in a ratio of 1:2.

The best equation that can be written for this reaction is therefore

$$Mg(s) + F_2(g) \rightarrow Mg^{2+}, 2F^-(s)$$

Exercise 4B

Reaction of potassium with chlorine

The equations using Lewis symbols provide us with more information than the usual equations for chemical reactions. The reaction of potassium metal with chlorine gas is shown in Fig. 4.1.

(i) Write an equation, using Lewis symbols, that shows the changes in electronic structure that occur when potassium atoms react with chlorine atoms.

(ii) Write the balanced equation for the reaction of potassium with chlorine gas (Cl_2).

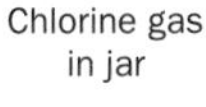

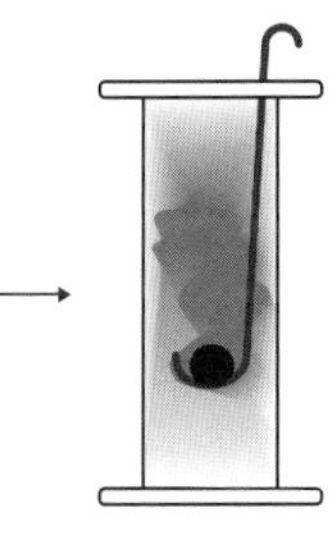

Fig. 4.1 The reaction of potassium metal with chlorine gas.

Exercise 4C

Writing equations for the formation of ionic compounds using Lewis symbols

Write equations for the formation of the following compounds, using Lewis symbols. Use Table 3.3 for electronic arrangements:

(i) magnesium oxide (MgO) from Mg and O atoms

(ii) calcium fluoride (CaF_2) from Ca and F atoms

(iii) aluminium chloride ($AlCl_3$) from Al and Cl atoms

(iv) lithium oxide (Li_2O) from Li and O atoms

(v) aluminium oxide (Al_2O_3) from Al and O atoms.

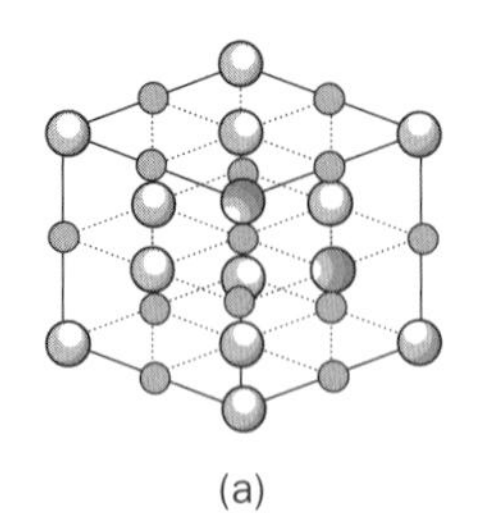

(a)

Arrangement of ions in sodium chloride

In a sodium chloride crystal, the Na^+ and Cl^- ions are arranged in a *giant lattice* structure. The 'building brick' of this structure is a **unit cell** is as shown in Fig 4.2. The larger spheres represent Cl^- ions, whereas the smaller spheres represent Na^+ ions. A crystal of sodium chloride consists of many billions of these unit cells stacked together in the lattice.

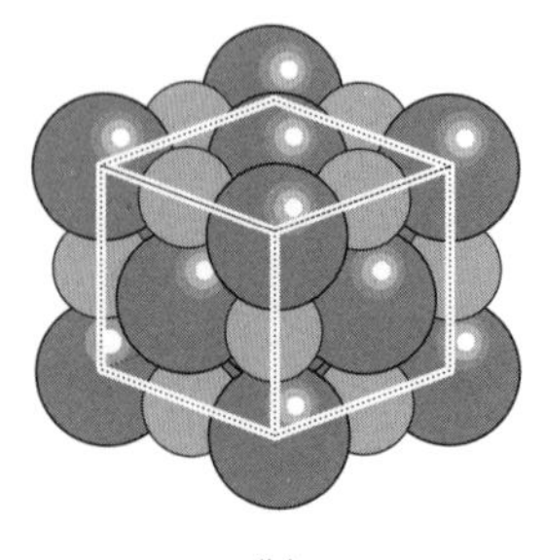

(b)

Fig. 4.2 The unit cell of sodium chloride: (a) ball-and-stick model; (b) space-filling model.

Cubic salt crystals. An orderly arrangement of Na^+ and Cl^- ions gives rise to this shape.

In the unit cell, each sodium ion is surrounded by six chloride ions and each chloride ion is surrounded by six sodium ions. The ions are held together by the attraction of their opposite charges (**electrostatic forces**). These forces are very strong. **X-ray diffraction** allows us to 'see' ions in an ionic compound, as described in Box 4.2.

All ionic substances consist of ions arranged in a giant lattice. Different ionic structures have their ions arranged in different patterns. The sodium chloride structure is one of the simplest arrangements.

BOX 4.2

Electron density map for sodium chloride

How do the ideas of Kossel and Lewis fit in with the modern view of the arrangement of electrons in an atom? The modern theory teaches that electrons have a wave-like nature. In regions where there is a high probability of finding the electron, there is a high concentration of electrical charge and we say that the **electron density** is large. By similar reasoning, electron density is small where there is a low probability of finding the electron. Through using a technique known as X-ray diffraction, it is possible to obtain electron density contour maps of ionic substances. Figure 4.3 shows such a map for sodium chloride – each line connects points with the same electron density, and the nuclei of each atom are represented by dots at the centre of each ion. The outermost contours have a squared shape because the ions are attracted to ions of opposite charge around them. Notice that there is a space between the nuclei where there are no contours; this is a characteristic feature of the electron density map for an ionic substance.

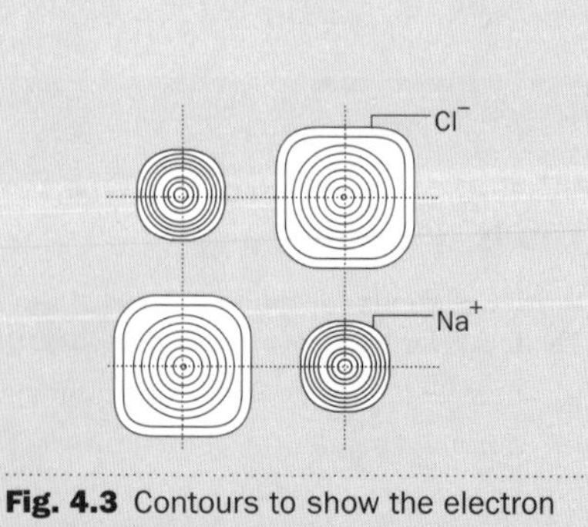

Fig. 4.3 Contours to show the electron density distribution around sodium and chloride ions.

Properties of ionic compounds

Ionic compounds have certain properties common to all of them:

1. High melting and boiling points (they are non-volatile)

They are solids at room temperature. When a substance melts, then boils, its particles separate. A great deal of energy is needed to break down the lattice structure because the ions are attracted together by strong forces. The structure must therefore be heated to a high temperature before it melts.

2. Soluble in water

Water contains molecules that have one end positively charged and the other end negatively charged (the charge separation is called a **dipole**). When ionic compounds are put into water, the water molecules are able to break down the crystal lattice by attracting the ions in the lattice and pulling them into the solution. In this way, the lattice breaks up and the ions mix in with the water molecules – the substance *dissolves*. Water is a **polar solvent** and tends to dissolve ionic compounds. The positive end of a water molecule can attach itself to an anion in the unit cell and pull it away from the structure. Similarly, the negative end of the water molecule can attract away a cation. This is shown in Fig. 4.4.

3. Conduct electricity when molten or in aqueous solution

In order for a substance to conduct electricity, it must possess **charged particles** that can *move* (in a metal, mobile electrons can carry the electric current). Solid ionic compounds *do not conduct electricity* – the ions are held firmly by strong electrostatic forces and cannot move. When the substance is melted or dissolved in water, however, the ions move freely and can carry an electric current. An aqueous solution, or melt, of an ionic substance that behaves in this way, is called an **electrolyte**.

4. Ionic crystals shatter easily

When a force is applied to the crystal, the layers of ions in the crystal structure can 'slip' so that similarly charged ions are next to one another. The like charges of the ions repel one another and the crystal structure shatters, see Fig. 4.5.

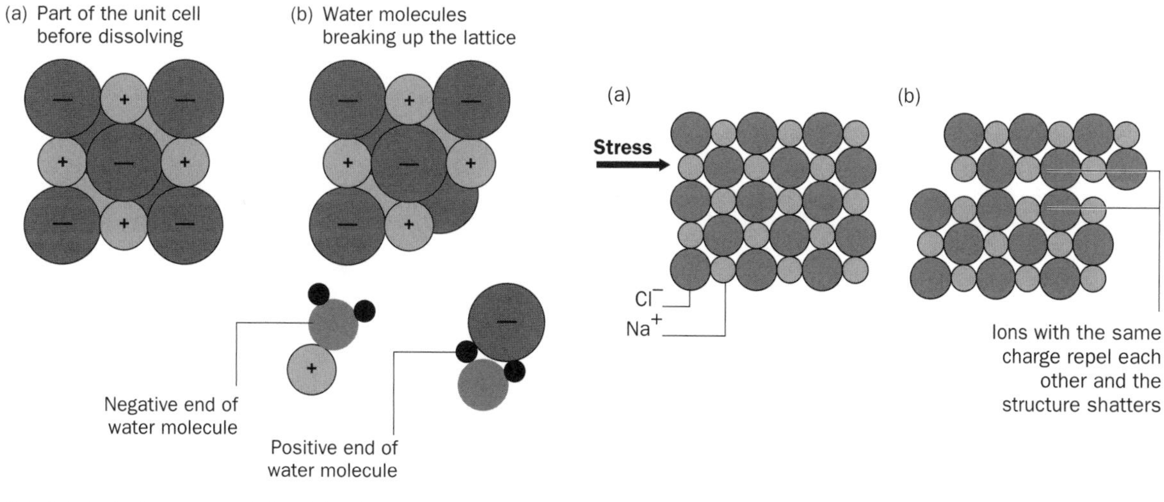

Fig. 4.4 An ionic compound dissolving in water.

Fig. 4.5 The shattering of an ionic crystal: (a) before, (b) after.

When ionic bonding is likely to occur

Ionic bonding occurs between elements when the energy required to remove the outer shell electrons (**the ionization energy**) of one of those elements is relatively low. Elements with such low ionization energies are *metals*; they generally contain *no more than three electrons* in the outer shell of their atoms. Therefore, compounds containing a metal tend to be ionic. Non-metals generally contain more than three electrons in the outer shells of their atoms. Note that, although hydrogen contains one electron in the first shell of its atoms, the ionization energy for that atom is very high because it is close to, and highly attracted to, the positive nucleus. Hydrogen is therefore classified as a non-metal and most of its compounds are not ionic.

BOX 4.3

Hydrogen ion

For the process

$$H(g) \rightarrow H^+(g) + e^-$$

the ionization energy is very high. Its ionization energy is 1310 kJ mol^{-1} (higher than the first ionization energy for xenon at 1170 kJ mol^{-1}!). The hydrogen ion can be formed only when it is **solvated**. Solvation occurs when the hydrogen ion attaches itself to a solvent molecule; the energy released when the hydrogen ion is solvated makes it 'worthwhile' ionizing hydrogen to H^+. The most common solvent with which to solvate H^+ is water; the hydrogen ion in water should be written as H_3O^+ or $H^+(aq)$ after the process

$$H^+(g) + H_2O(l) \rightarrow H_3O^+$$
$$\text{or } H^+(g) + H_2O(l) \rightarrow H^+(aq)$$

This process is known as **hydration** and the ion H_3O^+ is commonly called the **hydronium ion**. Remember that although you may see the symbol H^+, $H^+(aq)$ or a reference to the hydrogen ion, in water all of these terms refer to H_3O^+.

4.3 Covalent bonding

What type of bonding occurs if it takes too much energy to remove electrons from one of the atoms taking part in bonding? This situation occurs when two or more non-metals combine – their atoms have to *share* electrons to achieve the stable electronic configuration of a noble gas. This type of bonding is called **covalent** bonding.

Consider chlorine gas; the Lewis symbol for chlorine is:

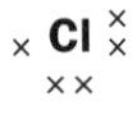

```
  ××
× Cl ×
     ×
  ××
2.8.7
```

If two chlorine atoms share one electron, they achieve the stable electron arrangement of argon. We can represent this on a diagram called a **Lewis structure**:

```
  ××   oo
× Cl × Cl o
×    o    o
  ××   oo
2.8.8 2.8.8
```

A **molecule** of chlorine is formed – there are no charged particles. The electrons of one chlorine atom are usually represented by dots and those of the other by crosses, to enable you to see where they come from. *Remember, though, that once the electrons are involved in the covalent bond there is no way to distinguish between them.* Chlorine gas therefore, chemical formula Cl_2, exists as a **diatomic** molecule because this allows its atoms to achieve the stable electronic arrangement of a noble gas. The **structural formula** of the molecule is written as

Cl——Cl

where '——' represents a pair of shared electrons, i.e. a **covalent bond**.

Example 4.2

Ammonia (NH_3) is a covalent substance. Draw a Lewis structure to represent the bonding in ammonia.

▶ Answer

Nitrogen, which has the electronic structure 2.5 shares its outer electrons with three hydrogens (electron arrangement 1). In this way nitrogen, and the three hydrogens, achieve the electronic structure of noble gases:

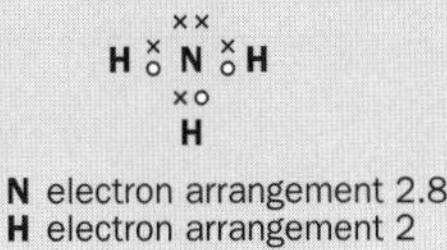

N electron arrangement 2.8
H electron arrangement 2

▶ Comment

The structural formula of ammonia is written:

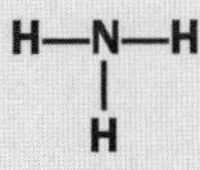

(remember, —— means 'a pair of shared electrons')

Example 4.3

Molecules of tetrachloromethane (often called carbon tetrachloride) have the formula CCl_4. What is the Lewis structure of the molecule?

▶ Answer

Carbon has the electronic structure 2.4 and chlorine the electronic structure 2.8.7. One carbon atom shares electrons with four chlorine atoms so that all atoms achieve the electronic structure of a noble gas:

```
         xx
        x Cl x
        x    x
   xx    xo    xx
 x Cl x  C  x Cl x
 x    o     o    x
   xx    xo    xx
        x Cl x
        x    x
         xx
```

C electron arrangement 2.8
Cl electron arrangement 2.8.8

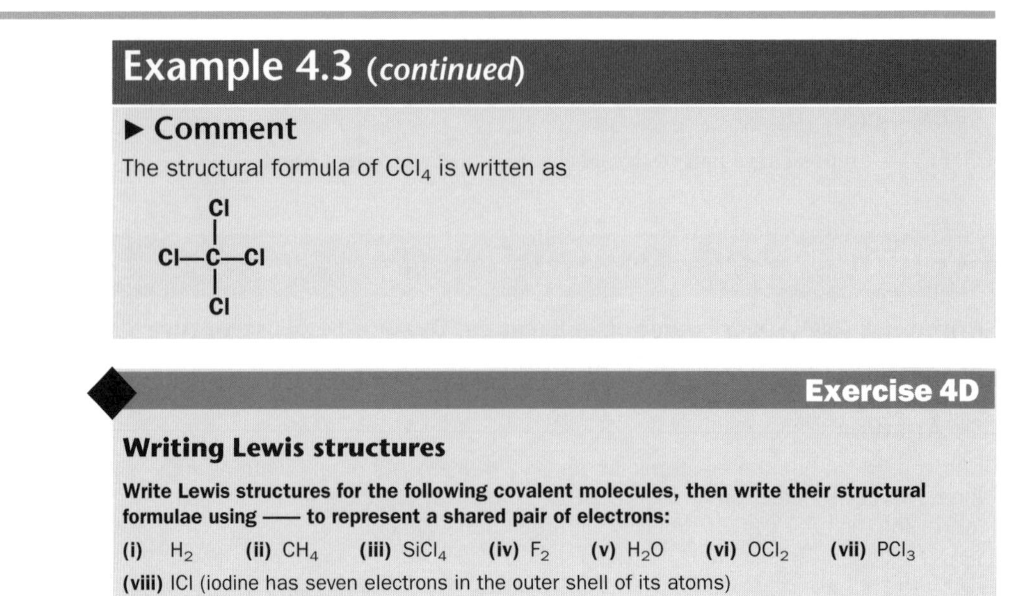

Example 4.3 (*continued*)

▶ Comment

The structural formula of CCl_4 is written as

```
     Cl
     |
Cl—C—Cl
     |
     Cl
```

Exercise 4D

Writing Lewis structures

Write Lewis structures for the following covalent molecules, then write their structural formulae using —— to represent a shared pair of electrons:

(i) H_2 **(ii)** CH_4 **(iii)** $SiCl_4$ **(iv)** F_2 **(v)** H_2O **(vi)** OCl_2 **(vii)** PCl_3

(viii) ICl (iodine has seven electrons in the outer shell of its atoms)

(ix) HBr **(x)** $CHCl_3$ (C is the central atom).

Valency of an atom

Can you now see where we got the idea of 'hooks' to describe combining power, or valency, in Chapter 2? *The number of 'hooks' that an atom or group possesses is actually a simple way of describing how many electrons that atom or group needs to give away, accept or share in order to obtain the stable configuration of a noble gas.* Carbon, for example, has the electronic structure 2.4. We describe it as having a valency four, or four 'hooks'; this is just another way of saying that it needs to share four more electrons to achieve the stable electronic structure of neon, namely 2.8.

Multiple bonds

Sometimes, more than one pair of electrons is shared. Consider oxygen gas (O_2):

```
××× ○○
O ○× O
××○ ○○
```

Here, each oxygen shares *two pairs* of electrons to achieve the electron configuration of neon. The structural formula of this oxygen molecule is written

O═O

A **double bond** exists between the oxygen atoms. In a structure where three pairs of electrons are shared, the bond is called a **triple bond**.

Exercise 4E

Covalent molecules with multiple bonds

Write Lewis structures for the following, then write their structural formulae:

(i) CO_2

(ii) N_2

(iii) $COCl_2$ (C is the central atom)

(iv) CS_2

(v) H_2CO (C is the central atom).

BOX 4.4

Families of carbon compounds

Carbon forms many, many covalent compounds in which the carbon atoms are covalently bonded to each other and to other atoms, such as hydrogen. Carbon has four outer shell electrons to share with other atoms and, in these examples, achieves the stable electronic structure of neon by sharing electrons with hydrogen or another carbon atom. Carbon atoms can share one, two or even three electrons with each other. Some examples are:

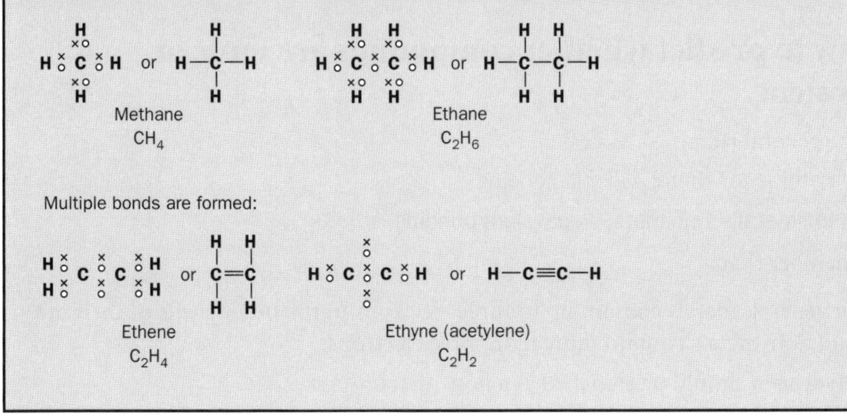

BOX 4.5

Electron density map for hydrogen (H_2)

The results of quantum mechanical calculations can be used to plot the electron density distribution in a hydrogen molecule. The contour lines join regions of the same electron density. Notice that there is a region of high electron density between the nuclei, which corresponds to the covalent bond in the Lewis structure. This electron density attracts the positive nuclei and prevents repulsion between the two positive nuclei from driving the atoms apart – it 'glues' the atoms together. A region of high electron density, holding atoms together in this way, is

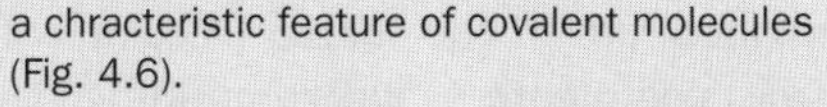

a chracteristic feature of covalent molecules (Fig. 4.6).

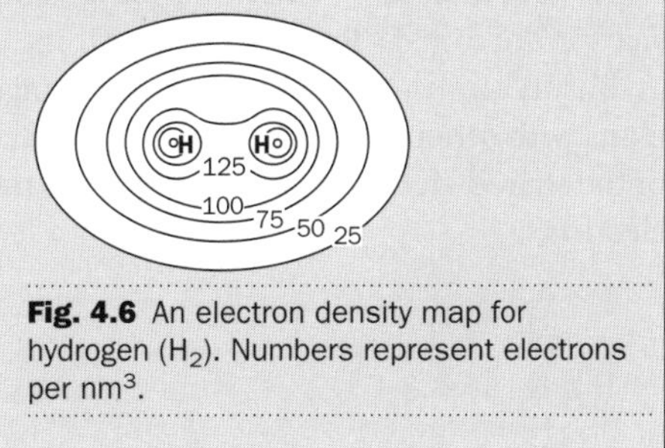

Fig. 4.6 An electron density map for hydrogen (H_2). Numbers represent electrons per nm^3.

Properties of covalent compounds

Covalent compounds have the following characteristic properties:

1. Low melting points and boiling points (they are volatile)

They are often liquids or gases at room temperature (think of O_2, H_2O, N_2 and CO_2). The molecules are not attracted towards each other by strong electrostatic forces – so they are pulled apart at relatively low temperatures. The forces that attract molecules together (the **intermolecular** forces) are called **van der Waals' forces** and are relatively weak.

2. Low solubility in water

Covalent compounds do not contain ions and therefore do not mix so readily with water molecules. Covalent compounds tend to dissolve readily in **non-polar organic solvents**, such as benzene, which also contain covalent molecules.

3. Do not conduct electricity

Covalent compounds cannot conduct electricity because they do not contain ions.

How to predict whether compounds are ionic or covalent

Some general rules:

metal + non-metal → ionic bonding

non-metal + non-metal → covalent bonding

Remember that:

- In general, metals contain up to three electrons in the outer shells of their atoms and non-metals contain more than three electrons.
- Hydrogen should be classified as a non-metal.

> **Exercise 4F**
>
> **Formation of ionic or covalent compounds**
>
> **What type of bonding do you think would be present in compounds formed by the combination of the following atoms?**
>
> **(i)** hydrogen and chlorine, which form hydrogen chloride
>
> **(ii)** potassium and fluorine, which form potassium fluoride
>
> **(iii)** silicon and hydrogen which form silane, SiH_4.

4.4 Coordinate bonding

Covalent bonds are sometimes formed between two atoms when one of the atoms provides *both* of the shared electrons. This type of covalent bonding is given the names **coordinate** or **dative** bonding. The two electrons that one atom donates for sharing are called a **lone pair** of electrons.

Boron atoms have the electron arrangement 2.3 and can only share three electrons with other atoms. In the compound NH_3BF_3, the nitrogen atom provides both of the shared electrons to boron, so that the boron atom is surrounded by an octet of electrons:

```
             ××
      H     ×F×
      •□     ○×   ××
  H□• N □□  B ×○ F ×
      •□     ○×   ××
      H     ×F×
             ××
```

To help you distinguish where the electrons come from:

● = electrons from the H atom
□ = electrons from the N atom
× = electrons from the F atom
○ = electrons from the B atom.

The structural formula of the molecule may be written

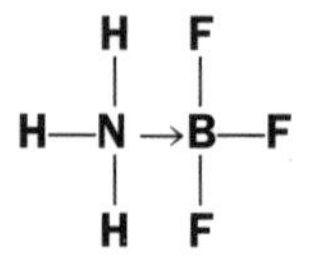

where '——' represents a covalent bond, where one electron has been donated from each atom, and '⟶' represents a coordinate covalent bond, where both electrons arise from one atom (in this case the nitrogen). In this way, all the atoms in the molecule end up with the electronic configuration of an inert gas. There is no difference in an 'ordinary covalent bond' and a coordinate bond once they are formed (*the electrons no longer 'belong' to a particular atom taking part in the covalent bond*) and the above structure is often represented as

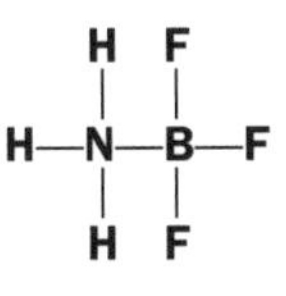

Exercise 4G

Coordinate bonding

(i) Draw a Lewis structure for carbon monoxide (CO), bearing mind it contains a multiple bond and a coordinate bond. Draw the structural formula of the molecule.

(ii) Draw a Lewis structure for the hydrated hydrogen ion, or hydronium ion H_3O^+.

(iii) Draw a Lewis structure for the ammonium ion, NH_4^+.

4.5 Ionic and covalent compounds – two extremes

Although sodium chloride is regarded by chemists as an ionic compound and the hydrogen molecule considered to be 'completely covalent', the vast majority of chemical compounds fall between these two extremes. Water, for example, is regarded as a covalent substance, and its molecules are neutral overall, but the molecules do have a slight positive charge at one end and an equal negative charge at the other end, i.e. there is a slight degree of ionic character. By the same token some compounds which are generally regarded as ionic, such as calcium iodide (CaI_2), have some covalent character.

Polar covalent molecules

The electrons in a bond between two identical atoms, such as that present in H_2, can be considered to be equally shared between both of them. If the atoms are not identical, however, the electrons may be more attracted to one atom than the other. For example in hydrogen fluoride, HF, the Lewis structure might be more accurately written as

××
H ×○ F ××
××

Fluorine attracts electrons in a covalent bond much more than hydrogen – it has a high **electronegativity**. Because the shared electrons are more associated with the fluorine atom rather than the hydrogen atom, the structural formula of the molecule is sometimes written as

$H^{\delta+}$—$F^{\delta-}$

Here, δ means 'slightly': the hydrogen end of the bond is slightly positively charged and the other end of the bond is slightly negatively charged. The molecule is *neutral overall* because the charges at either end balance out. A covalent bond which has a positive end and a negative end is called a **polar covalent bond** and a molecule which has a positive end and a negative end is said to possess a **dipole** (two poles).

Exercise 4H

Comparing the polarity of bonds

Work out the differences in electronegativities between the elements in the following covalent bonds, rank them in order of increasing polarity, and show the direction of each dipole:

(i) Be—H

(ii) O—H

(iii) C—H

(iv) C—Cl

(v) C—O.

BOX 4.6

Pauling's electronegativity values

The ability of an atom to attract a pair of shared electrons in a covalent bond is called the **electronegativity** of an atom.

Linus Pauling calculated relative electronegativity values for a number of elements. He assigned a value of 4 to fluorine, the most electronegative element. Electronegativity values are shown below:

H 2.1						
Li 1.0	Be 1.5	B 2.0	C 2.5	N 3.0	O 3.5	F 4.0
Na 0.9	Mg 1.2	Al 1.5	Si 1.8	P 2.1	S 2.5	Cl 3.0
K 0.8	Ca 1.0				Se 2.5	Br 2.8

If two elements combine, the difference in their electronegativity values can help decide whether the bonds holding them together are ionic or covalent:

1. The difference between the electronegativity values for sodium and chlorine is

$3.0 - 0.9 = 2.1.$

This is a large difference and, in general, differences greater than 1.7 indicate that compounds are ionic.

2. The difference in electronegativity values between the two atoms in a hydrogen molecule is

$2.1 - 2.1 = 0.0$

A value of 0 shows that the compound is completely covalent.

These examples are two extremes. By calculating these differences for covalent bonds between atoms, we can compare the **polarities** of covalent bonds. For example, the H–Cl bond (difference in electronegativities of H and Cl = 0.9) is much more polar than the H–Br bond (difference in electronegativities of the elements = 0.7).

Electronegativities can also show, for a polar covalent bond, which ends can be marked δ^+ and δ^-. The atom that has the highest electronegativity, or has the greatest attraction for the shared electrons, is marked δ^-. In the HBr molecule, Br has the largest electronegativity value (2.8) and so the polarity in the molecule is shown as

$\mathbf{H^{\delta+}—Br^{\delta-}}$

Exercise 4I

Electronegativity values

Box 4.6 gives you only a selection of elements together with their calculated electronegativity values. One way of determining electronegativities is to measure the polarities of bonds between various atoms. You would not be able to use this method, however, to measure the electronegativities of the noble gases He, Ne or Ar. Why do you think this is?

Polarization in ionic compounds

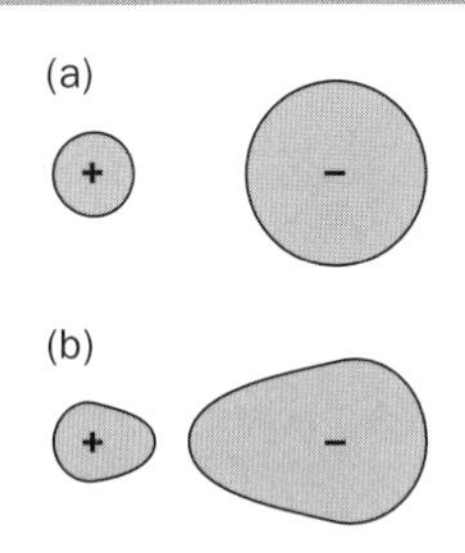

Fig. 4.7 Polarization: (a) purely ionic; (b) polarized.

A small, positively charged ion (**a cation**) in an ionic compound can attract the electrons of a neighbouring negatively charged ion (**an anion**) towards it and distort the anion. When this happens, the anion is said to be **polarized**. This distortion can be better represented by using the electron density model of an ionic compound, rather than using Lewis symbols. The outer electron density contours of a purely ionic bond, and an ionic bond which is polarized, are shown in Fig. 4.7.

Notice that, in the polarized case, more electron density is concentrated between the ions. A region of high electron density between atoms is characteristic of covalent compounds, so that the polarized compound in Fig. 4.7(b), although still mainly ionic, has some covalent character. When is significant polarization likely to occur? We can answer this question by applying the following rules.

The polarization of an ionic bond (i.e. its covalent character) is high if:

1. The cation is (a) small and (b) carries a high positive charge.

2. The anion is (a) large and (b) carries a high negative charge.

Compounds in which both conditions apply will have a great deal of covalent character. An example of a strongly polarizing cation is Al^{3+}, which is small and carries a large positive charge. The anion I^- is large; the degree of polarization in AlI_3 is so large, that it is considered to be a covalent compound. However, the compound AlF_3 has far more ionic character, because F^- is a much smaller anion and is not polarized to the same degree as I^-.

Exercise 4J

Polarization

The sizes of some cations and anions are listed below:

Cation	Radius/pm	Anion	Radius/pm
Li^+	60	F^-	136
Na^+	95	Cl^-	181
K^+	133	S^{2-}	184
Mg^{2+}	65		
Al^{3+}	53		

(i) In the series of ionic chlorides: LiCl, NaCl and KCl, which compound would have the most covalent character?

(ii) From the selection of anions and cations above, which compound would you expect to be the most ionic?

(iii) Which compound would you expect to have the most covalent character?

Ionic and covalent bonding: a summary

The overall picture is summed up in Fig. 4.8. Again, the outer contours of electron density maps have been used instead of Lewis symbols.

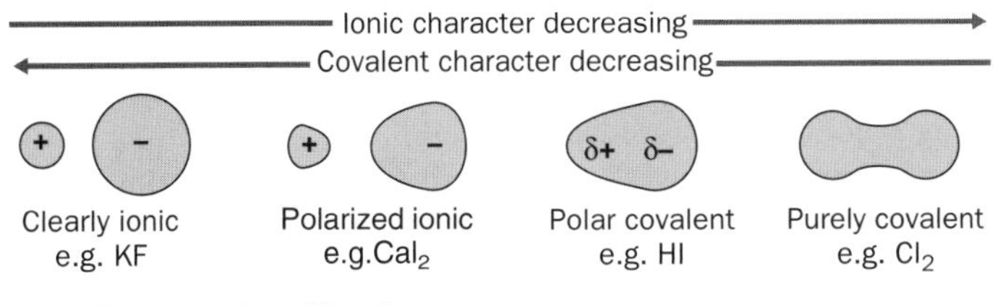

Fig. 4.8 An overview of bonding types.

Ionic compounds with polyatomic ions

Both ionic and covalent bonding exist within some compounds. Consider the deadly poisonous compound potassium cyanide (KCN). This is an ionic compound; it contains the ions K^+ and CN^-. A Lewis structure for the cyanide anion can be written as

$[\bullet\text{C}\ \text{N}]^-$

× C electrons
o N electrons
• electrons donated by potassium

The whole is written in square brackets with the negative charge outside to indicate that the negative charge is spread across the ion and does not belong to one particular atom – this 'spreading' of the negative charge or **delocalization** of the charge makes the ion more stable. The atoms C and N in the cyanide ion achieve the stable electronic arrangement of inert gases. They do this by covalent bonding *within* the ion. The structural formula of the ion is written:

$[C{\equiv}N]^-$

Ions such as CN^- that contain more than one atom are called **polyatomic ions**.

Other examples of polyatomic anions that form ionic compounds, but have covalent bonding *within* the anion are hydroxide (OH^-), nitrate (NO_3^-), sulfate (SO_4^{2-}), carbonate (CO_3^{2-}), hydrogencarbonate (HCO_3^-) and phosphate (PO_4^{3-}).

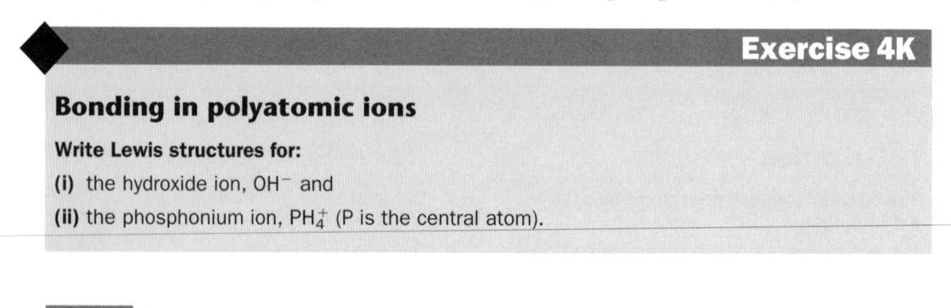

Exercise 4K

Bonding in polyatomic ions

Write Lewis structures for:

(i) the hydroxide ion, OH^- and

(ii) the phosphonium ion, PH_4^+ (P is the central atom).

4.6 Resonance structures

If you tried to write a structural formula for the carbonate ion you might have written:

$[(O)(O)C{=}O]^{2-}$

Or perhaps you might have come up with:

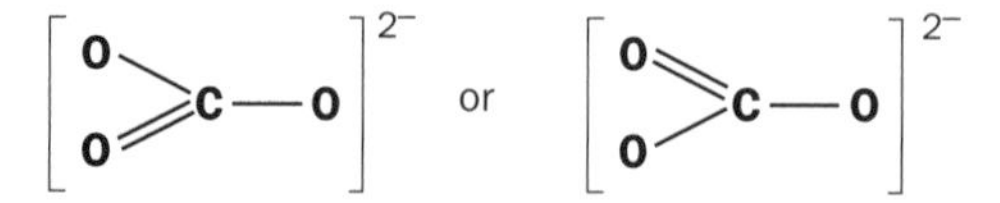

Which one is correct? Actually, none of them is exactly correct. In cases like this the Lewis model of bonding is not completely adequate.

It is possible for chemists to measure the lengths of the bonds in compounds. Double bonds between atoms are shorter than single bonds between the same atoms, so we might expect that measurement of the bond lengths in the carbonate ion would reveal two long bonds and one short one. In fact, all the bonds are *of the same length*; this bond length is shorter than would be expected for a C–O bond, but longer than that for C=O. The 'real' structure of the carbonate ion is a mixture of the three structures that we have already written and can be represented by:

This 'mixing' of structures is called **resonance** and the resulting structure, such as the one written above, is called a **resonance hybrid**. This idea is similar to the relationship between a donkey, a mule and a horse. The mule is the 'resonance hybrid' – it is neither a donkey nor a horse but has features of both of them.

Box 4.7 gives some historical information on scientists who have contributed to bonding theory.

Exercise 4L

Resonance Structures

(i) There are two possible structural formulae that might represent NO_2^-. Draw both of these.

(ii) Draw the best representation of the structural formula of the NO_2^- ion; it is a resonance hybrid.

BOX 4.7

G. N. Lewis and Linus Pauling

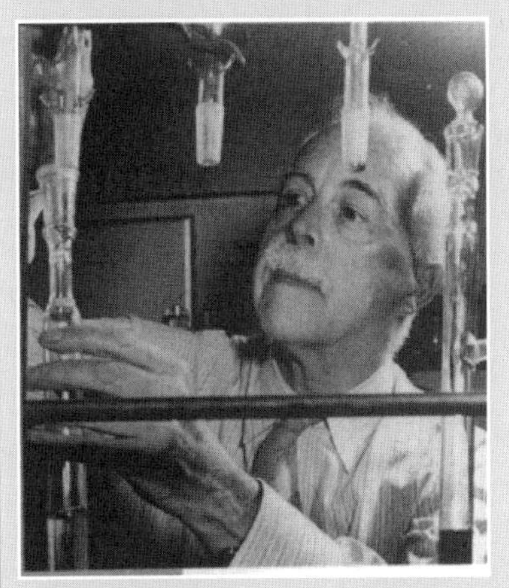

Gilbert Newton Lewis, an American chemist. His theories led to a clearer understanding of the concept of valency.

Early sketches of his ideas, by G. N. Lewis. He eventually became professor of chemistry at the University of California.

Linus Pauling, an American chemist and twice Nobel Prize winner. His book *The Nature of the Chemical Bond* is regarded as a classic by chemists.

Gilbert Lewis (1875–1946) was the first American to win the Nobel prize. This was awarded in 1914. Referring to covalent bonding, he wrote 'an electron may form part of the shell of two different atoms and cannot be said to belong to either one exclusively'. He was not without his critics, however. It was argued that this statement was the same as saying that if a couple have two dollars in a joint bank account and six dollars each besides (total 14 dollars), they both have eight dollars apiece (total 16 dollars). Lewis structures, however, still explain the properties of a great deal of covalent compounds today – the bank account analogy does not hold true in the world of chemistry!

Linus Pauling (1901–1994), who proposed electronegativity values, wrote *The Nature of the Chemical Bond* in 1939. It is still regarded as a classic chemistry textbook. He received not one, but **two** Nobel prizes: one for chemistry in 1954 and one for peace in 1962. The peace prize was awarded for his campaign against the testing of nuclear weapons. He did not take the courses required to graduate from high school and only received his diploma *after* he had gained two Nobel prizes! He advocated very large doses of vitamin C as a preventative treatment for colds and cancer. This treatment is still very controversial, but it cannot be denied that he lived to a ripe old age and was still directing scientific research at the end of his life.

Revision questions

4.1. Write Lewis symbols for the following atoms:

(i) Li **(ii)** B **(iii)** N **(iv)** O **(v)** Si.

4.2. Write Lewis symbols for the following ions:

(i) S^{2-} **(ii)** H^- **(iii)** P^{3-} **(iv)** Ca^{2+} **(v)** Br^-.

4.3. Use Lewis symbols to describe the reactions between the following elements:

(i) Mg and S **(ii)** Ca and Br **(iii)** K and S **(iv)** Mg and N.

4.4. Why does sodium not form the Na^{2+} ion in its compounds?

4.5. Draw Lewis structures for the following molecules:

(i) H_2S **(ii)** CF_4 **(iii)** I_2 **(iv)** PH_3.

4.6. Write a Lewis structure for F_2CO (C is the central atom).

4.7. The ion BF_4^- contains a coordinate bond. Write:

(i) the Lewis structure
(ii) the structural formula of the ion.

4.8. Draw Lewis structures for the following ions:

(i) ClO^- **(ii)** NO_2^+ **(iii)** NH_2^-

4.9. Use the electronegativity values in Box 4.6 to decide which of the following bonds is the most polar:

(i) N–O **(ii)** P–Cl **(iii)** Si–Br.

4.10. What is the electronegativity difference between the atoms Mg and N? Is the bond between these atoms likely to be ionic or covalent?

4.11. Two Lewis structures exist for ozone (O_3), each of which contains one double and one single bond:

(i) Write the structural formulae of both of the Lewis structures.
(ii) The actual structure of ozone is a resonance hybrid of both structures. Draw a diagram to show the structure of the resonance hybrid.

Extension material to support this unit is available on our website at **http://www.palgrave.com/foundations/lewis.**

UNIT 5

More about Bonding

Objectives

- ▶ Discusses exceptions to the octet rule
- ▶ Explains how we can predict the shapes of molecules
- ▶ Distinguishes between polar and non-polar molecules
- ▶ Describes the different types of intermolecular attractive forces

Contents

After reading this chapter you should realize how useful Lewis's bonding theory is! With small additions, the basic ideas can be used to explain a number of phenomena: how molecules get their shape; why metals conduct electricity; what makes diamond such a hard substance; and why water is a liquid under normal conditions. All these questions are answered in the following pages. First, however, we need to discuss how some exceptions to the ideas developed in the last chapter can be explained.

5.1 Exceptions to the octet rule

Compounds in which the central atom is surrounded by fewer than eight electrons

The idea that there is a stability associated with the electron arrangements of the atoms of noble gases is sometimes called the **octet rule**. This rule applies to many compounds, but it is not always obeyed (unfortunately scientific theories are often not as 'neat' as we would like them to be!). When does the octet rule break down?

Two important exceptions to remember are beryllium (Be, electronic structure 2.2) and boron (B, electronic structure 2.3). According to the reasoning in Chapter 4, they should react by losing two and three electrons, respectively. They do not react in this way, however, because their ionization energies (*the energies required to remove their outer electrons*) are very high. Because the atoms are small, the outer electrons in their atoms are close to the positive nucleus and are strongly attracted to it. Other atoms with two and three electrons in their outer shells, such as magnesium and aluminium, can lose these electrons when they form compounds because the electrons are farther away from the nucleus and more shielded from its attractive charge by full shells of electrons. Instead beryllium and boron choose to share electrons and form covalent compounds. Those compounds are stable enough to exist at room

Exercise 5A

Lithium

Beryllium and boron do not react by losing electrons, whereas lithium (Li, 2.1) does. Can you suggest a reason for this?

temperature, but the central atoms do not have a stable octet of electrons surrounding them.

For example, BF_3 is a covalent substance and its formula may be written:

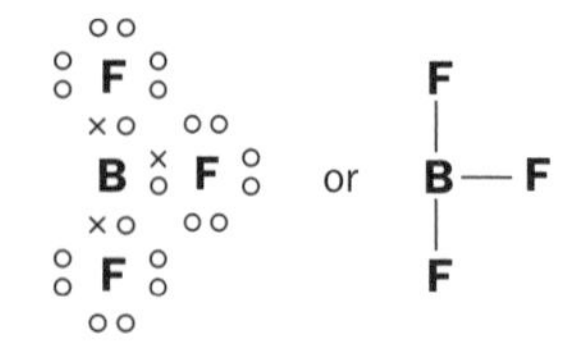

In this compound, boron is surrounded by only six electrons. The compound readily reacts with any substance that will give it the extra electrons it needs to obtain a stable octet. Two examples of such reactions are shown in the following equations:

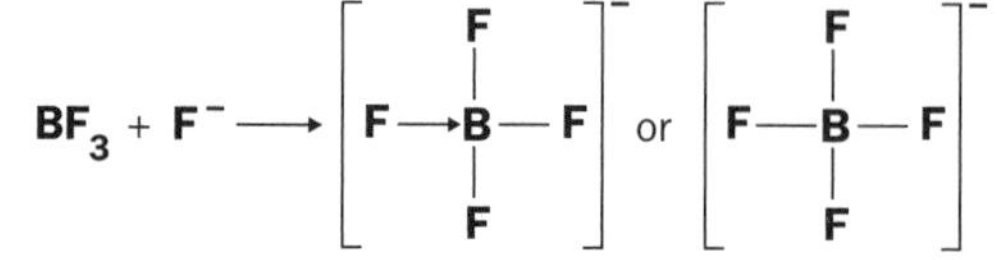

and

$BF_3 + NH_3 \longrightarrow$ H—N(H)(H)→B(F)(F)—F or H—N(H)(H)—B(F)(F)—F

Exercise 5B

More about beryllium compounds

(i) Draw a Lewis structure and also a structural formula for covalent $BeCl_2$.

(ii) In $BeCl_2$, how many electrons are in the outer shell of the Be atom?

(iii) $BeCl_2$ reacts with Cl^- ions to form the polyatomic ion $[BeCl_4]^{2-}$. Draw the structure of this ion.

(iv) How many electrons are in the outer shell of the Be atom in $[BeCl_4]^{2-}$?

(v) Which do you think is the more stable compound, $BeCl_2$ or $[BeCl_4]^{2-}$? Why?

Compounds which contain an atom with an expanded octet

The other class of compounds in which the octet rule is not obeyed is covalent compounds in which the central atom can accommodate more than eight electrons in its outer shell.

Remember that the first electron shell of an atom can hold up to two electrons ($1s^2$), the second shell can hold eight electrons ($2s^2\ 2p^6$) and the third shell can hold 18 electrons ($3s^2\ 3p^6\ 3d^{10}$). If the central atom in a covalent compound has electrons in the third (or a higher) shell, then it can use its empty d orbitals to hold extra electrons. In theory such elements can share up to nine pairs of electrons (making a total of 18), but the number of atoms around the central atom is restricted by the size of the central atom, it gets too 'crowded'. It is rare to find the central atom sharing more than six pairs of electrons.

Example 5.1

Phosphorus forms two fluorides PF_3 and PF_5:

In PF_3, the octet rule is obeyed:

But in PF_5, the central phosphorus atom has 10 electrons surrounding it:

Here, phosphorus uses its empty 3d orbitals to accommodate the extra electrons.

Exercise 5C

Covalently bonded atoms with an expanded octet

(i) Draw Lewis structures for the following molecules or ions:

(a) SF_6

(b) XeF_4

(c) BrF_3

(d) IF_5

(e) PCl_6^-

(f) ClF_4^+.

(ii) NF_3 exists, but NF_5 does not – suggest why.

(iii) IF_7 exists, but ClF_7 does not – suggest a reason for this.

5.2 Shapes of molecules

If we can draw Lewis structures for covalent molecules, we can predict their shapes by applying a few simple rules. The theory that accounts for the shapes of molecules is called **valence shell electron pair repulsion theory** (VSEPR theory) and is based on the reasoning that electron pairs attempt to get as far away from other electron pairs as possible because their negative charges repel each other.

Molecules with bonding electron pairs only

1. Consider the molecule $BeCl_2$

This can be represented as

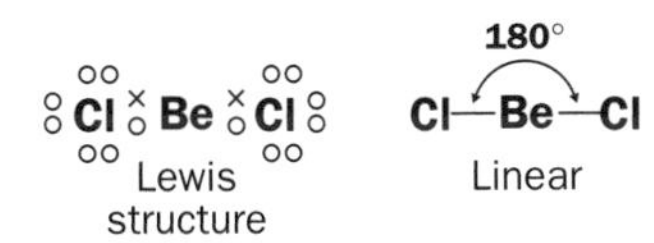

There are two shared electron pairs (**bonding pairs**) around the central beryllium atom. These two electron pairs arrange themselves as far apart from each other as they can, so that the angle between the two Be–Cl bonds is 180° and the molecule is **linear**.

2. The BF_3 molecule

Now let us apply this theory to a situation where there are three pairs of bonding electrons around the central atom, as in BF_3:

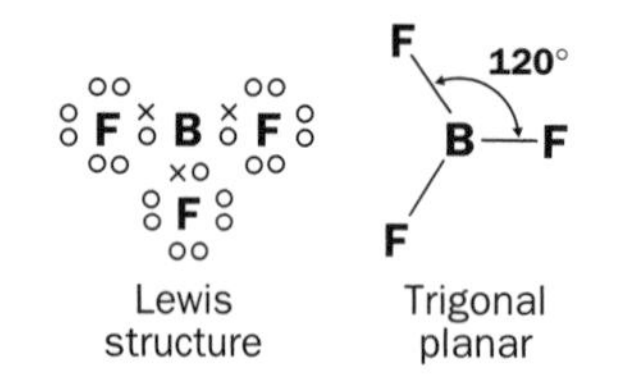

The three electron pairs around boron minimize the repulsion between them when the molecule adopts the **trigonal planar** shape. The fluorine atoms are in the same plane, at the three corners of a triangle drawn around the central atom. The three angles between the B–F bonds are all 120°.

3. Methane, CH_4

To minimize repulsion, the four electron pairs adopt a **tetrahedral** shape, with a bond angle of 109° 28′:

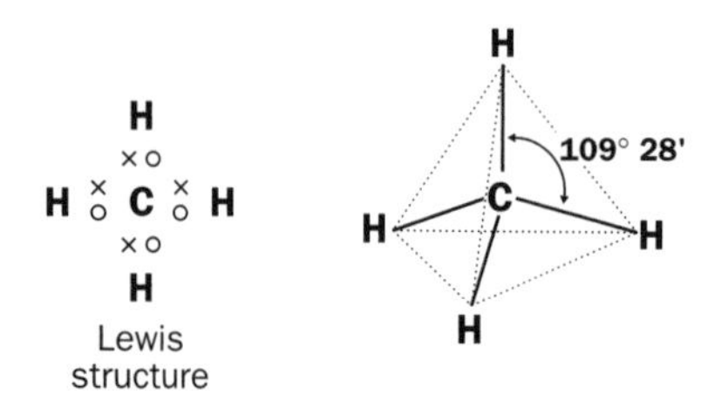

4. The PF_5 molecule

The five electron pairs adopt a **trigonal bipyramidal** shape, with bond angles of 120° and 90°:

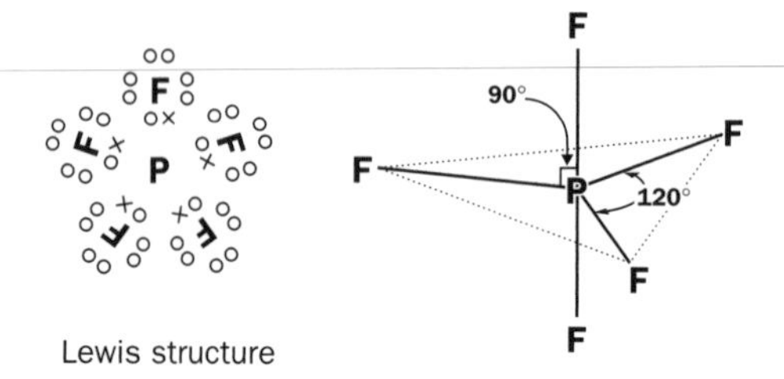

5. The SF_6 molecule

In this molecule the six electron pairs arrange themselves in an **octahedral** shape. All bond angles are 90°.

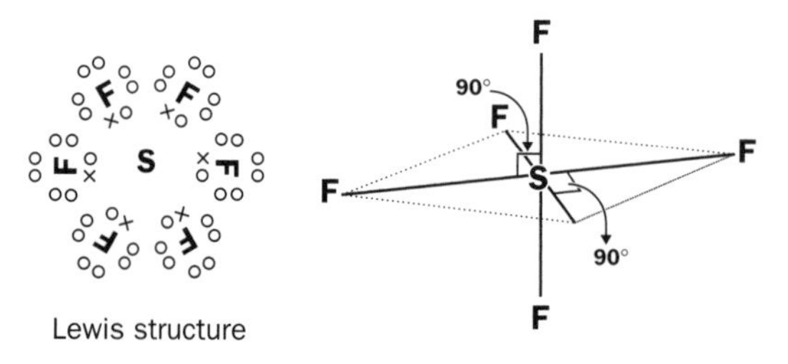

To summarize, Table 5.1 shows the shapes adopted by covalent compounds with different numbers of electron pairs around the central atom.

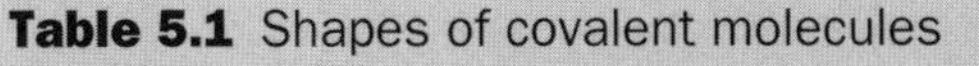

Table 5.1 Shapes of covalent molecules

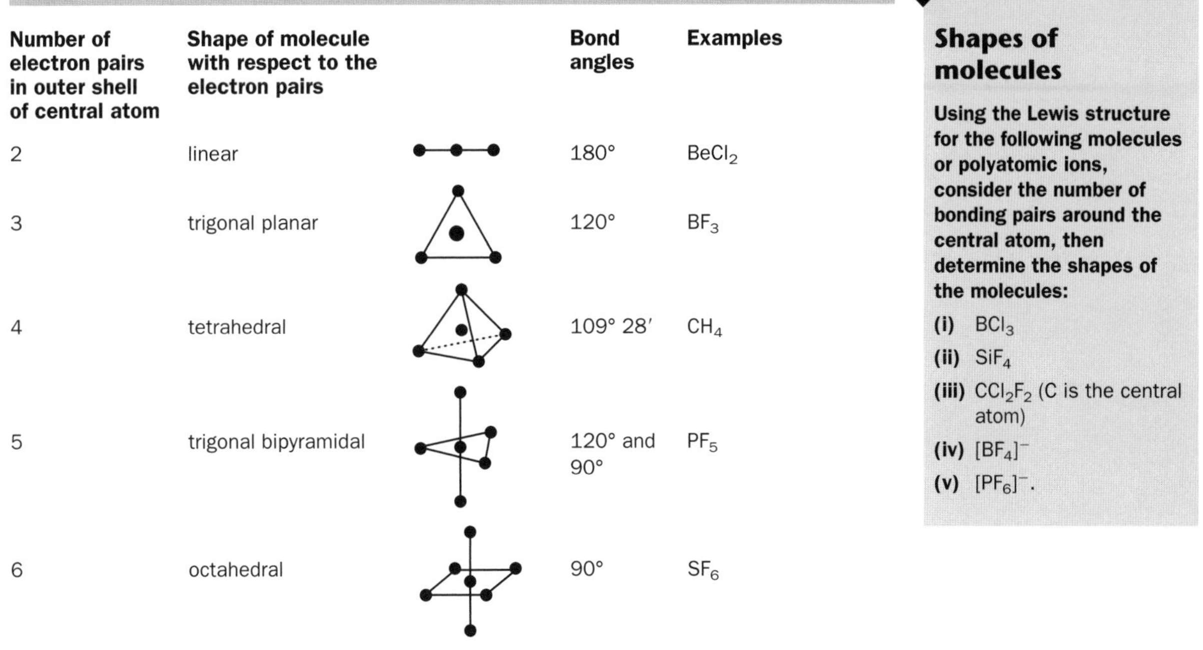

Number of electron pairs in outer shell of central atom	Shape of molecule with respect to the electron pairs	Bond angles	Examples
2	linear	180°	$BeCl_2$
3	trigonal planar	120°	BF_3
4	tetrahedral	109° 28′	CH_4
5	trigonal bipyramidal	120° and 90°	PF_5
6	octahedral	90°	SF_6

Remember that in all the shapes, the electron pairs are *as far apart from each other as possible.*

Molecules with lone pairs

In all the examples in Table 5.1, the electron pairs are shared between atoms – they are **bonding pairs**. Some molecules have pairs of electrons that are not shared by two atoms. These electrons are called **lone pairs**.

The order of repulsion between electron pairs is:

lone pairs repel lone pairs more than **lone pairs repel bonding pairs** more than **bonding pairs repel bonding pairs**

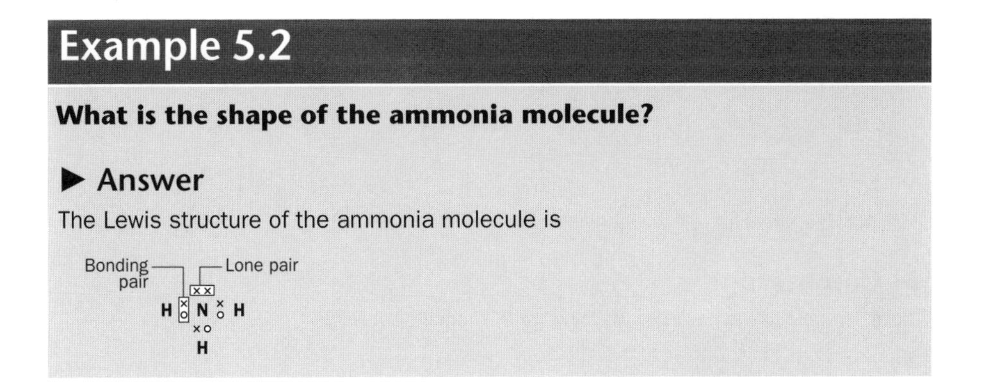

Example 5.2

What is the shape of the ammonia molecule?

▶ Answer

The Lewis structure of the ammonia molecule is

Exercise 5D

Shapes of molecules

Using the Lewis structure for the following molecules or polyatomic ions, consider the number of bonding pairs around the central atom, then determine the shapes of the molecules:

(i) BCl_3

(ii) SiF_4

(iii) CCl_2F_2 (C is the central atom)

(iv) $[BF_4]^-$

(v) $[PF_6]^-$.

Example 5.2 (*continued*)

Because lone pairs are not shared between two positive nuclei, they are closer to the central nucleus than bonding pairs and repel nearby electron pairs more than bonding pairs. In the ammonia example, there are four electron pairs around the central nitrogen atom but the angle between the N–H bonds is not 109°28′ but 107°. This lower angle between the N–H bonds is a result of the lone pair pushing the bonding pairs closer together; the lone pair repels the bonding pairs more than they repel each other. The shape of the ammonia molecule *with respect to the electron pairs* is a **distorted tetrahedron**. The structure is shown in Fig. 5.1.

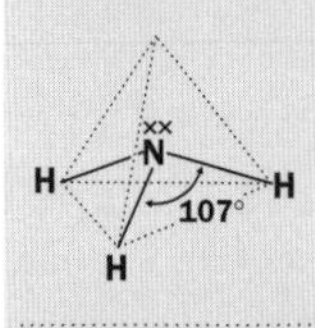

Fig. 5.1 The structure of the ammonia molecule.

▶ Comment

The ammonia molecule is also described as **trigonal pyramidal** *with respect to its N–H bonds* because it looks like a three-cornered pyramid.

Example 5.3

What is the shape of the water molecule?

▶ Answer

The Lewis structure for water is

```
  ××
× O × H
×   o
  o×
  H
```

There are four electron pairs around the central oxygen atom: two bonding pairs and two lone pairs. Again, the shape of the water molecule *with respect to the electron pairs is* **distorted tetrahedral**, but the repulsive effect of two lone pairs is to push the O–H bonds together so that the bond angle between them is 105°. The structure is shown in Fig. 5.2.

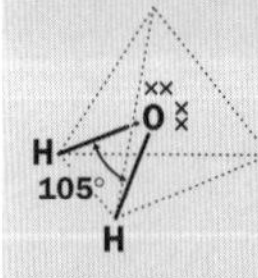

Fig. 5.2 The structure of the water molecule.

▶ Comment

The water molecule is described as **'bent' or 'V-shaped'** *with respect to its O–H bonds.*

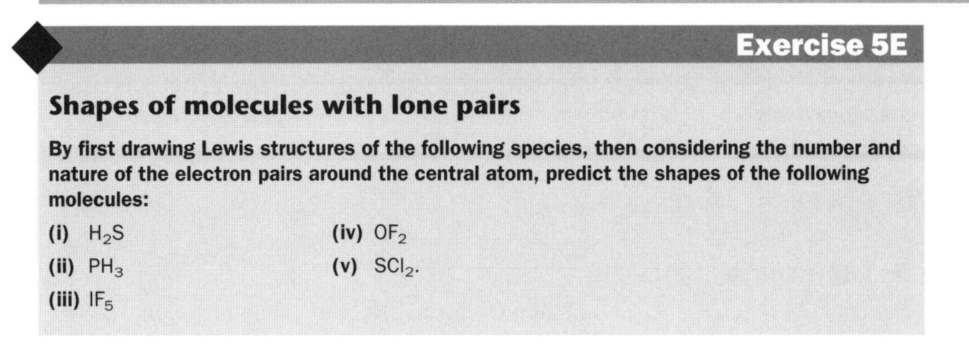

Exercise 5E

Shapes of molecules with lone pairs

By first drawing Lewis structures of the following species, then considering the number and nature of the electron pairs around the central atom, predict the shapes of the following molecules:

(i) H_2S
(ii) PH_3
(iii) IF_5
(iv) OF_2
(v) SCl_2.

5.3 Shapes of molecules with multiple bonds

How do we work out the shape of a molecule when it contains double, or even triple bonds? We can still use VSEPR theory, but a double or triple bond is treated as though it were a single bond, that is, one bonding pair of electrons. Carbon dioxide has the structural formula:

O=C=O

Although each double bond contains two shared pair of electrons, they are held together in the same place that a bonding pair of electrons would be held in a single bond. Therefore, if each double bond is treated as a single bond, the carbon atom has the equivalent of two bonding pairs of electrons around it and the shape of the molecule is **linear.**

Example 5.4

What is the shape of ethyne (common name acetylene)?

▶ Answer

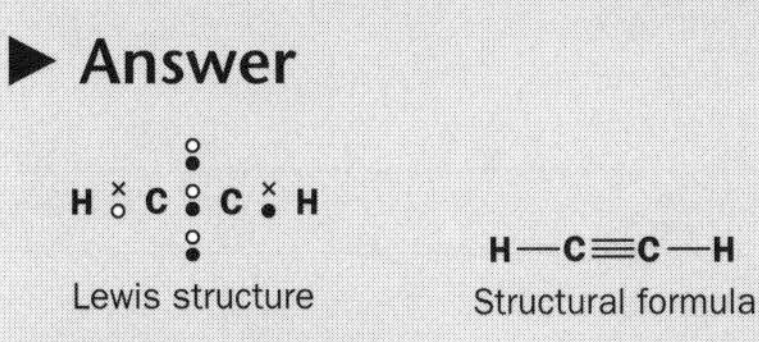

Lewis structure
Structural formula

Here the triple bond and single bond on either side of the carbon atoms are regarded as single bonds in order to work out the overall shape of the molecule. The molecule is **linear**.

Example 5.5

What is the shape of sulfur dioxide, SO_2?

▶ Answer

For the purposes of working out the shape, the central sulfur atom acts as if it has three electron pairs around it, one of which is a lone pair. The molecule is

Example 5.5 (*continued*)

therefore a distorted trigonal planar shape with respect to the electron pairs, or V-shaped with respect to the S=O bonds:

O S O or O=S=O

► Comment

The angle between the S=O bond, at 119.5°, is less than the usual trigonal planar angle of 120° because the S=O bonds are 'pushed' closer together by the repulsion of the lone pair.

Shapes of molecules with multiple bonds

What are the shapes of the following?

(i) the carbonate ion, CO_3^{2-}

(ii) sulfur trioxide, SO_3

(iii) carbon disulfide, CS_2 (draw the Lewis structure, then the structural formula first!)

(iv) the sulfate ion, SO_4^{2-}

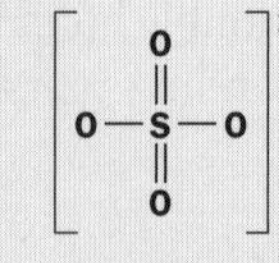

(v) the nitrate ion, NO_3^-

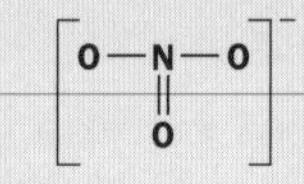

BOX 5.1

Finding out molecular shapes

How do we know that VSEPR theory is good for predicting the shapes of molecules? The predictions have been shown to be true by a number of techniques which can be used to gain information about the shapes of molecules:

1. X-ray diffraction X-ray diffraction can be used to determine the arrangement of atoms in a crystal. Simply put, this involves using X-rays (which are electromagnetic radiation with wavelengths about the same as the distances between neighbouring atoms in a crystal) to take 'pictures' of the crystal structure.

2. Infrared spectroscopy A pair of covalently bonded atoms vibrate when they absorb electromagnetic radiation. The frequency of this radiation depends on the atoms involved and upon the strength of the bond between them. The infrared spectrum of a molecule measures its absorbance over the infrared region of the electromagnetic spectrum and the spectrum depends, in part, on the geometry of the molecule.

3. Dipole moment measurements We have already discussed polar covalent bonds. Molecules which contain these bonds *may* have a positive end and a negative end (a dipole), depending on their geometry. Measurement of the magnitude of dipoles (**the dipole moment**), therefore, could give valuable information about the shape of a molecule. Dipole moments are discussed in the next section.

4. Microwave spectroscopy Molecules rotate in the gaseous state. Upon absorbing microwave radiation, the molecules rotate faster. The frequencies of radiation absorbed by molecules and the relative intensities of such absorptions depend upon the geometry of the molecules. A mathematical analysis of the microwave spectrum of a molecule can provide chemists with bond angles and bond lengths. This information is much more accurate than that obtained by infrared spectroscopy. Study of the effect of a magnetic field on a microwave spectrum allows the dipole moment of the molecule to be calculated.

5.4 Molecules with and without dipoles

In order to work out whether a molecule has a dipole or not we need to consider

1. whether the molecule contains polar covalent bonds;

2. the shape of the molecule.

Molecules without dipoles

Carbon dioxide is a linear molecule, containing carbon and oxygen atoms:

$O{=}C{=}O$

The electronegativity difference between carbon (electronegativity 2.5) and oxygen (electronegativity 3.5) is 1, so the C=O bond is **polar**:

$C^{\delta+}{=}O^{\delta-}$

The carbon dioxide molecule, however, *does not have an overall dipole* because it is linear:

$O^{\delta-}{=}C^{\delta+}{=}O^{\delta-}$

The dipoles of each polar C=O bond are equal in size but point in opposite directions and cancel each other out. So, the molecule does not have an overall dipole – *although it contains polar bonds, the molecule is non-polar overall.*

Carbon dioxide is an example of those molecules which, although they contain polar bonds, do not have an overall dipole because the geometry of the molecules allow the bond dipoles to cancel each other out. These geometries include trigonal planar, tetrahedral, trigonal bipyramidal and octahedral. Note that for this to be true, *all the covalent bonds must be the same.* Some examples are

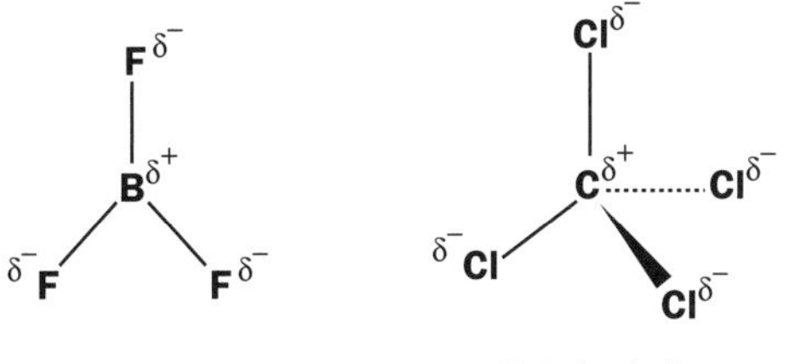

Molecules with dipoles

A water molecule is V-shaped and, because of the difference in electronegativities between oxygen and hydrogen; it contains polar covalent bonds:

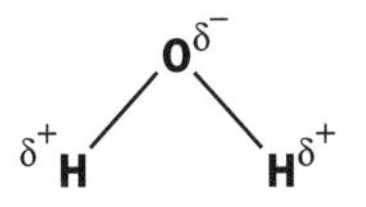

Because the molecule is not linear, it has a positive end and a negative end – it has a dipole and *is polar overall*, as shown in Fig. 5.3.

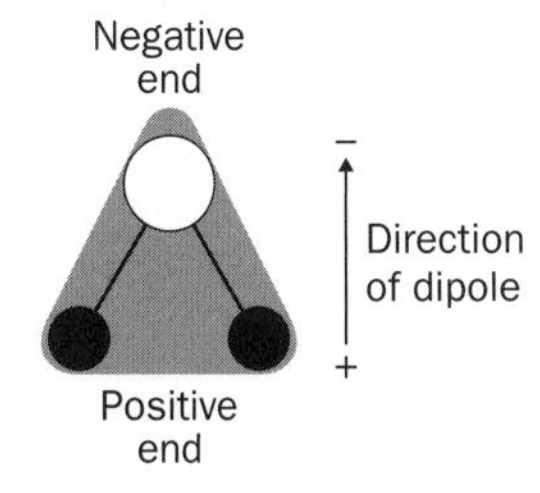

Fig. 5.3 A polar molecule.

Exercise 5G

Polar and non-polar molecules

1. Which of the following molecules have an overall dipole moment?

(i) CS_2 **(ii)** NH_3 **(iii)** $SiCl_4$
(iv) H_2S **(v)** $CHCl_3$.

2. An electron density map for hydrogen fluoride, HF, is shown below in Fig. 5.4.

(i) Does the molecule have a dipole?

(ii) In which direction is the dipole?

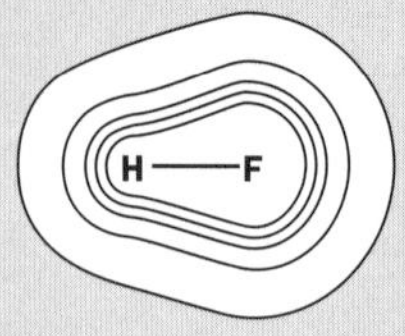

Fig. 5.4 Electron density map for hydrogen fluoride. (Note that the electron density increases going from the outer to the inner contours.)

BOX 5.2

Demonstrating the polarity of molecules

You can easily demonstrate that water molecules are polar by rubbing a plastic ruler with a dry duster, then bringing it close to a narrow stream of water flowing from a burette (*do not allow the ruler to touch the water!*). The stream of water will bend because the polar water molecules are attracted to the charged rod. Do the same experiment with hexane, instead of water. Hexane molecules are non-polar and the stream of hexane should not be deflected (Fig. 5.5).

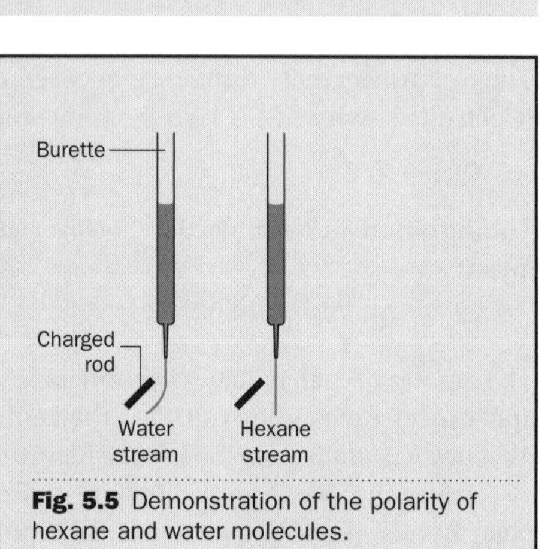

Fig. 5.5 Demonstration of the polarity of hexane and water molecules.

5.5 Metallic bonding

As we have seen, metals have very distinctive properties. Any successful model of the bonding in metals must account for these properties. Metals have a **crystalline** structure – a crystal can be regarded as a rigid substance in which particles are arranged in a repeating pattern. We can regard a metallic crystal as consisting of a lattice of positive ions, surrounded by a cloud of electrons which are free to move. The electrostatic attraction between the positive ions and the negatively charged cloud of electron density 'glues' the whole structure together. Each atom of the metal has given up its outer (i.e. **valence**) electrons to become a positive ion. These valence electrons form part of the electron cloud and, because they are free to move, they no longer 'belong' to the particular atom from which they have been released. This is illustrated in Fig. 5.6.

The model explains the properties of metals:

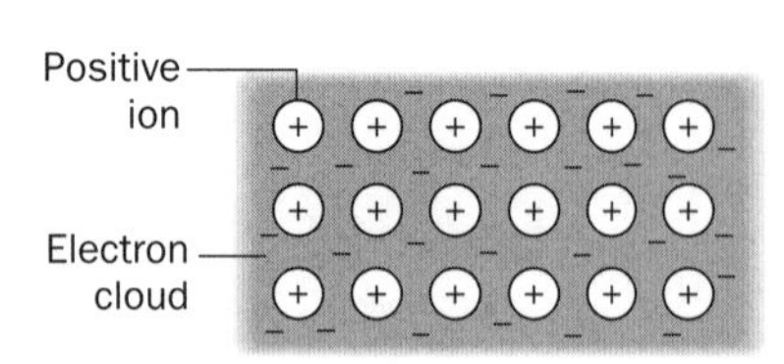

Fig. 5.6 Metallic bonding.

1. Metals are shiny because the free electrons absorb light that falls on the metal and then re-emit the light back. Most of the light that falls on the metal is re-emitted, so the metal appears shiny. This is why metals are used in mirrors. Most metals are a silvery colour – the red/gold colours of copper and gold occur because they absorb some wavelengths of visible light more than others.
2. The free electrons in the structure allow metals to conduct electricity

in the solid state (in contrast to ionic compounds). There is no chemical change to the metal when electricity is passed through it. If a potential difference is applied to the metal, the negative cloud of electrons will be attracted to the positive potential. These same electrons can transfer kinetic energy through the solid quickly, explaining why metals are good conductors of heat.

3. The structure can be distorted without breaking into pieces, because it is held together as a whole by the electron cloud. This is why metals are malleable and ductile. The strength of the metallic bond depends upon ratio of the number of free electrons in the electron cloud per ion. The more valence electrons an atom of a metal releases into the electron cloud, the stronger the metallic bond.

Exercise 5H

Metallic bonding

(i) How many electrons are there in the valence shell of magnesium?

(ii) Draw a diagram, similar to Fig. 5.6, to illustrate metallic bonding in magnesium.

(iii) How many electrons are there in the valence shell of aluminium?

(iv) Arrange the following metals in order of increasing strength of their metallic bonds: magnesium, aluminium and sodium.

(v) Which metal in **(iv)** would you expect to have the highest melting point and why?

5.6 Giant molecules

Allotropes of carbon

The element carbon exists in three solid forms, called **allotropes**, which differ in the way that the carbon atoms are bonded together. Allotropes are different forms of the same element; in each form the atoms are arranged in different ways. Three allotropes of carbon are: diamond, graphite and buckminsterfullerene.

BOX 5.3

Allotropes

Some other elements which exist as different allotropes are:

1. Oxygen consists of diatomic molecules in *oxygen gas* (O_2) and triatomic molecules in *ozone* (O_3).

2. Phosphorus can exist as three allotropes: *white phosphorus* (P_4, its structure is shown on page 19); *red phosphorus*, which is used in matches and consists of chains of P_4 units (Fig. 5.7); and *black phosphorus*, obtained by heating white phosphorus under pressure, which consists of sheets of phosphorus atoms.

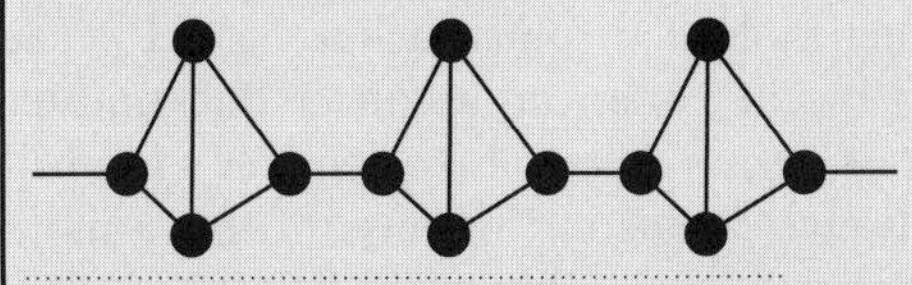

Fig. 5.7 The structure of red phosphorus.

3. Sulfur can exist in many forms including *rhombic sulfur* and *monoclinic sulfur*. They both contain S_8 molecules (page 19, but crystallize in different forms. *Plastic sulfur*, which consists of chains of sulfur atoms, is a form of sulfur that can be 'stretched'.

Diamond

Fig. 5.8 The arrangement of carbon atoms in diamond.

Diamond is a transparent solid which does not conduct electricity. In diamond, each carbon atom forms single covalent bonds to four other carbon atoms, which are at the corners of a tetrahedron. The four carbon atoms are, in turn, covalently bonded to four other carbon atoms and so on. This bonding exists throughout the entire crystal. A part of this arrangement is shown in Fig. 5.8. When all the atoms in a crystal are covalently bonded to one another throughout the whole crystal, the solid is termed a **network solid**.

Diamond is a very hard substance because the carbon–carbon single bonds are very strong and extend throughout the whole structure. In order to melt diamond, many of these bonds need to be broken and this requires a high temperature. This explains why diamond has a very high melting point (3823 K).

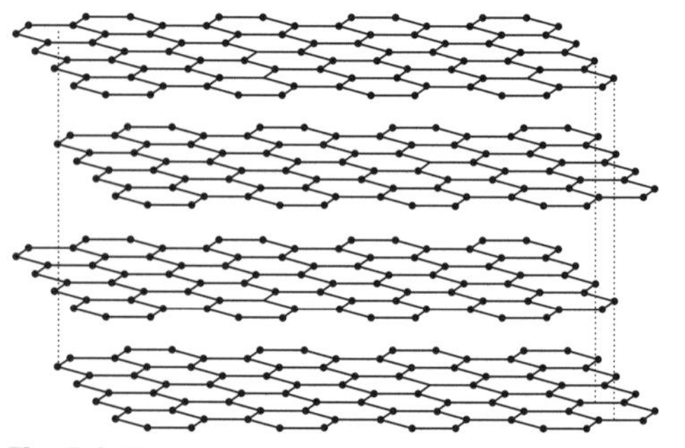

Fig. 5.9 The arrangements of carbon atoms in graphite.

Graphite

Graphite is another allotrope of carbon and another example of a network solid. Graphite is a soft, blackish solid that conducts electricity (unusual for a covalently bonded substance!). In graphite, there are layers of carbon atoms. Within each layer, each carbon atom is covalently bonded to three others in a trigonal planar arrangement (Fig. 5.9).

Although the covalent bonds between the carbon atoms within the layers are strong, the layers are held together by weak forces (**London dispersion forces**) and can slide over one another easily. Graphite therefore has a 'slippery' feel and is used as a lubricant and as the so-called 'lead' in pencils. The pencil marks paper because the layers of graphite are easily rubbed off. Each carbon atom uses only three of its valency electrons in bonding to three other carbon atoms and therefore has one valency electron left over. These electrons form a cloud above and below the layers, rather like that in the metallic bond. The free electrons allow graphite to conduct electricity and account for its shiny appearance. Soot consists of small crystals of graphite.

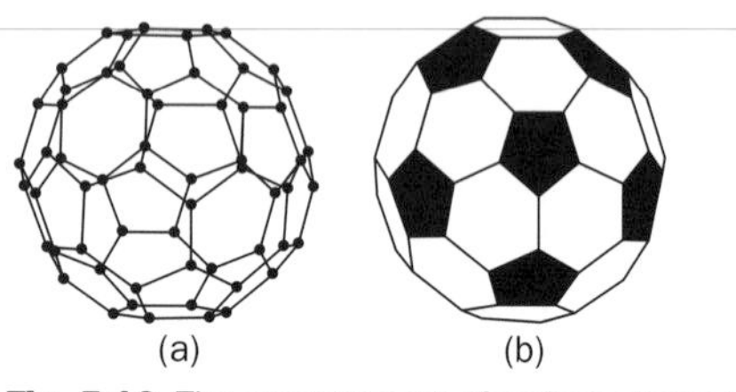

Fig. 5.10 The arrangement of carbon atoms in buckminsterfullerene: (a) is the arrangement of the carbon atoms; (b) is a football for comparison.

Buckminsterfullerene

Unlike diamond and graphite, which have been known for many centuries, this third allotrope of carbon has been discovered only recently (1985). The allotrope has the formula C_{60} and, although it cannot be classified as a giant molecule in the same way as diamond and graphite are giant molecules, it is still much larger than other molecular solids such as S_8 or P_4. The shape of C_{60} resembles domes designed by the American architect Buckminster Fuller, hence its name. The arrangement of carbon atoms also resembles a football (Fig. 5.10), with twenty hexagons and twelve pentagons; because of this a more informal name for the allotrope is sometimes used – **bucky ball**. Buckminsterfullerene is a dark solid at room temperature and dissolves in covalent solvents, such as benzene. Research still continues into its properties. Chemical derivatives of the allotrope are believed to have great potential for a number of applications, including lubricants, batteries and semiconductors. Three scientists – Harry Kroto, Richard

Smalley and Robert Curl – were jointly awarded the 1996 Nobel Prize for Chemistry for the discovery of C_{60}.

Exercise 5I

Another allotrope of carbon?

Chemists in the USA have very recently reported the discovery of yet another allotrope of carbon, called LAC. LAC consists of long chains of carbon atoms, where the alternate carbon–carbon bonds are of different lengths. Remembering that carbon atoms have four valence electrons, can you suggest a structure for LAC? *Hint*: LAC stands for linear acetylenic carbon.

5.7 Forces between covalent molecules

So far, we have discussed the bonding within covalent molecules (**intramolecular bonding**). In this section, we will describe the forces that hold covalent molecules (and some neutral atoms) together in the liquid and solid states (**intermolecular bonding**). Intermolecular forces have the general name **van der Waals' forces**, after the Dutch scientist Johannes van der Waals (1837–1923). The different types of intermolecular forces discussed are shown in Table 5.2.

Table 5.2 Types of intermolecular forces

Types of molecules	Types of intermolecular forces between molecules
Molecules with dipoles	(i) dipole–dipole attractions (ii) hydrogen bonding (iii) London dispersion forces
Molecules (or atoms) without dipoles	London dispersion forces only

Dipole–dipole interactions

Molecules which have dipoles, $X^{\delta+}–Y^{\delta-}$, tend to be attracted towards one another as shown in Fig. 5.11. The negative end of one molecule is attracted to the positive end of another and vice versa. Because these attractions are between *partial* charges ($\delta+$ and $\delta-$), they are much weaker electrostatic interactions than exist between

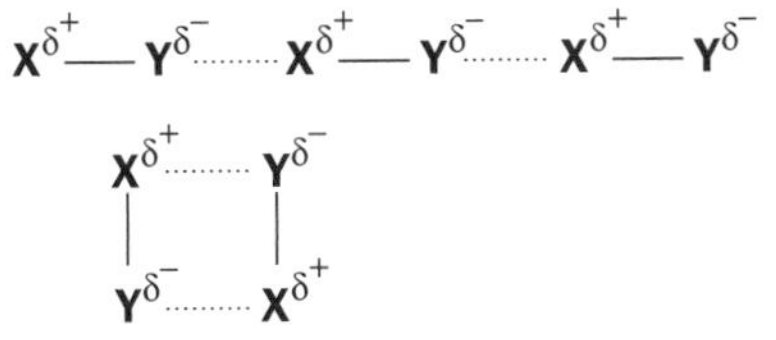

$X^{\delta+}—Y^{\delta-} \cdots X^{\delta+}—Y^{\delta-} \cdots X^{\delta+}—Y^{\delta-}$

$X^{\delta+} \cdots Y^{\delta-}$
$|\qquad |$
$Y^{\delta-} \cdots X^{\delta+}$

Fig. 5.11 The attraction between molecules with dipoles.

ions in ionic substances. The melting and boiling points of substances which consist of molecules with dipoles, therefore, tend to be much lower than those of ionic solids, but higher than those substances that consist of non-polar molecules of similar size.

London dispersion forces

London dispersion forces are another type of van der Waals' force. They are attractive forces between *all* molecules (or atoms). How do we know that attractive forces exist between non-polar molecules or neutral atoms? We know these forces exist because elements and compounds that contain such particles can be liquefied and/or solidified and this is only possible if attractive forces exist between particles in the liquid and solid states. Iodine, for example, is a solid at room temperature and it consists of I_2 molecules with no dipole moment. Similarly, although the smallest particles of the inert or noble gases are neutral atoms, they can be liquefied.

How do these attractive London forces arise? Although non-polar molecules and inert gas atoms have no dipole moment, the electron cloud in these particles, at one instant, may be denser on one side of the particle than the other. This causes a **temporary dipole**. The particle has, for a brief moment, one end with a small positive charge and the other end with an equally small negative charge. This temporary dipole causes (**induces**) another temporary dipole in a nearby particle – the positive end of the first particle attracts electron density from a neighbouring particle, making one end of the second particle slightly electron deficient and the other end electron rich. This process of particles inducing dipoles in nearby particles continues throughout. Weak attractive forces between the particles in the substance exist, at any particular instant. As the electron density in molecules or atoms shifts around, these dipoles change direction (fluctuate) but they are always present (Fig. 5.13).

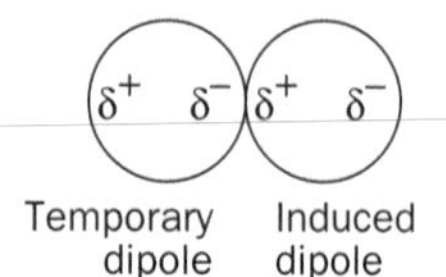

Fig. 5.13 Temporary dipole-induced dipole attractions.

The attractions due to London dispersion forces increase as the molecules or atoms get larger. The number of electrons increases and the fluctuations in the electron density produce bigger temporary dipoles. Molecular shape also influences the strength of the London interactions.

London dispersion forces are responsible for attractions between *all types of molecules*, both polar and non-polar, and can range from weak to quite strong.

BOX 5.4

Arrangement of molecules in solid iodine

Covalent compounds can form crystals, just like ionic substances. In black, shiny crystals of iodine, the iodine molecules (I_2) are arranged in a lattice structure but they are held together by London dispersion forces. Because these forces are weak the iodine is a volatile substance; the crystal structure can be pulled apart very easily. The arrangement is shown in Fig. 5.12.

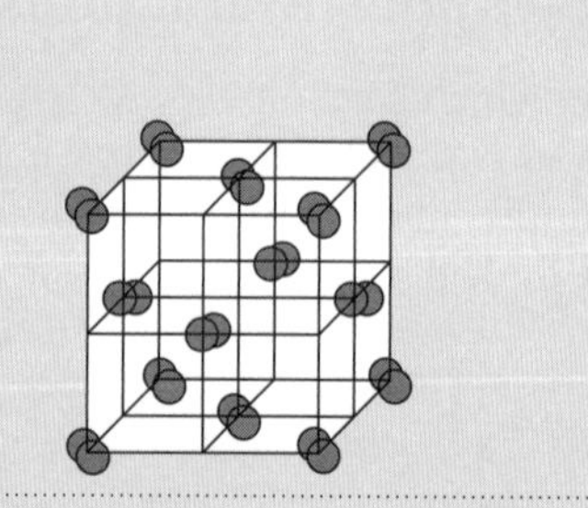

Fig. 5.12 The unit cell in solid iodine. The pairs of overlapping circles indicate the orientations of the I_2 molecules.

BOX 5.5

London dispersion forces and the shapes of molecules

Molecules of pentane (C_5H_{12}) exist with the same number of electrons, but very different shapes. One form of C_5H_{12} has molecules which have a cylindrical shape, whereas the other has molecules which are spherical. The cylindrical form has a boiling point of 36 °C, while the spherical form boils at 10 °C, suggesting that weaker London interactions exist between the spherical molecules. The valence electrons in the molecules with a spherical shape are 'hidden' deeper within the molecule than if the molecule is cylindrical, and cannot interact as strongly with neighbouring molecules (Fig. 5.14).

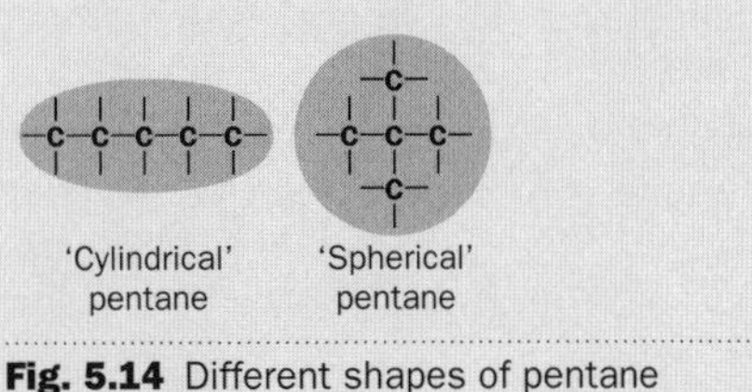

Fig. 5.14 Different shapes of pentane molecules.

Exercise 5J

London forces

(i) Arrange the following in order of decreasing boiling points. Explain your answer.

Inert gas	Diameter of atom/pm
He	100
Ne	130
Ar	190
Kr	220

(ii) At room temperature, F_2 and Cl_2 are gases, whereas Br_2 is a liquid and I_2 a solid. Explain these observations in terms of the strength of intermolecular forces within these substances.

Hydrogen bonding

Hydrogen bonding is a special type of dipole–dipole attraction *in which the hydrogen atom acts as a bridge between two electronegative atoms.* A hydrogen bond follows the general formula

$$A^{\delta-}\text{—}H^{\delta+} \cdots\cdots B^{\delta-}$$

Examples of hydrogen bonding are $F^{\delta-}$–$H^{\delta+} \cdots O^{\delta-}$, where A is F and B is O, and $F^{\delta-}$–$H^{\delta+} \cdots F^{\delta-}$, where A and B are F (see Example 5.6). These three atoms are usually bonded in a straight line. A and B are the electronegative atoms, such as F, O or N. Such atoms possess one or more lone pairs of electrons.

Example 5.6

A molecule of hydrogen fluoride has a large dipole moment. The partial positive charge on the hydrogen is attracted to the lone pair of electrons on the partially negatively charged fluorine atom of a neighbouring molecule – it forms a **hydrogen bond** with the fluorine. Liquid hydrogen fluoride contains zigzag chains of molecules of HF joined together by hydrogen bonds (Fig. 5.15).

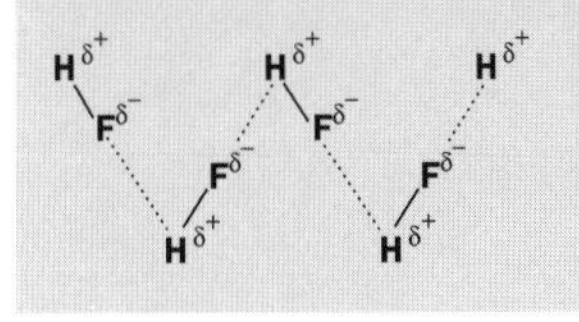

Fig. 5.15 Hydrogen bonding between molecules of hydrogen fluoride (hydrogen bonds are usually represented by dotted lines).

Example 5.7

Hydrogen bonds also exist between water molecules in liquid water and in ice. Because each oxygen atom in the water molecule has two lone pairs of electrons, it can attract two hydrogen atoms from other water molecules (Fig. 5.16).

Icebergs float on water. The open arrangement of water molecules in ice partially collapses when the substance melts, so that the liquid has a higher density up to about 4 °C.

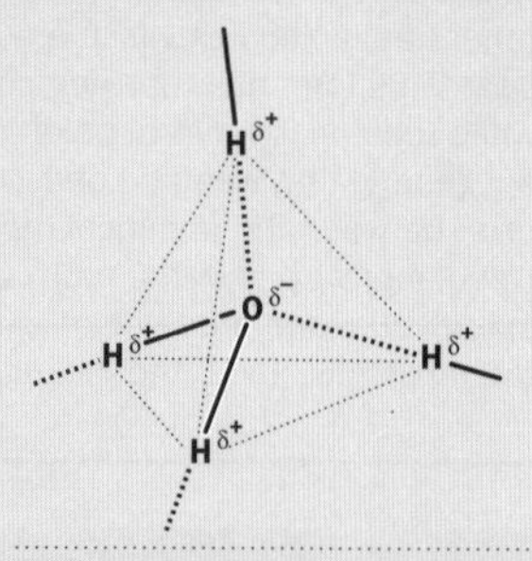

Fig. 5.16 Hydrogen bonding in water. Note that the O···H bonds are longer and weaker than the O–H bonds.

► Comment

Hydrogen bonding has a pronounced effect on the properties of water:

1. Water has a much higher melting or boiling point than compounds of similar formula containing hydrogen. For example H_2S, H_2Se and H_2Te have the same general formulae but are all gases at room temperature. Water, of course, is a liquid at room temperature because more energy is needed to separate H_2O molecules. Water molecules have much stronger attractive forces between them, due to hydrogen bonding, than do H_2S, H_2Se and H_2Te molecules.

2. Water has a high **surface tension**. The surface of water forms a kind of 'skin' on which insects can walk, or a needle can float. The high surface tension of water is due to hydrogen bonding – the intermolecular forces within the water pull surface molecules inwards.

3. A remarkable consequence of hydrogen bonding is the low density of ice. A substance is generally more dense in the solid state than the liquid state, but ice is *less dense* than water at 0°C (ice floats on water). The tetrahedral arrangement of water molecules in ice creates a very open structure (Fig. 5.17). When ice melts this structure breaks up and the water molecules approach each other more closely, so that the liquid has a higher density.

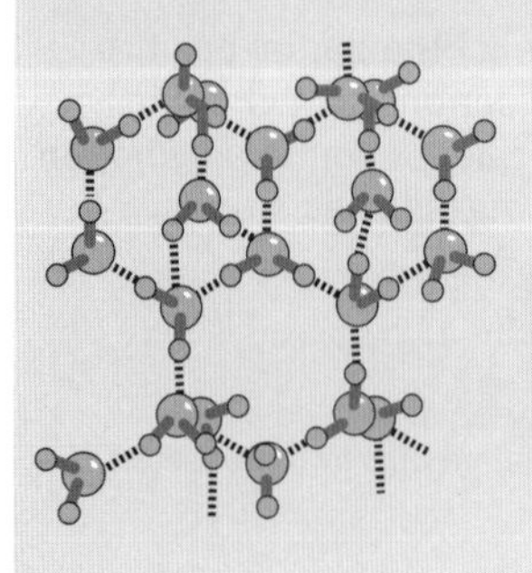

Fig. 5.17 Arrangement of water molecules in ice (dotted lines show hydrogen bonds).

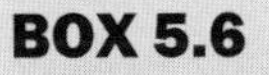

BOX 5.6

The boiling points of hydrides

The graph in Figure 5.18 shows the boiling points of four sets of hydrides with similar formulae. Note that for the set CH_4, SiH_4, GeH_4 and SnH_4 the boiling points increase with increasing mass of the molecules. The hydrides NH_3, H_2O and HF, however, have boiling points much higher than would be expected from the trends shown by the other members of their set because the process of boiling requires the breaking of hydrogen bonds. These compounds also have abnormally high melting points, latent heats of melting and vaporization because of hydrogen bonding between their molecules.

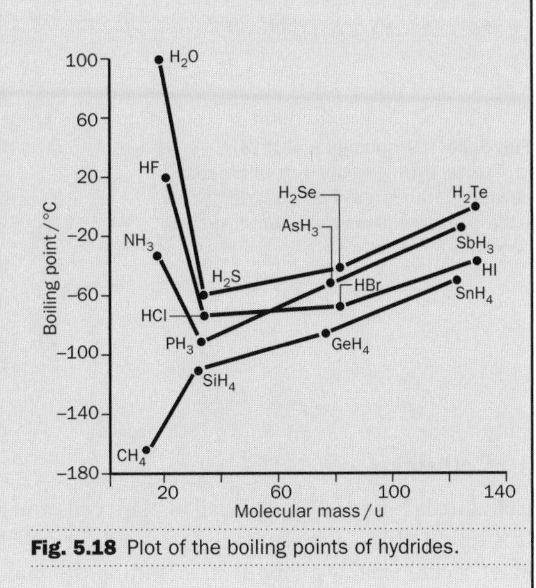

Fig. 5.18 Plot of the boiling points of hydrides.

Hydrogen bonding in biology

1. Proteins

Proteins are the basic 'building blocks' of living tissue. Protein molecules consist of long chains of atoms, containing polar >C=O and H–N< bonds. Hydrogen bonding can occur between these two groups:

$$>C^{\delta+}=O^{\delta-}\cdots\cdots H^{\delta+}-N^{\delta-}<$$

Many protein molecules have the chains of atoms twisted or coiled, and these coils are held together by many hydrogen bonds. These hydrogen bonds help to give the protein its shape (Fig. 5.19).

Because the hydrogen bonds holding the protein molecule together are weak, relative to the covalent bonds in the long chain of the molecule, they can be broken by heating. *Clear* egg white contains proteins suspended in water. When the egg white is heated, the increased vibrations of the protein molecules break the hydrogen bonds giving them their shape and they unravel. The protein turns the familiar colour of cooked egg white. When a protein loses its shape (and function) in this way it is said to be **denatured**.

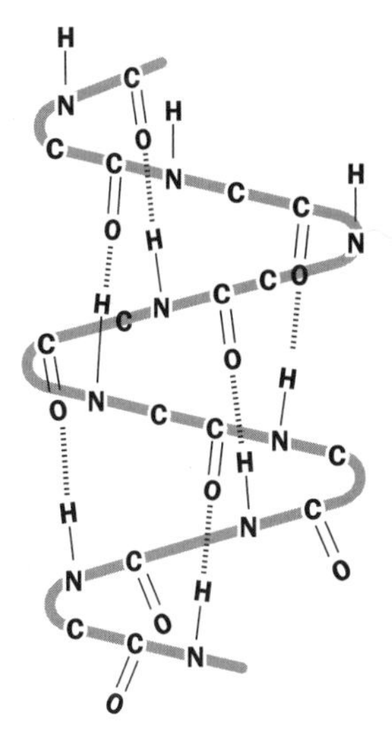

Fig. 5.19 Structure of a protein.

2. DNA

Deoxyribonucleic acid (DNA) stores the genetic information present in living cells. It allows the cell to make proteins according to a definite sequence of atoms. The DNA molecule consists of two long chains molecules twisted around each other and held together by hydrogen bonds This structure is called a **double helix** (Fig. 5.20). For this discovery, F. Crick, M. Wilkins and J. Watson gained the Nobel Prize for Medicine in 1962.

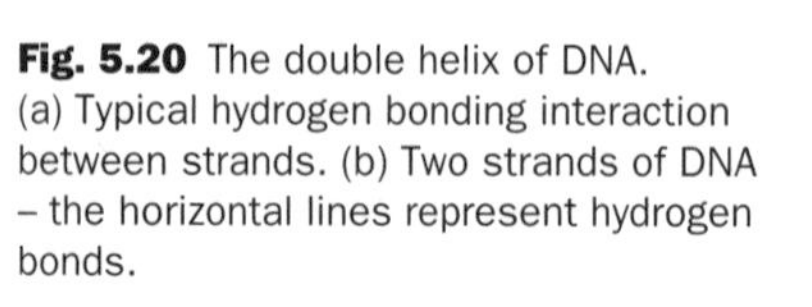

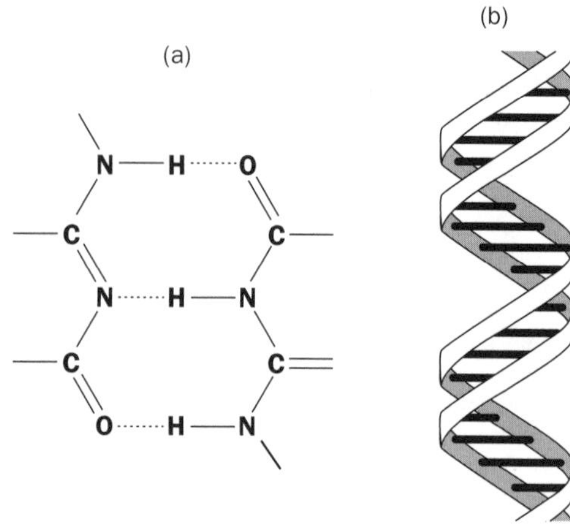

Fig. 5.20 The double helix of DNA. (a) Typical hydrogen bonding interaction between strands. (b) Two strands of DNA – the horizontal lines represent hydrogen bonds.

3. Cellulose

Cellulose is found in the cell wall of plant cells and helps to give plants their structure. It is found in large amounts in trees and cotton fibres (which are nearly pure cellulose). Cellulose has many uses, including the making of paper. Cellulose molecules have many O–H bonds and the strength of wood is due, in part, to hydrogen bonding between nearby molecules.

Strength of bonds: an overview

The last two chapters have discussed the different ways in which atoms bond together and the various attractions between molecules. Table 5.3 gives an idea of the relative strengths of some of these interactions.

Table 5.3 Examples of the approximate relative strengths of interparticle attractions

Interacting species	Description	Typical energies associated with the attractions ($kJ\,mol^{-1}$)
Intramolecular bonding		
Ions e.g. Na^+ and Cl^-	Ionic bonding – attraction between oppositely charged ions	500–4000
Atoms sharing electron pairs e.g. Cl—Cl	A covalent bond – two nuclei share an electron pair	200–1100
Metal atoms e.g. Cu	Metallic bonding – metal cations surrounded by an electron cloud	100–1000
Intermolecular bonding		
Polar molecules e.g. H—Br · · · H—Br	Dipole–dipole – attraction between dipoles	5–25
Non-polar and polar molecules and/or atoms e.g. I_2 · · · Ar	London forces between molecules and/or atoms caused by temporary dipoles in electron clouds	>0–40
Covalently bonded H and covalently bonded O, F or N e.g. O—H · · · F—H	A hydrogen bond – a 'special' dipole–dipole attraction	10–50

BOX 5.7

Relative strengths of London forces and hydrogen bonds

Are hydrogen bonds stronger than London forces? *Not necessarily*, despite the fact that the figures in Table 5.3 suggest this. The dispersion forces between many very *large* molecules, because they contain a large number of electrons, cause those large molecules to have a greater intermolecular attraction than that between small molecules such as water, even though hydrogen bonds are present between water molecules.

Cooking oil contains molecules with high molecular masses and therefore many electrons. The London forces between the molecules are so large that the oil has a higher boiling point than water. Battered fish and chips get their distinctive taste because vaporized water makes holes in the batter and potatoes. If oil did not have a much higher boiling point than water, then battered fish and chips would not have such a crispy texture.

Revision questions

5.1. Draw Lewis structures for the following:

(i) ICl_3 **(ii)** BeF_2 **(iii)** $BFCl_2$ **(iv)** XeF_2 **(v)** CH_2Cl_2.

5.2. Determine the shapes of the molecules in Question 5.1.

5.3. Draw a Lewis structure for covalent HCN. Determine the shape of the molecule.

5.4. Which of the following molecules have a dipole moment?

(i) SF_6 **(ii)** XeF_2 **(iii)** SO_3 **(iv)** SO_2 **(v)** CH_2Cl_2.

5.5. What physical evidence is there for the existence of London dispersion forces?

5.6. Which metal would you expect to have the highest melting point K or Ca? Explain your reasoning.

5.7. Explain why diamond does not conduct electricity, but graphite does.

5.8. Explain the type of intermolecular bonding that exists in solid carbon dioxide (**dry ice**).

5.9. Water molecules have stronger hydrogen bonding between them than ammonia molecules. Why?

5.10. Propanone (old name acetone) has the structure

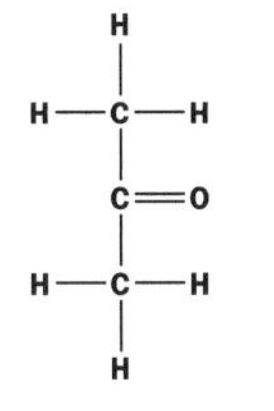

Is it possible for molecules of propanone to hydrogen bond together in the liquid state? Explain your reasoning

5.11. Since the discovery of buckminsterfullerene, a number of carbon molecules with a similar spherical structure (called fullerenes) have been made. It was once believed that any fullerene smaller than C_{60} would be too unstable to isolate in bulk, but in 1998 the Zettl Research Group announced the preparation of C_{36}. As with other fullerenes, the spherical molecule of C_{36} is composed of pentagons and hexagons of carbon atoms. Using a model kit (or plasticine and straws) can you make a model of the molecule and decide how many hexagons and pentagons it contains?

Extension material to support this unit is available on our website at **http://www.palgrave.com/foundations/lewis.**

Reactions of Ions in Solution

Contents

Objectives

- ▶Explains how to construct ionic equations
- ▶Discusses the reactions of HCl(g), H_2SO_4(l) and NH_3(g) with water
- ▶Lists the reactions of acids and bases
- ▶Outlines how simple tests may be used to identify common anions and cations

In Chapter 4 we saw that ions are atoms (or groups of atoms bonded together) which have lost or gained electrons. Loss of electrons produces positively charged ions (**cations**), while the gain of electrons produces negatively charged ions (**anions**). This chapter reviews some reactions of common ions. Such reactions are responsible for most of the coloured products and bubbling gases often associated with the chemistry laboratory.

6.1 Dissolution of salts in water

Ions produced when salts dissolve

Water is a covalently bonded substance, and if we want substantial numbers of ions in water we have to introduce them ourselves. The easiest way is to add a soluble salt such as sodium chloride.

Dissolving sodium chloride destroys its crystal structure and frees the Na^+ and Cl^- ions which can then disperse through the body of the water. This process is called **ionic dissociation**. The ions are loosely attracted to the surrounding water molecules and are said to be **hydrated**. We represent this by 'aq', an abbreviation for 'aqueous' which means watery. In our example we write Na^+(aq) and Cl^-(aq). The sum of the dissociation and hydration steps is called **dissolution**. The dissolution may be summarized by the equation:

$$Na^+,Cl^-(s) \xrightarrow{H_2O} Na^+(aq) + Cl^-(aq)$$

The table of valencies for common ions on page 431 lists many of the ions met in the college or school laboratory. (You are advised to learn this table.) We can use the table to work out the ions produced when we dissolve soluble salts in water.

Example 6.1

What ions are produced when magnesium nitrate dissolves in water?

▶ Answer

The formula of magnesium nitrate is $Mg(NO_3)_2$. Its ionic formula is $Mg^{2+},2NO_3^-$. This generates magnesium ($Mg^{2+}(aq)$) and nitrate ($NO_3^-(aq)$) ions in water:

$$Mg^{2+},2NO_3^-(s) \xrightarrow{H_2O} Mg^{2+}(aq) + 2NO_3^-(aq)$$

BOX 6.1

Composition of natural waters

It has been estimated that there is 1.4×10^{21} dm^3 of water on the Earth, with 97% of the water being present in the oceans. About 8×10^{18} dm^3 is tied up as fresh water rivers and 3×10^{19} dm^3 as ice in ice-caps and the glaciers. Table 6.1 shows the average concentrations of ions in rain, riverwater and ocean water. The units of concentration are milligrams per dm^3.

The main ions to be found in the oceans are sodium and chloride, i.e. seawater contains a relatively high concentration of common salt. Note that fluoride ions are not found in fresh water.

Table 6.1 Average composition (mg dm^{-3}) of rain, river water and sea water

Ion	Rain	River water	Sea water
Na^+	2.0	6.3	10 770
K^+	0.3	2.3	398
Mg^{2+}	0.3	4.1	1 290
Ca^{2+}	0.6	15.0	412
Cl^-	3.8	7.8	19 500
SO_4^{2-}	2.0	11.2	900
HCO_3^-	0.1	58.4	28
F^-	0.0	0.0	1.3

Source: *Encyclopaedia of Physical Science and Technology*, Academic Press (1987) Vol 5, p. 260.

Exercise 6A

Dissolution of salts in water

Write equations to show the dissolution in water of:

(i) calcium nitrate

(ii) potassium sulfate

(iii) sodium carbonate decahydrate ($Na_2CO_3 \cdot 10H_2O$)

6.2 Ionic equations

Reaction of zinc metal with copper ions

The reaction between copper(II) sulfate solution and zinc metal makes zinc(II) sulfate solution and copper metal:

$$CuSO_4(aq) + Zn(s) \rightarrow ZnSO_4(aq) + Cu(s)$$

During the reaction, the blue copper(II) sulfate solution changes to colourless zinc(II) sulfate solution and the copper metal settles out.

The equation suggests that zinc atoms react directly with copper(II) sulfate. However, this is misleading – copper(II) sulfate is a soluble salt and is therefore fully dissociated into ions in solution. We can get a better picture of the reaction that occurs if we write out the ions that are present as reactants and as products. In doing so, we will be writing out an **ionic equation**.

How to assemble the ionic equation for the reaction between copper(II) sulfate and zinc metal

1. Look at the table of valencies and symbols of ions (page 431). This shows that copper(II) ions are symbolized Cu^{2+} and that sulfate ions are symbolized SO_4^{2-}. In solution, copper(II) sulfate consists of separate copper(II) and sulfate ions. In all ionic equations, solids (whether metals or insoluble compounds) are represented by their chemical formulae. Here, we represent the zinc metal as Zn(s). The left-hand side of the chemical equation may now be written as

   $$Cu^{2+}(aq) + SO_4^{2-}(aq) + Zn(s) \rightarrow$$

2. Use of the valence table allows us to write $ZnSO_4(aq)$ as $Zn^{2+}(aq)$ and $SO_4^{2-}(aq)$, respectively. The copper metal is symbolized as Cu(s). The right-hand side of the equation becomes:

   $$\rightarrow Zn^{2+}(aq) + SO_4^{2-}(aq) + Cu(s)$$

3. The two halves are now written together:

   $$Cu^{2+}(aq) + SO_4^{2-}(aq) + Zn(s) \rightarrow Zn^{2+}(aq) + SO_4^{2-}(aq) + Cu(s)$$

4. The sulfate ion appears unchanged on both sides of the equation. This shows that the sulfate ion is not involved in the chemical reaction. The sulfate ion is termed a **spectator ion** and may be eliminated in the same way as identical terms may be cancelled on either side of a mathematical equation:

   $$Cu^{2+}(aq) + \cancel{SO_4^{2-}}(aq) + Zn(s) \rightarrow Zn^{2+}(aq) + \cancel{SO_4^{2-}}(aq) + Cu(s)$$

   giving

   $$Cu^{2+}(aq) + Zn(s) \rightarrow Zn^{2+}(aq) + Cu(s)$$

The last equation shows that the reaction does not involve sulfate ions, but that the products are made as a result of reaction between copper(II) ions and zinc atoms. In Fig. 6.1(a) all the ions present in the reactants and products are shown. In Fig. 6.1(b), the sulfate spectator ions have been removed and it is now easier to see that only the copper ions and zinc atoms are involved in the reaction.

The reaction of $Cu^{2+}(aq)$ and Zn(s) is also an example of a *redox reaction* (see page 100).

Exercise 6B

Ionic equations

Assemble the ionic equation for the reaction of copper(II) nitrate with zinc metal:

$$Cu(NO_3)_2(aq) + Zn(s) \rightarrow Zn(NO_3)_2(aq) + Cu(s)$$

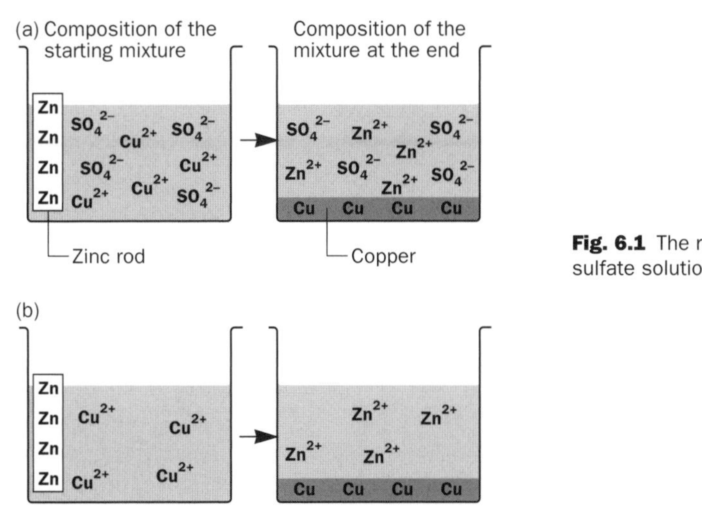

Fig. 6.1 The reaction of copper(II) sulfate solution with zinc metal.

Solubilities of ionic compounds

Table 6.2 shows the solubilities of common ionic compounds. If an ionic compound is insoluble, then that substance will fall out of (or **precipitate** from) solution when the appropriate ions are present. For example, if copper(II) sulfate solution is added to sodium carbonate solution, Table 6.2 shows that the combination of copper(II) and carbonate ions forms a precipitate of copper(II) carbonate:

$$Cu^{2+}(aq) + CO_3^{2-}(aq) \rightarrow \underset{\text{copper carbonate}}{CuCO_3(s)}$$

Where the compound is listed in Table 6.2 as slightly soluble, the precipitate may appear only as a milkiness in the solution.

Table 6.2 The colours of common cations and the solubilities of common salts. With three exceptions, all the cations are colourless in solution. All the anions are colourless in solution.

Cation	Cl^-, Br^-, I^-	SO_4^{2-}	CO_3^{2-}	S^{2-}	OH^-	NO_3^-	CH_3COO^-
Na^+, K^+, NH_4^+							
Mg^{2+}			•	d	•		
Ca^{2+}		•	•	d	○		
Ba^{2+}		•	•	d	○		
Al^{3+}			–	d	•		
Zn^{2+}			•	•	•		
Cu^{2+} (blue)			•	•	•		
Fe^{2+} (pale green)			•	○	•		
Fe^{3+} (yellow brown)			–	d	•		
Pb^{2+}	•	•	•	•	•		
Ag^+	•	○	•	•	–		

A blank indicates that the salt is soluble, • = salt is insoluble, ○ = salt is slightly soluble, d = decomposes in water, – = does not exist

Exercise 6C

Solubilities

(i) Which zinc salts are insoluble in water?

(ii) Name two compounds which when dissolved in water will produce a precipitate of barium sulfate.

(iii) Name two compounds which when dissolved in water will produce a precipitate of iron(II) hydroxide.

6.3 Producing ions in water by chemical reaction

The dissolution of an ionic salt is an example of physical change, since the ions are present in both solid salt and in solution. By definition, covalent substances contain no ions, but some covalent substances may *produce* ions as a result of their chemical reaction with water. The presence of such ions may be confirmed by testing the solution to see if it conducts electricity, but it is the new reactions of the solution – the reactions of the ions produced – that conclusively prove that chemical change has taken place. We now look at three important examples: the reactions with water of hydrogen chloride, pure sulfuric acid and ammonia.

Reaction of hydrogen chloride with water

Hydrogen chloride (HCl) is a gas at room temperature. The HCl molecule is covalent but it reacts with water producing a mixture of **hydronium** and chloride ions that is commonly referred to as hydrochloric acid (Fig. 6.2):

$$HCl(g) + H_2O(l) \rightarrow \underset{\text{hydronium ion}}{H_3O^+(aq)} + Cl^-(aq)$$

The hydronium ion is a **hydrated hydrogen ion**, and is often abbreviated as $H^+(aq)$. The reaction of HCl(g) with water may alternatively be written as

$$HCl(g) \xrightarrow{H_2O} H^+(aq) + Cl^-(aq)$$

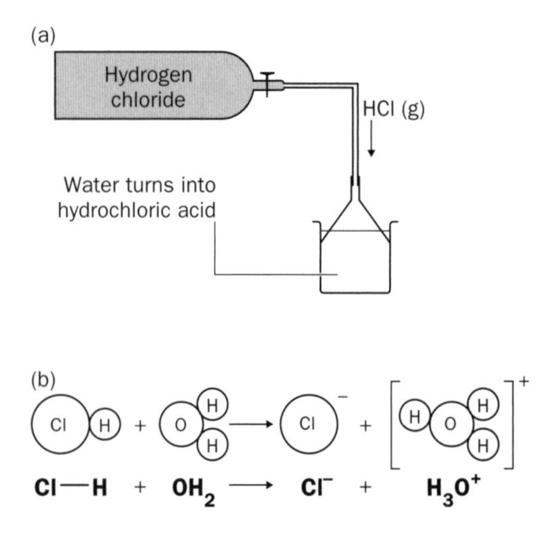

Fig. 6.2 Making hydrochloric acid. (a) Hydrogen chloride gas from a cylinder is brought into contact with water. A funnel maximizes the contact between gas and liquid. At room temperature, 100 g of water will hold up to 37 g of HCl(g). (b) Shows the reaction of water and HCl(g) on a molecular scale.

The reaction between HCl gas and water makes hydrogen chloride extremely soluble in water (see the fountain experiment, page 186). A considerable amount of heat is also given out, although this is quickly absorbed by the water.

Reactions of hydrochloric acid

The reactions of hydrochloric acid are simply the reactions of the hydronium and chloride ions. Concentrated and dilute hydrochloric acid display similar reactions, but the concentrated acid reacts at a faster rate.

Exercise 6D

Hydronium ion

In the reaction between HCl(g) and $H_2O(l)$, a proton (H^+) bonds to the water molecule by sharing a pair of electrons from the oxygen atom in the water molecule. Do you remember the name of this type of covalent bond?

Exercise 6E

Reaction of hydrogen bromide and water

Hydrogen bromide, HBr(g), dissolves in water in a similar way to HCl(g) to make hydrobromic acid, HBr(aq). Represent this by an ionic equation.

The H^+(aq) ion is the ion that is responsible for the *reactions of all acids*. Such reactions are known as **acidic properties**. One of the best known reactions of H^+(aq) ions and, therefore, of acids in general, is their ability to cause **acid–base indicators** to change colour. The commonest indicator is litmus (litmus is usually absorbed in paper) which turns red in an acidic solution and blue in basic (alkaline) solution. It follows that dry HCl gas does not affect dry blue litmus paper, but if the HCl gas or the litmus is damp, then the litmus will turn red.

The presence of the chloride ion in hydrochloric acid may be confirmed by adding silver ions, when a white precipitate of silver chloride (AgCl) is seen:

$$Ag^+(aq) + Cl^-(aq) \rightarrow AgCl(s)$$

Silver nitrate solution is usually used as the source of silver ions. A precipitate is obtained by mixing silver ions with *any* chloride and this is used as the basis for a test to detect the presence of a chloride. In testing unknown solutions using silver nitrate, it is usual to add dilute nitric acid to prevent the precipitation of other insoluble silver compounds, such as silver carbonate and silver sulfite.

Before moving on, it is interesting to note that hydrogen chloride gas dissolves in benzene (a non-polar solvent) without producing any ions. Such a solution contains hydrogen chloride molecules weakly attracted (solvated) to the surrounding benzene molecules. As we might expect, a solution of HCl in benzene does not conduct electricity and it does not show any of the reactions of hydrochloric acid.

Bags of fertilizer. Fertilizers have greatly increased the productivity of crop farmers, and may have prevented large-scale starvation. Many fertilizers are made from ammonia (NH_3), sulfuric acid (H_2SO_4) and phosphoric acid (H_3PO_4).

Exercise 6F

Reactions of silver nitrate

(i) Which of the following would give a white precipitate when added to a mixture of silver nitrate and nitric acid?

(a) potassium sulfate

(b) magnesium nitrate

(c) calcium chloride.

(ii) If a solution containing bromide ions (Br^-) is added to silver nitrate solution, a cream precipitate is produced. If a solution containing iodide ions (I^-) is added to silver nitrate solution, a pale yellow precipitate is produced. What do you think these precipitates are? Write an ionic equation for each precipitation reaction.

Reaction of pure sulfuric acid with water

Sulfuric acid (H_2SO_4) is usually sold as concentrated sulfuric acid, which contains 98% sulfuric acid and 2% water by mass: this is sufficiently pure to justify the name pure sulfuric acid.

Pure sulfuric acid (like hydrogen chloride) is covalent. Unlike HCl(g), however, it is a liquid at room temperature. Pure sulfuric acid is a colourless oily liquid that looks and pours like glycerine, but there the similarity ends. If pure sulfuric acid is added to water, a violent reaction occurs. Much heat is produced. The resulting colourless solution is called dilute sulfuric acid. The overall equation for the reaction is:

$$H_2SO_4(l) \xrightarrow{H_2O} 2H^+(aq) + SO_4^{2-}(aq)$$

Fig. 6.3 Always add acid to water, never water to acid.

The reaction of pure sulfuric acid with water is so violent that dilute sulfuric acid should always be prepared by slowly adding the pure acid to water. Under no circumstances should water be added to the pure acid. So much heat would be generated where the two liquids mix that some of the mixture might spit out into your face (Fig. 6.3).

Pure sulfuric acid is so greedy for water that it is commonly used as a drying agent. As most gases are only slightly soluble in the pure acid, they may be dried by bubbling the gas through the acid contained in a glass bottle. However, the greed of the acid for water also means that it is extremely corrosive to the skin and eyes.

Use of concentrated sulfuric acid in the brown ring test for nitrates

Concentrated sulfuric acid is used to identify the nitrate ion (NO_3^-), in a test called the **brown ring test**. (Another test for nitrates is described on page 116.) This involves the careful addition of concentrated sulfuric acid to a solution of the suspected nitrate. A solution of iron(II) sulfate is then slowly added to the mixture. If a nitrate is present, the mixture produces nitric oxide gas (NO) which reacts with $Fe^{2+}(aq)$ to produce a brown ring of $Fe(NO)^{2+}$ at the junction of the acid mixture and the $FeSO_4$ solution:

$$Fe^{2+}(aq) + NO(aq) \rightarrow Fe(NO)^{2+}(aq)$$

Reactions of dilute sulfuric acid

The reactions of dilute sulfuric acid are really the reactions of the $H^+(aq)$ and $SO_4^{2-}(aq)$ ions. The $H^+(aq)$ ions provide the acidic properties. The presence of a sulfate ion is confirmed by adding barium ($Ba^{2+}(aq)$) ions, when a dense white precipitate of barium sulfate is observed:

$$Ba^{2+}(aq) + SO_4^{2-}(aq) \rightarrow BaSO_4(s)$$

The usual source of barium ions is barium chloride solution. In order to make this test specific for $SO_4^{2-}(aq)$ ions, hydrochloric acid is added to prevent barium sulfite or barium carbonate from precipitating. (See Exercise 6G.)

Exercise 6G

Reactions of barium chloride

Which of the following would give a white precipitate when added to a mixture of barium chloride and hydrochloric acid?

(i) potassium sulfate

(ii) magnesium carbonate

(iii) sodium chloride.

Reaction of ammonia gas with water

Ammonia gas ($NH_3(g)$) is extremely soluble in water. The reaction is

$$NH_3(g) + H_2O(l) \rightleftharpoons \underbrace{NH_4^+(aq)}_{\text{ammonium ion}} + OH^-(aq)$$

Ammonia solution is sometimes (incorrectly) called ammonium hydroxide. A saturated solution at room temperature contains about 28% of ammonia by mass. The density of such a solution is $0.88\,g\,cm^{-3}$. For this reason a saturated ammonia solution is referred to as '88' ammonia.

The $OH^-(aq)$ ion is the **hydroxide ion**. It is this ion that gives solutions of bases their characteristic reactions; for example, it causes red litmus to turn blue.

The $\rightleftharpoons$ symbol reflects the fact that it is impossible to convert all the ammonia molecules to ammonium and hydroxide ions. In other words, the reaction is a **reversible** or **equilibrium** reaction.

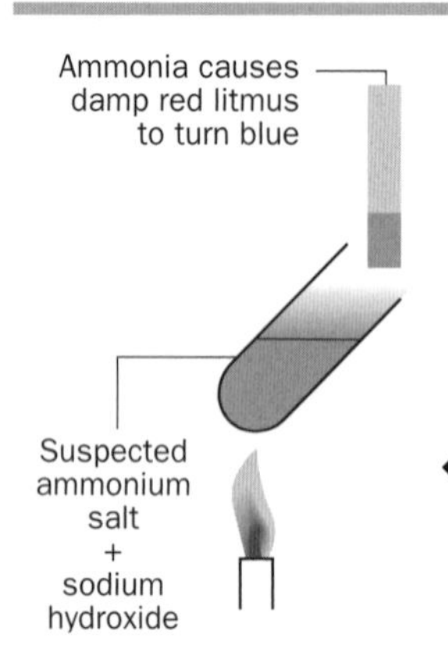

Fig. 6.4 Testing a solution to see if it contains ammonium (NH_4^+) ions. If ammonia is detected, the presence of the ammonium ions is confirmed.

Heating a mixture of ammonium and hydroxide ions causes them to recombine making ammonia gas and water. This works whatever the source of the ammonium ions. For example, heating ammonium chloride with sodium hydroxide solution produces ammonia. This is used as a laboratory test to confirm the presence of the ammonium ion. The ammonia gas is detected by its very characteristic smell (usually described as pungent) and by the fact that it turns damp red litmus blue (Fig. 6.4 and Exercise 6H).

Exercise 6H

Reactions that produce ammonia gas

Which of the following would be expected to produce ammonia gas when heated with sodium hydroxide solution?

(i) lead(II) nitrate **(ii)** ammonium carbonate **(iii)** calcium carbonate.

Write an ionic equation for the reaction(s) that are involved.

6.4 Acids and bases

Acids

An acid is a substance that produces $H^+(aq)$ ions when dissolved in water.

The commonest acids in the laboratory are sulfuric acid (H_2SO_4), hydrochloric acid (HCl) and nitric acid (HNO_3). These are known as the mineral acids.

Another common acid is ethanoic acid (acetic acid, CH_3COOH), a smelly liquid which boils at 118 °C. Ethanoic acid is the chemical that gives vinegar its sour taste. Its full structure is as follows:

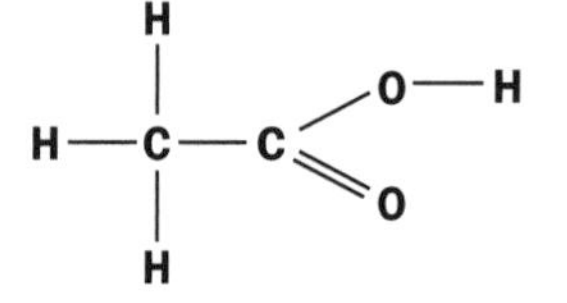

Only one of the hydrogen atoms in this molecule makes a hydrogen ion in solution, and for this reason the hydrogen atom is called an **acidic hydrogen**. In ethanoic acid, the acid hydrogen is the one bonded to the oxygen atom. The ionization of ethanoic acid may then be represented by the equation

$$CH_3COOH(l) \overset{H_2O}{\rightleftharpoons} \underset{\text{ethanoate (or acetate) ion}}{CH_3COO^-(aq)} + H^+(aq)$$

or,

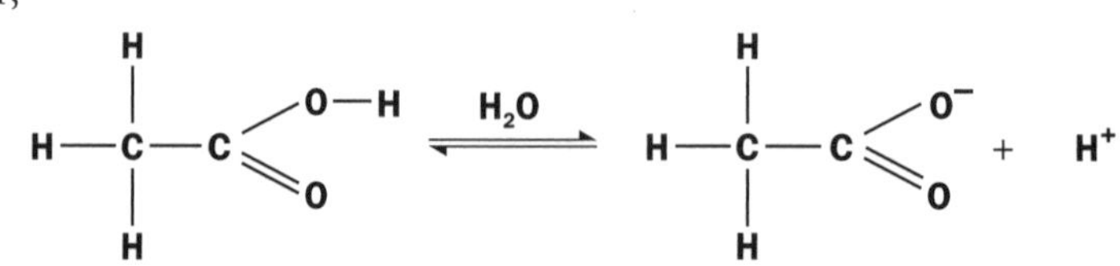

The $\rightleftharpoons$ sign shows that only some of the ethanoic acid molecules are ionized in solution.

Exercise 6I

Number of acidic hydrogen atoms

1. How many acidic hydrogen atoms do molecules of **(i)** sulfuric acid and **(ii)** hydrogen chloride possess?
2. Hydrochloric acid forms salts known as chlorides. What are names of the types of salts produced from **(i)** ethanoic acid and **(ii)** nitric acid?

Bases and alkalis

A base is a substance that reacts with an acid in solution producing a salt and water only.

Generalizing:

acid + base → salt + water

This reaction is known as **neutralization.**

An alkali is a base that dissolves in water (Fig. 6.5). A solution of an alkali contains the hydroxide ion, $OH^-(aq)$.

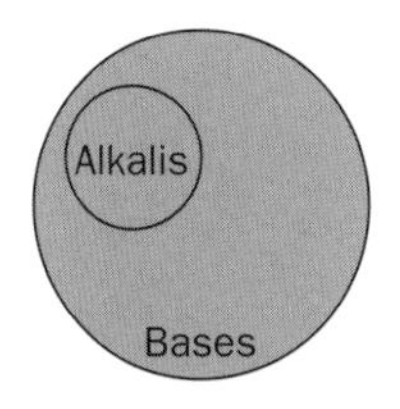

Fig. 6.5 Bases and alkalis – although all alkalis are bases, not all bases are alkalis.

Alkalis are usually hydroxides of metals. The common alkalis are the hydroxides of calcium, potassium and sodium. They are all are ionic solids which completely dissociate into ions in water, for instance:

$$\underset{\text{sodium hydroxide}}{Na^+,OH^-(s)} \xrightarrow{H_2O} Na^+(aq) + \underset{\text{hydroxide ion}}{OH^-(aq)}$$

Ammonia solution is also regarded as an alkali because it contains the hydroxide ion.

Bases which are insoluble in water include the oxides of metals such as magnesium oxide (Mg^{2+},O^{2-}) and copper(II) oxide (Cu^{2+},O^{2-}) and organic compounds (compounds based on carbon) which contain nitrogen atoms such as propylamine ($C_3H_7NH_2$).

Exercise 6J

Dissolution of potassium and calcium hydroxides

Write an equation showing the dissolution of potassium and calcium hydroxides in water.

6.5 Reactions of acids

The reactions of acids are the reactions of the $H^+(aq)$ ion).

1. Effect on acid–base indicators

Taking litmus as our example:

$$\text{red coloured litmus} \underset{H^+(aq)}{\overset{OH^-(aq)}{\rightleftharpoons}} \text{blue coloured litmus}$$

2. Reaction with bases (neutralization)

Acids neutralize bases, producing a salt and water only. Examples of reactions of alkalis with acids include

$$NaOH(aq) + HCl(aq) \rightarrow NaCl(aq) + H_2O(l)$$

$$2KOH(aq) + H_2SO_4(aq) \rightarrow K_2SO_4(aq) + 2H_2O(l)$$

Both neutralization reactions reduce to the ionic equation

$$H_3O^+(aq) + OH^-(aq) \rightarrow 2H_2O(l)$$

or, more simply, as

$$H^+(aq) + OH^-(aq) \rightarrow H_2O(l)$$

Two examples of the reactions of insoluble bases with acids are

$$CuO(s) + H_2SO_4 \rightarrow CuSO_4(aq) + H_2O(l)$$

$$MgO(s) + 2HNO_3(aq) \rightarrow Mg(NO_3)_2(aq) + H_2O(l)$$

Both reactions reduce to the ionic equation

$$O^{2-}(s) + 2H^+(aq) \rightarrow H_2O(l)$$

where $O^{2-}(s)$ is the oxide ion.

A dramatic example of the reaction between H_2 and O_2. The airship *Hindenberg* contained 200 000 m^3 of hydrogen in gas bags. A mixture of air (oxygen) and hydrogen is extremely flammable, and a stray spark caused the *Hindenberg* to explode as it was landing in the USA in 1937. The gas helium is also less dense than air, and since it does not burn is much safer to use in airships than hydrogen. However, helium is very much more expensive.

The reaction of organic bases such as propylamine $C_3H_7NH_2$ with acids also produces salts:

$$C_3H_7NH_2(l) + HCl(aq) \rightarrow C_3H_7NH_3 \cdot Cl(aq)$$

($C_3H_7NH_3 \cdot Cl$ is an ionic salt similar to NaCl, and is better represented as $C_3H_7NH_3^+,Cl^-$. It dissociates in aqueous solution forming $C_3H_5NH_3^+(aq)$ and $Cl^-(aq)$ ions.)

Organic acids containing long carbon chains are known as 'fatty acids'. The neutralization of fatty acids by alkalis produces salts which are soapy to the feel. Fatty acids are present on our skin, and this explains why alkalis feel soapy. Sodium stearate (common soap) is produced by neutralising stearic acid with sodium hydroxide. (For more about soap, see page 175.)

3. Reaction with metals

Solutions of acids react with some metals, making hydrogen gas and a salt. For example, magnesium ribbon reacts with hydrochloric acid, producing magnesium chloride and hydrogen gas. The hydrogen gas is produced so quickly that the solution fizzes:

$$Mg(s) + 2HCl(aq) \rightarrow MgCl_2(aq) + H_2(g)$$

The ionic equation for this reaction is:

$$Mg(s) + 2H^+(aq) \rightarrow Mg^{2+}(aq) + H_2(g)$$

We may test for hydrogen on a **small** scale by exploding a hydrogen–air mixture in a flame (Fig. 6.6):

$$2H_2(g) + O_2(g) \rightarrow 2H_2O(l)$$

Other metals that react with $H^+(aq)$ include calcium, zinc, aluminium and iron. Potassium and sodium react with explosive violence. Copper, silver and gold do not react.

Exercise 6K

Reaction of iron with dilute sulfuric acid

Write **(i)** a chemical equation and **(ii)** an ionic equation, for the reaction of iron with dilute sulfuric acid to produce iron(II) sulfate and hydrogen gas.

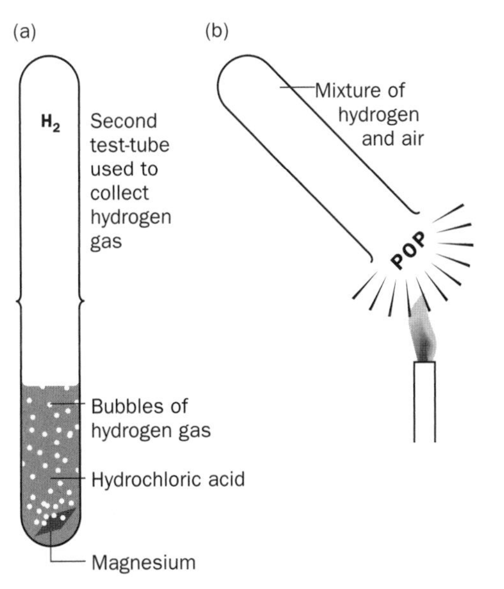

Fig. 6.6 Testing for hydrogen gas. An inverted test-tube is used to collect the hydrogen (which rises because it is less dense than air). After about 20 seconds, the open end of the test-tube is exposed to a flame. A 'pop' confirms the presence of hydrogen.

4. Reaction with carbonates and hydrogencarbonates

All acids react with carbonates to produce carbon dioxide gas. For example, the ethanoic acid in vinegar will fizz with sodium carbonate. The reaction is

$$\underset{\text{ethanoic acid in vinegar}}{2CH_3COOH(aq)} + Na_2CO_3(aq) \rightarrow \underset{\text{sodium ethanoate}}{2CH_3COONa(aq)} + CO_2(g) + H_2O(l)$$

The ionic equation for this reaction is

$$\underset{\text{any acid}}{2H^+(aq)} + \underset{\text{carbonate ion}}{CO_3^{2-}(aq)} \rightarrow CO_2(g) + H_2O(l)$$

Hydrogencarbonates (also called bicarbonates) also fizz with acid:

$$H^+(aq) + \underset{\text{hydrogencarbonate ion}}{HCO_3^-(aq)} \rightarrow CO_2(g) + H_2O(l)$$

(Note that hydrogencarbonates decompose upon *heating* giving $CO_2(g)$:

$$2HCO_3^-(aq) \rightarrow CO_2(g) + H_2O(l)$$

The decomposition of solid sodium hydrogencarbonate ('kitchen bicarb') is utilized in cooking, with the CO_2 gas causing cake mixtures to rise.)

5. Reaction with sulfites

Sulfites contain the ion SO_3^{2-}. (Don't confuse this ion with the sulfate ion (SO_4^{2-}), which does not react with acids.) Acids react with sulfites making the sharp-smelling gas sulfur dioxide (SO_2). The ionic equation is

$$\underset{\text{sulfite ion}}{SO_3^{2-}(aq)} + 2H^+(aq) \rightarrow SO_2(g) + H_2O(l)$$

Hydrogensulfites (also called bisulfites) also react with acid:

$$H^+(aq) + \underset{\text{hydrogensulfite ion}}{HSO_3^-(aq)} \rightarrow SO_2(g) + H_2O(l)$$

6. Reaction with sulfides

The reaction of acids with sulfides (compounds containing the sulfide ion, S^{2-}) produces the foul-smelling gas hydrogen sulfide – one of the products of rotten eggs. The ionic equation for the reaction is

$$\underset{\text{sulfide ion}}{S^{2-}(s)} + 2H^+(aq) \rightarrow H_2S(g)$$

Metal ions often react with sulfide ions to produce insoluble sulfides. For example,

$$Cu^{2+}(aq) + S^{2-}(aq) \rightarrow \underset{\text{black}}{CuS(s)}$$

Sodium sulfide solution or hydrogen sulfide gas are used as the source of sulfide ions. Solutions of Na^+, Ca^{2+} and K^+ will not produce precipitates because their sulfides are water soluble. Zinc sulfide (ZnS) is white, while lead(II) and silver sulfides (PbS and Ag_2S) are black.

6.6 Acids produced when gases CO_2, SO_2 and NO_2 dissolve in water

Carbon dioxide and sulfur dioxide gases are both water soluble. Some of the dissolved gas molecules then react with water producing hydrogencarbonate and hydrogensulfite ions, respectively. The reactions are:

$$CO_2(aq) + H_2O(l) \rightleftharpoons \underset{\text{hydrogencarbonate ion}}{HCO_3^-(aq)} + H^+(aq)$$

$$SO_2(aq) + H_2O(l) \rightleftharpoons \underset{\text{hydrogensulfite ion}}{HSO_3^-(aq)} + H^+(aq)$$

The $\rightleftharpoons$ sign shows that both reactions do not go to completion, i.e. attempts to react the gases completely with water will always produce an equilibrium mixture of unreacted gas and hydrogencarbonate or hydrogensulfite and hydronium ions. Under the same conditions, more SO_2 molecules react with water than do CO_2 molecules. This, and the fact that $SO_2(g)$ is more soluble in water than $CO_2(g)$, means that saturated solutions of SO_2 are more acidic (contain a higher H^+concentration) than solutions of CO_2. This explains why $CO_2(g)$ only turns damp blue litmus paper a purple/red colour, whereas $SO_2(g)$ produces a much more definite red colour.

In the presence of oxygen, a solution of sulfur dioxide is very slowly converted to dilute sulfuric acid at room temperature:

$$SO_2(aq) + \tfrac{1}{2}O_2(aq) + H_2O(l) \rightarrow SO_4^{2-}(aq) + 2H^+(aq)$$

In the atmosphere, the conversion of $SO_2(aq)$ to sulfuric acid within clouds occurs faster, and is the main source of acid rain (see page 415).

If CO_2 or SO_2 are dissolved in *alkaline* solutions, all the hydrogencarbonate and hydrogensulfite ions are converted to carbonates and sulfites, respectively:

$$HCO_3^-(aq) + OH^-(aq) \rightarrow CO_3^{2-}(aq) + H_2O(l)$$

$$HSO_3^-(aq) + OH^-(aq) \rightarrow SO_3^{2-}(aq) + H_2O(l)$$

Summarizing, when CO_2 and SO_2 gases are bubbled into a beaker of water, a mixture of undissolved gas and hydrogencarbonate or hydrogensulfite ions are present. However, in the presence of $OH^-(aq)$, the only species present are carbonate or sulfite ions.

Nitrogen dioxide ($NO_2(g)$), is a toxic brown gas which dissolves in water producing a highly acidic cocktail of hydronium ions, nitrate ions ($NO_3^-(aq)$) and nitric oxide gas (NO):

$$3NO_2(g) + H_2O(l) \rightarrow 2H^+(aq) + NO(g) + 2NO_3^-(aq)$$

6.7 Reactions of the hydroxide ion

The hydroxide ion in solution is normally written as $OH^-(aq)$.

1. Effect on litmus

Litmus turns blue in basic solution.

2. Neutralization of acids

The $OH^-(aq)$ ion neutralizes acids giving a salt and water only.

3. Reaction with NH_4^+ ion

Ammonia gas is produced:

$$NH_4^+(aq) + OH^-(aq) \xrightarrow{\text{heat}} H_2O(l) + NH_3(g)$$

4. Reaction with some metal ions in solution: the use of precipitation in analysis

The hydroxide ion causes insoluble metal hydroxides to precipitate out of solution. For example, if we add a few drops of sodium hydroxide solution to (blue) copper(II) sulfate solution, a beautiful blue precipitate of copper(II) hydroxide is observed:

$$Cu^{2+}(aq) + 2OH^-(aq) \rightarrow \underset{\text{blue precipitate}}{Cu(OH)_2(s)}$$

The $Cu^{2+}(aq)$ comes from the copper(II) sulfate and the $OH^-(aq)$ comes from the sodium hydroxide. Similar reactions take place with other metal ions:

$$\underset{\text{zinc ion (colourless)}}{Zn^{2+}(aq)} + 2OH^-(aq) \rightarrow \underset{\text{white precipitate}}{Zn(OH)_2(s)}$$

$$\underset{\text{aluminium ion (colourless)}}{Al^{3+}(aq)} + 3OH^-(aq) \rightarrow \underset{\text{white precipitate}}{Al(OH)_3(s)}$$

$$\underset{\text{iron(II) ion (pale-green colour)}}{Fe^{2+}(aq)} + 2OH^-(aq) \rightarrow \underset{\text{white precipitate which almost instantaneously turns green}}{Fe(OH)_2(s)}$$

$$\underset{\text{iron(III) ion (yellow/brown colour)}}{Fe^{3+}(aq)} + 3OH^-(aq) \rightarrow \underset{\text{reddish-brown precipitate}}{Fe(OH)_3(s)}$$

$$\underset{\text{lead(II) ion (colourless)}}{Pb^{2+}(aq)} + 2OH^-(aq) \rightarrow \underset{\text{white precipitate}}{Pb(OH)_2(s)}$$

$$\underset{\text{magnesium ion (colourless)}}{Mg^{2+}(aq)} + 2OH^-(aq) \rightarrow \underset{\text{white precipitate}}{Mg(OH)_2(s)}$$

Zinc, aluminium and lead(II) hydroxides will dissolve to make clear solutions if we add excess sodium hydroxide solution. This happens because all three hydroxides are **amphoteric** (see Box 6.2). This effect is useful because it helps to distinguish between magnesium hydroxide (whose precipitate is insoluble in excess hydroxide) and zinc, aluminium and lead(II) hydroxides (whose precipitates are soluble in excess hydroxide).

Since zinc, aluminium and lead(II) hydroxides are white, additional tests are needed to distinguish between them. A quick identification test for lead(II) ions depends upon the fact that lead(II) sulfate is insoluble in water, whereas aluminium and zinc sulfates are soluble. Addition of dilute sulfuric acid to lead(II) ions will therefore precipitate the lead(II) sulfate:

$$Pb^{2+}(aq) + SO_4^{2-}(aq) \rightarrow \underset{\text{white precipitate}}{PbSO_4(s)}$$

BOX 6.2

Amphoteric hydroxides

Hydroxides (like all bases) react with acids. For example, aluminium hydroxide dissolves in acid giving $Al^{3+}(aq)$:

$$Al(OH)_3(s) + 3H^+(aq) \rightarrow \underset{\text{colourless}}{Al^{3+}(aq)} + 3H_2O(l)$$

However, some hydroxides will also react with hydroxide ion if it is present in high enough concentration. For example, with aluminium hydroxide the reaction is:

$$Al(OH)_3(s) + OH^-(aq) \rightarrow \underset{\text{colourless}}{[Al(OH)_4]^-(aq)}$$

In this reaction, aluminium hydroxide is acting as an acid. Aluminium hydroxide is said to be **amphoteric**, meaning that it acts like a base or acid depending on whether excess $H^+(aq)$ or $OH^-(aq)$ is present. Zinc and lead(II) hydroxides are also amphoteric and both react in a similar way to aluminium hydroxide.

Zinc and aluminium ions may be distinguished by heating a small amount of their hydroxides on a metal spatula. Aluminium hydroxide decomposes to form white aluminium oxide. Zinc hydroxide also decomposes into the oxide, but zinc oxide (although white when cold) is yellow when hot.

Silver hydroxide does not exist. When hydroxide ions are added to silver ions, a dark brown precipitate of silver oxide is formed:

$$2Ag^+(aq) + 2OH^-(aq) \rightarrow Ag_2O(s) + H_2O(l)$$

Because potassium and sodium hydroxides are very soluble, they do not precipitate when OH^- ions are added to sodium or potassium salts. Calcium hydroxide is not very soluble in water, and appears as a cloudiness when a concentrated solution of OH^- is mixed with a concentrated solution of calcium salts. $Na^+(aq)$, $K^+(aq)$ and $Ca^{2+}(aq)$ may be identified using **flame tests** (see pages 195 and 197).

6.8 Use of reactions in the identification of ions in solution

We have looked at many ion reactions in this chapter, and we shall now look at how simple test-tube experiments may be used to identify common ions. A list of the ions which are relevant to the problems below is given in Table 6.3.

Example 6.1

A solution of a simple salt (AB, where A is the cation and B is the anion) is pale green. Addition of a mixture of barium chloride solution and dilute hydrochloric acid produces a white precipitate. Addition of sodium hydroxide solution to AB produces a green precipitate. Identify A and B.

▶ **Answer**

The colour of the solution suggests that the iron(II) ion is present. This is confirmed by the green precipitate with hydroxide ion. The precipitate with $BaCl_2$ + HCl confirms the presence of a sulfate.

$A = Fe^{2+}(aq)$; $B = SO_4^{2-}(aq)$; $AB = FeSO_4$

Table 6.3 Ions discussed in this chapter

Cations	Anions
H^+, NH_4^+, K^+, Na^+, Ca^{2+}, Cu^{2+}, Fe^{2+}, Fe^{3+}, Pb^{2+}, Al^{3+}, Zn^{2+}, Mg^{2+}, Ba^{2+}, Ag^+	OH^-, Cl^-, SO_4^{2-}, Br^-, I^-, SO_3^{2-}, S^{2-}, CO_3^{2-}, NO_3^-

Example 6.2

A solution, containing a cation Z and an anion Y, fizzed with sodium carbonate solution and gave a cream coloured precipitate when added to a mixture of silver nitrate solution and dilute nitric acid. Identify Z and Y.

▶ Answer

The fizzing suggests that Z is probably $H^+(aq)$. The cream coloured precipitate is probably silver bromide. Therefore, the unknown is a solution of hydrobromic acid.

$Z = H^+(aq)$; $Y = Br^-(aq)$; $ZY = HBr(aq)$

Exercise 6L

Identification of unknowns

(i) A colourless solution CD contains an anion C and a cation D. Addition of sodium hydroxide solution to CD gave a white precipitate H which dissolved in excess hydroxide to produce a clear solution. H turns yellow when strongly heated in a flame. Addition of CD to a mixture of silver nitrate solution and dilute nitric acid gave a white precipitate. Identify C and D.

(ii) A colourless solution is either magnesium carbonate, sodium nitrate or dilute sulfuric acid. Addition of sodium sulfite solution to the unknown (with warming) produces a sharp smelling gas. What is the identity of the colourless solution? Suggest a test which would confirm your answer.

(iii) The labels have fallen off four bottles. One of the bottles contains copper(II) sulfate solution, another sodium hydroxide solution, the third sodium carbonate solution and the fourth dilute hydrochloric acid. Using only these bottles (and clean test tubes) plan tests which would enable you to identify the contents of each bottle.

Revision questions

6.1. **(i)** Pure nitric acid (HNO_3) is a covalent liquid which does not react with dry sodium carbonate. Yet, dilute nitric acid does. Explain this.

(ii) Explain why ammonia gas turns damp red litmus paper blue.

(iii) Why should concentrated sulfuric acid not be used as a drying agent for ammonia gas?

6.2. Write equations showing the ions formed when the following dissolve in water:

(i) copper(II) sulfate crystals ($CuSO_4 \cdot 5H_2O$)

(ii) lithium hydroxide (LiOH, i.e. Li^+,OH^-).

6.3. Oxalic acid is a toxic acid found in rhubarb leaves. It possesses the structure

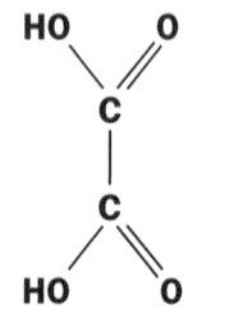

which may also be written as $(COOH)_2$. Both hydrogen atoms are acidic. Write down a chemical and ionic equation for the reaction of oxalic acid with: **(i)** potassium hydroxide solution, **(ii)** zinc metal.

6.4. Write chemical *and* ionic equations for the reaction of:

(i) magnesium carbonate with dilute sulfuric acid

(ii) ammonium sulfate and hot sodium hydroxide

(iii) sodium sulfide and dilute nitric acid

(iv) sodium sulfite (a preservative) and vinegar.

6.5. (i) Much more $CO_2(g)$ is produced annually than $SO_2(g)$, yet $SO_2(g)$ causes greater damage to limestone buildings than CO_2. Why?

(ii) The reaction of calcium hydroxide solution (limewater) with carbon dioxide is the basis of the **limewater test** for CO_2. During the reaction, a white precipitate is produced. Name the precipitate and write an ionic equation for the reaction.

6.6. An insoluble blue solid, containing an anion J and a cation K, fizzed violently with dilute sulfuric acid, producing a clear blue solution B. The gas produced had no smell and solution B gave a blue precipitate with sodium hydroxide solution. Identify J and K. Explain your reasoning.

6.7. A solution of a simple salt GH contains a cation G and an anion H. GH gave a red-brown precipitate when mixed with sodium hydroxide solution. GH gave a brown ring with concentrated H_2SO_4 and $FeSO_4$. Identify the cation and anion present.

6.8. A simple salt ST (containing an anion S and a cation T) is insoluble in water. ST produced hydrogen sulfide when warmed with hydrochloric acid. The resulting solution was divided into two. One half gave a white precipitate with sodium sulfate solution. The other half produced a white precipitate with sodium hydroxide solution which was soluble in excess NaOH. Identify S and T. Explain your reasoning and include ionic equations for any reactions that occur.

6.9. The detrimental effect of acid rain upon fish is partly due to the release of toxic aluminium ions ($Al^{3+}(aq)$) under acidic conditions. Explain this by writing ionic equations which show what happens to $Al^{3+}(aq)$ ions in **(i)** an alkaline environment, **(ii)** in a previously alkaline environment which has been overwhelmed with acid rain.

Extension material to support this unit is available on our website at **http://www.palgrave.com/foundations/lewis.**

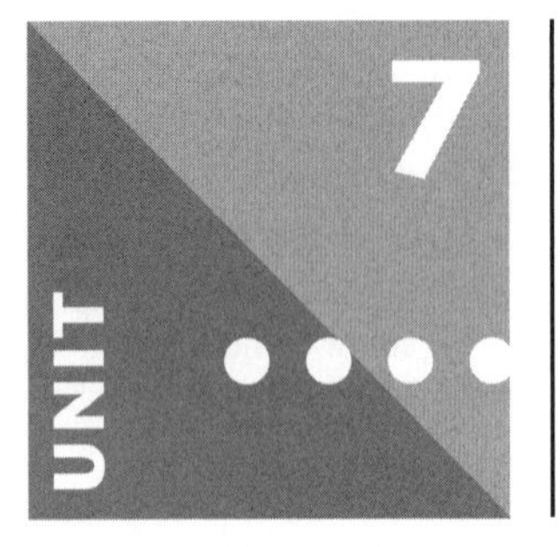

Oxidation and Reduction

Contents

Objectives

- ▶Defines oxidation, reduction and redox reaction
- ▶Explains how to work out oxidation numbers and write redox equations
- ▶Shows you how to use standard reduction potentials to decide if reactions can occur
- ▶Discusses corrosion and redox reactions in nature

7.1 Redox reactions

A **redox** reaction is an important type of chemical reaction. In such a reaction, one reactant is **oxidized** and another reactant is **reduced**.

Definitions of oxidation and reduction

Oxidation was once regarded as a chemical reaction in which oxygen was added to a substance, whereas reduction occurred when oxygen was lost.

For example, in the reaction

$$CO_2(g) + C(s) \rightarrow 2CO(g)$$

carbon is oxidized to carbon monoxide (an atom of carbon gains an oxygen atom), while carbon dioxide is reduced to carbon monoxide (a molecule of carbon dioxide loses an atom of oxygen).

The definition was later broadened to include hydrogen: oxidation was removal of hydrogen from a substance, while reduction took place when hydrogen was added.

These definitions, however, do not help when oxygen and hydrogen are not involved. Consider a reaction such as

$$Cu(s) + S(s) \rightarrow CuS(s)$$

Oxygen and sulfur both have six electrons in the outer shell of their atoms and their reactions tend to be very similar. It is reasonable to expect that copper undergoes much the same process here, as it does when it reacts with oxygen.

When copper reacts with oxygen (**is oxidized**) to form copper(II) oxide, the chemical equation may be written as

$$2Cu(s) + O_2(g) \rightarrow 2CuO(s)$$

In this reaction, a copper atom loses two electrons:

$$Cu \rightarrow Cu^{2+} + 2e^-$$

Each oxygen atom in an oxygen molecule gains two electrons from the copper atom (**the oxygen is reduced**):

$$O + 2e^- \rightarrow O^{2-}$$

Or, since oxygen gas reacts with copper,

$$O_2 + 4e^- \rightarrow 2O^{2-}$$

The overall equation can be obtained by adding the two 'half-reactions' together, so that the electrons cancel out on each side:

1. First multiply the equation for the oxidation of copper by two, so that this half-reaction contains the same number of electrons as the half equation for the reduction of O_2:

$$2Cu \rightarrow 2Cu^{2+} + 4e^-$$

$$O_2 + 4e^- \rightarrow 2O^{2-}$$

2. Then add the two half-reactions and the number of electrons on each side cancels out:

$$2Cu + O_2 + \cancel{4e^-} \rightarrow 2Cu^{2+} + 2O^{2-} + \cancel{4e^-}$$

3. The overall equation is

$$2Cu(s) + O_2(g) \rightarrow 2Cu^{2+}(s) + 2O^{2-}(s)$$

Similar equations can be written for the reaction of copper with sulfur:

$$Cu \rightarrow Cu^{2+} + 2e^- \text{ (oxidation)}$$

$$S + 2e^- \rightarrow S^{2-} \text{ (reduction)}$$

Adding and cancelling electrons gives

$$Cu(s) + S(s) \rightarrow Cu^{2+}(s) + S^{2-}(s)$$

Here, a copper atom loses two electrons just as it did in the reaction with oxygen – **it is oxidized**. Sulfur gains two electrons and **is reduced**.

Reactions such as these, led to a far more general definition of the terms oxidation and reduction:

Oxidation occurs when electrons are lost, reduction occurs when electrons are gained.

To help you remember:

OILRIG
Oxidation **i**s electron **l**oss; **r**eduction **i**s electron **g**ain

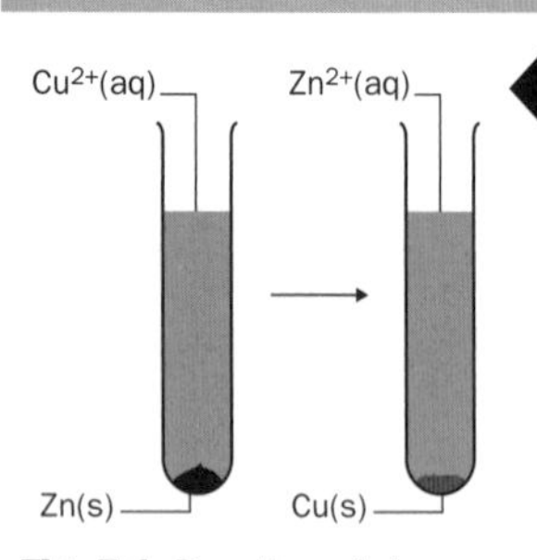

Fig. 7.1 Reaction of zinc with copper(II) sulfate solution.

Exercise 7A

The reaction of zinc with aqueous copper(II) sulfate

When solid zinc is added to blue copper(II) sulfate solution, the solution gradually becomes paler and a red-brown solid (copper) is deposited at the bottom of the test tube. This is shown in Fig. 7.1.

In this reaction, the zinc metal supplies two electrons to the copper ions and copper is precipitated.

The two half-reactions are:

$Zn(s) \rightarrow Zn^{2+}(aq) + 2e^-$

$Cu^{2+}(aq) + 2e^- \rightarrow Cu(s)$

(i) Which half-reaction represents reduction?

(ii) Which half-reaction represents oxidation?

(iii) Write the overall redox equation for the reaction.

(iv) The sulfate ions are not included in the redox equation. Why?

Exercise 7B

Oxidation and reduction

In the following reactions, which substances are oxidized and which are reduced?

(i) $2Zn(s) + O_2(g) \rightarrow 2ZnO(s)$

(ii) $Mg(s) + Cu^{2+}(aq) \rightarrow Mg^{2+}(aq) + Cu(s)$

(iii) $Cl_2(g) + 2I^-(aq) \rightarrow 2Cl^-(aq) + I_2(aq)$

(iv) $2Na(s) + Cl_2(g) \rightarrow 2NaCl(s)$

(v) $H_2(g) + Cl_2(g) \rightarrow 2HCl(g)$

7.2 Oxidation numbers

The **oxidation number** or **oxidation state** of an atom is a positive or negative number which is decided using agreed rules (see Box 7.1). Oxidation numbers are used to widen the definition of oxidation and reduction, with **oxidation being defined as an increase in oxidation number and reduction as a decrease in oxidation number.**

Note that the oxidation number system is just designed to help you work out whether substances are oxidized or reduced. *It does not tell you anything about the bonding in a compound.* For example, the fact that carbon has an oxidation number of +4 in some of its compounds does *not* mean that it exists as +4 ions in these compounds.

BOX 7.1

Oxidation numbers

The following elements nearly always have the same oxidation number in their compounds. They are used as 'standards' to help assign the oxidation numbers of other elements.

There are important exceptions to the elements marked with an asterisk (*), but they will not be dealt with here. Different oxidation numbers of Cl and O will be dealt with at a later stage in the chapter.

Element	Oxidation number
K^+ Na^+	+1
Mg^{2+} Ca^{2+}	+2
Al^{3+}	+3
H^+ (or covalent H)*	+1
F^- (or covalent F)	−1
Cl^- (or covalent Cl)*	−1
O^{2-} (or covalent O)*	−2

Rules for working out oxidation numbers

1. The oxidation number of an atom of a *free element* is *zero*. For example the oxidation number of nitrogen in N_2 or calcium in Ca is 0.
2. The oxidation number of the ion of an element is equal to its charge. For example the oxidation number of copper in Cu^{2+} is $+2$ and the oxidation number of oxygen in O^{2-} is -2.
3. The *algebraic sum* of the oxidation numbers of the atoms in the formula of an electrically neutral compound is zero. For $Ca(OH)_2$, for example, the sum of the oxidation numbers is

   [+2 for Ca] + [2(−2) for O] + [2(+1) for H] = 0
4. For convenience, shared electrons in covalent compounds are assigned to the element having the greater electronegativity. For example, the oxidation number for phosphorus in PCl_3 is $+3$ (chlorine is -1), since phosphorus is less electronegative than chlorine.
5. The *algebraic sum* of the oxidation numbers of all the atoms of an ion is equal to the charge on the ion. In SO_4^{2-}, for example, the sum of the oxidation numbers is

   [+6 for S] + [4(−2) for O] = −2

Remember that:

during chemical reactions an increase in oxidation number signifies OXIDATION; a decrease REDUCTION.

Exercise 7C

Assigning oxidation numbers

Assign oxidation numbers to the underlined atoms in the following examples:

(i) $\underline{S}O_2$
(ii) $\underline{N}H_3$
(iii) $\underline{S}_8$
(iv) $\underline{Cr}Cl_5$
(v) $\underline{N}O_3^-$
(vi) $\underline{Mn}O_4^-$
(vii) $\underline{Xe}F_2$
(viii) $Ca(\underline{V}O_3)_2$
(ix) $\underline{Bi}O_3^-$
(x) $\underline{Cr}_2O_7^{2-}$.

Exercise 7D

Oxidation ladders

Some elements have many oxidation states. You can construct redox 'ladders' like the one shown below for sulfur:

(i) What is the oxidation state of sulfur on each rung of the 'ladder'?
(ii) SO_2 reacts with water to form the *sulfite* ion SO_3^{2-}. Is this a redox reaction?
(iii) H_2S reacts with SO_2 to precipitate sulfur:

$$2H_2S(g) + SO_2(g) \rightarrow 2H_2O(l) + 3S(s)$$

Which species is reduced? Which species is oxidized?

SO_4^{2-}	SO_3
SO_3^{2-}	SO_2
S	
H_2S	

Exercise 7E

Redox ladder for chlorine

Chlorine can have different oxidation states. Using the following species, construct a redox ladder for chlorine:

Cl_2, HCl, ClO_3^-, ClO^-, Cl^-, $HClO_4$

Note that those species with the lowest oxidation number for chlorine should be at the bottom of the ladder, whereas those with the highest oxidation number for chlorine are at the top.

Exercise 7F

Oxidizing and reducing agents

Look back at the redox ladder for sulfur in Exercise 7D:

(i) Which sulfur species can act as both reducing and oxidizing agents in different chemical reactions?

(ii) Which sulfur species can only act as oxidizing agents and why?

(iii) Which sulfur species can only act as reducing agents and why?

7.3 Oxidizing and reducing agents

An **oxidizing agent** is a substance that takes up electrons during a chemical reaction and, in doing so, becomes **reduced**. A **reducing agent** supplies the electrons in this process and so becomes **oxidized**.

Example 7.1

When iron is heated in chlorine gas, iron(III) chloride is produced. The 'ordinary' equation is:

$$2Fe(s) + 3Cl_2(g) \rightarrow 2FeCl_3(s)$$

The redox half-equations are

$$Fe(s) \rightarrow Fe^{3+}(s) + 3e^-$$

$$Cl_2(g) + 2e^- \rightarrow 2Cl^-(s)$$

Here, iron is acting as a *reducing agent* because it supplies electrons to chlorine and is *oxidized.*
Chlorine is acting as an *oxidizing agent* because it accepts electrons from iron and is *reduced.*

7.4 Writing and balancing redox equations

In order to write a redox equation, it is first necessary to write the two half-reactions that identify the oxidation and reduction processes taking place. The overall redox equation is then obtained by adding these two half-reactions together, so that the electrons in each half reaction cancel out.

To write the half reactions, follow these simple rules:

1. Identify the atoms that are oxidized and reduced, using the oxidation number method.

Most of the reactions that you will come across at this stage, will occur in neutral or acid solution, and step 2 applies.

2. Balance the half-reactions:

(i) Make sure that there are the same number of atoms of the element that is oxidized (or reduced) on each side of the half-reaction.

(ii) If there are any oxygen atoms present, balance them by adding water molecules to the other side of the half-reaction.

(iii) If there are hydrogen atoms present, balance them by adding hydrogen ions on the other side of the half-reaction.

(iv) Make sure that the half-reactions have the same overall charge on each side by adding electrons.

Note that the rules are slightly different if the reaction occurs in basic solution: hydrogen atoms are balanced using H_2O molecules and then the same number of OH^- ions are added to the opposite side of the equation to balance the oxygens. Carry on as before, adding electrons to balance the charges.

Example 7.2

Write the overall redox equation for the oxidation of iron(II) ions to iron(III) ions by the manganate(VII) or permanganate ion, MnO_4^-, in acid solution. The manganate(VII) ion reacts to form the manganese(II) ion, Mn^{2+}.

▶ Answer

The oxidation half-reaction

The question tells you that iron(II) is oxidized, but you can check on this by working out the oxidation numbers of iron before and after the reaction:

$Fe^{2+} \rightarrow Fe^{3+}$; by oxidation numbers: $+2 \rightarrow +3$

therefore, an increase in oxidation number, i.e. oxidation, has occurred.

There are the same number of atoms on each side of this half-reaction, and no oxygen or hydrogen atoms, so all that remains to be done is to make sure that the overall charge is the same on each side of the half-equation. By adding one electron to the right-hand side, the overall charge on each side of the equation becomes +2 (*think of an electron as a unit negative charge, which will cancel out a positive charge*):

$$Fe^{2+} \rightarrow Fe^{3+} + e^-$$

The half-equation above represents the oxidation reaction.

The reduction half-reaction

The manganate(VII) ion reacts to form manganese(II). Check that this is a reduction by working out the oxidation numbers of manganese before and after the reaction:

$MnO_4^- \rightarrow Mn^{2+}$; by oxidation numbers: $+7 \rightarrow +2$

therefore, a decrease in oxidation number, i.e. reduction, has occurred.

There are the same number of manganese atoms on each side of the half-reaction, but the oxygen atoms need to be balanced:

1. Balance the oxygens with water molecules:
$$MnO_4^- \rightarrow Mn^{2+} + 4H_2O$$

2. Now H atoms have been introduced, so they need to be balanced with H^+ ions:
$$MnO_4^- + 8H^+ \rightarrow Mn^{2+} + 4H_2O$$

3. Balance the charges on each side of the half-reaction by adding electrons:
$$MnO_4^- + 8H^+ + 5e^- \rightarrow Mn^{2+} + 4H_2O$$

Each side of the equation now has an overall charge of +2. The above equation represents the reduction reaction. To construct the overall redox equation, the two half-reactions must be added together so that the electrons cancel out:

$$Fe^{2+} \rightarrow Fe^{3+} + e^- \quad \text{(oxidation)}$$
$$MnO_4^- + 8H^+ + 5e^- \rightarrow Mn^{2+} + 4H_2O \quad \text{(reduction)}$$

The first half-reaction must be multiplied by five, then the equations added together:

$$5Fe^{2+} \rightarrow 5Fe^{3+} + 5e^- \quad \text{(oxidation)}$$
$$MnO_4^- + 8H^+ + 5e^- \rightarrow Mn^{2+} + 4H_2O \quad \text{(reduction)}$$
$$MnO_4^- + 8H^+ + \cancel{5e^-} + 5Fe^{2+} \rightarrow Mn^{2+} + 4H_2O + 5Fe^{3+} + \cancel{5e^-}$$

The overall redox equation is therefore

$$MnO_4^-(aq) + 8H^+(aq) + 5Fe^{2+}(aq) \rightarrow Mn^{2+}(aq) + 4H_2O(l) + 5Fe^{3+}(aq)$$

Exercise 7G

Writing and balancing redox equations

1. Write half-equations for:

(i) the oxidation of I^- to IO_3^-

(ii) the reduction of NO_3^- to NO in acidic solution

(iii) the reduction of $Cr_2O_7^{2-}$ to Cr^{3+} in acidic solution

(iv) the reduction of H_2O_2 to H_2O in acidic solution

(v) the oxidation of S^{2-} to SO_4^{2-} in *basic* solution.

2. Write an overall redox equation for:

(i) The reaction of copper with silver nitrate, in aqueous solution, to produce copper(II) nitrate and silver.

(ii) The oxidation of copper metal to copper(II) nitrate by concentrated nitric acid (*regard the reduction of nitric acid as the ion NO_3^- being reduced to brown NO_2 gas*).

(iii) The reduction of I_2 by thiosulfate, $S_2O_3^{2-}$, to I^- in aqueous solution. Thiosulfate is oxidized to tetrathionate, $S_4O_6^{2-}$.

(iv) The oxidation of I^- to I_2 by H_2O_2 in acid solution (H_2O_2 is reduced to H_2O). (Note that in H_2O_2, the oxidation number of oxygen is -1.)

(v) The oxidation of H_2O_2 to O_2 by ClO_2^- (ClO_2^- is reduced to Cl^-).

BOX 7.2

The Breathalyser

The first type of breathalyser, still in use today, was a disposable device consisting of a plastic tube packed with yellow crystals. The person was required to blow through the tube into a 1 dm^3 plastic bag for about 15 s. Alcohol vapour in the subject's breath reacted with the yellow crystals and turned them green. If the green stain extended beyond the red line drawn on the tube, then the subject had failed the test (Fig. 7.2).

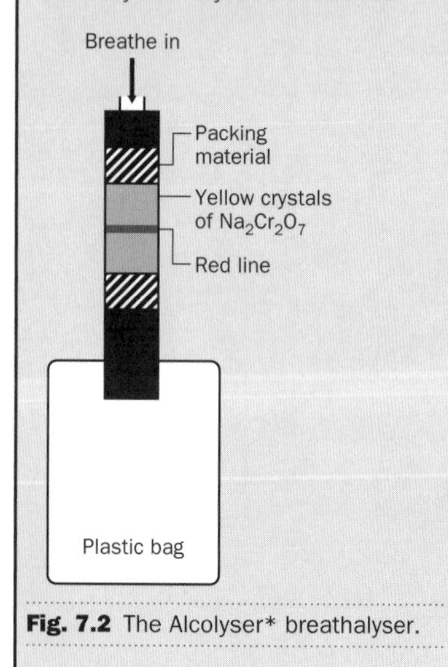

Fig. 7.2 The Alcolyser* breathalyser.

The yellow crystals are sodium dichromate, $Na_2Cr_2O_7$ and the redox reaction taking place is the reduction of the dichromate(VI) ion, $Cr_2O_7^{2-}$, to the green Cr^{3+} ion by alcohol, under *acidic* conditions:

$$\underset{\text{yellow}}{Cr_2O_7^{2-}} + 14H^+ + 6e^- \rightarrow \underset{\text{green}}{2Cr^{3+}} + 7H_2O$$

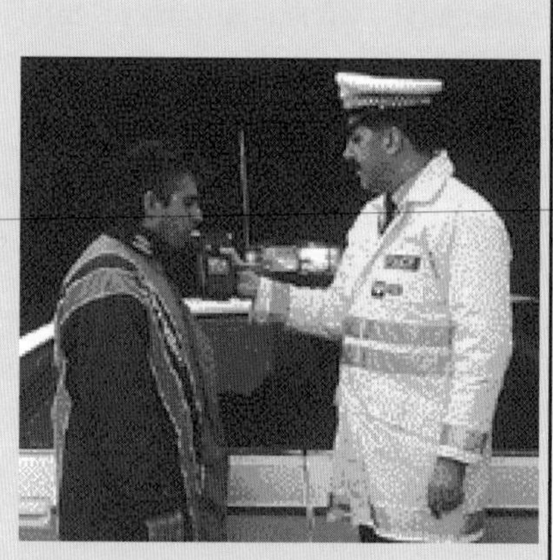

Many police forces now use an Alcolmeter* instead of the disposable dichromate tubes. The Alcometer is a small instrument with disposable mouthpiece tubes. The subject blows into the instrument and their alcohol level can be read from a digital display. The instrument is an electrochemical cell known as a fuel cell (see Chapter 22) which generates a voltage in proportion to the alcohol vapour concentration in the breath.

If a person is 'over the limit' for either of these roadside tests, they are taken to the police station, where a second instrument (a simple infrared spectrometer (see Chapter 20)) is used to measure accurately the alcohol concentration.

*Trademarks of Lion Laboratories Ltd, Ty Verlon Industrial Estate, Barry, South Wales.

7.5 Redox couples

Think of a piece of zinc metal, Zn(s), dipped in a solution of silver ions, Ag^+(aq). A reaction is seen to occur, and solid silver is produced:

$$Zn(s) + 2Ag^+(aq) \rightarrow Zn^{2+}(aq) + 2Ag(s)$$

The half-reactions that make up this overall reaction are:

$$Zn(s) \rightarrow Zn^{2+}(aq) + 2e^-$$

and

$$2Ag^+(aq) + 2e^- \rightarrow 2Ag(s)$$

The fact that there is a reaction between Zn(s) and Ag^+(aq) suggests that the silver ions want to accept electrons more than the zinc atoms want to keep them.

No reaction takes place when zinc metal is added to magnesium ions, Mg^{2+}(aq). This suggests that magnesium ions in solution have a smaller tendency to accept electrons, and form Mg(s), than do silver ions, forming Ag(s).

The pairs of species involved here, Ag^+(aq)/Ag(s) and Mg^{2+}(aq)/Mg(s) are known as **redox couples**. Therefore, our experiments suggest that the Ag^+(aq)/Ag(s) redox couple is a *stronger oxidizing agent* than the Mg^{2+}(aq)/Mg(s) redox couple. Since a strong oxidizing agent may be described as a weak reducing agent, we could also say that the Mg^{2+}(aq)/Mg(s) redox couple is a *stronger reducing agent* than the Ag^+(aq)/Ag(s) redox couple.

The relative oxidizing or reducing strengths of redox couples are expressed in terms of their **standard electrode potentials**, E°, which have the units of volts. It is impossible to measure the standard electrode potentials of redox couples in isolation, without introducing other metals into the electrical circuit. In practice, the redox couple under study is connected to a **reference redox couple** in an **electrochemical cell**. Voltage measurements then give a relative potential difference. The reference redox couple used is H^+(aq)/H_2(g), which (under agreed standard conditions) is given a E° value of zero.

The standard hydrogen electrode (SHE)

Experimentally, the H^+(aq)/H_2(g) couple is arranged in the form of a **hydrogen electrode**. A hydrogen electrode is made up of a piece of platinum foil, coated with fine particles of platinum, dipped into a solution of hydrogen ions, with hydrogen gas bubbling over the surface of the platinum (Fig. 7.3).

The platinum provides a surface upon which either of the following reactions may occur:

$$2H^+(aq) + 2e^- \rightarrow H_2(g)$$

or

$$H_2(g) \rightarrow 2H^+(aq) + 2e^-$$

These two possibilities are summarized by the equation

$$2H^+(aq) + 2e^- \rightleftharpoons H_2(g)$$

or it could also be written as

$$2H_3O^+(aq) + 2e^- \rightleftharpoons H_2(g) + 2H_2O(l)$$

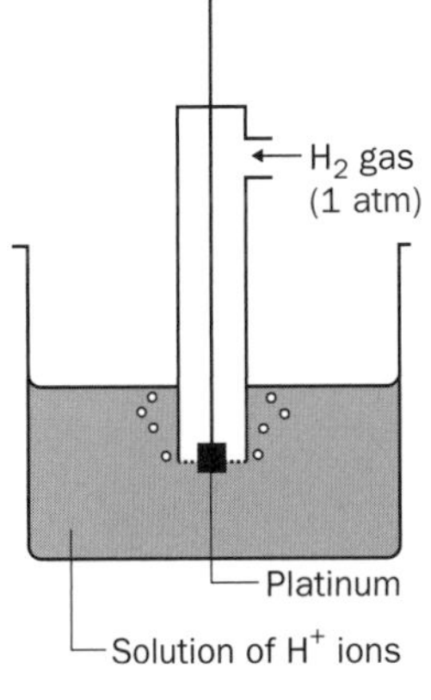

Fig. 7.3 The standard hydrogen electrode.

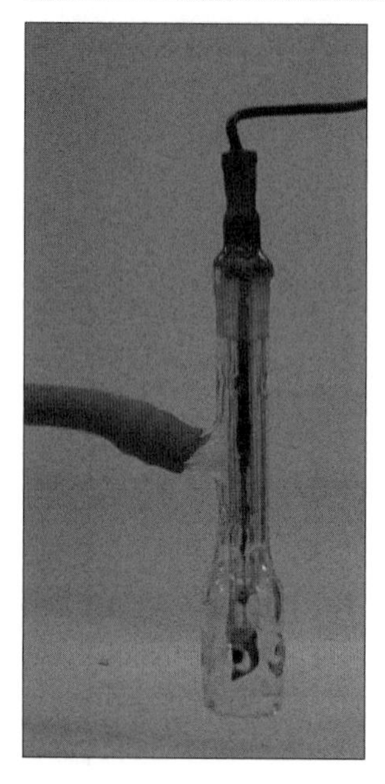

Photograph of a standard hydrogen electrode

The **standard hydrogen electrode** (SHE; Fig. 7.3) is a hydrogen electrode where $H^+(aq)$ and $H_2(g)$ are in their **standard states** (see page 217). This means that the concentration of $H^+(aq)$ is $1\ mol\ dm^{-3}$, and the hydrogen gas is at a pressure of 1 atm. By agreement, the potential difference between the $H^+(aq)$ solution and the Pt electrode in the SHE is set to zero at all temperatures. This is written as:

$$E^{\ominus}\ (H^+(aq),H_2(g)) = 0$$

where the use of the symbol $\ominus$ shows that the substances are in their standard states.

Other standard electrodes

The electrode of the $Ag^+(aq)/Ag(s)$ redox couple consists of a rod of silver dipped in a solution of silver ions. The $Ag^+(aq)/Ag(s)$ electrode becomes the **standard** $Ag^+(aq)/Ag(s)$ electrode if the silver is pure, and the silver ion concentration is $1\ mol\ dm^{-3}$.

The electrode of the $Fe^{3+}(aq)/Fe^{2+}(aq)$ redox couple has a different construction. It consists of a piece of platinum dipped in a mixture of both ions. In an electrochemical cell, the Pt connects the electrode to the rest of the cell. The Pt also provides a surface upon which electrode reactions may occur. The **standard** $Fe^{3+}(aq)/Fe^{2+}(aq)$ electrode contains both ions at a concentration of $1\ mol\ dm^{-3}$.

Measuring $E^{\ominus}$ values by making an electrochemical cell

1. Measuring $E^{\ominus}(Ag^+(aq)/Ag(s))$

We will now see how the standard electrode potential of a redox couple may be found by experiment. Figure 7.4 shows the electrochemical cell composed of the standard hydrogen and standard silver ion/silver electrodes. All measurements are made at 25 °C. The two electrodes are connected by:

1. a wire through which electrons can pass;

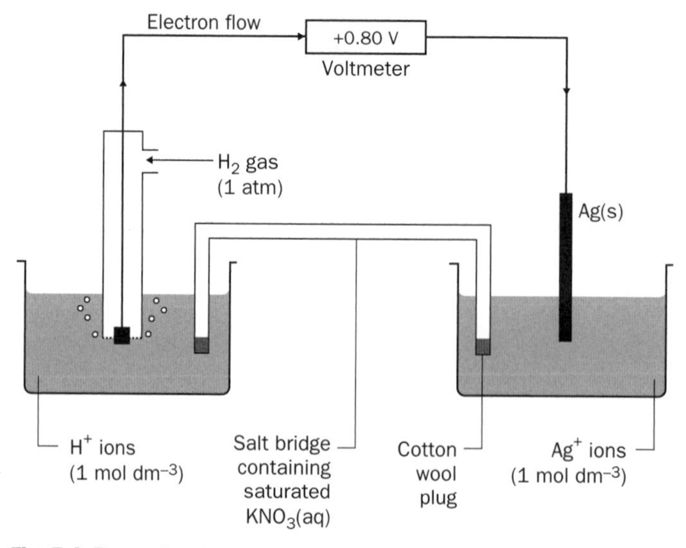

Fig. 7.4 Electrochemical cell used in the determination of $E^{\ominus}(Ag^+/Ag(s))$.

2. a **salt bridge** (a glass tube that contains a saturated solution of an ionic salt, such as KNO_3) which completes the electrical circuit.

If we measure the voltage of the cell (the **cell potential**) using a high-resistance voltmeter, we will not be using up any electrical current produced by the cell, and very little chemical reaction takes place. Under these conditions, the measured voltage does not fall during the measurement.

Which electrode is connected to the positive terminal of the voltmeter? By trial and error, it is found that only if the $Ag^+(aq)/Ag(s)$ electrode is connected to the positive terminal is the cell potential **positive**. (A voltage of $+0.80$ V is actually obtained; if the connections were reversed, the voltage reading would be -0.80 V.) Because of this, the $Ag^+(aq)/Ag(s)$ electrode is said to be the cell **cathode** (+) and the hydrogen electrode the cell **anode** (−).

Electrons flow from the anode to the cathode. This means that the hydrogen electrode is losing electrons (oxidation); this is only possible if the $H^+(aq)/H_2(g)$ redox couple is undergoing the reaction:

$$H_2(g) \rightarrow 2H^+(aq) + 2e^-$$

The electrons then pass through the circuit and move to the cathode where they cause the $Ag^+(aq)/Ag(s)$ redox couple to be reduced within the $Ag^+(aq)/Ag(s)$ electrode:

$$Ag^+(aq) + e^- \rightarrow Ag(s)$$

The overall cell reaction is:

$$H_2(g) + 2Ag^+(aq) \rightarrow 2H^+(aq) + 2Ag(s)$$

(The same reaction occurs outside a cell, if hydrogen gas is bubbled into a solution containing silver ions. However, in that case heat – not electricity – is made.)

The **cell diagram** for any electrochemical cell follows the pattern:

reactant → product (−) **anode** (in which oxidation takes place)	‖	reactant → product (+) **cathode** (in which reduction takes place)

where ‖ symbolizes the salt bridge. In our case:

$$\mathrm{Pt}\mid \underbrace{H_2(g) \mid H^+(aq)}_{\text{goes to}} \parallel \underbrace{Ag^+(aq) \mid Ag(s)}_{\text{goes to}}$$

which shows that $H_2(g)$ is oxidized to $H^+(aq)$ (the product at the anode), while $Ag^+(aq)$ is reduced to $Ag(s)$ (the product at the cathode).

The standard potential of the cell, $E^\ominus = +0.80$ V, is related to the standard electrode potentials for each redox couple by the following equation:

$$E^\ominus = E^\ominus_R - E^\ominus_L$$

which applies to all cells, and in which $E^\ominus_R$ is the standard electrode potential of the right-hand electrode (cathode) and $E^\ominus_L$ the standard electrode potential of the left-hand electrode (anode) as they appear in the cell diagram. Here, $E^\ominus_R = E^\ominus(Ag^+(aq)/Ag(s))$ and $E^\ominus_L = E^\ominus(H^+(aq),H_2(g)) = 0$ (by definition). Thus:

$$+0.80\,V = E^\ominus(Ag^+/Ag(s)) - 0$$

and $E^\ominus(Ag^+/Ag(s)) = +0.80$ V.

2. Finding $E^\ominus(Zn^{2+}/Zn(s))$

Suppose we wanted to measure $E^\ominus(Zn^{2+}(aq)/Zn(s))$. Our first step is to set up the electrochemical cell, consisting of the SHE and standard $(Zn^{2+}(aq)/Zn(s))$ electrode (consisting of pure Zn metal dipped into 1 mol dm^{-3} $Zn^{2+}(aq)$) connected by a salt bridge.

If the standard $Zn^{2+}(aq)/Zn(s)$ electrode is connected to the negative pole of the voltmeter, the voltage reading is found to be +0.76 V. The positive voltage shows that the $Zn^{2+}(aq)/Zn(s)$ electrode is the anode. At the anode, the $Zn^{2+}(aq)/Zn(s)$ redox couple undergoes oxidation:

$$Zn(s) \rightarrow Zn^{2+}(aq) + 2e^-$$

The electrons cause the $H^+(aq)/H_2(g)$ redox couple to be reduced within the cathode of the SHE:

$$2H^+(aq) + 2e^- \rightarrow H_2(g)$$

The overall cell reaction is

$$Zn(s) + 2H^+(aq) \rightarrow Zn^{2+}(aq) + H_2(g)$$

The cell diagram is

$$Zn(s) \mid Zn^{2+}(aq) \parallel Pt \mid H^+(aq) \mid H_2(g)$$

The cell potential is +0.76 V:

$$E^\ominus = E^\ominus_R - E^\ominus_L$$
$$0.76 = E^\ominus(H^+(aq),H_2(g)) - E^\ominus(Zn^{2+}(aq)/Zn(s))$$
$$0.76 = 0 - E^\ominus(Zn^{2+}(aq)/Zn(s))$$

or,

$$E^\ominus(Zn^{2+}(aq)/Zn(s)) = -0.76\,V$$

We have now looked at two electrochemical cells. In the first, where the $E^\ominus$ of the redox couple under investigation proved to be positive, the $H^+(aq)/H_2(g)$ couple undergoes oxidation (Fig. 7.5(a)). In the second, where the $E^\ominus$ of the redox couple under investigation proved to be negative, the $H^+(aq)/H_2(g)$ couple undergoes reduction (Fig. 7.5(b)). Generalizing:

1. *A negative $E^\ominus$ means that a redox couple is a stronger reducing agent than the $H^+(aq)/H_2(g)$ couple.*

2. *A positive $E^\ominus$ means that a redox couple is a weaker reducing agent than the $H^+(aq)/H_2(g)$ couple.*

The reverse statements apply to the oxidizing power of these redox couples.

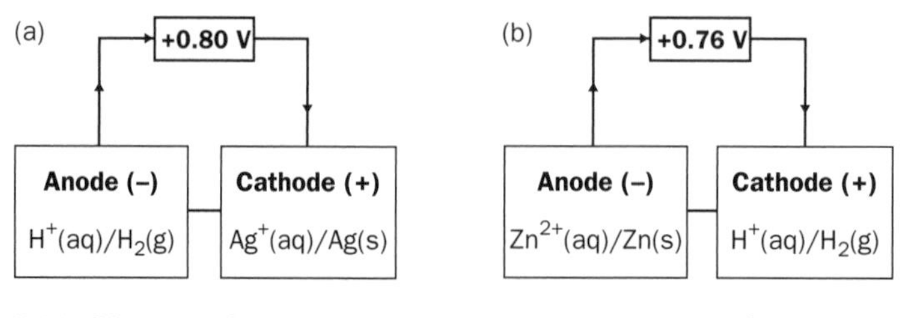

Fig. 7.5 The flow of electrons in cells involving the SHE in which (a) $Zn^{2+}(aq)/Zn(s)$, and (b) $Ag^+(aq)/Ag(s)$ form the other electrode.

Table 7.1 Standard electrode potentials

	Redox couple	Reaction equation when redox couple is reduced	$E^\ominus$ at 25°C/V
Strong	$F_2(g)/F^-(aq)$	$F_2(g) + 2e^- \rightarrow 2F^-(aq)$	+2.87
Oxidising	$Cl_2(g)/Cl^-(aq)$	$Cl_2(g) + 2e^- \rightarrow 2Cl^-(aq)$	+1.36
Agents	$Br_2(l)/Br^-(aq)$	$Br_2(l) + 2e^- \rightarrow 2Br^-(aq)$	+1.09
	$Ag^+(aq)/Ag(s)$	$Ag^+(aq) + e^- \rightarrow Ag(s)$	+0.80
	$Fe^{3+}(aq)/Fe^{2+}(aq)$	$Fe^{3+}(aq) + e^- \rightarrow Fe^{2+}(aq)$	+0.77
	$I_2(s)/I^-(aq)$	$I_2(s) + 2e^- \rightarrow 2I^-(aq)$	+0.54
	$Cu^{2+}(aq)/Cu(s)$	$Cu^{2+}(aq) + 2e^- \rightarrow Cu(s)$	+0.34
	$H^+(aq)/H_2(g)$	$2H^+(aq) + 2e^- \rightarrow H_2(g)$	0 (by definition)
	$Fe^{2+}(aq)/Fe(s)$	$Fe^{2+}(aq) + 2e^- \rightarrow Fe(s)$	−0.44
	$Cr^{3+}(aq)/Cr(s)$	$Cr^{3+}(aq) + 3e^- \rightarrow Cr(s)$	−0.74
Strong	$Zn^{2+}(aq)/Zn(s)$	$Zn^{2+}(aq) + 2e^- \rightarrow Zn(s)$	−0.76
Reducing	$Mg^{2+}(aq)/Mg(s)$	$Mg^{2+}(aq) + 2e^- \rightarrow Mg(s)$	−2.37
Agents	$Na^+(aq)/Na(s)$	$Na^+(aq) + e^- \rightarrow Na(s)$	−2.71

Exercise 7H

Electrochemical cells and $E^\ominus$

A cell, consisting of the SHE (as anode) and the standard $Cu^{2+}(aq)/Cu(s)$ electrode, gives a standard cell potential of +0.34 V at 25 °C.

(i) Write down the cell diagram.

(ii) Calculate $E^\ominus(Cu^{2+}(aq)/Cu(s))$.

(iii) Write down the overall reaction that occurs in the cell.

Key points about the standard electrode potentials of redox couples

1. $E^\ominus$ values are also known as **standard reduction potentials**. Table 7.1 shows selected $E^\ominus$ values which were measured at 25°C. Note that the redox couples are listed with the oxidized species first (e.g. $Na^+(aq)/Na(s)$ not $Na(s)/Na^+(aq)$).
2. Measurements of $E^\ominus$ may be carried out at other temperatures. $E^\ominus$ values do vary with temperature, apart from $E^\ominus$ ($H^+(aq)/H_2$), which by agreement is set to zero at all temperatures.

3. Some $E^\ominus$ values, such as $E^\ominus(Na^+(aq)/Na(s))$, cannot be found using simple electrochemical cells, and are calculated from other data.
4. Strictly speaking, $E^\ominus$ values apply to couples, and not to single species. For example, we cannot speak of $E^\ominus$(Na). The fact that $E^\ominus(Na^+(aq)/Na(s)) = -2.71$ V, shows specifically that sodium has a strong tendency to form Na^+ ions in solution (and not to form, say, Na^-).
5. $E^\ominus$ values may be used to predict whether or not redox reactions are allowed to occur. (Using the language of Chapter 15, we would say that a reaction that is allowed to occur has a large value for the equilibrium constant K_c.) This is discussed in the next section, but note that $E^\ominus$ values do not permit any predictions to be made about the rate of a reaction.

BOX 7.3

Dry cells

The trouble with 'wet' cells, is that the solutions (*the electrolyte*) might leak out. In the dry cell, the electrolyte is a damp paste. This type of cell is used in torches, walkmans and clocks. Many dry cells use the reactions of zinc and manganese(IV) oxide:

$Zn(s) \rightarrow Zn^{2+}(aq) + 2e^-$ (oxidation)

$MnO_2(s) + H_2O(l) + e^- \rightarrow MnO(OH)(s) + OH^-(aq)$ (reduction)

Other reactions occur, but we will not discuss them here. A diagram of a dry cell is shown in Fig. 7.6.

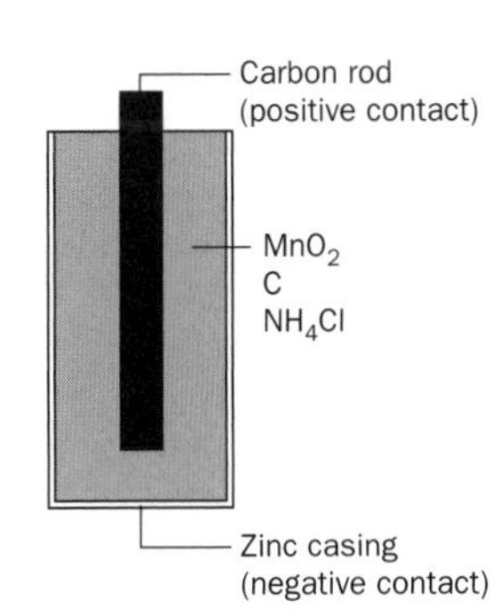

Fig. 7.6 A dry cell.

Predicting whether reactions can occur

Standard electrode potentials (E° values) may be used to decide whether redox reactions are allowed to occur. The key point to remember here is that the half-reaction having the more positive electrode potential occurs as a reduction, while the half-reaction having the more negative electrode potential occurs as an oxidation.

To help you apply this easily:
Write down both half-reactions, with the half-reaction possessing the *more negative (or least positive)* E° value at the top. Then draw *anticlockwise arrows* to predict whether the overall reaction can occur.

Example 7.3

Will hydrogen gas reduce $Fe^{3+}(aq)$ to $Fe^{2+}(aq)$?

▶ Answer

Write down both half-reactions, with the half-reaction possessing the more negative (or least positive) potential at the top:

$2H^+(aq) + 2e^- \rightarrow H_2(g) \qquad E^{\circ} = 0.00\,V$

$Fe^{3+}(aq) + e^- \rightarrow Fe^{2+}(aq) \qquad E^{\circ} = +0.77\,V$

Draw anticlockwise arrows:

$2H^+(aq) + 2e^- \rightarrow H_2(g) \qquad E^{\circ} = 0.00\,V$

$Fe^{3+}(aq) + e^- \rightarrow Fe^{2+}(aq) \qquad E^{\circ} = +0.77\,V$

The direction in which the arrows point show that H_2 changes to H^+ and that Fe^{3+} changes to Fe^{2+}. The predicted overall reaction is therefore:

$H_2(g) + 2Fe^{3+}(aq) \rightarrow 2Fe^{2+}(aq) + 2H^+(aq)$

▶ Comments

The conclusion that we can draw from this is that hydrogen gas is capable of reducing $Fe^{3+}(aq)$ to $Fe^{2+}(aq)$ at room temperature.

Note, however, *that we cannot predict how fast (or slow!) the reaction will be*. The reaction may be so slow that nothing will appear to happen.

In addition, our predictions only apply to solutions, since E° values cannot be obtained for dry gases or solids.

Example 7.4

Will bromine water ($Br_2(aq)$) react with an aqueous solution of potassium chloride ($Cl^-(aq)$)?

▶ Answer

Following the same procedure as above:

Example 7.4 *(continued)*

$Br_2(aq) + 2e^- \rightarrow 2Br^-(aq)$ $E^\ominus = +1.09\,V$

$Cl_2(aq) + 2e^- \rightarrow 2Cl^-(aq)$ $E^\ominus = +1.36\,V$

In this case, the direction of the arrows show that $Br_2(aq)$ and $Cl^-(aq)$ *stay as they are*. The conclusion is that *there is no reaction* at room temperature.

Exercise 7I

Predicting reactions

Using the standard reduction potentials in Table 7.1, determine whether the following reactions may occur:

(i) $Br_2(aq) + I^-(aq) \rightarrow I_2(aq) + 2Br^-(aq)$

(ii) $Cu(s) + 2H^+(aq) \rightarrow Cu^{2+}(aq) + H_2(g)$

(iii) $Zn(s) + 2Fe^{3+}(aq) \rightarrow Zn^{2+}(aq) + 2Fe^{2+}(aq)$

(iv) $2Fe^{3+}(aq) + 2I^-(aq) \rightarrow 2Fe^{2+}(aq) + I_2(aq)$.

7.6 Activity series of metals

The more negative the standard electrode potential, $E^\ominus$, of a redox couple $M^{n+}(aq)/M(s)$, the more powerful a reducing agent is that metal. This means it is a more reactive metal because it loses electrons more easily. We can arrange the metals in order of reducing power, producing an **activity series** of metals (Table 7.2). The metals at the top of the series are more reactive than those below.

A metal in solution can react with a metal ion that appears below it in the table. Hydrogen is included so that the reactivity of metals with aqueous acids may be worked out; when a metal above hydrogen is added to an acidic solution, a reaction will occur and hydrogen gas is given off.

Never mix strongly reducing metals, such as potassium, with acid solutions – the reactions are extremely violent.

Table 7.2 The activity series of metals

Reduction half-reaction	$E^\ominus\ (M^{n+}(aq)/M(s))/V$
$K^+(aq) + e^- \rightarrow K(s)$	−2.92
$Ca^{2+}(aq) + 2e^- \rightarrow Ca(s)$	−2.87
$Na^+(aq) + e^- \rightarrow Na(s)$	−2.71
$Mg^{2+}(aq) + 2e^- \rightarrow Mg(s)$	−2.37
$Al^{3+}(aq) + 3e^- \rightarrow Al(s)$	−1.67
$Zn^{2+}(aq) + 2e^- \rightarrow Zn(s)$	−0.76
$Fe^{2+}(aq) + 2e^- \rightarrow Fe(s)$	−0.44
$Ni^{2+}(aq) + 2e^- \rightarrow Ni(s)$	−0.25
$Sn^{2+}(aq) + 2e^- \rightarrow Sn(s)$	−0.14
$Pb^{2+}(aq) + 2e^- \rightarrow Pb(s)$	−0.13
$2H^+(aq) + 2e^- \rightarrow H_2(g)$	0
$Cu^{2+}(aq) + 2e^- \rightarrow Cu(s)$	+0.34
$Ag^+(aq) + e^- \rightarrow Ag(s)$	+0.80
$Au^+(aq) + e^- \rightarrow Au(s)$	+1.68

Exercise 7J

Activity series of metals

Write redox equations for any reactions that might occur when the following substances are mixed:

(i) magnesium and silver nitrate solution

(ii) copper and lead(II) nitrate solution

(iii) nickel and copper(II) sulfate solution

(iv) zinc and dilute hydrochloric acid ($H^+(aq)$)

(v) copper and dilute sulfuric acid ($H^+(aq)$).

7.7 Corrosion of iron

Rusting

Many metals, including iron, react with air and/or water (**corrode**). The reactivity of metals with air or water can be predicted using reduction potentials. For a neutral solution:

$$2H_2O(l) + 2e^- \rightarrow H_2(g) + 2OH^-(aq) \qquad E^\circ = -0.42\ V$$

In the case of iron:

$$Fe^{2+}(aq) + 2e^- \rightarrow Fe(s) \qquad E^\circ = -0.44\ V$$

$$Fe^{3+}(aq) + e^- \rightarrow Fe^{2+}(aq) \qquad E^\circ = +0.77\ V$$

Therefore pure water has only a slight tendency to oxidize Fe(s) to $Fe^{2+}(aq)$.

However, if oxygen is present, the following half-reaction can occur in neutral solution (pH = 7; see page 151)

$$O_2(g) + 4H^+(aq) + 4e^- \rightarrow 2H_2O(l) \qquad E^\circ = +0.82$$

Check, by using anticlockwise arrows, that this couple can oxidize Fe to Fe^{2+} and then from Fe^{2+} to Fe^{3+}.

Both oxygen and water can therefore oxidize Fe to Fe^{2+} and oxygen is further capable of oxidizing Fe^{2+} to Fe^{3+}. These reactions are known as **rusting**.

Rust is a brown, insoluble compound of formula $Fe_2O_3 \cdot xH_2O$ (the x in the formula indicates that it has a variable composition) which is formed when iron reacts with air and water. Rusting requires the presence of oxygen, water and ionic substances dissolved in the water (**electrolytes**). In the absence of any one of these, little rusting will occur.

Exercise 7K

Rusting

(i) Explain why iron nails do not rust if they are placed in a sealed tube containing boiled water.

(ii) Why do cars that are kept by the sea rust more rapidly?

(iii) Rusting occurs much more quickly in areas that are polluted. Can you give a reason for this?

Rust prevention

The following methods may be adopted in order to prevent corrosion of iron or steel:

1. Protection of the surface of the metal from air and water by painting, oiling, greasing or coating with a plastic.
2. Coat the metal with a more reactive metal (one with a more negative value for E°) so that, even if the coating is scratched, the more reactive metal will lose electrons in preference to the metal that has been coated. Zinc is often used to coat metals such as iron; this is achieved by dipping the iron in molten zinc, or by **electroplating** the iron. Zinc exposed to the air becomes covered with a film of

zinc oxide, which protects it from further corrosion – the zinc becomes **passive**. Covering a metal with a zinc layer is known as **galvanizing**.

3. It is not practical to galvanize large objects, such as a ships or pipes. Instead a block of a reactive metal, such as magnesium (or zinc), is attached to the large object and, again, preferentially loses electrons to oxygen. This method of protecting the metal is known as **sacrificial protection**.

Exercise 7L

Tinning

Steel food cans coated with tin used to be common. Unfortunately, when damaged they corrode more rapidly than if the coating was not present – explain this phenomenon.

7.8 Redox reactions in nature

Nitrogen fixation

Living organisms need nitrogen to make proteins, the 'building bricks' of plants and animals. Although nitrogen gas is abundant, making up approximately four-fifths of the atmosphere, most living organisms cannot obtain it directly from the air because it is an unreactive gas. Nitrogen becomes available to plants and animals via the **nitrogen cycle** (Fig. 7.7).

Atmospheric nitrogen is particularly unreactive; however, it is oxidized to NO when lightning flashes. Oxygen in the atmosphere oxidizes NO to NO_2, which reacts with rain water to form the acids HNO_2 and HNO_3. These acids react with metal oxides and carbonates in the soil, to form nitrate and nitrite salts.

Plants obtain nitrogen in the form of nitrate and ammonium ions. Nitrates are very soluble in water and reach the roots of plants easily; the nitrate ion is then taken up and reduced to ammonia by the plant. Nitrogen-fixing bacteria that live in the soil, or in **nodules** on the roots of plants, convert (i.e. **biologically fix**) atmospheric nitrogen into ammonium salts. Animals eat plants and, when both of these organisms die, their organic matter eventually decays into ammonium compounds. Nitrifying and denitrifying bacteria convert ammonium compounds into NO_3^- and NO_2^-, then N_2O and N_2. In this way nitrogen is returned to the atmosphere and the cycle is complete.

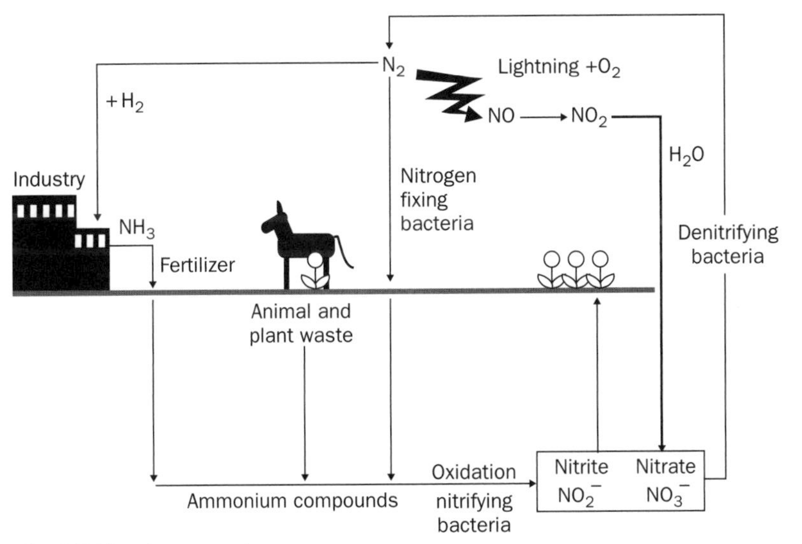

Fig. 7.7 The nitrogen cycle.

BOX 7.4

Another test for nitrates

If a substance is suspected to contain a nitrate ion (NO_3^-), the presence of the ion can be detected by heating the nitrate with sodium hydroxide solution and **Devarda's alloy**. Devarda's alloy is a roughly equal mixture of aluminium and copper, to which a little zinc is added. If a nitrate is present it is reduced and ammonia is evolved. Ammonia can be detected by its choking smell or, because it is an alkaline gas, it turns damp red litmus paper blue. The detection of ammonia confirms that a nitrate was present.

One redox equation for the reaction is

$$8Al(s) + 3NO_3^-(aq) + 5OH^-(aq) + 18H_2O(l) \rightarrow 8Al(OH)_4^-(aq) + 3NH_3(g)$$

See also the brown ring test on page 89.

The availability of nitrogen to plants (and therefore to animals) is vital for producing food. This is why *fertilizers*, ammonium and nitrate salts, are produced industrially to help supply nitrogen to crops.

The **Haber–Bosch process** is an industrial process whereby nitrogen gas is reduced to ammonia using an iron catalyst and is the first step in the production of fertilizers:

$$3H_2(g) + N_2(g) \rightleftharpoons 2NH_3(g)$$

This is a very expensive process, requiring high temperatures and pressures and the expense is one of the reasons why Third World countries find difficulty in feeding their populations. Biological nitrogen fixation, however, requires only solar energy as its power source and chemists are currently trying to find catalysts that can act in the same way as nitrogen-fixing bacteria.

Exercise 7M

Nitrogen cycle

Work out the oxidation numbers of all the nitrogen-containing species in the previous section. Construct an 'oxidation state' ladder for nitrogen, using these species.

Electron transport in living systems

Both **photosynthesis** and **respiration** are energy conversion processes and involve redox reactions.

Photosynthesis is a process that occurs when plants take in carbon dioxide and water to make sugar. Photosynthesis needs solar energy and is catalysed by **chlorophyll**, the green substance in leaves; oxygen is also produced:

$$6CO_2(g) + 6H_2O(l) \rightarrow C_6H_{12}O_6(s) + 6O_2(g)$$

For more about photosynthesis, see page 395.

Respiration is a reaction that supplies living organisms with the energy they need. It is the opposite reaction to photosynthesis:

$$C_6H_{12}O_6(s) + 6O_2(g) \rightarrow 6CO_2(g) + 6H_2O(l)$$

Note that energy is stored in glucose during photosynthesis and released again during respiration.

Respiration involves a series of redox reactions, some of which involve the participation of iron-containing substances called **cytochromes**. Cytochromes are electron carriers. They accept electrons from better reducing agents and give electrons to better oxidizing agents. The iron 'flips' back and forth between Fe^{3+} (the oxidized form) and Fe^{2+} (the reduced form). A similar set of cytochromes are used for the transport of electrons when green plants undergo photosynthesis.

Revision questions

7.1. What are the oxidation numbers of the named element in the following compounds?
(i) C in CF_4 **(ii)** Mn in MnO_2 **(iii)** S in SO_4^{2-}
(iv) Sn in $SnCl_4$ **(v)** N in Mg_3N_2.

7.2. Use standard electrode potentials to predict whether copper metal could reduce aqueous iron(III) ions to iron(II) at room temperature.

7.3. Using standard electrode potentials, predict if reactions could occur in the following situations. Write balanced redox equations for those reactions that could occur at room temperature.
(i) An iron nail is placed in an aqueous solution of copper(II) sulfate ($CuSO_4(aq)$).
(ii) A silver ring is placed in an aqueous solution of zinc nitrate ($Zn(NO_3)_2(aq)$).

7.4. In the equation

$$MnO_4^-(aq) + 8H_3O^+(aq) + 5Au(s) \rightarrow Mn^{2+}(aq) + 12H_2O(l) + 5Au^+(aq)$$

(i) Which species is oxidized and what is its change in oxidation number?
(ii) Which species is reduced?
(iii) Which species is acting as an oxidizing agent?

7.5. When an aqueous solution of iron(III) ions is reacted with sulfur dioxide gas (this forms an acid solution), aqueous iron(II) ions and the sulfate ion (SO_4^{2-}) are produced. Write a balanced overall redox equation for the reaction.

7.6. In the following cell, all solutions are of concentration 1 mol dm^{-3} and at 25 °C:

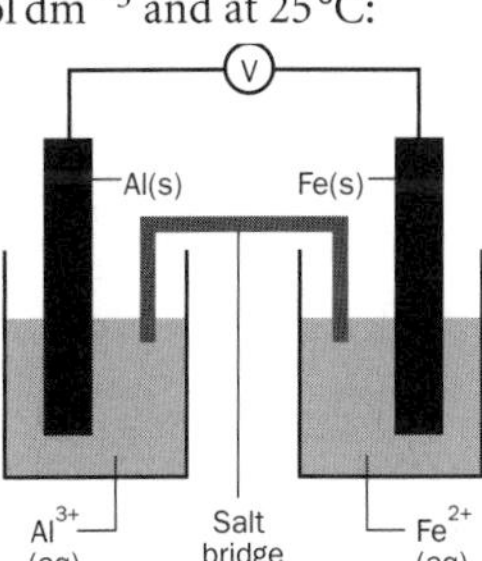

(i) Mark in the direction of electron flow.
(ii) Give equations for the processes occurring at each of the electrodes.
(iii) Calculate E_{cell}.
(iv) Draw a cell diagram for the system.

7.7. Tarnished silver jewellery occurs because the silver metal reacts with hydrogen sulfide in the presence of air to form silver sulfide (Ag_2S). The silver object can be 'cleaned' by dipping it into an aqueous solution of sodium chloride, contained in an aluminium metal container.
(i) Write down symbols for the ions contained in silver sulfide.
(ii) Using the standard electrode potentials for silver and aluminium, work out if a reaction occurs between the species present in the container.
(iii) Why is this a more desirable way of cleaning the silver than using silver polish?

7.8. Write a balanced redox equation for the oxidation of concentrated hydrochloric acid (H^+, Cl^-) to chlorine (Cl_2) by solid manganese(IV) oxide (MnO_2). The manganese(IV) oxide is reduced to $Mn^{2+}(aq)$.

7.9. In basic solution, zinc metal is oxidized by the nitrate ion ($NO_3^-(aq)$) to $[Zn(OH)_4]^{2-}(aq)$. The nitrate ion is reduced to ammonia. Write a balanced ionic equation for this reaction.

7.10. In the reaction

$$Ni^{2+}(aq) + Zn(s) \rightarrow Ni(s) + Zn^{2+}(aq)$$

which species is **(i)** the oxidizing agent **(ii)** the reducing agent?

7.11. Chromium plated steel (steel is mostly iron) is often used in motor car bumpers. If the bumper is scratched, will the chromium continue to protect the bumper against rusting?

Extension material to support this unit is available on our website at **http://www.palgrave.com/foundations/lewis.**

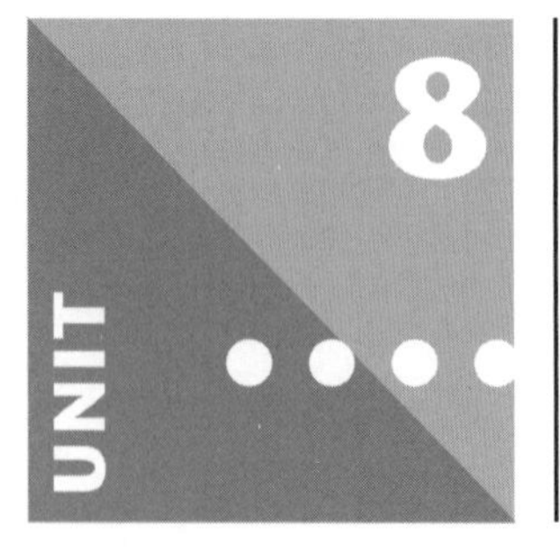

The Mole

Contents

Objectives

- ▶ Explains why chemists work in moles
- ▶ Describes how to perform chemical calculations, involving masses and gas volumes using moles
- ▶ Defines 'percentage yield'
- ▶ Explains how to work out which substance is the limiting reagent in a chemical reaction

8.1 Molecular mass

The idea of **molecular mass** was introduced in Chapter 3. This is the mass of one molecule of substance on the atomic mass scale. Molecular masses are calculated using the atomic masses of the constituent atoms. A list of approximate atomic masses for atoms of elements is shown in Table 8.1. You should use these for calculations, unless you are instructed otherwise.

Examples of calculations of molecular mass

$$m(H_2O) = 1 + 1 + 16 = 18\,u$$

$$m(C_6H_5Cl) = (6 \times 12) + (5 \times 1) + (35.5) = 112.5\,u$$

$$m(H_2SO_4) = (2 \times 1) + (32) + (4 \times 16) = 98\,u$$

Substances such as sodium chloride (Na^+, Cl^-) and copper(II) nitrate (Cu^{2+}, $2NO_3^-$) consist of ions (not molecules) and the term 'molecular mass' is not strictly appropriate. Nevertheless the 'formula mass' of these substances are calculated in a similar way to neutral molecules. Examples are

$$m(Na^+, Cl^-) = 23 + 35.5 = 58.5\,u$$

$$m(Cu^{2+}, 2NO_3^-) = (63.5) + 2(14 + 16 + 16 + 16) = 187.5\,u$$

Table 8.1 Approximate atomic masses of selected elements

Element	Symbol	Approximate atomic mass/u	Element	Symbol	Approximate atomic mass/u
Hydrogen	H	1	Calcium	Ca	40
Helium	He	4	Iron	Fe	56
Carbon	C	12	Nickel	Ni	59
Nitrogen	N	14	Copper	Cu	63.5
Oxygen	O	16	Zinc	Zn	65
Fluorine	F	19	Bromine	Br	80
Neon	Ne	20	Silver	Ag	108
Sodium	Na	23	Tin	Sn	119
Magnesium	Mg	24	Iodine	I	127
Aluminium	Al	27	Barium	Ba	137
Phosphorus	P	31	Gold	Au	197
Sulfur	S	32	Mercury	Hg	201
Chlorine	Cl	35.5	Lead	Pb	207
Potassium	K	39	Uranium	U	238

Exercise 8A

Calculating the molecular and formula mass

Write down the mass (u) of the following:

(i) nitric acid, HNO_3

(ii) magnesium sulfate, $MgSO_4$

(iii) ethyne, C_2H_2

(iv) sulfur molecules, S_8

(v) ethyl ethanoate, $CH_3COOC_2H_5$

(vi) hydrated iron(III) nitrate, $Fe(NO_3)_3\ 9H_2O$

(vii) one atom of neon.

8.2 Moles

The mole

How do you identify a chemist? One simple way is to ask the question: 'What is a mole?' A keen gardener will answer 'A small furry animal that digs holes in the lawn'; a doctor or nurse will answer 'a dark spot on skin'; and the head of a company may answer 'a spy'. A chemist will **always** answer 'a pile of atoms' or, being more specific, 'just over six hundred thousand trillion atoms'. When chemists have to calculate amounts of reacting substances they constantly work (and think!) in moles; that is why they will not hesitate to give the answer described above.

Why do chemists have to work in moles? Consider a reaction

$$A + B \rightarrow AB$$

You have already learned that the equation tells us that one particle of A reacts with one particle of B to form one particle of the compound AB. If a chemist wishes to get an exact amount of A to react with an exact amount of B, and not have excess A or B left over, then equal numbers of the particles of A and B must be reacted together.

The particles might be atoms, molecules or ions. Such particles are very small, so in order that the amounts involved are practical to work with, a great many particles of A must be added to the same number of particles of B.

Balances in the laboratory are used to weigh amounts of substance in grams and not atoms, molecules or ions, so it is *not* useful to weigh out say 1 g of A to react with 1 g of B – A and B are different substances so their particles *do not have the same mass*. Equal masses of A and B *will not contain the same number of particles*. Chemists work in moles, because *one mole of any substance contains the same number of particles*. Note that the *name* of the amount is **mole** and the symbol for the *unit* is **mol**.

Avogadro's constant

The number of particles (atoms, molecules or ions) in one mole of a substance is defined as being

the number of atoms contained in exactly 12 grams of carbon-12.

This number is very large and has been found by experiment to be

602 200 000 000 000 000 000 000 or 6.022×10^{23},
often approximated to 6×10^{23}

This number is called **Avogadro's constant** and is symbolized N_A.
From Table 8.1,

$$m(C) = 12\,u$$
$$m(He) = 4\,u$$
$$m(H_2O) = 18\,u$$
$$m(H_2) = 2\,u$$

and we can reason as follows:

- One atom of carbon is three times as heavy as one atom of helium. Therefore, one-thousand atoms of carbon are three times as heavy as one-thousand atoms of helium. Therefore, if a sample of carbon has three times the mass of a sample of helium they must have the same number of atoms, so that 12 g of carbon and 4 g of helium must each contain the same number of atoms – this number is numerically equal to N_A. Similar reasoning produces the following conclusions:
- One molecule of water is 4.5 times as heavy as one atom of helium, so 18 g of water contains N_A *molecules* of water, whereas 4 g of helium contains N_A *atoms* of helium.
- One atom of helium is twice as heavy as one molecule of hydrogen, so 4 g of helium contains N_A *atoms* of helium and 2 g of hydrogen contains N_A *molecules of hydrogen*.

Generalizing, the mass of a substance that contains N_A *particles is numerically equal to the atomic mass or molecular mass of that substance expressed in grams.* This mass is called the **molar mass** of the substance because it is the mass of one mole of that substance.

The symbol for molar mass is M. Its units are normally grams per mole ($g\,mol^{-1}$). For example,

$$M(H_2O) = 18\,g\,mol^{-1}$$

See Table 8.2 and Fig. 8.1.

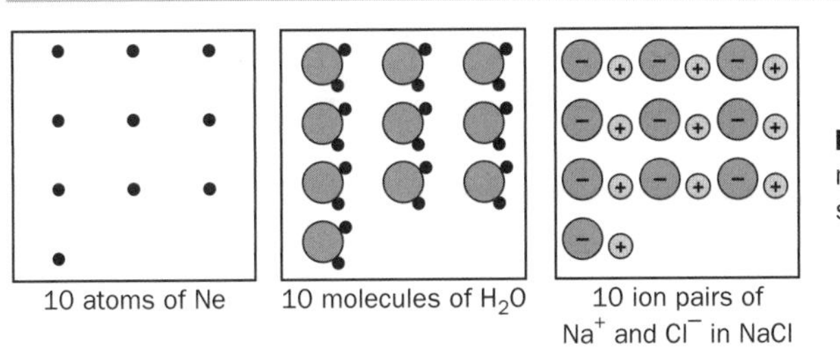

Fig. 8.1 An equal number of moles of neon, water and sodium chloride.

Table 8.2 The relationship between mass, molar mass and the number of moles for selected substances. (For simplicity N_A has been taken as exactly 6×10^{23})

Substance	Mass of substance /g	Formula of substance	Molar mass /g mol^{-1}	Number of moles of substance in the stated mass	Number of particles in the stated mass
Carbon	12	C	12	1	6×10^{23}
Carbon	120	C	12	10	6×10^{24}
Oxygen gas	32	O_2	32	1	6×10^{23}
Oxygen gas	3.2	O_2	32	0.1	6×10^{22}
Sodium chloride	58.5	NaCl	58.5	1	6×10^{23}*
Sodium chloride	0.0585	NaCl	58.5	0.001	6×10^{20}*
Benzene	78	C_6H_6	78	1	6×10^{23}
Benzene	7800	C_6H_6	78	100	6×10^{25}

* This is the number of Na^+,Cl^- ion pairs.

Exercise 8B

Moles, molar mass and Avogadro's constant

(i) Neon gas consists of single atoms. What mass of neon contains 6×10^{23} atoms?

(ii) How many potassium ions are there in 94 g of potassium oxide, $2K^+$, O^{2-}?

(iii) Magnesium metal consists of magnesium atoms. What mass of magnesium contains 6×10^{22} atoms?

(iv) What is the molar mass of copper(II) sulfate, $CuSO_4$?

Relationship between moles, molar mass and the mass of substance

To convert a mass of substance into moles the following equation can be used:

$$\text{amount of substance in moles} = \frac{\text{mass}}{\text{molar mass}}$$

Notice that units of molar mass are grams per mol, then the units of mass must be grams.

The rearranged forms of this equation are also useful:

$$\text{molar mass} = \frac{\text{mass}}{\text{amount of substance in moles}}$$

and

$$\text{mass} = \text{amount of substance in moles} \times \text{molar mass}$$

Example 8.1

How many moles of hydrated copper(II) sulfate ($CuSO_4 \cdot 5H_2O$) are contained in 5.00 g of the substance?

▶ Answer

$m(CuSO_4 \cdot 5H_2O) = (63.5 + 32 + 4(16) + 5(18)) = 249.5\,u$

therefore,

$M(CuSO_4 \cdot 5H_2O) = 249.5\,g\,mol^{-1}$.

$$\text{The amount of copper sulfate in moles} = \frac{\text{mass}}{\text{molar mass}} = \frac{5.00}{249.5} = 0.0200\,\text{mol}$$

Example 8.2

0.059 mol of calcium carbonate ($CaCO_3$) is required in an experiment. What mass of $CaCO_3$ needs to be weighed out?

▶ Answer

$M(CaCO_3) = 100\,g\,mol^{-1}$

Therefore, mass of calcium carbonate

= amount of calcium carbonate in moles × molar mass

= 0.059 × 100 = 5.9 g

Example 8.3

0.20 mol of a compound containing carbon and hydrogen has a mass of 3.2 g. What is the molecular mass of the compound? What is its likely formula?

▶ Answer

$$\text{molar mass} = \frac{\text{mass}}{\text{amount of substance in moles}} = \frac{3.2}{0.20} = 16\,g\,mol^{-1}$$

Since $M(\text{compound}) = 16\,g\,mol^{-1}$, $m(\text{compound}) = 16\,u$.

Carbon has an atomic mass of 12. The molecule is likely to be methane (CH_4) with

$m(CH_4) = 12 + 1 + 1 + 1 + 1 = 16\,u$

▶ Comment

Where there could be confusion, you should always be careful to specify *exactly* the substance you are referring to. For example the statement '1 mol of hydrogen' could mean 1 mol of H atoms *or* 1 mol of H_2 molecules. These have different molar masses of $1\,g\,mol^{-1}$ and $2\,g\,mol^{-1}$, respectively.

Exercise 8C

More calculations involving moles, mass, and molar mass

1. What is the mass of:

(i) 1 mol of N atoms

(ii) 4 mol of Fe atoms

(iii) 1.50 mol of Cl^- ions

(iv) 20 mol of Na_2SO_3?

2. How many moles of substance are present in:

(i) 40 g of calcium metal (Ca)

(i) 123.2 g of CCl_4

(iii) 0.49 g of SO_4^{2-}

(iv) 14 g of N_2?

3. A pure solid consists of either (i) sodium chloride (NaCl) (ii) copper carbonate ($CuCO_3$) or (iii) sodium carbonate (Na_2CO_3). 1.06 g of the solid contains 0.01 mol of compound. What is the chemical identity of the solid?

4. Calculate:

(i) the number of molecules of sulfur (S_8) in 16 g of solid sulfur.

(ii) the number of aluminium ions in 0.056 g of aluminium oxide ($2Al^{3+}$, $3O^{2-}$).

($N_A = 6 \times 10^{23}\,mol^{-1}$).

Simple experiment to estimate the Avogadro constant

Oil spreads out on the surface of water as far as it can. Theoretically, this could be until a layer one molecule thick is present. The oil used in this experiment is oleic acid, $C_{18}H_{34}O_2$ (Fig 8.2).

If one small drop of a solution of oil is dropped on to the surface of some water in a suitable container it spreads out in a circular layer. If the surface of the water is previously covered with a fine layer of talc, the diameter of the oil layer can be estimated with a ruler. Assuming the layer of oil is one molecule thick, the approximate thickness of an oil molecule and an estimate of the size of the Avogadro constant, N_A can be calculated. Specimen results are shown below.

To find the thickness of an oil molecule

Volume of oil in drop of solution = $y\,cm^3$.

Density of oil = $0.891\,g\,cm^{-3}$.

Diameter of oil layer on water = d cm.

Area of circular oil layer on water = $\pi r^2 = \pi \times (d/2)^2\,cm^2$.

Volume of oil layer = $\pi r^2 h = \pi \times (d/2)^2 \times h$

(think of the layer as a cylinder, one molecule thick and the thickness of the molecules = h).

Since the volume of the layer must be equal to the volume of the drop,

$$y = \pi \times (d/2)^2 \times h$$

So h, the thickness of a molecule of oil, can be calculated.

To estimate the value of the Avogadro constant N_A

Assume that the molecules are cubes, of side h.

Then volume of a molecule = h^3.

The molecular mass of the oil = 282

$$\text{Volume of 1 mol of oil} = \frac{\text{molar mass}}{\text{density}} = \frac{282}{0.891}$$

$$\text{Then } N_A = \frac{\text{volume of 1 mol of molecules}}{\text{volume of 1 molecule}}$$

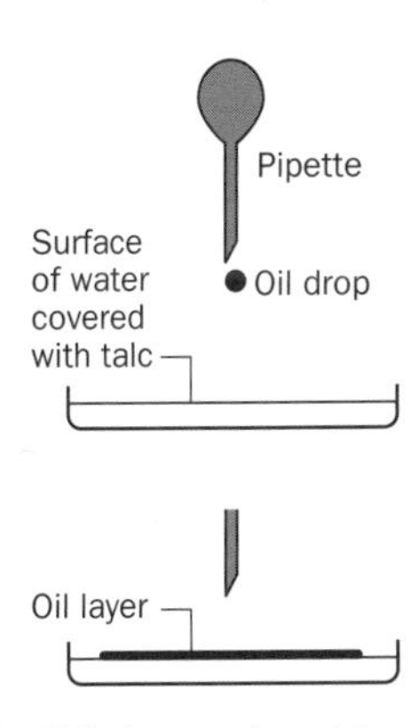

Fig. 8.2 An experiment to estimate the Avogadro constant.

The answer you obtain from this calculation should only give you some idea of the **magnitude** of N_A. There are a number of oversimplifications in the above reasoning, especially the assumption that the layer of oil is only one molecule thick.

Exercise 8D

Estimating N_A by experiment

Some typical experimental results from an experiment to estimate the Avogado constant are given below. Using these results, calculate a value for N_A.

(i) Volume of oil in drop = 2.54×10^{-5} cm³.

(ii) Diameter of oil layer on water = 20 cm.

Determining formulae by experiment

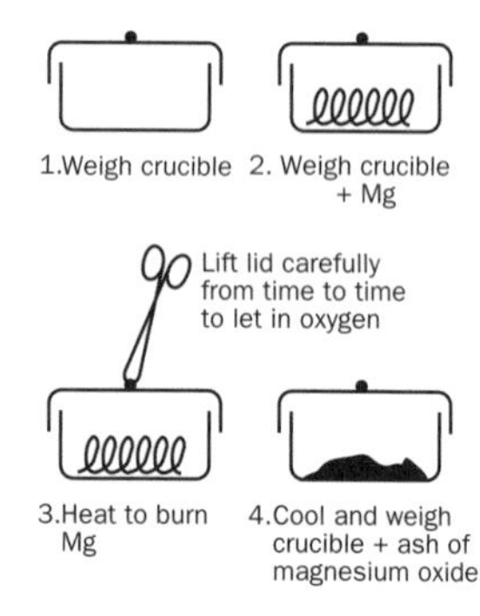

Fig. 8.3 Experiment to determine the formula of magnesium oxide.

To work out a formula by experiment, it is necessary to find out how many moles of each element combine together to make the compound. Magnesium metal, in the form of a strip (*magnesium ribbon*) is heated in air until all the magnesium is combined with oxygen. The diagrams in Fig. 8.3 show different steps in the experiment, in order.

Some sample results are:

Mass of crucible = 20.40 g
Mass of crucible + Mg = 22.26 g
Mass of crucible + magnesium oxide = 23.52 g

The formula of magnesium oxide can be worked out by answering the following questions (answers in brackets):

1. Find the mass of magnesium used in the experiment (22.26 – 20.40 = 1.86 g).

2. Calculate how many moles of magnesium atoms the answer to **1.** represents (1.86/24 = 0.078).

3. Find the mass of magnesium oxide formed (23.52 – 20.40 = 3.12 g).

4. Find the mass of oxygen that combines with magnesium (3.12 – 1.86 = 1.26 g).

5. How many moles of oxygen atoms does the answer to **3.** represent? (1.26/16 = 0.079)

6. What is the ratio of the number of moles of magnesium atoms to number of moles of oxygen atoms (0.078:0.079) as a whole number? (almost exactly 1:1).

7. The formula of magnesium oxide is therefore MgO.

8.3 Percentage composition by mass

The percentage of an element in a compound is found by using the formula of the compound and the atomic masses of the elements in the compound. It is more correctly called the **percentage composition by mass** of the element in that compound.

The percentage composition by mass of an element is given by the formula

$$\frac{\text{molar mass of element} \times \text{number of atoms of the element in the formula of the compound}}{\text{molar mass of compound}} \times 100$$

Example 8.4

Calculate the percentage composition by mass of nitrogen in ammonium nitrate, NH_4NO_3.

▶ Answer

$m(N) = 14\,u$

$m(NH_4NO_3) = 14 + (4 \times 1) + 14 + (3 \times 16) = 80\,u$

number of atoms of N = 2

$$\text{percentage composition by mass} = \frac{14 \times 2 \times 100}{80} = 35\%$$

Simplest formula

Just as the percentage composition by mass can be calculated from the formula of a compound, the simplest formula of a compound can be found from the percentage composition by mass of each element. The simplest formula is also known as the **empirical formula**. Here, 'empirical' means 'obtained from experimental results'. To find the simplest formula:

1. Write down the symbols of the elements involved in the compound.
2. Write the percentage composition of each element under its symbol.
3. Divide each percentage by the molar mass of that element ('the atomic mass in grams') to obtain a molar ratio.
4. Find the simplest, whole number, molar ratio to work out the formula of the compound.

Example 8.5

Calculate the simplest, or empirical, formula for a compound that contains 27.3% C and 72.7% O by mass.

▶ Answer

Symbols of elements involved	C	O
Percentage composition	27.3	72.7
Divide by molar masses	$\frac{27.3}{12}$	$\frac{72.7}{16}$
Molar ratio	2.3	4.5
Simplest whole number ratio (this can often be obtained by dividing through by the smallest number)	$\frac{2.3}{2.3}$	$\frac{4.5}{2.3}$
This is approximately	1	2

Therefore, the simplest formula of the compound is CO_2.

▶ Comment

Sometimes the composition of elements in the compound is given as mass of elements in a given mass of compound. In this case, the procedure is the same, but the mass of each element is substituted for the percentage composition by mass.

Exercise 8E

Calculating percentage composition by mass

Find the percentage composition by mass of:

(i) sulfur in Na_2SO_4

(ii) sodium in $Na_2CO_3 \cdot 10H_2O$

(iii) oxygen in $Al(NO_3)_3$

(iv) bromine in $CaBr_2$

(v) sulfate in $(NH_4)_2SO_4$.

Exercise 8F

Calculating simplest (or empirical) formulae

(i) When methane is analysed it is found to contain 75% C and 25% H. What is its empirical formula?

(ii) 15.3 g of aluminium atoms combine with 13.6 g of oxygen atoms to form an oxide. What is the simplest formula of the oxide?

(iii) 0.4764 g of an oxide of iron contained 0.3450 g of iron. What is the empirical formula of the oxide?

(iv) A compound contains sodium (32.3%), sulfur (22.5%) and oxygen only. What is its empirical formula?

Finding molecular formulae

The simplest formula is not always the 'true' or molecular formula: it is only the simplest ratio of the atoms contained in the substance. If, for example, you calculated the simplest formula of hydrazine (N_2H_4) from its percentage composition by mass of nitrogen and hydrogen, you would obtain NH_2. In order to establish the molecular formula, you need the molecular mass as well as the empirical formula.

Example 8.6

Dichloroethane has a molecular mass of 99 u. Analysis of a sample shows that it contains 24.3% carbon, 4.1% hydrogen and 71.6% chlorine. What is its molecular formula?

▶ Answer

Calculate the simplest formula:

Element	C	H	Cl
Percentage composition	24.3	4.1	71.6
Divide by molar masses	$\frac{24.3}{12}$	$\frac{4.1}{1}$	$\frac{71.6}{35.5}$
Molar ratio	1	2	1

Therefore, the simplest formula is CH_2Cl.

The molecular mass of $CH_2Cl = 12 + (2 \times 1) + 35.5 = 49.5\,u$, so it is not the molecular formula of dichloroethane.

Multiply the number of each atom by two, to maintain the same ratio, and you get $C_2H_4Cl_2$ which has a molecular mass of

$$(2 \times 12) + (4 \times 1) + (2 \times 35.5) = 99\,u$$

So, the molecular formula of dichloroethane is $C_2H_4Cl_2$.

Exercise 8G

Determining molecular formulae

(i) Work out the empirical and molecular formulae for a compound that has a molar mass of 30 g mol^{-1} and is of percentage composition 80% carbon and 20% hydrogen.

(ii) Nicotine contains 74.9% carbon, 8.7% hydrogen and 17.3% nitrogen. The compound contains two nitrogen atoms per molecule. What are the empirical and molecular formulae of nicotine?

8.4 Water of crystallization

The formulae of some ionic compounds may be written as $AB \cdot nH_2O$, where '$\cdot nH_2O$' indicates that n molecules of water are associated with each AB unit within the crystal lattice. Because the water is within the lattice, the substance is perfectly dry to the touch but it contains **water of crystallization** and is said to be *hydrated*. The water may be driven off by strong heating (**dehydration**) and the residue, that now does not contain water of crystallization, is termed **anhydrous**.

Blue copper(II) sulfate crystals ($CuSO_4 \cdot 5H_2O$) become white, anhydrous copper(II) sulfate on heating:

$$\underset{\text{blue}}{CuSO_4 \cdot 5H_2O(s)} \rightarrow \underset{\text{white}}{CuSO_4(s)} + 5H_2O(l)$$

If water is added to anhydrous copper(II) sulfate, it turns blue again.

Other common substances that contain water of crystallization are

sodium carbonate decahydrate (washing soda): $Na_2CO_3 \cdot 10H_2O$

magnesium sulfate heptahydrate (Epsom salts): $MgSO_4 \cdot 7H_2O$

Exercise 8H

Substances that contain water of crystallization

The dehydration of copper(II) sulfate by heating is a 'stepwise' process

(i) First, the pentahydrate loses water to become first the trihydrate.

(ii) Then the monohydrate is formed.

(iii) Finally the anhydrous salt remains.

(iv) If the anhydrous salt is heated very strongly, it decomposes to copper(II) oxide and sulfur trioxide.

Write balanced equations for all of these reactions.

8.5 Calculating amounts from equations

Calculating masses

You have already seen that chemical equations give the following information:

1. The chemical composition of the reactants and products in a reaction.
2. The simplest whole-number relationship of the molecules involved in the reaction.

Using the mole concept, together with the atomic masses of the elements involved, we are now in a position to work out the masses of substances involved in chemical reactions.

Example 8.7

What mass of iodine will react completely with 10 g of aluminium to form aluminium iodide (Al_2I_6)?

▶ Answer

Write the balanced equation:

$$2Al(s) + 3I_2(s) \rightarrow Al_2I_6(s)$$

Work out the relationship between the molecules involved:

$$2 \text{ atoms Al} + 3 \text{ molecules } I_2 \rightarrow 1 \text{ molecule } Al_2I_6$$

You could also write:

$$4 \text{ atoms Al} + 6 \text{ molecules } I_2 \rightarrow 2 \text{ molecules } Al_2I_6$$

or

$$200 \text{ atoms Al} + 300 \text{ molecules } I_2 \rightarrow 100 \text{ molecules } Al_2I_6$$

or even

$$2 \times 6.022 \times 10^{23} \text{ atoms Al} + 3 \times 6.022 \times 10^{23} \text{ molecules } I_2 \rightarrow 1 \times 6.022 \times 10^{23} \text{ molecules } Al_2I_6$$

The last line can also be written as

$$2 \text{ mol Al} + 3 \text{ mol } I_2 \rightarrow 1 \text{ mol } Al_2I_6$$

Convert the numbers of moles to grams using molar masses, $M(\text{Al}) = 27\,\text{g mol}^{-1}$ and $M(\text{I}) = 127\,\text{g mol}^{-1}$:

$$2 \times 27\,\text{g Al} + 3 \times 254\,\text{g } I_2 \rightarrow 1 \times 816\,\text{g } Al_2I_6$$

or

$$54\,\text{g Al} + 762\,\text{g } I_2 \rightarrow 816\,\text{g } Al_2I_6$$

We now have the ratio between the quantities of reactants and products in grams.

In the question, you start with 10 g of aluminium, so what masses of iodine and aluminium iodide are involved if the above ratio is to apply?

For 1 g of aluminium:

$$1\,\text{g Al} + \frac{762}{54}\,\text{g } I_2 \rightarrow \frac{816}{54}\,\text{g } Al_2I_6$$

Therefore, for 10 g aluminium:

$$10\,\text{g Al} + \frac{762 \times 10}{54}\,\text{g } I_2 \rightarrow \frac{816 \times 10}{54}\,\text{g } Al_2I_6$$

or

$$10\,\text{g Al} + 141\,\text{g } I_2 \rightarrow 151\,\text{g } Al_2I_6$$

So, 10 g of aluminium reacts with 141 g I_2.

▶ Comment

The calculation is written in simple steps to help you understand. In practice, you can leave out some of the more obvious steps and ignore the calculations involving Al_2I_6, since they are not asked for in the question – the question only asks for the relationship between aluminium and iodine.

Exercise 8I

Calculations based on equations

(i) What mass of magnesium oxide is formed by burning 1.2 g of magnesium metal in oxygen?

$2Mg(s) + O_2(g) \rightarrow 2MgO(s)$

(ii) What mass of oxygen is formed when 4.34 g of mercury(II) oxide is completely decomposed by heating?

$2HgO(s) \rightarrow 2Hg(l) + O_2(g)$

(iii) What mass of copper(II) nitrate is needed to produce 16 g of copper(II) oxide upon decomposition?

$2Cu(NO_3)_2(s) \rightarrow 2CuO(s) + 4NO_2(g) + O_2(g)$

(iv) How much graphite (carbon, C) must be burned in oxygen gas to produce 2.2 g of carbon dioxide gas? (You provide the balanced equation.)

(v) Solid copper(II) carbonate decomposes on heating to solid copper(II) oxide and carbon dioxide gas. How much copper(II) oxide is obtained when 1.28 g of the carbonate is heated? (Again, first provide the balanced equation.)

8.6 Calculating gas volumes

When gases are involved in reactions, it is easier to measure their volumes rather than weigh them and find their masses. Although we can calculate the masses of gases used up or produced in chemical reactions using the methods described in the previous section, it is often more convenient to calculate the volumes involved.

At this point, it is worth remembering that

One mole of ANY gas occupies a volume of 24 dm^3 at room temperature and pressure.

'Room temperature' is taken as 20°C or 293 K. 'Normal' atmospheric pressure is 1 atm, i.e. 101 325 Pa. (The pascal (Pa) is the SI unit of pressure – see page 157.)

Example 8.8

What volume of carbon dioxide can be obtained by decomposing 50 g of calcium carbonate at room temperature and pressure?

▶ Answer

Write the balanced equation:

$CaCO_3(s) \rightarrow CaO(s) + CO_2(g)$

Work out the mole relationship between the substances you are considering:

1 molecule $CaCO_3 \rightarrow$ 1 molecule CO_2

Therefore,

1 mol $CaCO_3 \equiv$ 1 mol CO_2

Note that '$\rightarrow$' can be replaced by '$\equiv$' which can be taken to mean 'reacts with' or 'reacts to form' according to the amounts shown by the equation.
Work out the mole relationship between the substances you are considering:

Example 8.8 (*continued*)

Convert moles to grams in the case of calcium carbonate and moles to cubic decimetres in the case of carbon dioxide, because the question asks about the volume of the gas:

$$100\,g\ CaCO_3 \equiv 24\,dm^3\ CO_2$$

Therefore,

$$50\,g\ CaCO_3 \equiv 12\,dm^3\ CO_2$$

So, $12\,dm^3$ CO_2 are produced.

Exercise 8J

Calculating gas volumes from equations

All gas volumes are measured at room temperature and pressure

(i) What is the volume occupied by 0.25 mol of chlorine gas?

(ii) What is the mass of $600\,cm^3$ of sulfur dioxide gas?

(iii) What volume of hydrogen is obtained from the reaction of 4.8 g of magnesium metal with excess hydrochloric acid?

$Mg(s) + 2HCl(aq) \rightarrow MgCl_2(aq) + H_2(g)$

(iv) What volume of oxygen is needed for complete combustion of $2\,dm^3$ of propane (C_3H_8).

$C_3H_8(g) + 5O_2(g) \rightarrow 3CO_2(g) + 4H_2O(l)$

(v) What volume of hydrogen chloride gas is produced by the reaction of $40\,cm^3$ of chlorine with hydrogen? (First write a balanced equation.)

(vi) What mass of water is made when $10\,dm^3$ of hydrogen gas burns completely in excess air? (First write a balanced equation.)

8.7 Percentage yield

The previous calculations can be used to find out how much product you would expect, if you carried out a reaction with a measured amount of reactants in the laboratory. In practice it is often extremely difficult to obtain the calculated yield, especially in organic chemistry, because of many reasons including:

1. Loss of reactants and/or product during experimental manipulations.
2. Competing side reactions, where other products are formed.
3. As you will see later, in some reactions (**reversible** or **equilibrium reactions**), the reactants do not completely change into the products (*go to completion*).

The yield of the product obtained by experiment, therefore, may be lower than expected; sometimes it is a great deal lower. It is usual to report your **percentage yield** when you are doing preparative chemistry:

$$\text{percentage yield} = \frac{\text{experimental yield}}{\text{theoretical yield}} \times 100\%$$

where the experimental yield is the mass of product obtained and the theoretical yield is the mass of product you calculated you would get from the balanced equation.

Example 8.9

Methane is converted to carbon dioxide and water when burned in a plentiful supply of oxygen (*complete combustion*):

$$CH_4(g) + 2O_2(g) \rightarrow CO_2(g) + 2H_2O(l)$$

If 10 g of CO_2 were obtained when 16 g of CH_4 were burned in a limited supply of oxygen gas, what would be the percentage yield of carbon dioxide?

▶ Answer

If 16 g of methane is burned in oxygen and 10 g of carbon dioxide are formed, then the theoretical yield of carbon dioxide formed can be found from the equation:

$$CH_4(g) + 2O_2(g) \rightarrow CO_2(g) + 2H_2O(l)$$

1 mol $CH_4(g)$ produces 1 mol $CO_2(g)$

16 g $CH_4(g)$ produces 44 g $CO_2(g)$

$$\text{percentage yield of } CO_2 = \frac{\text{experimental yield}}{\text{theoretical yield}} \times 100$$

$$= \frac{10}{44} \times 100$$

$$= 23\%$$

▶ Comment

In a limited supply of oxygen, not all of the methane would be converted to carbon dioxide – carbon monoxide and soot (carbon) would also be formed. In this case, *incomplete combustion* occurs.

Exercise 8K

Percentage yields

A student prepared solid magnesium oxide by burning 4.8 g of magnesium metal in air. Unfortunately, much of the white magnesium oxide produced in this way formed a white smoke which dispersed into the air of the laboratory. The student recovered 4.5 g of magnesium oxide. What was the percentage yield of magnesium oxide?

8.8 Limiting reagents

We have very often specified that a particular reagent is **in excess**; this means that there are more than enough moles of it to react with the other reactant. Under these circumstances it is the reagent that is *not* in excess that controls the amount of product. This reagent is called the **limiting reagent**.

Example 8.10

In order to understand the idea of a limiting reagent, imagine that you are packing biscuits. If each packet can hold ten biscuits, the 'reaction' is

1 empty packet + 10 loose biscuits → packet of biscuits

You have 25 empty packets and 200 loose biscuits. How many packets of biscuits can you make?

▶ Answer

20 with 5 empty packets left over. Here, you do not have enough biscuits to fill all the packets. The number of loose biscuits (*the limiting reagent*) controls the number of packets of biscuits you can make.

▶ Comment

If you have 12 empty packets and 150 biscuits, you can make 12 packets of biscuits and 30 biscuits remain. This time it is the number of empty packets that is the limiting reagent.

In a chemical reaction, unless the exact amounts of each reagent (as specified by the balanced equation for the reaction) are reacted together, then one reagent is always limiting. The limiting reagent can be spotted by comparing the ratio of the amounts of reactants available with the ratio of the amounts of reactants obtained from the balanced equation for the reaction.

Example 8.11

Copper reacts with silver nitrate solution according to the equation

$$Cu(s) + 2AgNO_3(aq) \rightarrow Cu(NO_3)_2(aq) + 2Ag(s)$$

If 0.50 mol of copper is added to 1.5 mol of silver nitrate, which is the limiting reagent and how many moles of silver are formed?

▶ Answer

Decide which is the limiting reagent

According to the equation, 1 mol Cu ≡ 2 mol $AgNO_3$, so

$$0.50 \text{ mol Cu} \equiv 2 \times 0.50 = 1.0 \text{ mol } AgNO_3$$

but there are 0.50 mol Cu and 1.5 mol $AgNO_3$. Therefore, $AgNO_3$ is present in excess and Cu is the limiting reagent.

Calculate how many moles of silver are formed

Use the amount of the limiting reagent to find the amount of product. According to the equation, 1 mol Cu ≡ 2 mol Ag. Therefore,

$$0.50 \text{ mol Cu} \equiv 2 \times 0.50 = 1.0 \text{ mol Ag}$$

So 1.0 mol Ag is formed.

Exercise 8L

Limiting reagents

(i) What is the limiting reagent when 0.5 mol of solid phosphorus and 0.5 mol of oxygen react according to the equation

$P_4(s) + 5O_2(g) \rightarrow P_4O_{10}(s)$

and how many moles of P_4O_{10} are formed?

(ii) What is the limiting reagent when 1.5 g of magnesium and 1.5 g of nitrogen combine according to the equation

$3Mg(s) + N_2(g) \rightarrow Mg_3N_2(s)$

and how many grams of Mg_3N_2 are formed?

(iii) Zinc reacts with heated copper(II) oxide to form zinc oxide and copper metal. If 3.0 g of zinc are reacted with 3.0 g copper(II) oxide, which is the limiting reagent? What is the mass of copper metal formed?

BOX 8.1

Avogadro

Lorenzo Romano Amedeo Carlo Avogadro de Quaregnae di Cerreto (1776–1856) had a very impressive name – and an equally impressive intellect. He graduated in law at 16 and received a doctorate at 20. He practised law for a while, but became interested in science and mathematics and eventually gave up law to teach these subjects.

He had read that one volume of oxygen gas always reacted with twice the volume of hydrogen gas to form water. If the temperature was high enough, then the water was formed as steam and exactly two volumes of steam were produced. He concluded that two *particles* of hydrogen gas must react with one *particle* of oxygen gas to produce two *particles* of water. After looking at the results of the experiments of another famous chemist, Gay-Lussac, in which reacting gas volumes had been measured, Avogadro came up with the following law (see page 159):

Equal volumes of different gases (at the same temperature and pressure) contain the same number of particles

He called the gas particles **molecules**. If Avogadro's law is applied to volumes of reacting gases, it is possible to deduce their formulae. The law was discounted by the scientific community until very late in his lifetime.

Revision questions

8.1. Calculate the molecular mass of:
(i) nitrogen gas (N_2)
(ii) copper(II) chloride
(iii) phosphoric acid (H_3PO_4)
(iv) calcium nitrate
(v) sodium sulfate decahydrate.

8.2. What is the mass of one mole of:
(i) chlorine atoms (Cl)
(ii) chlorine molecules (Cl_2)
(iii) phosphorus molecules (P_4)
(iv) iodide ions (I^-)
(v) buckminsterfullerene molecules (C_{60}).

8.3. Calculate the number of moles in:
(i) 30 g of oxygen molecules (O_2)
(ii) 31 g of phosphorus molecules (P_4)
(iii) 25 g of calcium carbonate
(iv) 3.28 g of sulfur dioxide
(v) 980 g of sulfuric acid.

8.4. Calculate the mass of each of the following:
(i) 0.500 mol of NaCl
(ii) 0.250 mol of CO_2
(iii) 2 mol of H_2
(iv) 2.50 mol of $CuSO_4 \cdot 5H_2O$
(v) 100 mol of HNO_3.

8.5. A sample of ammonia weighs 2.00 g. What mass of sulfur dioxide contains the same number of molecules as are in 2.00 g of ammonia?

8.6. What mass of magnesium metal would react completely with 8 g of sulfur?

$Mg(s) + S(s) \rightarrow MgS(s)$

8.7. **(i)** What *mass* of oxygen would be produced by heating 4.25 g of sodium nitrate?

$$2NaNO_3(s) \rightarrow 2NaNO_2(s) + O_2(g)$$

(ii) What *volume* of oxygen, at room temperature and pressure, does the answer to **(i)** represent?

8.8. When sodium hydrogen carbonate ($NaHCO_3$) is heated, the following change occurs:

$$NaHCO_3(s) \rightarrow Na_2CO_3(s) + H_2O(l) + CO_2(g)$$

(i) Balance the equation written above.
(ii) What is the mass of two moles of $NaHCO_3$?
(iii) What is the mass of CO_2 produced when 2.0 mol of $NaHCO_3$ are decomposed? What is the volume of this mass at room temperature and pressure?
(iv) What volume of carbon dioxide, at room temperature and pressure, would be obtained if 4.2 g of $NaHCO_3$ were completely decomposed?

8.9. What percentage by mass is water in the salt $MgSO_4 \cdot 7H_2O$?

8.10. When the gas ethene is analysed, it is found to contain 85.72% carbon and 14.28% hydrogen by mass. The molar mass of ethene is 28 g mol^{-1}. What is its molecular formula?

8.11. Aluminium was burned in a stream of chlorine so that it was all converted to aluminium chloride. The following results were obtained:
Mass of aluminium at the start of the experiment = 13.50 g.
Mass of aluminium chloride at the end of the experiment = 66.75 g.
From these results, determine the formula of aluminium chloride.

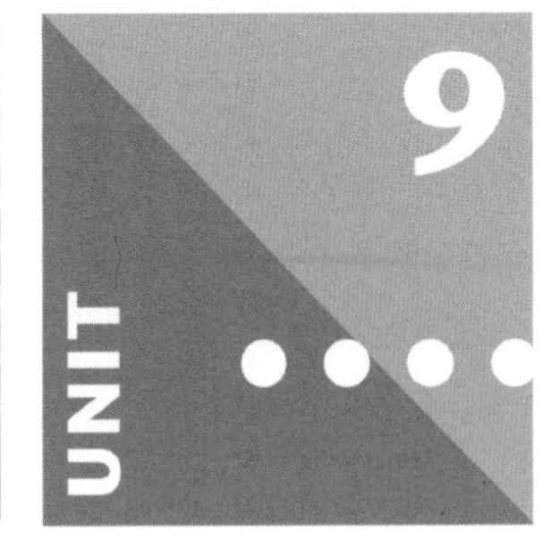

Calculating Concentrations

Objectives

- ▶ Describes how the concentrations of solutions can be expressed in appropriate units
- ▶ Gives examples of how volumetric analysis can be used to find the concentration of solutions
- ▶ Explains what is meant by 'pH'

Contents

9.1 Concentration of solutions

We have already discussed how to calculate masses of solids or volumes of gases, for the reactants or products in balanced equations. What if solutions are involved in the reaction? An aqueous solution may be concentrated or dilute. That is, it may contain a large or small amount of solid dissolved in a given amount of water as shown in Fig. 9.1.

How can we express the concentrations of the solutions?

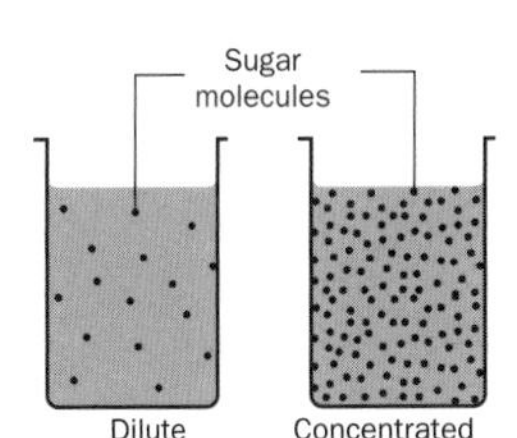

Fig. 9.1 Sugar solutions at two different concentrations.

Molar concentration

It is common to measure the concentration of a solution in *moles per cubic decimetre* (**mol dm**$^{-3}$). For example, a solution of sodium chloride, of concentration 1 mol dm^{-3}, contains 1 mol (or a mass of 23 + 35.5 = 58.5 g) of NaCl dissolved in 1 dm^3 of *solution. Always use the term 'solution' and not water.* A mixture of 58.5 g of NaCl dissolved in 1 dm^3 of water is *not* quite the same thing.

Different textbooks, according to their age and their country of origin, may express concentration as M (molar) or mol L^{-1} (moles per litre). Both these units may be taken to mean the same as mol dm^{-3}.

Some examples of molar concentration are:

1. A solution of sodium chloride of concentration 2.00 mol dm^{-3} contains 2.00 mol (or $2.00 \times 58.5 = 117$ g) of NaCl dissolved in 1 dm^3 of solution.
2. Sodium chloride solution of concentration 0.500 mol dm^{-3} contains 0.500 mol (or $0.500 \times 58.5 = 29.3$ g) of NaCl dissolved in 1 dm^3 of solution.

In general for solutions,

amount of substance in mol = volume × molar concentration

When using this equation, volume is expressed in dm^3 *and concentration is in* $mol\,dm^{-3}$.

Rearranged forms of this equation may be needed:

$$\text{volume} = \frac{\text{amount in mol}}{\text{molar concentration}}$$

or

$$\text{molar concentration} = \frac{\text{amount in mol}}{\text{volume}}$$

Example 9.1

What is the concentration of a solution of sodium hydroxide (NaOH) that contains 0.1 mol of NaOH dissolved in 250 cm^3 of solution?

▶ Answer

amount of NaOH in mol = volume × molar concentration

Substituting, and remembering that 1 dm^3 = 1000 cm^3, so we must divide 250 cm^3 by 1000 to convert it to dm^3,

$$0.1 = \frac{250}{1000} \times \text{molar concentration}$$

Therefore the concentration of the NaOH solution is

$$\frac{1000 \times 0.1}{250} = 0.4\,\text{mol}\,\text{dm}^{-3}$$

Example 9.2

How many grams of sodium chloride are there in 500 cm^3 of a solution which has a concentration of 0.500 $mol\,dm^{-3}$?

▶ Answer

$$\begin{aligned}\text{amount of NaCl in mol} &= \text{volume} \times \text{molar colcentration} \\ &= \frac{500}{1000} \times 0.500 \\ &= 0.250\,\text{mol}\end{aligned}$$

To convert moles into grams:

$$\text{amount of NaCl in mol} = \frac{\text{mass NaCl}}{\text{molar mass NaCl}}$$

$$0.250 = \frac{\text{mass NaCl}}{58.5}$$

which gives

mass of NaCl = 0.250 × 58.5 = 14.6 g

Therefore, 14.6 g of NaCl are dissolved in 500 cm^3 of solution.

BOX 9.1

Concentrations of pure solids and pure liquids

Concentration is measured in mol dm^{-3}. This is a measure of the number of particles per unit volume of the substance. In pure solids or pure liquids the number of particles in a given volume can be calculated from the density and this is constant, at constant temperature. Since the density of a given pure solid or pure liquid is constant, at constant temperature, **so is its concentration**.

1. The density of copper metal at 298 K is 8.92 g cm^{-3}. This is equivalent to 8920 g dm^{-3}, or a concentration of 8920/63.5 mol dm^{-3}, which approximates to 140 mol dm^{-3}.
2. The density of pure water at 298 K is 1 g cm^{-3}, equivalent to 1000 g dm^{-3} or a concentration of 1000/18 mol dm^{-3}. This approximates to 56 mol dm^{-3}.

Exercise 9A

Concentration of solutions

(i) Sodium chloride (58.5 g) is dissolved in water. More water is then added so that the total volume is 100 cm^3. What is the concentration of the solution in mol dm^{-3}?

(ii) Calculate the concentration (in mol dm^{-3}) of hydrated copper(II) sulfate ($CuSO_4 \cdot 5H_2O$) in a solution originally prepared by dissolving 0.1 g of the solid in water and making up to 50 cm^3 with water.

(iii) What volume of sodium chloride solution of concentration 0.2 mol dm^{-3} contains 10 mol of salt?

(iv) What mass of hydrated sodium carbonate ($Na_2CO_3 \cdot 10H_2O$) should be dissolved in water and made up to exactly 1 dm^3 in order to make the solution a concentration of 0.100 mol dm^{-3}?

Ion concentrations

When ionic solids are dissolved in water, they are regarded as completely ionized in solution. It is sometimes necessary to express the solution in terms of the concentrations of the ions present, as shown in the examples below:

1. An aqueous NaCl solution, of concentration 1 mol dm^{-3}, contains $Na^+(aq)$ and $Cl^-(aq)$ ions. Each NaCl unit is dissociated in solution as follows:

   $NaCl(aq) \rightarrow Na^+(aq) + Cl^-(aq)$

   As 1 mol NaCl produces 1 mol Na^+ and 1 mol Cl^-, a solution of NaCl of concentration 1 mol dm^{-3}, contains a concentration of 1 mol dm^{-3} $Na^+(aq)$ ions and 1 mol dm^{-3} $Cl^-(aq)$ ions.

2. An aqueous $CaCl_2$ solution, of concentration 1 mol dm^{-3}, contains $Ca^{2+}(aq)$ and $Cl^-(aq)$ ions. Each $CaCl_2$ unit is dissociated in solution as follows:

   $CaCl_2(aq) \rightarrow Ca^{2+}(aq) + 2Cl^-(aq)$

   A solution of $CaCl_2$ of concentration 1 mol dm^{-3}, contains a concentration of 1 mol dm^{-3} $Ca^{2+}(aq)$ and contains a concentration of 2 mol dm^{-3} $Cl^-(aq)$.

3. Similarly, a 0.1 mol dm^{-3} solution of $Na_3PO_4(aq)$ contains $Na^+(aq)$ at a concentration of 0.3 mol dm^{-3} and $PO_4^{3-}(aq)$ at a concentration of 0.1 mol dm^{-3}.

Exercise 9B

Assuming the following compounds are completely ionized in solution, identify the ions that exist in the following aqueous solutions and state the concentration of each ion.

(i) Potassium chloride, KCl (0.1 mol dm^{-3})

(ii) Sodium carbonate, Na_2CO_3 (0.125 mol dm^{-3}.

(iii) Ammonium sulfate, $(NH_4)_2SO_4$ (0.25 mol dm^{-3}).

(iv) Copper nitrate, $Cu(NO_3)_2$ (0.15 mol dm^{-3}).

9.2 Standard solutions

A **standard solution** is a solution of known concentration. **Primary standards** are substances that are used to make up standard solutions, by accurately weighing out a given mass of the standard, then dissolving the standard in deionized (i.e. pure) water to make up a known volume of solution. Primary standards should be:

1. chemically stable in the solid and aqueous states
2. pure
3. soluble in water.

Anhydrous sodium carbonate is an example of a primary standard. Standard solutions of primary standards can be used to standardize other solutions, as you will see in the next section.

Exercise 9C

Calculating the concentration of standard solutions

An aqueous solution containing 0.10 mol of anhydrous sodium carbonate (Na_2CO_3) is placed in each of the volumetric flasks, and carefully diluted with deionized water to each of the volumes shown. Calculate the concentration of each solution.

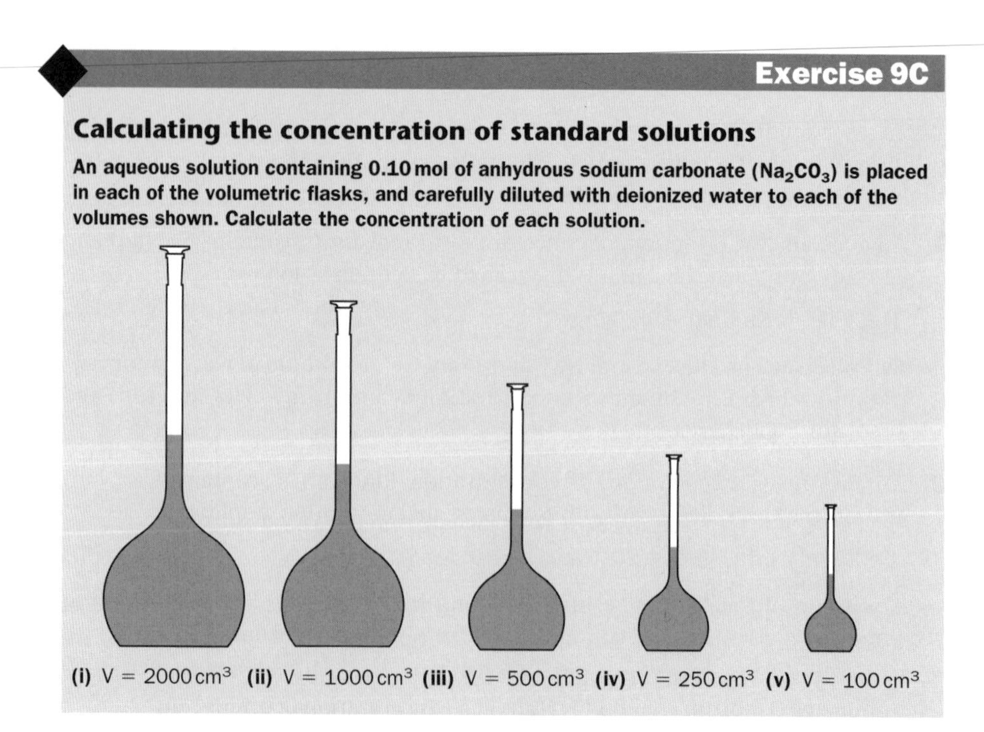

Standard solutions are made up in special flasks, called **volumetric flasks**, which have been calibrated to contain a definite volume of liquid. A graduation line is marked on the neck of the flask. When the flask contains solution to this level it contains the volume of solution marked on the front of the flask. Some volumetric flasks are shown in Exercise 9C.

BOX 9.2

Making a standard solution

A standard solution of, for example, sodium carbonate, is made by weighing out the required mass of sodium carbonate using a very sensitive balance called an **analytical balance**. The mass is generally measured to the nearest 0.0001 g. The solid is then *completely* dissolved in deionized water in a beaker and the resulting solution transferred to a volumetric flask. The beaker is rinsed with deionized water a few times and the washings also transferred to the flask, so that *all* the sodium carbonate solution is transferred. The flask is then carefully topped up to the graduation mark on the flask with deionized water, so that the bottom of the solution meniscus is 'sitting on' the graduation line when viewed at eye level. Finally, the flask is shaken well, so that the concentration of the solution it contains is homogeneous.

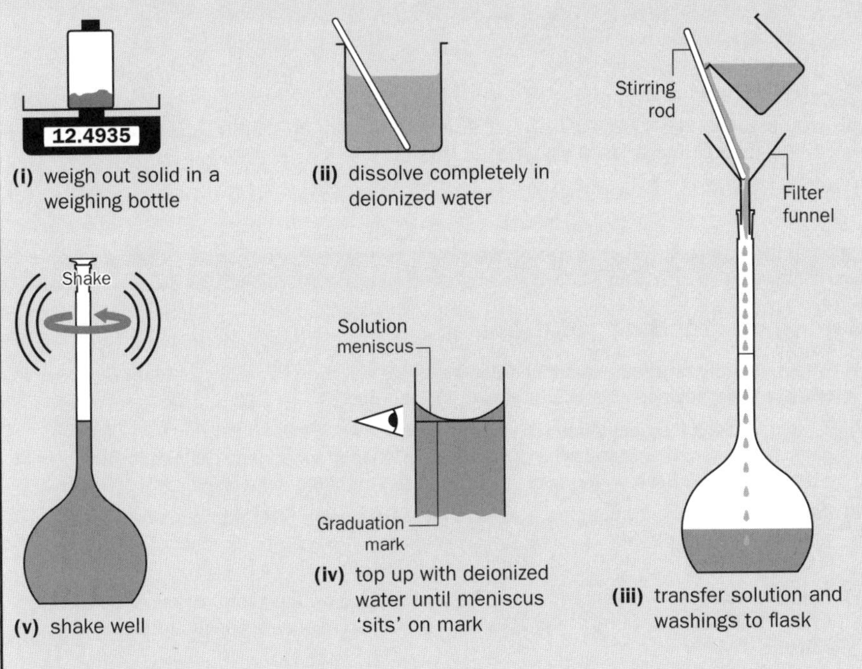

Diluting standard solutions

A dilute standard solution can also be made by carefully diluting a more concentrated standard solution. In this instance, a known volume of the concentrated standard solution is delivered into a standard volumetric flask, and the solution diluted to the graduation line with deionized water. **Pipettes** are designed to deliver specific volumes of liquid. They can contain various volumes (from 0.50 to 200 cm^3) and the volume of the pipette is marked on its bulb or shown by graduation marks on the side of the pipette (Fig. 9.2).

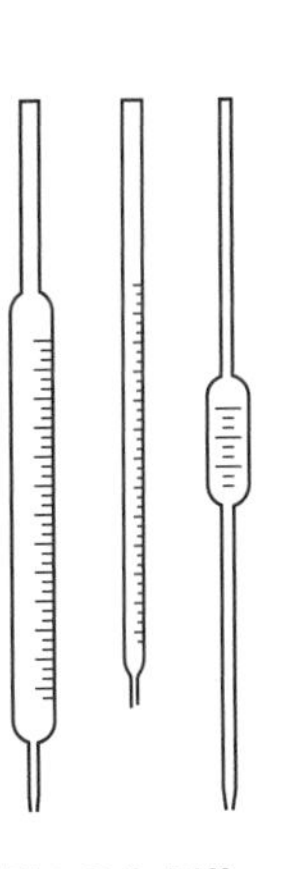

Fig. 9.2 Different types of pipette.

Example 9.3

How much of a sodium chloride solution, of concentration 0.500 mol dm^{-3}, would you take and dilute in order to make exactly 500 cm^3 of a solution of concentration 0.100 mol dm^{-3}?

▶ Answer

First, work out how many moles of sodium chloride you would end up with.

Since the amount of NaCl in moles = volume × the molar concentration, 500 cm^3 of a solution of NaCl of concentration 0.100 mol dm^{-3} contains

$$\frac{500}{1000} \times 0.100 = 0.0500\ \text{mol}$$

What volume of the solution of concentration 0.500 mol dm^{-3} does this correspond to?

$$\text{volume} = \frac{\text{amount of NaCl in mol}}{\text{molar concentration}}$$

$$= \frac{0.0500}{0.500} = 0.100\ \text{dm}^3 \text{ or } 100\ \text{cm}^3$$

▶ Comment

You would therefore take 100 cm^3 of solution of concentration 0.500 mol dm^{-3} and dilute it to 500 cm^3 in a volumetric flask.

Exercise 9D

Making up standard solutions

(i) What mass of anhydrous sodium carbonate would you have to weigh out to make 500 cm^3 of a standard solution of concentration 0.010 mol dm^{-3}?

(ii) If, when making the solution in **(i)**, you did not add enough water to bring the volume of the solution completely up to the 500 cm^3 graduation mark on your volumetric flask, what effect would that have on the concentration of the standard solution?

(iii) Calculate the concentration of a solution that results when 50 cm^3 of a sodium hydroxide solution, of concentration 0.1 mol dm^{-3}, is pipetted into a 250 cm^3 volumetric flask and made up to the graduation mark with deionized water.

(iv) You have some acid of concentration 16.0 mol dm^{-3}. How much of the concentrated acid would you dilute in order to make 1000 cm^3 of acid solution at a concentration of 2.00 mol dm^{-3}?

(v) How much of a copper(II) sulfate solution, of concentration 2 mol dm^{-3}, would you need to take and dilute to 200 cm^3 in order to end up with a solution of concentration 0.5 mol dm^{-3}?

9.3 Volumetric analysis

The procedure involved in finding the concentration of a solution is called **volumetric analysis.** It involves reacting a solution of *known* concentration (a standard solution) with one of *unknown* concentration, in order to determine the **equivalence point.**

The equivalence point is reached when the reactants have reacted together exact ratio of quantities, as given in the balanced equation for the reaction example, in an acid–base titration, the equivalence point involves addition of e enough base to neutralize any acid present.

The technique whereby one reactant is slowly added to a second reagent unt equivalence point is reached is called a **titration**. **Chemical indicators** are subst that change colour at a particular ratio of reactant concentrations called the **point**. In a titration, a chemical indicator is chosen so that the endpoint and eq lence point are the same.

By measuring the volumes of the solutions that have reacted together, and the balanced equation for the reaction, the unknown solution concentration can be determined. If carried out carefully, volumetric analysis is quite an accurate technique and a skilled worker should determine an unknown concentration within an error of no more than 0.2%.

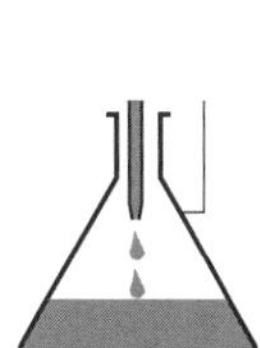

Fig. 9.3 Delivering a measured volume of acid.

An acid–base titration

The (previously unknown) concentration of an acid can be determined by titrating it with a standard alkaline solution. The acid is neutralized by the alkali. The procedure is as follows:

1. A measured volume of the acid is delivered into a conical flask, using a pipette as shown in Fig. 9.3.
2. A few drops of a suitable indicator are added to the acid solution.
3. The standard alkaline solution is added from a **burette**. A burette is a long glass tube with graduation marks on it, with a stopcock at the bottom end. It allows the user to measure the volume of solution delivered through the stopcock. The alkaline solution is added until the endpoint is reached and the indicator changes colour; the total volume of alkaline solution added is then noted. A diagram of this procedure is shown in Fig. 9.4.

Some typical results from such a titration are shown in the following example.

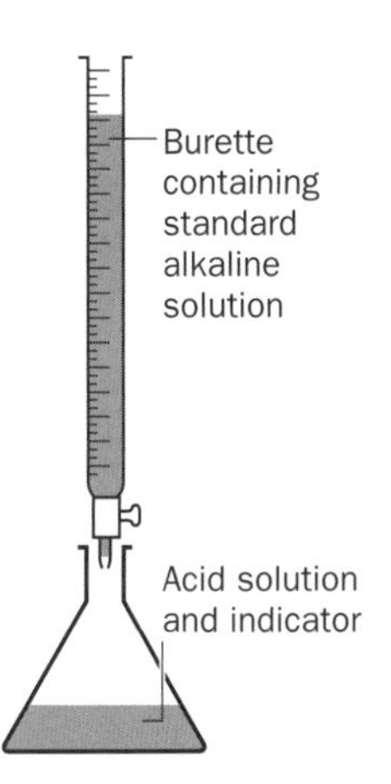

Fig. 9.4 Addition of standard alkaline solution.

Example 9.4

Hydrochloric acid, of unknown concentration, was titrated with a standard solution of potassium hydroxide of concentration 0.500 mol dm^{-3}. It was found that 25.0 cm^3 of the acid reacted with 37.5 cm^3 of the potassium hydroxide solution. Calculate the concentration of the hydrochloric acid.

▶ Answer

Write the balanced equation for the reaction:

$$HCl(aq) + KOH(aq) \rightarrow KCl(aq) + H_2O(l)$$

Work out the molar relationship between the two reactants:

$$1\ \text{mol HCl} \equiv 1\ \text{mol KOH}$$

Example 9.4 *(continued)*

Express this ratio as a fraction:

$$\frac{\text{amount of HCl in moles}}{\text{amount of KOH in moles}} = \frac{1}{1}$$

Since the amount of substance in moles = volume × molar concentration, we can substitute:

$$\frac{\text{vol} \times \text{molar conc HCl}}{\text{vol} \times \text{molar conc KOH}} = \frac{1}{1}$$

Substitute the values given in the equation:

$$\frac{25.0/1000 \times \text{molar concentration of HCl}}{37.5/1000 \times 0.500} = \frac{1}{1}$$

Therefore, the molar concentration of HCl is

$$\frac{37.5 \times 0.500}{25} = 0.750\ \text{mol dm}^{-3}$$

Exercise 9E

Calculation of unknown concentrations by titration

Use approximate atomic masses for these calculations.

(i) How much dilute sulfuric acid of concentration 1.0 mol dm^{-3} will react completely with 25 cm^3 of potassium hydroxide solution of concentration 4.0 mol dm^{-3}?

$$H_2SO_4(aq) + 2KOH(aq) \rightarrow K_2SO_4(aq) + 2H_2O(l)$$

(ii) A solution of sodium hydroxide contains 4.0 g dm^{-3}. An endpoint is reached when 25 cm^3 of this solution react with 35 cm^3 of dilute nitric acid. Calculate the concentration of the nitric acid.

$$NaOH(aq) + HNO_3(aq) \rightarrow NaNO_3(aq) + H_2O(l)$$

(iii) A standard solution was prepared by dissolving 2.6061 g of anhydrous sodium carbonate in deionized water and the solution diluted to 250 cm^3. A 25.0 cm^3 portion of this solution was titrated against hydrochloric acid, using a suitable indicator. The endpoint was reached after 18.7 cm^3 of acid had been added. Calculate the concentration of the acid.

$$Na_2CO_3(aq) + 2HCl(aq) \rightarrow 2NaCl(aq) + H_2O(l) + CO_2(g)$$

(iv) How many cubic centimetres of a solution of hydrochloric acid, of concentration 2.0 mol dm^{-3}, are required to react completely with 50 cm^3 of a solution of $Ba(OH)_2$ that contains 68.4 g dm^{-3}?

$$2HCl(aq) + Ba(OH)_2(aq) \rightarrow BaCl_2(aq) + 2H_2O(l)$$

Redox titrations

The technique of titration may also be used to determine the concentration of an unknown solution, when the solutions react together in a **redox** reaction. Potassium permanganate (or potassium manganate(VII), $KMnO_4$) is a powerful oxidizing agent and is used to estimate the concentration of many reducing agents. Potassium permanganate solution is purple in colour, but is virtually colourless when reduced to Mn^{2+} – it can therefore act as a **self-indicator**. As it is added to the reducing solution it turns colourless, but at the endpoint a very small amount of excess permanganate will be present. The excess is shown when the reacting mixture turns pale pink. The half-reaction for the reduction in *acid solution* is:

$$MnO_4^-(aq) + 8H^+(aq) + 5e^- \rightarrow Mn^{2+}(aq) + 4H_2O(l)$$

pink colourless

Two common reactions involving oxidation by potassium permanganate are:

1. Oxidation of iron(II) to iron(III):

$$Fe^{2+}(aq) \rightarrow Fe^{3+}(aq) + e^-$$

2. Oxidation of oxalates to carbon dioxide. This reaction needs to be carried out at 70 °C, because it is too slow at room temperature:

$$C_2O_4^{2-}(aq) \rightarrow 2CO_2(g) + 2e^-$$

Potassium dichromate (or potassium dichromate(VI), $K_2Cr_2O_7$) is also used as an oxidizing agent in similar reactions to those described for potassium permanganate. In acid solution, it reacts according to the following half-reaction:

$$Cr_2O_7^{2-}(aq) + 14H^+(aq) + 6e^- \rightarrow 2Cr^{3+}(aq) + 7H_2O(l)$$

orange green

The green colour of the chromium(III) ion 'masks' the orange colour of the dichromate ion, so the oxidizing agent cannot act as a self-indicator. The solution is used together with a few drops of a **redox indicator**, such as diphenylamine. This indicator is oxidized to an intense blue compound in the presence of a slight excess of dichromate.

Example 9.5

Iron(II) sulfate is oxidized by potassium permanganate in acid solution. The overall ionic equation is

$$5Fe^{2+}(aq) + MnO_4^-(aq) + 8H^+(aq) \rightarrow Mn^{2+}(aq) + 4H_2O(l) + 5Fe^{3+}(aq)$$

What volume of 0.010 mol dm^{-3} iron(II) sulfate will be oxidized by 25.00 cm^3 of 0.020 mol dm^{-3} permanganate solution?

▶ Answer

Work out the molar relationship between the two reactants:

$$5\text{ mol } Fe^{2+} \equiv 1\text{ mol } MnO_4^-$$

Express this ratio as a fraction:

$$\frac{\text{amount of } Fe^{2+} \text{ in moles}}{\text{amount of } MnO_4^- \text{ in moles}} = \frac{5}{1}$$

Since the *amount of substance in moles* = *volume* × *molar concentration*, we can substitute:

$$\frac{\text{vol} \times \text{molar conc } Fe^{2+}}{\text{vol} \times \text{molar conc } MnO_4^-} = \frac{5}{1}$$

Substituting values into the equation:

$$\frac{\text{vol} \times 0.010}{25/1000 \times 0.020} = \frac{5}{1}$$

Therefore, the volume of Fe^{2+} = 0.250 dm^3 = 250 cm^3.

Exercise 9F

Redox titrations

(i) 25 cm³ of a solution of an iron(II) salt needed 13.5 cm³ of a 0.020 mol dm^{-3} solution of potassium dichromate for complete oxidation to iron(III). Write the balanced overall redox equation for the reaction and calculate the concentration of Fe^{2+}(aq) in the original solution.

(ii) Cerium(IV) ions are reduced by oxalate ions in solution:

$2Ce^{4+}(aq) + C_2O_4^{2-}(aq) \rightarrow 2Ce^{3+}(aq) + 2CO_2(g)$

Ferroin indicator is used in the titration. Calculate the volume of 0.050 mol dm^{-3} cerium(IV) solution needed to exactly react with 25 cm³ of 0.050 mol dm^{-3} oxalate ion solution.

(iii) Calculate the volume of a 0.020 mol dm^{-3} solution of potassium permanganate that is needed for the complete oxidation of 0.0070 mol of iron(II) oxalate solution. (*Be careful! – there are two oxidations occurring here – the oxidation of iron(II) to iron(III) and the oxidation of the oxalate ion to carbon dioxide.*)

Reactions of solids with solutions

Calculations involving the reaction of a solid with a solution can be worked out in a similar manner.

Example 9.6

What mass of zinc will react completely with 50.0 cm³ of 0.100 mol dm^{-3} sulfuric acid?

▶ Answer

Write the balanced equation for the reaction:

$$Zn(s) + H_2SO_4(aq) \rightarrow ZnSO_4(aq) + H_2(g)$$

Work out the molar relationship between the two reactants:

$$1\ \text{mol Zn} \equiv 1\ \text{mol } H_2SO_4$$

Express this ratio as a fraction:

$$\frac{\text{amount of Zn in moles}}{\text{amount of } H_2SO_4 \text{ in moles}} = \frac{1}{1}$$

Since, for a solution, the number of moles = volume × molar concentration, we can substitute this expression for H_2SO_4 **only** (zinc is a solid, not a solution):

$$\frac{\text{amount of Zn in moles}}{\text{vol} \times \text{molar concentration } H_2SO_4} = \frac{1}{1}$$

Substitute the values given in the equation:

$$\frac{\text{amount of Zn in moles}}{50.0/1000 \times 0.100} = \frac{1}{1}$$

Therefore, the moles of zinc needed is

$$\frac{50.0}{1000} \times 0.100 = 0.00500\ \text{mol}$$

and the mass of zinc required = 0.00500 × 65 = 0.325 g.

Exercise 9G

Problems involving the reactions of solids with solutions

(i) What is the volume of sulfuric acid of concentration 1.50 mol dm^{-3} which would be needed to react with 7.95 g of copper(II) oxide?

$H_2SO_4(aq) + CuO(s) \rightarrow CuSO_4(aq) + H_2O(l)$

(ii) Magnesium hydroxide is a compound used to neutralize stomach acid (dilute hydrochloric acid) and so cure acid indigestion. The reaction is

$Mg(OH)_2(s) + 2HCl(aq) \rightarrow MgCl_2(aq) + 2H_2O(l)$

The magnesium hydroxide in an indigestion tablet was found to react with 20 cm^3 of 0.50 mol dm^{-3} HCl. How many grams of $Mg(OH)_2$ were in the tablet?

(iii) Carbon dioxide gas is prepared by the action of 2.00 mol dm^{-3} hydrochloric acid on marble:

$CaCO_3(s) + 2HCl(aq) \rightarrow CaCl_2(aq) + H_2O(l) + CO_2(g)$

What volume of acid is required to give 10.0 dm^3 of carbon dioxide at room temperature and pressure?

(iv) What mass of solid copper(II) carbonate is needed to react completely with 50 cm^3 of sulfuric acid of concentration 0.50 mol dm^{-3}?

Gravimetric analysis

When two solutions react together to form an insoluble product, the concentration of one of the solutions can be worked out from the mass of precipitate formed in the reaction. The process is called **gravimetric analysis**. The precipitate is filtered off and dried thoroughly. The precipitate is heated and weighed several times until the mass of the product does not change (**heated to constant mass**); this process ensures that the insoluble product is thoroughly dried.

Example 9.7

Barium chloride solution reacts with the sulfate ion in aqueous solution to form insoluble barium sulfate. The ionic equation for the reaction is as follows:

$Ba^{2+}(aq) + SO_4^{2-}(aq) \rightarrow BaSO_4(s)$

If excess barium chloride solution was added to a 250.0 cm^3 sample of an unknown solution to precipitate all the sulfate ions present as barium sulfate, and 1.2210 g of dry $BaSO_4$ precipitated, what was the original concentration of sulfate ion in the unknown sample?

▶ Answer

$M(BaSO_4) = 137.327 + 32.066 + (4 \times 15.999) = 233.389\,g\,mol^{-1}$

(Do not use approximate atomic masses in calculations of gravimetric determinations; use the atomic masses in the Periodic Table.)

amount of $BaSO_4 = 1.2210/233.389 = 5.2316 \times 10^{-3}$ mol

Since 1 mol of $BaSO_4$ contains 1 mol SO_4^{2-} ions:

amount of SO_4^{2-} present $= 5.2316 \times 10^{-3}$ mol

This amount of SO_4^{2-} was originally present in 250.0 cm^3 solution, therefore the concentration of SO_4^{2-} ions in original solution is

$5.2316 \times 10^{-3} \times 4 = 0.02093\,mol\,dm^{-3}$

Exercise 9H

Gravimetric analysis

The concentration of nickel ions in a solution can be determined by precipitation of scarlet nickel dimethylglyoxime, $Ni(DMG)_2$. The ionic equation for the reaction is as follows:

$$Ni^{2+}(aq) + 2DMGH \rightarrow Ni(DMG)_2(s) + 2H^+(aq)$$

where DMGH = $C_4H_7O_2N_2$.

A student used a 100.0 cm³ sample of a 2000 cm³ solution and precipitated all of the nickel in the sample as $Ni(DMG)_2$. The precipitate was filtered off and heated to constant mass. The results were:

mass of container	30.8749 g
mass of container + precipitate (1st weighing)	30.9885 g
mass of container + precipitate (2nd weighing)	30.9877 g
mass of container + precipitate (3rd weighing)	30.9877 g

(i) What was the mass of the dry precipitate?

(ii) Why were the precipitates heated to constant mass?

(iii) Calculate the concentration of nickel ions in the solution.

(iv) What was the mass of nickel in the original 2000 cm³ of solution?

(use m(Ni) = 58.69 u, m(C) = 12.011 u, m(H) = 1.008 u, m(O) = 15.999 u, m(N) = 14.007 u)

Exercise 9I

Parts per million by mass

(i) What total mass of mercury is present in a fish of mass of 2.0 kg, which has been found to contain mercury at a concentration of 0.60 ppm (or 0.60 mg kg^{-1})?

(ii) A 5.00 kg sample of soil is found to have a total lead content of 2.00 g. What is the concentration of lead in the soil, in ppm (mg kg^{-1})?

9.4 Other units of concentration

Parts per million

Very low concentrations of substances are usually expressed in terms of parts per million (ppm) or even parts per billion (ppb) by mass, where:

$$\text{ppm} = \frac{\text{mass of trace substance}}{\text{mass of sample}} \times 10^6; \quad \text{ppb} = \frac{\text{mass of trace substance}}{\text{mass of sample}} \times 10^9$$

1 ppm of a species is often expressed in the units:

1 mg of the species per kg of a sample

or

1 ppm is equivalent to 1 mg kg^{-1}

See Box 9.3 for examples.

Trace concentrations in gaseous mixtures

If we are dealing with gaseous mixtures, it is often more convenient to express ppm and ppb by *volume*, rather than by mass:

$$1 \text{ ppm by volume} = \frac{\text{volume of trace substance}}{\text{total volume of sample}} \times 10^6$$

or

1 ppm by volume is equivalent to 1 cm³ in 10³ dm³ of sample

These units are useful for describing the levels of pollutants in air. For example, the concentration of NO_2 in the atmosphere around a large city was 15 ppm; this corresponds to 15 cm³ of the gas in a sample of 10³ dm³ of air.

ppm in aqueous solution

For dilute aqueous solutions at room temperature, the density of the solution is close to the density of water (1 g cm^{-3}), so the volume of the solution in cubic centimetres may be used instead of the mass of the solution in grams. For dilute aqueous solutions, therefore, we can use:

1 ppm = 1 mg of substance dissolved in 1 dm^3 of water

and

1 ppb = 1 μg of substance dissolved in 1 dm^3 of water

These units are more convenient for low concentrations because they avoid using the very low numbers that would be involved if the concentrations were to be expressed in mol dm^{-3}; water pollutants are often given in ppm units. Also, we may use these units to describe trace concentrations of metals when we are not sure of the exact nature of the species involved. For example, drinking water may be found to have a total aluminium content at a concentration of 3 ppm. The nature and concentration of the different aluminium compounds present in a sample of drinking water is dependent on pH and can be a very complex problem to solve; the ions Al^{3+}, $AlOH^{2+}$ and $Al(OH)_2^+$ are but a few species that may be found. Without exact knowledge of the species present, calculations of molar concentration cannot be made, but it is always possible to express concentration in terms of the total mass of metal present.

BOX 9.3

LGC

How little is 'little'?
One lump of sugar (6g) dissolved in:

Setting standards in analytical science

Tea pot	0.6 dm^3	1% = 1 per cent	10 g/kg	10 mg/g
Bucket	6 dm^3	0.1% = per thousand	1g/kg	1 mg/g 1 milligram = 0.001 g
Tank lorry	6000 dm^3	1ppm = 1 per million	1 mg/kg	1 μg/g 1 microgram = 0.000 001 g
Super-tanker	6 million dm^3	1ppb = 1 per billion	1 μg/kg	1 ng/g 1 nanogram = 0.000 000 001 g
Reservoir	6 billion dm^3	1ppt = 1 per trillion	1 ng/kg	1 pg/g 1 picogram = 0.000 000 000 001 g
Bay	6 trillion dm^3	1ppq = 1 per quadrillion	1 pg/kg	1 fg/g 1 femtogram = 0.000 000 000 000 001g

Reproduced courtesy of LGC, www.lgc.co.uk

Example 9.8

If a sample of drinking water contains aluminium ions at a concentration of 3.0 ppm, how many mol dm^{-3} of aluminium ions does this represent?

▶ Answer

The sample contains 3.0 mg of Al^{3+} in 1 dm^3, or

0.0030 g of Al^{3+} in 1 dm^3

or

$$\frac{0.0030}{27} \text{ mol of } Al^{3+} \text{ in } 1\,dm^3 = 1.1 \times 10^{-4}\,mol\,dm^{-3} \text{ of } Al^{3+}$$

Example 9.9

If 20 cm^3 of a standard solution of magnesium ions (Mg^{2+}) of concentration 100 ppm was diluted to 200 cm^3, what would be the concentration of the new solution?

▶ Answer

First, work out the factor by which you have diluted the solution by dividing the new volume by the old volume. The 100 ppm solution of Mg^{2+} has been diluted from 20 cm^3 to 200 cm^3, i.e. by a factor of

200/20 = 10

The new concentration is worked out by dividing the old concentration by the dilution factor:

$$\text{concentration of the new diluted solution} = \frac{100}{10} = 10 \text{ ppm of } Mg^{2+}$$

Example 9.10

A solution of NaCl has a concentration of 0.100 mol dm^{-3}. What is its concentration in ppm (mg dm^{-3})?

▶ Answer

1.00 mol of NaCl has a mass of (23 + 35.5) = 58.5 g. Therefore a 0.100 mol dm^{-3} solution contains 5.85 g dm^{-3} or 5850 mg dm^{-3} of NaCl. This is also 5850 ppm.

▶ Comment

When the salt dissolves in water 1 mol of NaCl produces 1 mol of Na^+ ions. A 0.1 mol dm^{-3} solution contains 2.3 g dm^{-3} of Na^+, or 2300 ppm (mg dm^{-3}) Na^+, and 3.55 g dm^{-3} or 3550 ppm (mg dm^{-3}) Cl^-.

BOX 9.4

Measuring the alcohol content in spirits

An old-fashioned way of measuring the alcohol (the chemical name is **ethanol**) content in spirits is to quote the concentration of alcohol in terms of **degrees proof**. The measurement was made by pouring the spirit over gunpowder. If the gunpowder would not burn, the spirit was 'under-proof' – it contained too much water. The spirit that had a concentration of ethanol that would *just* allow gunpowder to burn was termed 100 proof spirit, written 100°. A sample that is 100° contains 50% ethanol (v/v), so 70° whisky contains about 35% ethanol.

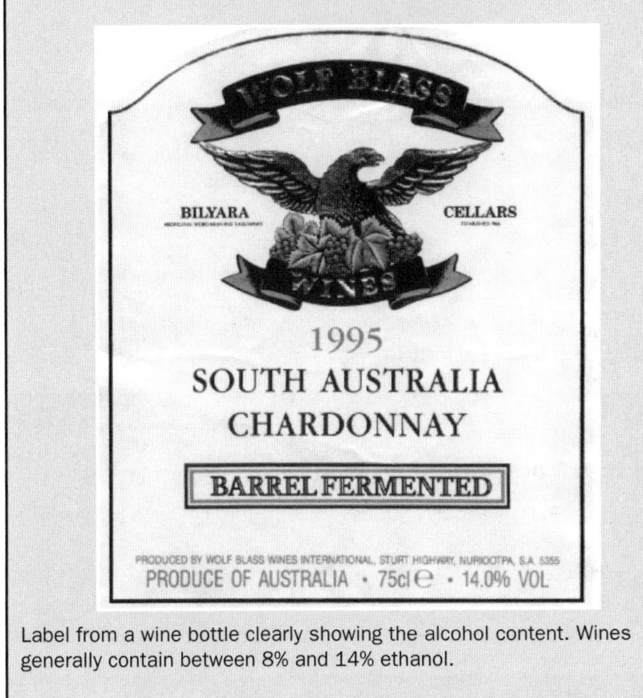

Label from a wine bottle clearly showing the alcohol content. Wines generally contain between 8% and 14% ethanol.

Exercise 9J

Aqueous concentrations in ppm and ppb

(i) Drinking water is considered acceptable if it contains less than 500 ppm of chloride ions (Cl^-). What is this concentration in $mol\,dm^{-3}$?

(ii) A standard solution contains 1000 ppm of chloride ions. If a 50 cm^3 sample of this standard solution were diluted to 250 cm^3, what would be the concentration of the new solution?

(iii) An aqueous solution of silver ions (Ag^+) is of concentration 0.0005 ppm. Express this concentration in ppb.

Concentration as percentage composition

Many solutions in industry are expressed in one of two ways:

1. Percent by mass (or 'weight'), often written as percentage (w/w):

percentage composition by mass

$$= \frac{\text{mass of solute}}{\text{mass of solution}} \times 100$$

2. Percent by volume, often written as percentage (v/v):

percentage composition by volume

$$= \frac{\text{volume of liquid solute}}{\text{volume of solution}} \times 100$$

Exercise 9K

Percentage compositions

(i) Glucose sugar has the formula $C_6H_{12}O_6$. If you dissolved 15.5 g of glucose in 100 g of water, what would be the concentration of glucose, expressed as a percentage composition by mass?

(ii) A concentrated ammonia solution is labelled as 28% (w/w). How many grams of ammonia are in 150 g of the solution?

(iii) A wine bottle has 12% (v/v) written on the label, this describes the ethanol content in the wine. What volume of ethanol is in a 1.5 L bottle of wine?

(iv) Strong vodka is 85° proof. What volume of ethanol does a 1 L bottle contain?

Concentration expressed as mole fraction

Sometimes the physical properties of solutions, such as **vapour pressure** (a measure of the number of molecules escaping from the surface of the liquid, see page 165), depend upon the relative numbers of particles of the components of the solution. In these cases, the composition of a solution can be described in terms of the **mole fraction** of each of the components. The mole fraction of any component is calculated by dividing the number of moles of that component by the total number of moles present in the mixture.

For example, in a mixture of A and B, the mole fraction of A (symbol X_A), is defined as follows:

$$X_A = \frac{n_A}{n_A + n_B}$$

where n_A is number of moles of A present and n_B is number of moles of B present.

Similarly, for B,

$$X_B = \frac{n_B}{n_A + n_B}$$

(Note that $X_A + X_B = 1$.)

Also note that:

1. A mole fraction has no units, it is a ratio.

2. A mole fraction has a value between 0 and 1.

Exercise 9L

Mole fractions in a four-component mixture

(i) In a four-component mixture of P, Q, R and S, write an expression for the mole fraction of R, X_R.

(ii) Show that $X_P + X_Q + X_R + X_S = 1$.

(iii) What is the composition of the mixture if $X_P = 1$?

Exercise 9M

Calculating mole fractions

(i) A mixture consists of 1 mol of ethanol (C_2H_5OH), 3 mol of methanol (CH_3OH) and 6 mol of water. What is the mole fraction of methanol present?

(ii) A mixture consists of 37.25 g of potassium chloride dissolved in 1000 g of water. What is the mole fraction of potassium chloride present?

9.5 pH scale

The pH scale is a way of expressing the $[H_3O^+]$, or $[H^+(aq)]$, as a *small* number. Note that the square brackets, [ ], stand for 'the concentration of' (in mol dm^{-3}). The pH of a solution can be calculated using the equation

$$pH = -\log [H^+(aq)]$$

Alternatively, the hydrogen ion concentration can be calculated using the equation

$$[H^+(aq)] = 10^{-pH}$$

The pH is a measure of the *acidity* of a solution. A low pH corresponds to a high value of $[H_3O^+]$ or $[H^+(aq)]$, showing the solution to be strongly acidic Table 9.1.

The values of $[H^+(aq)]$, and thus pH, can vary depending on the acidity or basicity of the solution. The pH scale below shows the range of common pH values.

0 1 2 3 4 5 6 | 7 | 8 9 10 11 12 13 14

←increasingly acidic   increasingly basic→

Acidic solutions have pH values below 7. Solutions with pH values above 7 are basic. A pH of 7 at 25 °C (the pH of pure water) is neutral. A **pH meter** can be used to measure the pH of a solution. It consists of an electrode which is dipped into the solution and connected to a device which gives a digital readout. Alternatively, **universal indicator paper** may be used. It contains a mixture of dyes that changes to different colours depending on the pH of the solution (see page 301).

Table 9.1 Approximate pH values of common substances

Substance	pH
Gastric juice	1.2–3.0
Vinegar	3.0
Lemonade	3.0–3.5
Tomato juice	4.2
Coffee	5.0
Saliva	6.6
Milk	6.7
Pure water	7.0
Blood	7.4
Eggs	7.8

Example 9.11

What is the pH of a solution of 0.020 mol dm^{-3} nitric acid, HNO_3 at 25 °C?

► Answer

As the acid may be assumed to be completely dissociated (the concentration of the acid equals the concentration of hydrogen ions),

$[H^+(aq)] = 0.020$

$pH = -\log[H^+(aq)] = -\log[0.020] = 1.70$

Since pH is a logarithmic quantity, it has no units. (For an explanation of significant figures see Unit 1, page 10.)

Example 9.12

The pH of blood is 7.4. What is the hydrogen ion concentration in blood?

► Answer

$pH = -\log[H^+(aq)] = 7.4$

$[H^+(aq)] = 10^{-7.4}$

$[H^+(aq)] = 4 \times 10^{-8}$ mol dm^{-3}

Exercise 9N

The pH scale

Classify the following as strongly acidic, weakly acidic, strongly basic or weakly basic:

(i) lemon juice (pH 4)

(ii) rain water (pH 6 because it contains dissolved carbon dioxide)

(iii) bicarbonate of soda (or sodium hydrogencarbonate, pH 12)

(iv) sodium hydroxide of concentration 0.1 mol dm^{-3}

(v) soap solution (pH 9)

(vi) wine (pH 3.5)

(vii) household ammonia (pH 12).

Revision questions

9.1. What mass of solute must be used in order to prepare the following solutions?

(i) 250 cm^3 of 0.200 mol dm^{-3} sodium hydrogencarbonate ($NaHCO_3(aq)$)?

(ii) 2.50 dm^3 of 0.500 mol dm^{-3} potassium nitrate ($KNO_3(aq)$)?

(iii) 750 cm^3 of 0.050 mol dm^{-3} oxalic acid ($H_2C_2O_4 \cdot 2H_2O(aq)$)?

9.2. Standard solutions of sodium hydroxide are usually obtained in the laboratory by dilution of a commercial standard solution, or they are standardized by acid–base titrations.

Why do you think that solid sodium hydroxide is not very suitable to use as a primary standard? (**Hint** what happens to solid sodium hydroxide in air?)

9.3. What volumes of the following solutions will neutralize 50 cm^3 of 0.10 mol dm^{-3} sulfuric acid?

(i) 0.20 mol dm^{-3} potassium hydroxide solution

(ii) 0.05 mol dm^{-3} of sodium carbonate solution.

9.4. Lead(II) nitrate solution ($Pb(NO_3)_2$) reacts with potassium iodide solution (KI) according to the following ionic equation:

$$Pb^{2+}(aq) + 2I^-(aq) \rightarrow PbI_2(s)$$

(i) What volume of 1.00 mol dm^{-3} KI(aq) is needed to react exactly with 25.0 cm^3 of $Pb(NO_3)_2(aq)$ of concentration 2.00 mol dm^{-3}?

(ii) This reaction can be used for the gravimetric determination of lead(II) ions in solution. What mass of $PbI_2(s)$ would be precipitated from the lead(II) nitrate solution in **(i)**? (Use accurate atomic masses in this calculation.)

9.5. What mass of the following will react with 25 cm^3 of 0.250 mol dm^{-3} nitric acid?

(i) solid potassium hydroxide (KOH)

(ii) solid copper carbonate ($CuCO_3$).

9.6. Sodium thiosulfate ($Na_2S_2O_3 \cdot 5H_2O$) solution reacts with aqueous iodine in a redox reaction according to the equation

$$2S_2O_3^{2-}(aq) + I_2(aq) \rightarrow S_4O_6^{2-}(aq) + 2I^-(aq)$$

A solution, of volume 500 cm^3, contains 12.400 g of dissolved hydrated sodium thiosulfate.

(i) What is the concentration of the solution in mol dm^{-3}?

(ii) 11.8 cm^3 of this solution reacted exactly with 25 cm^3 of a solution of aqueous iodine. What is the concentration of the iodine solution in g dm^{-3}?

9.7. How much of a standard solution of $Ca^{2+}(aq)$ of concentration 1000 ppm should you take and dilute in order to form 2 dm^3 of a solution of concentration 50 ppm?

9.8. A bottle of aqueous hydrochloric acid (HCl) has 35% (w/w) on the label. What mass of solution contains 7.0 g of HCl?

9.9. Complete this table:

Substance	Molar concentration /mol dm^{-3}	Metal ion concentration /ppm or mg dm^{-3}
KBr	0.050	
Na_2SO_4	0.010	
$Pb(NO_3)_2$	0.0060	

9.10. Calculate the mass (in grams) of lead in 200 cm^3 of a solution of lead nitrate that contains 100 mg of lead nitrate per dm^3 of solution.

9.11. What mass of $Na_2CO_3 \cdot 10H_2O$ is required to make 1 dm^3 of a 20 ppm solution of $Na^+(aq)$?

9.12. **(i)** Calculate the pH of a solution whose $[H^+(aq)] = 4.48 \times 10^{-9}$ mol dm^{-3}.

(ii) Calculate the hydrogen ion concentration (in mol dm^{-3}) for a solution with a pH value of 5.93.

Extension material to support this unit is available on our website at **http://www.palgrave.com/foundations/lewis.**

Gases, Liquids and Solids

Objectives

- ▶ Defines heat and temperature
- ▶ Looks at changes of state
- ▶ Introduces the kinetic molecular theory of matter
- ▶ Discusses the gas laws and the idea of vapour pressure
- ▶ Shows how to use the ideal gas equation in calculations

Contents

Although people have always recognized the distinction between liquids and solids, the idea of a gas as we know it today only began to develop in the 1760s. The absence of such a concept severely hindered chemists in their attempt to make sense of the burning of fuels and of the tarnishing of metals in air, both of which we now know to be chemical reactions involving gaseous oxygen.

10.1 Heat and temperature

The hotness or coldness of an object is measured by its **temperature** (Box 10.1). If we make an object hot, we have transferred energy to that object. This energy is called **heat**.

10.2 Changes in the state of matter

Everyday observations show that solids have a definite shape and volume. Liquids possess the shape of their container but their volume is fixed and does not depend upon the volume of their container. Gases take up the shape and volume of their container (Fig. 10.2(a)).

Kinetic molecular theory of matter

The idea that matter consists of moving particles (molecules, atoms or ions) is the basis of the **kinetic molecular theory** (or simply, **kinetic theory**). Its two main assumptions are:

BOX 10.1

Temperature scales

The units of temperature are degrees Celsius (°C) or kelvin (K), (Fig. 10.1). The size of one kelvin and one degree Celsius is the same, and the two are interconverted using the expression

$$T(\text{K}) = T(°\text{C}) + 273.15$$

Although there is no theoretical limit to the upper temperature of matter, the lowest temperature that is permitted in nature is −273.15 °C (0 K): this temperature is known as **absolute zero**. The limitation is similar to that found in mechanics, where no object may travel faster than the speed of light. (NB Absolute zero is often approximated to −273 °C.)

Even at 0 K all motion does not cease. Particles in a solid at 0 K vibrate very slightly and so matter still possesses some energy, even at absolute zero. This energy is called the **zero point vibrational energy**.

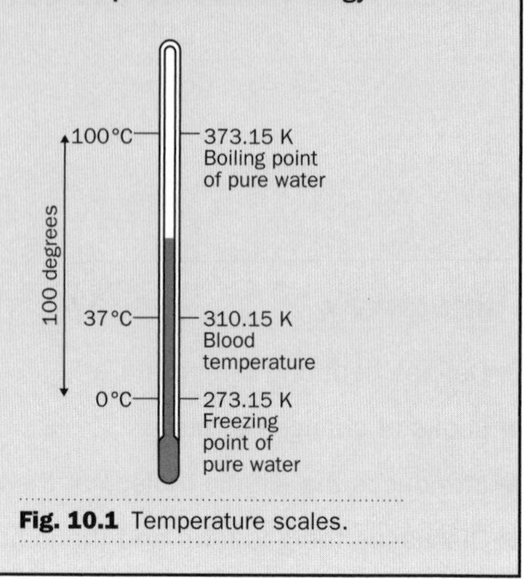

Fig. 10.1 Temperature scales.

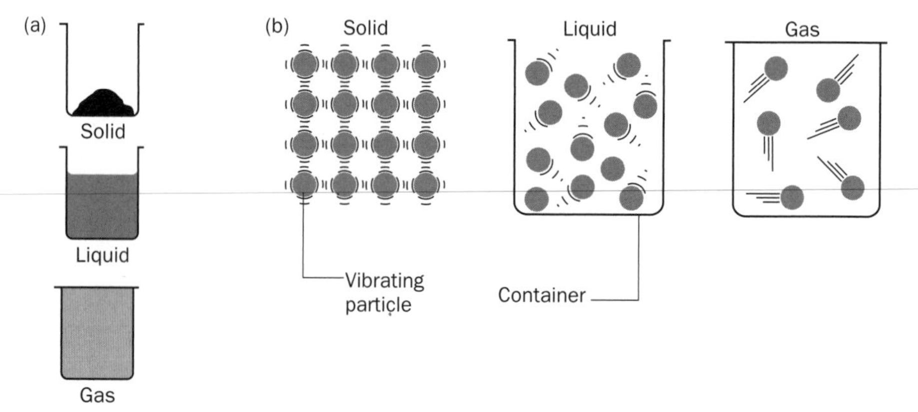

Fig. 10.2 Gases, liquids and solids.
(a) The bulk appearance of the three states of matter.
(b) The molecular picture.

1. The higher the temperature of a collection of particles, the higher the *average* kinetic energy of those particles. On average, particles move *faster* at higher temperature.
2. There is a stickiness (attractive force) between all particles.

The attractive forces between particles in a material pull the particles together, whereas the movement of the particles (the kinetic energy of the particles) pushes them apart. Changes of state occur when *one of these factors dominate*. For example, if we raise the temperature of a solid, the particles will eventually possess enough energy to partially break away from each other, and melting takes place producing a liquid state. Further heating causes the particles to move so fast that the attractive forces in the liquid are unable to hold them even loosely together and they become well separated

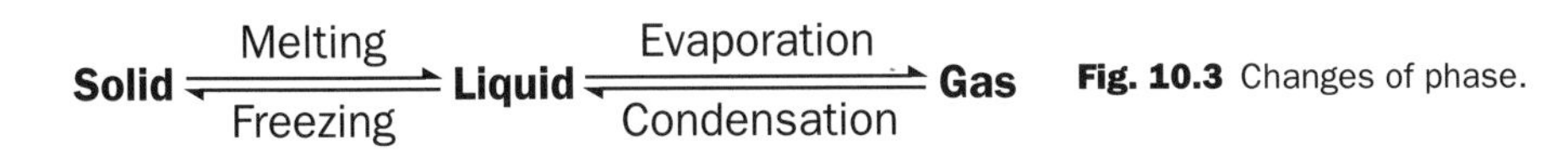

Fig. 10.3 Changes of phase.

from each other. This process is called evaporation, and the resulting gas consists of particles which (at low gas pressure) move entirely independently of each other.

Similar explanations apply (but in reverse) to the other changes of state. These 'phase changes' are shown in Fig. 10.3.

The 'molecular picture' of the three states of matter is summarized in Fig. 10.2(b), with the particles in a solid vibrating, the particles in a liquid sliding over each other (explaining why liquids take up the shape of their container), and the particles in a gas moving so rapidly that they take up the shape *and* volume of their container.

The position of atoms (and ions) in solids and liquids may be studied using **X-ray diffraction**. This technique confirms that atoms in a liquid are in a state of orderliness which is intermediate between that of gases and solids.

Exercise 10A

Changes of phase

What is the significance of the double pointed arrows ($\rightleftharpoons$) in Fig. 10.3?

Explanations based upon the kinetic molecular theory

The kinetic theory of matter explains many everyday observations including:

1. The relative density of solids, liquids and gases

The density of a material tells us how closely packed the particles are. Experiments show that solids usually possess the greatest density, and gases (in which the molecules are moving fastest) the least. Gases have low densities because the gaps between the particles are relatively great. For this reason, gases have a volume which may easily be reduced by applying pressure. This is what we mean when we say that gases are easily compressed.

2. The variation of gas pressure with temperature

Gas pressure results from the bombardment of the container walls by molecules of the gas (Fig. 10.2(b)). At high temperatures, the gas molecules are moving faster and both the number of collisions per second, and the energy of the colliding molecules, increases. This causes an increase in gas pressure.

3. The evaporation of liquids

A drop of alcohol in a dish slowly vanishes. This 'vanishing act' is no magic, because the alcohol has simply become alcohol gas (alcohol *vapour*). Evaporation of liquid molecules from the liquid surface occurs at all temperatures. Only fast-moving molecules are able to escape from the attractive forces of the other molecules in the liquid. Since the number of fast-moving molecules increases as the average kinetic energy of the sample of molecules increases, the rate of evaporation of a liquid increases with temperature.

4. Diffusion

The spreading out of particles is called **diffusion**. For example, a gas (or vapour) will fill up its container because the gas molecules are in constant and random movement, and a crystal of copper(II) sulfate will slowly dissolve in water (even without agitation) because the water molecules are continually striking the crystal and the individual copper and sulfate ions disperse into the water.

Exercise 10B

Kinetic theory

Use the kinetic theory to explain the following:

(i) Two beakers are heated with identical bunsen flames for the same time. Beaker A contains less water than beaker B. Why does the temperature of the water in beaker A rise faster than that of beaker B?

(ii) The drying of clothes on a line is an example of evaporation. Explain why clothes dry faster on hot, windy days.

(iii) A smell of a stink bomb released in a corner of a room soon spreads to the rest of the room.

Heating and cooling curves

The phase changes shown in Fig. 10.3 may be represented using a **heating curve** for the substance under study (Fig. 10.4).

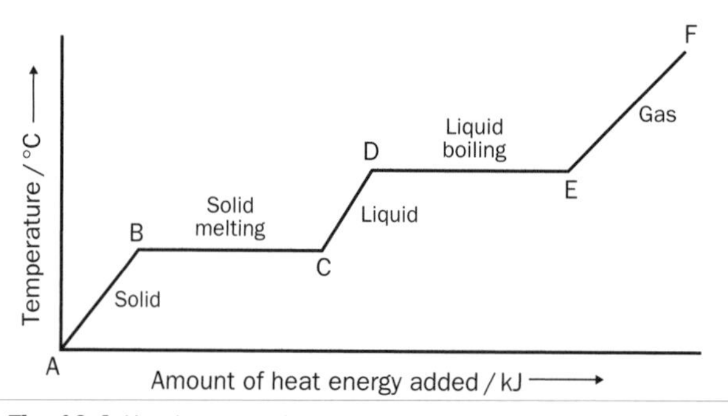

Fig. 10.4 Heating curve for a pure substance.

A heating curve is a plot of the temperature of a substance as it is being heated (i.e. as heat energy is added to the substance). The general shape of heating curves is the same for most pure substances.

Melting

As we start to heat the solid substance slowly, its temperature rises (from A to B). The slope of the line AB depends upon the **specific heat capacity** of the solid. As soon as the first drop of liquid is produced (at B), the temperature of the liquid–solid mixture remains constant (from B to C). This constant temperature is the **melting** or **freezing point** (or **melting temperature**) of the substance. The constant temperature arises because all the added heat is used to melt the solid: no energy is used to raise the temperature of the substance. At point C, all the solid has melted. The melting point of water at a surrounding air pressure of 101 kPa (1 atm) is 0 °C.

Boiling

As we continue heating, the liquid becomes hotter (C to D). The slope of the line CD depends upon the specific heat capacity of the liquid. At D, the liquid begins to boil.

When the liquid boils, its temperature does not rise – all the added heat is used to boil the liquid. This constant temperature is called the **boiling point** (or **boiling temperature**) of the substance. The boiling point of water at an air pressure of one atmosphere is 100 °C. (At this temperature, the vapour pressure of water equals one atmosphere, see page 166.)

At E, all the substance is present as a gas. If we contained the gas in a sealed vessel and then applied more heat, its temperature would continue to rise. The slope of EF depends upon the specific heat capacity of the gas.

If a substance is cooled (e.g. by putting it in a freezer) the resulting **cooling curve** is the mirror image of Fig. 10.4. The energy changes of Fig. 10.4 are also reversed. For example, steam condensing to liquid water and liquid water that is freezing both **give out** heat. Now try Exercise 10C.

Exercise 10C

Boiling

When cooking vegetables by boiling them in water on a cooker, does the speed at which the vegetables cook depend upon the rate at which the water is boiled?

10.3 Gas laws

Think of a gas in a sealed container. We can measure the number of moles n of gas inside, the pressure P of the gas (Box 10.2), the volume V of the gas (which equals the volume of the container) and the temperature T of the gas. The observed relationships between P, V, n and T are summarized by four laws: Charles' law, Boyle's law, Avogadro's law and Dalton's law. Gases that obey these laws perfectly are known as **ideal gases**. Real gases behave approximately as ideal gases at room temperature and at atmospheric pressure. Gases become less ideal at higher pressures ($>$ 1 atm) and at low temperatures (especially close to their boiling point).

If a gas is at 25 °C and at a pressure of exactly 100 kPa (1 bar), the gas is said to be at **standard ambient temperature and pressure** (symbolized SATP). An earlier set of conditions, known as **standard temperature and pressure** (STP) refers to a gas at 0 °C and exactly 1 atm pressure.

Exercise 10D

Pressure

Use the information in Box 10.2 to convert the following to Pascals:

(i) 20 torr
(ii) 10 atmospheres
(iii) 107 000 N m^{-2}
(iv) 1 kPa.

BOX 10.2

Gas pressure

Pressure is defined as force per unit area:

$$\text{Pressure} = \frac{\text{force}}{\text{area}}$$

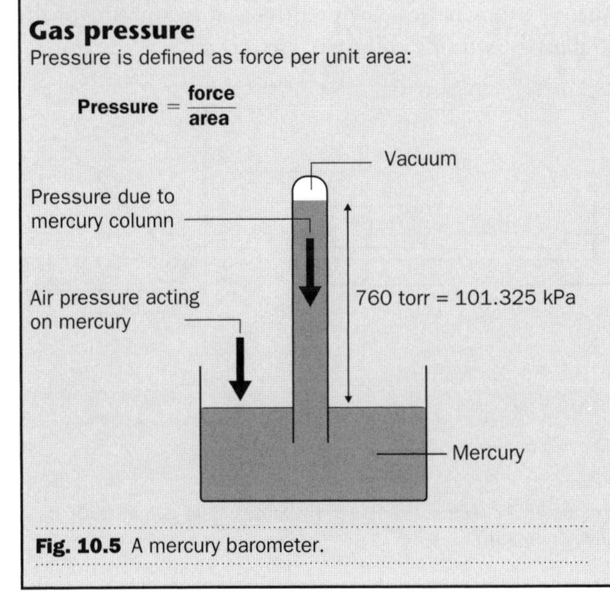

Fig. 10.5 A mercury barometer.

The SI unit of force is the newton (N), and the SI unit of area is the square metre (m^2). So, the SI unit of pressure is newton per square metre (N m^{-2}) also known as the **pascal** (Pa).

The pressure of air at sea level is said to be one 'standard atmosphere' (often simply called 'one atmosphere'). The symbol for (standard) atmosphere is **atm**:

$$1\text{ atm} = 101\,325\text{ N m}^{-2} = 101.325\text{ kPa}$$

$$1\text{ atm} \approx 1 \times 10^5\text{ N m}^{-2} \approx 101\text{ kPa}$$

Exactly 100 kPa is called 1 bar:

$$100\text{ kPa} = 1\text{ bar}$$

The 'torr' is still used as a unit of pressure in some laboratories. One torr is the pressure of air which supports 1 mm of mercury in a barometer at sea level (Fig. 10.5). There are 760 torr in 1 atmosphere:

$$1\text{ atm} = 760\text{ torr} = 101.325\text{ kPa}$$

Charles' law

The expansion of a fixed mass of gas at constant pressure as its temperature is raised is predicted by **Charles' law**:

The volume of a gas is directly proportional to its temperature in kelvin.

One way of allowing a sealed mass of gas to expand is to use a piston (Fig. 10.6). Since the piston handle is free to move, the pressure of the gas inside the piston will adjust itself so that it equals the pressure on the piston handle (the 'external pressure'). For example, if the external pressure (due to the air) is 1 atm, the gas inside the piston will remain at a pressure of 1 atm throughout the expansion.

Mathematically, Charles' law may be expressed as

$$V \propto T$$

where V is the volume of gas and T is its temperature in kelvin. The symbol $\propto$ means 'proportional to' and shows (for example) that if T doubles, V also doubles.

Charles' law can be used only where the volume of a gas is *allowed* to change. For example, heating a sample of gas in a sealed metal tube will not alter the volume of the gas because the rigid walls of the container prevent expansion.

Suppose a piston contains 100 cm^3 of hydrogen gas at 300 K (the temperature of the laboratory) and is under an external air pressure of 1 atm. Imagine that the piston is heated so that the gas eventually reaches a constant temperature of 600 K. Heating causes the temperature of the gas to rise. This causes the gas molecules to collide more often and more violently with the inner wall of the piston handle, resulting in a momentary increase in hydrogen gas pressure above 1 atm. Since the pressure of the hydrogen gas is greater than the pressure on the piston handle, the gas forces the piston handle back. In this way the gas increases its volume, but since the same number of molecules of hydrogen gas are now spread out in a larger space, the number of collisions per second on the inner wall of the piston handle falls and the pressure of the hydrogen gas begins to fall back towards 1 atm. The piston handle stops moving when the volume of the gas in the piston has been increased so that the pressure of the hydrogen equals 1 atm. Charles' law predicts that a doubling of the hydrogen gas temperature doubles the volume of the gas – see Fig. 10.6.

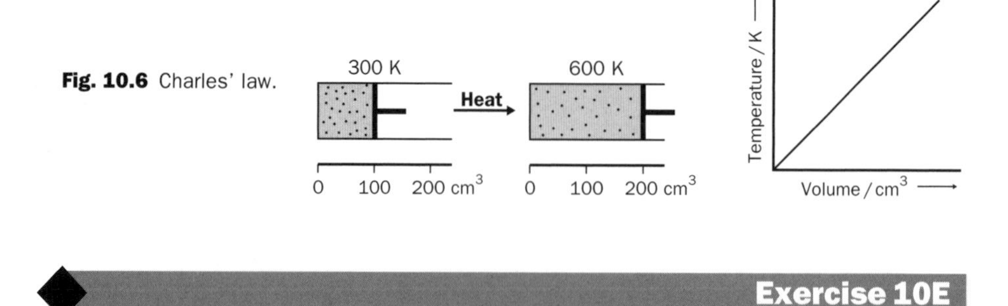

Fig. 10.6 Charles' law.

Exercise 10E

Charles' law

A sample of nitrogen at 500 K occupies 73.0 cm^3. Predict the volume of the gas at 2500 K. (Assume that its pressure has not changed.)

If the piston is allowed to cool, the reverse effect occurs, and the piston handle moves inwards so reducing the volume of gas inside.

Boyle's law

Think of a sample of gas of volume *V* contained in a piston (Fig. 10.7). If we push the piston handle inwards, the pressure on the gas exceeds atmospheric pressure and the volume of the gas drops. Boyle's law states that for a fixed mass of gas at a particular temperature

The volume (*V*) of a gas is inversely proportional to the pressure (*P*) of the gas.

Mathematically, Boyle's law is expressed as

$$V \propto \frac{1}{P}$$

Figure 10.7 illustrates the case where the applied pressure increases from 1 atm to 2 atm. This halves the gas volume, as predicted by the above equation.

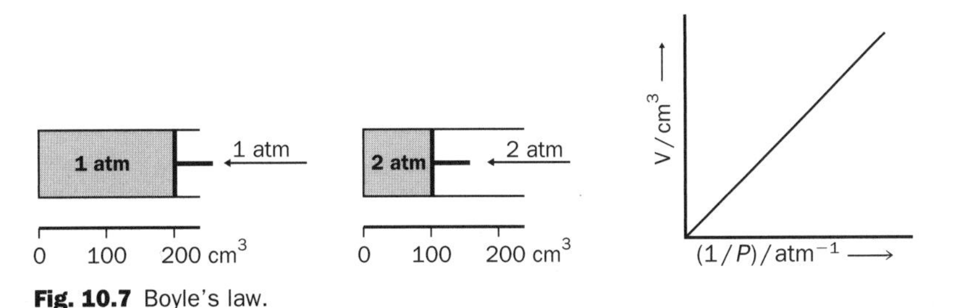

Fig. 10.7 Boyle's law.

Exercise 10F

Boyle's law

A gas contained in a piston at a pressure of 30.0 kPa occupies a volume of 200.0 cm^3 at 300 K. The pressure of the gas is slowly increased to 90.0 kPa without a change in temperature. What is the new volume of the gas?

Avogadro's law

Avogadro's law states that at a fixed pressure and temperature

The volume of a gas is proportional to the number of moles (or molecules) of gas present.

Mathematically we can write

$$V \propto n$$

where *n* is the number of moles of gas.

The following statement follows from Avogadro's law:

Equal volumes of all gases at the same temperature and pressure contain the same number of molecules.

It also follows that one mole of *any* gas occupies the same volume at the same temperature and pressure. This volume is known as the **molar volume** of the gas. Experiments show that the molar volume of an ideal gas at SATP is 24.79 dm^3. The molar volume of an ideal gas at 1 atm pressure and 20 °C is 24 dm^3 (Fig. 10.8), and it is this volume that we will use in routine calculations.

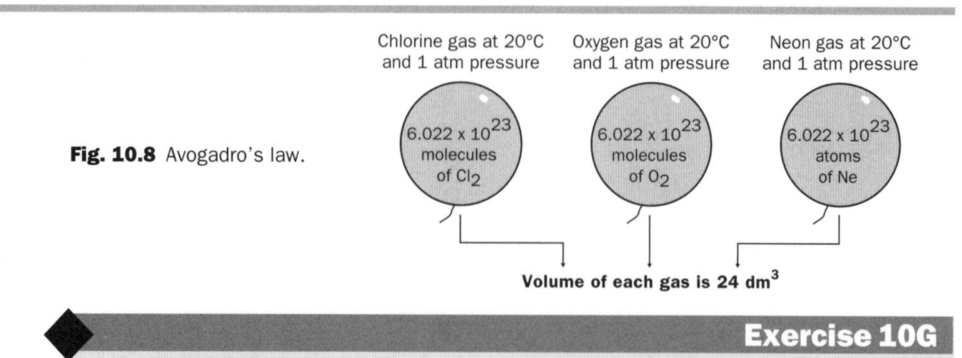

Fig. 10.8 Avogadro's law.

Exercise 10G

Avogadro's law

0.50 g of zinc reacts completely with excess dilute sulfuric acid at 20 °C. Calculate the volume of hydrogen produced at this temperature. (The molar volume of a gas at 20 °C and 1 atm pressure is 24 dm^3; m(Zn) = 65 u.)

Dalton's law of partial pressures

This law applies to gases which do not react chemically. It states that

> **The total pressure of a gas mixture is the sum of the partial pressures of the individual gases in the mixture.**

The **partial pressure** of a gas in a mixture is defined as the **pressure that gas would exert if it alone occupied the container.**

Mathematically, Dalton's law may be expressed as

$$P_T = p_A + p_B + p_C \dots$$

where P_T is the total pressure of the mixture and p_A, p_B, p_C . . . are the partial pressures of the gases A, B, C . . . in the mixture.

An example will show how easy it is to use Dalton's law. Suppose we add nitrogen to an initially empty gas cell so that the pressure of the nitrogen in the cell becomes 1.0 atm. If we now add an amount of oxygen (such that the pressure of oxygen if it alone occupied the gas cell would be 0.5 atm), a measurement would show that the total pressure of the mixture is 1.5 atm. This is predicted by Dalton's law:

$$P_T = p_{N_2} + p_{O_2} = 1.0 + 0.5 = 1.5\,\text{atm}$$

Dalton's law 'works' because the gaps between gas molecules are so large that the molecules of the constituent gases do not interfere with each other. The total pressure of the mixture is then the sum of the pressures exerted by the individual gases.

The most important gas mixture is, of course, the air around us (Box 10.3).

10.4 Kinetic molecular theory of gases

Assumptions of the kinetic molecular theory of gases

The kinetic theory of gases is a physical model that may be expressed in mathematical form. It assumes:

1. Gases consist of tiny particles (molecules or atoms) which are in chaotic and random motion.

BOX 10.3

Air

Table 10.1 shows the composition of air by volume and by mass. Water is also present in air, but the percentage varies with the location and with local temperature.

Table 10.1 Composition of dry air at sea level

Gas	Average percentage (by volume)	Average percentage (by mass)
Nitrogen N_2	78.05	75.50
Oxygen O_2	21.00	23.21
Argon Ar	0.94	1.28
Carbon dioxide CO_2	0.03	0.04
Neon Ne	0.0015	0.0011

The percentage composition of air remains almost constant at all altitudes, but the air *pressure* rapidly decreases from its sea level value of 1 atm according to the equation:

$$P = e^{-(34h/T)}$$

where P is the air pressure (in atm) at altitude h kilometres above sea level, and where the temperature is T kelvin. For example, the air pressure on a mountain 5 km above sea level where the temperature is −30 °C (= 243 K) is:

$$P = e^{-(34 \times 5/243)} = e^{-(0.70)} = 0.50\ \text{atm}$$

Exercise 10H

Air

Use the data in Box 10.3 to calculate the volume of argon in 1.0 dm³ of air? What mass of argon is contained in 1.0 kg of air? (Assume the air is dry and at sea level.)

2. The particles *do not* attract each other.

3. The collision of two molecules does not alter the overall energy of both molecules.

4. The volume of the particles is negligible compared with the volume of the container.

As we have already noted, assumption 1 immediately provides us with an explanation of gas pressure. The particles frequently collide with the molecules which make up the container walls: it is the force of these collisions which (on a gigantic scale) is responsible for the pressure of a gas.

Assumption 2 is a simplifying feature which applies *only* to gases at low pressures where the particles are so far apart from each other that they behave as independent 'mathematical points'. Attractive forces *cannot* be ignored in the solid and liquid state nor in explaining why gases condense to form liquids. This is one reason why solids and liquids are theoretically more complicated to deal with than gases, and why there is no simple equivalent of the ideal gas equation (discussed below) for these states of matter.

Assumption 3 is another way of stating that the average kinetic energy of particles in a gas remains fixed at a particular temperature. If the energy of particles were lost in collisions (either wall–gas particle or gas particle–gas particle collisions) the total energy of the gas would be continuously draining away, and (contrary to experimental observation) the pressure of a gas would fall with time.

Kinetic energy of gas molecules

Since the molecules of a gas are in constant motion, they all possess **kinetic energy**. The greater the kinetic energy of a gas molecule, the greater is its speed. Experiments show that molecules in a sample of gas are not all travelling at the same speed. Figure 10.9 shows the spread of molecular speeds of nitrogen at 300 K and 3000 K, and of chlorine gas at 300 K.

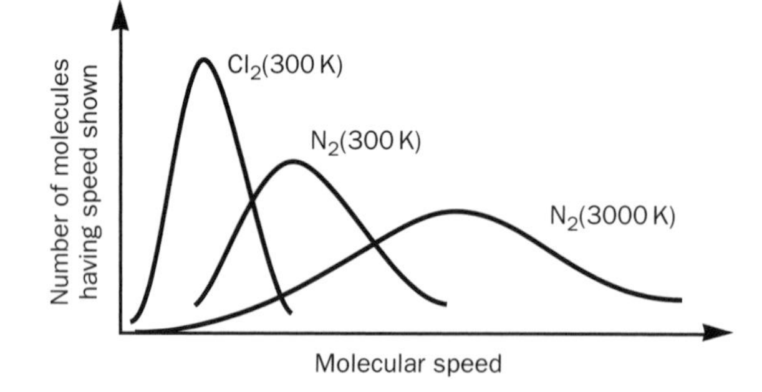

Fig. 10.9 The spread of molecular speeds in chlorine and nitrogen gases.

1. The **average speed** of gas particles at a particular temperature depends upon the mass of the gas molecules, with smaller molecules (such as N_2) moving faster than heavier ones (like Cl_2). Even relatively heavy molecules travel at average speeds of several hundred metres per second at room temperature!
2. The average speed of molecules in a gas also *increases* with the temperature of the gas, but the spread of speeds is greater at higher temperatures.

10.5 Ideal gas equation

The kinetic theory of gases reduces to a remarkably simple mathematical form, known as the **ideal gas equation.**

$$PV = nRT$$

where P is the pressure of a gas (Pa), V is the volume of gas (m^3), T the gas temperature (K), and n the number of moles of gas. R is the 'universal gas constant' with a value of $8.3145\,J\,mol^{-1}\,K^{-1}$.

Key points about the ideal gas equation

1. The ideal gas equation shows that P, V, n and T are not independent. These variables cannot take on any values we like – nature has dictated that if three of them are known, then the fourth is fixed.
2. The equation contains no factor which is *specific* to any gas. The type of gas (helium, chlorine, oxygen etc.) is irrelevant.
3. The rearranged forms of the ideal gas equation are often used in calculations:

$$V = \frac{nRT}{P}; \quad T = \frac{PV}{nR}; \quad P = \frac{nRT}{V}; \quad n = \frac{PV}{RT}$$

4. The inter-relationships between P, V, n and T, summarized in the gas laws of Charles, Boyle and Avogadro, are 'inbuilt' into the ideal gas equation, and the equation makes the separate use of these laws in calculations redundant. For example, the ideal gas equation shows that if the pressure and the number of moles of gas does not change, then V is proportional to T. In other words, it predicts Charles' law.

Examples in the use of the ideal gas equation

Note that, to make calculations easier, always express T in kelvin, V in cubic metres, P in pascals, and n in moles.

Example 10.1

Calculate the volume of 1.0000 mol of gas at exactly 20 °C at a pressure of 101.325 kPa.

▶ **Answer**

$T = 20 + 273.15 = 293.15\,\text{K}$

Rearranging the ideal gas equation,

$$V = \frac{nRT}{P} = \frac{1.0000 \times 8.3145 \times 293.15}{101\,325} = 0.024\,055\,\text{m}^3 \text{ (to five significant figures)}$$

Remembering that $1\,\text{m}^3 = 1000\,\text{dm}^3$,

volume $= 0.024\,055 \times 1000 = 24.055\,\text{dm}^3$

▶ **Comment**

This is the molar volume of a gas at 'room temperature and pressure'.

Example 10.2

Calculate the number of molecules of methane in 0.50 m³ of the gas at a pressure of 2.0 × 10² kPa and a temperature of exactly 300 K.

▶ **Answer**

$2.0 \times 10^2\,\text{kPa} = 2.0 \times 10^5\,\text{Pa}$

$$n = \frac{PV}{RT} = \frac{2.0 \times 10^5 \times 0.50}{8.3145 \times 300} = 40 \text{ (to two significant figures)}$$

To convert moles to molecules, we multiply by Avogadro's constant N_A:

number of molecules $= 6.022 \times 10^{23} \times 40 = 2.4 \times 10^{25}$

Exercise 10I

Ideal gas law

Calculate the pressure of 1.0 mol of helium in a 2.0 dm³ container at 20.0 °C.

Experimental measurements of the deviations of real gases from ideal behaviour

An ideal gas is one that exactly follows the ideal gas equation. Rearranging this equation,

$$n = \frac{PV}{RT}$$

If we use 1 mole of a gas (*any* gas) then the ratio PV/RT is predicted to have a numerical value of 1 *at all pressures*. If a gas does not obey the ideal gas law, the ratio will be either greater than 1 or less than 1.

Experiments show that at pressures below 1 atm (101 kPa), the ratio PV/RT does not deviate appreciably from 1 for most gases. This gives us a rough guide – **most real gases behave nearly ideally if their pressure is below 1 atm**. Even at a pressure of several atmospheres, many gases behave sufficiently ideally that the use of the ideal gas equation remains justified.

A plot of PV/RT against P for three gases is shown in Fig. 10.10. Deviations from ideal behaviour occur because, above 1 atm pressure, the attractive forces between the gas molecules can no longer be neglected. A single particle striking the wall of the containing vessel will be slightly attracted (and therefore 'held back') by other molecules in the bulk of the gas. As a consequence, the force with which the molecule strikes the container wall is *reduced*. On a bulk scale this causes the pressure of the gas to be *lower* than that of an ideal gas. Because P is lower than that for an ideal gas, the ratio PV/RT is smaller than it would be for an ideal gas. This is evident at moderately high pressures in Fig. 10.10. For example, PV/RT for CO_2 is less than 1 between 10 and 600 atm.

Exercise 10J

Deviation from ideal gas behaviour

Why would you expect carbon dioxide to deviate from ideal behaviour more than hydrogen or nitrogen?

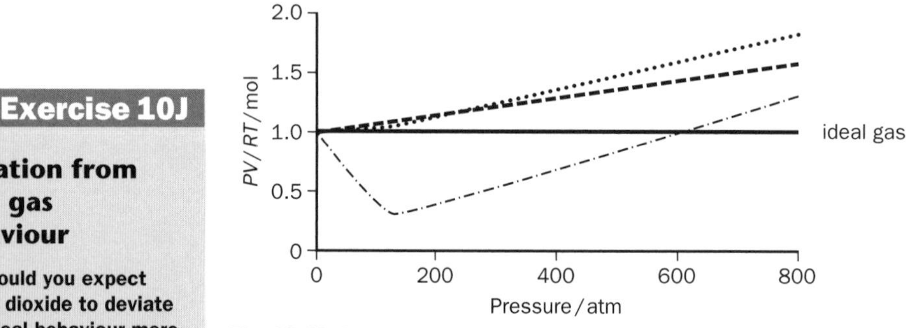

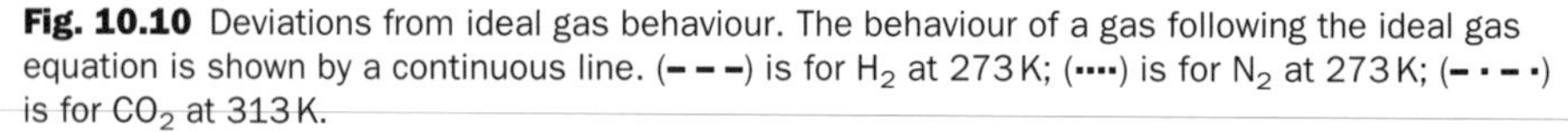

Fig. 10.10 Deviations from ideal gas behaviour. The behaviour of a gas following the ideal gas equation is shown by a continuous line. (– – –) is for H_2 at 273 K; (····) is for N_2 at 273 K; (– · – ·) is for CO_2 at 313 K.

At yet higher pressures, it is no longer reasonable to assume that the volume of the gas molecules is negligible compared with the container volume. This makes PV/RT greater than 1 because the volume V is greater than that for an ideal gas. For CO_2, this is observed at pressures above 600 atm.

10.6 Adsorption of gases on solids

Charcoal is used in gas masks to remove toxic gases. This is one example of a general phenomenon known as **adsorption**.

In adsorption, a substance (the **adsorbate**) sticks to the *surface* of another material (the **adsorbent**). This is distinct from **absorption**, in which a substance penetrates another material. The difference between the two phenomena may be remembered by thinking of a bath sponge. If dust sticks to the surface of the dry sponge then adsorption has occurred, but if water is taken into the sponge, absorption has occurred.

It is often impossible to decide whether adsorption, absorption or a mixture of both are occurring. For this reason, the noncommittal term **sorption** is used for both processes. An example of this difficulty is found in chromatography, where a

Gas masks contain activated charcoal which absorbs toxic gases in the surrounding air. This photograph of a family out shopping in London was taken during a tear gas test in 1941 (World War II).

combination of solute absorption and adsorption may bring about the separation of a mixture on a chromatographic column (Chapter 19).

The forces that bind the adsorber to adsorbent may be physical (i.e. intermolecular forces) or chemical (when chemical bonds are formed). The adsorption of gases by charcoal is physical, whereas the adsorption of gases on some catalysts is chemical. Toxic gases adsorb to charcoal better than the oxygen and nitrogen of the air because they are often relatively large molecules and are frequently polar. This increases the intermolecular forces between adsorbate and adsorbent.

The charcoal used in gas masks is called 'activated charcoal', and is made by heating wood or coal in carbon dioxide, water vapour or in a limited supply of air. Activated charcoal has an enormous surface area – crucial to adsorption – as high as 1000 m^2 of gas per gram of charcoal! At a given temperature and pressure, a sample of activated charcoal will adsorb only a fixed mass of adsorbate, but the mass increases as the pressure of adsorbate gas is increased or if the temperature is lowered. The charcoal may be regenerated by heating, which drives off the adsorbate.
Adsorption will also occur in solution. For example, charcoal beds are used to adsorb trace organic compounds (such as pesticides and dyes) from water.

10.7 Vapour pressure

Meaning of vapour pressure

A vapour is a gas in contact with a liquid of the same substance. For example, the water 'gas' above the surface of liquid water is described as water vapour. The gas pressure of the water vapour is known as its **vapour pressure**.

Suppose that liquid water is placed into a container and that the gas (mainly air) above the water is pumped away using a powerful pump, and the container sealed without allowing air to re-enter. Some of the liquid water then evaporates so that the only gas in the space above the liquid water is water vapour. Figure 10.11(a) shows how a pressure measuring device might be used to measure the vapour pressure of the water.

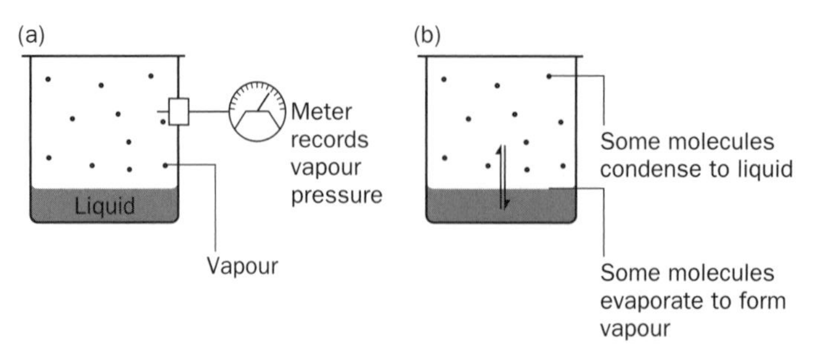

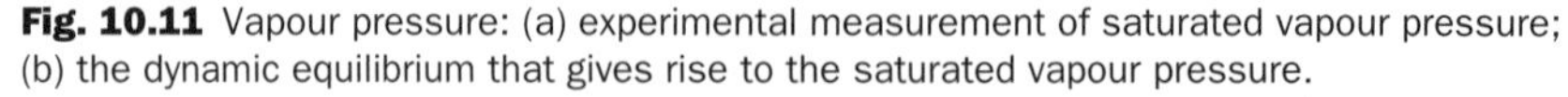

Fig. 10.11 Vapour pressure: (a) experimental measurement of saturated vapour pressure; (b) the dynamic equilibrium that gives rise to the saturated vapour pressure.

The vapour pressure increases as soon as the container is sealed, but the rate of increase slows down after a few minutes because, although water molecules continue to evaporate, some water molecules also condense back to form liquid water. After a few more minutes the vapour pressure levels off and reaches a maximum called the **saturated vapour pressure** – so-called because the air is saturated with water vapour at that temperature. For pure water at 20 °C, the saturated vapour pressure is 2.33 kPa. Even at this stage, molecules continue to evaporate and condense, but a **dynamic equilibrium** has been achieved in which the number of molecules evaporating per second equals the number of molecules which are condensing per second (Fig. 10.11(b)). For this reason, saturated vapour pressure is often called **equilibrium vapour pressure**. We represent the equilibrium by the equation

$$\text{liquid} \rightleftharpoons \text{vapour}$$

If the container was open to the air, the water vapour would continually diffuse away and the vapour pressure would not reach its equilibrium value. More and more water would slowly evaporate into the air until (after several days) no more water was left in the container.

Figure 10.12 shows the saturated vapour pressure of water at various temperatures. The figure also shows the way that the saturated vapour pressure of ethanol varies with temperature. The vapour pressure of water and ethanol (like that of all liquids) increases with temperature. This is expected, since at the higher temperature more liquid molecules have sufficient energy to escape the attractive forces holding molecules in the liquid.

At the boiling point of a liquid, the saturated vapour pressure *equals* the external air pressure. At sea level, the air pressure is 1 atm (101 kPa). The boiling point of a liquid at 1 atm external pressure is called the **normal boiling point** of the liquid. The normal boiling points of water and ethanol are 100 °C and 78 °C, respectively.

Exercise 10K

Boiling point

Explain the following:

(i) The boiling point of water drops with increasing altitude.

(ii) Salty water boils at a higher temperature than pure water.

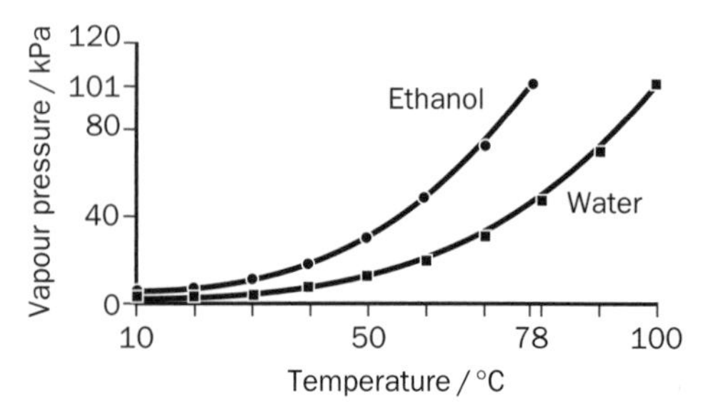

Fig. 10.12 The vapour pressure of ethanol and water between 10 °C and 100 °C.

Vapour pressure and total gas pressure

If a *sealed vessel* contains a liquid X (such as ethanol or water) and a gas Y (such as nitrogen), it is usually safe to assume that the gas Y is saturated with the vapour of X. As expected from Dalton's law, the total gas pressure above the liquid is the sum of the gas and vapour pressures. This is illustrated by the following example.

Example 10.3

A sealed bottle contains carbon monoxide gas (CO) and about 3 cm^3 of liquid water at 20°C. A pressure gauge measured the total gas pressure in the bottle as 81.2 kPa. What is the partial pressure of carbon monoxide in the bottle? The saturated vapour pressure of water at 20°C is 2.33 kPa.

▶ Answer

The presence of liquid water in the bottle shows that there is enough water present to saturate the CO with water vapour (Fig. 10.13). Since

$$\text{total gas pressure} = 81.2\,\text{kPa} = p_{CO} + p_{H_2O}$$

where p represents the vapour pressure contributions from both substances. Since $p_{H_2O} = 2.33\,\text{kPa}$,

$$p_{CO} = 81.2 - 2.33 = 78.9\,\text{kPa}$$

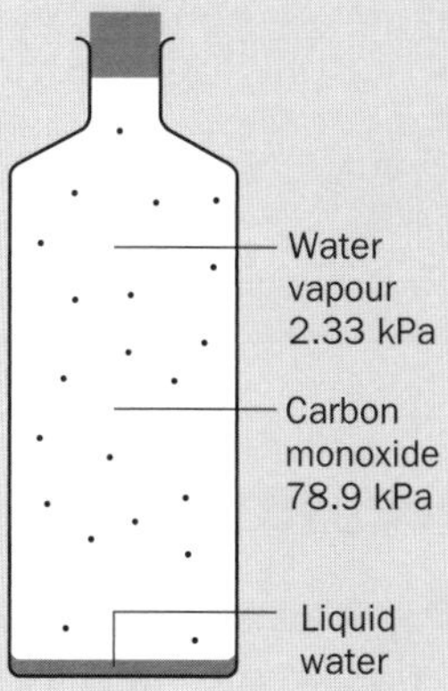

Water vapour 2.33 kPa

Carbon monoxide 78.9 kPa

Liquid water

Fig. 10.13 Vapour pressure and total gas pressure.

Exercise 10L

Vapour pressure calculations

A corked test-tube at 25°C contains air and a few drops of liquid ethanol. The total gas pressure in the tube is 101 kPa. What is the partial pressure of air in the tube? (Saturated vapour pressure of ethanol at 25°C is 7.9 kPa.)

Volatility of solvents

Ethanol evaporates much more readily than water; therefore, we say that ethanol is more **volatile**. This is reflected in Fig. 10.12, where the vapour pressure of ethanol exceeds the vapour pressure of water at every temperature.

One way to estimate the volatility of solvents at room temperature is to saturate pieces of filter paper with solvent, and time how long it takes for all the solvent to evaporate. Alternatively, the boiling point of a solvent may be taken as a rough indication of its volatility. One of the most volatile solvents available in the laboratory is ethoxyethane (diethyl ether), with a normal boiling point of 34.5°C (see Box 10.4).

BOX 10.4

Flash point of solvents

Many solvents are good fuels, and will readily burn in air. This can be a safety hazard in laboratories and factories. The temperature at which a solvent catches fire in the *absence* of a flame is called its **auto-ignition temperature**. This is surprisingly high for most solvents (Table 10.2). The temperature at which the vapour pressure of a pure solvent in air is sufficient to be ignited by a naked flame is called the **flash point** of the solvent (Fig. 10.14). This can be very low (Table 10.2). For example, the flash point of petroleum ether (a fraction obtained from the distillation of crude oil and which boils between 30 and 40 °C) is −51 °C. This means that petroleum ether stored in an open (or leaking) container will still represent a serious fire hazard even if stored in a freezer operating at −51 °C. Flash points are used to classify the **flammability** of solvents.

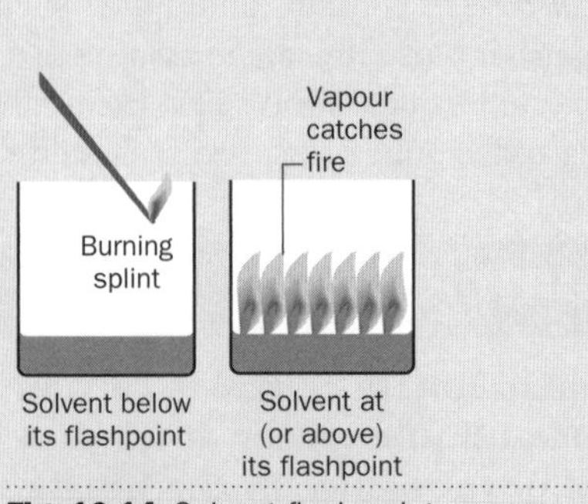

Fig. 10.14 Solvent flash point.

Solvents with a flash point below 0 °C (and with boiling points below 36 °C) are described as **extremely flammable**. Solvents with a flash point above 0 °C but below 21 °C are described as **highly flammable**. Solvents with a flash point between 21 and 55 °C are described as **flammable**.

Table 10.2 Flash points and ignition temperatures of common solvents

Solvent	Flash point/°C	Boiling point/°C	Auto-ignition temperature/°C
Propanone (acetone)	−20	56	538
Benzene	−11	80	536
Ethoxyethane (diethyl ether)	−40	34.5	170
Ethanol	13	78	425
30–40 petroleum spirit	−51	30–40	230
Butan-1-ol	29	118	365

Exercise 10M

Critical temperature and pressure

(i) A mixture of hydrogen and oxygen at −119 °C is subjected to a very high pressure. Would you expect liquid hydrogen and liquid oxygen to condense?

(ii) Historically, hydrogen was one of the last gases to be liquefied in the laboratory – why? (**Hint** What is the boiling point of $H_2(l)$?)

10.8 Critical temperature and pressure

If the volume of a gas (or vapour) is reduced by compressing it (as in a piston), or cooling it, or both, the gas may condense to a liquid. Experiments show that a gas *cannot be liquefied* by pressure alone unless it is at (or below) a temperature known as the **critical temperature**. Put another way, the critical temperature is the highest temperature at which a liquid can exist.

At a molecular level, molecules in a sample at a temperature above the critical temperature simply have too much energy to 'stick together' – no matter what the applied pressure.

The pressure needed to liquefy a gas at its critical temperature is called its **critical pressure**. The critical temperatures and pressures for some gases are given in Table 10.3. For example, if ethanol vapour is *above* 243 °C, no amount of pressure will convert the vapour to liquid. If the ethanol vapour were exactly at a temperature of 243 °C, a pressure of 63 atm would need to be applied to the vapour in order to force it to condense to a liquid.

Table 10.3 Critical temperatures and pressures for common gases

Gas	Normal boiling point/°C	Critical temperature/°C	Critical pressure/atm
Ammonia	−33	132	112
Carbon dioxide	−78 (sublimes)	31	73
Ethanol	78	243	63
Hydrogen	−252	−240	13
Oxygen	−183	−119	50
Nitrogen	−196	−147	34
Water	100	374	218

Revision questions

10.1. Convert:
(i) 0.124 torr to standard atmospheres
(ii) exactly 1.5 atm to pascals
(iii) 89 kPa to newtons per square metre ($N\,m^{-2}$)
(iv) 101.325 kPa to bar
(v) −100 °C to kelvin.

10.2. In an experiment (in which all gases were at the same temperature and pressure before and after reaction) it was found that 1.21 dm^3 of hydrogen gas reacted completely with 1.21 dm^3 of chlorine gas to produce 2.42 dm^3 of hydrogen chloride. Using Avogadro's law, explain whether or not these volumes confirm the equation

$$H_2(g) + Cl_2(g) \rightarrow 2HCl(g)$$

10.3. Potassium chlorate decomposes upon heating in the presence of a catalyst to give oxygen gas:

$$2KClO_3(s) \rightarrow 2KCl(s) + 3O_2(g)$$

Calculate the volume of oxygen gas made at room temperature and pressure (20 °C and 1 atm) when 10.0 g of chlorate is completely decomposed.

10.4. 25.0 kPa of chlorine gas is fed into a previously evacuated cell with a volume of 0.0100 m^3. Oxygen gas was then added to the cell so that its partial pressure was 50.0 kPa. The cell temperature is exactly 20 °C. Calculate:
(i) the total pressure of gas in the cell in pascals
(ii) the mole fraction of oxygen gas in the gas mixture, where

$$\text{mole fraction of } O_2 = \frac{\text{number of moles of } O_2}{\text{total number of moles of gas}}$$

10.5. A metal can of volume 2000.0 cm^3 contains 2.0 g of nitrogen gas and 6.0 g of hydrogen gas at exactly 200 °C. What is the total pressure of gas in the can?

10.6. A flexible leak-proof balloon of volume 1.00 m^3 was filled with helium to a pressure of 1.00×10^5 Pa ($N\,m^{-2}$) at 293 K. When allowed to ascend to an altitude where the helium pressure was 5.00×10^4 Pa, the balloon volume had increased to 1.66 m^3. Use the ideal gas equation to calculate the number of moles of He in the balloon at 293 K. Then calculate the atmospheric temperature at the new altitude.

10.7. Experiments show that the molar volumes of hydrogen, carbon dioxide and ammonia at 1 atm pressure and 273 K are H_2 22.4 dm^3, CO_2 22.3 dm^3, NH_3 22.1 dm^3. What do these values suggest about the degree to which these gases deviate from the ideal gas equation? Explain the trend. (Start this problem by calculating the molar volume of an ideal gas under these conditions.)

10.8. Calculate the total pressure of the air in the atmosphere at an altitude of 80 km, where $T = -100$ °C (see Box 10.3).

10.9. A stoppered 2.0 dm^3 flask was half-filled with water at exactly 40 °C. Calculate the number of water molecules in the water vapour. (The saturated vapour pressure of water at 40 °C is 7370 Pa.)

10.10. Why do solvents with high vapour pressures at room temperature usually possess low flash points?

Extension material to support this unit is available on our website at **http://www.palgrave.com/foundations/lewis.**

Solutions and Solubility

Contents

Objectives

- ▶ Examines solvent miscibility and immiscibility
- ▶ Explains the idea of solubility product
- ▶ Looks at distribution ratios and gas solubility
- ▶ Discusses osmosis and its applications
- ▶ Introduces colloids

11.1 Solubility

A **solution** is a mixture consisting of a **solvent** (the 'dissolver') and the **solute** (the substance that is being dissolved). For example, if we dissolve sugar in water, the water is the solvent, the sugar the solute and the sugary water is the solution. If we keep adding sugar to some water, a point will be reached when the water will not be able to hold any more sugar. The solution is now said to be **saturated**. Adding more sugar simply results in sugar settling on the bottom of the container. Raising the temperature of the solution allows the water to hold more sugar before it becomes saturated. Many solids, like sugar, are more soluble at higher temperatures, although the reverse usually applies to gases, which are less soluble in hot water than in cold water.

Rules of solubility

The word 'polar' was introduced in Unit 5 (see page 71). A polar substance is a substance that contains ions or consists of polar molecules. A polar solvent is a solvent which consists of polar molecules.

We start by reminding ourselves of the following:

1. If a *polar substance* dissolves, it dissolves only in *polar solvents*.
2. If a *non-polar substance* dissolves, it dissolves only in *non-polar solvents*.

These generalizations are summarized in the rule *like dissolves like*. Solvents may be placed in order of polarity by testing their solubility in each other. The order of

Table 11.1 Polarity of common solvents – in order of increasing polarity with heptane the least polar and water the most polar

Solvent	Formula	Density at 25°C/g cm^{-3}
Heptane	$CH_3(CH_2)_5CH_3$	0.68
Hexane	$CH_3(CH_2)_4CH_3$	0.66
Cyclohexane	C_6H_{12}	0.77
Tetrachloromethane[1]	CCl_4	1.58
Methyl benzene[2]	$C_6H_5CH_3$	0.86
Ethoxyethane[3]	$C_2H_5OC_2H_5$	0.71
Dichloromethane	CH_2Cl_2	1.32
Propan-2-ol	$CH_3CH(OH)CH_3$	0.78
Tetrahydrofuran	C_4H_8O	0.89
Trichloromethane[4]	$CHCl_3$	1.48
Ethanol[5] (absolute)	CH_3CH_2OH	0.79
Ethyl ethanoate[6]	$CH_3COOC_2H_5$	0.90
Propanone[7]	CH_3COCH_3	0.79
Methanol[8]	CH_3OH	0.79
Ethanenitrile[9]	CH_3CN	0.78
Dimethyl sulfoxide	CH_3SOCH_3	1.10
Water	H_2O	1.00

Alternative names: [1]Carbon tetrachloride, [2]Toluene, [3]Diethyl ether, [4]Chloroform, [5]Ethyl alcohol, [6]Ethyl acetate, [7]Acetone, [8]Methyl alcohol, [9]Acetonitrile.

solvents in Table 11.1 was obtained in this way. Of the common solvents, water is the most polar and the hydrocarbons heptane and hexane the least polar.

Miscibility

If, when two solvents are mixed, a single layer (consisting of a solution of the two solvents) is produced, the solvents are said to be **miscible**. If two layers are produced and both layers consist of pure solvent, the liquids are said to be **immiscible** (Fig. 11.1). If two layers are produced, the solvent with the lowest density floats on the top.

The word 'layer' is often replaced by the word **phase**. Thus, a mixture of hexane and water produces two phases.

Table 11.2 shows which pairs of common solvents are miscible, with ● denoting immiscibility. For example, the table shows that water is immiscible with trichloromethane and with ethyl ethanoate.

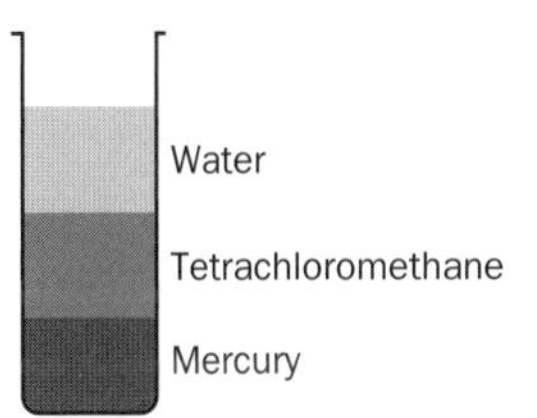

Fig. 11.1 Three immiscible liquids—tetrachloromethane, mercury and water: mercury (density 13.6 g cm^{-3} at 25°C) sinks to the bottom; tetrachloromethane (density 1.6 g cm^{-3}) occupies the middle position; and water (density 1.0 g cm^{-3}) floats on top.

Partially miscible solvents

Few solvents are truly immiscible, and even though two liquids may not appear to mix, there will still be a tiny amount of each solvent present in the other layer. Table 11.3 shows the solubilities of organic solvents in water, and of water in organic solvents. The units of the solubilities are grams of organic solvent per 100 g of saturated water, and grams of water per 100 g of saturated organic solvent.

Exercise 11A

The captain's bet

The captain of a sailing ship challenged the leader of his mutinous crew to a bet, with the winner to take the ship as prize. The captain mixed oil and beer in a large earthenware jug and immediately poured the mixture into two pewter tankards. The captain quickly picked up a tankard and drank down its contents in one go, but the crewman was unable to drink more than a mouthful. Explain how the captain won his bet.

Table 11.2 Miscibility of common solvents. (Adapted, with permission of Fisher Scientific UK from their 1996 Chemicals Catalogue)

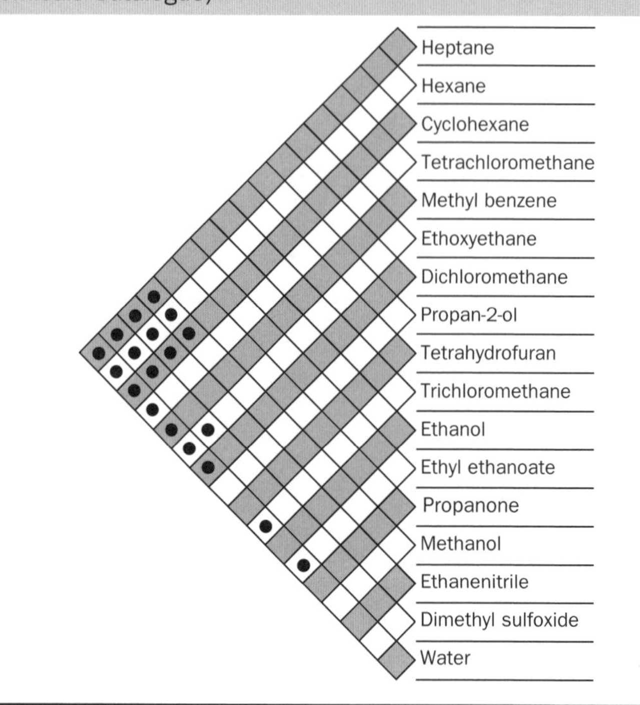

Table 11.3 Approximate solubilities of organic liquids in water, and of water in organic liquids at 25 °C

Solvent	Concentration g per 100 g: Organic liquid in water phase	Concentration g per 100 g: Water in organic phase
Hexane	0.001	0.01
Cyclohexane	0.01	0.01
Tetrachloromethane	0.08	0.008
Methyl benzene	0.074	0.03
Ethoxyethane	6.9	1.3
Trichloromethane	0.82	0.056
Ethyl ethanoate	8.7	3.3

Exercise 11B

Concentrations in two phases

Cyclohexane is mixed with pure water, producing two layers.

(i) Which liquid forms the top layer?

(ii) How much water is contained in 130 cm^3 of the cyclohexane layer?

An oil slick. Crude oil is largely immiscible with water and floats on its surface. It then washes ashore. It is particularly dangerous to birds since it is miscible with the oils on a bird's body.

Suppose, for example, that 100 g of hexane is mixed with 100 g of water at room temperature. The liquids are almost immiscible, but the hexane layer would actually contain about 0.01 g of water, and the water layer would contain about 0.001 g of hexane.

Justification of the rule that 'like dissolves like'

The information contained in Table 11.2 confirms the rule that 'like dissolves like'. This rule is based upon experimental observation. In order to explain the rule, we rely upon a knowledge of the forces between ions, atoms and molecules:

1. Inorganic salts (such as Na^+, $Cl^-(s)$) are polar by definition. The solubility of sodium chloride in water is due to the strong attractive forces between the Na^+ and Cl^- ions, and the polar ends of the water molecules. Salts such as sodium chloride are also fairly soluble in ethanol and methanol (which, although less polar than water, still give rise to fairly strong ion–water molecule forces).

2. Salts are immiscible with non-polar solvents such as hexane and tetrachloromethane. The forces between ions (such as Na^+ and Cl^-) and molecules of non-polar solvents are London dispersion forces. Such forces are much weaker than the attractions between opposite ions which hold the sodium chloride lattice together.

3. Solvents such as hexane and tetrachloromethane are soluble in similar solvents (such as hydrocarbons) because the London dispersion forces between the two different kinds of molecules are roughly equal to (or greater than) the London dispersion forces between the same molecules. For example, hexane dissolves in heptane because the forces between the heptane and hexane molecules are at least equal to those between heptane and heptane or between hexane and hexane molecules. If water is added to heptane or hexane, the water molecules 'stick together' and form a separate phase because the hydrogen bonds between the water molecules are much stronger than the forces between the water and hydrocarbon molecules.

4. Even large covalent molecules may be soluble in water if they contain sufficiently polar bonds. This is the case for sucrose (table sugar, $C_{12}H_{22}O_{11}$) which contains eight polar $O^{\delta-}$–$H^{\delta+}$ groups per molecule and so is able to link with the water molecules using hydrogen bonding. However, the general rule is that within a series of organic compounds, solubility in water decreases as the number of carbon atoms in the molecular chain increases. For example, consider ethanoic and hexanoic acids:

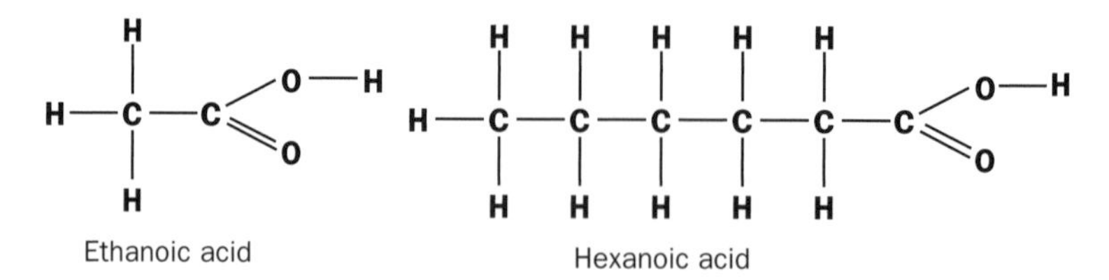

Ethanoic acid is completely miscible with water in all proportions. This means that ethanoic acid will never form two layers with water no matter how much acid is present. However, hexanoic acid is virtually insoluble in water. Although both acids contain the highly polar C=O and –O–H groups, the long carbon chain of the hexanoic acid molecule increases the size of the London dispersion forces between the acid molecules; this causes the acid molecules to stick together. Even the possibility of hydrogen bonding between acid and water molecules is not enough to make hexanoic acid soluble in water, and the only hydrogen bonding present occurs between acid molecules.

Exercise 11C

Effect of chain length upon solubility

Look at this list of alcohols and their formulae:

butan-1-ol	$CH_3CH_2CH_2CH_2OH$
hexan-1-ol	$CH_3CH_2CH_2CH_2CH_2CH_2OH$
nonan-1-ol	$CH_3CH_2CH_2CH_2CH_2CH_2CH_2CH_2CH_2OH$
octan-1-ol	$CH_3CH_2CH_2CH_2CH_2CH_2CH_2CH_2OH$

Which of these alcohols is the most soluble in water?

Hydrophilic and hydrophobic ends of large organic molecules

We can look at large organic molecules in an even simpler way, by thinking of organic molecules such as hexanoic acid as consisting of a polar 'head' (the –COOH part) and a zigzag hydrocarbon 'tail' (the $-CH_2-$ chain), as shown in Fig. 11.2. The heads are attracted to water and are said to be **hydrophilic** (water loving), whereas the tails attract each other (and so pull away from water molecules) and are said to be **hydrophobic** (water hating).

The application of these ideas to soap action (**detergency**) is discussed in Box 11.1. Detergency is an example of **emulsification** (or **solubilization**) in which fats are prevented from forming two layers in water-based mixtures.

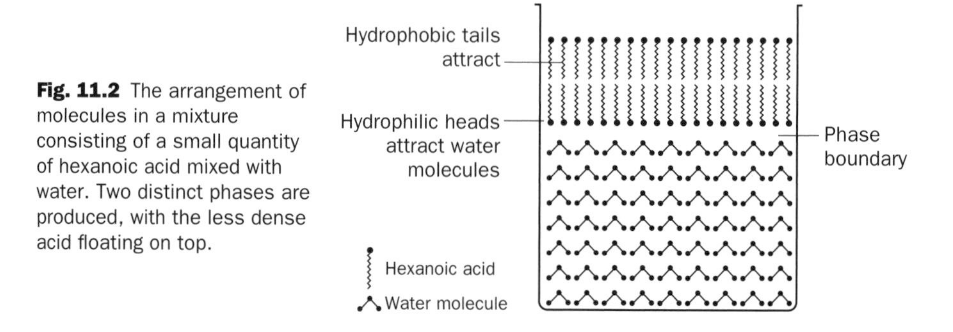

Fig. 11.2 The arrangement of molecules in a mixture consisting of a small quantity of hexanoic acid mixed with water. Two distinct phases are produced, with the less dense acid floating on top.

BOX 11.1

Soaps and detergents

What are soaps?

Soaps are salts of fatty acids. The commonest soap is sodium stearate, made by reacting sodium hydroxide (caustic soda) with stearic acid:

$$\underset{\text{sodium hydroxide}}{NaOH} + \underset{\text{stearic acid}}{C_{17}H_{35}COOH} \rightarrow \underset{\text{sodium stearate}}{C_{17}H_{35}CO_2^-,Na^+} + H_2O$$

The sodium stearate dissociates into sodium and stearate ions in solution.

The $C_{17}H_{35}-$ part of the soap consists of a long chain of carbon atoms, with the COO^-Na^+ forming an ionic 'end' to the molecule. The full structure of soap may be written as

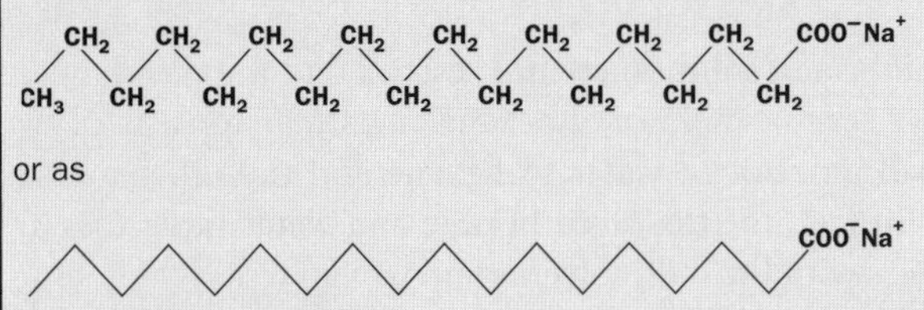

What are detergents?

Detergents are soap substitutes made in the oil industry. They possess similar molecules to soaps, with the sulfonate group (OSO_3^-) being the commonest ionic group. For example, the detergent sodium lauryl sulfate ($C_{17}H_{35}OSO_3^-,Na^+$) has the following structure:

The sodium lauryl sulfate dissociates into sodium and lauryl sulfate ions in solution.

Cleaning action of soaps and detergents

Dirt is a greasy material in which waste solid particles get stuck. Water cannot dislodge dirt (Fig. 11.3(a)). The cleaning action of soap and detergents is identical and involves the loosening of the solid particles from surfaces (normally of clothes or some plates or dishes) and can be thought of as taking place in three steps:

1. The hydrocarbon end (the hydrophobic end) of the stearate or detergent ion is attracted to the grease in the dirt. The ionic (hydrophilic end) of the soap or detergent is attracted to the water (Fig. 11.3(b)).

2. The stearate or detergent ion reduces the stickiness (the so-called **interfacial tension**) between the dirt and the surfaces of the clothes or dishes. This is done by forming a layer of stearate or detergent ions on the clothes or dishes (Figs 11.3(c) and (d)). As a result, grease droplets are encouraged to enter the water phase.

3. The stearate or detergent ions surround small droplets of grease and form tiny clumps (containing typically 50–100 ions) called **ionic micelles** (Fig. 11.3(d)). The overall result is a colloidal dispersion of oil droplets which, together with the dirt particles, can be physically washed away.

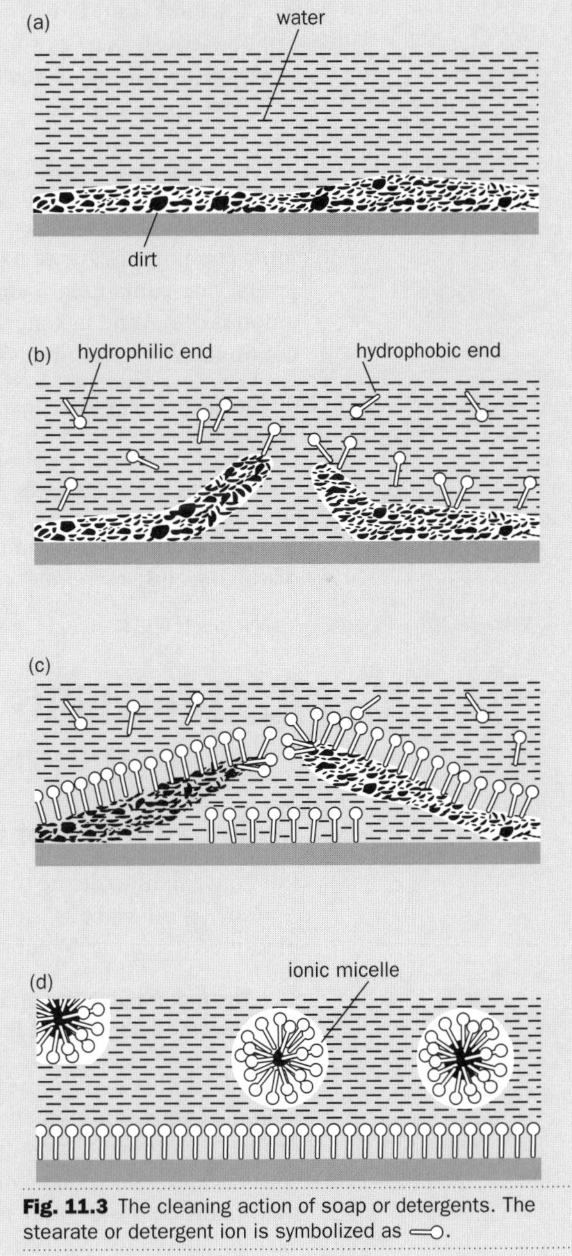

Fig. 11.3 The cleaning action of soap or detergents. The stearate or detergent ion is symbolized as ⊸○.

11.2 Dynamic nature of dissolution

Equilibrium involved in dissolving

Think of a beaker containing saturated potassium iodide solution, with undissolved (solid) K^+,I^- at the bottom of the beaker. Experiments show that although the concentration of dissolved KI remains fixed at that temperature, the ions in the solution and in the solid are continuously exchanging places. Suppose at an instant, a pair of $K^+(aq)$ and $I^-(aq)$ ions come together and precipitate at the bottom of the beaker. Simultaneously, a pair of K^+ and I^- ions in the solid break up into the solution. In other words, a **dynamic equilibrium** exists between the iodide in the solid and in solution. We represent this as:

$$K^+,I^-(s) \rightleftharpoons K^+(aq) + I^-(aq)$$

Examples of dynamic equilibria in solution are not restricted to the dissolving of ionic compounds. As we have seen, a mixture of hexane and water consists of two layers, one containing a small amount of water in hexane and the other a small amount of hexane in water. In fact, the dissolved hexane and water molecules are continuously exchanging places with the bulk solvents and we can write:

water in water layer $\rightleftharpoons$ water in hexane layer

also

hexane in hexane layer $\rightleftharpoons$ hexane in water layer

Other examples of dynamic equilibria we shall discuss in this chapter include the dissolving of gases in water and the distribution of a solid between two solvents.

11.3 Solubility of sparingly soluble ionic compounds

Molar solubility of ionic compounds

It is convenient to define the solubility of ionic substances in units of g per 1000 g of saturated *solution*, or as the number of moles of substance needed to produce 1 dm^3 of saturated *solution*. **Molar solubility** is defined as follows:

The molar solubility of a compound is its concentration ($mol\,dm^{-3}$) in a saturated solution at that temperature.

Molar solubility is given the symbol *s*.

Ionic compounds which are *only slightly soluble* in water are said to be **sparingly soluble.** Examples include silver chloride (Ag^+,Cl^-) and barium sulfate (Ba^{2+},SO_4^{2-}). Since such compounds are only slightly soluble, the volume of water used to dissolve a salt may be taken to be equal to the volume of the solution. (For example, if we dissolve 0.0010 g of silver chloride in 1000 cm^3 of water, we may safely assume that the volume of the solution is also 1000 cm^3.) This approximation does not hold for very soluble salts, where the presence of appreciable concentrations of salt in the water causes a slight change in the liquid volume.

One reminder – although many ionic substances are only sparingly soluble in water, any ionic substance that *does* dissolve completely dissociates into separate ions.

Example 11.1

The solubility of silver chloride is 1.8×10^{-3} g of AgCl per 1000 g of saturated AgCl solution at 25 °C. Express this solubility in mol of AgCl per dm^{-3} of AgCl solution. What is the concentration of $Ag^+(aq)$ and $Cl^-(aq)$ (in mol dm^{-3}) in a saturated solution of AgCl? (Assume that 1 cm^3 of solution has a mass of exactly 1 g at 25 °C.)

▶ Answer

The molar mass of AgCl = M(AgCl) = 143.5 g mol^{-1}. Therefore, a mass of 1.8×10^{-3} g contains

$$1.8 \times 10^{-3}/143.5 = 1.3 \times 10^{-5} \text{ mol AgCl}$$

The solution, which contains 1000 g of water, has a volume of 1000 cm^3 = 1.000 dm^3. Therefore, the molar solubility of AgCl is

$$s = \frac{\text{amount of substance (in moles)}}{\text{volume}}$$

$$= \frac{1.3 \times 10^{-5}}{1.000} = 1.3 \times 10^{-5} \text{ mol dm}^{-3}$$

The dissolution of silver chloride is represented by the equation

$$\underset{(\text{i.e. } Ag^+,Cl^-)}{AgCl(s)} \rightleftharpoons Ag^+(aq) + Cl^-(aq)$$

The Ag^+,Cl^- pairs that do dissolve, dissociate completely into ions. Since one pair of dissolved Ag^+,Cl^- ions produces one Ag^+ ion and one Cl^- ion, the concentration of these ions is also 1.3×10^{-5} mol dm^{-3}:

$$[Cl^-(aq)] = [Ag^+(aq)] = 1.3 \times 10^{-5} \text{ mol dm}^{-3}$$

Exercise 11D

Molar solubility

Lead(II) chloride ($PbCl_2$) has a molar solubility at 25 °C of 1.60×10^{-2} mol dm^{-3} of solution. What mass of $PbCl_2$ does 100 cm^3 of water at 25 °C which is saturated with $PbCl_2$ contain? What is the concentration of lead and chloride ions (in mol dm^{-3}) in the saturated solution?

Solubility product of sparingly soluble salts

Experiments show that, for a saturated solution of silver chloride,

$$[Ag^+(aq)] \times [Cl^-(aq)] = \text{a constant at that temperature}$$

where the square brackets signify equilibrium concentrations in mol dm^{-3}. This is the **solubility product expression** for silver chloride. The constant is known as the **solubility product** or **solubility constant** (symbolized K_s) for AgCl(s) at that temperature. At 25 °C, $K_s(AgCl) = 1.6 \times 10^{-10}\ mol^2\ dm^{-6}$, i.e.

$$[Ag^+(aq)] \times [Cl^-(aq)] = K_s(AgCl) = 1.6 \times 10^{-10}\ mol^2\ dm^{-6}$$

We can put this into words as follows: 'In a solution saturated with AgCl at 25 °C, the concentrations of silver and chloride ions when multiplied together must equal $1.6 \times 10^{-10}\ mol^2\ dm^{-6}$'.

We can define the solubility products of other sparingly soluble ionic substances in the same way:

1. For barium sulfate the equilibrium is

$$BaSO_4(s) \rightleftharpoons Ba^{2+}(aq) + SO_4^{2-}(aq)$$

$$K_s(BaSO_4) = [Ba^{2+}(aq)] \times [SO_4^{2-}(aq)] = 1.1 \times 10^{-10}\ mol^2\ dm^{-6}$$

2. For magnesium fluoride the equilibrium is

$$MgF_2(s) \rightleftharpoons Mg^{2+}(aq) + 2F^-(aq)$$

$$K_s(MgF_2) = [Mg^{2+}(aq)] \times [F^-(aq)]^2 = 6.4 \times 10^{-9}\ mol^3\ dm^{-9}$$

Note that the fluoride ion concentration is *squared* in the last solubility product expression because two F^- ions are produced when one MgF_2(s) dissociates in water.

The solubility products of selected compounds are listed in Table 11.4. Solubility product expressions can be used only for sparingly soluble salts. They cannot be used (without modification) for very soluble salts (like sodium chloride) because the concentration of ions is so high that the ions influence each other and the effective concentration of ions is lower than their concentration in mol dm^{-3}. The errors in using similar equations for slightly soluble ionic substances (such as calcium hydroxide) are smaller, but are still significant in accurate work. (Now try Exercise 11E.)

Table 11.4 Solubility products of sparingly soluble ionic substances. The units of concentration are mol dm^{-3}.

Salt	K_s at 25 °C
Silver chloride AgCl	1.6×10^{-10}
Barium sulfate $BaSO_4$	1.1×10^{-10}
Magnesium fluoride MgF_2	6.4×10^{-9}
Calcium carbonate $CaCO_3$	8.7×10^{-9}
Copper(I) iodide CuI	5.1×10^{-12}
Iron(II) hydroxide $Fe(OH)_2$	1.6×10^{-14}
Aluminium hydroxide $Al(OH)_3$	1.0×10^{-33}
Zinc hydroxide $Zn(OH)_2$	2.0×10^{-27}
Calcium sulfate $CaSO_4$	2.4×10^{-5}
Lead(II) chloride $PbCl_2$	1.6×10^{-5}

Exercise 11E

Solubility product

Write down solubility product expressions for the following sparingly soluble compounds and state their units:

(i) copper(I) iodide

(ii) aluminium hydroxide.

Relationship between solubility product and molar solubility

The dissolution of silver chloride,

$$AgCl(s) \rightleftharpoons Ag^{+}(aq) + Cl^{-}(aq)$$

shows that the molar solubility of AgCl(s), *s*(AgCl), equals either of the ion concentrations:

$$[Cl^{-}(aq)] = [Ag^{+}(aq)] = s(AgCl)$$

or,

$$K_s(AgCl) = [Ag^{+}(aq)] \times [Cl^{-}(aq)] = s(AgCl) \times s(AgCl) = s(AgCl)^2$$

Rearranging for *s*(AgCl) gives

$$s(AgCl) = \sqrt{K_s(AgCl)}\ \text{mol dm}^{-3}$$

At 25 °C, $K_s(AgCl) = 1.6 \times 10^{-10}\ \text{mol}^2\ \text{dm}^{-6}$, so that:

$$s(AgCl) = \sqrt{1.6 \times 10^{-10}}\ \text{mol dm}^{-3}$$
$$= 1.3 \times 10^{-5}\ \text{mol dm}^{3}$$

(compare Example 11.1).

Exercise 11F

Solubility product and molar solubility

Use the data in Table 11.4 to calculate the molar solubility of:

(i) barium sulfate and

(ii) magnesium fluoride at 25 °C.

Use of solubility products in predicting whether or not precipitation will take place: the common ion effect

For AgCl(s) at 25 °C,

$$[Ag^{+}(aq)] \times [Cl^{-}(aq)] = K_s(AgCl) = 1.6 \times 10^{-10}\ \text{mol}^2\ \text{dm}^{-6}$$

This expression tells us that in a saturated solution of AgCl(s) the silver and chloride ion concentrations can take on any values **provided** that when these concentrations are multiplied together they equal 1.6×10^{-10}.

We can decide whether or not a salt precipitates in water by substituting the **actual** ion concentrations into the solubility product expression for that salt:

1. If the product of the actual ion concentrations is (temporarily) **greater** than the solubility product, some of the salt precipitates out in order to restore the product of ion concentrations to the value of K_s (here, 1.6×10^{-10}).
2. If the product of the ion concentrations **exactly equals** the solubility product, the solution is saturated with salt, but no precipitation takes place.
3. If the product of the ion concentrations is **less than** the solubility product, the ions remain in solution, the solution is unsaturated and no precipitation takes place.

Suppose we add a minute grain of silver chloride to some water so that there is 1.0×10^{-5} mol of AgCl(s) in 1 dm^{-3} of water. Will all the silver chloride dissolve?

The easiest way to approach this question is to *suppose* that all the AgCl(s) *did* completely dissolve and then to decide whether or not the resulting ion concentrations would exceed the solubility product of AgCl(s).

One AgCl(s) dissociates into one $Ag^{+}(aq)$ and one $Cl^{-}(aq)$ ion, so that

$$[Ag^{+}(aq)] = [Cl^{-}(aq)] = 1.0 \times 10^{-5}\ \text{mol dm}^{-3}$$

The product of the *actual* ion concentrations would be

$$[Ag^{+}(aq)] \times [Cl^{-}(aq)] = 1.0 \times 10^{-5} \times 1.0 \times 10^{-5} = 1.0 \times 10^{-10}\ \text{mol}^2\ \text{dm}^{-6}$$

At 25°C, $K_s(AgCl) = 1.6 \times 10^{-10}\,mol^2\,dm^{-6}$, so that the product of the ion concentration is *smaller* than K_s. No precipitation takes place, and so we conclude that all the AgCl(s)remains in solution. In other words, the grain of silver chloride completely dissolves in water.

Now suppose that we add a few crystals of solid sodium chloride (NaCl) to the same solution of AgCl. The chloride ions from the sodium chloride flood into the solution. Suppose that the concentration of Cl^- in the mixture from the NaCl is $1.0 \times 10^{-2}\,mol\,dm^{-3}$. The product of the ion concentrations for silver chloride is now

$$[Ag^+(aq)] \times [Cl^-(aq)] = \underbrace{(1.0 \times 10^{-5})}_{\text{from AgCl only}}\,\underbrace{(1.0 \times 10^{-5} + 1.0 \times 10^{-2})}_{\text{from AgCl plus NaCl}}$$

Compared with 1.0×10^{-2}, 1.0×10^{-5} is small, and makes a negligible contribution to the chloride ion concentration. We therefore take $1.0 \times 10^{-5} + 1.0 \times 10^{-2}$ as being equal to 1.0×10^{-2}, and

$$[Ag^+(aq)] \times [Cl^-(aq)] = (1.0 \times 10^{-5})(1.0 \times 10^{-2}) = 1.0 \times 10^{-7}$$

1.0×10^{-7} is greater than 1.6×10^{-10}. Because the ion product exceeds K_s, virtually all the silver ions precipitate out as silver chloride.

The effect of swamping a solution with one of the ions in the solubility product expression is called the **common ion effect**. The addition of a common ion *reduces the molar solubility* of the sparingly soluble salt and so most of the salt precipitates out. In our example, we can prove that the molar solubility of AgCl has fallen, as follows.

We start by remembering that, at 25°C,

$$[Ag^+(aq)] \times [Cl^-(aq)] = 1.6 \times 10^{-10}\,mol^2\,dm^{-6}$$

Upon the addition of NaCl,

$$[Cl^-(aq)] = 1.0 \times 10^{-2}\,mol\,dm^{-3}$$

We then write

$$[Ag^+(aq)] \times (1.0 \times 10^{-2}) = 1.6 \times 10^{-10}\,mol^2\,dm^{-6}$$

or,

$$[Ag^+(aq)] = \frac{1.6 \times 10^{-10}}{1.0 \times 10^{-2}} = 1.6 \times 10^{-8}\,mol\,dm^{-3} = s$$

so that the molar solubility of AgCl is now $1.6 \times 10^{-8}\,mol\,dm^{-3}$. Before adding the NaCl it was $1.3 \times 10^{-5}\,mol\,dm^{-3}$ (page 179). In other words, AgCl is about 800 times less soluble in $0.01\,mol\,dm^{-3}$ NaCl solution than in pure water.

The common ion effect is also observed with very soluble salts, although the arithmetic is more complicated because the concentrations in the solubility product expression need to be modified in order to apply in these cases. A dramatic example of the common ion effect is seen by making a saturated solution of sodium chloride and adding a few drops of concentrated hydrochloric acid. The chloride ions from the hydrochloric acid flood into the salt solution, the solubility product of NaCl is exceeded, and salt appears as a white precipitate (Fig. 11.4).

Example 11.2 considers a slightly different case, where the relevant ions are introduced by adding two very soluble ionic substances (iron(II) sulfate and sodium hydroxide) to water.

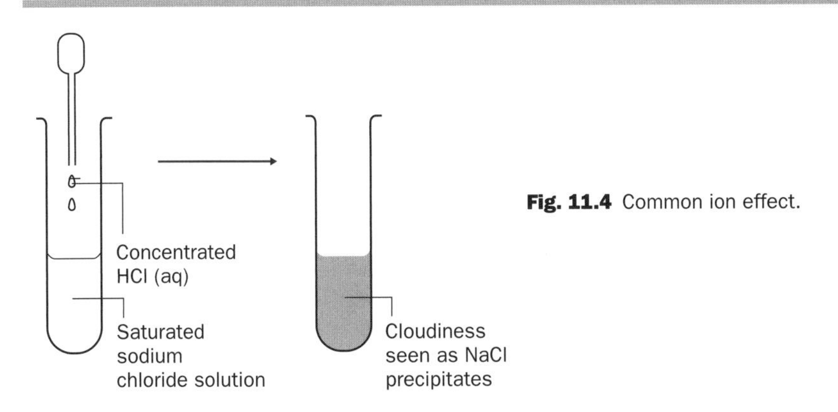

Fig. 11.4 Common ion effect.

Example 11.2

25.0 cm³ of a 0.0020 mol dm⁻³ solution of iron(II) sulfate was added to 25.0 cm³ of a 0.0040 mol dm⁻³ solution of sodium hydroxide. Will iron(II) hydroxide precipitate?

▶ Answer

For iron(II) hydroxide

$$K_s(Fe(OH)_2) = [Fe^{2+}(aq)] \times [OH^-(aq)]^2$$

(Notice that the hydroxide ion concentration is squared because iron(II) hydroxide produces two hydroxide ions when it dissociates.)

total volume of solution = $25.0\,cm^3 + 25.0\,cm^3 = 50.0\,cm^3$

Iron(II) sulfate = $FeSO_4$; the concentration of Fe^{2+}(aq) ion will be $0.0010\,mol\,dm^{-3}$ because of the dilution from $25.0\,cm^3$ to $50.0\,cm^3$.

Sodium hydroxide = NaOH; the concentration of OH^-(aq) ion will be $0.0020\,mol\,dm^{-3}$ because of the dilution from $25\,cm^3$ to $50\,cm^3$.

The product of the ion concentrations is then

$$[Fe^{2+}(aq)] \times [OH^-(aq)]^2 = (0.0010) \times (0.0020)^2 = 4.0 \times 10^{-9}\,mol^3\,dm^{-9}$$

This is very much greater than the solubility product of $Fe(OH)_2$ ($1.6 \times 10^{-14}\,mol^3\,dm^{-9}$, Table 11.4) and so iron(II) hydroxide will precipitate out.

Exercise 11G

Precipitation

(i) A solution of volume $10.0\,cm^3$ contains magnesium ions at a concentration of $0.0010\,mol\,dm^{-3}$, and fluoride ions at a concentration of $0.0030\,mol\,dm^{-3}$. Will magnesium fluoride precipitate? What happens if $90\,cm^3$ of pure water is then added to the solution?

(ii) $5.0\,cm^3$ of $2.0 \times 10^{-4}\,mol\,dm^{-3}$ barium chloride is mixed with $5.0\,cm^3$ of $4.0 \times 10^{-4}\,mol\,dm^{-3}$ sodium sulfate solution. Will barium sulfate precipitate out?

11.4 Distribution of a solute between two solvents

Distribution ratio

If a solute is soluble in two solvents, and the solvents are immiscible, it is found that the concentration of solute in the one layer is always bigger than the concentration of solute in the other layer by a fixed amount. This ratio (symbolized K_d and known as the **distribution** or **partition ratio**) is a constant at a particular temperature:

$$K_{d(T)} = \frac{\text{concentration of solute in solvent A}}{\text{concentration of solute in solvent B}}$$

where the subscript T reminds us that the ratio is fixed only at that particular temperature. This equation does not hold if the solute dissociates, associates (such as forming pairs of hydrogen bonded molecules) or otherwise chemically reacts with the solvents present.

Our first example is the distribution of solid iodine (the solute) between water and tetrachloromethane (the solvents). If we add a few crystals of iodine to some water and tetrachloromethane (CCl_4) in a flask and shake the mixture well, the iodine colours the water a pale brown colour and the tetrachloromethane a purple colour. There is more iodine in the organic layer (the tetrachloromethane layer) than in the aqueous layer (the water layer). At 25 °C experiments show that

$$K_d(I_2) = \frac{\text{concentration of } I_2 \text{ in } CCl_4}{\text{concentration of } I_2 \text{ in } H_2O} = 85$$

K_d is unitless, because the units of concentration cancel out.

It is important to realize that the value of K_d at one temperature is the same whether we add 0.1 g, 1 g, or 2 g (etc.) of iodine to the mixture of solvents. This is because the partitioning of a solute between two solvents is another example of a dynamic equilibrium in which the relative concentration of the iodine in the two phases is constantly being maintained:

$$I_2(\text{in } H_2O) \rightleftharpoons I_2(\text{in } CCl_4)$$

If we add iodine to *one* layer and shake, the iodine concentration in both layers alters so as to maintain the value of K_d.

At its simplest, the distribution ratio allows us to calculate the concentration of solute in one layer, knowing the concentration of solute in the other. For example, if the concentration of iodine in the aqueous layer was determined by titration with

Exercise 11H

Distribution ratio

The distribution ratio of the gas sulfur dioxide (SO_2) between water and trichloromethane (chloroform) is about 0.90 at room temperature.

In an experiment, sulfur dioxide was bubbled through a mixture of chloroform and water. At equilibrium the concentration of SO_2 in the chloroform layer was 0.10 mol dm^{-3}. What is the concentration of sulfur dioxide in the water layer?

thiosulfate ion and found to be 1.0×10^{-3} mol dm^{-3}, then the concentration of iodine in the tetrachloromethane layer must be 85 times as great:

$$\text{concentration of } I_2 \text{ in } CCl_4 \text{ layer} = 85 \times 1.0 \times 10^{-3} = 8.5 \times 10^{-2}\,\text{mol dm}^{-3}$$

Solvent extraction

Frequently in organic chemistry, it is necessary to separate an organic compound from a watery mixture. This is done by adding an organic solvent (such as dichloromethane or ethoxyethane) and shaking – a process called **solvent extraction**. The organic solvent contains most of the desired organic compound (the solute) and the compound is obtained by evaporation of the organic layer. The arithmetic of solvent extraction is discussed in Chapter 19.

Chromatography

One of the most important ways of separating mixtures, known as **chromatography**, is really a series of solvent extractions. More details are given in Chapter 19.

11.5 Solubility of gases in water

The solubilities of common gases are given in Table 11.5. Some gases (e.g. H_2, O_2, N_2, CH_4) are only slightly soluble in water. The dissolution of these gases may be represented by the equation

$$\text{gas molecules} \xrightleftharpoons{\text{water}} \text{dissolved molecules}$$

The dissolved gas may be driven out of solution by heating.

Other gases (e.g. NH_3, HCl, NO_2) **chemically react** with water. These gases are generally more soluble than H_2, O_2, etc. The dissolution of such gases may be thought of as occurring in two stages:

$$\text{gas molecules} \xrightleftharpoons{\text{water}} \text{dissolved molecules} \xrightleftharpoons{\text{water}} \text{products}$$

Table 11.5 Solubilities of common gases at 0 °C. The pressure of the gases in contact with the water is 1 atm

Gas	Solubility of gas/m^3 of gas per m^3 of water
Helium He	0.0094
Hydrogen H_2	0.021
Nitrogen N_2	0.024
Carbon monoxide CO	0.035
Oxygen O_2	0.049
Methane CH_4	0.054
Argon Ar	0.056
Carbon dioxide CO_2	1.7
Chlorine Cl_2	4.6
Hydrogen sulfide H_2S	4.7
Sulfur dioxide SO_2	80
Hydrogen chloride HCl	506
Ammonia NH_3	1300

for example,

$$NH_3(g) \xrightleftharpoons{\text{water}} NH_3(aq)$$

followed by

$$NH_3(aq) + H_2O(l) \rightleftharpoons NH_4^+(aq) + OH^-(aq)$$

Variation of the solubility of gases with the partial pressure of the gas

How does the solubility of gases vary with the pressure of the dissolving gas? Because a gas at high pressure will be 'pushed' more into the water than a gas at low pressure, we predict that the greater the (partial) pressure of the dissolving gas the greater the concentration of gas in solution. This turns out to be the case for all gases:

The solubility of a gas increases with its partial pressure.

The data in Table 11.5 refer to the case when the partial pressure of the dissolving gas is 1 atm (101 kPa). Notice that ammonia is the most soluble of all the gases listed.

Measurements of the partial pressure of a gas and its equilibrium concentration in solution may be plotted graphically. Some gases (Fig. 11.5) give straight line plots and obey the equation

$$c = K_H \times p$$

where c is the concentration of gas in solution (in mol dm^{-3}), p is the partial pressure of the dissolving gas (in atm) and K_H is a constant (with the units mol dm^{-3} atm^{-1}). This equation is the mathematical expression of **Henry's law** which may be stated as follows:

The equilibrium concentration of a gas in a solvent at a particular temperature is proportional to the partial pressure of the dissolving gas.

Table 11.6 Selected Henry's law constants for water at 25 °C

Gas	K_H/mol dm^{-3} atm^{-1}
O_2	1.28×10^{-3}
CO_2	3.38×10^{-2}
H_2	7.90×10^{-4}
CH_4	1.34×10^{-3}
N_2	6.48×10^{-4}
NO	2.0×10^{-4}

Source: *Environmental Chemistry*, S.E. Manahan

K_H values are called Henry's law constants (Table 11.6). They vary with temperature.

Most gases which chemically react with water only follow Henry's law at *very* low partial pressures (for SO_2, less than 0.001 atm). However, since calculations involving gas solubilities often involve gases (such as SO_2 and CO_2) which occur at trace levels in the atmosphere, Henry's law constants are still useful.

The solubility of a gas is unaffected by the presence of other gases unless the total pressure of gas (i.e. dissolving gas added to other gases) is above about 1 atm (101 kPa) pressure. Above this pressure, *all* gases show significant deviations from Henry's law because the gas mixture behaves nonideally.

For examples of calculations involving Henry's Law see Appendix 11 in the website.

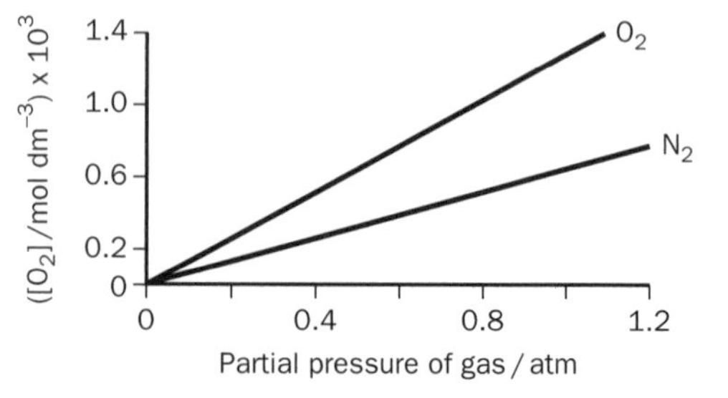

Fig. 11.5 The solubilities of O_2 and N_2 at different partial pressures at 25 °C. The slopes are equal to the Henry's law constants for these gases.

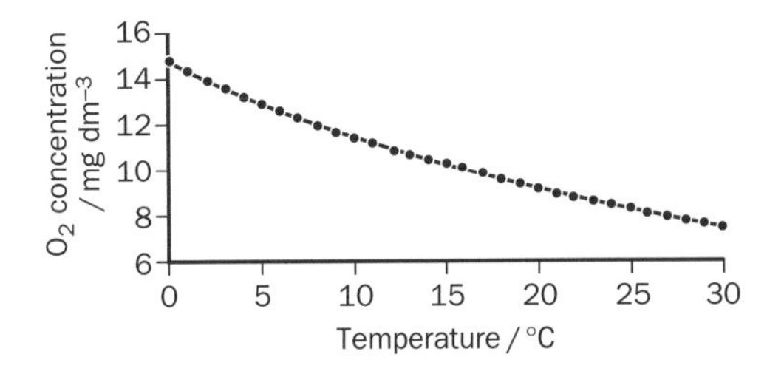

Fig. 11.6 Variation of the concentration of dissolved oxygen in pure water with temperature.

Variation of the solubility of gases with the temperature of the solution

Figure 11.6 shows the concentration of dissolved oxygen in water (in equilibrium with dry air containing 20.9% O_2 by volume at 1 atm total pressure) at different temperatures. As expected, the concentration of dissolved O_2 falls with increasing temperature. The concentrations vary from 14.6 to 7.6 mg dm^{-3} between 0 and 30 °C. This means that over this temperature range, 1 dm^3 of water contains between 10 and 6 cm^3 of dissolved oxygen gas.

The presence of dissolved solids in water further reduces the solubility of oxygen. Sea water contains roughly 10 000 mg dm^{-3} NaCl. At 25 °C, this reduces the solubility of oxygen from 8.4 to 7.6 mg dm^{-3}. Typically, fish growth is inhibited below about 6 mg dm^{-3} of dissolved oxygen. (See page 418.)

Demonstrations involving gas solubility

1. Lemonade bottle

A bottle of lemonade is essentially a solution of carbon dioxide in flavoured water and so Henry's law applies:

$$c(CO_2) = K_H \times p_{CO_2}$$

where $c(CO_2)$ is the concentration of CO_2 dissolved in the water. When a new bottle of lemonade is opened, some of the carbon dioxide gas above the liquid escapes. Since p_{CO_2} falls, so does the concentration of CO_2 in the water – the CO_2 is lost as bubbles which rush out of the water.

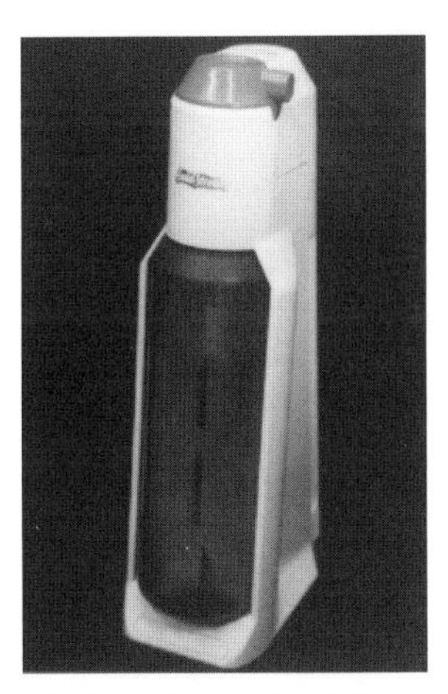

CO_2 fizzy drinks maker. CO_2 is injected from the gas cylinder at above atmospheric pressure. The dissolved CO_2 makes the drinks fizzy, and contributes to the tangy taste.

2. Carbonating water

If we try and carbonate a bottle of water which contains drink concentrate it will spray everywhere. This is because the presence of concentrate in the water reduces the solubility of the carbon dioxide – the excess CO_2 has nowhere to go and so it forces some of the liquid out from the bottle. A less dramatic effect is observed if you carbonate water first and add concentrate later – but note the release of CO_2 as the concentrate is added.

3. Diving

As divers descend in water, the pressure from the column of sea water above them increases. To counteract this, the divers' regulators provide them with air from cylinders at a pressure equal to the pressure from the surrounding sea. This means that scuba divers breathing from a cylinder of air (containing 21% O_2 and the rest N_2) will be breathing higher partial pressures of oxygen underwater than they would be at sea level. This can cause CO_2 poisoning because the higher oxygen pressure triggers less frequent breathing which allows CO_2 levels in the diver's blood to build up. To match this, divers use 'air' mixtures which contain smaller percentages of oxygen, the exact percentage depending on the depth at which they operate.

Operating underwater (at higher pressure) causes more gas (notably nitrogen) to dissolve in the blood. If the diver returns to the surface too quickly, the nitrogen bubbles out from the blood, and the diver suffers 'the bends'. The bends can be reduced by replacing the nitrogen in the diver's cylinder with helium gas which (as shown in Table 11.5) is less soluble in water. Even so, divers often have to allow their bodies to slowly equilibrate with the gases they breathe. This is accomplished by slowly descending or ascending.

4. Floating candle

If we float a candle on water and cover the candle with a jam jar, the candle eventually goes out. During this period, the water level rises (Fig. 11.7).

The burning of the candle consumes oxygen. Therefore, we might expect the volume of air to be reduced by about 21%, i.e. for the water level to rise by about one-fifth. The observed reduction in volume is much less than this because (i) the candle goes out before the oxygen gas contained in the jar is exhausted, and (ii) the carbon dioxide that is produced in the burning is only slightly soluble in water. (If CO_2 were *completely insoluble*, the production of one mole of CO_2 for every mole of O_2 consumed would produce no change in the overall volume of gas in the jar.)

5. Fountain experiment

Very soluble gases (such as ammonia and hydrogen chloride) can be made to perform the 'fountain experiment'. Figure 11.8 shows the experiment carried out with HCl(g). A flask is fitted with a cork and tube and filled with dry hydrogen chloride. Without delay, the flask is lowered into a deep bucket of water. So much gas dissolves in the water rising up the tube that a partial vacuum is created in the flask. This draws

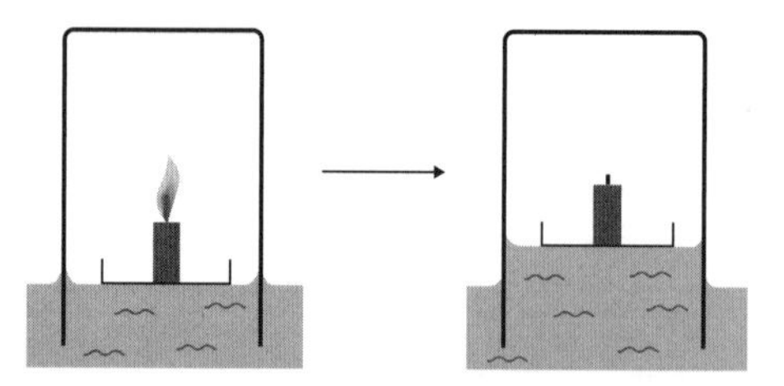

Fig. 11.7 A burning candle consumes oxygen.

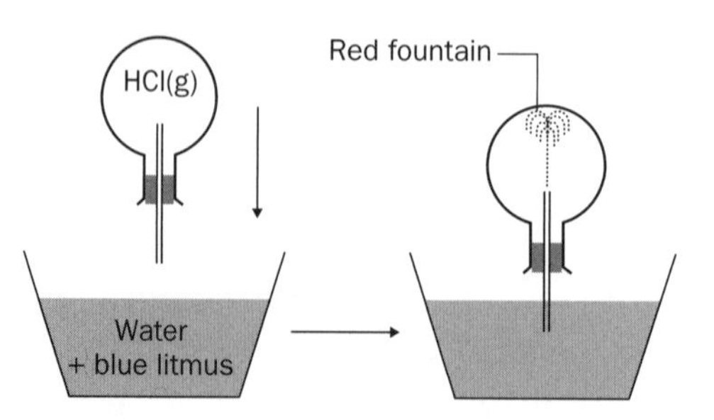

Fig. 11.8 The fountain experiment.

up water into the flask so fast that it sprays like a fountain. If blue litmus is added to the bucket, the water entering the flask turns red because of the hydrochloric acid formed and a red fountain is observed. If the experiment is repeated using ammonia, red litmus is used and the colour change is red to blue.

11.6 Osmosis

Example of osmosis

When red blood cells are put into water they burst. What is happening? The cell membrane allows only water to pass through. The fluid in the red cells contains a mixture of organic compounds and inorganic ions, and is therefore more concentrated than the water outside. Water enters through the cell walls in an attempt to dilute the solution within the cell. Eventually, so much water enters that the cell wall bursts apart.

If fresh red blood cells are placed in a 2% solution of salt, the cells shrink. In this case, the water within the cells enters the more concentrated solution outside the cell, and the cells dehydrate.

Key points about osmosis

1. Osmosis is the passage of a *solvent* (usually water) from a zone of low concentration to one of high concentration. The solution of higher concentration is said to be **hypertonic** ('hyper-' mean more), whereas the solution of lower concentration is said to be **hypotonic** ('hypo-' means less). After osmosis is complete, the solutions are equally concentrated and are said to be **isotonic** ('iso-' means the same). An example of osmosis is shown in Fig. 11.9.
2. The membrane allows solvent to pass through but not solute; it is called a **semipermeable membrane**.
3. The movement of solvent from low to high concentration is similar to the spreading out of a gas (diffusion) from high gas pressure to low pressure. The idea of pressure is also applicable to osmosis, and the **osmotic pressure** may be thought of as the force acting per unit area of membrane which drives solvent molecules through the membrane.

Semipermeable membranes may be natural or artificial. No membrane is semipermeable to all solutes. Some membranes keep out ions (such as Na^+ and Cl^-)

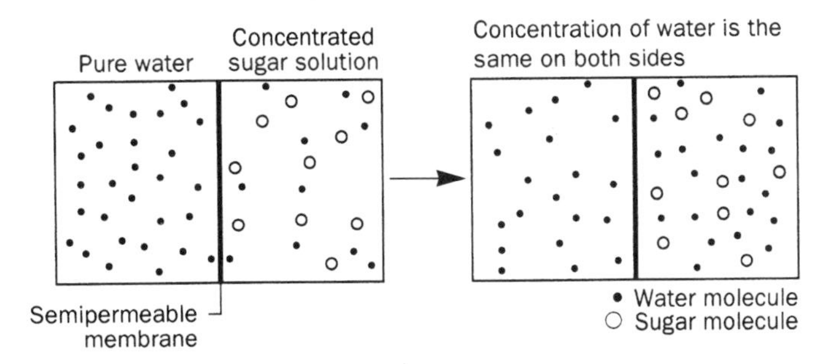

Fig. 11.9 Osmosis involves the equalization of concentrations across a semipermeable membrane.

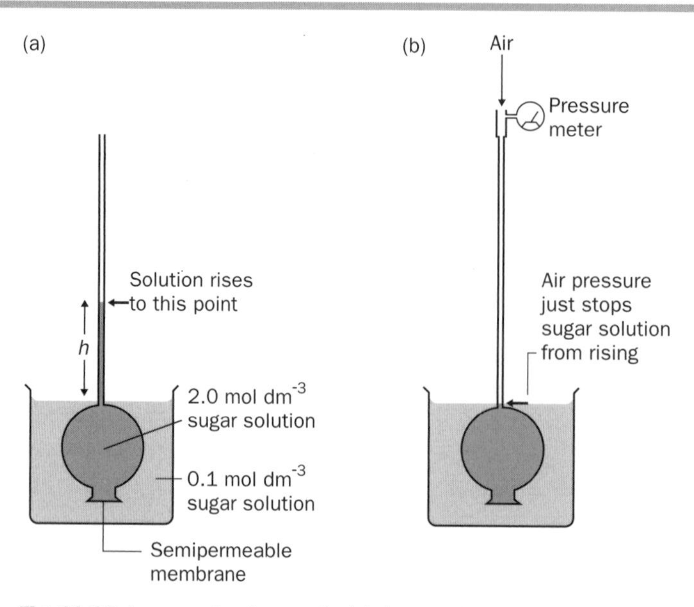

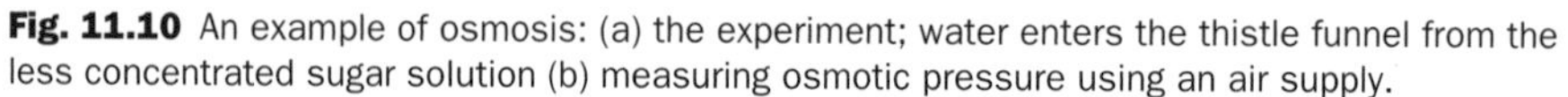

Fig. 11.10 An example of osmosis: (a) the experiment; water enters the thistle funnel from the less concentrated sugar solution (b) measuring osmotic pressure using an air supply.

whereas others will keep out only larger molecules (such as metal-complex ions or sugars). The exact way that semipermeable membranes work is still poorly understood. A common semipermeable membrane is cellulose acetate, which will not allow sugar molecules or hydrated ions (made bulky by being attached to water molecules) to pass through.

More about osmotic pressure

Osmotic pressure will be greatest at the start of an osmosis experiment, when the concentration difference across the membrane is greatest. We should use the phrase 'initial osmotic pressure', but since the initial osmotic pressure is the only one in which we are interested we shall drop the prefix 'initial'.

Figure 11.10(a) illustrates a simple laboratory osmosis experiment. At the end of the osmosis process the solution inside the thistle funnel has stopped rising.

We cannot easily use the height the sugar solution has risen during the osmosis as a measure of osmotic pressure. This is because the solution in the thistle funnel is less concentrated at the end of the experiment than at the beginning. Since a less concentrated solution is a less dense one, a simple measurement of h overestimates the osmotic pressure.

A better way of measuring osmotic pressure is to apply a known amount of external pressure until the osmosis just stops. The applied pressure then exactly equals the osmotic pressure of the solution. The *principle* of this type of measurement is shown in Fig. 11.10(b). We start by connecting the top of the thistle funnel to a high pressure air supply. The pressure of the air supply is slowly increased so as to just stop the osmosis – this air pressure is equal to the osmotic pressure across the membrane.

Calculating osmotic pressure

The osmotic pressure π may be calculated using the equation

$$\pi = \Delta c \times RT$$

where Δc is the *difference* in the concentration of solute between the two solutions used in the experiment, R is the universal gas constant (present in the expression even though we are talking about solutions!), and T the temperature of the solutions. Solutions which obey this equation are said to be behaving ideally, and this is usually true when the solute concentrations are low.

The key points about this equation are:

1. The osmotic pressure of a solution is unaffected by the membrane used. The only assumption we make is that the membrane allows solvent to pass but not solute. However, the equation says nothing about the speed (rate) at which the equalization of concentration occurs. The speed will be dependent upon the type and dimensions of the membrane.
2. The osmotic pressure π depends upon the difference in *molar* concentration, i.e. the difference in the numbers of solute molecules in equal volumes of both solutions. The type of solute (sugar, salt, or a particular protein) is irrelevant. This reminds us of the ideas behind the model of ideal gases, where gas pressure is independent of the type of gas. However, the equation does assume that the solute does not dissociate or otherwise react with the solvent, since this would alter the number of particles per unit volume of solution.
3. Units – as with gases, we use pressure in units of Pa ($\mathrm{N\,m^{-2}}$), and c will have units of moles per cubic metre ($\mathrm{mol\,m^{-3}}$). Remember that $1000\,\mathrm{mol\,m^{-3}} \equiv 1\,\mathrm{mol\,dm^{-3}}$.
4. Although water is usually the solvent, the osmotic pressure does not depend on the type of solvent used.

Example 11.3

A solution of sugar of concentration $0.30\,\mathrm{mol\,dm^{-3}}$ is separated from pure water by a semipermeable membrane.

(i) Calculate the osmotic pressure at 20.0 °C.
(ii) What would be the osmotic pressure if the water was replaced by a sugar solution of concentration $0.10\,\mathrm{mol\,dm^{-3}}$?

▶ Answer

(i) The difference in concentration is

$$\Delta c = 0.30 - 0.00\ (\text{water}) = 0.30\,\mathrm{mol\,dm^{-3}} = 0.30 \times 10^3\,\mathrm{mol\,m^{-3}}$$

$$\pi = \Delta cRT = 0.30 \times 10^3\,\mathrm{mol\,m^{-3}} \times 8.3145\,\mathrm{J\,mol^{-1}\,K^{-1}} \times 293\,\mathrm{K}$$

$$= 7.3 \times 10^5\,\mathrm{Pa}\ (\approx 7.3\,\mathrm{atm})$$

(ii) The difference in concentration is

$$\Delta c = 0.30 - 0.10 = 0.20\,\mathrm{mol\,dm^{-3}} = 0.20 \times 10^3\,\mathrm{mol\,m^{-3}}$$

$$\pi = \Delta cRT = 0.20 \times 10^3\,\mathrm{mol\,m^{-3}} \times 8.3145\,\mathrm{J\,mol^{-1}\,K^{-1}} \times 293\,\mathrm{K}$$

$$= 4.9 \times 10^5\,\mathrm{Pa}\ (\approx 4.9\,\mathrm{atm})$$

Example 11.3 *(continued)*

▶ Comment

The osmotic pressures developed by these solutions are surprisingly high. This means that the osmotic pressure of very dilute solutions may be measured. This is useful in determining the molecular mass of polymers and large biologically important molecules which are not usually very soluble in any solvent.

Exercise 11I

Osmotic pressure

Calculate the average molecular mass of a protein given that a solution containing 0.25 g of the protein in 1.0 dm^3 of water produced an osmotic pressure (taken against pure water) of 91 Pa at exactly 300 K.

11.7 Colloids

What is a colloid?

A colloid is a mixture of solute and solvent in which the particles of solute are intermediate in size between those found in true solutions (such as sugar in water) and suspensions (such as chalk or sand in water).

Examples of colloids include tea, starch and milk in water, gelatine (a protein) in water, aluminium hydroxide 'gel', and soapy water. Colloids are sometimes called **colloidal solutions**, **colloidal suspensions** or **colloidal dispersions**.

Key points about colloids

1. The particles in a colloid have diameters in the range 1–1000 nm. Such particles are too small to be seen by optical microscopes. The particles are either very large molecules (such as proteins) or else aggregates of molecules or ions.
2. The particles of colloids do not settle at the bottom of the container in the same way that suspensions do, nor can they be filtered off using ordinary filter papers. However, colloids can be separated from true solutions using a special membrane – such a separation is called **dialysis**.
3. Colloids can be distinguished from true solutions by shining light through the solution. The relatively large particles of a colloid scatter the light. This is known as the **Tyndall effect** (Fig. 11.11). Car headlights in fog is a familiar example of this effect.

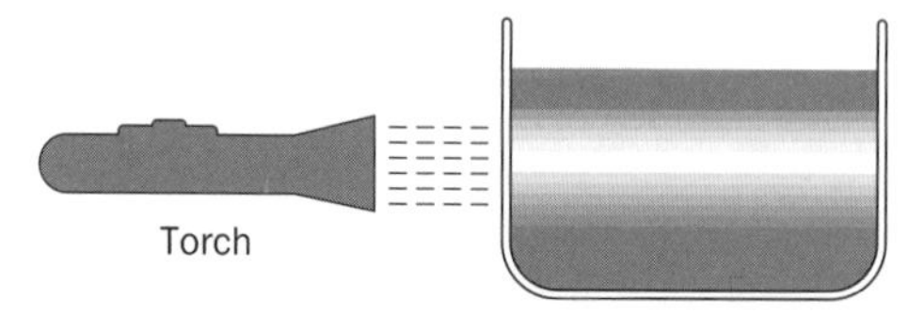

Fig. 11.11 The Tyndall effect.

Table 11.7 Classification of common types of colloid

Colloid type	Composition	Examples
Sol	solid particles dispersed in a liquid	most paints, starch and tea in water, inks, soapy water, many proteins in water
Gel (or solid emulsion)	continuous network of solid particles dispersed in a liquid	jellies, gelatine
Aerosol	solid or liquid dispersed in a gas	smoke in a room, fog
Foam	gas dispersed in liquid	beer foam, soapsuds, whipped cream and shaving foam
Emulsion	liquid dispersed in liquid	milk, mayonnaise, butter

Classification of colloids

Colloids do not only involve solids and liquids. The classification of common types of colloid is shown in Table 11.7.

Emulsions

Dispersions of liquids in liquids are important commercially, and occur in foods and in living organisms.

Suppose we violently shake 1 cm^3 of oil with 10 cm^3 of water for a few seconds. Upon stopping shaking, the oil remains in the form of tiny droplets for a fraction of a second before the mixture reverts to two layers again. The purpose of **emulsification** (also called **solubilization**) is to ensure that the solute (here the oil) remains dispersed permanently in the form of tiny droplets. To achieve this, we add an **emulsifying agent**.

Milk is an emulsion of butterfat droplets in an aqueous mixture of sugars, mineral salts and proteins. Ice cream is another emulsion with a similar composition. The emulsifying agents in both emulsions are proteins with smaller amounts of phospholipids, mainly lecithin. In egg yolk, the emulsifying agents are phospholipids and cholesterol.

Emulsions play an important role in the digestion of fats in the intestines. In the small intestine, fats are emulsified by the alkaline secretions of the pancreas and the liver. Why is it necessary to emulsify fats before they can be digested? The answer is that the rate at which fats are digested depends upon the surface area of fat 'solution' exposed to the intestines. For example, if 1 cm^3 of oil is dispersed to make oil droplets of 5 nm diameter, the total area of all the drops is about 1200 m^2! Emulsification spreads out the oil so effectively that enzymes can break down fats more rapidly.

Soap

The colloidal particles of soap and other detergents are called **ionic micelles**. Such micelles typically contain 50–100 ions stuck together (see page 175). Ionic micelles are important in explaining the emulsification of fats and oils by soaps and detergents. Apart from its use in washing clothes, soaps are used as emulsifying agents in many cosmetics.

Revision questions

11.1. Which of the following are true?

(i) Water is more soluble in hexane than hexane is soluble in water.

(ii) Long-chained organic molecules are usually more soluble in water than shorter-chained molecules.

(iii) The osmotic pressure of a solution depends upon the molecular mass of the solute.

(iv) The hydrophobic end of a molecule is attracted to water molecules.

(v) An aerosol is defined as a dispersion of a solid in a gas.

11.2. Use Table 11.2 to find out whether the following pairs of liquids are immiscible or miscible, in each case explaining your answer by considering the intermolecular forces involved:

(i) Propan-2-ol and water.

(ii) Propanone and methyl benzene.

(The molecular formulae are listed in Table 11.1.)

11.3. The solubility of sodium chloride (in grams NaCl per 100 g solvent) at 25 °C in three solvents (A, B and C) is (A) 36.1 g, (B) 1.3 g and (C) 0.1 g. Which solvent is likely to be the least polar? Explain your answer.

11.4. $K_s(CaCO_3) = 8.7 \times 10^{-9}\,mol^2\,dm^{-6}$ at 25 °C.

(i) Calculate **(a)** the molar solubility and **(b)** the solubility in $g\,dm^{-3}$ of solution, of calcium carbonate at 25 °C.

(ii) A test-tube containing 2.0×10^{-5} g of calcium chloride in $10.0\,cm^3$ of aqueous solution was added to $10.0\,cm^3$ of $2.0 \times 10^{-5}\,mol\,dm^{-3}$ sodium carbonate solution at 25 °C. Will calcium carbonate precipitate?

11.5. A saturated solution of calcium oxalate ($Ca(C_2O_4)$) was found to contain $4.47 \times 10^{-5}\,mol\,dm^{-3}$ of calcium ion at 25 °C. What is the solubility product of calcium oxalate at this temperature? (The oxalate ion is $C_2O_4^{2-}$.)

11.6. The distribution of a drug between the oil octanol and water is sometimes used to mimic the absorption of drugs from the aqueous solution in the stomach into the (lipid) stomach wall. If the K_d (octanol/water) for a drug was 1000 at 25 °C, and the concentration of drug in the stomach (i.e. the aqueous layer) was $0.01\,mg\,dm^{-3}$, estimate the equilibrium concentration of drug in the stomach wall.

11.7. The solubility of chlorine gas at 0 °C and an air pressure of 1 atm is about $4.6\,m^3$ of gas per cubic metre of water. Assuming no subsequent reaction, estimate the concentration of dissolved gas (in $mol\,dm^{-3}$) in a saturated solution under these conditions. (The molar volume of an ideal gas at 0 °C and 1 atm pressure is $22.4\,dm^3$.)

11.8. The solubility of oxygen gas in water at a pressure of 1 atm is 14.6 and $7.6\,mg\,dm^{-3}$ at 0 °C and 30 °C, respectively. Show that these concentrations correspond to about 10 and $6\,cm^3$ of dissolved oxygen gas per dm^3 of water. (Take the molar volume of oxygen as $22.4\,dm^3$ at 0 °C and as $24.9\,dm^3$ at 30 °C.)

11.9. (i) Calculate the osmotic pressure at exactly 25 °C produced by a $0.0050\,mol\,dm^{-3}$ aqueous solution of a protein separated from pure water by a semipermeable membrane.

(ii) Use the equation $PV = nRT$ to calculate the gas pressure exerted by a sample of gas of concentration $0.0050\,mol\,dm^{-3}$ at exactly 25 °C. Compare your answer with that obtained in **(i)**.

11.10. (i) 5.0 g of polystyrene was dissolved in exactly $200\,cm^3$ methyl benzene (toluene) and its osmotic pressure measured (against pure methyl benzene) as 0.0175 atm at 25 °C. Calculate the average molecular mass of the polystyrene polymer.

(ii) Calculate the osmotic pressure (against pure water) of a solution of a polypeptide of average molecular mass 100 000 u, which contains 0.100 g of polypeptide dissolved in $1.00\,dm^3$ of water at 20 °C.

11.11. Light from a torch is spread out when passed through soapy water – why? Could the particles causing this effect be filtered off with standard laboratory filter paper?

11.12. The shell of an egg is mainly calcium carbonate, and is easily dissolved by soaking the egg in vinegar. If a raw egg is left in a jar of vinegar for 3 days, it swells up. If the egg is now placed in syrup (concentrated sugar solution) it shrinks. Explain these observations.

Extension material to support this unit is available on our website at **http://www.palgrave.com/foundations/lewis.**

Chemical Families

Objectives

- ▶Takes you on a guided tour of the Periodic Table
- ▶Describes the similarities and changes in the properties of the members of selected groups of the table
- ▶Introduces the chemistry of the d-block elements transition elements
- ▶Looks at trends in the properties of the elements across a period of the table

Contents

12.1 Periodic Table

At first glance, the known elements present a bewildering array of different properties. Sodium, for example, is a very reactive solid that has a shiny appearance and conducts electricity, whereas neon is a non-conducting gas that does not react with anything. How do chemists organise the elements so that they can explain the properties of known elements and predict the properties of, as yet, undiscovered elements?

The Russian chemist Dimitri Mendeleev made a major contribution to chemistry when he presented his Periodic Table 1869. The arrangement of the elements in that Periodic Table arose from observations that certain groups of elements show similar properties. For example F_2, Cl_2, Br_2 and I_2 are all coloured, volatile, poisonous and reactive non-metals, whereas Li, Na and K are all soft, light metals that tarnish in air.

The modern-day Periodic Table, shown on the inside front cover of this book, has the atoms of elements listed in order of increasing atomic number. The numbered vertical columns are called **groups** (or **families**) and elements in the same group often have similar properties. Some of the groups have special names:

1. Group 1 elements (Li, Na, K, Rb, Cs and Fr) are called the **alkali metals.**
2. Group 2 elements (Be, Mg, Ca, Sr, Ba and Ra) are the **alkaline earth metals.**
3. Group 15 elements (N, P, As, Sb and Bi) are (rarely) called the **pnicogens.**
4. Group 16 elements (O, S, Se, Te and Po) are sometimes referred to as the **chalcogens.**
5. Group 17 elements (F, Cl, Br, I and At) are commonly known as the **halogens.**
6. Group 18 elements (He, Ne, Ar, Kr, Xe and Rn) are the **noble gases.**

Exercise 12A

Electron arrangements of elements in a group

(i) Atoms of Li (atomic number 3) have the electronic structure 2.1. Write down the electron arrangements (Bohr model and s,p,d,f structures) for the other members of Group 1 – Na, K and Rb. Can you give a general expression for the electron arrangements of the elements in Group 1?

(ii) Write down the electronic structure (Bohr model and s,p,d,f arrangements) for the elements in Group 17 – F, Cl, Br, I. Give a general expression for the electron arrangements of the elements in this group.

An older numbering system for the groups is also shown on the Periodic Table, whereby Groups 13, 14, 15, 16, 17 and 18 are shown as Groups 3, 4, 5, 6, 7 and 0, respectively. This system does not number the block of elements known as the d-block elements (transition elements).

Other points to note about the Periodic Table:

- The block of elements between Groups 2 and 13 is named the **d-block elements** or transition elements.
- The horizontal rows in the table are called **periods**. Period I consists only of the two elements H and He. Period 2 consists of the elements Li, Be, B, C, N, O, F and Ne. The zigzag line in the table separates the metals from the non-metals. The metals are on the left and you can see that there are a great deal many more metals in the table than non-metals.
- Hydrogen does not really belong in any group of the table, so it is usually placed at the top, on its own.

In order that the Periodic Table should fit on a single page, the **lanthanoids** (elements 57–70) and **actinoids** (elements 89–102) series are usually written below the table. We will not concern ourselves with the chemistry of these elements in this book.

After completing Exercise 12A, you should have gained a valuable clue as to why elements in the same group often have similar physical and chemical properties:

All the elements in a given group have the same number of electrons in the outermost shell of their atoms.

Since the type of bonding in compounds of an element is dependent on its ionization energy, which is related to the number of electrons in the outermost shell of its atoms, then we might expect those elements with similar electronic configurations to exhibit similar properties. Although this is generally true of the elements in a group, there are gradations (gradual changes) of the properties of the elements within a group.

If you write the electronic configuration of the elements in s,p,d,f notation, you should be able to see that:

- Group 1 and Group 2 elements are often called the **s-block elements** because they have s-orbital electrons in the outermost shell of their atoms.
- Groups 13 to 18 elements are called the **p-block elements** because they have p-orbital electrons in the outermost shell of their atoms.
- The block of elements *between* Groups 2 and 13 is known as the **d-block elements**, because d electrons generally play an important part in determining their chemistry.

12.2 Group 1 elements – the alkali metals: Li, Na, K, Rb, Cs, Fr

Electronic configuration

These elements have one electron in the outermost shell of their atoms; their chemistry is dominated by the tendency to lose this electron and they are good **reducing agents**. The general ionic equation for this loss of an electron can be written as

$$M \rightarrow M^{+} + e^{-}$$

Most of the compounds of the alkali metals are therefore ionic, and the alkali metals have an oxidation state of +1 in their compounds.

Exercise 12B

Ionization energies of the alkali metals

Whereas the first ionization energies (*the energy to remove one electron*) of the alkali metals are relatively low, the second ionization energies are very high.

Metal	First ionization energy/kJ mol^{-1}	Second ionization energy/kJ mol^{-1}
Li	520	7298
Na	494	4563
K	419	3051
Rb	403	2632
Cs	376	2420

(i) Would you expect Na to form an Na^{2+} ion?

(ii) Why do you think K^+ is much more stable than K^{2+}?

(iii) Suggest reasons why the second ionization energy of the metals is a great deal higher than the first.

Reactions of the alkali metals

1. They are all soft metals which tarnish in air, and for this reason they are generally stored under oil. The products obtained on burning the metals with oxygen, in the air, depend upon the metal:

   $4Li(s) + O_2(g) \rightarrow 2Li_2O(s)$ the **normal oxide** is formed

   $2Na(s) + O_2(g) \rightarrow Na_2O_2(s)$ a **peroxide** is formed

   $K(s) + O_2(g) \rightarrow KO_2(s)$ for K, Rb and Cs a **superoxide** is formed, M^+,O_2^-

2. They react readily with water, to form the corresponding alkali and hydrogen. The reaction becomes very violent as the group is descended. A general equation for this reaction is

   $2M(s) + 2H_2O(l) \rightarrow 2MOH(aq) + H_2(g)$

   The ionic equation is

   $2M(s) + 2H_2O(l) \rightarrow 2M^+(aq) + 2OH^-(aq) + H_2(g)$

   All the Group 1 hydroxides are very soluble in water.

3. The metals and their compounds display characteristic flame colours. If a moistened platinum wire is dipped into a substance containing the alkali metal and introduced into a Bunsen flame, the flame becomes coloured. The colours are shown in Table 12.1.

4. They form water-soluble, thermally stable carbonates (but note that Li_2CO_3 can be decomposed with strong heating in a Bunsen flame).

5. With the exception of $LiHCO_3$, which exists only in solution, they form solid hydrogencarbonates. Sodium hydrogencarbonate (baking powder) decomposes on heating:

   $2NaHCO_3(s) \rightarrow Na_2CO_3(s) + H_2O(l) + CO_2(g)$

Table 12.1 Flame colours of the alkali metals and their compounds

Alkali metals	Flame colour
Lithium	red
Sodium	yellow
Potassium	lilac (crimson through blue glass)
Rubidium	red
Caesium	blue

Exercise 12C

Trends in the physical properties of the alkali metals

The boiling points of the metals are as follows:

Metal	Boiling point/K
Li	1615
Na	1156
K	1033
Rb	–
Cs	942

(i) Explain the trend shown by the boiling point values as the group is descended. (**Hint** consider the strength of the metallic bonding in the elements.)

(ii) Predict a boiling point value for Rb.

(iii) The alkali metals are so soft that they can all be cut with a pen-knife. They become softer as the group is descended. Can you think of an explanation for this trend?

(iv) The metals have low densities (Li, Na and K float on water). Why do you think this is?

6. They form nitrates that decompose to nitrites and oxygen on strong heating (but note that $LiNO_3$ decomposes to the oxide). The general equation for decomposition of the nitrates is

$$2MNO_3(s) \rightarrow 2MNO_2(s) + O_2(g)$$

7. Their salts (usually colourless or white) are ionic and *not hydrolysed* in solution (they do not react with water, they only dissolve in water).

The elements become *more reactive* as the group is descended. The atomic radius of the elements increases as the group is descended and the single electron in the outer shell of the atoms becomes farther away and more shielded from the pull of the nucleus by full shells of inner electrons. The outer electrons therefore become less strongly bound down the group and the ionization energy of the elements decreases. This is shown by the values given in Exercise 12B.

Trends in the chemical properties of the alkali metals and their compounds are summarized in Table 12.2.

The chemistry of francium (Fr) is not as well studied as that for the rest of the alkali metals, because it is radioactive and scarce. The longest-lived isotope of the element has a half-life of about 20 min.

Exercise 12D

Chemistry of francium

From your knowledge of the alkali metals and the trends in their properties, predict:

(i) The reactivity of francium towards air and water. Write balanced equations for the reactions that occur.

(ii) The formula, solubility and thermal stability of francium carbonate.

(iii) The effect of heat on francium nitrate. Write an equation for the reaction.

Table 12.2 Trends in the chemical properties of the alkali metals and their compounds

Metal	Reactivity with air and water	Thermal stability of carbonates in a bunsen flame	Thermal stability of hydroxides	Thermal stability of nitrates	Solubilities of carbonates
Li	↓	decomposes	↓	decomposes to oxide	↓
Na, K, Rb, Cs	more vigorous ↓	stable and increasing stability ↓	stable and increasing stability ↓	decompose to nitrites increasing stability ↓	soluble and increasing solubility ↓

12.3 Group 2 elements – the alkaline earth metals: Be, Mg, Ca, Sr, Ba, Ra

The chemistry of these elements (except Be) is dominated by their tendency to lose two electrons and their compounds are generally ionic. The general equation for this reaction is

$$M \rightarrow M^{2+} + 2e^-$$

The metals tend to be less reactive than their corresponding alkali metals because removal of two electrons requires more energy. The alkaline earth metals also have higher densities and melting points than the alkali metals. Reactivity of the Group 2 metals increases as the group is descended (Table 12.3).

Some chemical properties of Group 2 metals (with the exception of Be, which forms mostly covalent compounds) are:

1. They burn in air to form normal oxides (barium forms some peroxide). The general equation is
   $$2M(s) + O_2(g) \rightarrow 2MO(s)$$
2. The alkaline earth metals, when heated, also react with nitrogen to form ionic nitrides:
   $$3M(s) + N_2(g) \rightarrow M_3N_2(s)$$
3. Ca, Sr and Ba react with water to form the hydroxide and hydrogen gas:
   $$M(s) + 2H_2O(l) \rightarrow M(OH)_2(aq) + H_2(g)$$
   Magnesium reacts with cold water only slowly, but rapidly reacts to form the oxide when heated in steam:
   $$Mg(s) + H_2O(l) \rightarrow MgO(s) + H_2(g)$$
4. The metals react with aqueous acids, liberating hydrogen:
   $$M(s) + 2HCl(aq) \rightarrow MgCl_2(aq) + H_2(g)$$
5. Some alkaline earth metals have characteristic flame colours, shown in Table 12.4.
6. Their carbonates decompose on strong heating into carbon dioxide and the oxide:
   $$MCO_3(s) \rightarrow MO(s) + CO_2(g)$$
7. The Group 2 nitrates decompose into the oxide, brown gaseous nitrogen dioxide and oxygen:
   $$2M(NO_3)_2(s) \rightarrow 2MO(s) + 4NO_2(g) + O_2(g)$$
8. Their hydroxides are not as thermally stable as those of the alkali metals. Group 2 hydroxides decompose on heating:
   $$M(OH)_2(s) \rightarrow MO(s) + H_2O(l)$$

Exercise 12E

Physical properties of the alkaline earth metals

Can you explain why the melting points of the alkaline earth metals are higher than those of the corresponding alkali metals?

Table 12.3 Trends in the properties of the alkaline earth metals and their compounds

Metal	Reactivity with air, water and acids	Solubility of sulfates	Solubility of carbonates	Basic strength of hydroxides	Solubility of hydroxides
Be	poor		does not exist	amphoteric	insoluble
Mg		soluble		weak	insoluble
Ca	increases	insoluble			
Sr		increasingly	increasingly	increasingly	increasingly
Ba	↓	↓ insoluble	↓ insoluble	↓ strong	↓ soluble

Table 12.4 Flame colours of the alkaline earth metals and their compounds

Element	Flame colour
Calcium	brick red
Strontium	scarlet
Barium	apple green

BOX 12.1

Hardness of water

Hardness in water is caused by the presence of dissolved calcium and magnesium compounds, such as calcium hydrogencarbonate or magnesium sulfate. The hardness of the water can be gauged by its ability to form a thick lather with soap. Water that lathers easily is called **soft**, whereas water that produces very little lather and a scum, is termed **hard**. Soap (see page 175) is sodium stearate and scum forms as a result of the precipitation of an insoluble stearate:

$$Ca^{2+}(aq) + 2St^{-}(aq) \rightarrow CaSt_2(s)$$

where St^{-} = stearate ($C_{18}H_{35}O_2^{-}$)

Lather cannot form until all the calcium or magnesium ions have been precipitated. Note that soapless detergents, which are an ingredient of many washing-up liquids and detergents, do not form a scum with hard water.

Hard water is produced when rainwater, which is weakly acidic because of the presence of dissolved carbon dioxide, falls upon limestone or chalk which are both forms of calcium carbonate. The acidic solution dissolves calcium carbonate producing a solution of calcium hydrogencarbonate, or hard water:

$$H_2O(l) + CO_2(g) + CaCO_3(s) \rightleftharpoons Ca(HCO_3)_2(aq)$$

This type of hard water is called temporary hard water, because it can be removed by boiling. Permanently hard water contains calcium and/or magnesium salts which are not decomposed when the water is boiled. Calcium hydrogencarbonate solution decomposes on boiling to calcium carbonate (scale), which can block pipes or fur kettles in hard water districts:

$$Ca(HCO_3)_2(aq) \rightleftharpoons H_2O(l) + CO_2(g) + CaCO_3(s)$$

Stalagtites and stalagmites are also found in hard water districts. When drops of calcium hydrogencarbonate solution collect on the roof of a cave, the solution decomposes, some of the water evaporates, carbon dioxide is lost and a tiny precipitate of calcium carbonate is left behind. Over many years these specks of calcium carbonate accumulate to form a **stalactite**. Solution that drips on to the floor undergoes the same decomposition, forming a **stalagmite** (Fig. 12.1).

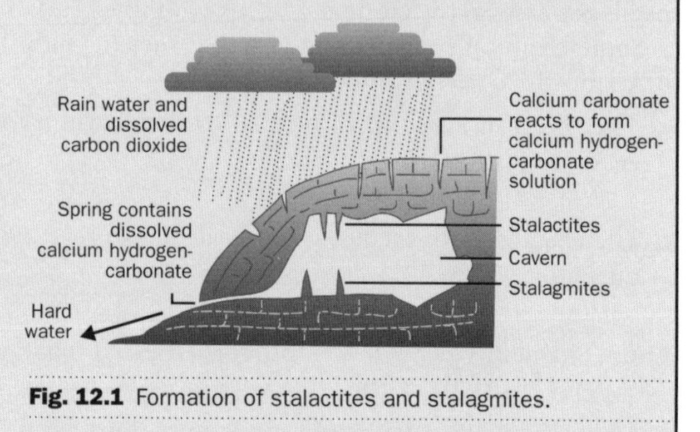

Fig. 12.1 Formation of stalactites and stalagmites.

Temporary or permanent hard water are commonly softened by **ion exchangers**, which remove the Group 2 metal ions and replace them with sodium ions. Although hard water has disadvantages, it does have a pleasant taste and the calcium compounds it contains are good for bones and teeth.

Exercise 12F

Atomic and ionic radii of the s-block elements

The atomic and ionic radii of the elements of Groups 1 and 2 are shown below:

Element	Atomic radius /pm	Ionic radius /pm	Element	Atomic radius /pm	Ionic radius /pm
Li	145	60	Be	105	31
Na	180	95	Mg	150	65
K	220	133	Ca	180	99
Rb	235	148	Sr	200	113
Cs	266	169	Ba	215	135

(i) For each element, why is its ionic radius much smaller than its atomic radius?

(ii) Why do the atomic radii of the elements increase as the groups are descended?

(iii) Although they are in different groups, the chemistry of Li is quite similar to that of Mg. Why do you think this is?

12.4 Group 14 elements: C, Si, Ge, Sn, Pb

All these elements have four electrons in the outermost shell of their atoms (or the general outer electronic structure ns^2np^2, where n is a whole number greater than one). The elements form compounds with similar formulae but they show marked changes in properties, as the group is descended, from carbon to lead.

About the elements

The element carbon exists mainly in two allotropic forms, diamond and graphite, and has a very large branch of chemistry (organic chemistry) concerned with the compounds that it forms because of its ability to form long chains by bonding with atoms of itself. The ability of the atoms of an element to covalently bond with themselves is called **catenation**.

Silicon and germanium are metalloids, whereas tin and lead are metals. Silicon and germanium have structures similar to that of diamond, tin exists in allotropic forms, but lead exists in only one metallic form.

BOX 12.2

Allotropic forms of tin

Tin has three crystalline forms:

$$\alpha\text{-tin} \underset{}{\overset{13\,°\text{C}}{\rightleftharpoons}} \beta\text{-tin} \overset{161\,°\text{C}}{\rightleftharpoons} \gamma\text{-tin}$$

α-tin (grey tin) has a structure like diamond; β-tin (white tin) metallic; γ-tin metallic

Below 13 °C, powdery grey α-tin forms on the white, metallic β-tin allotrope and the metal crumbles. Napoleon's soldiers had tin buttons fastening their jackets and they used tin pots and pans to cook with. During the winter invasion of Russia, in 1812, their buttons and pots crumbled and it was said that this contributed to their defeat.

Catenation and multiple bonds

Apart from carbon, catenation does not occur to any great degree in the chemistry of the other elements. Si and Ge do form hydrides that might be compared with the lighter hydrocarbons, but they are not as stable. Catenation in these elements involves the formation of Si–Si and Ge–Ge covalent bonds, which are longer and therefore weaker than C–C bonds.

Carbon is also the only member of the group which is able to form multiple bonds with itself or other elements, such as oxygen. Although the dioxides of carbon and silicon have similar formulae (CO_2 and SiO_2), they have very different physical and chemical properties, as shown in Table 12.5.

Carbon dioxide molecules are simple, linear, non-polar molecules with the structure

$$O^{\delta-}{=}C^{\delta+}{=}O^{\delta-}$$

Silica does not have the same structure even though it has the formula SiO_2. Silicon cannot form multiple bonds to oxygen and forms single covalent bonds instead. Silica occurs in several forms, but in all of them silicon has covalent single bonds to four oxygen atoms in a tetrahedral arrangement. Many units are bonded this

Table 12.5 Properties of carbon and silicon dioxides

Compound	Carbon dioxide (CO_2)	Silicon dioxide or silica (SiO_2)
Form under normal conditions	colourless gas	sand or quartz
Chemical properties		
	weakly acidic gas, reacts with bases to form salts (carbonates)	unreactive
	slightly soluble in water: the weakly acid solution formed is carbonic acid	insoluble in water
	in soft drinks under pressure it carbonates them or makes them fizzy	
	used in fire extinguishers because it does not support combustion	

way to form a network solid (not unlike the diamond structure), resulting in a high-melting-point solid. The formula of silica is SiO_2 because each silicon atom has a 'half-share' of four oxygen atoms, as shown in Fig. 12.2.

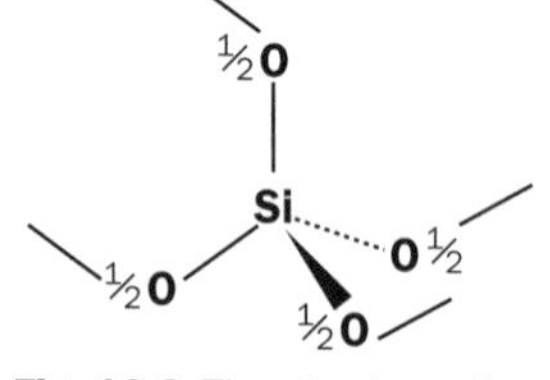

Fig. 12.2 The structure of silica.

Exercise 12G

Silicon and germanium hydrides

(i) Write a structural formula for the covalent compound Si_6H_{14}.

(ii) Describe the bonding and shape of the molecule Ge_2Cl_6.

BOX 12.3

Dry ice

Solid carbon dioxide (dry ice) is obtained by releasing CO_2 from cylinders so that it expands and cools. A 'snow' is produced that is pressed into blocks. The solid carbon dioxide is used to refrigerate ice cream, meat and other foodstuffs. Solid carbon dioxide **sublimes** (changes directly from a solid to a gas) into carbon dioxide vapour at normal atmospheric pressure so that it is liquid-free when used as a refrigerant – hence the name **dry ice.**

Oxidation states

The elements of Group 14 combine with oxidation number +2 and +4. As the group is descended, there is an *increased tendency* for the elements to combine with other elements using the oxidation number +2 instead of +4. Carbon generally exhibits oxidation state four in its compounds, whereas the other elements in the group can form tetravalent and divalent compounds. The most important oxidation states for the elements are shown in Table 12.6. Lead(IV) compounds however, are oxidizing agents, because lead(II) is the most stable oxidation state for the element. This gradual preference of the valency (or oxidation number/state) of 2 instead of 4 can be explained by the **inert pair effect**, described in Box 12.4.

Most of the compounds of Group 4 elements are covalent, with the exception of compounds of tin and lead in oxidation state +2, which are normally described as ionic.

Table 12.6 Oxidation states of Group 14 elements. Very unusual oxidation states are in brackets, preferred oxidation states are in boxes

Element	Oxidation states	
Carbon C	(2)	[4] e.g. CCl_4
Silicon Si	(2)	[4] e.g. SiO_2
Germanium Ge	2	[4] e.g. $GeCl_4$
Tin Sn	2	[4] e.g. SnO_2
Lead Pb	[2] e.g. PbO	4

Exercise 12H

Oxidation numbers (states) of Group 14 elements

(i) Using Lewis symbols, describe the bonding in $SnCl_4$ and $SnCl_2$.

(ii) PbO_2 is a strong oxidizing agent, whereas PbO is not. Can you offer an explanation?

(iii) Whereas CH_4 is a very stable compound, PbH_4 is not. Why do you think this is?

(iv) Can you think of a compound where carbon (very unusually) has oxidation state +2?

(v) Write the formulae for two oxides of germanium. Which would you expect to be the more stable?

BOX 12.4

Inert pair effect

Carbon has the electronic structure 2.4 or $1s^22s^22p^2$. If we use the 'electrons in boxes' model the electronic structure can be written as

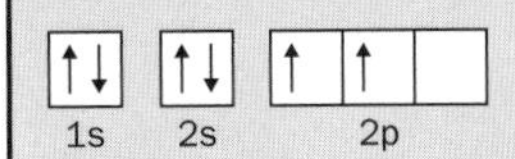

If an electron is promoted from a 2s orbital to a 2p orbital, the electronic structure will be

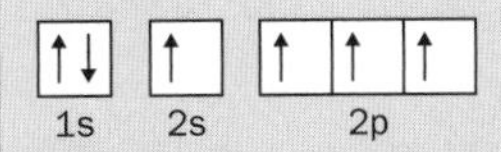

Now the carbon can combine with a covalency of four, because it has four unpaired electrons which can combine with the unpaired electrons of other elements, or groups, to make four covalent bonds. Remember that when covalent bonds are formed, energy is released and the formation of four covalent bonds will release more energy than the formation of two covalent bonds. Energy is required to promote an electron from the 2s orbital to the 2p orbital, but this is more than 'paid back' in the case of carbon, when the element forms four covalent bonds. Because carbon is a small atom, it forms short and strong covalent bonds – a great deal of energy is released when they are formed. As Group 14 is descended the covalent bonds formed by the elements become longer and weaker (because their atoms are larger) and the energy release gained by the formation of four covalent bonds does not become 'worth' the energy of promotion of *n*s electrons into *n*p orbitals. This is why the oxidation state of +2 becomes more favourable as compared with that of +4 and lead(IV) compounds are good oxidizing agents.

Hydrolysis of tetrachlorides

The tetrachlorides of Group 14 (XCl_4) are liquids with a tetrahedral structure. All, *except* CCl_4, react with water (or *hydrolyse*) to form HCl and the oxides (XO_2).

Exercise 12I

Hydrolysis of Group 14 chlorides

(i) Write a balanced equation for the hydrolysis of $SiCl_4$.

(ii) In the hydrolysis of $SiCl_4$, Si–Cl bonds must be broken. Why is the Si–Cl bond easier to break than a C–Cl bond?

Exercise 12J

Semiconductors

(i) Diamond is a very good insulator because the valence electrons around its carbon atoms tend to stay firmly in place. Why do the electrons in Si–Si covalent bonds delocalize much more easily?

(ii) In a sample of silicon, if electrons from Si–Si covalent bonds are set free, what stops the structure from collapsing?

Semiconductors

Silicon and germanium are used as **semiconductors**. Semiconductors have electrical resistances somewhere between those for conductors and insulators and the electrical resistance of a semiconductor decreases as the temperature gets higher. In metals, which are good conductors, the valence electrons are able to break free from the individual atoms and move through the metal when a potential difference is applied across it. In insulators, the valence electrons cannot move freely around the material in the same way. The valence electrons in semiconductors, however, can break free of their atoms if a little energy is applied. As the temperature rises more valence electrons are able to break away (**become delocalized**) and the electrical conduction of the substance increases. The semiconducting properties of silicon are used in integrated circuits known as **silicon chips**. The element silicon contains many silicon atoms covalently bonded to each other in tetrahedral arrangements so that a giant molecule is built up. At very low temperatures, the electrons in the covalent bonds that hold the silicon atoms in place tend to stay associated with their 'parent' atoms. As silicon is heated, however, more of these electrons gain sufficient energy to delocalize and the resistance of the silicon decreases.

BOX 12.5

Doping semiconductors

The electrical properties of semiconductors can be changed by adding small amounts of other substances. This process is known as **doping**. In ***n*-type** semiconductors, a small number of Group 15 element atoms, such as arsenic, are spread out in a sample of silicon. The arsenic atoms use four of their five valence electrons to bond with surrounding silicon atoms and the fifth electron is delocalised in the structure. This makes the substance a much better electrical conductor than pure silicon. The name *n*-type is derived from the fact that the doped material has an excess of negative charge. This is shown in Fig. 12.3

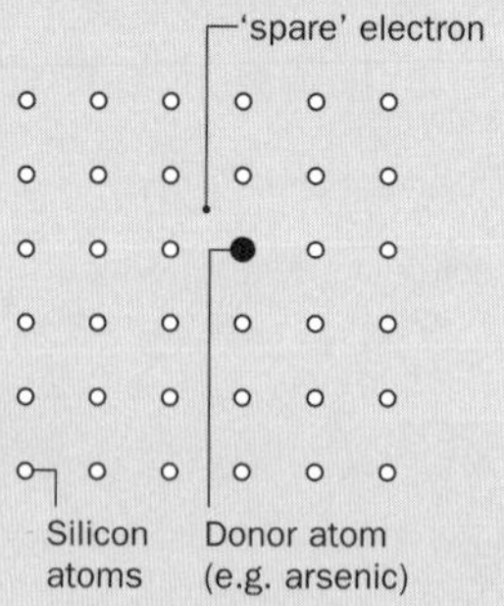

'spare' electron

Silicon atoms

Donor atom (e.g. arsenic)

Fig. 12.3 An *n*-type semiconductor.

If small traces of a Group 13 element, such as boron, are added to silicon then a deficiency of electrons is created. Boron has only three valence electrons to bond with four surrounding silicon atoms. The incomplete covalent bond created contains a 'hole' where an electron should be. A neighbouring valence electron from a silicon atom can jump into this hole, creating a new hole ready to be occupied by another electron, and so on. In this way electrons move throughout the sample and, again, the doped substance is a better conductor of electric current. These type of doped semiconductors are called ***p*-type**; the *p* refers to an excess of positive charge.

12.5 Group 17 elements – the halogens: F, Cl, Br, I, At

The halogens have seven electrons in the outermost shell of their atoms (with the general outer electronic structure ns^2np^5, where n is greater than one) and gain the stable electronic configuration of a noble gas by either accepting an electron to form a negatively charged ion, or by sharing an electron to form a covalent bond. Some of their common properties are:

1. The non-metallic elements exist as stable diatomic molecules: F_2, Cl_2, Br_2 and I_2.
2. They are volatile elements: fluorine and chlorine are gases at room temperature, whereas bromine and iodine are a liquid and volatile solid respectively. Solid iodine sublimes when warmed.
3. They are coloured; the colours of the elements are shown in Table 12.7.
4. Halogens combine directly with most metals to form ionic solids:

   $$2Na(s) + Cl_2(g) \rightarrow 2NaCl(s)$$

   Some iodides have covalent character, however, because of the high polarizability of the iodide ion.
5. They form gaseous (with the exception of HF, bp 20 °C), covalent hydrides with the general formula HX. The thermal stability of these hydrides decreases down the group. The hydrides form acidic aqueous solutions.
6. Their silver salts become increasingly covalent from AgF to AgI. A test for aqueous halides is the precipitation of the silver salt using silver nitrate solution. The general ionic equation is

   $$Ag^+(aq) + X^-(aq) \rightarrow AgX(s)$$

   The halogen can be identified from the colour of the precipitate formed: AgCl is white, AgBr is cream, and AgI is pale yellow.

   When exposed to light, silver chloride and bromide decompose into silver and the halogen. The finely divided silver appears black. This is why silver halides are used in photographic films.
7. The reactivity of the halogens decreases as the group is descended. Chlorine will displace bromine and iodine from bromides and iodides, respectively, and bromine will displace iodine from iodides.

   $$Cl_2(aq) + 2Br^-(aq) \rightarrow 2Cl^-(aq) + Br_2(aq)$$

   In the reaction above, chlorine is acting as an oxidizing agent, the bromide ion is acting as the reducing agent.

The trends apparent within the group are shown in Table 12.8.

Table 12.7 Colours of the halogens

Element	Colour
Fluorine	pale yellow gas
Chlorine	pale green gas
Bromine	red/brown liquid
Iodine	black solid, purple in organic solvents or vapour phase

Table 12.8 Trends in the properties of the halogens and their compounds

Element	Colour	Melting and boiling points	Stability of hydrides	Colour of silver salt	Oxidizing ability
F	getting darker ↓	increasing ↓	decreasing ↓		decreases ↓
Cl				white	
Br				cream	
I				pale yellow	

Exercise 12K

Halogens

Astatine, the fifth halogen, is little studied because its isotopes are radioactive and short-lived. From your knowledge of the physical and chemical properties of the halogens, predict the physical and chemical properties of the element. Include:

(i) the colour of the element

(ii) its state at room temperature

(iii) the stability of hydrogen astatide

(iv) its strength as an oxidizing agent

(v) the nature of the bonding present in its silver salt

(vi) the colour of AgAt.

12.6 Group 18 elements – the noble gases: He, Ne, Ar, Kr, Xe, Rn

These elements, with the exception of helium, have the outer electronic configuration ns^2np^6. There is a stable octet of electrons in the outer shells of their atoms.

Group 18 elements have the following properties in common:

1. The elements are all colourless gases.
2. The lighter members of the group do not combine with any substance, xenon reacts directly only with highly reactive fluorine:

$$Xe(g) + F_2(g) \rightarrow XeF_2(s)$$

Although they are all gases, the elements become more dense as the group is descended.

Exercise 12L

Compounds of xenon

(i) Write a balanced equation for the reaction of xenon with fluorine to form xenon tetrafluoride.

(ii) Draw a Lewis structure for the product of the reaction in **(i)**. What sort of shape would the molecule have?

(iii) Why are all the fluoride compounds of xenon strong oxidizing agents? What is the reduction product in these reactions?

BOX 12.6

Discovery of the noble gases

In 1888 Lord Rayleigh (1842–1915) found that the density of nitrogen, obtained from the air, had a higher value than that obtained for nitrogen which had been prepared by chemical means. Together with William Ramsay (1852–1916), he sought an explanation. The two scientists eventually concluded that a new gas was present in samples of nitrogen obtained from the air. Because Ramsay, despite many chemical experiments, could not get this new gas to combine with another substance, the gas was named **argon** (from the Greek 'idle'). The gas was subsequently found to be monoatomic, with a molar mass of about 40 g mol^{-1}. The two scientists realized that there was room for another group of elements in the Periodic Table that had been proposed by Mendeleev and the search began for the other elements of the group.

Helium had been named some years earlier when unknown spectral lines were observed in the spectrum of the sun. It had been suspected that the lines were caused by the presence of an unknown element. Ramsay now managed to isolate helium from a mineral called cleveite, and show that it had similar properties to argon.

In 1898 Ramsay isolated three more gases in the group from liquid air – neon, krypton and xenon – all found to be chemically inert. The group was complete a year later, when a gas was found to be coming from samples of the Group 2 element radium. The new inert gas was first called 'emanation' (Em) and finally radon (Rn). Ramsay received a Nobel prize for the work in 1904.

Chemists were intrigued by the apparent unreactivity of the gases and attempts followed to react them with other substances. Argon was found not to react with the most reactive element of all, fluorine, yet H. G. Wells wrote that the Martians in the *War of the Worlds* attacked Earth with a poisonous argon compound! In the early 20th century, when the electronic configurations of the elements were known, the fact that the atoms of the inert gases had completed shells of electrons was accepted as a reason for their lack of reactivity. However, the story was not finished: in the early 1960s XeF_4 was prepared. Although the ionization energies of the outer electrons in argon were too high for the atoms to act as electron donors in reaction with fluorine, xenon atoms are larger and the outer electrons are more shielded from their nuclei. In fact, the first ionization energy for xenon is very close to that obtained for oxygen. Xenon and the heavier inert gases can therefore form compounds under the right conditions.

12.7 Elements of the first transition series

The block of elements between Group 2 and Group 13 of the Periodic Table are known as the **transition elements** or **d-block elements** (Sc to Zn and the elements below them). The **elements of the first transition series** are those elements that have partly filled d orbitals in any of their common oxidation states, which are the block of elements headed by Ti to Cu. Here, we will look mainly at the properties of the **first transition series**: Ti, V, Cr, Mn, Fe, Co, Ni and Cu. These elements are typical metals and are often referred to as the transition metals. They have very similar physical properties. The changes in the atomic radii and first ionization energies across the first transition series are small, because each increase in nuclear charge is well shielded by the inner 3d electrons and only a small increased attraction is noticed by the outer electrons in the 4s subshell. See Box 12.7.

The transition metals have the following properties in common:

1. They are hard and have high melting points, both indications that strong metallic bonding exists within the metals.
2. They form alloys with one another and with other metals. Transition metals can form alloys with each other because their atoms are similar in size – the atoms of one metal can occupy positions in another metal's lattice. (Steel is an alloy of iron with other transition metals such as chromium.)

Exercise 12M

Magnetic properties of the d-block elements

Write down the electronic configurations ('electrons in boxes') of the following metals and decide whether the metals are paramagnetic or diamagnetic:

(i) V **(ii)** Cr **(iii)** Zn.

Which metal do you think shows the strongest paramagnetism?

3. Metallic Fe, Co and Ni are **ferromagnetic** – they are able to become permanently magnetized. Other metals and their compounds are **paramagnetic** – they are weakly attracted into a magnetic field. Paramagnetism is caused by the presence of **unpaired** spinning electrons. Two electrons paired up, in the same orbital, spin in opposite directions and their moments cancel each other out. If a substance has all its electrons paired up it is termed **diamagnetic**, and the substance is not attracted into a magnetic field.

BOX 12.7

Changes across the first transition series

Element	Ti	V	Cr	Mn	Fe	Co	Ni	Cu
Electronic structure	$[Ar]3d^24s^2$	$[Ar]3d^34s^2$	$[Ar]3d^54s^1$	$[Ar]3d^54s^2$	$[Ar]3d^64s^2$	$[Ar]3d^74s^2$	$[Ar]3d^84s^2$	$[Ar]3d^{10}4s^1$
Atomic radius/pm	146	131	125	129	126	125	124	128
First ionization energy /kJ mol^{-1}	658	650	653	717	759	758	737	746
Melting point/K	1930	2160	2130	1520	1800	1770	1730	1360

4. Many of the metals can exist in a wide range of oxidation states. These are shown in Box 12.8. The 3d electrons are very close in energy to the 4s electrons and both types of electrons can be involved in bonding.

Notice that the later metals in the series do not display anywhere near the maximum oxidation states possible. This is because the nuclear charge increases across the series and the outer electrons are held more tightly to the atoms.

BOX 12.8

Oxidation numbers of the d-block elements

Metal	Sc	Ti	V	Cr	Mn	Fe	Co	Ni	Cu	Zn
					7					
				6	6	6				
Oxidation states			**5**	5	5	5	5			
		4	**4**	4	**4**	4	4	4		
	3	**3**	**3**	**3**	3	**3**	**3**	3	3	
		2	2	**2**	**2**	**2**	**2**	**2**	**2**	**2**
		1	1	1	1	1	1	1	1	

(Important oxidation numbers are in bold type.)

5. Many transition metals, or their compounds, are important catalysts. For example, Fe is used as a catalyst in the Haber process to synthesize ammonia:

$$N_2(g) + 3H_2(g) \rightleftharpoons 2NH_3(g)$$

When the substance acts as a catalyst, bonds are formed between atoms of the reactants and the catalyst surface, so weakening the bonds in the reactants. A transition metal's ability to exist in different oxidation states allows it to easily make and break bonds with other species, so acting as an efficient catalyst.

6. The transition metals form **complex ions**. A complex ion consists of a central metal ion with molecules or ions attached to the metal ion by coordinate bonding. Some main group metals, such as Al, can form complex ions, but the transition metals form a much greater variety of complexes. Transition metal complexes are described in more detail in the next section.
7. Compounds of the transition metals tend to be coloured. Although the d electrons in the atoms of transition metals are **degenerate** (of the same energy), in complex ions of the transition metals, the d electrons become split in energy, some at a lower energy than others. The difference between the energies of the d electrons is of the same order as the energies of light photons in the visible region of the spectrum. If white light is passed through a solution of a transition metal complex, d electrons of lower energy may absorb some of the light and be promoted to a higher energy level, thereby absorbing some of the visible light. The unabsorbed light is responsible for the colour observed.

Common colours of transition metal ions in aqueous solution are:

Cu^{2+}	sky blue	Cr^{3+}	violet
Fe^{2+}	pale green	Co^{2+}	pink
Fe^{3+}	yellow brown	Mn^{2+}	pale pink
Ni^{2+}	green		

Exercise 12N

Oxidation states of the transition metals

Transition metals achieve their oxidation states *first* by the loss of the outer s electrons and *then* by the loss of the underlying d electrons, which are very close in energy.

(i) What is the electronic structure of atoms of Mn in oxidation state +2?

(ii) What is the electronic structure of atoms of Mn in oxidation state +3?

(iii) What do you think is the maximum oxidation state of Mn? Can you think of a common ion that has Mn in this oxidation state?

Exercise 12O

d-block compounds and their colours

(i) Why is the complex ion $[Sc(H_2O)_6]^{3+}$ colourless?

(ii) Why are aqueous solutions of zinc compounds colourless?

Transition metal complexes

In transition metal complexes the ions or groups attached to the metal cation are called **ligands**. Copper(II) sulfate solution, for example, contains the complex ion $[Cu(H_2O)_6]^{2+}$ in which six water ligands are attached to the central copper(II) ion by coordinate bonds. This complex is often simply written as $Cu^{2+}(aq)$.

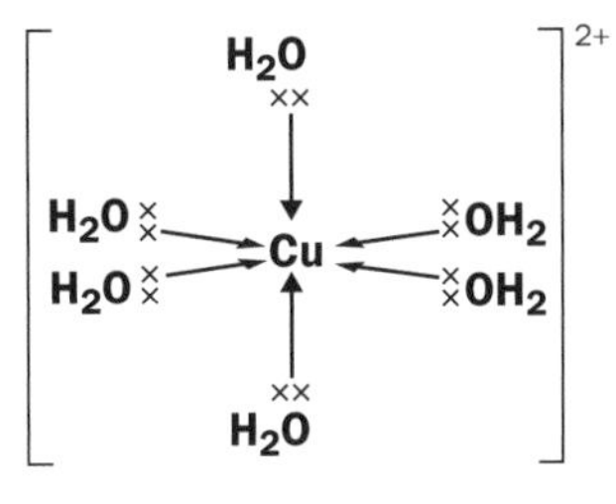

Notice that a transition metal complex ion is written in *square* brackets and the overall charge on the ion is written *outside* the brackets. In this example, the copper ion carries two positive charges and the water molecules are neutral, so the overall charge on the ion is 2+. The number of atoms directly attached to the central metal atom (in this example, six) is called the **coordination number**.

BOX 12.9

Understanding the names of complex ions

Cationic (and neutral) complexes

Cationic complex ions are named by indicating the name and number of the ligands, followed by the name and oxidation state of the central metal ion:

$[Cu(H_2O)_6]^{2+}$ is the hexaaquacopper(II) ion
$[Ag(NH_3)_2]^{+}$ is the diamminesilver(I) ion
$[FeCl(H_2O)_5]^{+}$ is the pentaaquachloroiron(II) ion

Notice that water and ammonia produce 'aqua' and 'ammine' complexes. Negatively charged ligands, such as Cl^- (chloro) end in 'o'. The prefixes 'di', 'tri', 'tetra', 'penta' and 'hexa' are used to show the number of ligands present and ligands are named in alphabetical order.

Anionic complexes

If the complex ion is **anionic**, the rules are followed as for cationic complexes but 'ate' is added to the metal's name:

$[PtCl_4]^{2-}$ is the tetrachloroplatinate(II) ion
$[Ni(CN)_4]^{2-}$ is the tetracyanonickelate(II) ion
$[Zn(OH_4]^{2-}$ is the tetrahydroxozincate(II) ion

Notice that 'ate' is added to the Latin name of a metal, if it exists. Anionic complexes of copper therefore end in 'cuprate' (from *cuprum*), whereas those of iron end in 'ferrate' (from *ferrum*).

Exercise 12P

Complex ions

Four Cl^- ions can act as ligands to form a complex ion with Cu^{2+}:

(i) What is the overall charge of the complex ion?
(ii) Write the formula for the complex ion.
(iii) What is the coordination number of the complex ion?
(iv) Name the complex.

Exercise 12Q

Names of complex ions

1. Name the following ions:

(i) $[Fe(CN)_6]^{3-}$
(ii) $[Zn(NH_3)_4]^{2+}$
(iii) $[Ti(H_2O)_6]^{3+}$
(iv) $[Cu(NH_3)_4(H_2O)_2]^{2+}$
(v) $[Co(NH_3)_5Br]^{2+}$
(vi) $[Pt(NH_3)_2Cl_4]$.

2. What are the formulae of the following ions?

(i) hexachloroplatinate(IV)
(ii) tetraaquadichlorochromium(III)
(iii) hexacyanopalladate(IV)
(iv) tetrabromocobaltate(II)
(v) hexaaquairon(III).

Cordination numbers two, four and six are found in transition metal complexes, with coordination number six commonly found.

Complexes with coordination number two tend to be **linear**, for example

$[Ag(NH_3)_2]^+$ $\quad [H_3N \rightarrow Ag \leftarrow NH_3]^+$

Those with coordination number four tend to be **square planar** or **tetrahedral**, for example

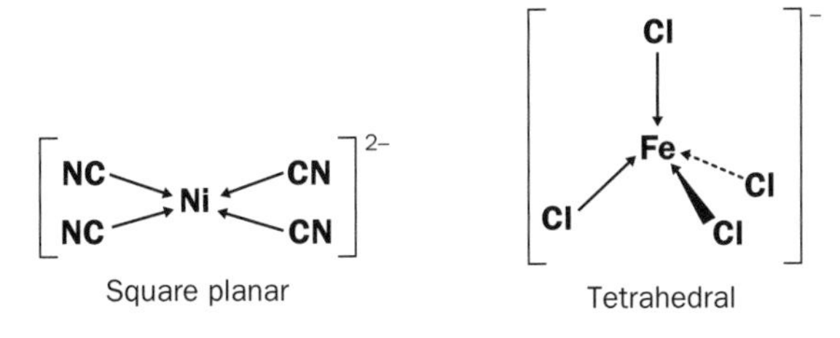

Square planar Tetrahedral

Complexes with coordination number six tend to be **octahedral**, for example

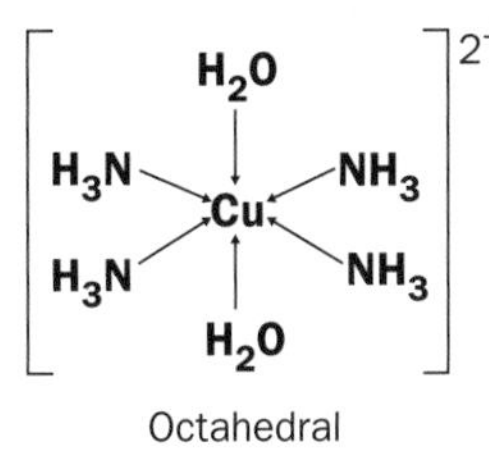

Octahedral

The ligands shown in the previous examples only donate one electron pair to the central ion. They are **monodentate** (or 'one tooth') ligands. Ligands that can donate more than one electron pair to the central metal ion are **polydentate** ('many toothed') ligands. Thus a ligand that donates two electron pairs is **bidentate**, whereas one that donates three electron pairs is **tridentate** and so on. A ligand which has the common name ethylenediamine (abbreviation 'en') has the structure

$\overset{\times\times}{N}H_2 — CH_2 — CH_2 — \overset{\times\times}{N}H_2$

This has two electron pairs to donate to the metal and acts as a bidentate ligand, for example in the octahedral complex $[Co(en)_3]^{3+}$:

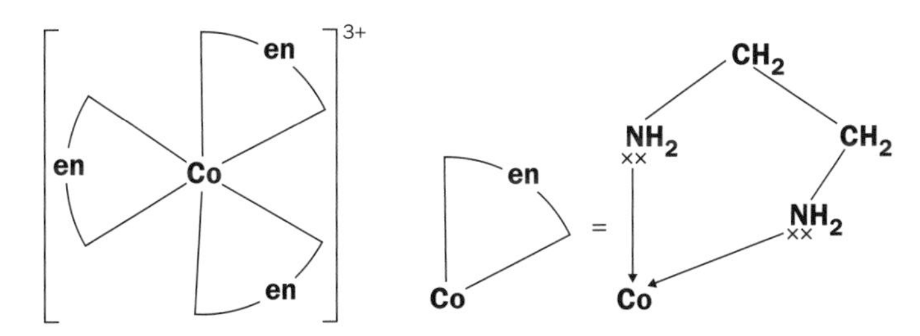

Because the ligand gives both electron pairs to the same metal, it is also known as a **chelate** ('claw-like') or **chelating ligand**. Notice that rings are formed that include the metal and the ligand and the coordination number of the metal is six.

Exercise 12R

Polydentate ligands

The ethylenediaminetetraacetate ion ($EDTA^{4-}$) is a polydentate ligand. Its structure is as follows:

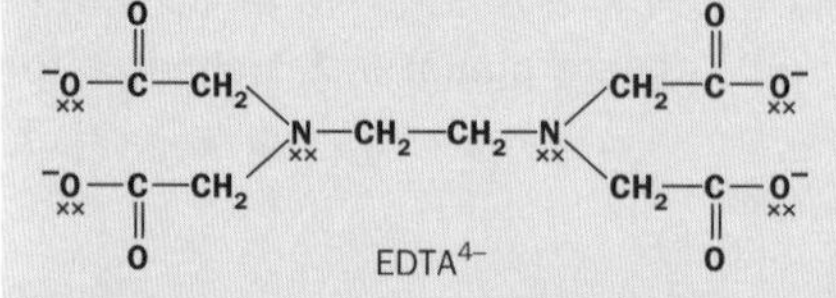

The lone pairs that are available for donation to a metal ion are shown on the diagram. $EDTA^{4-}$ also coordinates with Ca^{2+} or Mg^{2+} and is used to remove these ions from water (to 'soften' water) and is therefore often an ingredient of shampoos.

(i) How many lone pairs can $EDTA^{4-}$ coordinate to a metal ion?

(ii) One molecule of an hydrated metal ion reacts with $EDTA^{4-}$, so that all of the water molecules are replaced by one molecule of $EDTA^{4-}$:

$[M(H_2O)_6]^{2+} + EDTA^{4-}$

What is the overall charge of the new complex ion formed?

12.8 Variation of properties of elements within groups and periods

Changes in properties within groups

By now, you will have noticed that, although elements in the same group have similar properties, there are definite variations in the properties of the elements and their compounds as any group in the Periodic Table is descended.

As any group in the Periodic Table is descended, the general trends are as follows:

1. The elements become *more metallic*

On descending a group, the atoms become larger and the outer electrons become further away and more shielded from the positive charge of the nucleus that binds them to the atom. The ionization energies of the outer electrons therefore decrease. Metals react by losing electrons to form positive ions. The 'ease' of this reaction is related to the energy required to take away the outer shell electrons. Therefore, as the ionization energies of the elements decrease, the metallic character becomes more pronounced. This trend may be difficult to see in Groups 1 or 2, because the elements are all metallic, but the trend is very apparent in Group 14.

Exercise 12S

Atomic and ionic radii

(i) Explain why the radius of a sodium atom is *larger* than that of a sodium ion, whereas that of a chlorine atom is *smaller* than a chloride ion.

(ii) The ions Na^+ and Mg^{2+} are **isoelectronic** (contain the same number of electrons), but Mg^{2+} is a much smaller ion. Explain why this should be so.

(iii) N^{3-} and F^- are also isoelectronic, but N^{3-} is the larger ion – why?

2. Atomic and ionic radii *increase*

Examples from Group 1 and 2 elements are shown back in Exercise 12F. Atomic and ionic radii for the halogens are shown below:

Symbol	Atomic radius/pm	Symbol	Ionic radius/pm
F	50	F^-	136
Cl	99	Cl^-	181
Br	114	Br^-	195
I	133	I^-	215

3. The chlorides and oxides of the elements become *more ionic*

These properties are related to the metallic nature of the element – metallic chlorides tend to be *ionic*, whereas non-metallic chlorides are **covalent**. Also, metallic oxides tend to be *ionic and basic*. Highly basic oxides such as Na_2O form *hydroxides* when dissolved in water:

$$Na_2O(s) + H_2O(l) \rightarrow 2NaOH(aq)$$

Non-metallic oxides tend to form covalent, *acidic oxides*. These oxides produce acids when dissolved in water, for example:

$$SO_2(g) + H_2O(l) \rightarrow H_2SO_3(aq)$$

4. The elements become *less electronegative*

The ability of an atom in a molecule to attract electron density to itself is related to the size of the atom. The further away the electrons in the bond are from the nucleus, the weaker they are attracted to the atom. In Group 17, therefore, fluorine is the most electronegative element and iodine is the least electronegative.

Changes in properties across a period

Consider the elements of the second period, excluding Ne:

Element	Li	Be	B	C	N	O	F
Electronic structure	2.1	2.2	2.3	2.4	2.5	2.6	2.7
s,p,d,f structure	$1s^22s^1$	$1s^22s^2$	$1s^22s^22p^1$	$1s^22s^22p^2$	$1s^22s^22p^3$	$1s^22s^22p^4$	$1s^22s^22p^5$

As we cross the period, from Li to F, the second shell of electrons is being filled. Each addition of one electron is accompanied by the addition of one proton into the nucleus, since the atoms in the elements are electrically neutral. Electrons in the *same shell* have little effect in shielding each other from nuclear charge, so that the *effective nuclear charge* felt by the outer electrons increases from left to right across the period. As a result of this increased 'pull' on the outer electrons by the nucleus:

1. The atomic radii of the elements *decreases*

In period 2, for example, the atomic radii decrease from Li at 145 pm to F with an atomic radius of 50 pm.

2. The first ionization energy *increases*

The energy required to remove one electron (the first ionization energy) increases as the effective nuclear charge felt by the electron increases. The variation of first ionization energy across the second period is shown in Figure 12.4.

Notice there is a *general increase* in ionization energy from left to right in the period. The 'saw-tooth' appearance of the plot is due to the slight differences in energies involved in removing electrons from filled or exactly half-filled subshells. A similar pattern is seen if the first ionization energies are plotted for the third period (Na to Ar). Note that sodium atoms have the lowest first ionization energy and argon atoms the highest.

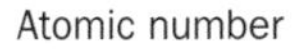

Fig. 12.4 First ionization energies of the elements of period 2.

3. The elements become *less metallic*

The tendency of atoms of the elements to lose electrons decreases from left to right across a period, so the elements gradually become less metallic. On the extreme right-hand side the halogens react, as typical non-metals, by *gaining electrons*. For example, for chlorine in the reaction of sodium metal with chlorine gas,

$$Cl_2(g) + 2e^-(g) \rightarrow 2Cl^-(s)$$

The halogens also react by sharing electrons to form covalent bonds. For example, in the reaction of hydrogen gas with chlorine gas,

$$H_2(g) + Cl_2(g) \rightarrow 2HCl(g)$$

4. Character of chlorides and oxides

The fact that the elements become less metallic, from left to right across a period, is shown by the characters of their chlorides and oxides. The chlorides of metals on the left-hand side of a period are *ionic* and *dissolve in water* to form *neutral solutions* whereas the chlorides of non-metals on the right-hand side of the table are *covalent and react with water*. There is a gradual change between these two extremes across the period. Consider the chlorides of period 3:

$NaCl(s)$	$MgCl_2(s)$	$AlCl_3(s)$	$SiCl_4(l)$	$PCl_3(l)$	$S_2Cl_2(l)$	$Cl_2(g)$
← white ionic → solids, dissolve in water		← covalent compounds, hydrolyse to → produce acidic solutions of HCl(aq), e.g. $SiCl_4(l) + 4H_2O(l) \rightarrow SiO_2 \cdot 2H_2O(s) + 4HCl(aq)$				

The oxides of period 33 change from *ionic to covalent* across the period. Oxides of the left-hand side metals are basic, whereas non-metallic oxides of the right-hand side elements are acidic. There is a gradual change from *basic to acidic* character of the oxides across the period:

$Na_2O(s)$	$MgO(s)$	$Al_2O_3(s)$	$SiO_2(s)$	$P_4O_6(s)$	$SO_2(g)$ $SO_3(l)$	$Cl_2O(g)$
←ionic solids→			←covalent→			
←basic→		amphoteric	←acidic→			

For example MgO(s) is basic, it *reacts with acids to give a salt and water*:

$$MgO(s) + 2HCl(aq) \rightarrow MgCl_2(aq) + H_2O(l)$$

The oxide $Al_2O_3(s)$ is **amphoteric**, it reacts with strong acids to produce salts:

$$Al_2O_3(s) + 3H_2SO_4(aq) \rightarrow Al_2(SO_4)_3(aq) + 3H_2O(l)$$

It will also react with a strong alkali to produce a salt:

$$Al_2O_3(s) + 2NaOH(aq) + 3H_2O(l) \rightarrow \underset{\text{sodium aluminate}}{2Na[Al(OH)_4](aq)}$$

The sulfur oxides are acidic; they react with water to form acids:

$$SO_2(g) + H_2O(l) \rightarrow H_2SO_3(aq)$$
$$SO_3(l) + H_2O(l) \rightarrow H_2SO_4(aq)$$

BOX 12.10

Distribution of metals and non-metals within the periodic table

The trends in metallic and non-metallic character of the elements within groups and periods within the periodic table may be summarized in the following diagram:

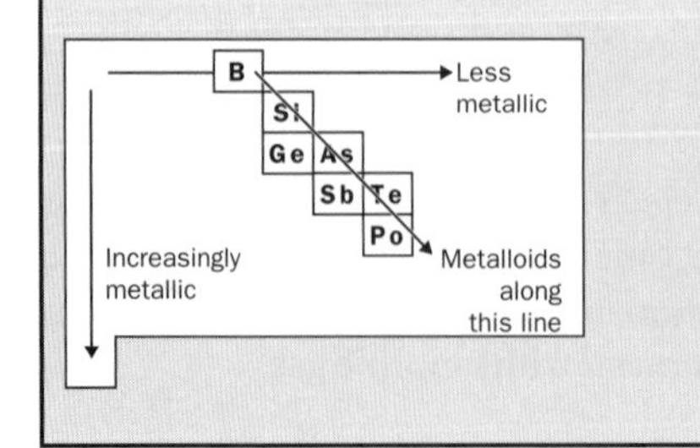

BOX 12.11

Mendeleev's periodic table

Dmitri Mendeleev, a Russian chemist and professor at the University of St. Petersburg. He was known to cut his hair only once a year, in the Spring!

Dimitri Mendeleev (1834–1907), a Siberian chemist, was not the first scientist to notice similarities between some groups of elements, but he was the first to organize the elements into a Periodic Table and use the concept to successfully predict the properties of then undiscovered, elements.

Convinced that the atomic weights (now atomic masses) of the elements were related to their properties and anxious to present some sort of 'ordering' system to the chemistry of the elements he published his **Periodic Law**:

Elements placed according to the value of their atomic weights present a clear periodicity of their properties.

Periodicity means 'reoccurring at regular intervals'. The order of atomic weights (now called atomic masses) did not always preserve the law and Mendeleev had to deviate from a strict order where elements were obviously misplaced. Iodine, for example, obviously belonged with the halogens even though the order that Mendeleev worked with did not predict this. Mendeleev realized that some elements had yet to be discovered, so he left gaps for those elements and dared to forecast some of their properties. Later, when these elements were discovered, some of his predictions were found to be astonishingly accurate and led to acceptance of the law and great excitement amongst the scientific community. Consider his predictions, for example, for the element gallium, which he named 'Eka-aluminium':

Predicted in 1871	discovered 1875
Eka-aluminium	Gallium
Atomic weight 68	atomic weight 69.9
Atomic volume 11.5	atomic volume 11.7
Specific gravity 6.0	specific gravity 5.96

The classification of elements in the Periodic Table is one of the greatest contributions made to modern chemistry. Today's improved Periodic Table positions the elements in order of atomic number rather than atomic weight because scientists have realized that, and to some extent why, the properties of elements depend on their atomic number (which gives us the number and arrangements of the electrons in their atoms).

The hunt for new elements continues. To date, many elements between 93 and 118 have been synthesized artificially by nuclear reactions. The heaviest elements are very unstable and cannot be of any practical use, but scientists predicted that an 'island of stability' exists at around element 114 (one atom of this element was reported in January 1999, and it does seem to be more stable than other similarly heavy elements). Attempts have already been made to predict the properties of, as yet, undiscovered elements using Periodic trends.

	Group							
Period	I	II	III	IV	V	VI	VII	VIII
1	H							
2	Li	Be	B	C	N	O	F	
3	Na	Mg	Al	Si	P	S	Cl	
	K	Ca	*	Ti	V	Cr	Mn	Fe Co Ni
4	Cu	Zn	*	*	As	Se	Br	
	Rb	Sr	Y	Zr	Nb	Mo	*	Ru Rh Pd
5	Ag	Cd	In	Sn	Sb	Te	I	

The asterisk * marked spaces for elements which had yet to be discovered. They were scandium, gallium, germanium and technetium.

Revision questions

12.1. Using only the symbols shown in this outline of the Periodic Table

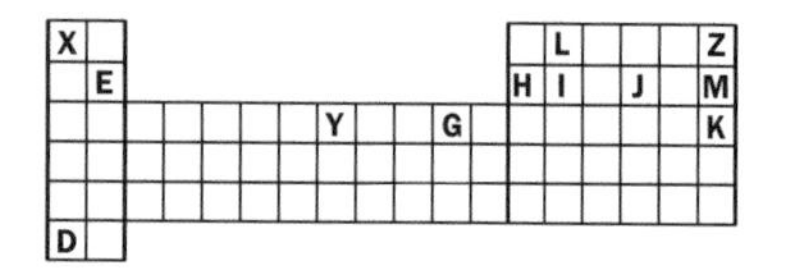

give the symbols for:

(i) an element in Group 13 of the Periodic Table
(ii) an element in period 2 of the table
(iii) a transition metal
(iv) an alkali metal
(v) a noble gas
(vi) an element the atoms of which have the electron arrangement 2.8.4.
(vii) an element which forms an ion with a 2− charge
(viii) an element which forms an ion with a 3+ charge
(ix) the most reactive metal
(x) an element that forms a great many compounds in which it exhibits catenation.

12.2. Radium is an alkaline earth metal.
(i) How would you expect it to behave when reacting with water? Write a balanced equation for this reaction.
(ii) Comment on the thermal stability and solubility of radium carbonate.

12.3. Tin-plated metal cans are generally not used in very cold countries – why?

12.4. What is the formula of a simple compound of carbon and sulfur in which carbon has oxidation state +2? (The compound is not very stable.)

12.5. Consider the following oxides formed by some elements in period 2 of the Periodic Table:

Li_2O, BeO, B_2O_3, CO_2, NO_2

(i) Which oxide would you expect to be the most basic? Write an equation for its reaction with water.
(ii) Which oxide would you expect to be the most acidic?
(iii) Give the formula of an amphoteric oxide.

12.6. What type of bonding exists in **(i)** RbCl **(ii)** ICl?

12.7. The graph shows the boiling points of the first 18 elements:

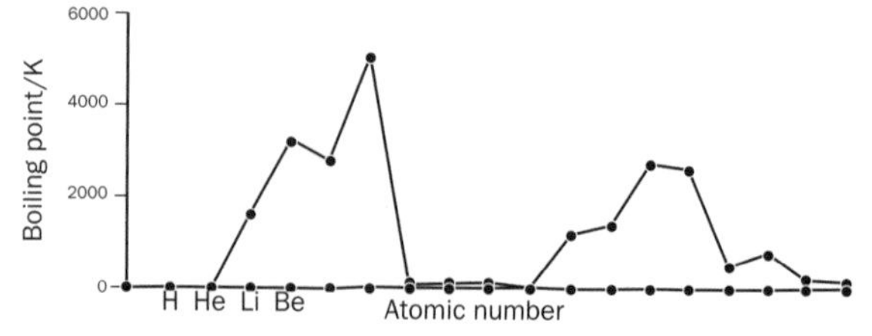

(i) Complete the graph by labelling the elements.
(ii) Which two elements have the highest boiling points? Describe the bonding in these elements.
(iii) Name two elements with very low boiling points. Describe the bonding in these elements.
(iv) Name two elements which exhibit metallic bonding. Describe metallic bonding.
(v) Is there a periodic variation in boiling point among the elements of the second and third periods? Explain.

12.8. Consider the complex compound $[Ni(H_2O)_4Cl_2]$.
(i) What is the coordination number of the central metal atom?
(ii) What is the oxidation state of the nickel?
(iii) Name the compound.

12.9. Write formulae for:
(i) the tetraaquadichlorochromium(III) ion
(ii) the hexabromoplatinate(IV) ion.

12.10. Sketch and describe the structures of:
(i) $[Cu(NH_3)_2]^+$
(ii) $[Co(NH_3)_6]^{3+}$

Extension material to support this unit is available on our website at **http://www.palgrave.com/foundations/lewis.**

Energy Changes in Chemical Reactions

Objectives

- ▶ Defines different types of standard enthalpy change
- ▶ Explains Hess's law
- ▶ Introduces calculations involving enthalpy changes
- ▶ Shows how enthalpy changes may be measured
- ▶ Discusses fuels, nutrition and explosives

Contents

13.1 Conservation of energy

Following the experiments of J. P. Joule (1818–1889), it was concluded that energy is neither created or destroyed but is merely converted from one form to another. For example, a dynamo converts mechanical energy to electrical energy. If the dynamo does 100 J of work, then the sum of the electrical energy produced and the energy lost as friction also equals 100 J. This preservation (or 'conservation') of the total amount of energy is called the **law of conservation of energy**.

In a chemical reaction, new substances are made. The law of conservation of energy tells us that the total energy of the reactants must equal the total energy of the products and any energy lost to (or gained from) the surroundings:

total energy of reactants = total energy of products + energy lost (or gained)

Enthalpy

The energy of a substance under constant atmospheric pressure (such as in an open test tube or beaker) is called its **enthalpy** (symbol, H). The enthalpy of elements and compounds cannot be measured or calculated, but differences in enthalpy (**enthalpy changes**) are easily measured in the laboratory. If a chemical or physical change is carried out at constant pressure, the amount of heat energy absorbed or produced in the change equals the change in enthalpy that has taken place.

More about enthalpy changes

The enthalpy change ΔH of a reaction

reactants → products

is defined by the equation

ΔH = (sum of enthalpies of products) − (sum of enthalpies of reactants)

Using the mathematical symbol Σ for sum, and the letters R and P for reactants and products, we can write this as as follows:

$$\Delta H = \Sigma H_{\mathrm{P}} - \Sigma H_{\mathrm{R}} \qquad \textbf{(13.1)}$$

Suppose that heat is given out during a chemical reaction (this is the case for the burning of fuels). Such a reaction is said to be **exothermic**, and the container holding the reaction mixture will be warm or hot to the touch. This loss of energy shows that the reactant molecules were at a higher level of enthalpy than the product molecules. Therefore, for exothermic reactions, ΔH (as defined by equation (13.1)) is **negative.** This is represented in Fig. 13.1(a), which shows the total enthalpy of the substances before the reaction starts and after reaction is complete. In this sense, the *x* axis (labelled 'progress of reaction') may be thought of as a time axis.

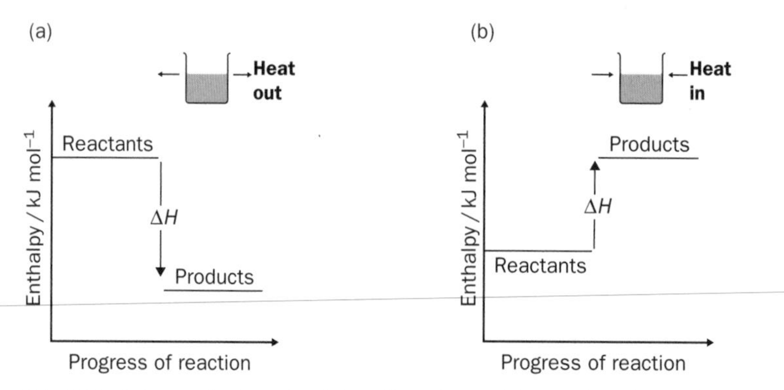

Fig. 13.1 Enthalpy diagrams for (a) exothermic and (b) endothermic reactions.

Figure 13.1(b) shows the change of enthalpy for a reaction in which heat is taken in (an **endothermic reaction**). The container holding the reaction mixture will be cool or cold to the touch. In this case, the reactant molecules are at a lower level of enthalpy than the product molecules. Therefore, for endothermic reactions ΔH is **positive.**

Once the enthalpy change of a reaction is known, we may write it down on the right-hand side of the chemical equation to which it applies. For example, the reaction between ethene and hydrogen is exothermic with $\Delta H = -137$ kJ per mol of reacted ethene at 25 °C. We would write this as

$$CH_2{=}CH_2(g) + H_2(g) \rightarrow CH_3CH_3(g) \qquad \Delta H = -137\ \mathrm{kJ\,mol^{-1}}$$

The combination of a chemical equation and its ΔH value is known as a **thermochemical equation.**

One application of reactions that take in or give out heat is in cooling and heating packs (Box 13.1).

BOX 13.1

Cooling and heating packs

Instant ice packs are used to cool food and also to numb sports injuries on the field of play. The ice packs do not require refrigeration and can be activated simply by striking them with a fist.

The ice packs consist of a thin plastic bag containing water, which is contained in a larger outer bag containing ammonium nitrate. When the pack is thumped, the bag containing the water breaks and the ammonium nitrate dissolves in the water. Ammonium nitrate is used because when it dissolves, heat is taken in

$$NH_4NO_3(s) \rightarrow NH_4^+(aq) + NO_3^-(aq) \qquad \Delta H \approx +26\,kJ\,mol^{-1}$$

A mixture containing 50% (by mass) of water and ammonium nitrate reaches a temperature of about −15 °C.

Heat packs use several types of reaction. In one type, the reaction of damp iron powder (in the pack) and oxygen (in the air) produces enough heat to warm up food. This type of heat pack cannot be reused.

A second type of heat pack, used as a hand warmer in winter, depends upon the fact that the crystallization of some salts from solution (the reverse of dissolution) is exothermic. Sodium thiosulfate solution is chosen because it can be made to hold more than the equilibrium concentration of solute (sodium thiosulfate is said to make a **supersaturated solution**) and so there are more crystals available to crystallise out. The sodium thiosulfate is contained in a plastic bag and the crystallization started by pushing in a steel disc upon whose surface crystallization occurs.

The thiosulfate heat pack can be reused because the crystallization is reversible. Immersing the bag of thiosulfate in hot water dissolves the crystals and regenerates the supersaturated solution.

The burning of acetylene (HC≡CH) in oxygen produces so much heat that part of the flame reaches 3000 °C. This is high enough to melt most metals (iron melts at 1535 °C) and is used to melt metals together, an operation known as gas welding.

Standard enthalpy changes of reaction ($\Delta H^\ominus$)

The enthalpy change of a reaction when the reactants and products are in their *standard states* is called the **standard enthalpy change of reaction**, $\Delta H^\ominus$ (with the superscript $\ominus$ sign being read as 'standard'):

$$\Delta H^\ominus = \Sigma H_P^\ominus - \Sigma H_R^\ominus \qquad \textbf{(13.2)}$$

What is a standard state? **A substance is said to be in its standard state when it is in its pure form at an atmospheric pressure of 1 atm** (101 kPa). Temperature is not part of the definition of a standard state and (for example) the standard state of iron is pure iron at a pressure of 1 atm whatever its temperature, although it happens that at 25 °C the iron would be a solid and at 1600 °C the iron would be a liquid.

The idea of a standard state is equally applicable to compounds. For instance, the standard state of sodium chloride would be pure sodium chloride at a pressure of 1 atm.

Where elements may exist in several allotropic forms, one form is chosen as the **reference state**. For example, carbon exists in two common forms, graphite and diamond, but graphite has been chosen as the reference state for the element carbon. Similarly dioxygen (O_2, but not ozone (O_3)), has been chosen as the reference state for the element oxygen. Usually, reference states are the most stable forms of an element. An exception to this rule is that white phosphorus (and not

Exercise 13A

Enthalpy change

A few reactions, such as the decomposition of silver chlorate(III) upon heating,

$$2AgClO_2(s) \rightarrow 2Ag(s) + Cl_2(g) + 2O_2(g) \qquad \Delta H \approx 0$$

involve changes in enthalpies which are so near zero that they cannot be measured. Sketch an enthalpy diagram (following the pattern in Fig. 13.1) for such a reaction.

the more stable red phosphorus), has been chosen as the reference state for the element phosphorus.

It is most common for standard enthalpy changes to be measured at 25 °C (more precisely, at 298.15 K), or else for standard enthalpy changes obtained at other temperatures to be theoretically corrected to 25 °C. Such enthalpy changes are represented by the notation $\Delta H^{\ominus}(298\,\text{K})$. Unless otherwise stated, we will assume that all standard enthalpy changes refer to a reaction where the temperature at the beginning and end of the reaction is 298 °C, and we shall therefore take $\Delta H^{\ominus}$ to mean $\Delta H^{\ominus}(298\,\text{K})$. (In fact, provided there are no changes in phase, the variation of $\Delta H^{\ominus}$ with *small* changes in temperature is often negligible, and $\Delta H^{\ominus}$ may then be considered as independent of temperature.)

13.2 Key points about enthalpy changes

1. ΔH values are meaningless unless we state how many moles of product are made, or else include a chemical equation to which that enthalpy change applies.

The enthalpy change of a reaction is an **extensive property**. This means that the enthalpy change depends upon the quantities of reactants which, in turn, control the quantities of product made. In the absence of any other information, we always assume that the ΔH value for a reaction is produced when the number of moles of reactants that combine are those indicated by the chemical equation. So, when we write the thermochemical equation

$$N_2(g) + 3H_2(g) \rightarrow 2NH_3(g) \qquad \Delta H^{\ominus} = -92.22\,\text{kJ mol}^{-1}$$

we immediately understand that the stated enthalpy change arises when three moles of hydrogen gas react with one mole of nitrogen gas to make two moles of ammonia. If we use up a different number of moles of reactants, the enthalpy change is worked out by simple proportion, as in the following example.

Example 13.1

Calculate the amount of heat produced when 0.32 g of methanol is completely burned in excess oxygen at 25 °C.

$$2CH_3OH(l) + 3O_2(g) \rightarrow 2CO_2(g) + 4H_2O(g) \qquad \Delta H^{\ominus} = -726\,\text{kJ mol}^{-1}$$

▶ **Answer**

$$\text{amount of methanol (in mol)} = \frac{0.32\,\text{g}}{32\,\text{g mol}^{-1}} = 0.010\,\text{mol}$$

The thermochemical equation shows that two mol of methanol burn to produce 726 kJ of heat. Therefore, the enthalpy change that accompanies the burning of 0.010 mol of methanol is

$$\frac{0.010}{2} \times 726 = -3.6\,\text{kJ}$$

2. The physical states of the reactants and products must be clearly shown.

The importance of specifying the physical states of the substances involved in a reaction may be illustrated by referring to the reaction of hydrogen and oxygen gases, with gaseous water as the product:

$$2H_2(g) + O_2(g) \rightarrow 2H_2O(g)$$

$\Delta H^\ominus$ for this reaction is $-483.64\ \text{kJ mol}^{-1}$. Experiments show that $\Delta H^\ominus$ for the reaction

$$2H_2(g) + O_2(g) \rightarrow 2H_2O(l)$$

(in which 2 mol of *liquid* water is made) is $-571.66\ \text{kJ mol}^{-1}$. The difference in the standard enthalpy changes for these reactions,

$$-571.66 - (-483.64) = -88.02\ \text{kJ mol}^{-1}$$

is the energy released when 2 mol of $H_2O(g)$ condense to form 2 mol of $H_2O(l)$.

Exercise 13B

Thermochemical equations

(i) Explain why the statement 'for the reaction of sodium with chlorine, $\Delta H^\ominus = -411.15\ \text{kJ mol}^{-1}$' is ambiguous. What extra information would make sense of this statement?

(ii) Given that

$$C(s) + O_2(g) \rightarrow CO_2(g) \qquad \Delta H^\ominus = -393.15\ \text{kJ mol}^{-1}$$

calculate the amount of heat given out when 120 g of carbon is completely burned in excess air.

3. Reverse reactions possess ΔH values which are equal in size, but opposite in sign, to the forward reaction.

What is $\Delta H^\ominus$ for the following reaction?

$$2H_2O(g) \rightarrow 2H_2(g) + O_2(g)$$

The answer is simply that since the reaction is *exactly* the reverse of the equation

$$2H_2(g) + O_2(g) \rightarrow 2H_2O(g)$$

then the enthalpy change is equal in size but opposite in sign, i.e. $+483.64\ \text{kJ mol}^{-1}$.

We summarize this in the following rule: **reactions which are exothermic in the forward direction are endothermic in the reverse direction.**

This rule applies to both physical and chemical changes. For example, heat is absorbed in vaporization because liquid molecules require energy to escape from their neighbours into the gas phase. Therefore, vaporization is always endothermic. The **standard enthalpy of vaporization** of acetone at 25 °C is +29.1 kJ per mol of acetone:

$$CH_3COCH_3(l) \rightarrow CH_3COCH_3(g) \qquad \Delta H^\ominus_{vap} = +29.1\ \text{kJ mol}^{-1}$$

It follows that 29.1 kJ of heat energy is given out when 1 mol of acetone *condenses* into acetone liquid:

$$CH_3COCH_3(g) \rightarrow CH_3COCH_3(l) \qquad \Delta H^\ominus = -29.1\ \text{kJ mol}^{-1}$$

We can also look at these relationships at a molecular level. Suppose that we were able magically to video the collision of two molecules to produce a new molecule in an exothermic reaction. If we were to play back the video in reverse, we would see that heat is absorbed just before the new molecule began breaking apart. The principle that if reactions are reversed the energy changes are equal in size but opposite in sign is important because if this were not so, it would be possible to destroy or create energy, and the law of conservation of energy would be broken.

Exercise 13C

Enthalpy changes of back reactions

The formation of magnesium oxide from its elements at 298 K obeys the thermochemical equation

$2Mg(s) + O_2(g) \rightarrow 2MgO(s)$

$\Delta H^\ominus = -601.7\ kJ\ mol^{-1}$

How much energy would be required to decompose one mol of magnesium oxide into oxygen gas and magnesium metal?

4. If a reaction can occur by more than one route, the enthalpy change is the same, whichever route is followed.

This is a statement of **Hess's law**. As an example, we shall look at the production of hydrogen iodide gas from iodine and hydrogen. The reaction can occur in one stage in which hydrogen iodide gas is made directly:

$$H_2(g) + I_2(s) \rightarrow 2HI(g) \qquad \Delta H^\ominus = +53\ kJ\ mol^{-1} \qquad \textbf{(13.3)}$$

We call this route A.

Alternatively, we can carry out this reaction in *two* stages:

$$I_2(s) \rightarrow I_2(g) \qquad \Delta H^\ominus = +62\ kJ\ mol^{-1} \qquad \textbf{(13.4)}$$

followed by

$$I_2(g) + H_2(g) \rightarrow 2HI(g) \qquad \Delta H^\ominus = -9\ kJ\ mol^{-1} \qquad \textbf{(13.5)}$$

We shall call reactions (13.4) and (13.5) route B.

Both routes are illustrated in Fig. 13.2(a), which shows the enthalpy changes for the reaction in which 1 mol of reactants is converted into 2 mol of HI. The diagram

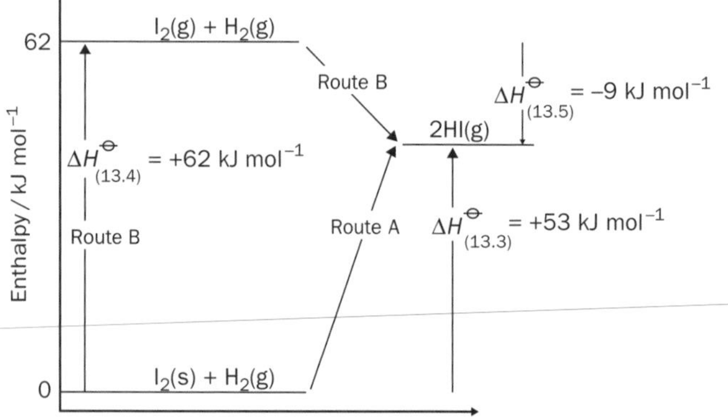

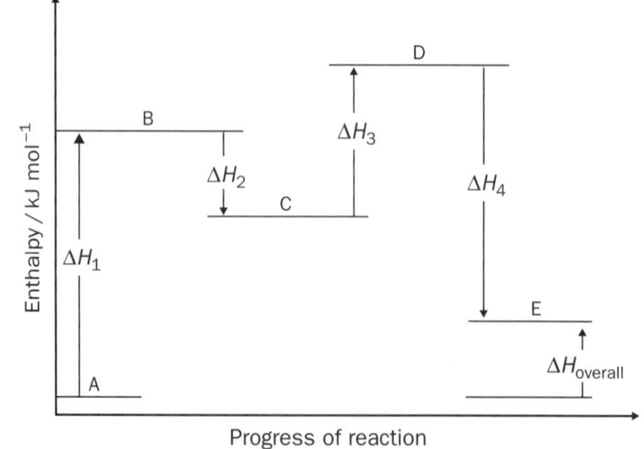

Fig. 13.2 Hess's law: (a) the manufacture of hydrogen iodide by two routes; (b) the general case.

shows that standard enthalpy change for reaction (13.3) is the *sum* of the enthalpy changes for reactions (13.4) and (13.5):

$$\Delta H^\ominus_{(13.3)} = \Delta H^\ominus_{(13.4)} + \Delta H^\ominus_{(13.5)}$$
$$= (+62) + (-9) = +53\,\text{kJ mol}^{-1}$$

Generalizing, if a reaction A $\rightarrow$ E is carried out in several stages,

$$\text{A} \xrightarrow{1} \text{B} \xrightarrow{2} \text{C} \xrightarrow{3} \text{D} \xrightarrow{4} \text{E}$$

then the enthalpy change for the overall reaction,

$$\Delta H(\text{A} \rightarrow \text{E})$$

is the sum of the individual enthalpy changes:

$$\Delta H(\text{A} \rightarrow \text{E}) = \Delta H(\text{A} \rightarrow \text{B}) + \Delta H(\text{B} \rightarrow \text{C}) + \Delta H(\text{C} \rightarrow \text{D}) + \Delta H(\text{D} \rightarrow \text{E})$$

or,

$$\Delta H_{\text{overall}} = \Delta H_1 + \Delta H_2 + \Delta H_3 + \Delta H_4$$

as illustrated in Fig. 13.2(b).

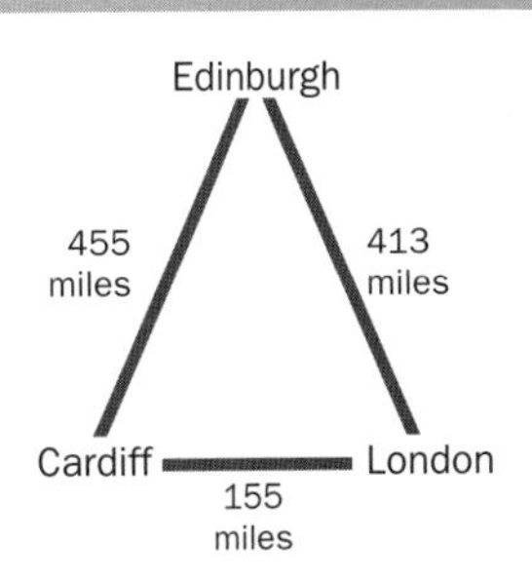

Fig. 13.3 A geographical analogy of Hess's law.

Hess's law: a geographical analogy

Cardiff is 155 miles from London. Suppose that instead of travelling direct from Cardiff to London, you travelled first from Cardiff to Edinburgh (455 miles) and then from Edinburgh to London (413 miles) as in Fig. 13.3. Whichever route you select, Cardiff is still only 155 miles from London. Similarly, whatever the chemical route taken in Fig. 13.2(a), the *energy difference* between I_2(s) + H_2(g) and 2HI(g) remains at 53 kJ mol^{-1}.

Use of Hess's law to calculate enthalpy changes

Hess's law allows us to work out a ΔH value for a reaction without carrying out that reaction in the laboratory. This is achieved by algebraically manipulating (adding, subtracting, or multiplying by a factor) the chemical equations of related reactions so that they produce the chemical equation for the reaction whose enthalpy change is required. The value for ΔH is then obtained by applying the same operations to the enthalpy changes of the related reactions. This procedure is illustrated in Example 13.2.

Example 13.2

From the thermochemical equations

$H_2(g) + Cl_2(g) \rightarrow 2HCl(g)$	$\Delta H^\ominus_{(13.6)} = -184.6\,\text{kJ mol}^{-1}$	**(13.6)**
$2NH_3(g) \rightarrow 3H_2(g) + N_2(g)$	$\Delta H^\ominus_{(13.7)} = +92.2\,\text{kJ mol}^{-1}$	**(13.7)**
$\frac{1}{2}N_2(g) + 2H_2(g) + \frac{1}{2}Cl_2(g) \rightarrow NH_4Cl(s)$	$\Delta H^\ominus_{(13.8)} = -314.4\,\text{kJ mol}^{-1}$	**(13.8)**

calculate the enthalpy change for the reaction

$HCl(g) + NH_3(g) \rightarrow NH_4Cl(s)$	$\Delta H^\ominus_{(13.9)} = ?$	**(13.9)**

Example 13.2 *(continued)*

▶ Answer

We start by multiplying equation (13.6) by $-\frac{1}{2}$, giving

$$-\tfrac{1}{2}H_2(g) + -\tfrac{1}{2}Cl_2(g) \rightarrow -HCl(g)$$

The arrow in this equation may be likened to the equals sign (=) in a mathematical equation, and the chemical equation may be re-arranged in a similar way to algebraic equations giving:

$$HCl(g) \rightarrow \tfrac{1}{2}H_2(g) + \tfrac{1}{2}Cl_2(g) \qquad \textbf{(13.10)}$$

(Compare this to the rearrangement of $-x + -z = -y$, to give $y = x + z$). The rearrangement puts HCl(g) on the left-hand side of an equation, as in equation (13.9).

Dividing equation (13.7) throughout by 2 gives

$$NH_3(g) \rightarrow \tfrac{3}{2}\ H_2(g) + \tfrac{1}{2}N_2(g) \qquad \textbf{(13.11)}$$

which keeps $NH_3(g)$ on the left-hand side, as in equation (13.9).

Adding equations (13.10) and (13.11) to equation (13.8) gives

$$HCl(g) + NH_3(g) + \tfrac{1}{2}N_2(g) + 2H_2(g) + \tfrac{1}{2}Cl_2(g)$$
$$\rightarrow \tfrac{1}{2}H_2(g) + \tfrac{1}{2}Cl_2(g) + \tfrac{3}{2}\ H_2(g) + \tfrac{1}{2}N_2(g) + NH_4Cl(s)$$

Collection and cancellation of like terms gives:

$$HCl(g) + NH_3(g) \rightarrow NH_4Cl(s)$$

which we recognize as equation (13.9). The *enthalpy change* is now obtained by carrying out similar operations to the *ΔH values* as follows:

$$\Delta H^{\circ}_{(13.9)} = -\tfrac{1}{2}\Delta H^{\circ}_{(13.6)} + \tfrac{1}{2}\Delta H^{\circ}_{(13.7)} + \Delta H^{\circ}_{(13.8)}$$
$$= -\tfrac{1}{2}(-184.6) + \tfrac{1}{2}(92.2) + (-314.4) = -176.0\ \text{kJ mol}^{-1}$$

Exercise 13D

Application of Hess's law

Given the following standard enthalpy changes at 298 K,

$$S(s) + O_2(g) \rightarrow SO_2(g) \qquad \Delta H^{\circ} = -296.8\ \text{kJ mol}^{-1} \qquad \textbf{(13.12)}$$

$$SO_2(g) + \tfrac{1}{2}O_2(g) \rightarrow SO_3(g) \qquad \Delta H^{\circ} = -98.9\ \text{kJ mol}^{-1} \qquad \textbf{(13.13)}$$

calculate the standard enthalpy change for the reaction:

$$S(s) + \tfrac{3}{2}O_2(g) \rightarrow SO_3(g) \qquad \Delta H^{\circ} = ? \qquad \textbf{(13.14)}$$

13.3 Determination of ΔH in the laboratory

The following section gives an example of a laboratory determination of the enthalpy change for a chemical reaction in which the reactants are mixed in water. After the reaction has taken place the final mixture consists of products, unused reactants and water. The principles of such measurements are as follows.

- The reaction is carried out in an insulated container which ideally prevents the reaction mixture from losing heat to (or gaining heat from) the surroundings.

- The chemical reaction should be rapid, so that the energy change is complete in a short period of time. This is achieved by rapidly mixing the reactants.
- An exothermic chemical reaction increases the amount of heat energy contained in the reaction mixture. This raises the temperature of the final mixture by ΔT°C, defined as $T_{\text{final}} - T_{\text{initial}}$. An endothermic reaction decreases the amount of heat energy contained in the reaction mixture, so lowering the temperature of the final mixture.
- The change in the amount of heat energy due to chemical reaction, q joules, is calculated using the equation

  $$q = -m \times C \times \Delta T$$

  (note the negative sign) where m is the mass and C the average specific heat capacity of the final mixture (mainly water) whose temperature is being measured. The value of q is negative if the chemical reaction is exothermic (when ΔT is positive); the value of q is positive if the chemical reaction is endothermic (when ΔT is negative).
- At constant pressure, the heat change undergone produced by the chemical reaction is equal to the enthalpy change of the reaction, i.e. $q = \Delta H$. Note that q and ΔH have the same sign.
- The calculations assume that the reaction 'goes to completion'.

Experimental details

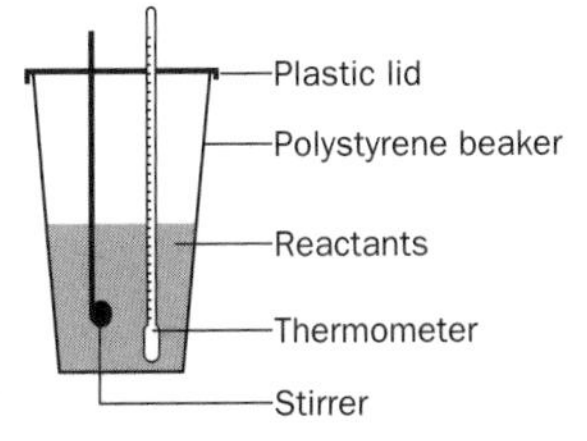

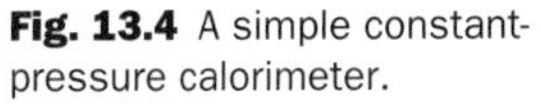
Fig. 13.4 A simple constant-pressure calorimeter.

1. The reactants are contained in an insulated container vented to the atmosphere so as to keep the contents at atmospheric pressure. Such a container is known as a **constant-pressure calorimeter**, and at its simplest the container would be a polystyrene coffee cup (Fig. 13.4). Temperature changes are measured using a thermometer.

2. A known quantity of each of the reactants is placed in separate containers (in the following example the reactants are powdered zinc and lead(II) nitrate solution) and allowed 10 min to achieve the constant temperature, T_{initial}, of the laboratory. This 'equilibration time' corresponds to the points between A and B in Fig. 13.5. If we want to measure $\Delta H^{\ominus}$ (298 K), T_{initial} should be 298 K and the air pressure should be exactly 1 atm.

3. The reactants are mixed at time t seconds and rapidly stirred so that reaction is complete within 30 s of mixing. As the reaction mixture can never be truly insulated, the container and its contents immediately start to lose heat (points C to D in an exothermic reaction, Fig. 13.5(a)) or gain heat (points E to F in an endothermic reaction, Fig. 13.5(b)). The temperature of the mixture changes too rapidly for a single definitive measurement to be taken immediately after mixing. Instead, tempera-

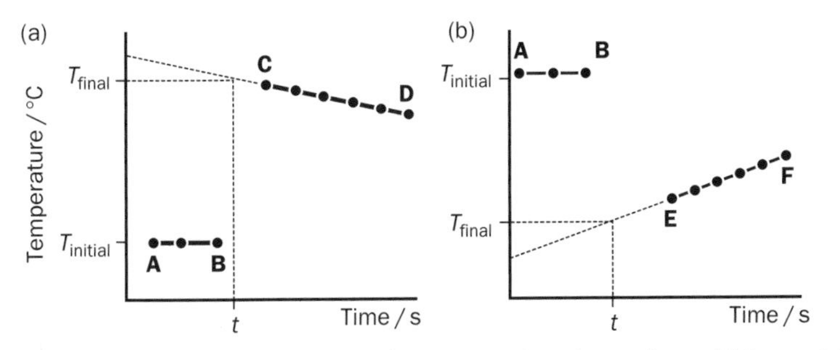

Fig. 13.5 Temperature–time profiles for (a) an exothermic reaction and (b) an endothermic reaction.

ture measurements are recorded at regular intervals (say every 5 s) after reaction completion and extrapolation of the resulting graph (shown by the dotted lines in Fig. 13.5) is used to give the temperature change ΔT at mixing.

Example 13.3

Find the enthalpy change for the exothermic reaction

$$Zn(s) + Pb(NO_3)_2(aq) \rightarrow Zn(NO_3)_2(aq) + Pb(s)$$

The ionic reaction involved is

$$Zn(s) + Pb^{2+}(aq) \rightarrow Zn^{2+}(aq) + Pb(s)$$

In an experiment, 0.500 g of finely powdered zinc was rapidly stirred into 100.0 cm^3 of 0.100 mol dm^{-3} lead(II) nitrate solution (this is an excess of lead(II) nitrate solution) in a polystyrene cup. The recorded temperature of the mixture rose from 20.0 °C (before mixing) to 22.6°C so that

$$\Delta T = T_{final} - T_{initial} = 22.6 - 20.0 = +2.6°C.$$

We make the approximation that the heat capacity of the polystyrene cup and of the reactants and products is relatively small and may be ignored. The lead(II) nitrate solution is so dilute that its specific heat capacity may be taken to be the same as that of pure water (4.18 J $g^{-1}°C^{-1}$). In effect, we are assuming that the heat produced by the reaction is entirely absorbed by the 100 g of water present.

The change in heat energy due to the chemical reaction, q, is calculated as follows:

$$q = -m \times C \times \Delta T = -100.0 \times 4.18 \times 2.6 = -1087\ \text{J} = -1.087\ \text{kJ}$$

(units: g J $g^{-1}°C^{-1}$ °C = J)

The experiment involves the conversion of 0.500 g of Zn to $Zn^{2+}(aq)$:

$$\text{amount of Zn} = \frac{0.500\ \text{g}}{65.39\ \text{g mol}^{-1}} = 0.007\,65\ \text{mol}$$

Therefore, 1.087 kJ of heat is produced by the chemical reaction (and subsequently absorbed by the water) when 0.007 65 mol of Zn is used up. If one mol of Zn(s) were consumed in the reaction, the heat change would be

$$q\,(\text{per mol}) = -1.087 \times \frac{1}{0.007\,65} = -142\ \text{kJ mol}^{-1}$$

As we have only measured ΔT to two significant figures, q should be rounded down to -140 kJ mol^{-1}. Finally, as q and ΔH are equal,

$$\Delta H = -140\ \text{kJ mol}^{-1}$$

This will be approximately equal to $\Delta H^{\ominus}$(298 K). The accepted value of $\Delta H^{\ominus}$ (298 K) is -152 kJ mol^{-1}, the difference between the experimental and literature values being mainly due to excessive heat losses by the mixture in the polystyrene cup and by the approximations made in the calculation. More accurate experimental determinations rely upon more efficient insulation and a more rapid (electronic) means of recording temperature. In these cases, the heat capacities of the container and of the reactants and products are also taken into account.

Exercise 13E

Determining enthalpy changes 1

In Example 13.3

(i) Why was the zinc powdered?

(ii) Calculate the number of moles of lead nitrate in the solution, and hence show that the lead nitrate is in excess.

Exercise 13F

Determining enthalpy changes 2

50.0 cm³ of 1.00 mol dm⁻³ HCl and 50.0 cm³ of 2.00 mol dm⁻³ NaOH (both at an initial temperature of 17.40 °C) were rapidly mixed in a vacuum flask and the maximum temperature recorded electronically as 23.60 °C. Write down an ionic equation for this reaction and estimate the enthalpy change of reaction per mol of water formed (the literature value for $\Delta H^{\ominus}$ is −57.3 kJ mol⁻¹).

13.4 Special kinds of standard enthalpy change

Some examples of special standard enthalpy changes are listed in Table 13.1.

Table 13.1 Examples of special standard enthalpy changes

Name	Symbol	Description	Example of process
Enthalpy of formation	$\Delta H_f^{\ominus}$	one mole of compound made from its elements	$K(s) + \frac{1}{2}Cl_2(g) \rightarrow KCl(s)$
Enthalpy of combustion	$\Delta H_c^{\ominus}$	one mole of a fuel burned in excess oxygen	$CH_4(g) + 2O_2(g) \rightarrow CO_2(g) + 2H_2O(l)$
Lattice enthalpy	$\Delta H_L^{\ominus}$	one mole of a crystal completely separated into isolated particles in the gas phase	$K^+,Cl^-(s) \rightarrow K^+(g) + Cl^-(g)$
Enthalpy of neutralization	$\Delta H_N^{\ominus}$	one mole of water formed by the neutralization of an acid by a base	$CH_3COOH(aq) + NaOH(aq) \rightarrow CH_3COONa(aq) + H_2O(l)$
Bond dissociation enthalpy (of a specific bond A–B)	$\Delta H^{\ominus}$(A–B)	one mole of bonds broken: all species in the gas phase	$HCl(g) \rightarrow H(g) + Cl(g)$
Ionization energy (also known as **ionization enthalpy**)	$\Delta H_{ion}^{\ominus}$ or I	one mole of atoms (or molecules) ionized: all species in the gas phase	$Na(g) \rightarrow Na^+(g) + e^-$ (first ionization energy)
Enthalpy of electron gain ($-\Delta H_{eg}^{\ominus}$ is called **electron affinity**)	$\Delta H_{eg}^{\ominus}$	one mole of anions being formed: all species in the gas phase	$F(g) + e^- \rightarrow F^-(g)$
Enthalpy of vaporization	$\Delta H_{vap}^{\ominus}$	one mole of vapour being formed from liquid without a change in temperature	$H_2O(l) \rightarrow H_2O(g)$
Enthalpy of fusion	$\Delta H_{fus}^{\ominus}$	one mole of liquid being formed from the solid without a change in temperature	$H_2O(s) \rightarrow H_2O(l)$
Enthalpy of atomization	$\Delta H_{atom}^{\ominus}$	one mole of substance broken up into its isolated atoms in the gas phase	$Ca(s) \rightarrow Ca(g)$

13.5 Standard enthalpy of formation

The standard enthalpy of formation of a compound $\Delta H_f^{\ominus}$ is the standard enthalpy change at a particular temperature when *one mole of compound is made from its elements.*

The standard enthalpies of formation used in this book are those which have been measured (or calculated) at 25 °C (298 K), and we shall take $\Delta H_f^{\ominus}$ as meaning $\Delta H_f^{\ominus}$(298 K). The units of $\Delta H_f^{\ominus}$ are kJ *per mole of compound* made.

Key points about standard enthalpies of formation

- $\Delta H_f^{\ominus}$ is a *standard* enthalpy change, and therefore applies to a reaction in which both the elements and compound involved are present in their pure forms at an atmospheric pressure of 1 atm.
- $\Delta H_f^{\ominus}$ at a particular temperature is a constant for a compound. Extensive compilations of $\Delta H_f^{\ominus}$ values are available. Selected values are shown in Table 13.2.

Table 13.2 Standard enthalpies of formation of selected elements and their compounds at 298 K

Substance	Formula and state	$\Delta H_f^{\ominus}$/kJ mol^{-1}
Aluminium	$Al(s)$	0
Aluminium oxide	$Al_2O_3(s)$	−1675.7
Bromine	$Br_2(l)$	0
Bromine	$Br_2(g)$	+30.91
Hydrogen bromide	$HBr(g)$	−36.40
Calcium	$Ca(s)$	0
Calcium oxide	$CaO(s)$	−635.09
Calcium chloride	$CaCl_2(s)$	−795.8
Carbon (graphite)	$C(s)$ (graphite)	0
Carbon (diamond)	$C(s)$ (diamond)	+1.895
Carbon monoxide	$CO(g)$	−110.53
Carbon dioxide	$CO_2(g)$	−393.51
Tetrachloromethane	$CCl_4(l)$	−135.44
Hydrogen cyanide	$HCN(g)$	+135.1
Methane	$CH_4(g)$	−74.81
Ethyne (acetylene)	$C_2H_2(g)$	+226.73
Butane	$C_4H_{10}(g)$	−126.15
Cyclohexane	$C_6H_{12}(l)$	−156.4
Benzene	$C_6H_6(l)$	+49.0
Ethanol	$C_2H_5OH(l)$	−277.69
Sucrose	$C_{12}H_{22}O_{11}(s)$	−2222
Urea	$NH_2CONH_2(s)$	−333.51
Chlorine	$Cl_2(g)$	0
Chlorine atom	$Cl(g)$	+121.68
Hydrochloric acid	$HCl(aq)$	−167.16
Hydrogen chloride	$HCl(g)$	−92.31
Copper	$Cu(s)$	0
Copper(II) oxide	$CuO(s)$	−157.3
Copper(II) sulfate	$CuSO_4(s)$	−771.36
Copper(II) sulfate	$CuSO_4 \cdot 5H_2O(s)$	−2279.7
Fluorine	$F_2(g)$	0
Fluorine ion	$F^-(aq)$	−332.63
Hydrogen fluoride	$HF(g)$	−271.1
Hydrogen	$H_2(g)$	0
Hydrogen atom	$H(g)$	+217.97
Deuterium	$D_2(g)$	0
Water	$H_2O(l)$	−285.83
Water	$H_2O(g)$	−241.82
Deuterium oxide	$D_2O(l)$	−294.60
Hydrogen ion	$H^+(aq)$	0
Iodine	$I_2(s)$	0
Iodide ion	$I^-(aq)$	−55.19
Hydrogen iodide	$HI(g)$	+26.48
Iron	$Fe(s)$	0
Iron(III) oxide	Fe_2O_3 (haematite)	−824.2
Nitrogen	$N_2(g)$	0
Nitrogen(II) oxide	$NO(g)$	+90.25
Nitrogen(IV) dioxide	$NO_2(g)$	+33.18
Dinitrogen tetroxide	$N_2O_4(g)$	+9.16
Ammonia	$NH_3(g)$	−46.11
Ammonium chloride	$NH_4Cl(s)$	−314.43
Ammonium chlorate(VII)	$NH_4ClO_4(s)$	−295.31
Oxygen	$O_2(g)$	0
Ozone	$O_3(g)$	+142.7
Hydroxide ion	$OH^-(aq)$	−229.99
Phosphorus (white)	P (white)	0
Phosphorus (red)	P (red)	−18
Phosphorus(III) chloride	$PCl_3(l)$	−319.7
Potassium	$K(s)$	0
Potassium hydroxide	$KOH(s)$	−424.76
Potassium chloride	$KCl(s)$	−436.75
Sodium	$Na(s)$	0
Sodium chloride	$NaCl(s)$	−411.15
Sulfur (rhombic)	$S(s)$ (rhombic)	0
Sulfur (monoclinic)	$S(s)$ (monoclinic)	+0.33
Sulfuric acid	$H_2SO_4(l)$	−813.99
Sulfur dioxide	$SO_2(g)$	−296.83
Sulfur trioxide	$SO_3(g)$	−395.72
Hydrogen sulfide	$H_2S(g)$	−20.63

- If an element exists in more than one allotrope, $\Delta H_f^\ominus$ refers to the compound producing reaction in which that element is present in its reference state.

Think of water. According to the definition, the standard enthalpy of formation of water is the standard enthalpy change involved when pure hydrogen and pure oxygen react to give one mole of pure water – all at a pressure of 1 atm. At our chosen temperature of 25 °C, both hydrogen and oxygen will be gases and water will be a liquid. The reaction will then be represented as

$$H_2(g) + \tfrac{1}{2}O_2(g) \rightarrow H_2O(l) \qquad \Delta H_f^\ominus = -285.83\,\text{kJ mol}^{-1}$$

The standard enthalpy change, represented by the equation

$$2H_2(g) + O_2(g) \rightarrow 2H_2O(l) \qquad \Delta H^\ominus = -571.66\,\text{kJ mol}^{-1}$$

does not equal the standard enthalpy change of formation, since the equation indicates that two moles of water are made.

Although $\Delta H_f^\ominus(H_2O)$ may be measured experimentally, by studying the reaction of H_2 and O_2 in the laboratory, for many compounds direct measurement of $\Delta H_f^\ominus$ is impossible. For example, consider the thermochemical equation showing the formation of methanoic (formic) acid:

$$H_2(g) + O_2(g) + C(s) \rightarrow HCO_2H(l) \qquad \Delta H_f^\ominus(298\text{ K}) = -402.1\,\text{kJ mol}^{-1}$$

If we heat hydrogen, oxygen and graphite together, so little methanoic acid is produced that it is impossible to directly measure its enthalpy change of formation. For this reason, the enthalpy change of formation is measured indirectly, using Hess's law.

Examples of thermochemical equations where the standard enthalpy changes are equal to the standard enthalpy of formation of compounds are

1. *ammonia*

$$\tfrac{1}{2}N_2(g) + \tfrac{3}{2}H_2(g) \rightarrow NH_3(g) \qquad \Delta H_f^\ominus = -46.1\,\text{kJ mol}^{-1}$$

2. *benzoic acid*

$$7C(s) + 3H_2(g) + O_2(g) \rightarrow C_6H_5CO_2H(s) \qquad \Delta H_f^\ominus = -385.1\,\text{kJ mol}^{-1}$$

3. *hydrogen iodide*

$$\tfrac{1}{2}H_2(g) + \tfrac{1}{2}I_2(s) \rightarrow HI(g) \qquad \Delta H_f^\ominus = +26.48\,\text{kJ mol}^{-1}$$

Standard enthalpy of formation of elements

The standard enthalpy of formation of diamond refers to the process

$$C_{(s)\,(graphite)} \rightarrow C_{(s)\,(diamond)} \qquad \Delta H_f^\ominus = +1.895\,\text{kJ mol}^{-1}$$

Not surprisingly, the standard enthalpy of formation of an element in its **reference state** is zero, since the reaction involves no change:

$$C_{(s)\,(graphite)} \rightarrow C_{(s)\,(graphite)} \qquad \Delta H_f^\ominus = 0.000\,\text{kJ mol}^{-1}$$

This might appear trivial, but it will be important to keep this in mind when we use $\Delta H_f^\ominus$ values in calculations later.

Exercise 13G

Enthalpy of formation

Complete the following thermochemical equations which refer to the formation of one mole of the substances shown at 298 K:

.... $\rightarrow H_2SO_4(l)$ $\Delta H_f^\ominus = -814\,kJ\,mol^{-1}$

.... $\rightarrow C_6H_{12}O_6(s)$ $\Delta H_f^\ominus = -1268\,kJ\,mol^{-1}$

Calculation of $\Delta H^\ominus$ using standard enthalpies of formation

On page 222, we used Hess's law to calculate $\Delta H^\ominus$ for the reaction

$$HCl(g) + NH_3(g) \rightarrow NH_4Cl(s)$$

obtaining $\Delta H^\ominus = -176.0\,kJ\,mol^{-1}$. The same answer may be obtained by adding the standard enthalpies of formation of HCl(g) and $NH_3(g)$ together, and subtracting this sum from the standard enthalpy of formation of ammonium chloride:

$$\Delta H^\ominus = \Delta H_f^\ominus(NH_4Cl(s)) - [\Delta H_f^\ominus(HCl(g)) + \Delta H_f^\ominus(NH_3(g))]$$

The $\Delta H_f^\ominus$ values ($kJ\,mol^{-1}$) are

$$\Delta H_f^\ominus(NH_4Cl(s)) = -314.4; \quad \Delta H_f^\ominus(HCl(g)) = -92.3; \quad \Delta H_f^\ominus(NH_3(g))] = -46.1$$

so that

$$\Delta H^\ominus = (-314.4) - (-92.3 + (-)\,46.1) = -176.0\,kJ\,mol^{-1}$$

This method is easier to use than the algebraic method based upon Hess's law, and we will now generalize the calculation so that it applies to all reactions:

- To work out the standard enthalpy change ($\Delta H^\ominus$) of any reaction from $\Delta H_f^\ominus$ values, substitute into the following equation:

  $$\Delta H^\ominus = \Sigma\Delta H_f^\ominus(\text{products}) - \Delta\Sigma H_f^\ominus(\text{reactants}) \quad \textbf{(13.15)}$$

  where $\Sigma\Delta H_f^\ominus$(products) is the sum of the standard enthalpies of formation of the products in the reaction, and $\Delta\Sigma H_f^\ominus$(reactants) is the sum of the standard enthalpies of formation of the reactants.
- When summing the enthalpies of formation, the enthalpy of formation of each substance must be multiplied by its **coefficient** in the balanced equation for the reaction. This will be seen in Example 13.4.

Example 13.4

Calculate $\Delta H^\ominus$ at 298 K for the production of three moles of copper by the reaction:

$$3CuO(s) + 2NH_3(g) \rightarrow N_2(g) + 3H_2O(l) + 3Cu(s) \quad \textbf{(A)}$$

▶ Answer

$$\Delta H^\ominus = \Sigma\Delta H_f^\ominus(\text{products}) - \Sigma\Delta H_f^\ominus(\text{reactants})$$

$$\Delta H^\ominus =$$

$$[\Delta H_f^\ominus(N_2(g)) + \underset{\uparrow}{3}\Delta H_f^\ominus(H_2O(l)) + \underset{\uparrow}{3}\Delta H_f^\ominus(Cu(s))] - [\underset{\uparrow}{3}\Delta H_f^\ominus(CuO(s)) + \underset{\uparrow}{2}\Delta H_f^\ominus(NH_3(g))]$$

The numbers indicated by arrows are the coefficients in the chemical equation. Using the standard enthalpy of formation data in Table 13.2,

Example 13.4 *(continued)*

$\Delta H_f^\ominus(H_2O(l)) = -285.83\,kJ\,mol^{-1}$
$\Delta H_f^\ominus(CuO(s)) = -157.3\,kJ\,mol^{-1}$
$\Delta H_f^\ominus(NH_3(g)) = -46.11\,kJ\,mol^{-1}$

and remembering that the standard enthalpy of formation of the elements N_2 and Cu is zero

$$\Delta H^\ominus = [0 + 3(-285.83) + 3(0)] - [3(-157.3) + 2(-46.11)]$$
$$= -293.4\,kJ\,mol^{-1}$$

Therefore the standard enthalpy change for reaction **(A)** is $-293.4\,kJ\,mol^{-1}$.

Exercise 13H

Standard enthalpy change

Use the $\Delta H_f^\ominus$ data in Table 13.2 to calculate the standard enthalpy change at 298 K of the following reactions:

(i) The burning of two moles of carbon monoxide gas:

$2CO(g) + O_2(g) \rightarrow 2CO_2(g)$

(ii) the reduction of one mole of iron(III) oxide by carbon monoxide:

$Fe_2O_3(s) + 3CO(g) \rightarrow 2Fe(s) + 3CO_2(g)$

(iii) the decomposition of one mole of ammonia gas:

$NH_3(g) \rightarrow \frac{1}{2}N_2(g) + \frac{3}{2}H_2(g)$

You might be wondering why the calculation of $\Delta H^\ominus$ from $\Delta H_f^\ominus$ values by this method works. The answer lies in the definition of standard enthalpy of formation, which requires $\Delta H_f^\ominus$ values for elements in their reference states to be zero. Although the standard enthalpies (i.e. energies) of compounds and elements cannot be measured or calculated, the *difference* between the standard enthalpies of two substances is equal to the difference in the corresponding standard enthalpies of formation. This is the reason why equation (13.15) strongly resembles equation (13.2).

Standard enthalpies of formation and compound stability

The value of $\Delta H_f^\ominus$ is an approximate measure of the stability of a substance relative to the elements from which it is made. The standard enthalpies of formation of graphite, diamond, water, ethyne (acetylene, C_2H_2), ammonia and sodium chloride are shown in Fig. 13.6. The reference states of elements define an energy baseline or 'sea level'. Compounds such as ethyne, for which $\Delta H_f^\ominus$ is positive, and which therefore possess a greater enthalpy than their constituent elements, appear above 'sea level' and are called **endothermic compounds**. Compounds such as water, ammonia and sodium chloride, for which $\Delta H_f^\ominus$ is negative and which therefore possess a lower enthalpy than their constituent elements, appear below 'sea level' and are called **exothermic compounds**.

Since $\Delta H_f^\ominus$ values represent the difference in enthalpy between compounds and the elements from which they are made, it is unhelpful to compare the $\Delta H_f^\ominus$ values of

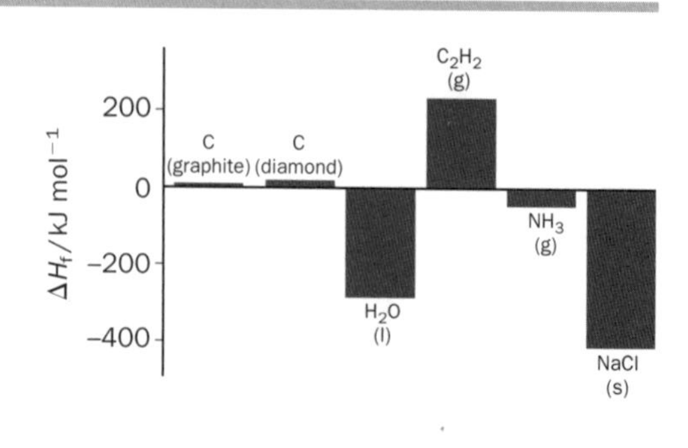

Fig. 13.6 A comparison of the standard enthalpies of formation of six substances at 25 °C. (The value for diamond has been multiplied by five; the value for graphite is zero.).

say CH_4(g) and SO_2(g), since neither compound contains an element common to both. However, a comparison of $\Delta H_f^\ominus$ values (kJ mol^{-1}) for compounds in the series:

HF(g)	HCl(g)	HBr(g)	HI(g)
−271	−92	−36	+27

is useful since it suggests a gradual reduction in stability from left to right. This trend fits in with experimental observations which show that HI(g) is the least stable of the series, with HI(g) being the only hydride in the series that decomposes at room temperature. However, $\Delta H_f^\ominus$ values cannot be used to predict *how fast* such a decomposition will occur.

13.6 Standard enthalpy of combustion

The standard enthalpy of combustion $\Delta H_c^\ominus$ is the

standard enthalpy change when one mole of substance is burned in oxygen.

For example, methane (the main constitutent of natural gas) burns as follows:

$$CH_4(g) + 2O_2(g) \rightarrow CO_2(g) + 2H_2O(l) \qquad \Delta H_c^\ominus \text{ (298 K)} = -890 \text{ kJ mol}^{-1}$$

(From now on we shall assume that all $\Delta H_c^\ominus$ values apply at 298 K.)

The heating power of fuels is often expressed as the **energy value**, defined as the heat produced when one gram of fuel burns completely. Its units are kilojoules per gram (kJ g^{-1}).

We work out the energy value of methane as follows. The molar mass of methane is 16.0 g mol^{-1}. When 16.0 g of methane burns, 890 kJ of heat is produced. When 1 g of methane is burned, 890/16.0 ≈ 56 kJ of heat are produced, so that the energy value of methane is 56 kJ g^{-1}. The energy values of common fuels are shown in Table 13.3.

Table 13.3 Energy values

Fuel	Energy value/ kJ g^{-1}
Hydrogen (H_2(g))	142
Methane (CH_4(g))	56
Ethyne (acetylene, C_2H_2(g))	50
Octane (C_8H_{18}(l))	48
Coal (bituminous)	≈30
Pine wood	18
Methanol (CH_3OH(l))	23
Apples	2
White bread	11
Butter	30
Cheese	4–18
Hard boiled eggs	7
Full cream milk	3
Baked potatoes	4
Chips	11
Nitroglycerine	7
TNT	5
Ammonium nitrate	2

Determination of enthalpy changes of combustion in the laboratory

The determination of ΔH_c is, in principle, similar to that for other enthalpy changes. In practice, the fuel (or food) is mixed with an oxidizing agent (such as solid sodium peroxide (Na_2O_2) or pure oxygen gas) and sealed in a steel-walled vessel called a **bomb calorimeter**. The mixture is ignited by an electric current and the temperature rise recorded. The combustion reaction in a bomb calorimeter is not taking place at constant pressure, but this is easily allowed for in the final calculations.

Exercise 13I

Energy from fuels

(i) Given the thermochemical equation

$C_{12}H_{22}O_{11}(s) + 12O_2(g) \rightarrow 12CO_2(g) + 11H_2O(l)$

$\Delta H_c^{\ominus} = -5644\,kJ\,mol^{-1}$

calculate the energy value of sucrose in $kJ\,g^{-1}$.

(ii) From Table 13.3, work out the volume of hydrogen gas which, when burned completely, produces 100 kJ of heat (assume the molar volume of a gas at room temperature and pressure is $24\,dm^3$).

13.7 Nutrition

Energy value of food

Table 13.3 includes the energy value of common foodstuffs, since these can be thought of as 'human fuel'. If 1 g of apple were completely burned in a flame, it would give out about 2 kJ of heat energy. Energy is also produced when foods are digested in the body, but there are three important differences between the breakdown of foodstuffs (more properly called **food metabolism**) and the burning of a fuel in a flame:

1. In the human body, foodstuffs are consumed slowly in a complicated series of chemical reactions. About 40% of the enthalpy change of combustion of a food is available to the body to be used in exercise.

2. Oxygen gas is not used as the oxidizing agent in the metabolism of food. The oxidizing agent used in the cell is called **nicotinamide adenine dinucleotide** (NAD).

3. Some of the energy released in the metabolism of foods is not given out as heat, but is used to make a substance called **adenosine triphosphate** (ATP). ATP may be converted to **adenosine diphosphate** (ADP) when required, and it is the ATP $\rightarrow$ ADP reaction that produces the energy upon which we depend for the synthesis of biologically important molecules, for the transport of vital nutrients within the body and for muscle action (see Appendix 13 in the website).

Exercise 13J

Nutrition

Diet plans often tabulate the energy value of food in kilocalories (symbolized Cal) with 1 Cal = 4.18 kJ. A 20 g bar of chocolate possessed a calorific value of 120 Cal. What is the energy value of the chocolate, in $kJ\,g^{-1}$?

Measuring energy use during bodily exercise

The metabolism of fats and glucose during prolonged exercise requires oxygen and produces carbon dioxide and water. The energy consumption of an individual during exercise can be estimated by measuring the ratio of the volume of CO_2 produced to the volume of O_2 used up. This ratio is called the **respiratory exchange ratio** (RER):

$$RER = \frac{\text{volume of } CO_2 \text{ produced per minute}}{\text{volume of } O_2 \text{ consumed per minute}}$$

Experimentally, the RER is obtained by analysis of the inhaled and exhaled air of the person exercising.

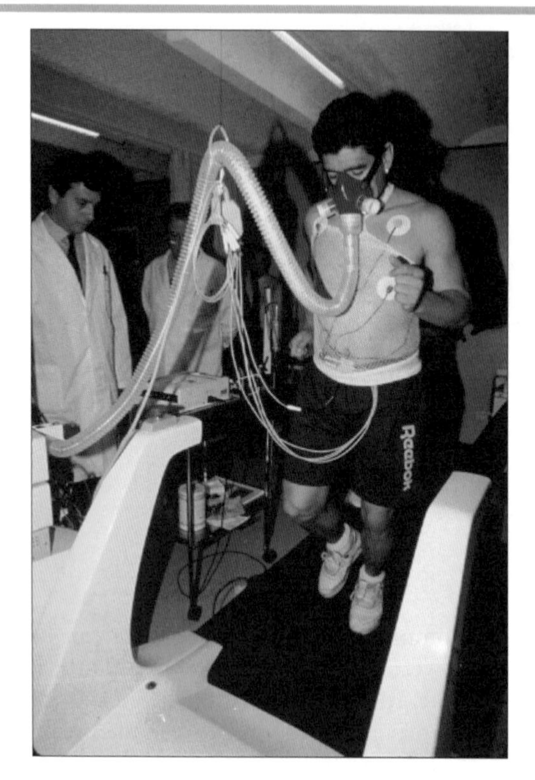

A footballer on a treadmill, showing the equipment used to measure the respiratory exchange of oxygen and carbon dioxide.

If a person's body is using only carbohydrates to provide energy, the RER is about 1.0. If only fats are being used as the source of energy, the ratio is about 0.7. In both cases, about 20 kJ of energy is being produced per dm^3 of oxygen consumed.

For an average person at rest (e.g. sitting in a chair) who lives on a mixed diet, the RER is about 0.8. Typically, an average male at rest consumes about 0.30 dm^3 O_2 per minute, or $0.30 \times 60 \times 24 \approx 430$ dm^3 per day. (The man would breath in much more oxygen gas than this, but only a small fraction of the inhaled oxygen would be actually used up in the metabolism of foodstuffs.) Every day the typical man would produce $20 \times 430 \approx 8600$ kJ (or 2060 Cal) per day. This would also be the minimum daily energy requirement required for such a man – if his diet was of a lower energy value his body would have to draw upon its food reserves. If the man engaged in walking, reading or writing, more than 8600 kJ day^{-1} would be required. In extreme cases, athletes can consume over 50 000 kJ per day! To lose weight, a person persistently ingests food with a calorific value which is less than their daily requirement – the practice is more difficult than the theory!

Exercise 13K

Energy consumption

Estimate the minimum energy consumption per day of an average woman. Assume that her oxygen consumption is 270 cm^3 min^{-1}.

One commercially useful group of materials which produce heat energy extremely rapidly are **explosives**. These are discussed in Box 13.2.

BOX 13.2

Explosives

An explosive is a substance which undergoes a very rapid chemical reaction in which gas is produced. The volume of gas produced is large compared with the bulk of the explosive, and the gas escapes so rapidly that a pressure wave is set up in the surrounding air – this causes the bang. For example, the explosive nitroglycerine (Fig. 13.7) decomposes according to the overall equation

$$4C_3H_5(NO_3)_3(l) \rightarrow 12CO_2(g) + 10H_2O(g) + 6N_2(g) + O_2$$

Although the earliest explosive was gunpowder (black powder), consisting of sulfur, potassium nitrate (as oxidizing agent) and charcoal, it was the discovery of nitroglycerine (in 1847), of trinitrotoluene (TNT, in 1863) and of dynamite (made in 1866 by Alfred Nobel by absorbing nitroglycerine into finely powdered earth) that revolutionized the explosives industry. The early investigations of explosives caused many fatalities, including Alfred Nobel's brother and father. Although we associate explosives with their military applications, most explosives are used in quarries, mines and for clearing land in road construction. The commonest explosive is called **ANFO** and consists of about 95% by mass of **a**mmonium **n**itrate ($NH_4NO_3(s)$, as oxidising agent) and 5% **f**uel **o**il.

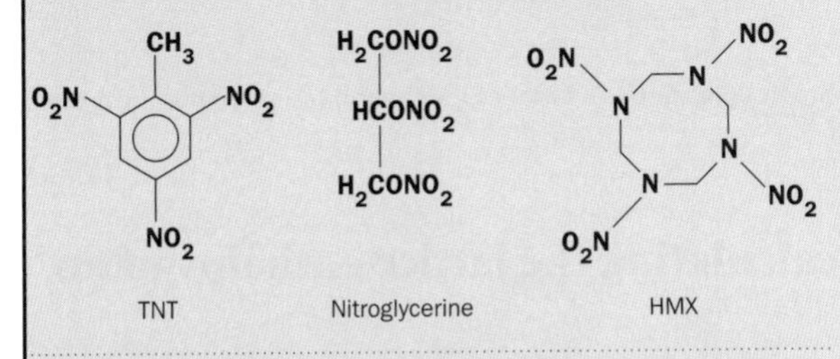

Fig. 13.7 Structures of the explosives TNT, nitroglycerine and HMX (tetramethylenetetranitramine). Nitroglycerine and HMX contain sufficient oxygen to burn completely, but TNT is blended with 80% ammonium nitrate to ensure complete oxidation during the explosion.

All common explosives contain nitrogen, and nitrogen gas is always produced as one of the gaseous products. The explosion reaction is over in a very short time. For example, although the burning of nitroglycerine produces only 7 kJ g^{-1} of heat (Table 13.3), this heat energy is produced in less than 0.001 s. This is equivalent to a power of 7000 kW, and the temperature of the escaping gas may reach 3000 °C and speeds of up to 7000 m s^{-1}!

To illustrate the pressures involved, suppose that 5 g of nitroglycerine is exploded in a sealed container of volume 10 cm^3 (10^{-5} m^3). About 0.16 mol of gases are produced at 3000 °C. We can crudely estimate the pressure of the gas at the instant of the explosion using the ideal gas equation

$$P \approx \frac{nRT}{V} \approx \frac{0.16 \times 8.3145 \times (273 + 3000)}{10^{-5}} \approx 4 \times 10^8\ \text{Pa}\ (4000\ \text{atm})$$

This is an enormous pressure (roughly 40 times the gas pressure in a large commercial gas cylinder) and explains the ability of explosives to smash rocks and metal.

13.8 Lattice enthalpy

The standard enthalpy change in which 1 mol of a crystal lattice is broken up into isolated gaseous particles is called the **lattice enthalpy** ($\Delta H_L^\ominus$) of the lattice. For example, 776 kJ of energy is needed to break up 1 mol of pure sodium chloride crystal lattice into gaseous sodium and chloride ions at 298 K and 1 atm pressure:

$$Na^+,Cl^-(s) \rightarrow Na^+(g) + Cl^-(g) \qquad \Delta H_L^\ominus = +771\ \text{kJ mol}^{-1}$$

When 1 mol of gaseous sodium ions condenses with 1 mol of chloride ions, under the same conditions, 1 mol of sodium chloride crystals are made:

$$Na^+(g) + Cl^-(g) \rightarrow Na^+,Cl^-(s) \qquad \Delta H^\ominus = -\Delta H_L^\ominus = -771\ \text{kJ mol}^{-1}$$

Generalizing:

Heat needs to be absorbed in order to break down a crystal lattice

Heat is given out when a crystal lattice is formed

Lattice enthalpy as a measure of the strength of ionic bonding

The lattice enthalpy of NaCl(s) is an index of the strength of the bonding between Na^+ and Cl^- ions in the sodium chloride crystal. However, ions in a sodium chloride crystal lattice are not paired, and each sodium (or chloride) ion is attracted by six oppositely charged neighbouring ions and (less strongly) by more distant ions. This means that $\Delta H_L^\ominus$ is a measure of the *total* attractive force between sodium and chloride ions within the lattice, and does not simply reflect the force of attraction between isolated *pairs* of Na^+,Cl^-.

The 'MX'-type lattices NaCl(s), NaI(s) and NaBr(s) possess the following lattice enthalpies: +771, +684 and +731 kJ mol^{-1} respectively. Of these crystals, NaCl possesses the most positive $\Delta H_L^\ominus$ and is said to be the 'most stable' of the series.

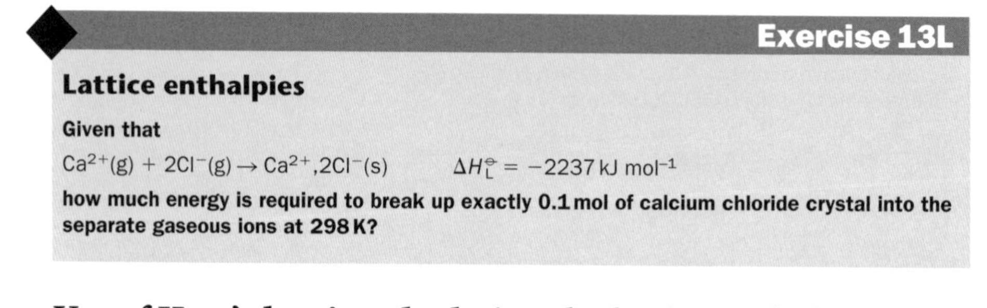

Exercise 13L

Lattice enthalpies

Given that

$Ca^{2+}(g) + 2Cl^-(g) \rightarrow Ca^{2+},2Cl^-(s)$ $\qquad \Delta H_L^\ominus = -2237$ kJ mol^{-1}

how much energy is required to break up exactly 0.1 mol of calcium chloride crystal into the separate gaseous ions at 298 K?

Use of Hess's law in calculating the lattice enthalpy of an ionic crystal

The thermochemical equation showing the reaction in which sodium chloride is formed from its elements at 298 K,

$$Na(s) + \tfrac{1}{2}Cl_2(g) \rightarrow Na^+,Cl^-(s) \qquad \Delta H_f^\ominus = -411 \text{ kJ mol}^{-1} \qquad \textbf{(13.16)}$$

may be split up into several imaginary steps known as the **Born–Haber cycle** (Fig. 13.8). The enthalpy change for each step has been defined in Table 13.1. As always,

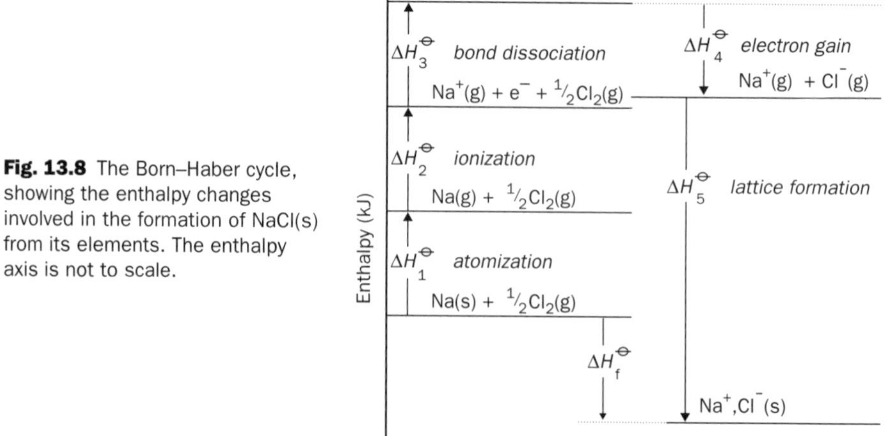

Fig. 13.8 The Born–Haber cycle, showing the enthalpy changes involved in the formation of NaCl(s) from its elements. The enthalpy axis is not to scale.

the endothermic steps cause the total enthalpy to increase (i.e. to move up the figure), whereas the exothermic steps cause the total enthalpy to decrease (i.e. to move down the figure). The stages are:

1. The conversion of solid sodium to gaseous sodium (*atomization*):

$Na(s) \rightarrow Na(g) \qquad \Delta H_1^\ominus = +109\ \text{kJ mol}^{-1}$

(i.e. the standard enthalpy of atomization of sodium).

2. The ionization of sodium atoms in the vapour:

$Na(g) \rightarrow Na^+(g) + e^- \qquad \Delta H_2^\ominus = +494\ \text{kJ mol}^{-1}$

(i.e. the standard first ionization energy of sodium).

3. The breaking of the Cl–Cl bond:

$\tfrac{1}{2}Cl_2(g) \rightarrow Cl^\bullet(g) \qquad \Delta H_3^\ominus = +121\ \text{kJ mol}^{-1}$

i.e. half the standard bond dissociation enthalpy of the Cl–Cl bond. The enthalpy change for this process is **half** the energy required to break one Cl–Cl bond because we only require one $Cl^\bullet$ atom to react with each Na. The standard bond dissociation enthalpy of Cl–Cl is +242 kJ mol^{-1}, so that $\Delta H_3^\ominus = 242/2 = 121$ kJ mol^{-1}.

4. The gain of one electron by the chlorine atom:

$Cl^\bullet(g) + e^- \rightarrow Cl^-(g) \qquad \Delta H_4^\ominus = -364\ \text{kJ mol}^{-1}$

i.e. the standard enthalpy of electron gain of the chlorine atom. The $Cl^\bullet$ atom is reaching a more stable configuration in gaining an electron.

5. The attraction of the sodium and chloride ions to make crystalline sodium chloride:

$Na^+(g) + Cl^-(g) \rightarrow Na^+,Cl^-(s) \qquad \Delta H_5^\ominus = -\Delta H_L^\ominus$

Note that addition of the equations involving the $\Delta H_1^\ominus$, $\Delta H_2^\ominus$, $\Delta H_3^\ominus$, $\Delta H_4^\ominus$ and $\Delta H_5^\ominus$ terms gives equation (13.16). According to Hess's law,

$$\Delta H_f^\ominus = \Delta H_1^\ominus + \Delta H_2^\ominus + \Delta H_3^\ominus + \Delta H_4^\ominus + \Delta H_5^\ominus \qquad \textbf{(13.17)}$$

The stages shown in Fig. 13.8 are not meant to indicate the way that sodium and chlorine actually react, i.e. they do not represent the *mechanism* of the reaction.

Similar energy cycles may be set up for other ionic compounds, such as Ca^{2+},O^{2-}. The importance of equation (13.17) is that we can calculate one of the enthalpy terms knowing all the others. The unknown is usually the lattice enthalpy (which equals $-\Delta H_5^\ominus$), and which cannot be measured directly by experiment.

Rearrangement of equation (13.17) gives

$$\Delta H_5^\ominus = \Delta H_f^\ominus - \Delta H_1^\ominus - \Delta H_2^\ominus - \Delta H_3^\ominus - \Delta H_4^\ominus$$

For example, for the sodium and chlorine cycle,

$$\Delta H_5^\ominus = -411 - (+109) - (+494) - (+121) - (-364) = -771\ \text{kJ mol}^{-1}$$

Exercise 13M

Enthalpy changes

(i) What is the standard enthalpy change for the process $Na(s) \rightarrow Na^+(g) + e^-$?

(ii) Estimate the amount of energy required for the following (hypothetical) process under standard conditions:

$Na^+,Cl^-(s) \rightarrow Na(g) + \tfrac{1}{2}Cl_2(g)$

Since

$$\Delta H_5^{\ominus} = -\Delta H_L^{\ominus}(NaCl)$$

$$\Delta H_L^{\ominus}(NaCl) = -\Delta H_5^{\ominus} = -(-771) = +771\,kJ\,mol^{-1}$$

Exercise 13N

Lattice enthalpy of magnesium fluoride

Calculate the lattice enthalpy of magnesium fluoride given the following standard enthalpy data:

enthalpy of atomization of magnesium, $\Delta H_{atom}^{\ominus}(Mg(s)) = +148\,kJ\,mol^{-1}$

first ionization energy of magnesium, $\Delta H_{ion}^{\ominus}(Mg(g)) = +744\,kJ\,mol^{-1}$

Second ionization energy of magnesium, $\Delta H_{ion}^{\ominus}(Mg^+(g)) = +1456\,kJ\,mol^{-1}$

bond dissociation enthalpy of $F_2(g)$, $\Delta H^{\ominus}(F\text{–}F) = +158\,kJ\,mol^{-1}$

enthalpy of electron gain of F(g), $\Delta H_{eg}^{\ominus} = -339\,kJ\,mol^{-1}$

enthalpy of formation of $MgF_2(s)$, $\Delta H_f^{\ominus}(MgF_2(s)) = -1121\,kJ\,mol^{-1}$.

13.9 Energetics of bond breaking and bond making

Bond dissociation enthalpies

If a covalent bond exists between two atoms, it will take energy to pull those atoms apart. For example, the dissociation of the A–B bond in the molecule AB may be represented by the equation

$$A\text{–}B(g) \rightarrow A(g) + B(g)$$

(Note that both reactants and products are gaseous.) The amount of energy required to break a particular bond in a gaseous molecule under standard conditions is called the **(standard) bond dissociation enthalpy** of that bond, symbolised $\Delta H_{A\text{–}B}^{\ominus}$. For example, the standard enthalpy change for the dissociation of chlorine molecules at 298 K,

$$Cl_2(g) \rightarrow Cl^{\bullet}(g) + Cl^{\bullet}(g)$$

is 242 kJ per mole of Cl–Cl bonds. Therefore, the bond dissociation enthalpy is 242 kJ mol^{-1}:

$$\Delta H^{\ominus}(Cl\text{–}Cl) = 242\,kJ\,mol^{-1}$$

Selected bond dissociation enthalpies are shown in Table 13.4. Covalent bonds where $\Delta H_{A\text{–}B}^{\ominus}$ is greater than 400 kJ mol^{-1} are usually described as strong bonds.

When a reaction is reversed the energy changes are also reversed. We predict that the reaction

$$Cl(g) + Cl(g) \rightarrow Cl_2(g)$$

will involve a standard enthalpy change at 298 K of -242 kJ mol^{-1}, the released heat energy reflecting the fact that in this reaction the two chlorine atoms are achieving a stable electronic configuration through the formation of a covalent bond.

Bond dissociation enthalpies of polyatomic molecules

The standard enthalpy change for the reaction:

$$\text{H–O–H(g)} \rightarrow \text{H(g)} + \text{OH(g)} \qquad \Delta H^{\ominus} = +499\ \text{kJ mol}^{-1}$$

shows that the bond dissociation enthalpy of the HO–H bond, symbolized $\Delta H^{\ominus}_{\text{HO–H}}$, is 499 kJ per mol of O–H bonds. However, if we now break the O–H bond in OH(g), only 428 kJ of energy per mol of O–H is required for complete dissociation:

$$\text{O–H(g)} \rightarrow \text{O(g)} + \text{H(g)} \qquad \Delta H^{\ominus} = \Delta H^{\ominus}_{\text{O–H}} = +428\ \text{kJ mol}^{-1}$$

The **mean bond enthalpy** for the stepwise dissociation of both O–H bonds in water is $(499 + 428)/2 = 463.5\ \text{kJ mol}^{-1}$.

This data does not mean that the O–H bonds in water are different. On the contrary, they are identical in every respect. The different values of bond enthalpies for the O–H bonds arise because they apply to very different species, H_2O and OH.

Table 13.4 Standard bond enthalpies

Bond	$\Delta H^{\ominus}_{\text{A–B}}$ (298K) /kJ mol^{-1}
Bond dissociation enthalpies	
H–H	436
N≡N	945
O=O	497
F–F	158
Cl–Cl	242
H–F	565
H–Cl	431
H–Br	366
H–I	299
O**C=O**	531
H_3**C–Cl**	339
H_2**C=C**H_2	699
Mean bond enthalpies	
C–H	412
C–C	348
C=C	612
C=O	743
O–H	463
N–N	163
N–H	388
C–Cl	338

The atoms on either side of the bond are displayed in bold.

Mean bond enthalpies

The bond dissociation enthalpies of bonds such as C–H, C–Cl, C=O, N=N and O–H are approximately the same in different molecules. If the values of bond dissociation enthalpy for a bond between two atoms (A and B) in several different molecules are averaged, the resulting value is called the **mean bond enthalpy** (Table 13.4). Mean bond enthalpies are useful in estimating enthalpy changes for reactions for which standard enthalpies of formation are unavailable, but the fact that such values may have been obtained by averaging bond dissociation enthalpies from different types of molecule may lead to substantial errors in the calculated value of $\Delta H^{\ominus}$.

Exercise 13O

Mean bond enthalpy of C=O in CO_2

The carbon dioxide molecule (O=C=O) contains two identical C=O bonds. If the bonds in CO_2 are broken in sequence it is found that $\Delta H^{\ominus}_{\text{OC=O}} = 531\ \text{kJ mol}^{-1}$ and $\Delta H^{\ominus}_{\text{C=O}} = 1075\ \text{kJ mol}^{-1}$. What is the mean bond enthalpy of the C=O bond in CO_2?

Exercise 13P

Bond enthalpies 1 (use of data in Table 13.4)

(i) Table 13.4 shows that $\Delta H^{\ominus}_{\text{C–H}} = 412\ \text{kJ mol}^{-1}$. Estimate the enthalpy change for the following reaction under standard conditions:

$CH_3(g) + H(g) \rightarrow CH_4(g)$

(ii) The bond lengths (in pm) for the hydrogen halides are H–F(92), H–Cl(128), H–Br(141) and H–I(160). Do these figures support the idea that long bonds are weak bonds?

(iii) Does the bond enthalpy data in Table 13.4 suggest that double and triple bonds are stronger than single bonds?

Use of bond enthalpies: fuels – where does the heat come from?

The burning of methane in air follows the equation:

$$CH_4(g) + 2O_2(g) \rightarrow CO_2(g) + 2H_2O(g) \qquad \textbf{(13.18)}$$

We know that heat is given out during this reaction. But where exactly does the heat energy come from? We will now calculate the enthalpy change for this reaction using mean bond enthalpies. To do this we carry out the following operations:

1. We start by writing out the molecular structures for the reactants and products in full (Fig. 13.9). We see that going from left to right in the equation, the reaction involves the breaking of four C–H bonds and two O=O bonds and the making of two C=O bonds and of four O–H bonds.

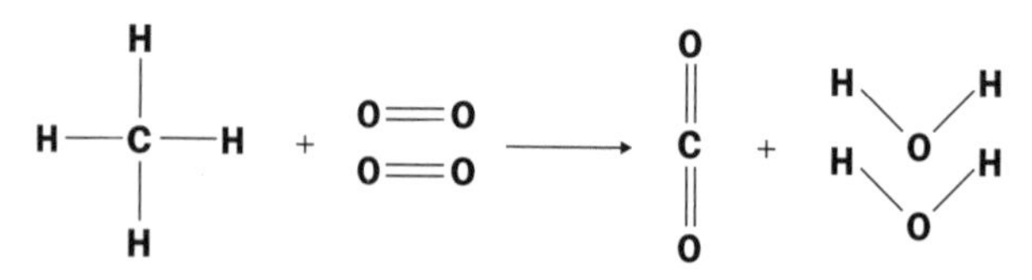

Fig. 13.9 The burning of methane in air.

2. We imagine that the oxygen and methane molecules are *completely broken up into atoms* during the reaction, and that the resulting isolated atoms are then *completely assembled* into the product water and carbon dioxide molecules. (It does not matter whether or not the reaction really follows this mechanism because, according to Hess's law, the results of 'our energy count' are independent of how the molecules actually react or whether or not the products are made in several intermediate steps.)

3. The overall enthalpy change for reaction (13.18) is obtained by adding the enthalpy changes involved in the breaking up of the reactant molecules and in the subsequent formation of the product molecules. The breaking up of molecules involves positive enthalpy changes because energy needs to be absorbed in order to break apart atoms forming a covalent bond. The forming of bonds involves negative enthalpy changes because of the extra stability achieved by the atoms after combination.

Table 13.5 shows the arithmetic involved in using bond enthalpies to work out whether, overall, heat is taken in or given out in reaction (13.18). With the exception

Table 13.5 Estimating the enthalpy change for the burning of methane

Bonds	Energy change	Totals*/kJ mol^{-1}
Four C–H bonds broken	4 × 412 = 1648 kJ taken in	+1648
Two O=O bonds broken	2 × 497 kJ taken in	+994
Two C=O bonds made	2 × 743 kJ given out	−1486
Four O–H bonds made	4 × 463 kJ given out	−1852
	Estimated enthalpy change $\Delta H^{\ominus}$ =	−696

* Positive value indicates heat taken in, negative value indicates heat given out.

of the O=O bond, mean bond enthalpies have been used in this calculation. Overall, *we predict* that the reaction involves the production of 696 kJ of heat per mol of methane burned. The reason that heat is given out is simply that, on balance, *the product molecules possess stronger bonds than the reactants.* The arithmetic shows that if the C–H bonds and O=O bonds were stronger, little or no heat would be given out!

We may generalize our results to all reactions:

- **If the sum of the bond enthalpies of the product molecules is greater than the sum of the bond enthalpies for the reactants, the reaction will be exothermic.**
- **If the sum of the bond enthalpies of the product molecules is less than the sum of the bond enthalpies for the reactants, the reaction will be endothermic.**

Whereas the calculated enthalpy change for reaction (13.18) is $-696\,\text{kJ mol}^{-1}$, the experimentally measured value is $\Delta H_c^{\ominus} = -802\,\text{kJ mol}^{-1}$. The main error in our calculation lies in the use of the mean bond energy for the C=O bond, which for CO_2 is $803\,\text{kJ mol}^{-1}$ – not $743\,\text{kJ mol}^{-1}$ (see Exercise 13O). Use of a value of $803\,\text{kJ mol}^{-1}$ in Table 13.5 yields a predicted enthalpy change of -816 kJ, much closer to the observed value (check this yourself). This example serves to remind us that the choice of bond enthalpies in such calculations should be carefully reviewed beforehand.

Exercise 13Q

Bond enthalpies 2

Estimate the standard enthalpy change ($\Delta H^{\ominus}$) for the following reactions at 298 K:

(i) $CH_2{=}CH_2(g) + Cl_2(g) \rightarrow CH_2Cl{-}CH_2Cl(g)$

(ii) $2H_2O(l) \rightarrow 2H_2(g) + O_2(g)$ (Note: $H_2O(l) \rightarrow H_2O(g)$ $\Delta H^{\ominus} = +44\,\text{kJ mol}^{-1}$).

Revision questions

13.1. Which of the following is false?

(i) For exothermic reactions, $\Delta H^{\ominus}$ is always positive

(ii) $\Delta H_c^{\ominus}(H_2(g)) = \Delta H_f^{\ominus}(H_2(g))$

(iii) $\Delta H_c^{\ominus}(C(s)\text{ graphite}) = \Delta H_f^{\ominus}(CO_2(g))$

(iv) $\Delta H^{\ominus}$ for the reaction $2S(s) + 3O_2(g) \rightarrow 2SO_3(g)$ is the same as $\Delta H_f^{\ominus}(SO_3(g))$.

(v) $\Delta H_f^{\ominus}(C(s)\text{ graphite}) = 0$

(vi) $\Delta H_c^{\ominus}(C(s)\text{ graphite}) = \Delta H_c^{\ominus}(C(s)\text{ diamond})$

(vii) The O–H bonds in the water molecule are equally strong.

13.2. **(i)** Define 'standard enthalpy of combustion'.

(ii) Calculate the standard enthalpy of combustion of methyl benzene (toluene), $C_7H_8(l)$, given that $\Delta H_f^{\ominus}$ (methyl benzene) $= +12.0\,\text{kJ mol}^{-1}$. (For other data, use Table 13.2.)

(iii) Calculate the energy value of toluene fuel in units of kJ g^{-1}, and kJ cm^{-3} (the density of methyl benzene at room temperature is $0.86\,\text{g cm}^{-3}$).

13.3. Use the data in Table 13.2 to calculate the standard enthalpy changes for the following reactions at 298 K:

(i) P(s) (white) → P(s) (red)

(ii) the production of ethyl ethanoate from ethanoic acid and ethanol:

$$CH_3COOH(l) + C_2H_5OH(l) \rightarrow CH_3COOC_2H_5(l) + H_2O(l)$$

using $\Delta H_f^{\ominus}(CH_3COOH(l)) \approx -485\,\text{kJ mol}^{-1}$, and $\Delta H_f^{\ominus}(CH_3COOC_2H_5(l)) \approx -481\,\text{kJ mol}^{-1}$.

(iii) $6NH_4ClO_4(s) + 10Al(s) \rightarrow 3N_2(g) + 9H_2O(g) + 5Al_2O_3(s) + 6HCl(g)$ (This is the reaction used in the booster rockets of the space shuttle – the huge billowing clouds are made of $Al_2O_3(s)$.)

13.4. 50 cm^3 of 0.050 mol dm^{-3} $Ag^+(aq)$ and 50 cm^3 of 0.050 mol dm^{-3} $Cl^-(aq)$ were rapidly mixed in a simple calorimeter. The temperature rose by 0.41 °C. Estimate the enthalpy change of the reaction

$$Ag^+(aq) + Cl^-(aq) \rightarrow AgCl(s)$$

13.5. $\Delta H_f^\ominus$ for methanoic acid,

$$H_2(g) + O_2(g) + C(s) \rightarrow HCOOH(l) \qquad \Delta H_f^\ominus = ?$$

cannot be determined directly. However, since methanoic acid readily burns in air, its enthalpy of combustion can be measured accurately. It is found that $\Delta H_c^\ominus = -277.2$ kJ mol^{-1}. Using this value of $\Delta H_c^\ominus$ (and the $\Delta H_f^\ominus$ data for CO_2 and H_2O contained in Table 13.2), calculate $\Delta H_f^\ominus$ (HCOOH(l)).

13.6. Glucose ($C_6H_{12}O_6(s)$) is an important 'fuel' in the body.

(i) Write a balanced equation for the complete combustion of glucose in air to make liquid water and carbon dioxide gas.

(ii) Calculate the amount of energy generated in the body (assume that only 40% of the enthalpy of combustion of the fuel is utilised) from the metabolism of 10 g of glucose ($\Delta H_f^\ominus$ ($C_6H_{12}O_6(s)$) = −2816 kJ mol^{-1}).

13.7. **(i)** The third standard ionization energy of aluminium is 2751 kJ mol^{-1}. Write a thermochemical equation to which this enthalpy term applies.

(ii) The first, second and third standard ionization energies of aluminium are 584, 1823 and 2751 kJ mol^{-1}, respectively. The standard enthalpy of atomization of aluminium is 326 kJ mol^{-1}. Calculate the enthalpy change for the process

$$Al(s) \rightarrow Al^{3+}(g) + 3e^-$$

13.8. The following standard enthalpy data (at 298 K) are provided:

(a) enthalpy change of atomization of calcium metal, $\Delta H_{atom}^\ominus$ (Ca(s)), = +193 kJ mol^{-1}

(b) first ionization energy of calcium, $\Delta H_{ion}^\ominus(Ca(g))$ = +590 kJ mol^{-1}

(c) second ionization energy of calcium, $\Delta H_{ion}^\ominus(Ca^+(g))$ = +1150 kJ mol^{-1}

(d) bond dissociation enthalpy of $O_2(g)$, $\Delta H^\ominus(O{=}O)$ = +497 kJ mol^{-1}

(e) enthalpy of double electron gain of the oxygen atom (forming $O^{2-}(g)$), $\Delta H_{eg}^\ominus$, = +703 kJ mol^{-1}.

(f) lattice enthalpy of CaO, $\Delta H_L^\ominus$ = 3513 kJ mol^{-1}.

(i) Write down thermochemical equations that apply to the above enthalpy terms.

(ii) Use the data to calculate the standard enthalpy change of formation ($\Delta H_f^\ominus$) of calcium oxide (CaO).

13.9. The standard enthalpy change for the formation of sulfur tetrafluoride (SF_4) from the gaseous elements at 298 K follows the thermochemical equation

$$2F_2(g) + S(g) \rightarrow SF_4(g) \qquad \Delta H^\ominus = -994 \text{ kJ mol}^{-1}$$

Given that $\Delta H_{F-F}^\ominus$ = 158 kJ mol^{-1}, estimate $\Delta H_{S-F}^\ominus$ in SF_4.

Extension material to support this unit is available on our website at **http://www.palgrave.com/foundations/lewis.**

Speed of Chemical Reactions

UNIT 14

Objectives

- Discusses the factors affecting reaction rate
- Defines the terms *rate constant* and *order of reaction*
- Introduces *activation energy* and *reaction mechanism*
- Looks at first-order reactions in detail
- Explains the role of catalysts

Contents

14.1 Reaction rate

Chemical reactions take place at different speeds. Rusting, the reaction of iron with oxygen and water is a slow process, whereas the reaction of potassium metal with water is explosively fast. Reaction speeds are more properly referred to as 'reaction rates'. The study of the rate of reactions (and of the factors controlling reaction rates) is known as **chemical kinetics**.

A knowledge of the rate at which a reaction takes place is often crucial. In industry, reactions are economically profitable only if the yield of product is sufficient **and** if the products are made in a short enough time. In the chemistry of pollution, the rates at which a pollutant is formed and destroyed are important factors in assessing the hazard posed by the pollutant.

Definition of reaction rate

The word 'rate' is part of everyday language (Table 14.1). The average rate of the chemical reaction:

$$M + N \rightarrow MN$$

(where M and N are reactants and MN is the product) is defined as the rate at which the *reactant* concentration *falls* between two times:

$$\text{rate of reaction} = \frac{\text{change in concentration of reactant}}{\text{change in time}}$$

$$= -\frac{\Delta[M]}{\Delta t}$$

Table 14.1 Examples of 'rates'

Rate	Units
Oil production rate	barrels day^{-1}
Speed	$m\,s^{-1}$
Plant growth	cm $month^{-1}$
Population growth	people $year^{-1}$
Reaction rate	concentration s^{-1}

Example 14.1

The reaction of hydrogen and iodine to make hydrogen iodide at a particular temperature,

$$H_2(g) + I_2(g) \rightarrow 2HI(g)$$

was studied at various times by stopping the reaction (by cooling the mixture) and titrating any unreacted iodine with sodium thiosulfate. At 100.0 s after the start of the reaction, the iodine concentration had fallen from 0.010 mol dm^{-3} to 0.0080 mol dm^{-3}. What is the average rate of reaction during this period?

▶ Answer

average rate = $(0.010 - 0.0080)/100.0 = 2.0 \times 10^{-5}$ mol dm^{-3} s^{-1}

Therefore, the average rate of reaction in the first 100 s was 2.0×10^{-5} mol dm^{-3} s^{-1}.

where $\Delta[M]$ ('delta bracket M') is read as 'change in concentration of M' and Δt ('delta t') as 'change in time'. The minus sign shows that [M] is falling.

The dimensions of reaction rate are

$$\frac{\text{units of concentration}}{\text{units of time}}$$

The usual units of reaction rate used in this book are mol dm^{-3} s^{-1}.

Alternatively, the average rate of a reaction may be defined as the rate at which the *product* concentration *rises* over a period of time:

$$\text{rate of reaction} = \frac{\text{change in concentration of product}}{\text{change in time}}$$

$$= + \frac{\Delta[MN]}{\Delta t}$$

Summarizing, the rate of reaction for reaction M + N → MN is given by the expressions

$$-\frac{\Delta[M]}{\Delta t} = -\frac{\Delta[N]}{\Delta t} = +\frac{\Delta[MN]}{\Delta t}$$

where the minus sign signifies that the concentration of M (and N) falls as time passes, and the plus sign signifies that the concentration of MN increases as time passes.

These equations need to be modified when we are dealing with chemical reactions in which the ratio of reactants to products is not 1:1. For example, consider the decomposition of dinitrogen pentoxide (N_2O_5) into nitrogen dioxide (NO_2) and oxygen:

$$2N_2O_5(g) \rightarrow 4NO_2(g) + O_2(g)$$

The equation shows that for every two molecules of N_2O_5 that decompose, four NO_2 molecules and one oxygen molecule are produced. This means that NO_2 is produced at *twice* the rate at which N_2O_5 decomposes, and also that O_2 is produced at *half* the rate at which N_2O_5 is destroyed.

Suppose that under a particular temperature and set of concentrations, the rate at which N_2O_5 decomposes at an instant was found to be 1×10^{-4} mol dm^{-3} s^{-1}. The rate of NO_2 and O_2 production at the same instant is 2×10^{-4} and 0.5×10^{-4} mol dm^{-3} s^{-1}, respectively. To avoid ambiguity in such cases, we should state to which substance a measured rate refers.

Exercise 14A

Rate of reaction

The reaction between nitric oxide (NO) and hydrogen to make nitrous oxide (N_2O) and water obeys the equation:

$H_2(g) + 2NO(g) \rightarrow N_2O(g) + H_2O(g)$

If the rate (at a particular instant) at which hydrogen disappears was found to be 0.01 mol dm^{-3} s^{-1}, what is the simultaneous rate at which **(i)** nitric oxide disappears, **(ii)** nitrous oxide appears, **(iii)** water appears?

Experimental measurements of the rate of a chemical reaction

In order to measure reaction rates experimentally, we need to monitor the concentration of reactants or products with time. Such concentrations are measured using standard analytical techniques such as titration, those depending upon the degree of light absorption of a substance, or by changes in the electrical conductivity of the reaction mixture.

Occasionally we are able to follow the rate of a reaction because one of the products or reactants is coloured. An example is the reaction between bromine solution and methanoic acid:

$$\underset{\text{dark red}}{Br_2(aq)} + HCOOH(aq) \xrightarrow{H^+(aq)} 2Br^-(aq) + 2H^+(aq) + CO_2(g)$$

(Note that H^+ is used as a catalyst.) It happens that bromine is the only coloured species involved in the reaction, and as the reaction proceeds its colour fades (Fig. 14.1). The intensity of the colour is proportional to the concentration of bromine.

Change of rate of reaction with time

When two reactants are mixed together (for example, in a beaker) the rate of reaction is greatest at the start of the reaction. *As the reaction continues, the rate continues to fall.* If the concentration of reactant (such as bromine, Fig. 14.1) or product is plotted against time, a graph like Fig. 14.2(a) is produced. (If the reaction is an equilibrium reaction, the concentration of reactants in Fig. 14.2(a) will level out when equilibrium has been achieved, but will not fall to zero.)

The rate of reaction at an instant t seconds after the reactants are mixed, is calculated by working out the slope of the *tangent* of the concentration–time graph at that time. As time increases, the slope of the tangents at each instant of time decreases, confirming that the reaction rate is continuously falling.

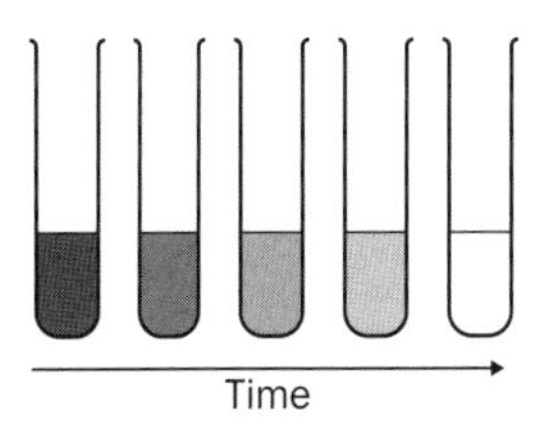

Fig. 14.1 The reaction between bromine and methanoic acid. As the reaction continues, the red colour of bromine fades.

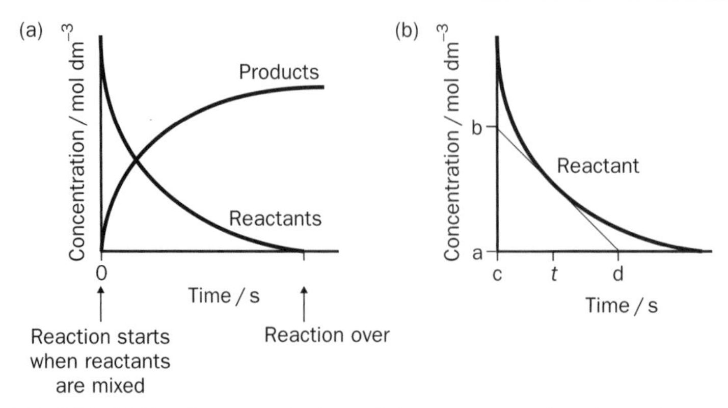

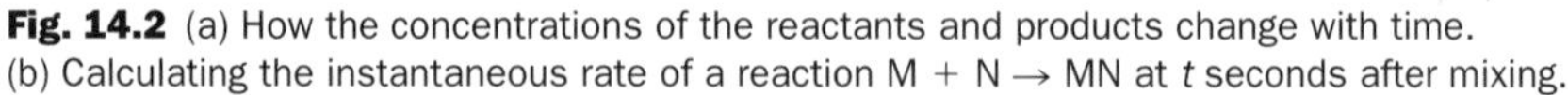

Fig. 14.2 (a) How the concentrations of the reactants and products change with time. (b) Calculating the instantaneous rate of a reaction M + N → MN at *t* seconds after mixing.

Figure 14.2(b) shows the tangent to the reactant curve at time *t*. The rate of reaction at this instant is

$$\frac{\text{change in concentration}}{\text{change in time}} = \frac{b - a}{d - c} \text{ mol dm}^{-3}\text{ s}^{-1}$$

This equation may be used with the concentration of reactants (Exercise 14B) or products (Exercise 14C).

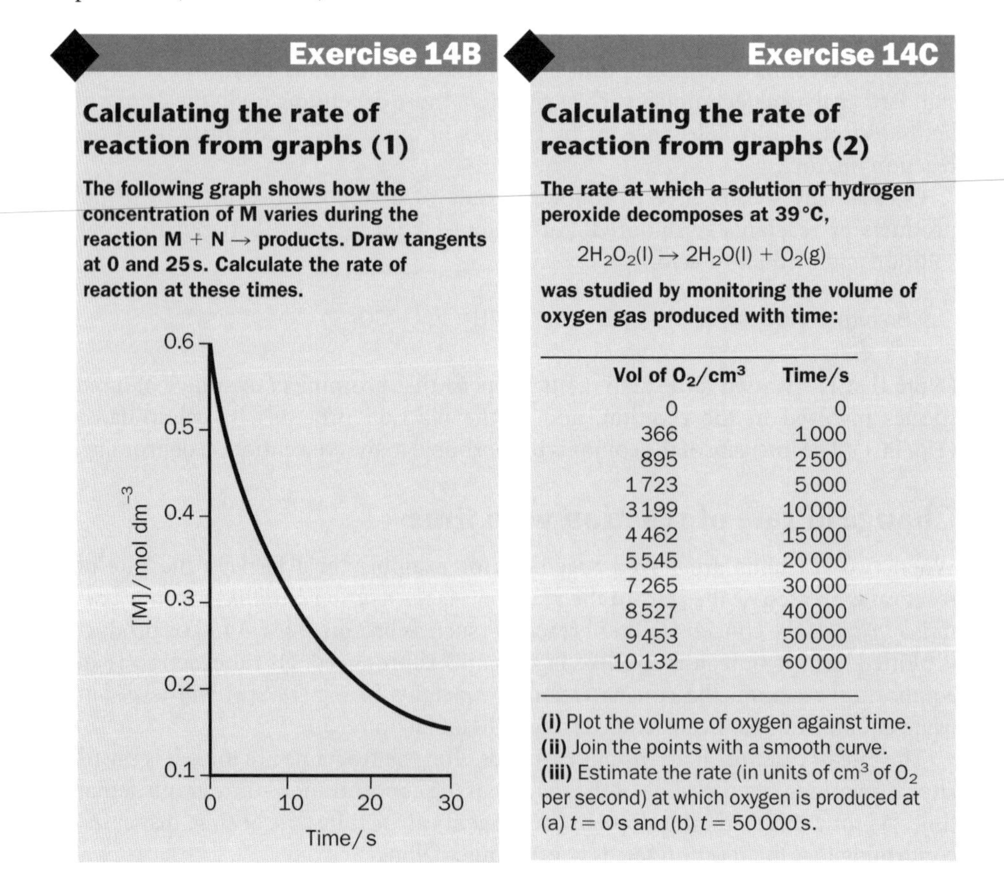

Exercise 14B

Calculating the rate of reaction from graphs (1)

The following graph shows how the concentration of M varies during the reaction M + N → products. Draw tangents at 0 and 25 s. Calculate the rate of reaction at these times.

Exercise 14C

Calculating the rate of reaction from graphs (2)

The rate at which a solution of hydrogen peroxide decomposes at 39 °C,

$$2H_2O_2(l) \rightarrow 2H_2O(l) + O_2(g)$$

was studied by monitoring the volume of oxygen gas produced with time:

Vol of O_2/cm^3	Time/s
0	0
366	1000
895	2500
1723	5000
3199	10000
4462	15000
5545	20000
7265	30000
8527	40000
9453	50000
10132	60000

(i) Plot the volume of oxygen against time.
(ii) Join the points with a smooth curve.
(iii) Estimate the rate (in units of cm^3 of O_2 per second) at which oxygen is produced at (a) $t = 0$ s and (b) $t = 50\,000$ s.

Why reaction rate falls with time

So far we have not explained why reaction rate should begin to fall as soon as the reactants are mixed. Our explanation depends upon two assumptions. The first is that in order for a reaction to take place, collisions must occur between reactant molecules (say, M and N to give the product molecule MN). The second, is that the number of collisions per second between M and N falls as the concentration of M and N falls. There are no product molecules (MN) at the start of the reaction, and the concentration of M and N is at a maximum. As soon as reaction starts, M and N molecules start to combine. The concentrations of M and N begin to fall (and of MN rises), so reducing the number of collisions per second between unreacted M and N molecules. This reduces the reaction rate (Fig. 14.3).

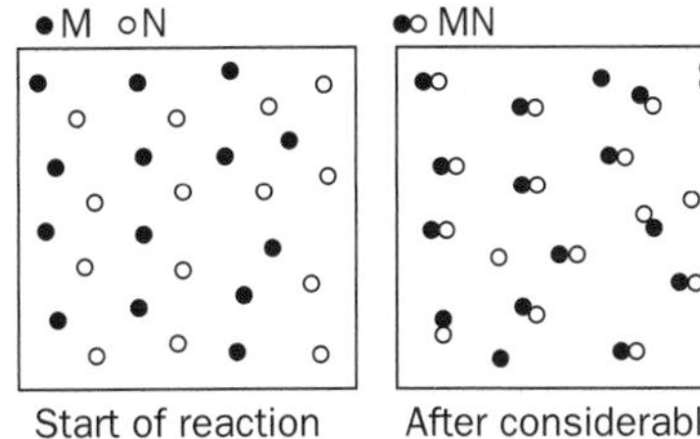

Fig. 14.3 For the general reaction, M + N → MN, reaction can only occur if molecules M and N collide. The chance of collisions between M and N molecules falls as more product (MN) molecules are made.

14.2 Factors affecting reaction rate

1. Nature of the reaction

Some reactions are naturally faster than others. Compare the reaction of hydronium and hydroxide ions,

$$H_3O^+(aq) + OH^-(aq) \rightarrow 2H_2O(l)$$

with the hydrolysis of bromoethane,

$$C_2H_5Br(l) + OH^-(aq) \rightarrow C_2H_5OH(aq) + Br^-(aq)$$

If in an experiment we kept the initial reactant concentrations in both reactions the same, and carried out the experiment at room temperature, the initial rate of the first reaction would be about 1 000 000 000 000 000 times faster than the second reaction. In fact, the reaction between H_3O^+ and OH^- is so fast that it appears to be instantaneous. Special equipment is needed to study such extremely fast reactions.

It is sometimes possible to predict whether reactions will be fast by simply looking at the reactants. Reactions in which the reactants are oppositely charged ions are usually very fast because the charges pull the ions towards each other. This is the case with the reaction,

$$Ag^+(aq) + Cl^-(aq) \rightarrow AgCl(s)$$

and many other common precipitation reactions. Reactants which are atoms or molecules with odd electrons (known as **free radicals**, such as $H^\bullet$ and $CH_3^\bullet$) also react rapidly in an effort to use the odd electron to form a covalent bond. For example, the reaction,

$$CH_3^{\bullet}(g) + CH_3^{\bullet}(g) \rightarrow C_2H_6(g)$$

is naturally very fast.

2. Concentration (or for gases, pressure)

Increasing the concentration of reactants increases the chances of collisions between reactant molecules. In other words, increasing concentration increases the number of collisions between molecules *per second*. This increases the reaction rate. The concentration of a gas increases with its partial pressure. So, we can increase the rate of gaseous reactions by increasing the partial pressure of the reactants.

Where one of the reactants is a solid, increasing the surface area of the solid also increases the chance of the particles of reactants colliding. For example, powdered carbon burns in oxygen faster than lumps of carbon:

$$C(s) + O_2(g) \rightarrow CO_2(g)$$

This is why, in power stations, coal is pulverized before allowing it to burn. Fine powders can sometimes react extremely rapidly. For example, stringent safety regulations need to be applied to grain and flour factories, since fine grain and flour can burn explosively fast in air (Exercise 14D).

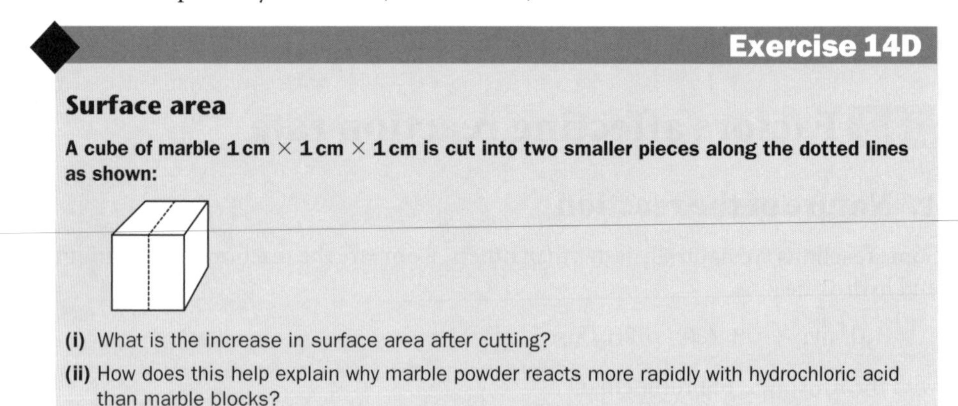

Exercise 14D

Surface area

A cube of marble 1 cm × 1 cm × 1 cm is cut into two smaller pieces along the dotted lines as shown:

(i) What is the increase in surface area after cutting?

(ii) How does this help explain why marble powder reacts more rapidly with hydrochloric acid than marble blocks?

3. Temperature

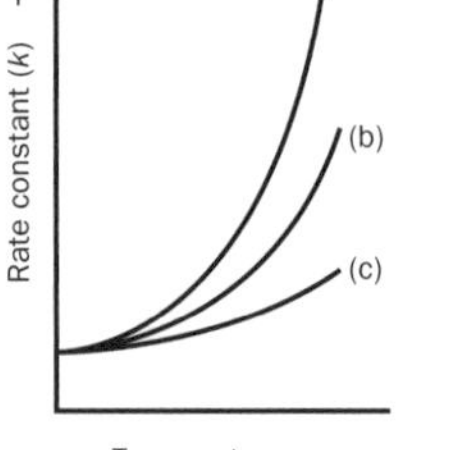

Fig. 14.4 Variation of reaction rate with temperature for reactions which are (a) very sensitive, (b) fairly sensitive, (c) insensitive to changes in temperature. (The rate constant is the reaction rate with the reactants at a concentration of 1.0 mol dm^{-3}.)

Increasing the temperature of a reaction mixture increases the rate of the chemical reaction. (This happens whether the reaction is exothermic or endothermic.) This increase is often dramatic (Fig. 14.4).

Although the number of collisions per second between reactant molecules rises with temperature, calculations show that this makes only a small contribution to the increase of reaction rate with temperature. The accepted explanation for the increase in reaction rate involves a key idea in chemical kinetics, that of **activation energy**.

Think of a chemical reaction in which molecules of M and N form the product molecule MN in a single step. In order for a collision between molecules to result in chemical change, old bonds need to be broken and new ones formed. Bond destruction and bond formation require a redistribution of the bonding electrons within

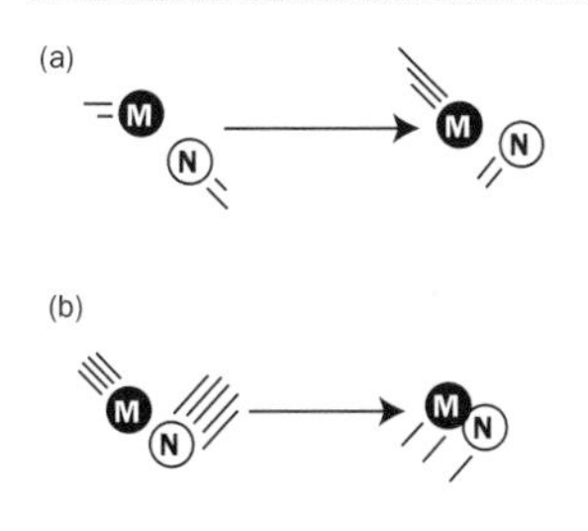

Fig. 14.5 (a) Unsuccessful collision – the kinetic energy of M and N is less than E_A and so the reactants bounce apart.
(b) Successful collision – the kinetic energy of M and N is greater or equal to E_A and a bond is formed between M and N.

the reactant molecules. However, as the molecules of M and N approach each other their outer electrons repel, so preventing the reactants from getting close enough for the bonding electrons to be rearranged. The reactant molecules then simply 'bounce off' each other, and the collision does not result in chemical change.

In order to overcome the repulsive forces between the molecules, the colliding molecules must between them possess a minimum value of kinetic energy, termed the **activation energy** (E_A). Unless the sum of the energies possessed by the reactant molecules involved in the collision equals (or exceeds) this minimum energy, no reaction can take place (Fig. 14.5).

The activation energy for a reaction *does not itself* vary with temperature, but increasing the temperature of a material does increase the average energy of its molecules. *It follows that the higher the temperature, the greater is the fraction of colliding molecules which possess at least the activation energy* (Fig. 14.6). We then expect more collisions to result in chemical reaction at higher temperatures, and this explains the experimental observation of an increased reaction rate.

The rate of some reactions changes more rapidly with temperature than others (see Fig. 14.4), and we are now in a position to explain this observation by assuming that *reactions whose rates are very sensitive to temperature possess relatively high activation energies*:

- The speed of a reaction which possesses a large activation energy changes rapidly with temperature because few of the reacting molecules will possess an energy equal to (or greater than) E_A at room temperature. Therefore, an increase in temperature will drastically increase the number of collisions which result in chemical change.

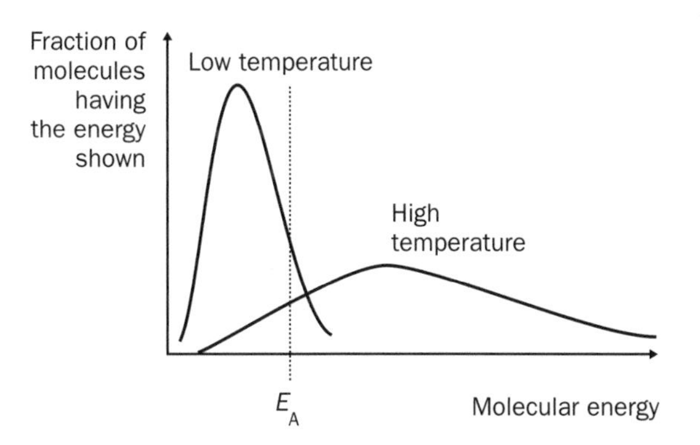

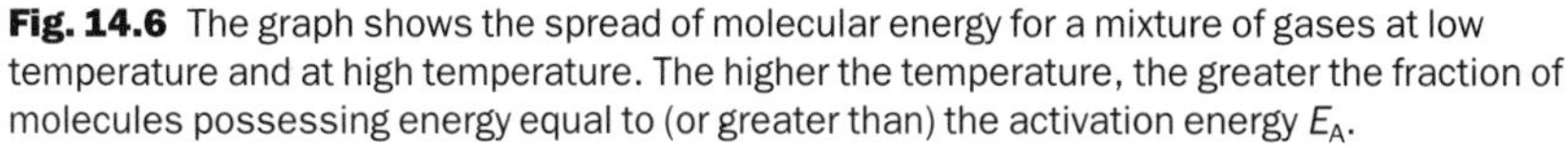

Fig. 14.6 The graph shows the spread of molecular energy for a mixture of gases at low temperature and at high temperature. The higher the temperature, the greater the fraction of molecules possessing energy equal to (or greater than) the activation energy E_A.

Pressurized steam has a temperature *above* 100 °C. A pressure cooker allows the safe production of pressurized steam, which then cooks food faster than the conventional method in which food is simply immersed in boiling water at 100 °C in an open pot.

- A reaction which possesses a low E_A is relatively insensitive to temperature because most of the reacting molecules already possess the activation energy at room temperature.
- Reactions which are naturally very fast (such as $CH_3^{\bullet}(g) + CH_3^{\bullet}(g) \rightarrow C_2H_6(g)$) possess activation energies which are virtually zero. As a consequence, the rates of such reactions show very little dependence upon temperature.

Exercise 14E

Increase in reaction rate at higher temperature

(i) The chemical reactions involved in the hardening ('boiling') of an egg roughly double in rate for every 10 °C rise in temperature. How much faster will these reactions be at 100 °C than at 20 °C?

(ii) Very exothermic reactions are sometimes faster after the reaction has started, even though the initial reactant concentrations have fallen – why ?

Exercise 14F

Factors affecting reaction rate

Zinc reacts with hydrochloric acid to produce hydrogen gas:

$$Zn(s) + 2HCl(aq) \rightarrow ZnCl_2(aq) + H_2(g)$$

Experiments show that the rate of reaction decreases **(i)** if zinc powder is replaced by zinc nails and **(ii)** if the following are added to the reaction mixture: (a) water, (b) ice and (c) sodium carbonate. Carefully explain these effects.

BOX 14.1

Transition states and activation energy

As the bonds in reacting molecules are rearranging during a collision, the energy of the molecules rises to a maximum above the starting energy of the reactants (Fig. 14.7). This maximum is the **activation energy** E_A (in kJ mol^{-1}) of the (forward) reaction. The arrangement (or **configuration**) of atoms at this energy maximum is called the **transition state**. The transition state is a 'half-way house' between reactants and products. It can never be isolated and exists for only a vanishingly short period of time. Once formed, the transition state immediately 'falls back' to the reactants or else 'turns' into products.

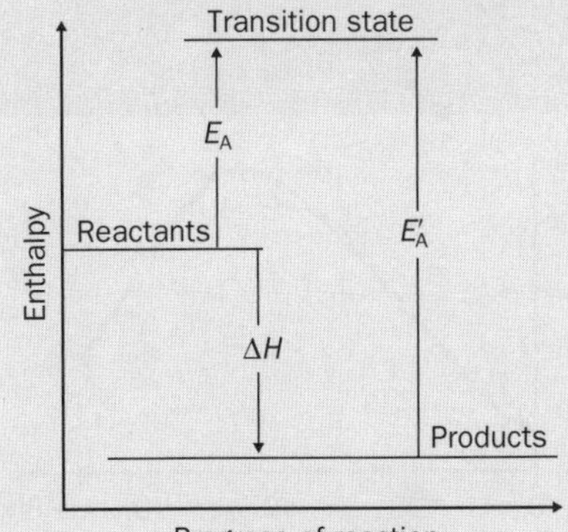

Fig. 14.7 Connection between reaction activation energy E_A and the reaction enthalpy change ΔH for the exothermic single-step reaction M + N → MN.

Figure 14.7 also shows the activation energy for the **reverse** reaction (MN → M + N), symbolised E'_A. When the forward reaction is exothermic, the activation energy of the forward reaction is always smaller than the activation energy for the reverse reaction, i.e. $E_A < E'_A$. (The reverse is true for an endothermic reaction.) If a chemical reaction occurs in several steps, each step possesses its own transition state, with its own forward and back activation energies.

The reaction between 1-bromopropane and hydroxide ion:

$$RCH_2Br(l) + OH^-(aq) \rightarrow RCH_2OH(aq) + Br^-(aq)$$

($R = CH_3CH_2$) occurs in one stage. The probable structure of the transition state is shown in Fig. 14.8 with the negative charge of the OH^- being spread out over the structure.

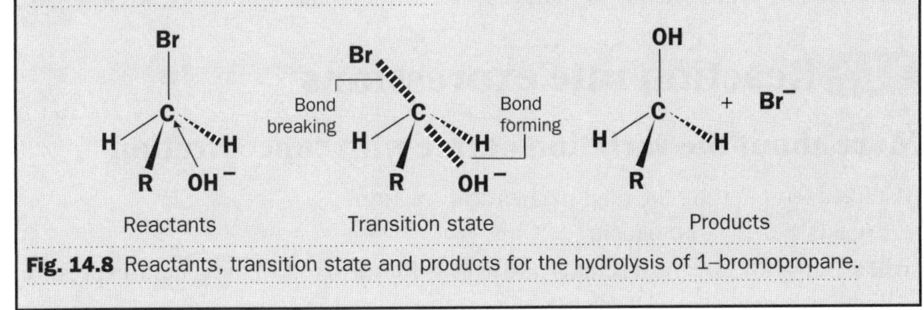

Fig. 14.8 Reactants, transition state and products for the hydrolysis of 1-bromopropane.

4. Catalysts

Catalysts are used to speed up reactions. Their general characteristics are:

- Catalysts take part in chemical reactions but are regenerated during the reaction.
- Catalysts do not affect the enthalpy change or equilibrium constant of the reaction.
- Catalysts are often specific, i.e. they will only affect the rate of a particular reaction. This is particularly true of some biological catalysts (**enzymes**). Enzymes are **proteins**. They are often deactivated (or even destroyed) if the temperature rises much above 40 °C (Fig. 14.9). Most enzymes operate in a narrow band of pH, and many require the presence of a metal ion or another complex organic molecule as **cofactor**. For more about enzymes, see Case Study 1 on the website.

Rusting is the reaction of iron with oxygen in the presence of water. The rate of rusting is controlled by the surface area of the iron exposed to the oxygen, by the concentration of water vapour in the air, and by temperature. Rusting proceeds slower at lower temperatures, but salt (often used on the roads in cold climates) acts as a catalyst and causes cars to rust faster.

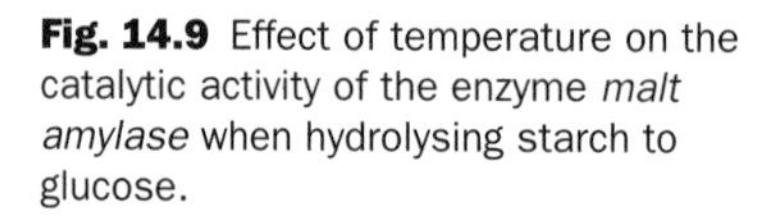

Fig. 14.9 Effect of temperature on the catalytic activity of the enzyme *malt amylase* when hydrolysing starch to glucose.

Catalysts are important in industry. For example, iron is used to catalyse the production of ammonia from nitrogen and hydrogen, and nickel is used to speed up the conversion of vegetable oils to margarines. For more about catalysis, see page 262.

14.3 Reaction rate expressions

More about the variation of rate with concentration

At a fixed temperature the rates of chemical reactions

$$A + B + C \ldots \rightarrow \text{products}$$

(where A, B and C are the reactants) are usually found to follow the general expression

$$\text{rate of reaction} = k[A]^x[B]^y[C]^z \ldots$$

where k is called the **rate constant** and the square brackets represent the concentrations (in mol dm^{-3}) of the reactants which are raised to the powers of $x, y, z \ldots$ The concentration of reactants will begin to fall as soon as the reactants are mixed. Therefore, this equation defines the rate of the reaction at *one instant* – as [A], [B] and [C] continuously fall, so does the rate of reaction.

The powers x, y and z are said to be the *individual orders* of reaction for reactants A, B and C, respectively. The *overall order* of reaction is the sum of the individual orders x, y, and $z \ldots$

$$\text{overall order} = x + y + z \ldots$$

Reactions with an overall order of one are said to be 'overall first-order'; reactions with an overall order of two are said to be 'overall second-order'; reactions with an overall order of zero are said to be 'overall zeroth order' and so on.

The values of x, y and z (etc.) are decided by experiment – they cannot be reliably predicted by simply looking at the chemical equation for the reaction.

The rate constant

- One way to look at k is to think of it as the rate of reaction when all of the reactant concentrations are exactly $1\ mol\ dm^{-3}$.
- k increases with the temperature of the reaction mixture. Therefore, the reaction rate also increases with temperature (Box 14.2).
- The size of k is decided by the activation energy for the reaction (a low E_A means a high k value and vice versa). Naturally fast reactions have high rate constants.
- The units of k depend upon the overall order of reaction (Table 14.2).

Table 14.2 Units of rate constants

Overall order of reaction	Units of rate constant
Zeroth	$mol\ dm^{-3}\ s^{-1}$
First	s^{-1}
Second	$mol^{-1}\ dm^3\ s^{-1}$
Third	$mol^{-2}\ dm^6\ s^{-1}$

Example 14.2

The reaction between A and B follows the rate expression

rate of reaction = k[A][B]

What is the overall order of reaction? What are the units of k?

▶ Answer

We can rewrite this as

Rate of reaction = $k[A]^x[B]^y$ where $x = y = 1$

The reaction is therefore first order with respect to the reactants A and B. Overall, the reaction order is $x + y = 1 + 1 = 2$, i.e. it is second order overall.

For the units of k, rearranging for k gives

$$k = \frac{\text{rate of reaction}}{[A][B]}$$

The units of k are therefore

$$\frac{mol\ dm^{-3}\ s^{-1}}{mol\ dm^{-3}\ mol\ dm^{-3}}$$

Cancelling

$$\frac{\cancel{mol}\ \cancel{dm}^{-3}\ s^{-1}}{\cancel{mol}\ \cancel{dm}^{-3}\ mol\ dm^{-3}} = mol^{-1}\ dm^3\ s^{-1}$$

A reaction which is *overall* second order would also be obtained if the rate expression was:

rate of reaction = $k[A]^2$ or rate of reaction = $k[B]^2$

Exercise 14G

Working out the overall order from the rate expression

What is the overall order of reactions possessing the following rate expressions?

(i) rate of reaction = $k[A][B]^0$

(ii) rate of reaction = $k[D]^2$

(iii) rate of reaction = $k[E]^{0.5}[G]^{0.5}$

(iv) rate of reaction = k[B][J][N].

BOX 14.2

Dependence of the rate constant k upon temperature: the Arrhenius equation

Experiments show that the way that the *rate constant* for a chemical reaction varies with temperature can be described by the Arrhenius equation:

$$k = Ae^{-E/RT}$$

where k is the rate constant at temperature T (K), A is a constant for the reaction, R is the gas constant (8.3145 J mol^{-1} K^{-1}) and E (in J mol^{-1}) is the so-called **Arrhenius activation energy** for the reaction. The Arrhenius equation is useful because it is found to apply even if the reaction under study does not occur in one simple step.

Taking natural logs (ln) to both sides of the Arrhenius equation gives an equation of the form

$$y = mx + c,$$

thus

$$\ln k = \ln A - \left(\frac{E}{R}\right)\frac{1}{T}$$

If we have measured the rate constants for a named reaction at various temperatures, we can plot ln k against $1/T$, giving a straight line graph of slope $-E/R$. This is the way that Arrhenius activation energies are calculated from experimental data (see the website).

The Arrhenius activation energy E is **not** the same as the reaction activation energy E_A (which is only meaningful for single-step reactions). It is better to look upon the Arrhenius activation energy simply as a useful experimental index which shows *how quickly* the rate constant of a reaction changes as the temperature is changed:

- If $E = 0$, the rate constant is independent of temperature.
- If E is small, the rate constant has only a slight dependence upon temperature (Fig. 14.4(c)).
- If E is large, the rate constant changes dramatically with changes in temperature (Fig. 14.4(a)).

If the rate constant of a reaction is k_1 at temperature T_1, and k_2 at temperature T_2, the Arrhenius equations for both temperatures may be combined to give the expression

$$\ln\left(\frac{k_1}{k_2}\right) = \frac{E}{R}\left(\frac{1}{T_2} - \frac{1}{T_1}\right)$$

Provided that we know the Arrhenius activation energy for the reaction, this equation may be used to calculate the rate constant at one temperature, knowing the rate constant at another temperature. For example, if for a particular chemical reaction,

$$k_1 = 1.0 \times 10^{-3}\,\text{mol}^{-1}\,\text{dm}^3\,\text{s}^{-1} \text{ at } 300\,\text{K}$$

and

$$E = 50\,\text{kJ mol}^{-1} \text{ (i.e. } 50\,000\,\text{J mol}^{-1}\text{)},$$

then, at 320 K,

$$\ln\left(\frac{10^{-3}}{k_2}\right) = \frac{50\,000}{8.3145}\left(\frac{1}{320} - \frac{1}{300}\right) = -1.25$$

or

$$\left(\frac{10^{-3}}{k_2}\right) = e^{-1.25} = 0.287$$

so that $k_2 = 10^{-3}/0.287 \approx 3.5 \times 10^{-3}$ mol^{-1} dm^3 s^{-1}. In other words, the rate constant has increased by a factor of 3.5 for an increase in temperature of only 20 K.

14.4 Examples of rate expressions found by experiment

Decomposition of ethanal

Ethanal (CH_3CHO) molecules may be broken up by heating:

$$CH_3CHO(g) \rightarrow CH_4(g) + CO(g)$$

Experiments show the rate expression to be

$$\text{rate of reaction} = k[CH_3CHO]^{1.5}$$

i.e., the order of reaction with respect to ethanal is 1.5. Since ethanal is the only reactant, the overall order is also 1.5. The order of reaction is something of a surprise, and we could certainly not predict it from the chemical equation for the reaction.

Hydrolysis of methyl ethanoate

The equation for hydrolysis of methyl ethanoate ester by sodium hydroxide is

$$CH_3COOCH_3(l) + OH^-(aq) \rightarrow CH_3COO^-(aq) + CH_3OH(aq)$$

The rate expression is found to be

$$\text{rate of reaction} = k[CH_3COOCH_3(l)][OH^-(aq)]$$

The order of reaction is therefore one (i.e. first order) with respect to both methyl ethanoate and sodium hydroxide and the overall order of reaction is two (i.e. second order).

If the sodium hydroxide is in excess (at least 10 times the concentration of ester), the *consumption* of sodium hydroxide during the reaction will be very small and may be neglected. This means that $[OH^-]$, the hydroxide ion concentration at an instant, may be replaced by the initial hydroxide concentration $[OH^-]_0$. k and $[OH^-]_0$ may now be lumped together as a new constant k'. Under these conditions the rate expression becomes

$$\text{rate of reaction} = k'[CH_3COOCH_3(l)]$$

where $k' = [OH^-]_0 \times k$. Experimentally, the reaction is now first order. We say that the reaction is **pseudo first order** and that k' is the pseudo-first-order rate constant. Here, the word 'pseudo' means 'falsely'. Deliberately making one of the reactant concentrations much higher than the other(s) is a useful strategy for forcing a reaction to show first-order kinetics.

14.5 Calculations using rate expressions

Calculating the rate constant from the initial rate of reaction

What is the rate of reaction at the start of a reaction? Immediately after mixing, the concentrations of reactants will not have changed very much from their starting values. We then define the *initial rate* of the reaction as

$$\text{initial rate} = k[A]_0^x[B]_0^y[C]_0^z \ldots$$

where $[A]_0$, $[B]_0$, and $[C]_0$ are the initial concentrations of reactants A, B and C.

The rate constant k at a specific temperature may now be calculated by rearrangement:

$$k = \frac{\text{initial rate}}{[A]_0^x[B]_0^y[C]_0^z \ldots}$$

Example 14.3

The formation of nitrosyl chloride (NOCl) at 400 K

$NO(g) + Cl_2(g) \rightarrow NOCl(g) + Cl(g)$

follows the rate expression

rate of reaction = $k[NO(g)][Cl_2(g)]$

The initial rate of reaction (with both reactants at an initial concentration of 0.010 mol dm^{-3}) was found to be 3.2 × 10^{-6} mol dm^{-3} s^{-1}.

(a) What is the order of reaction?

(b) What is the value of *k* at 400 K?

▶ Answer

(a) The reaction is first order with respect to nitric oxide, first order with respect to chlorine, and second order overall.

(b)

$$k = \frac{\text{initial rate of reaction}}{[NO(g)]_0[Cl_2(g)]_0}$$

Therefore,

$$k = (3.2 \times 10^{-6})/(0.010 \times 0.010) = 3.2 \times 10^{-2}\ \text{mol}^{-1}\,\text{dm}^3\,\text{s}^{-1}.$$

Exercise 14H

Calculating the rate of reaction

The reaction between nitric oxide and oxygen gas,

$2NO(g) + O_2(g) \rightarrow 2NO_2(g)$

is found to follow the rate expression

rate of reaction = $k[NO]^2[O_2]$

(i) What is the overall order of reaction?

(ii) At 300 °C the rate constant *k* is 7000 mol^{-2} dm^6 s^{-1}. What is the rate of reaction at the instant when [NO] and $[O_2]$ are both 0.060 mol dm^{-3}?

Exercise 14I

Calculation of rate constant

The hydrolysis of sucrose in the presence of acid catalyst is often represented as

$$C_{12}H_{22}O_{11} + H_2O \xrightarrow{[H^+(aq)]} 2C_6H_{12}O_6$$

The rate expression is found to be

rate of reaction = $k[C_{12}H_{22}O_{11}][H^+(aq)]$

(Notice that in this example, the concentration of catalyst appears in the rate expression.) A reaction mixture at 298 K initially contained 0.20 mol dm^{-3} sucrose and 0.10 mol dm^{-3} H$^+$. The initial rate of reaction was found to be 4.0 × 10^{-6} mol dm^{-3} s^{-1}.

(i) What is the overall order of the reaction?

(ii) Calculate *k* at 298 K and state its units.

Finding the order of a reaction using the initial rates method

We now look at the use of initial rates of reaction in helping us find the individual orders in a rate expression.

Suppose that we wish to find the overall order of the reaction

$$A + B \rightarrow \text{products}$$

This is another way of saying that we want to find out the individual orders (the values of x and y) in the rate expression

$$\text{rate of reaction} = k[A]^x[B]^y$$

The initial rate of this reaction is given by the expression

$$\text{initial rate} = k[A]_0^x[B]_0^y$$

where $[A]_0$ and $[B]_0$ are the initial concentrations of A and B.

Now suppose that we carry out two experiments at the same temperature (so that k remains fixed) and using the same initial concentration of B but different initial concentrations (0.1 mol dm^{-3} and 0.2 mol dm^{-3} respectively) of A.

For the first experiment,

$$\text{initial rate}_1 = k[0.1]^x \times [B]_0^y$$

For the second experiment

$$\text{initial rate}_2 = k[0.2]^x \times [B]_0^y$$

Both k and $[B]_0^y$ feature in both expressions. The initial rates in these experiments will be different only because the initial concentrations of A are different. We can see this more easily by writing the above equations in the form:

$$\text{initial rate}_1 = \text{constant} \times [0.1]^x$$

$$\text{initial rate}_2 = \text{constant} \times [0.2]^x$$

where

$$\text{constant} = k[B]_0^y$$

What if the initial rate doubles as [A] doubles?

If, in doubling [A], experiments show that the initial rate doubles, so that

$$\text{initial rate}_2 = 2 \times \text{initial rate}_1$$

i.e.

$$\text{constant} \times [0.2]^x = 2 \times \text{constant} \times [0.1]^x$$

then the order x *must* be one. This means that the reaction is first order with respect to A.

What if the initial rate increases by a factor of four as [A] doubles?

If, in doubling [A], the *initial rate* is found to have quadrupled, so that

$$\text{initial rate}_2 = 4 \times \text{initial rate}_1$$

i.e.

$$\text{constant} \times [0.2]^x = 4 \times \text{constant} \times [0.1]^x$$

then the order x *must* be two. This means that the reaction is second order with respect to A.

Example 14.4

The reaction of benzaldehyde with cyanide ions (as catalyst) at 50 °C was studied by the initial rates method:

$$2C_6H_5CHO(aq) \xrightarrow{CN^-(aq)} C_6H_5CH(OH)COC_6H_5(aq)$$

The results were:

Experiment number	Initial concentration/mol dm^{-3}		Initial rate of reaction /mol dm^{-3} s^{-1}
	$[C_6H_5CHO(aq)]_0$	$[CN^-(aq)]_0$	
1	0.500	0.500	1.00×10^{-4}
2	1.00	0.500	4.01×10^{-4}
3	1.00	1.00	7.98×10^{-4}

What is the overall order of reaction? What is the rate constant at this temperature?

▶ Answer

The rate expression is:

$$\text{rate of reaction} = k[C_6H_5CHO(aq)]^x[CN^-(aq)]^y$$

Experiment 1 is used as a 'norm'. Doubling the concentration of benzaldehyde (experiment 2) quadruples the initial rate. This means that $x = 2$.

Doubling the concentration of cyanide ions (experiment 3) produces an initial rate which is double that of experiment 2. Therefore, $y = 1$ so that

$$\text{overall order} = x + y = 2 + 1 = 3$$

The reaction is third order overall and the rate expression becomes

$$\text{rate of reaction} = k[C_6H_5CHO(aq)]^2[CN^-(aq)]$$

Rearranging this equation and considering the initial rate:

$$k = \frac{\text{initial rate}}{[C_6H_5CHO(aq)]_0^2\,[CN^-(aq)]_0}$$

For experiment 1,

$$k = \frac{1.00 \times 10^{-4}}{[0.500]^2\,[0.500]} = 8.00 \times 10^{-4}\,\text{mol}^{-2}\,\text{dm}^{-6}\,\text{s}^{-1}$$

For experiment 2,

$$k = \frac{4.01 \times 10^{-4}}{[1.00]^2\,[0.500]} = 8.02 \times 10^{-4}\,\text{mol}^{-2}\,\text{dm}^{-6}\,\text{s}^{-1}$$

For experiment 3,

$$k = \frac{7.98 \times 10^{-4}}{[1.00]^2\,[1.00]} = 7.98 \times 10^{-4}\,\text{mol}^{-2}\,\text{dm}^{-6}\,\text{s}^{-1}$$

The average rate constant at 50 °C is

$$k = \frac{(8.00 + 8.02 + 7.98) \times 10^{-4}}{3} = 8.00 \times 10^{-4}\,\text{mol}^{-2}\,\text{dm}^{-6}\,\text{s}^{-1}$$

What if there is no change in initial rate as [A] doubles?

If experiments show that the initial rate of reaction remains unchanged, even though [A] in experiment 2 is double [A] in experiment 1, then

$$\text{initial rate}_2 = \text{initial rate}_1$$

i.e.

$$\text{constant} \times [0.2]^x = \text{constant} \times [0.1]^x$$

This can be true only if $x = 0$, since 'anything to the power of zero is one'. This means that the reaction is zeroth order with respect to reactant A.

Similar arguments can be applied to reactant B, by carrying out experiments in which $[A]_0$ remains constant and in which the concentration of B is varied from experiment to experiment.

Exercise 14J

Initial rates, 1

The reaction:

$$M + Y = \text{products}$$

was studied by the initial rates method at 25 °C. It was found that the initial rate of reaction halved when [M] was halved and that doubling [Y] had no effect upon the initial rate. Write down the rate expression for the reaction.

Exercise 14K

Initial rates, 2

The kinetics of the reaction between bromide ($Br^-(aq)$), bromate ($BrO_3^-(aq)$) and $H^+(aq)$ ions:

$$5Br^-(aq) + BrO_3^-(aq) + 6H^+(aq) \rightarrow 3Br_2(l) + 3H_2O(l)$$

was studied by the initial rates method and the following data obtained:

	Initial concentration/mol dm^{-3}			
Experiment	**$[Br^-(aq)]_0$**	**$[BrO_3^-(aq)]_0$**	**$[H^+(aq)]_0$**	**Relative initial rate of reaction**
1	0.01	0.01	0.01	1.00
2	0.02	0.01	0.01	2.00
3	0.01	0.04	0.01	4.00
4	0.01	0.01	0.02	4.00

Write down a rate expression for the reaction.

14.6 More about first-order reactions

If the simple reaction A → B is first order, then

$$\text{rate of reaction} = k[A]$$

For such a first-order reaction, the concentration of A at time t seconds after the start of the reaction, symbolized $[A]_t$, is related to the initial concentration of A, i.e. $[A]_0$, by the equation

$$[A]_t = [A]_0\, e^{-kt}$$

where k is the first-order rate constant. Taking natural logs on both sides of this equation, followed by rearrangement, gives

$$\ln\left(\frac{[A]_0}{[A]_t}\right) = k \times t \qquad \textbf{(14.1)}$$

If the time at which exactly half of the concentration of A has disappeared is symbolised as $t_{1/2}$, then

$$[A]_{t_{1/2}} = [A]_0/2 \qquad \textbf{(14.2)}$$

Substituting equation (14.2) into equation (14.1) gives

$$\ln\left(\frac{[A]_0}{[A]_0/2}\right) = \ln 2 = k \times t_{1/2}$$

and, since $\ln 2 = 0.693$,

$$k \times t_{1/2} = 0.693$$

Note that:

- These equations apply to any first-order reaction of the type A $\rightarrow$ B.
- The formal definition for the **half-life** (or **reactant half-life**) of a reaction, symbolised $t_{1/2}$, is that it is the time taken for the concentration of a reactant to fall by half.
- The half-life of a first-order reaction does not depend upon the initial concentration of the reactant.

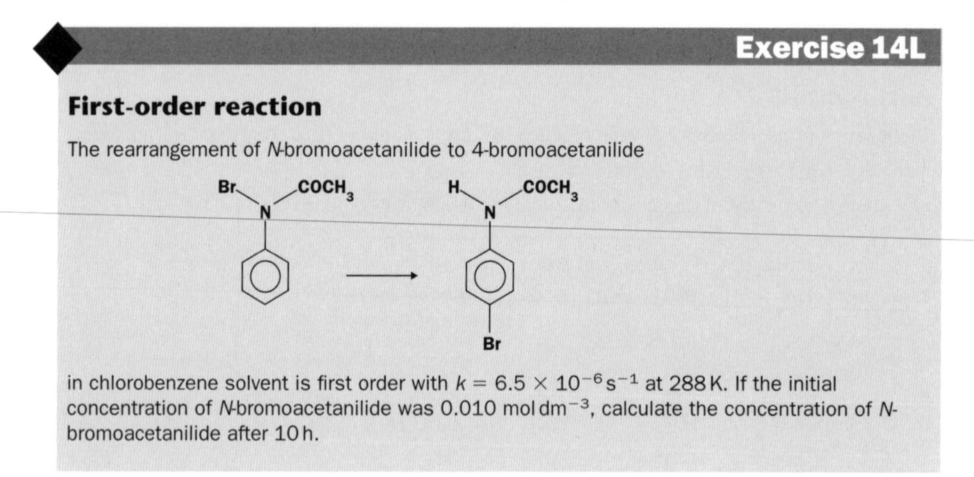

Exercise 14L

First-order reaction

The rearrangement of *N*-bromoacetanilide to 4-bromoacetanilide

in chlorobenzene solvent is first order with $k = 6.5 \times 10^{-6}\,s^{-1}$ at 288 K. If the initial concentration of *N*-bromoacetanilide was 0.010 mol dm^{-3}, calculate the concentration of *N*-bromoacetanilide after 10 h.

Example of half-life – the decomposition of azomethane

The half-life for the decomposition of azomethane,

$$CH_3N_2CH_3(g) \rightarrow CH_3CH_3(g) + N_2(g)$$

where

$$\text{rate of reaction} = k[CH_3N_2CH_3(g)]$$

is about 2000 s at 180 °C. This means that if the initial concentration of azomethane was 0.1 mol dm^{-3}, after 2000 s the concentration would have fallen to 0.05 mol dm^{-3}, after another 2000 s to 0.025 mol dm^{-3}, after a further 2000 s to 0.0125 mol dm^{-3}, and so on.

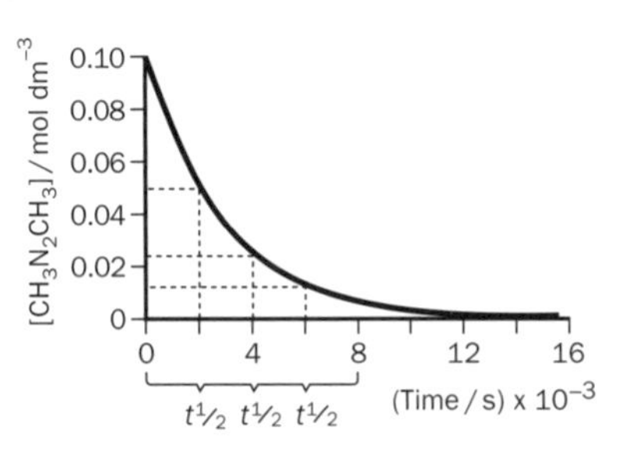

Fig. 14.10 The fall in azomethane concentration at 180 °C. Because the reaction is first order, its half-life does not depend upon the initial concentration of azomethane.

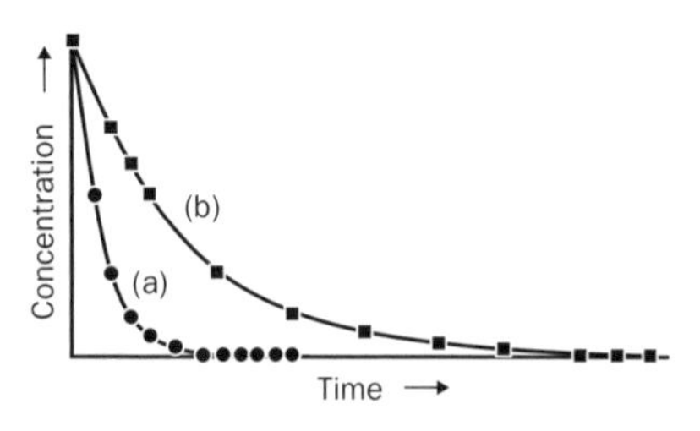

Fig. 14.11 The exponential decay of a reactant in a first-order reaction where (a) k is large (b) k is small. The larger the value of k, the faster is the decay.

Figure 14.10 shows a plot of $[CH_3N_2CH_3(g)]$ against t. The gradual fall in azomethane concentration follows equation (14.1) and is said to be *exponential*. The curvature of the plot is controlled by the value of k – the greater is k, the quicker the reactant concentration falls (Fig. 14.11).

The rate constant for the decomposition of $CH_3N_2CH_3(g)$ may be calculated directly from the half-life:

$$k = \frac{0.693}{t_{1/2}} = 0.693/2000 = 3.5 \times 10^{-4}\ s^{-1}$$

Exercise 14M

Calculating half-life from *k*

Calculate the half-life of the first-order reaction

cis-CHCl=CHCl → *trans*-CHCl=CHCl

at a temperature of 825 K given that $k = 9.6 \times 10^{-3}\ s^{-1}$.

Exercise 14N

Half-lives of pesticides

The chemical degradation of a pesticide (P) in a field

P → byproducts

was found to be roughly first order:

rate of reaction = k[P]

with $k = 5.73 \times 10^{-7}\ s^{-1}$. After spraying, a pesticide was washed off plants into a pond. The initial concentration of pesticide in the pond was 0.10 mg dm^{-3}. What concentration of pesticide would be left in the pond after 42 days? (**Hint** start by calculating the half-life of the pesticide.)

Examples of the use of half-lives

1. Radioactive decay

The decay of elements by radioactivity, although not a chemical reaction, follows first-order kinetics. For example, the half-life of $^{238}_{92}U$ is 4.5×10^9 years. This means that 4.5 billion years ago (when the earth was formed), there was double the amount of uranium-238 present than is the case now.

2. Degradation of organic compounds in the environment

Organic compounds (including solvents and pesticides) break up in the environment as a result of hydrolysis (reaction with water), photolysis (breakdown by light) or by the action of bacteria. These reactions are usually first order (or pseudo first order).

Ideally, toxic organic compounds in waste should have a low half-life so that they break down quickly. The insecticide **parathion** slowly hydrolyses at room temperature:

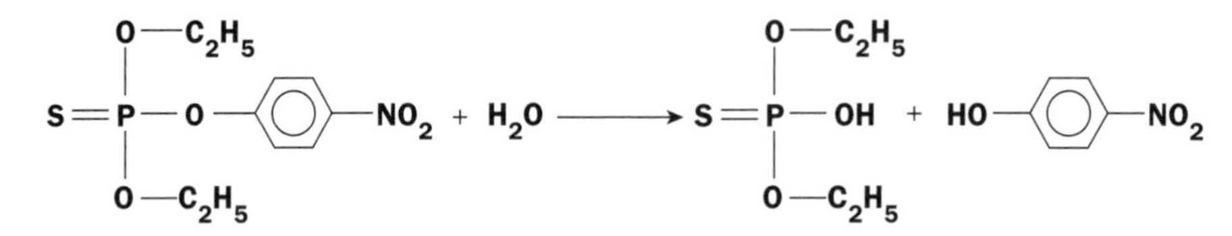

The $t_{1/2}$ of parathion in water due to hydrolysis is about 25 days – sufficiently long for it to pose a hazard.

14.7 Reaction mechanisms

Chlorination of methane

Chloromethane (CH_3Cl) is made by reacting chlorine and methane. The overall equation for the reaction is

$$Cl_2(g) + CH_4(g) \rightarrow CH_3Cl(g) + HCl(g)$$

A spark or a flash of light is needed to start the reaction. In either case the reaction can be explosively violent.

The overall equation suggests that chloromethane is produced on a molecular scale by the collision of chlorine and methane molecules. Experimental evidence shows that this is not the case: the production of chloromethane occurs in several (simple) steps known as **elementary reactions**. The sequence of steps is called the **reaction mechanism.** For the chlorination of methane the steps include

$$Cl_2(g) + \text{heat/light} \rightarrow 2Cl^\bullet(g) \quad \textbf{(14.3)}$$

$$Cl^\bullet(g) + CH_4(g) \rightarrow CH_3^\bullet(g) + HCl(g) \quad \textbf{(14.4)}$$

$$CH_3^\bullet(g) + Cl_2(g) \rightarrow CH_3Cl(g) + Cl^\bullet(g) \quad \textbf{(14.5)}$$

The spark (or flash of light) supplies the energy for reaction (14.3). Reactions (14.4) and (14.5) involve chlorine and methyl **free radicals** ($Cl^\bullet$ and $CH_3^\bullet$) as the **reaction intermediates**. Free radicals are chemically very reactive because they are eager to use their odd electrons (symbolized $^\bullet$) to covalently bond with other atoms.

Note that chlorine radicals are **regenerated** in reaction (14.5) and that these then react with more methane (reaction (14.4)). In this way the reaction gets faster and faster (a **chain reaction**) which may cause the reaction mixture to explode. But the reaction cannot go on indefinitely and eventually **termination reactions**, such as

$$CH_3^\bullet(g) + CH_3^\bullet(g) \rightarrow C_2H_6(g)$$

or

$$Cl^\bullet(g) + Cl^\bullet(g) \rightarrow Cl_2(g)$$

use up the free radicals and stop the overall reaction.

Chemical reactions which (like the chlorination of methane) occur in several steps are said to be **complex**.

Reaction intermediates are molecules, radicals or ions which are made (and then used up) during the reaction mechanism. Although they are often so reactive that they last for only a short time (or, are present in very small concentrations at any moment during the reaction) they may often be detected spectroscopically, i.e. by the characteristic frequencies of light which they absorb.

The use of reaction kinetics in helping to establish the mechanism of a reaction

The rate expression for a reaction often gives valuable information about the mechanism of a reaction. When we introduced rate expressions (page 250), we made the point that we could not reliably predict the rate expression from the equation for the reaction. The reason for this uncertainty is that we cannot usually say whether or not the chemical equation in question is elementary, i.e. whether or not it occurs in one or several steps.

It *is* possible to predict the rate expression for an *elementary reaction* from its chemical equation. For example, consider the elementary reaction represented by the equation

$$A + B \rightarrow C$$

The rate at which molecules A and B collide is proportional to their concentration. The rate expression is therefore

$$\text{rate} = k[A][B]$$

In practice, we usually look at the experimentally determined rate equation and see whether or not it is *consistent* with a proposed mechanism. As an example, consider the reaction of NO_2 and CO gases to produce NO(g) and CO_2. Suppose we *propose* that the reaction occurs in one step:

$$NO_2(g) + CO(g) \rightarrow NO(g) + CO_2(g)$$

The expected rate expression would then be

$$\text{rate} = k[NO_2(g)][CO(g)]$$

Experimentally it is found that the rate expression is

$$\text{rate} = k[NO_2(g)]^2$$

This provides evidence that the reaction of NO_2 and CO does not occur in one stage.

Now consider the reaction of 1-bromopropane with hydroxide ion. Let us propose that the reaction occurs in one step:

$$CH_3CH_2CH_2Br(l) + OH^-(aq) \rightarrow CH_3CH_2CH_2OH(l) + Br^-(aq)$$

Experiments show that the rate expression is:

$$\text{Rate} = k[CH_3CH_2CH_2Br(l)][OH^-(aq)]$$

Such a rate expression is consistent with (but does not absolutely prove) that the reaction occurs in one stage. (For more details of the bromopropane/hydroxide ion reaction, see page 249).

Complicated rate expressions (such as those involving fractional orders of reaction) always indicate that the mechanism of the reaction is complex. An example of this is the decomposition of ethanal (page 252).

14.8 Catalysis

How catalysts work

The idea of an overall reaction as consisting of several stages helps to explain how catalysts work. Catalysts cause the reactants to make products through a *different* mechanism in which the *slowest* step in the new mechanism is faster (i.e. possesses a lower activation energy) than the slowest step in the mechanism which applies to the uncatalysed reaction. This has the effect of speeding up the rate of the overall reaction.

Homogeneous catalysis

Catalysts which are in the same phase as the reactants (e.g. they are all in solution) are termed **homogeneous catalysts.** Examples of homogeneous catalysis include the hydrolysis of sugar (page 254), and the catalysis of ozone destruction by $Cl^{\bullet}$ (see page 394). Another example is the reaction of persulfate ($S_2O_8^{2-}$(aq)) and iodide (I^-(aq)) ions which obeys the overall reaction

$$2I^-(aq) + S_2O_8^{2-}(aq) \rightarrow I_2(aq) + 2SO_4^{2-}(aq) \qquad \textbf{(14.6)}$$

The direct reaction of I^- with $S_2O_8^{2-}$ possesses a high activation energy because the reacting ions are both negatively charged and so they repel each other. This makes the direct reaction slow, but the reaction is speeded up if Fe^{3+} ions are added. Although the reaction mechanism is complicated, the catalytic effect of Fe^{3+} ions probably involves the following (simplified) stages. First, the Fe^{3+} reacts with the iodide,

$$2I^-(aq) + 2Fe^{3+}(aq) \rightarrow I_2(aq) + 2Fe^{2+}(aq) \qquad \textbf{(14.7)}$$

Fe^{2+} then reacts with $S_2O_8^{2-}$,

$$S_2O_8^{2-}(aq) + 2Fe^{2+}(aq) \rightarrow 2SO_4^{2-}(aq) + 2Fe^{3+}(aq) \qquad \textbf{(14.8)}$$

regenerating the Fe^{3+}.

The catalytic steps are faster than direct reaction of I^- with $S_2O_8^{2-}$ because steps (14.7) and (14.8) involve two oppositely charged ions and so both reactions are expected to be very rapid. (Note that addition of equations (14.7) and (14.8) produces (14.6), the equation of the overall reaction.)

Heterogeneous catalysis

Catalysts which are in a different phase as the reactants (usually a solid catalyst speeding up reactants in a solution or gaseous mixture) are termed **heterogeneous catalysts**. Industrially important heterogeneous catalysts include solid nickel, platinium and rhodium. These catalysts work by allowing reactant molecules to stick (adsorb) to the surface of the metal catalyst, so causing the bonds between atoms in the reactant molecules to weaken, making molecular rearrangement easier.

As an example of heterogeneous catalysis, consider the reaction of ethene with hydrogen using a heated nickel catalyst:

$$\underset{\text{ethene}}{CH_2{=}CH_2(g)} + H_2(g) \xrightarrow[\text{Ni}]{400\,^{\circ}C} \underset{\text{ethane}}{C_2H_6(g)}$$

1. The starting materials, nickel, hydrogen and ethene, are mixed (Fig. 14.12(a)).
2. The ethene and hydrogen molecules **adsorb** on the surface of the surface of the nickel (Fig. 14.12(b)).

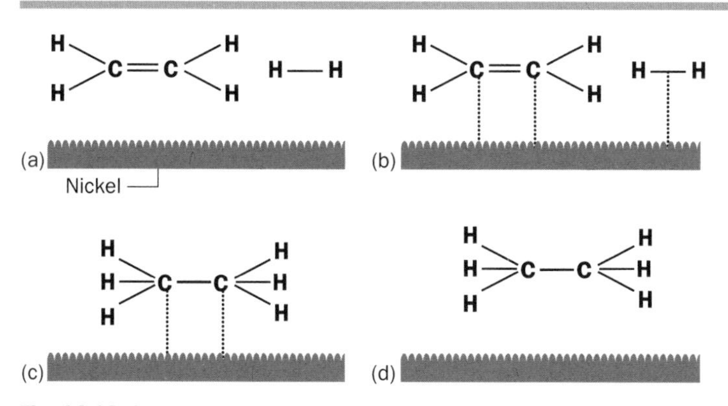

Fig. 14.12 Catalytic hydrogenation of ethene (adsorption is shown by dotted lines).

3. The binding of the ethene and hydrogen molecules to the metal surface weakens the bonds within the CH_2=CH_2 and H_2 molecules. This reduces the activation energy for the redistribution of bonding electrons involved in the hydrogenation reaction. This makes the hydrogenation much faster than the direct reaction of ethene and hydrogen in the *absence* of a catalyst. The product, ethane, is produced already adsorbed on the metal surface (Fig. 14.12(c)).
4. The ethane desorbs from the metal surface, and moves into the gas phase (Fig. 14.12(d)). The nickel is then able to adsorb fresh reactants.

Revision questions

14.1. Carefully explain the following key terms:
(i) rate of reaction **(ii)** transition state **(iii)** activation energy **(iv)** order of reaction **(v)** elementary reaction **(vi)** reaction mechanism **(vii)** reactant half-life.

14.2. Which of the following statements are true?

(i) Catalysts do not alter the enthalpy change (ΔH) of a reaction.

(ii) Reactions with high activation energies are usually very fast.

(iii) Activation energies vary with temperature.

(iv) Catalysts do not take part in the reactions they speed up.

(v) Reaction intermediates cannot be isolated

14.3. Which of the following diagrams shows the variation of reaction rate with temperature for **(i)** an enzyme-catalysed reaction and **(ii)** an uncatalysed reaction?

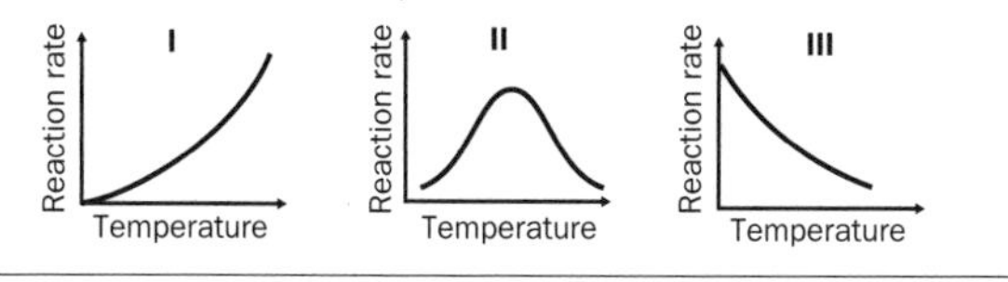

14.4. The decomposition of SO_2Cl_2,

$$SO_2Cl_2(g) \rightarrow SO_2(g) + Cl_2(g)$$

follows the rate expression

$$\text{rate of reaction} = k[SO_2Cl_2]$$

with $k = 2.00 \times 10^{-5}\,s^{-1}$ at 300 °C. In an experiment, SO_2Cl_2 was heated to 300 °C. The initial concentration of $SO_2Cl_2(g)$ was 0.100 mol dm^{-3}.

(i) What is the order of reaction?

(ii) What is the initial rate of reaction?

(iii) After 9 h the $SO_2Cl_2(g)$ was still decomposing, but its concentration has fallen to 0.0523 mol dm^{-3}. What is the rate of reaction at $t = 9$ h?

(iv) What is the half-life of $SO_2Cl_2(g)$ at 300 °C?

14.5. The decomposition of dinitrogen pentoxide at 65 °C:

$$2N_2O_5(g) \rightarrow 4NO_2(g) + O_2(g)$$

follows the rate expression

$$\text{rate of reaction} = k[N_2O_5(g)]$$

In an experiment carried out at 65 °C, the concentration of N_2O_5 changed with time as follows:

$[N_2O_5]$/mol dm^{-3}	Time after start of reaction/h
10×10^{-2}	0 (i.e. at start)
8.6×10^{-2}	0.50
7.3×10^{-2}	1.0
6.3×10^{-2}	1.5
5.4×10^{-2}	2.0
4.6×10^{-2}	2.5
3.9×10^{-2}	3.0
3.4×10^{-2}	3.5
2.9×10^{-2}	4.0

(i) What is the overall order of reaction?

(ii) Plot $[N_2O_5(g)]$ against time. Join the points with a smooth curve.

(iii) From your graph, find the initial rate of reaction.

(iv) Calculate the rate constant for the reaction at 65 °C and state its units.

(v) What is the half-life of N_2O_5 at this temperature?

14.6. The hydrolysis of bromomethane,

$$CH_3Br(l) + OH^-(aq) \rightarrow CH_3OH(aq) + H_2O(l)$$

follows the rate expression:

$$\text{rate of reaction} = k[CH_3Br(l)][OH^-(aq)]$$

The rate constant k was found to be 3.0×10^{-4} mol^{-1} dm^3 s^{-1} at 300 K.

(i) With excess $OH^-(aq)$, the reaction was found to be pseudo first order. What does this mean? What would be the reaction order if the bromomethane (and not the $OH^-(aq)$) was in excess?

(ii) In an experiment $[OH^-(aq)]_0 = 1.0 \times 10^{-2}$ mol dm^{-3} and $[CH_3Br(l)]_0 = 1.0 \times 10^{-4}$ mol dm^{-3}. What is the value of (a) the initial rate of reaction, (b) the pseudo-first-order rate constant and (c) the pseudo-reaction half-life, at 300 K at this hydroxide concentration?

14.7. The reaction of chlorine oxide ($ClO^\bullet$) radicals with nitrogen dioxide,

$$ClO^\bullet(g) + NO_2(g) + N_2(g) \rightarrow ClONO_2(g) + N_2(g)$$

is an important 'sink' for chlorine oxide radicals in the atmosphere. (The nitrogen absorbs the excess energy of colliding molecules and acts as a type of catalyst.) The reaction was studied in the laboratory and showed the following dependence upon the concentrations of the reactants at 298 K:

Experiment	Initial concentration/mol dm^{-3}			Initial rate of reaction/ mol dm^{-3} s^{-1}
	$[ClO(g)]_0$	$[NO_2(g)]_0$	$[N_2(g)]_0$	
1	1.0×10^{-5}	2.0×10^{-5}	3.0×10^{-5}	3.5×10^{-4}
2	0.5×10^{-5}	2.0×10^{-5}	3.0×10^{-5}	1.8×10^{-4}
3	1.0×10^{-5}	4.0×10^{-5}	3.0×10^{-5}	7.1×10^{-4}
4	0.5×10^{-5}	2.0×10^{-5}	6.0×10^{-5}	3.6×10^{-4}

(i) Write down the rate expression for the reaction.

(ii) Calculate an average value of k(298 K).

(iii) Is the rate expression consistent with the formation of chlorine nitrate being a single-stage reaction? Explain.

14.8. Draw an energy profile, similar to Fig. 14.7, for an **endothermic** reaction.

14.9. Show, starting from the equation $[A]_t = [A]_0 e^{-kt}$, that the time (symbolized $t_{0.99}$) it takes for a first-order reaction to use up 99% of the reactant is given by the expression

$$t_{0.99} = \frac{4.605}{k}$$

(**Hint** After $t_{0.99}$ seconds, the ratio of initial to actual concentrations is 1/100.)

14.10. The reaction:

$$2CO(g) + O_2(g) \rightarrow 2CO_2(g)$$

is catalysed by powdered platinum in 'catalytic converters'. Draw sketches (similar to those found in Fig. 14.12) which show a possible mechanism for the reaction. For simplicity, represent carbon monoxide as C=O.

Extension material to support this unit is available on our website at **http://www.palgrave.com/foundations/lewis.**

Dynamic Chemical Equilibria

Objectives

- ▶ Explains what is meant by dynamic chemical equilibrium
- ▶ Defines equilibrium constant
- ▶ Discusses factors which affect equilibrium concentrations
- ▶ Looks at calculations involving simple equilibria

Contents

15.1 Introduction

Chemical reactions are examples of dynamic equilibria. Equilibrium reactions are sometimes called **reversible reactions**. As an example of an equilibrium reaction, consider the reaction of hydrogen and iodine to make hydrogen iodide in a sealed container at high temperature.

If we heat pure hydrogen and iodine the *forward* reaction takes place:

$$H_2(g) + I_2(g) \rightarrow 2HI(g) \quad \textbf{(15.1)}$$

Almost as soon as HI is formed, some HI molecules decompose in the *back reaction*:

$$2HI(g) \rightarrow H_2(g) + I_2(g) \quad \textbf{(15.2)}$$

At **equilibrium** the *rates* of both reactions are *equal*. This does not mean the *concentrations* of the reactants and products are equal, merely that the 'status quo' (the equilibrium composition) is being maintained by two opposite processes which are proceeding at the same rate (Fig. 15.1). We summarise the forward and reverse reactions in one equation:

$$H_2(g) + I_2(g) \rightleftharpoons 2HI(g)$$

Figure 15.2 shows the general way that the concentrations of reactants and products varies with time for an equilibrium reaction. Equilibrium is achieved *t* seconds after the start of the reaction.

Athlete

Treadmill

Fig. 15.1 An illustration of a dynamic equilibrium – an athlete on a treadmill. The athlete does not fall off because he is moving at the *same* rate as the treadmill but in the *opposite* direction.

Picture of dynamic chemical equilibrium

The idea that although the *rates* of the forward and back reactions are equal at equilibrium, the *concentrations* of the reactants and products are not necessarily equal, is

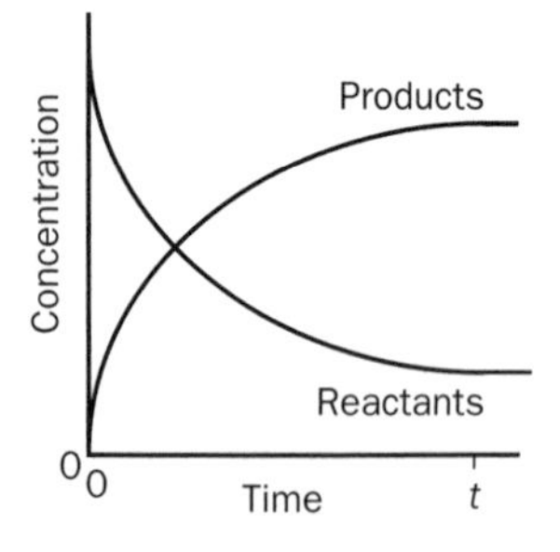

Fig. 15.2 Concentration–time profile for an equilibrium reaction.

often confusing to beginners in the subject and is worthy of further explanation. To do this, we consider the simplest kind of chemical reaction in which there is only one reactant (A) and one product (B). The reaction is then represented as

$$A \rightleftharpoons B$$

Figure 15.3 represents the reaction as consisting of a store of A and B molecules on either side of a barrier. The volume of the container is fixed, and so the numbers of molecules represents the concentrations of those molecules. In our model, reaction has taken place when a molecule passes over the barrier, and we show this by changing the colour of the circles (○→●) used to represent the molecules. We shall assume that the temperature is fixed.

In Fig. 15.3(a), we start with pure A. As the reaction proceeds, some of the A molecules change into B molecules, and soon after this some B molecules revert to A. At equilibrium (Fig. 15.3(b)) the composition of the mixture is fixed, but A molecules are still changing into B and vice versa. The key point though, is that at equilibrium the number of A molecules jumping over the barrier per second equals the number of B molecules jumping over the barrier per second. In our example, the equilibrium mixture contains 75% of A (12 out of 16 molecules) and only 25% of B. In chemical reactions, the composition of the reactant–product mixture at equilibrium depends upon the precise concentration of reactants and products (if any) at the start, and upon the temperature of the reaction mixture.

Our simple model may be used to illustrate another important feature of dynamic equilibria. Figure 15.3(c) shows the reaction starting with pure B with an initial concentration equal to that of A in (a). As reaction starts in Fig. 15.3(c), some B changes into A. Some A then reverts to B, and Fig. 15.3(d) shows the composition of the equilibrium mixture. It is evident that it is the same as Fig. 15.3(b), i.e. as was arrived at starting from pure A at the same temperature. We summarize this by saying that at the same temperature *the same equilibrium composition may be achieved from both forward and reverse directions.*

All reactions are, in principle, equilibrium reactions. However, for many reactions there is such a low concentration of limiting *reactants* at equilibrium that we may safely assume that the reactants have been completely converted to products, i.e. that the reaction 'has gone to completion'. We do not use the $\rightleftharpoons$ sign for reactions that go to completion, and the ordinary 'produces' sign ($\rightarrow$) is used instead.

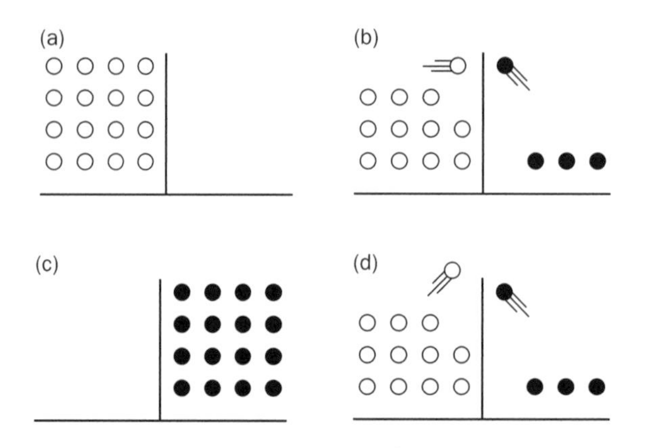

Fig. 15.3 The chemical equilibrium $A \rightleftharpoons B$: ○ represents A molecules; ● represents B molecules.

15.2 Equilibrium law and equilibrium constant

Equilibrium law

For any chemical reaction at equilibrium we define an **equilibrium constant** $K_{c(T)}$ in terms of concentration as follows. For the generalized reaction

$$aA + bB + \ldots \rightleftharpoons cC + dD + \ldots$$

$$K_{c(T)} = \frac{[C]^c[D]^d}{[A]^a[B]^b}\cdots \quad (15.3)$$

The fact that all equilibrium reactions can be fitted to this general equation is sometimes called the **equilibrium law**. The coefficients in the chemical equation are a, b, c, d, . . . and the square brackets represent the concentrations of the reactants and products *at equilibrium* (*not* the starting concentrations). The T in $K_{c(T)}$ reminds us that the equilibrium 'constant' depends upon temperature.

Applying the equilibrium law to the expression

$$H_2(g) + I_2(g) \rightleftharpoons 2HI(g)$$

gives

$$K_{c(T)} = \frac{[HI(g)]^2}{[H_2(g)][I_2(g)]} \quad (15.4)$$

Example 15.1

Write down an expression for the equilibrium constant for the reaction

$$2SO_2(g) + O_2(g) \rightleftharpoons 2SO_3(g) \quad (15.5)$$

and state its units.

▶ Answer

We might represent this equation more clearly as:

$$\mathbf{2}SO_2(g) + \mathbf{1}O_2(g) \rightleftharpoons \mathbf{2}SO_3(g)$$

in which the coefficients (2, 1 and 2, respectively) are displayed in bold. Comparison with equation (15.3) shows that, for this reaction,

$A = SO_2$ $B = O_2$ $C = SO_3$ $a = 2$ $b = 1$ $c = 2$

The expression for the equilibrium constant therefore becomes

$$K_{c(T)} = \frac{[SO_3(g)]^2}{[SO_2(g)]^2[O_2(g)]}$$

The units of an equilibrium constant depend upon the number of concentration terms in the expression. Here the units are

$$K_{c(T)} = \frac{(\text{mol dm}^{-3})^2}{(\text{mol dm}^{-3})^2(\text{mol dm}^{-3})} = \frac{\cancel{(\text{mol dm}^{-3})^2}}{\cancel{(\text{mol dm}^{-3})^2}(\text{mol dm}^{-3})}$$

i.e. $K_{c(T)}$ possesses the units of $\text{mol}^{-1}\,\text{dm}^3$.

$K_{c(T)}$ is practically important because it does not depend upon the concentrations of the reactants at the start of the reaction, and for a named reaction varies only with temperature of the equilibrium mixture. Remember,

The value of an equilibrium constant for a particular reaction changes only with temperature.

The equilibrium constants of reactions involving gases may also be expressed in terms of the partial pressures (in atm) of the reactants and products. The equilibrium constant is then symbolized $K_{p(T)}$. For example, for reaction (15.5),

$$K_{p(T)} = \frac{(p_{SO_3})^2}{(p_{SO_2})^2 \times (p_{O_2})^2} = 3.0 \times 10^4 \text{ atm}^{-1} \text{ at } 700\ K \quad \textbf{(15.6)}$$

(The conversion of $K_{c(T)}$ to $K_{p(T)}$ is described in the website, Appendix 15.)

Exercise 15A

Reactions that go to completion

Study the equations

$CO(g) + H_2(g) \rightleftharpoons H_2O(g) + C(s)$

and

$NaOH(aq) + HCl(aq) \rightarrow H_2O(l) + NaCl(aq)$

Which of the following statements are correct?

(a) If we started with 0.1 mol of H_2 and excess CO, we would expect to produce 0.1 mol of H_2O and 0.1 mol of C.

(b) If we started with 0.1 mol of NaOH and excess HCl, we would expect to produce 0.1 mol of H_2O and 0.1 mol of NaCl.

Exercise 15B

Equilibrium expressions

Write down expressions for the equilibrium constants of the following reactions:

(i) Oxidation of iron(II) ions by chlorine gas

$2Fe^{2+}(aq) + Cl_2(g) \rightleftharpoons 2Fe^{3+}(aq) + 2Cl^-(aq)$

(ii) Formation of chlorine nitrate gas

$ClO(g) + NO_2(g) \rightleftharpoons ClONO_2(g)$

(iii) Formation of nitrogen dioxide gas from dinitrogen tetroxide

$N_2O_4(g) \rightleftharpoons 2NO_2(g)$

What are the units of $K_{c(T)}$ in each case?

Equilibrium constants and rate constants

First, a reminder – 'big K' symbolizes an equilibrium constant, whereas 'little k' symbolizes a rate constant.

Experiments show that the rate of reaction (15.1) follows the rate expression

$$\text{rate of reaction} = k_f[H_2(g)]\,[I_2(g)]$$

where k_f (subscript f for 'forward') is the rate constant (see Chapter 14) and the brackets indicate the concentrations at that *instant*. Immediately after starting the reaction, the concentrations of hydrogen and iodine begin to fall and so the rate of the forward reaction falls.

The experimental rate expression for the *back reaction* (equation (15.2)) turns out to be equally straightforward:

$$\text{rate of reaction} = k_b[HI(g)]^2$$

As the concentration of HI builds up, the rate of the back reaction increases. At equilibrium, the rates of the forward and back reactions are equal:

$$k_b[HI(g)]^2 = k_f[H_2(g)]\,[I_2(g)]$$

Comparing this with equation (15.4) shows that

$$K_{c(T)} = \frac{k_f}{k_b}$$

- This equation applies to all equilibrium reactions (and may be used whether or not the reaction occurs in a single stage).
- Remember that a rate constant is the rate of reaction when the substances in the rate expression are at unit concentration, i.e. $1\,\text{mol}\,\text{dm}^{-3}$ (see page 251). The above equation shows that the size of the equilibrium constant depends simply upon how much faster the forward reaction is (at unit concentrations) than the back reaction (at unit concentrations) at that temperature. We need only to know the equilibrium constant and one of the rate constants in order to calculate the remaining rate constant. For example, for the reaction

  $$H_2(g) + I_2(g) \rightleftharpoons 2HI(g)$$

  $k_f = 1.66\,\text{mol}^{-1}\,\text{dm}^3\,\text{s}^{-1}$ and $K_{c(T)} = 45.6$ at 764 K. Therefore,

  $$k_b(764\,\text{K}) = \frac{1.66\,\text{mol}^{-1}\,\text{dm}^3\text{s}^{-1}}{45.6} = 0.0364\,\text{mol}^{-1}\,\text{dm}^3\,\text{s}^{-1}$$

The importance of stating the chemical equation to which the equilibrium constant applies

An equilibrium constant applies to a particular chemical equation. For example, whereas

$$K_{c(T)} = \frac{[SO_3(g)]^2}{[SO_2(g)]^2[O_2(g)]} \text{ for } 2SO_2(g) + O_2(g) \rightleftharpoons 2SO_3(g)$$

we also write

$$K_{c(T)} = \frac{[SO_3(g)]}{[SO_2(g)][O_2(g)]^{1/2}} \text{ for } SO_2(g) + \tfrac{1}{2}O_2(g) \rightleftharpoons SO_3(g)$$

The equilibrium constants for these expressions will have different numerical values. The moral is simple – never quote equilibrium constants unless you write down the chemical equation (or equilibrium expression) to which that value of $K_{c(T)}$ applies.

15.3 Meaning of equilibrium constants

Look back at Fig. 15.3(b). At equilibrium, the reaction mixture consisted of 12 molecules of A and 4 of B. Since, for a fixed volume, the molar concentration is proportional to the number of molecules, we can write

$$K_{c(T)} = \frac{[B]}{[A]} = \frac{4}{12} = 0.33$$

Suppose that for a different reaction, say C $\rightleftharpoons$ D, there were 15 molecules of C and 1 of D at equilibrium at a particular temperature. Then,

$$K_{c(T)} = \frac{[D]}{[C]} = \frac{1}{15} = 0.066$$

We see that the smaller the equilibrium constant, the smaller the relative number of molecules of product to reactant in the equilibrium mixture, and vice versa.

Where chemical equations contain more than one reactant or product molecule, the equilibrium constant for the reaction is not simply the ratio of a single product and single reactant concentration. (Equation (15.4), for example, contains a concentration which is squared.) Nevertheless, *large equilibrium constants always mean that the product molecules dominate the equilibrium mixture*, whereas small equilibrium constants show that reactant molecules dominate the equilibrium mixture.

As a rough guide, if $K_{c(T)}$ is greater than 1000, the product dominates the equilibrium mixture and we would say that (for most purposes) the reaction goes to completion at that temperature. If $K_{c(T)}$ is less than 0.001 we would say that (for most purposes) a negligible amount of product has been made and that 'no reaction has occurred' at that temperature. Summarising,

if $K_{c(T)} > 1000$ reaction goes to completion

if $K_{c(T)} < 0.001$ no reaction occurs

If $K_{c(T)}$ is in the range 1000–0.001, the number of concentrations being multiplied together in the equilibrium expression need to be carefully considered in deciding whether or not a $K_{c(T)}$ value implies a high or low concentration of products in the equilibrium mixture, or whether or not the reactants and product concentrations are roughly equal.

Reactions which go to completion include many of the reactions of inorganic ions we have already studied, for example

$Ag^+(aq) + Cl^-(aq) \rightarrow AgCl(s)$ $K_{c(25\,°C)} \gg 1000$

Most of the mole calculations discussed in previous units (including those involving titrations) have assumed that the reactions go to completion.

A list of $K_{c(T)}$ values for selected reactions is given in Table 15.1.

Table 15.1 Equilibrium constants

No	Reaction	$K_{c(T)}$	Temperature/K	Units of $K_{c(T)}$
1	$Cl_2(g) \rightleftharpoons Cl(g) + Cl(g)$	1.2×10^{-7}	1000	$mol\ dm^{-3}$
2(a)	$H_2(g) + Cl_2(g) \rightleftharpoons 2HCl(g)$	4.0×10^{31}	300	–
2(b)	$H_2(g) + Cl_2(g) \rightleftharpoons 2HCl(g)$	4.0×10^{18}	500	–
3	$2BrCl(g) \rightleftharpoons Br_2(g) + Cl_2(g)$	377	300	–
4	$CO_2(g) + H_2(g) \rightleftharpoons CO(g) + H_2O(g)$	1.56	1073	–
5	$HF(g) + HCN(g) \rightleftharpoons HCN\cdots HF(g)$	1.04	298	$mol^{-1}\ dm^3$
6	$I^-(aq) + I_2(l) \rightleftharpoons I_3^-(aq)$	7.1×10^{-2}	298	$mol^{-1} dm^3$
7	$Cu^{2+}(aq) + 4NH_3(aq) \rightleftharpoons Cu[(NH_3)_4]^{2+}(aq)$	1.4×10^{13}	298	$mol^{-4}\ dm^{12}$
8(a)	$N_2(g) + 3H_2(g) \rightleftharpoons 2NH_3(g)$	4×10^8	298	$mol^{-2}\ dm^6$
8(b)	$N_2(g) + 3H_2(g) \rightleftharpoons 2NH_3(g)$	2.2	623	$mol^{-2}\ dm^6$
9	$2NH_3(g) \rightleftharpoons N_2(g) + 3H_2(g)$	0.46	623	$mol^2\ dm^{-6}$
10	$Cu^{2+}(aq) + Zn(s) \rightleftharpoons Zn^{2+}(aq) + Cu(s)$	1×10^{37}	298	–

Equilibrium constants as indicators of whether or not a reaction is allowed to occur

Chemical reactions with large equilibrium constants are said to be 'thermodynamically allowed'. However, equilibrium constants tell us nothing about whether or not the reaction under consideration will occur at a *fast enough speed* to be detectable. Many reactions possess large $K_{c(T)}$ values, but are extremely slow in reaching equilibrium at room temperature, for example:

- The formation of ammonia gas from hydrogen and nitrogen gases, occurs at a negligible rate at room temperature.
- A mixture of hydrogen and oxygen gases at room temperature show no signs of producing water, no matter how long it is kept.

Fortunately, the rates of reactions can (within certain limits) be controlled by altering the experimental conditions, and reactions may often be accelerated using a combination of high temperature, high reactant concentration and a catalyst.

Exercise 15C

Equilibrium constants

(i) Which of the reactions shown in Table 15.1 go to completion at the temperatures listed?

(ii) The equilibrium constant for the reactions of sodium and magnesium ions with the complexing agent $EDTA^{4-}$ at room temperature are 50 $mol^{-1} dm^3$ and $5 \times 10^8 mol^{-1} dm^3$ respectively. Which ion cannot be satisfactorily analysed by titration with $EDTA^{4-}$, and why?

Values of equilibrium constants for reverse reactions

The equilibrium constant for the formation of hydrogen iodide,

$$H_2(g) + I_2(g) \rightleftharpoons 2HI(g)$$

is

$$K_{c(T)f} = \frac{[HI(g)]^2}{[H_2(g)][I_2(g)]}$$

where 'f' stands for 'forward reaction'. The equilibrium constant for the reverse or 'back' reaction,

$$2HI(g) \rightleftharpoons H_2(g) + I_2(g)$$

is

$$K_{c(T)b} = \frac{[H_2(g)][I_2(g)]}{[HI(g)]^2}$$

from which we see that

$$K_{c(T)b} = \frac{1}{K_{c(T)f}}$$

We conclude that *if the equilibrium constant for the forward reaction is large, then the equilibrium constant for the reverse process is small.*

Exercise 15D

Equilibrium constants for reverse reactions

The equilibrium constant for the reaction

$$N_2(g) + O_2(g) \rightleftharpoons 2NO(g)$$

is 7×10^{-9} at 1100 K. In an experiment, NO is heated to 1100 K. Will the NO decompose?

For example, $K_{c(T)}$ for the reaction

$$Cu^{2+}(aq) + Zn(s) \rightleftharpoons Zn^{2+}(aq) + Cu(s)$$

is 1×10^{37} at room temperature (Table 15.1). Thus $K_{c(T)}$ for the *reverse* reaction is 1×10^{-37}. The consequence of these values is clear cut: for all practical purposes the reaction of zinc metal and copper ions goes to completion, whereas the reaction of zinc ions with copper metal does not take place.

15.4 Effects of changing concentration, pressure and temperature upon equilibria

Detailed studies have been made of many different examples of equilibria. We look at two cases. In case 1, the hydrogen–iodine reaction, we confirm that $K_{c(T)}$ is independent of reactant concentrations. In case 2, an esterification reaction, we confirm that the equilibrium concentrations of reactants and products (the 'equilibrium composition') depends upon their starting concentrations.

Case 1: Use of the reaction between hydrogen and iodine gases to show that $K_{c(T)}$ is independent of concentration

The reaction is

$$H_2(g) + I_2(g) \rightleftharpoons 2HI(g)$$

Table 15.2 shows the initial (i.e. at $t = 0$) and equilibrium compositions of three different mixtures at 490 °C.

We start by checking whether or not the results obey the reaction equation. According to the equation, 1 mol of H_2 or I_2 produces 2 mol of HI. Therefore, the

Table 15.2 The composition of mixtures hydrogen containing hydrogen, iodine and iodide at 490 °C

Time	Concentrations/mol dm^{-3}		
	$[H_2]$	$[I_2]$	[HI]
Experiment 1			
$t = 0$	20.00	20.00	0
At equilibrium	4.58	4.58	30.86
Experiment 2			
$t = 0$	8.19	14.31	0
At equilibrium	0.72	6.84	14.94
Experiment 3			
$t = 0$	0	0	40.00
At equilibrium	4.58	4.58	30.86

concentration of HI in Table 15.2 should be twice the *loss* in concentration of H_2 or I_2 that takes place by the time equilibrium has been reached. Within experimental error, this is seen to be the case. For example, in experiment 1, the equilibrium concentration of HI is 30.86 mol dm^{-3}. The reduction in I_2 concentration is (20.00 − 4.58) = 15.42 mol dm^{-3}, almost half of 30.86.

Substitution of the equilibrium concentrations into equation (15.4) gives:

For experiments 1 and 3, $K_{c(T)} = (30.86)^2/(4.58)^2 = 45.4$

For experiment 2, $K_{c(T)} = (14.94)^2/(0.72) \times (6.84) = 45.3$

This confirms that although the starting concentrations of reactant are different, $K_{c(T)}$ remains the same (within experimental error) at that temperature.

Now look at experiment 3 more closely. Experiment 3 starts with 40.00 mol of pure HI. The HI decomposes into H_2 and I_2, and [HI] then falls until equilibrium is reached. 40.00 mol of HI happens to be the amount of HI that would be produced *if* we were able to convert all of the H_2 and I_2 in experiment 1 to HI. This is the reason that the equilibrium compositions of experiment 1 (in which we started with pure reactants) and experiment 3 (in which we started with pure product) are identical, confirming our earlier statement (page 266) that the same equilibrium composition may be achieved from both forward and reverse directions.

Exercise 15E

Calculations involving equilibrium concentrations

(i) Pure iodine and hydrogen gases were mixed and allowed to reach equilibrium. The equilibrium concentrations of iodine and hydrogen gases at 490 °C were each found to be 3.00 mol dm^{-3}. What is the equilibrium concentration of gaseous hydrogen iodide? What were the starting concentrations of reactants?

(ii) The equilibrium concentrations (in mol dm^{-3}) for a mixture of HI, I_2, and H_2 at 527 °C were [HI(g)] = 19.1, [I_2(g)] = 3.50 and [H_2(g)] = 2.50. Does $K_{c(T)}$ for the reaction increase or decrease with increasing temperature?

Case 2: Esterification of ethanol used to illustrate the effect of concentration on equilibrium composition

The reaction of ethanol with ethanoic acid (in the presence of sulfuric acid catalyst) produces a sweet-smelling compound known as an ester. The production of an ester is called **esterification**. The name of the particular ester produced in this reaction is ethyl ethanoate:

$$\underset{\text{ethanoic acid}}{CH_3COOH(l)} + \underset{\text{ethanol}}{C_2H_5OH(l)} \rightleftharpoons \underset{\text{ethyl ethanoate}}{CH_3COOC_2H_5(l)} + H_2O(l)$$

From this we obtain

$$K_{c(T)} = \frac{[CH_3COOC_2H_5(l)][H_2O(l)]}{[CH_3COOH(l)][C_2H_5OH(l)]} \quad \textbf{(15.7)}$$

Table 15.3 shows the composition of the equilibrium mixture for three initial mixtures of differing concentration. Substituting the equilibrium concentrations into equation (15.7) gives

Table 15.3 Equilibrium composition (at 25 °C) of mixtures containing ethanoic acid, ethanol, water and ethyl ethanoate

Time	Concentrations/mol dm^{-3}			
	$[CH_3COOH]$	$[C_2H_5OH]$	$[CH_3COOC_2H_5]$	$[H_2O]$
Experiment 1				
$t = 0$	1.00	1.00	0	0
At equilibrium	0.33	0.33	0.67	0.67
Experiment 2				
$t = 0$	4.00	1.00	0	0
At equilibrium	3.07	0.07	0.93	0.93
Experiment 3				
$t = 0$	4.00	1.00	1.00	0
At equilibrium	3.13	0.13	1.87	0.87

for experiment 1 $K_{c(T)} = 4.1$
for experiment 2 $K_{c(T)} = 4.0$
for experiment 3 $K_{c(T)} = 4.0$

Three features need commenting upon:

1. Constancy of $K_{c(T)}$

Within experimental error all three values of $K_{c(298\ K)}$ are the same.

2. Effect of higher concentration of reactant upon product concentration at equilibrium

In experiment 2 the initial mixture contains a higher concentration of acid than was used in 1. This causes the equilibrium concentration of ester and water to be greater than that observed in experiment 1. This effect is observed in other equilibria:

Increasing the concentration of reactant(s) increases the equilibrium concentration of products.

3. Effect of the presence of product upon the gain in product concentration in the equilibrium mixture

Although the equilibrium concentration of ester in experiment 3 is greater than in experiment 2, we need take into account that the initial reaction mixture in experiment 3 contained some ester. To do this, we define the **gain** in ester concentration (conc.):

gain in ester conc. = equilibrium ester conc. − initial ester conc.

The final concentration of ester in experiment 3 (1.87 mol dm^{-3}) represents a gain of $1.87 - 1.00 = 0.87$ mol dm^{-3} ester. This is **less** than that obtained with experiment 2, where the gain was 0.93 mol dm^{-3}. The same effect is observed in other equilibria.

Generalizing,

The presence of some product at the start of a reaction reduces the gain of product in the equilibrium mixture.

In principle, the effects of increased reactant concentration, and of the presence of product in the initial mixture, are applicable to all equilibria. However, if the equilibrium constant is very large (so that the reaction goes to completion) or very small (so that reaction hardly occurs), changes in concentration are found to have an insignificant effect upon equilibrium concentrations.

We could come to the same conclusions as in points **2** and **3** above by studying the expression for the equilibrium constant, equation (15.7). As an example, consider point **3**. Imagine that we add fresh ester to a mixture *already at equilibrium*. This causes $[CH_3COOC_2H_5(l)]$ to rise, and the ratio of concentrations,

$$\frac{[CH_3COOC_2H_5(l)][H_2O(l)]}{[CH_3COOH(l)][C_2H_5OH(l)]}$$

temporarily exceeds $K_{c(T)}$. The ratio eventually reduces to $K_{c(T)}$ because some ester decomposes into ethanoic acid and ethanol:

$$CH_3COOC_2H_5(l) + H_2O(l) \rightarrow CH_3COOH(l) + C_2H_5OH(l)$$

The equilibrium mixture now contains fewer moles of ester than the sum of the number of moles of added ester and the number of moles of ester at equilibrium before the disturbance took place. Note that it may take several minutes for the reaction to re-establish equilibrium, as shown in Fig. 15.4.

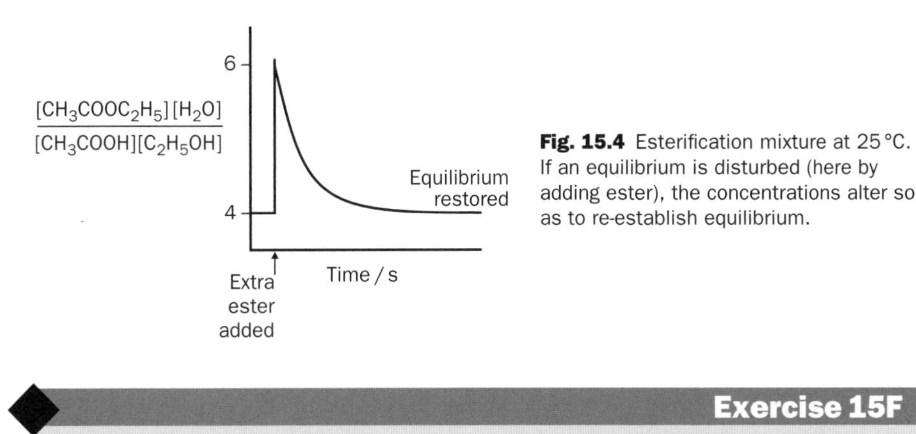

Fig. 15.4 Esterification mixture at 25 °C. If an equilibrium is disturbed (here by adding ester), the concentrations alter so as to re-establish equilibrium.

Exercise 15F

Working out whether or not a reaction has reached equilibrium

The decomposition of ammonia gas was studied at 623 K:

$$2NH_3(g) \rightleftharpoons N_2(g) + 3H_2(g)$$

It was found that $K_{c(T)} = 0.45\ mol^2\ dm^{-6}$. In a separate experiment, some ammonia gas was heated to 623 K and the composition of the mixture analysed after a set time. The concentrations were: $[N_2(g)] = 0.04\ mol\ dm^{-3}$, $H_2(g) = 0.03\ mol\ dm^{-3}$, and $NH_3(g) = 0.02\ mol\ dm^{-3}$. Has the reaction reached equilibrium?

Chemistry under non-equilibrium conditions

In the laboratory, and in industry, we often deliberately disturb a chemical equilibrium. For example, the percentage of the number of moles of ester at equilibrium in Experiment 2 in Table 15.3, is less than 50%. However, if the ester is *physically removed* from the reaction mixture by distillation, the composition of the mixture in the reaction vessel will adjust to give a higher yield of ester. The overall yield of ester (obtained by summing the ester from several distillations) is then very much greater.

Exercise 15G

Non-equilibrium conditions

$K_{c(T)}$ for the reaction of steam and heated solid iron,

$$2Fe(s) + 3H_2O(g) \rightleftharpoons Fe_2O_3(s) + 3H_2(g)$$

is very large at 600 K, yet it is quite easy to reverse this reaction by passing hydrogen gas over strongly heated iron oxide using the apparatus shown in Fig. 15.5. How is this possible?

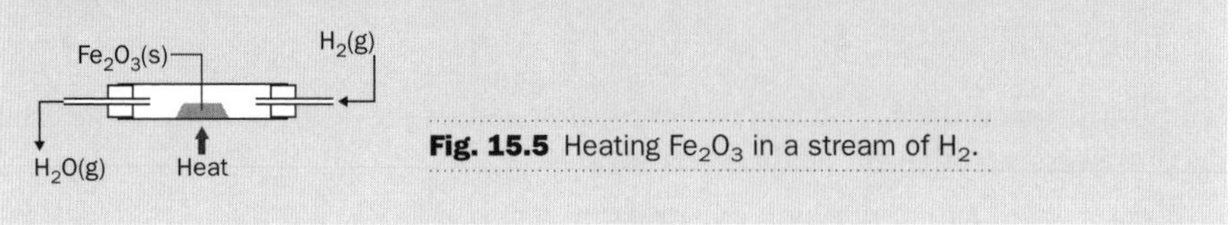

Fig. 15.5 Heating Fe_2O_3 in a stream of H_2.

Effect of temperature on equilibrium constants

The effect of temperature upon $K_{c(T)}$ is summarized as follows:

1. The equilibrium constant for an exothermic reaction decreases with increasing temperature.
2. The equilibrium constant for an endothermic reaction increases with increasing temperature.

For example, the formation of HI(g) from $H_2(g)$ and $I_2(g)$ is exothermic, and $K_{c(T)}$ for this reaction falls with increasing temperature (Fig. 15.6). This means that the percentage of HI(g) in the equilibrium mixture falls as the temperature is increased.

The variation of an equilibrium constant with temperature is often so dramatic that adjusting the operating temperature of a reaction is the most effective way of improving the yield of product. The following examples illustrate this.

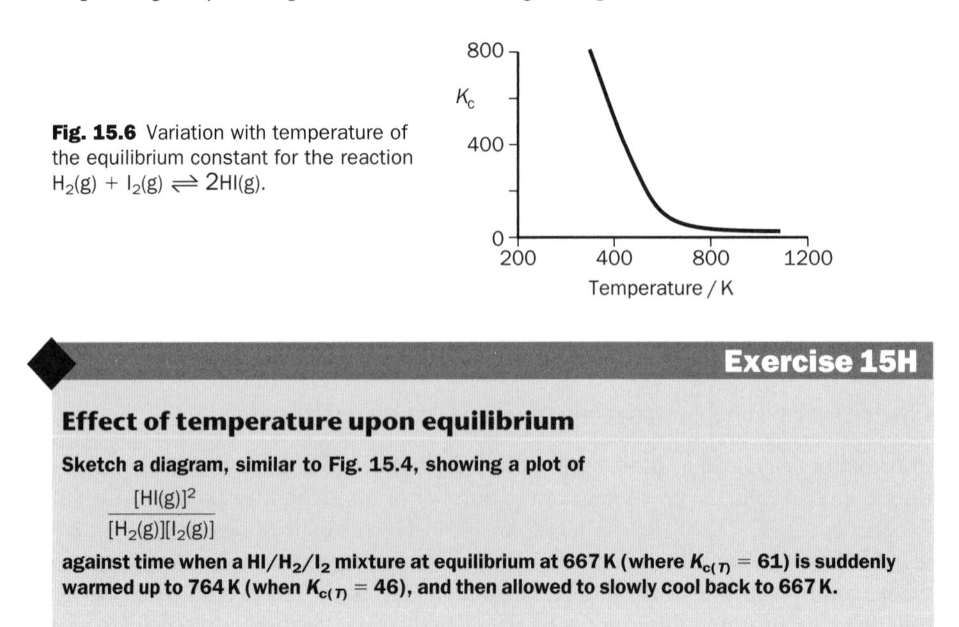

Fig. 15.6 Variation with temperature of the equilibrium constant for the reaction $H_2(g) + I_2(g) \rightleftharpoons 2HI(g)$.

Exercise 15H

Effect of temperature upon equilibrium

Sketch a diagram, similar to Fig. 15.4, showing a plot of

$$\frac{[HI(g)]^2}{[H_2(g)][I_2(g)]}$$

against time when a $HI/H_2/I_2$ mixture at equilibrium at 667 K (where $K_{c(T)} = 61$) is suddenly warmed up to 764 K (when $K_{c(T)} = 46$), and then allowed to slowly cool back to 667 K.

1. Formation of hydrogen-bonded compounds

The reactions in which *hydrogen-bonded compounds* are formed are exothermic and often possess very low equilibrium constants at room temperature. For example, $K_{c(T)}$ for the reaction between hydrogen fluoride and hydrogen cyanide gases,

$$\mathrm{HF(g) + HCN(g) \rightleftharpoons \underset{\text{hydrogen-bonded 'complex'}}{HCN\cdots HF(g)}}$$

is only 1.04 $mol^{-1}\,dm^3$ at 298 K. For a starting mixture of $2.7 \times 10^{-3}\,mol\,dm^{-3}$ HCN and $2.7 \times 10^{-3}\,mol\,dm^{-3}$ HF, this corresponds to an equilibrium concentration of HCN···HF complex of only $7 \times 10^{-6}\,mol\,dm^{-3}$! For this reason it is beneficial to study such complexes at lower temperatures when $K_{c(T)}$ is larger, as the increased concentration of complex makes them easier to detect.

However, since reaction rates generally decrease with decreasing temperature, it is important to remember that a lower temperature *always* means that a reaction takes longer to *reach equilibrium*. This may not be a severe hindrance in the laboratory, but it is frequently a very important economic consideration in the chemical and metallurgical industries.

Exercise 15I

Deducing whether a reaction is exothermic or endothermic from the effect of temperature upon its equilibrium constant

The decomposition of dinitrogen tetroxide into nitrogen dioxide produces a colour change:

$$\underset{\text{colourless gas}}{\mathrm{N_2O_4(g)}} \rightleftharpoons \underset{\text{brown gas}}{\mathrm{2NO_2(g)}}$$

A mixture of the gases was sealed in a steel tube containing a quartz window through which the colour of the gases could be observed. As the contents of the tube were heated, the mixture became darker brown. Is the reaction exothermic or endothermic?

2. Metal extraction

The extraction of a metal (M) from its oxide (MO) using carbon as a reducing agent is generally endothermic. As expected, experiments show that the equilibrium constants of the general reaction,

$$\mathrm{MO(s) + C(g) \rightleftharpoons CO(g) + M(s)}$$

increase with increasing temperature. This means that virtually *any* metal oxide may be reduced by carbon if the temperature is high enough. For example, in the reduction of zinc oxide,

$$\mathrm{ZnO(s) + C(g) \rightleftharpoons CO(g) + Zn(s)}$$

the equilibrium yield of the free metal is only high at temperatures above 800 °C. For aluminium, reduction takes place only above 2000 °C, and it is cheaper to produce this metal by electrolysis rather than attempt to maintain a furnace at such a high temperature.

BOX 15.1

Explaining the variation of $K_{c(T)}$ with temperature

Why do the equilibrium constants of reactions vary with temperature? The answer lies in the enthalpy–progress of reaction diagram discussed in Box 14.1 (page 249).

For a reaction that is *exothermic* from left to right, the activation energy of the forward reaction is *less* than the activation energy of the reverse reaction. The smaller the activation energy, the less sensitive is the associated rate constant to changes in temperature. Thus, for an exothermic reaction, k_f is less sensitive to temperature changes than k_b. Put another way, for a particular *increase* in temperature (say 50 °C), k_f increases by a smaller factor than k_b. Since $K_{c(T)} = k_f/k_b$, this causes $K_{c(T)}$ to decrease with temperature. For reactions which are endothermic, $E_{A(f)} > E_{A(b)}$ and so k_f changes faster with temperature than k_b. This causes $K_{c(T)}$ to increase with increasing temperature.

Catalysts and equilibria

A catalyst increases the rate constant for the forward and back reaction by the same factor. For example, it might double both rate constants. It follows from the equation

$$K_{c(T)} = k_f/k_b$$

that $K_{c(T)}$ is unaffected:

Catalysts do not alter the equilibrium constant of a reaction. They do not alter the equilibrium concentrations. They merely increase the speed at which equilibrium is achieved.

Le Chatelier's principle

The effects of changing conditions upon chemical equilibria are summarized in Table 15.4. One way of remembering (but not explaining) such effects is to think of the chemical reaction as actively opposing (almost as if it were alive!) any changes in conditions that have been made. This is the basis of Le Chatelier's principle:

The concentrations of reactants and products in an equilibrium mixture will alter so as to counteract any changes in pressure, temperature or concentration.

Predictions made using Le Chatelier's principle

1. If we add more reactant to an equilibrium mixture, the equilibrium composition shifts so that more reactants are used up, i.e. so that the equilibrium mixture contains a higher concentration of product. (We arrived at the same conclusion by looking at experimental data for the ethanoic acid–ethanol reaction.)

2. The effect of temperature can be predicted by thinking of the heat evolved as a special kind of *product*. Suppose we increase the temperature of an equilibrium mixture in which the reaction is exothermic:

$$aA + bB \rightleftharpoons cC + dD + \text{heat}$$

Le Chatelier's principle predicts that the equilibrium composition alters so that less 'heat product' will be made, i.e. it alters in the direction in which the temperature will be lowered. This is, of course, in the direction of the endothermic reaction. This means that the equilibrium concentration of reactants increases and that of the products falls. As we have seen, this is indeed what happens.

Table 15.4 Summary of the effect of varying conditions on the equilibrium $aA + bB \rightleftharpoons cC + dD$

Change	Effect on percentage of product at equilibrium	Effect on equilibrium constant
Increasing the concentration* of one or both reactants	**increases**	no change
Decreasing the concentration* of one or both reactants	**decreases**	no change
For gas reactions, increasing the pressure of the *reactants* and *products* (e.g. if the mixture is in a piston, we apply extra force to the piston handle)	**1. decreases** if there are fewer molecules on the left-hand side of the reaction equation, i.e. if $(a + b) < (c + d)$	no change
	2. increases if there are more molecules on the left-hand side of the reaction equation, i.e. if $(a + b) > (c + d)$	no change
For gas reactions, allowing the mixture to expand so reducing the pressure of the *reactants* and *products* (e.g. if the mixture is in a piston, we reduce the force on the piston handle)	**1. increases** if $(a + b) < (c + d)$	no change
	2. Decreases if $(a + b) > (c + d)$	no change
For gas reactions, injecting an inert gas into the mixture while keeping the total gas volume constant	no change. (This is true whatever the values of a, b, c and d).	no change
Increased temperature	**1. increases** for an endothermic reaction	**increases**
	2. decreases for an exothermic reaction	**decreases**
Catalyst	no change	no change

*For gases, the partial pressure.

3. Consider the gaseous reaction

$$H_2SO_4(g) \rightleftharpoons H_2O(g) + SO_3(g)$$

which we shall assume is occurring in a cylinder fitted with a piston heated to a constant temperature. There is an increase in the number of molecules going from left (reactants) to right (products) in this reaction, and therefore (at fixed volume) the pressure of the mixture increases as reaction takes place.

Imagine that the reaction mixture is allowed to reach equilibrium. Suppose we now increase the pressure of the equilibrium mixture by forcing the piston handle inwards. Le Chatelier's principle predicts that the reaction will attempt to resist such changes by shifting the equilibrium composition so that the pressure of the mixture decreases, i.e. so that fewer molecules are formed. Fewer molecules can only be made if the equilibrium concentration of product decreases. This is achieved by the combination of product molecules to form reactants.

If the reaction involves a reduction in the number of molecules going from left to right, Le Chatelier's principle correctly predicts that increasing the pressure of the mixture will cause the equilibrium concentration of product to increase.

If the reaction involves an equal number of molecules on the reactant and product sides of the equation, altering the total pressure of the equilibrium mixture will not affect the equilibrium concentration of reactants or products.

Finally, imagine that an equilibrium mixture is maintained at constant volume (for example, the reaction is carried out in a sealed and pistonless steel can) but that we subsequently increase the mixture pressure by injecting in an unreactive gas (such as argon). Because the partial pressures of the *reactants* and *products* are unchanged, Le Chatelier's principle successfully predicts that the equilibrium composition will be unaffected.

Example 15.2

Predict how the equilibrium yield of product(s) in the following reactions will be affected by (i) allowing the mixture to expand so that its pressure falls, and (ii) compressing the mixture so that its pressure increases:

$$H_2(g) + I_2(g) \rightleftharpoons 2HI(g) \qquad (A)$$

$$2NH_3(g) \rightleftharpoons 3H_2(g) + N_2(g) \qquad (B)$$

▶ Answer

We take 'equilibrium yield' to mean the concentration of products in the equilibrium mixture. Applying the rules discussed above:

Reaction	Pressure decreases	Pressure increases
(A)	no change	no change
(B)	$[H_{2(g)}]$ and $[N_{2(g)}]$ increases	$[H_{2(g)}]$ and $[N_{2(g)}]$ decreases

Reaction (A) is unaffected because there is no change in the number of molecules going from left to right.

Exercise 15J

Predicting the effects of changes on the equilibrium composition of a gaseous reaction

Ethene and hydrogen react in the presence of a catalyst, making ethane:

$$C_2H_4(g) + H_2(g) \xrightleftharpoons{\text{(Ni catalyst)}} C_2H_6(g) \qquad \Delta H^{\ominus} = -137\,\text{kJ}\,\text{mol}^{-1}$$

An equilibrium mixture at 300 K was subjected to the following changes in separate experiments:

(i) The mixture was compressed into a smaller volume.

(ii) Hydrogen gas was pumped into the mixture while keeping the volume constant.

(iii) The mixture was maintained at constant volume and nitrogen gas (which does not react with any of the reactants or products) was pumped into the mixture.

(iv) The mixture was cooled.

How will these changes affect (a) the equilibrium yield of ethane and (b) the equilibrium constant?

15.5 Production of ammonia by the Haber–Bosch process

A piece of history

Up until 1913, the bulk of the world's artificial nitrogenous fertilizers was made from sodium nitrate obtained from Chile. Calculations showed that this supply would soon be exhausted if demand continued to expand, and so Fritz Haber (1868–1934) attempted to make ammonia directly from nitrogen and hydrogen gases. After 8 years, Haber produced ammonia efficiently on a small scale. Collaboration with the brilliant chemical engineer Carl Bosch (1874–1940) enabled the production to be scaled up, and ammonia was first manufactured on a large scale in 1913. The world's production of ammonia now exceeds 100 million tonnes. About 80% of all ammonia produced is used to make fertilizers with another 5% being used to make nylon, and a further 5% used to make explosives.

Fritz Haber. Haber's initial results on the nitrogen–hydrogen reaction suggested that it was impossible to produce a significant yield of ammonia. To Haber's embarrassment, this was publicly challenged by the brilliant chemist Walter Nernst (widely regarded as the father of physical chemistry), who suggested that the yield might be significant if compressed gases were used. Haber followed up Nernst's suggestion and his efforts were so successful that he was awarded the Nobel Prize for chemistry in 1918.

Haber–Bosch process

Ammonia gas is produced on a massive scale using the **Haber–Bosch process**. The reaction is

$$N_2(g) + 3H_2(g) \rightleftharpoons 2NH_3(g) \qquad \Delta H^\ominus = -92\,\text{kJ mol}^{-1}$$

Nitrogen gas is obtained from air, while hydrogen is produced by the reaction of methane with steam (see page 313).

To make the Haber–Bosch process viable we require (1) a high equilibrium concentration of ammonia, and (2) that the equilibrium concentration of ammonia be produced in a short time (the 'kinetic' factor). In originally studying this reaction, Haber was aware of the following facts:

- Although K_c for this equilibrium is a massive $4 \times 10^8\,\text{mol}^{-2}\,\text{dm}^{-6}$ at 298 K, the concentration of ammonia in a mixture of nitrogen and hydrogen gases at room temperature is negligible. This is because the rates of the forward and back reactions are extremely slow so that dynamic equilibrium is never attained.

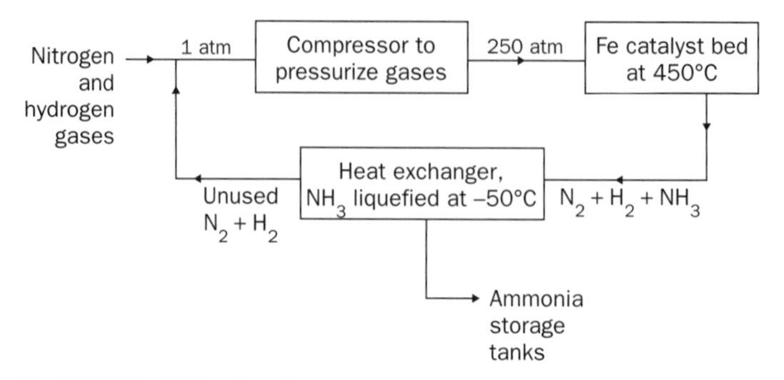

Fig. 15.7 The Haber–Bosch process.

- Operating at higher temperatures increases the reaction rate but reduces $K_{c(T)}$ enormously because the reaction is exothermic and the variation of equilibrium constant with temperature is similar to that shown in Fig. 15.6. (For the Haber–Bosch process, $K_c = 60\,\text{mol}^{-2}\,\text{dm}^{-6}$ at 227 °C and at $0.02\,\text{mol}^{-2}\,\text{dm}^{-6}$ at 527 °C.)
- Using higher partial pressures of hydrogen and nitrogen increases the equilibrium concentration of ammonia. For example, using 75 atm of $H_2(g)$ and 25 atm of $N_2(g)$ at 500 °C produces an equilibrium mixture in which 10% of all the molecules present are ammonia. The use of high partial pressures in this way also increases reaction rate, but (alone) is not enough to make the reaction fast enough.

Haber realized that a catalyst was needed to force the reaction to achieve equilibrium faster. In the commercial process, hydrogen and nitrogen in a 3:1 ratio and at 250 atm pressure are reacted in the presence of iron catalyst at a 'compromise' temperature of about 450 °C. Below 450 °C the reaction is too slow even with a catalyst. The use of pressures above 250 atm involves the use of very powerful compressor pumps which are too expensive to run.

Figure 15.7 shows the main stages of the commercial process. The reaction of nitrogen and hydrogen takes place in the Fe catalyst bed. The equilibrium mixture (containing ammonia and unused hydrogen and nitrogen) is passed into a heat exchanger which cools down the mixture so that the ammonia (but not the reactants) become liquefied. The reactants are then recirculated to the catalyst bed.

15.6 Heterogeneous equilibria

Equilibria in which the substances are all in the same phase (e.g. all gases) are known as **homogeneous equilibria.** Equilibria which involves substances in different phases are known as **heterogeneous equilibria.**

A simplification occurs with heterogeneous equilibria because the concentrations of pure solids and pure liquids are constant at a fixed temperature (see page 137). For example, the equilibrium expression for the production of carbon monoxide from coke and carbon dioxide,

$$C(s) + CO_2(g) \rightleftharpoons 2CO(g)$$

is

$$K_{c(T)} = \frac{[CO(g)]^2}{[CO_2(g)][C(s)]}$$

Since [C(s)] is constant,

$$K'_{c(T)} = \frac{[CO(g)]^2}{[CO_2(g)]}$$

where $K'_{c(T)} = K_{c(T)} \times [C(s)]$.

Application of the equilibrium law to physical equilibria

The equilibrium law also applies to physical processes. For example, the equilibrium law expression for the dissolution of silver chloride,

$$AgCl(s) \rightleftharpoons Ag^+(aq) + Cl^-(aq)$$

is

$$K_{(AgCl)} = \frac{[Ag^+(aq)][Cl^-(aq)]}{[AgCl(s)]} \quad (15.8)$$

However, since the concentration of a pure solid is constant

$$K_{(AgCl)} \times [AgCl(s)] = [Ag^+(aq)][Cl^-(aq)] = \text{constant}$$

This explains the expression

$$[Ag^+(aq)] \times [Cl^-(aq)] = \text{constant at that temperature} = K_s$$

discussed in Unit 11. The common ion effect (page 179) may now be explained using equation (15.8). If chloride ions are added to a saturated solution of AgCl, this equation predicts that $[Ag^+(aq)]$ must be reduced in order to keep $[Ag^+(aq)] \times [Cl^-(aq)]$ equal to $K_{(AgCl)}$. The reduction is achieved by precipitation of AgCl(s).

Exercise 15K

Application of the equilibrium law to vaporization

Show that application of the equilibrium law to the process

$$H_2O(l) \rightleftharpoons H_2O(g)$$

leads to the conclusion that the equilibrium vapour pressure of water is fixed at a particular temperature.

Revision questions

15.1. What is missing from the statement '$K_c = 3.5 \times 10^4$ for the reaction of hydrogen and bromine gases at 2040 °C'?

15.2. The table below shows the variation in the concentrations of $H_2(g)$, $I_2(g)$ and HI(g) with time when the following reaction was carried out at a fixed temperature:

$$H_2(g) + I_2(g) \rightleftharpoons 2HI(g)$$

	(Concentrations/mol dm^{-3})/10^{-3}			
Time/s	$H_2(g)$	$I_2(g)$	HI(g)	Q
0	1.000	1.000	0.000	
50	0.897	0.897	0.206	
100	0.813	0.813	0.374	
200	0.685	0.685	0.630	
300	0.592	0.592	0.817	
400	0.521	0.521	1.040	
500	0.465	0.465	1.069	
600	0.420	0.420	1.160	
700	0.383	0.383	1.234	
800	0.352	0.352	1.296	
850	0.333	0.333	1.334	
900	0.332	0.332	1.336	
950	0.331	0.331	1.338	

(Notice that the concentrations have been multiplied by 1000. For example, the concentration of HI at $t = 300$ s is $0.817 \times 10^{-3} = 8.17 \times 10^{-4}$ mol dm^{-3}.)

(i) Calculate the value of the expression

$$Q = \frac{[HI(g)]^2}{[H_2(g)][I_2(g)]}$$

at each time. (If possible, do this using a suitable computer spreadsheet such as *Microsoft Excel*.) Enter these values in the column labelled Q.

(ii) Plot $[H_2(g)]$ and [HI(g)] against time. Estimate the *initial rate* at which (a) H_2 is used up and (b) at which HI is produced. Comment upon your answers.

(iii) Plot Q versus t. At what time would you say that equilibrium has been reached? What is the value of the equilibrium constant for the reaction at this temperature?

(iv) Use Fig. 15.6 to estimate the temperature at which the experiments were carried out.

15.3. (i) Explain why the presence of water reduces the equilibrium yield of ester in the reaction

$$C_2H_5OH(l) + CH_3COOH(l) \rightleftharpoons CH_3COOC_2H_5(l) + H_2O(l)$$

(ii) In an experiment, ethanol and ethanoic acid were mixed and allowed to equilibrate at 25 °C. The equilibrium concentrations of ethanol, ethanoic acid and ester were found to be 2.67, 1.93 and 4.48 mol dm^{-3}, respectively. Calculate (a) the equilibrium constant for this reaction, and (b) the initial concentrations of ethanol and ethanoic acid.

(iii) The rate constant for the formation of ethyl ethanoate from ethanol at 25 °C is approximately 2.4 × 10^{-4} mol^{-1} dm^{-3} s^{-1}. Estimate the rate constant for the reverse reaction.

15.4. The formation of nitrosyl chloride from nitric oxide,

$$2NO(g) + O_2(g) \rightleftharpoons 2NOCl(g)$$

was investigated at a particular temperature. The pressures of gases at equilibrium were $p_{NO} = 0.22$ atm, $p_{O_2} = 0.11$ atm, and $p_{NOCl}) = 0.32$ atm. Calculate $K_{p(T)}$ and state its units.

15.5. K_c for the reaction of tin(II) ions with iron(III) ions,

$$Sn^{2+}(aq) + 2Fe^{3+}(aq) \rightleftharpoons Sn^{4+}(aq) + 2Fe^{2+}(aq)$$

is 1.0 × 10^{10} at 298 K. In an experiment at room temperature in which solutions of Sn^{2+}(aq) and Fe^{3+}(aq) were mixed, the equilibrium concentrations of tin ions were found to be

$$[Sn^{2+}(aq)] = 0.050 \text{ mol dm}^{-3}$$

$$[Sn^{4+}(aq)] = 0.040 \text{ mol dm}^{-3}$$

Calculate the concentrations of iron(III) and iron(II) ions at equilibrium.

15.6. The equilibrium constant for the decomposition of phosgene (a World War I poison gas) at 100°C is 2.2 × 10^{-10} mol dm^{-3}:

$$COCl_2(g) \rightleftharpoons CO(g) + Cl_2(g)$$

(Note how toxic the products are!) Phosgene is contained in a metal drum, whose outside surfaces are heated with steam. Would you expect all the phosgene to have decomposed once equilibrium had been achieved?

15.7. The exothermic reaction

$$CO(g) + 2H_2(g) \rightleftharpoons CH_3OH(g)$$

is allowed to reach equilibrium at 1000 K. How would the yield of methanol, and the value of the equilibrium constant, be affected by the following separate changes?

(i) Increasing the partial pressure of methanol by injecting some methanol into the mixture.

(ii) Compressing the mixture so that its pressure increases.

(iii) Allowing the mixture to expand so that its pressure decreases.

(iv) Pumping in an inert gas (such as argon) which increases the total pressure without a change in volume.

(v) Lowering the temperature of the equilibrium mixture.

15.8. The formation of brown nitrogen dioxide gas (NO_2) from nitrogen and oxygen, involves two stages:

$$N_2(g) + O_2(g) \rightleftharpoons 2NO(g) \quad \textbf{(1)}$$

$$2NO(g) + O_2(g) \rightleftharpoons 2NO_2(g) \quad \textbf{(2)}$$

The net reaction is

$$N_2(g) + 2O_2(g) \rightleftharpoons 2NO_2(g) \quad \textbf{(3)}$$

Show that $K_c(3) = K_c(1) \times K_c(2)$.

15.9. Show that:

(i) Henry's law, such as for the dissolution of oxygen in water,

$$\text{oxygen in air} \rightleftharpoons \text{oxygen in water}$$

and **(ii)** and the distribution law, such as for the solubility of I_2 between two solvents,

$$I_2 \text{ in } H_2O \rightleftharpoons I_2 \text{ in } CCl_4$$

both follow from an application of the equilibrium law.

Extension material to support this unit is available on our website at **http://www.palgrave.com/foundations/lewis.**

Acid–Base Equilibria

Objectives

- ▶ To use the ionic product constant of water in calculations
- ▶ Defines pH, pOH, pK_a and pK_b
- ▶ Discusses salt hydrolysis
- ▶ Looks at calculations involving weak acids and bases
- ▶ Explains the use of buffers and acid–base indicators

Contents

16.1 Ionic equilibria in water

Ionization of water

Pure water is a very poor conductor of electricity. This shows that there are virtually no ions present. However, there are some, and these play a very important role in the equilibria of aqueous solutions.

At room temperature about one water molecule in every 1 000 000 000 is ionized:

$$H_2O(l) + H_2O(l) \rightleftharpoons H_3O^+(aq) + OH^-(aq) \quad \textbf{(16.1)}$$

In this reaction, a proton is being transferred from one water molecule to another.

The equilibrium expression for this reaction is

$$K_{c(T)} = \frac{[H_3O^+(aq)][OH^-(aq)]}{[H_2O(l)]^2}$$

But $[H_2O(l)]$ is constant. This gives

$$K_{w(T)} = [H_3O^+(aq)][OH^-(aq)] \quad \textbf{(16.2)}$$

where $K_{w(T)}$, which equals $K_{c(T)} \times [H_2O(l)]^2$, is known as the **ionic product constant** (or **autoionization constant**) of water.

We can express equation (16.2) in words:

> **When the concentration of hydroxide and hydronium ions in water are multiplied together, the product is fixed at that temperature.**

Experiments show that equation (16.2) applies to water and also to *all* aqueous solutions, such as sodium chloride solution, hydrochloric acid and sodium hydroxide solution.

> **Exercise 16A**
>
> **Variation of K_w with temperature**
>
> **The data in Table 16.1 suggest that the dissociation of water is endothermic – Why?**

Can we be sure that reaction (16.1) has reached equilibrium at room temperature? Sophisticated kinetic experimental measurements have shown that reactions in which protons are lost or gained by molecules in solution are extremely fast. This means that we may safely assume that all the reactions discussed in this chapter have already reached equilibrium.

Table 16.1 shows the values of K_w at various temperatures.

Table 16.1 The ionic product constant of water at various temperatures

Temperature/°C	$10^{14} \times K_w$/mol^2 dm^{-6}
0	0.114
10	0.293
20	0.681
25	1.008
30	1.471
40	2.916
50	5.476
100	51.3

From Table 16.1, at 25 °C

$$K_w \approx 1.0 \times 10^{-14}\ \text{mol}^2\ \text{dm}^{-6}$$

(this is easy to remember) and we can now say that:

When the concentration of hydroxide and hydronium ions in water or in aqueous solutions are multiplied together, the product is always 1.0×10^{-14} mol^2 dm^{-6} at 25 °C.

Equation (16.1) shows that two water molecules produce one hydronium and one hydroxide ion; therefore, the concentration of these ions is equal. A solution containing equal numbers of each ion is said to be **neutral**. At 25 °C,

$$1.0 \times 10^{-14} = [H_3O^+(aq)][OH^-(aq)]$$

Therefore,

$$[H_3O^+(aq)] = [OH^-(aq)] = 1.0 \times 10^{-7}\ \text{mol dm}^{-3}$$

Calculations of pH were introduced in Unit 1 (page 13) and Unit 9 (page 150):

$$\text{pH} = -\log [H^+_{(aq)}],\ \text{or pH} = -\log[H_3O^+_{(aq)}]$$

We can now calculate the pH of pure water (water which contains no dissolved solids, liquids or gases):

$$\text{pH of pure water} = -\log(1.0 \times 10^{-7}) = 7.00$$

(Pure water is rarely encountered. Even distilled or deionized water contains dissolved carbon dioxide which gives it a pH of about 6. The pH of river water or lake water may be above 7 if it contains dissolved limestone. Acid rain has a pH as low as 4.)

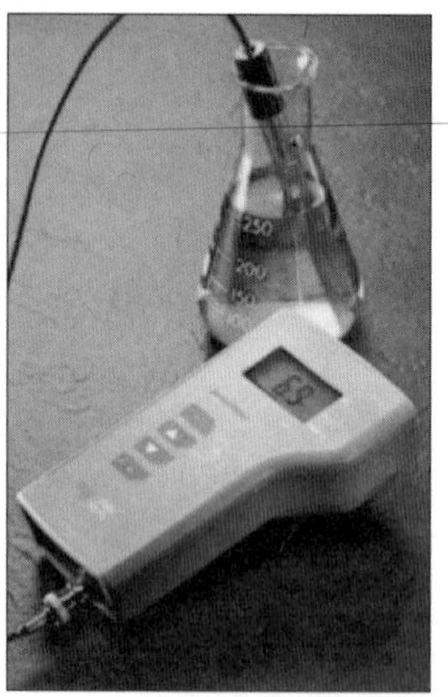

pH measurements require a meter, a measuring probe called a glass electrode, and a reference electrode. These are often packed together in a compact case. pH meters are used to provide rapid and accurate measurements of the pH of soils, of waste waters and of swimming pools.

Definitions of neutral, acidic and basic solutions

Neutral solution

A neutral solution is one where the concentrations of hydroxide and hydronium ions are equal. *At 25 °C, the pH of a neutral solution is 7.*

Acidic solution

An acidic solution is one where the concentration of hydronium ions is greater than the concentration of hydroxide ions. *At 25 °C, the pH of an acidic solution is less than 7.*

Basic solution

A basic (or alkaline) solution is one where the concentration of hydroxide ions is greater than the concentration of hydronium ions. *At 25 °C, the pH of a basic solution is greater than 7.*

Calculations using the ionic product constant of water

The ionic product constant of water (K_w) may be used to calculate the hydroxide ion concentration in solutions of acids. It may also be used to calculate the hydronium ion concentration in solutions of bases.

Example 16.1

A 0.456 mol dm^{-3} solution of hydrochloric acid (HCl(aq)) contains 0.456 mol dm^{-3} of hydronium ions. Calculate the pH of the solution of acid. What is the concentration of hydroxide ion in the acid? (T = 25 °C)

▶ **Answer**

First, calculate the pH:

$$pH = -\log(0.456) = -(0.341) = 0.341$$

Then, calculate the hydroxide ion concentration from K_w

$$[H_3O^+(aq) \times [OH^-(aq)] = 1.0 \times 10^{-14}\ mol^2\ dm^{-6}$$

or

$$[OH^-(aq)] = \frac{1.0 \times 10^{-14}}{[H_3O^+(aq)]} = \frac{1.0 \times 10^{-14}}{0.436} = 2.3 \times 10^{-14}\ mol\ dm^{-3}$$

▶ **Comment**

The number of decimal places in the calculated pH equals to the number of significant figures in $[H_3O^+(aq)]$, see Unit 1, page 13.

Exercise 16B

Calculations involving solutions of acids

(i) What is (a) the hydronium ion concentration and (b) the hydroxide ion concentration in a 0.10 mol dm^{-3} solution of nitric acid (HNO_3)? What is the pH of this solution? (Nitric acid is a strong acid and therefore completely dissociated into ions in solution.)

(ii) A solution of hydrochloric acid (a strong acid) has a pH of 2.20. What is the concentration of hydronium ions in the solution?

Example 16.2

A 0.020 mol dm^{-3} solution of sodium hydroxide at 25 °C contains 0.020 mol dm^{-3} of hydroxide ion. Calculate its pH.

▶ Answer

First, calculate the hydronium ion concentration:

$[OH^-(aq)] = 0.020$ mol dm^{-3}. We know that

$[H_3O^+(aq)] \times [OH^-(aq)] = 1.0 \times 10^{-14}$ mol^2 dm^{-3}

Rearranging,

$$[H_3O^+(aq)] = \frac{1.0 \times 10^{-14}}{[OH^-(aq)]} = \frac{1.0 \times 10^{-14}}{0.020} = 5.0 \times 10^{-13}\ \text{mol dm}^{-3}$$

To calculate the pH:

$pH = -\log [H_3O^+(aq)] = -\log (5.0 \times 10^{-13}) = 12.30$

▶ Comment

The pH of strongly basic solutions is usually above 11, so this is a reasonable answer. Now try Exercise 16C.

Exercise 16C

Calculations involving solutions of alkalis

(i) 0.10 g of calcium hydroxide ($Ca(OH)_2$), is dissolved in water and made up to 200 cm^3. Calculate the pH of the solution at 25 °C. (Calcium hydroxide is a 'strong base', i.e. it is completely dissociated into hydroxide and calcium ions in solution.)

(ii) A solution of sodium hydroxide (also a strong base) has a pH of 11.10. What is the concentration of (a) hydroxide ion and (b) hydronium ion in the solution? (Take the temperature to be 25 °C.)

pH, pOH and pK_w

Taking logs on both sides of equation (16.2) gives

$\log K_w = \log [H_3O^+(aq)] + \log [OH^-(aq)]$

Multiplying throughout by −1 gives

$-\log K_w = -\log [H_3O^+(aq)] - \log [OH^-(aq)]$

We know that $pH = -\log [H_3O^+(aq)]$. We similarly define pOH and pK_w as

$pOH = -\log[OH^-(aq)]$ and $pK_w = -\log K_w$

from which we see that

$pK_w = pH + pOH$

This expression is just another form of equation (16.2), and also applies to all aqueous solutions. At 25 °C, $pK_w = -\log (1.0 \times 10^{-14}) = 14$, so that

$pH + pOH = 14$

Now try Exercises 16D and 16E.

Exercise 16D

Relationship between pH and pOH

The pH of a solution is 5.80 at 25 °C. Calculate its pOH and $[OH^-(aq)]$.

Exercise 16E

pH and pOH scale

The following table shows the hydronium and hydroxide ion concentrations (in mol dm^{-3}) for 14 aqueous solutions – complete the table (the first entry has been done).

$[H_3O^+(aq)]$	$[OH^-(aq)]$	pH	pOH
10^{-14}	1	14	0
10^{-13}			
10^{-12}			
10^{-11}			
10^{-10}			
10^{-9}			
10^{-8}			
10^{-7}			
10^{-6}			
10^{-5}			
10^{-4}			
10^{-3}			
10^{-2}			
10^{-1}			

16.2 Acids and bases in aqueous solution

Strong acids

Hydrochloric acid is made by dissolving hydrogen chloride gas in water. The dissolved HCl (symbolized HCl(aq)) then reacts with water as follows:

$$HCl(aq) + H_2O(l) \rightarrow H_3O^+(aq) + Cl^-(aq)$$

Hydrochloric acid consists almost entirely of hydronium and chloride ions and the concentration of covalent acid molecules is negligible present. For this reason hydrochloric acid is said to be a **strong acid**:

a strong acid is completely ionized in solution

This means that if 0.1 mol of HCl(g) is dissolved in water and made up to 1 dm^3, the concentration of both H_3O^+ and Cl^- ions is also 0.1 mol dm^{-3}.

There are only a few strong acids. Apart from HCl(aq), the list includes:

- perchloric acid ($HClO_4$(aq)), and cheerfully referred to as the 'strongest acid known' in TV quizzes;
- nitric acid (HNO_3(aq)), whose ancient name, *aqua fortis*, means 'strong water';
- hydrobromic acid (HBr(aq)) and hydriodic acid (HI(aq)).

The case of sulfuric acid is discussed on page 294.

The term *strong acid* has nothing to do with the corrosive properties of an acid. It is also important to distinguish between acid strength (strong acid, weak acid) and the acidity of a solution. The *strength of an acid* in water is a property of the acid *molecule*, whereas the *acidity of a solution* is a qualitative desciption of its pH. A strongly acidic solution is usually regarded as one with a pH of below 3. A weakly acidic solution has a pH of 3–7. A low pH might result from a concentrated solution of a weak acid or from a very dilute solution of a strong acid. But hydrochloric acid is a strong acid whether its concentration is 0.5 mol dm^{-3} or 0.0005 mol dm^{-3}.

Weak acids

Ethanoic acid is only partially ionized in aqueous solution:

$$CH_3COOH(aq) + H_2O(l) \rightleftharpoons \underset{\text{ethanoate (or acetate) ion}}{CH_3COO^-(aq)} + H_3O^+(aq) \tag{16.3}$$

and is therefore termed a **weak acid**:

a weak acid is incompletely ionized in solution

If we apply the equilibrium law to equation (16.3) the equilibrium expression becomes

$$K_{a(T)} = \frac{[CH_3COO^-(aq)][H_3O^+(aq)]}{[CH_3COOH(aq)]}$$

where $K_{a(T)}$ is the equilibrium constant ('a' is for acid) at temperature T. As usual, the square brackets refer to the equilibrium (not starting) concentrations. Once again, the concentration of water (a constant) has been incorporated into $K_{a(T)}$ – this will be taken for granted in similar expressions below.

The term $K_{a(T)}$ is commonly known as the **acidity constant** (or the **dissociation constant**) of the acid. For ethanoic acid, $K_a(25\,°C) = 1.8 \times 10^{-5}$ mol dm^{-3}, the low value confirming that very few of the acid molecules have ionized. Hydrogen fluoride:

$$\underset{\text{hydrogen fluoride}}{HF(aq)} + H_2O(l) \rightleftharpoons \underbrace{F^-(aq) + H_3O^+(aq)}_{\text{hydrofluoric acid}}$$

is also a weak acid with $K_a(25\,°C) = 3.5 \times 10^{-4}$ mol dm^{-3}.

Table 16.2 shows the acidity constants for selected acids at 25 °C. Symbolizing the acids as AH, the acidity constants are the equilibrium constants for the general reaction

$$AH(aq) + H_2O(l) \rightleftharpoons A^-(aq) + H_3O^+(aq) \tag{16.4}$$

(If AH is a solid, liquid or gas at room temperature (e.g. $C_6H_5COOH(s)$, $CH_3COOH(l)$, or $HCl(g)$) we assume that AH first dissolves in water making AH(aq), which then reacts according to this equation.)

The higher the K_a value, the stronger the acid.

The acidity constants of strong acids are difficult to determine experimentally because the concentration of unionized molecules is minute, but there is no doubt that the acidity constants of strong acids are much higher than those of the weaker acids listed in Table 16.2. This reflects the fact that the acid molecules are almost entirely dissociated in solution.

We may also define pK_a, where

$$pK_a = -\log K_a$$

Table 16.2 K_a and K_b values for selected weak acids and weak bases in aqueous solution at 25 °C. These values change only slightly within the range 15–30 °C. The acidic hydrogen or basic nitrogen atom is highlighted in bold

Acids	K_a/mol dm^{-3}
Trichloroethanoic acid $CCl_3CO_2\mathbf{H}$	3.0×10^{-1}
Chloroethanoic acid $CH_2ClCO_2\mathbf{H}$	1.4×10^{-3}
Sulfur dioxide SO_2 (sulfurous acid $\mathbf{H}_2SO_3$*)	1.6×10^{-2}
Lactic acid $CH_3CH(OH)CO_2\mathbf{H}$	8.4×10^{-4}
Hydrogen fluoride **H**F (hydrofluoric acid)	3.5×10^{-4}
Benzoic acid $C_6H_5COO\mathbf{H}$	6.5×10^{-5}
Ethanoic acid $CH_3COO\mathbf{H}$	1.8×10^{-5}
Carbon dioxide CO_2 (carbonic acid $\mathbf{H}_2CO_3$*)	4.3×10^{-7}
Hydrogen cyanide **H**CN (hydrocyanic acid)	4.9×10^{-10}
Phenol (carbolic acid) $C_6H_5O\mathbf{H}$	1.3×10^{-10}

Bases	K_b/mol dm^{-3}
Urea $\mathbf{N}H_2CONH_2$*	1.3×10^{-14}
Phenylamine (aniline) $C_6H_5\mathbf{N}H_2$	4.3×10^{-10}
Morphine $C_{17}H_{19}O_3\mathbf{N}$	1.6×10^{-6}
Ammonia $\mathbf{N}H_3$	1.8×10^{-5}
Methylamine $CH_3\mathbf{N}H_2$	3.6×10^{-4}
Dimethylamine $(CH_3)_2\mathbf{N}H$	5.4×10^{-4}
Triethylamine $(C_2H_5)_3\mathbf{N}$	1.0×10^{-3}

* K_a or K_b given for first ionization.

For example, the pK_a of ethanoic acid at 25°C is $-\log(1.8 \times 10^{-5}) = 4.75$ and the pK_a of lactic acid is $-\log(8.4 \times 10^{-4}) = 3.08$.

The stronger the acid, the lower is its pK_a.

In equation (16.4), the acid is donating a hydrogen ion (a proton) to a water molecule to produce a H_3O^+(aq) ion. This means that we can look at a K_a value as an indicator of the **proton-donating power** of that acid molecule toward the water molecule:

The greater the K_a value, the greater the proton donor ability of the acid towards water.

The species A^-(aq) in equation (16.4) is referred to as the **conjugate base** of the acid AH(aq). This is because the A^-(aq) ion acts like a base in that it accepts a proton from H_3O^+(aq) in the reverse of reaction (16.4). For example, the conjugate base of ethanoic acid is the ethanoate ion CH_3COO^-(aq).

Calculating the pH of solutions of weak acids

We shall use ethanoic acid as an example of a weak acid which contains one acidic hydrogen atom. We will symbolize the initial concentration of ethanoic acid in a solution as C_A mol dm^{-3}. (For example, suppose we make a solution of ethanoic acid by dissolving 0.1 mol of pure ethanoic acid in water and making the solution up to 1 dm^3. Then $C_A = 0.1$ mol dm^{-3}.) The difficulty in calculating the pH of the resulting solution is that, unlike strong acids, the equilibrium concentration of hydronium ion in the solution is not the same as the concentration of the acid originally added, i.e. $[H_3O^+(aq)] \neq C_A$.

The $H_3O^+(aq)$ concentration of a very weak acid with one acidic hydrogen can be estimated by re-arranging the following equation:

$$K_{a(T)} \approx \frac{[H_3O^+(aq)]^2}{C_A} \quad \textbf{(16.5)}$$

This equation is derived in Box 16.1. The equation is a good approximation provided the percentage of acid molecules that are ionized,

$$\text{percentage of acid molecules ionized} = \frac{[H_3O^+(aq)]}{C_A} \times 100$$

does not exceed about 5%.

Example 16.3

Some white vinegar contains 0.50% by mass of ethanoic acid, i.e. 100 g of vinegar contains 0.50 g of acid. Estimate its pH at 25 °C. What percentage of the ethanoic acid molecules are ionized? (1.0 cm³ of vinegar has a mass of 1.0 g.)

▶ Answer

Consider exactly 1 dm³ of vinegar. 1 dm³ of vinegar contains 5.0 g CH_3COOH. The molar mass of ethanoic acid, $M(CH_3COOH)$, is 60 g mol⁻¹.

$$\text{number of moles of acid} = \frac{5.0\,\text{g}}{60\,\text{g mol}^{-1}} = 0.083$$

$$\text{concentration of acid} = \frac{0.083\,\text{mol}}{1\,\text{dm}^3} = 0.083\,\text{mol dm}^{-3}$$

Rearranging equation (16.5) gives

$$[H_3O^+(aq)] \approx \sqrt{(K_{a(T)} \times C_A)}$$

$C_A = 0.083\,\text{mol dm}^{-3}$ and $K_a = 1.8 \times 10^{-5}\,\text{mol dm}^{-3}$ (Table 16.2), so that

$$[H_3O^+(aq)] \approx \sqrt{(1.8 \times 10^{-5} \times 0.083)} = 1.2 \times 10^{-3}\,\text{mol dm}^{-3}$$

$$\text{pH} = -\log[H_3O^+(aq)] = -\log(1.2 \times 10^{-3}) = 2.92$$

The percentage of ethanoic acid molecules that are ionized in vinegar is calculated as

$$\frac{[H_3O^+(aq)]}{C_A} \times 100 = \frac{1.2 \times 10^{-3} \times 100}{0.083} = 1.4\%$$

▶ Comment

The percentage of ionized acid molecules is less than 5%, and this confirms that we are able to use equation (16.5) to calculate $[H_3O^+(aq)]$ for this acid.

Exercise 16F

Calculations involving weak acids

(i) Calculate the percentage of HCN molecules ionized in a HCN solution of concentration $0.020\,mol\,dm^{-3}$. What is the pH of the solution? (Use the data in Table 16.2.)

(ii) Benzoic acid is a natural bacteriostatic preservative, i.e. it helps prevent the growth of bacteria. Some fruit contains 0.050% by mass of benzoic acid. Estimate the pH of $1.0\,dm^3$ of the solution produced from 4.0 kg of the fruit. (**Hint** Start by calculating the concentration of benzoic acid.)

Exercise 16G

Weak acid equilibria

(i) $25.0\,cm^3$ of $0.10\,mol\,dm^{-3}$ HCl requires $25.0\,cm^3$ of $0.10\,mol\,dm^{-3}$ NaOH for neutralization. Even though only 1.3% of ethanoic acid molecules are ionized at room temperature, $25.0\,cm^3$ of $0.10\,mol\,dm^{-3}$ CH_3COOH also requires $25\,cm^3$ of $0.10\,mol\,dm^{-3}$ NaOH for neutralization. Explain this.

(ii) The enthalpy change of neutralization ($\Delta H^{\ominus}_{N}$) is the standard enthalpy change when one mol of water is made in the neutralization of an acid by a base. Values are:

Acid–base pair	$\Delta H^{\ominus}_{N}/kJ\,mol^{-1}$
HNO_3–KOH	−57.3
HCl–NaOH	−57.1
CH_3COOH–NaOH	−55.2

Why are the values for the first two pairs approximately the same? Why is $\Delta H^{\ominus}_{N}$ for the neutralization of ethanoic acid less negative than the others?

BOX 16.1

An expression to calculate the pH of a weak acid

We need the two following relationships to derive equation (16.5):

1. The equilibrium concentration of unionized ethanoic acid will be the difference between C_A and the equilibrium hydronium ion concentration:

$$[CH_3COOH(aq)] = C_A - [H_3O^+(aq)] \qquad \textbf{(16.6)}$$

2. For every molecule of ethanoic acid that ionizes, one hydronium and one ethanoate ion is produced. Thus,

$$[CH_3COO^-(aq)] = [H_3O^+(aq)] \qquad \textbf{(16.7)}$$

Using the equilibrium expression

$$K_{a(T)} = \frac{[CH_3COO^-(aq)][H_3O^+(aq)]}{[CH_3COOH(aq)]}$$

and substituting the right-hand sides of equations (16.6) and (16.7) gives

$$K_{a(T)} = \frac{[H_3O^+(aq)]^2}{C_A - [H_3O^+(aq)]}$$

For acids that are only slightly ionized ($< 5\%$) we are justified in assuming that $C_A - [H_3O^+(aq)]$ is approximately equal to C_A. We then obtain equation (16.5):

$$K_{a(T)} \approx \frac{[H_3O^+(aq)]^2}{C_A}$$

Sulfuric acid – a diprotic acid

Sulfuric acid possesses two acidic hydrogens and is said to be **diprotic**. There are two ionization steps, with sulfuric acid itself being the acid in the first step, and the hydrogensulfate ion being the acid in the second step:

First ionization

$$H_2SO_4(aq) + H_2O(l) \rightleftharpoons H_3O^+(aq) + \underset{\text{hydrogensulfate ion}}{HSO_4^-(aq)} \qquad K_a(298\,K) = \text{large}$$

Second ionization

$$HSO_4^-(aq) + H_2O(l) \rightleftharpoons H_3O^+(aq) + SO_4^{2-}(aq) \qquad K_a(298\,K) = 0.012\,\text{mol dm}^{-3}$$

The net reaction is the sum of both steps:

$$H_2SO_4(l) + 2H_2O(l) \rightleftharpoons 2H_3O^+(aq) + SO_4^{2-}(aq)$$

The K_a values show that sulfuric acid is a strong acid and that the hydrogensulfate ion is a relatively weak acid (although much stronger than most of the acids listed in Table 16.2).

Dilute sulfuric acid of concentration 0.0100 mol dm^{-3} contains an $H_3O^+(aq)$ concentration of 0.0145 mol dm^{-3}, giving a pH of 1.84. Of the total $[H_3O^+(aq)]$, 0.0100 mol dm^{-3} is contributed by the first ionization, and only 0.0045 mol dm^{-3} of the total $[H_3O^+(aq)]$ is provided by the second ionization. If both K_a values were large, each reaction would contribute a hydronium ion concentration of 0.0100 mol dm^{-3}, and the pH of the dilute acid would then be $-\log[0.0200] = 1.70$.

The HSO_4^- ion is more than strong enough to be used as an acid in its own right, and sodium hydrogensulfate crystals (Na^+,HSO_4^-) are used as a powerful disinfectant. They are also provided in chemistry sets as a safe substitute for sulfuric acid.

Strong and weak bases

The commonest strong bases are the water-soluble hydroxides of sodium, potassium, calcium, barium and lithium. These are all ionic solids. Since they are **strong bases** they are completely ionized in water. For example, NaOH(s) breaks up completely in water producing $Na^+(aq)$ and $OH^-(aq)$:

$$\underset{\text{sodium hydroxide}}{Na^+,OH^-(s)} \xrightarrow{H_2O} Na^+(aq) + \underset{\text{hydroxide ion}}{OH^-(aq)}$$

For example, a solution containing 0.5 mol of dissolved NaOH per dm^3 of solution contains 0.5 mol dm^{-3} of $OH^-(aq)$.

A second category of bases are those which produce hydroxide ions in solution by *reaction* with water. An example is ammonia:

$$NH_3(aq) + H_2O(l) \rightleftharpoons NH_4^+(aq) + OH^-(aq)$$

Ammonia is a **weak base** because it is incompletely ionized in solution, and

$$K_{b(T)} = \frac{[NH_4^+(aq)][OH^-(aq)]}{[NH_3(g)]}$$

where $K_{b(T)}$ is the **basicity constant** (or **dissociation constant**) of ammonia at temperature T. Generalizing, the basicity constant for a base B is the equilibrium constant for the reaction:

$$B(aq) + H_2O(l) \rightleftharpoons BH^+(aq) + OH^-(aq) \qquad \textbf{(16.8)}$$

Any solution which contains the hydroxide ion (such as NaOH(aq) or NH_3(aq)) will neutralize acids in the reaction:

$$H_3O^+(aq) + OH^-(aq) \rightarrow 2H_2O(l)$$

Selected K_b values are included in Table 16.2, with triethylamine being the strongest base listed because it has the *biggest K_b value*. By coincidence, K_b for ammonia (one of the commonest weak bases) is numerically equal to K_a for ethanoic acid at 25°C.

Books often tabulate pK_b values where

$$pK_b = -\log K_b$$

For example, the pK_b of ammonia is $-\log(1.8 \times 10^{-5}) = 4.74$. The stronger the base, the *lower* is its pK_b.

In reaction (16.8), the base is accepting a proton from a water molecule. This enables us to use $K_{b(T)}$ values as indicators of the *proton-accepting ability* of bases in water, in the same way that we used $K_{a(T)}$ values as indicators of the proton donating ability of acids in water. A large $K_{b(T)}$ value shows that the base is a strong proton acceptor.

In equation (16.8), the species BH^+(aq) is referred to as the **conjugate acid** of base B(aq). This is because BH^+(aq) donates a proton to the OH^-(aq) in the *reverse* of reaction (16.8).

The self-ionization of water,

$$H_2O(l) + H_2O(l) \rightleftharpoons H_3O^+(aq) + OH^-(aq)$$

involves one molecule of water donating a proton to another. This makes water both a proton donor and a proton acceptor.

When calculating the pH of a weak base, we follow the same pattern as for an acid,

$$K_{b(T)} \approx \frac{[OH^-(aq)]^2}{C_B}$$

where C_B is the initial molar concentration of base in the solution. If C_B and $K_{b(T)}$ are known, we can then calculate $[OH^-(aq)]$. $[H_3O^+(aq)]$ is then worked out using K_w. (Now try Exercise 16H.)

Exercise 16H

Calculations involving weak bases

(i) pK_b for nicotine, $C_{10}H_{14}N_2$, is 5.98. What is K_b?

(ii) Estimate the pH of a solution of ammonia of concentration 0.200 mol dm^{-3} at 25°C.

(iii) Write down an equation showing the reaction of phenylamine with water to produce hydroxide ions. What is the conjugate acid of phenylamine? Some phenylamine was dissolved in pure water. The pH of the solution was found to be 7.80 at 25°C. What was the concentration of phenylamine in the solution?

16.3 Hydrolysis of salts

pH of solutions of salts

A salt is produced when a base and acid neutralize each other. On this basis, four classes of salts are possible:

1. a salt of a strong acid and a strong base (SA–SB);
2. a salt of a weak acid and a strong base (WA–SB);

3. a salt of a strong acid and a weak base (SA–WB);

4. a salt of a weak acid and a weak base (WA–WB).

Sodium chloride solution (made with pure water), is neutral with a pH of 7 at 25 °C. This may lead us to suppose that solutions of all ionic salts are neutral. In fact, *only solutions of salts made from strong acids and strong bases are always neutral.* Solutions of other salts are usually either acidic or basic. The reasons for this behaviour are that:

1. These salts react with the water *producing a weak acid or a weak base.*
2. Since the weak acid or weak base is only partially ionized in water, formation of these molecules ties up hydroxide or hydronium ions.
3. This produces unequal concentrations of hydronium and hydroxide ions, and the resulting solution of salt is then acidic or basic.

Reactions with water are called **hydrolysis**, and the reactions of a salt with water are referred to as **salt hydrolysis**. We now look at one example of this effect.

Hydrolysis of a salt of a weak acid and a strong base

Using sodium ethanoate (CH_3COO^-,Na^+) as an example, the ions present in an aqueous solution of this salt are

$$CH_3COO^-,Na^+(s) \xrightarrow{H_2O} CH_3COO^-(aq) + Na^+(aq) \quad \text{(from the } \textit{salt}\text{)}$$

$$2H_2O(l) \rightleftharpoons H_3O^+(aq) + OH^-(aq) \quad \text{(from the } \textit{water}\text{)}$$

The potential products of any reaction between the ions in solution are NaOH (from the reaction of Na^+(aq) and OH^-(aq)) and CH_3COOH (from the reaction of CH_3COO^-(aq) and H_3O^+(aq)). The NaOH is fully ionized in solution, but the ethanoic acid is only partially ionized. This means that $[OH^-(aq)] > [H_3O^+(aq)]$ and the solution is basic.

Generalizing,

Solutions of salts of weak acids and strong bases are basic.

Consideration of the possible reactions in solutions of the other classes of salts leads to the conclusions given in Table 16.3.

Table 16.3 The pH of solutions of the four classes of salt

Class of salt	Example	Formula	Acidic	Basic	Neutral
SA–SB	sodium chloride	NaCl			✓
WA–SB	sodium ethanoate	CH_3COONa		✓	
SA–WB	ammonium chloride	NH_4Cl	✓		
WA–WB	ammonium cyanide	NH_4CN		✓*	

* If $K_a(WA) > K_b(WB)$ the solution is acidic. If $K_a(WA) < K_b(WB)$ the solution is basic. If $K_a(WA) = K_b(WB)$ the solution is neutral.

Exercise 16I

Salt hydrolysis

Use Table 16.3 to complete this table:

Salt	Formula	Parent acid and base	Acidic, basic or neutral?
Iron(III) nitrate			
Calcium chloride			
Sodium sulfate			
Ammonium benzoate			

16.4 Buffer solutions

Adding small amounts of acids or alkalis may lead to drastic changes in solution pH

The addition of even one drop of dilute hydrochloric acid to water drastically changes its pH (Box 16.2). Such changes in pH can be troublesome in the laboratory, and catastrophic in living cells. Changes in pH due to trace contamination with acids or bases can be prevented using a **buffer solution** (usually simply called a 'buffer' (Fig. 16.1)).

> **A buffer solution resists changes in pH when it is diluted or when acid or base is added.**

Figure 16.2 shows the pH values in separate experiments in which (a) 0.1 mol dm^{-3} HCl, (b) deionized water and (c) 0.1 mol dm^{-3} NaOH were added to a sodium ethanoate–ethanoic acid buffer. In this case, adding 3 cm^3 or less of (a) or (c) produces only a slight change in pH. If we add more acid or alkali, the buffer becomes exhausted and it can no longer resist changes in pH. However, dilution of the buffer by adding water causes virtually no change in buffer pH.

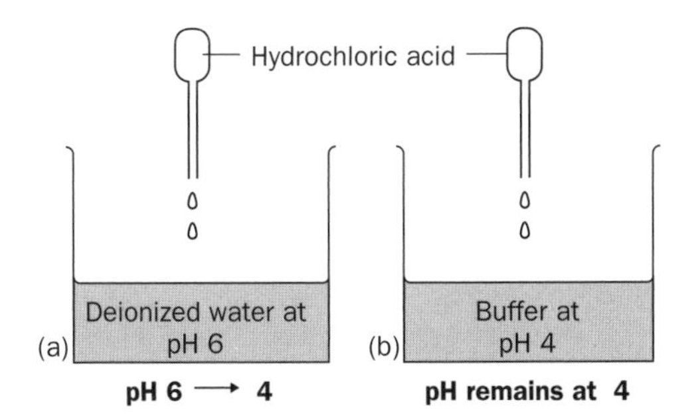

Fig. 16.1 A buffer in action. (a) When a drop of 2.0 mol dm^{-3} hydrochloric acid is added a water, the pH falls drastically. (b) There is no change in pH when an acid is added to the buffer solution.

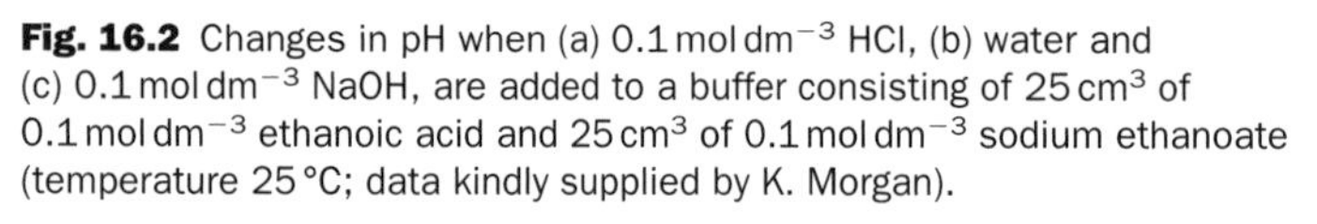

Fig. 16.2 Changes in pH when (a) 0.1 mol dm^{-3} HCl, (b) water and (c) 0.1 mol dm^{-3} NaOH, are added to a buffer consisting of 25 cm^3 of 0.1 mol dm^{-3} ethanoic acid and 25 cm^3 of 0.1 mol dm^{-3} sodium ethanoate (temperature 25 °C; data kindly supplied by K. Morgan).

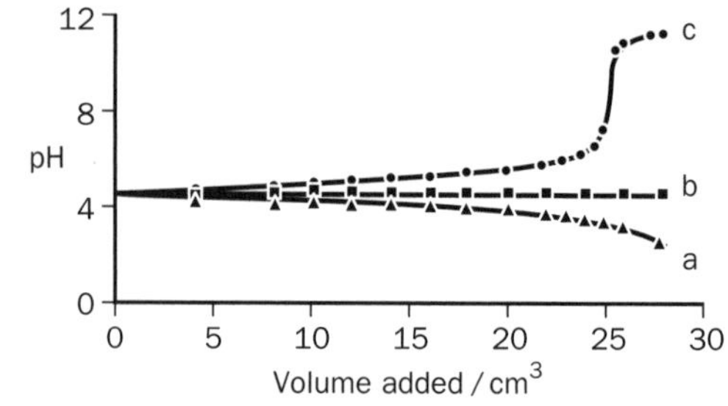

How buffers work

Buffers either consist of a weak base and one of its salts, or a weak acid and one of its salts. As an example, we look at a mixture containing ethanoic acid and sodium ethanoate. The ions present in such a mixture are shown by the following equations:

$$CH_3COO^-,Na^+(s) \xrightarrow{H_2O} CH_3COO^-(aq) + Na^+(aq)$$

$$CH_3COOH(aq) + H_2O(l) \rightleftharpoons CH_3COO^-(aq) + H_3O^+(aq) \quad \textbf{(16.9)}$$

The salt (since it is an ionic compound) is fully ionized, and generates a relatively high concentration of ethanoate ions.

Now consider the ionization of ethanoic acid in the presence of sodium ethanoate. The expression for the acidity constant of ethanoic acid is:

$$K_{a(T)} = \frac{[CH_3COO^-(aq)][H_3O^+(aq)]}{[CH_3COOH(aq)]} \quad \textbf{(16.10)}$$

If ethanoate ions are added to an aqueous solution of ethanoic acid, the equilibrium composition will shift in order to keep the right-hand side of this expression equal to $K_{a(T)}$. As a result of the shift, $[H_3O^+(aq)]$ becomes very low, meaning that very little

BOX 16.2

pH of a jug of water contaminated with a drop of dilute hydrochloric acid

Suppose we add 1 drop of 'bench hydrochloric acid' (concentration = 2.0 mol dm^{-3}) to 1 dm^3 of deionized water. What is the pH of the resulting solution?

The volume of the drop ($\approx$ 0.05 cm^3) is negligible compared with the water, so we may take the solution volume as 1 dm^3. The number of moles of HCl in the 0.05 cm^3 drop is

$$\frac{0.05}{1000} \times 2.0 = 1 \times 10^{-4}\,\text{mol}$$

and the concentration of HCl in the solution is therefore

$$\frac{1 \times 10^{-4}\,\text{mol}}{1\,\text{dm}^3} = 1 \times 10^{-4}\,\text{mol dm}^{-3}$$

The pH of the solution is calculated as follows:

$$\text{pH} = -\log[H_3O^+(aq)] = -\log[1 \times 10^{-4}] = 4.0$$

The pH of deionized water in the laboratory is about 6 because it contains dissolved carbon dioxide. Therefore,

the addition of only one drop of dilute acid has reduced its pH from 6 to 4.

ethanoic acid is now ionized. Calculations show that in the presence of sodium ethanoate, the ionization of ethanoic acid is so small that $[CH_3COO^-(aq)]$ may be taken to be equal to the initial sodium ethanoate salt solution, C_s.

How does the buffer mixture resist changes in pH? The explanation is as follows.

- Suppose we add some hydrochloric acid to the mixture. The equilibrium composition shifts, with ethanoate and hydronium ions combining together in the *reverse* of equation (16.9). This mops up the added $H_3O^+(aq)$ ions, and there is no change in pH. (The same conclusion is arrived at using Le Chatelier's principle.)
- If we add hydroxide ions (e.g. by adding sodium hydroxide solution), the tiny amount of $[H_3O^+(aq)]$ at equilibrium in the buffer mixture reacts with $OH^-(aq)$ in the neutralisation reaction

  $$H_3O^+(aq) + OH^-(aq) \rightarrow 2H_2O(l).$$

  More ethanoic acid ionizes to restore the equilibrium concentration of $H_3O^+(aq)$, and ionization continues until just enough extra $H_3O^+(aq)$ has been made to convert all the $OH^-(aq)$ to water. In this way, the addition of base does not lead to a change in the pH of the buffer.

Calculating the pH of a buffer solution

We can easily obtain an expression with which to estimate the pH of a buffer, starting with equation (16.10). Since virtually all the ethanoate ions in the buffer come from the sodium ethanoate, we are justified in substituting C_s for $[CH_3COO^-(aq)]$. Since so little ethanoic acid is ionized in the buffer, we are also justified in taking the equilibrium concentration of ethanoic acid, $[CH_3COOH(aq)]$, to be equal to the initial concentration of acid, C_A. Substitution of this information into equation (16.10) gives

$$K_{a(T)} \approx \frac{C_s \times [H_3O^+(aq)]}{C_A}$$

Rearranging,

$$[H_3O^+(aq)] \approx \frac{C_A \times K_{a(T)}}{C_s} \qquad \textbf{(16.11)}$$

This equation shows that dilution of the buffer mixture does not affect its pH because C_A and C_s are affected equally. For example, if we add 50 cm^3 of water to 50 cm^3 of buffer, *both* the acid (C_A) and salt (C_s) concentrations in the buffer solution are halved, and the pH predicted by equation (16.11) is unaffected by the dilution.

Exercise 16J

Calculating acidity constants

The acidity constant of an acid may be calculated from the way that the pH of a buffer changes as the concentration of acid and salt are changed. This involves using a *graphical* method. Can you see from equation (16.11) how this might be done?

Buffer solutions may also be prepared by mixing a weak base and one of its salts (such as ammonia solution and ammonium chloride). The $[OH^-(aq)]$ of such a mixture may be calculated using the expression

$$[OH^-(aq)] \approx \frac{C_B \times K_{b(T)}}{C_s}$$

where C_B is the concentration of base, C_s is the concentration of salt, and $K_{b(T)}$ is the basicity constant.

Data for the sodium ethanoate–ethanoic acid buffer

Let us use equation (16.11) to calculate the pH of an ethanoate–ethanoic acid buffer at 25 °C made by mixing 25 cm^3 of 0.100 mol dm^{-3} sodium ethanoate solution with 25 cm^3 of 0.100 mol dm^{-3} ethanoic acid solution. Since the volume doubles, the concentrations are halved and $C_A = C_s = 0.050$ mol dm^{-3}. K_a for ethanoic acid at 25 °C $= 1.8 \times 10^{-5}$ mol dm^{-3}. Substituting into equation (16.11) gives

$$[H_3O^+(aq)] = \frac{0.050 \times 1.8 \times 10^{-5}}{0.050} = 1.8 \times 10^{-5}\,\text{mol dm}^{-3}$$

The pH of the buffer mixture is

$$pH = -\log[1.8 \times 10^{-5}] = 4.74$$

(This is approximately the initial pH of the buffer in Fig. 16.2.) By using different acids (different K_a values) and their salts, buffers operating at different pH values may be obtained. To adjust the pH of a buffer more exactly, the ratio of the initial concentrations of acid and salt in the mixture are carefully controlled.

Buffer capacity

The amount of acid or alkali that needs to be added before the pH of a buffer changes is called the **buffer capacity** of the buffer. The buffer capacity of a buffer containing a relatively high number of moles of acid and salt is greater than the capacity of a buffer with a lower number of moles of acid and salt. As we have already noted, the buffer capacity of the buffer in Fig. 16.2 is equivalently to roughly 3 cm^3 of 0.1 mol dm^{-3} HCl or 3 cm^3 of 0.1 mol dm^{-3} NaOH.

Exercise 16K

Calculating the pH of buffer solutions

Investigate the effect of using different volumes of 0.100 mol dm^{-3} ethanoic acid and 0.100 mol dm^{-3} sodium ethanoate salt, by repeating the pH calculation starting with (i) 25 cm^3 of acid and 50 cm^3 of salt solution, and (ii) 50 cm^3 of acid and 25 cm^3 of salt solution.

Alternative form of equation (16.11) – the Henderson–Hasselbalch equation

Equation (16.11) is sometimes used in a different form. Taking logs on both sides gives

$$\log [H_3O^+(aq)] = \log K_{a(T)} + \log \frac{C_A}{C_s}$$

Multiplying by −1,

$$-\log [H_3O^+(aq)] = -\log K_{a(T)} - \log \frac{C_A}{C_s}$$

Remembering that

$$-\log \frac{C_A}{C_s} = \log \frac{C_s}{C_A}$$

we arrive at the relationship

$$pH = pK_a + \log \frac{C_s}{C_A}$$

which is known as the **Henderson–Hasselbalch** equation (Exercise 16L).

Exercise 16L

Using the Henderson–Hasselbalch equation

A buffer solution contains 0.10 mol of one of the acids listed in Table 16.2 mixed with 0.10 mol of its potassium salt. The observed pH of the buffer was 4.18. What is the name of the acid?

16.5 Acid–base indicators

Acid–base indicators, such as methyl orange, phenolphthalein and litmus, show two extreme colours, one at lower pH and the other at higher pH.

For example, in the case of methyl orange, the two extreme colours are red (lower pH) and yellow (higher pH). As with most indicators, the change in colour does not occur over a very small change of pH. The range of pH over which an indicator changes colour is called its **pH range**. The pH range of methyl orange is 3.2–4.4. Below pH 3.2 methyl orange is red, but as the pH is increased *above* 3.2, the observed colour contains an increasing amount of yellow until, beyond pH 4.4, it is entirely yellow.

Indicators are sometimes absorbed into paper strips (e.g. litmus paper). The indicator known as 'universal indicator' is a mixture of selected indicators which displays different colours at different pH values. (To demonstrate this, place a large crystal of tartaric acid in a small flask. Add dilute universal indicator solution and one drop of NaOH solution. Upon gently swirling the flask, the acid slowly dissolves, and the indicator passes through an impressive sequence of colours.)

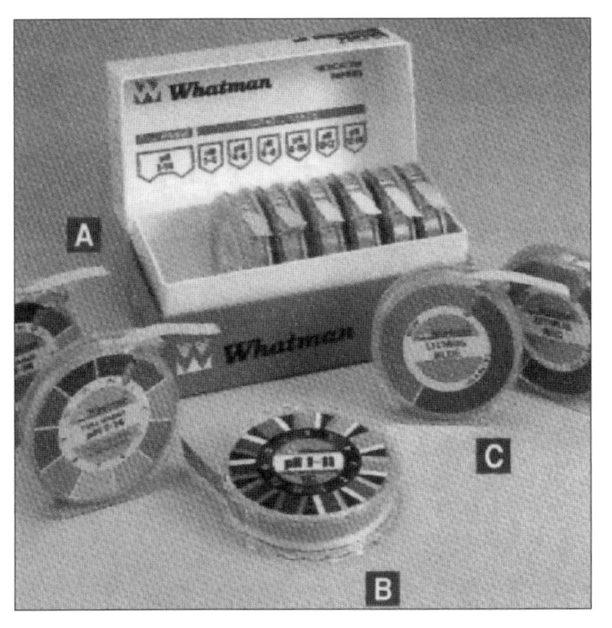

Acid–base indicators are available in easy to use strips.

Exercise 16M

Colours of acid–base indicators

Table 16.4 shows the pH range of selected acid–base indicators. Use this table to help you decide the colour of phenolphthalein and bromocresol green at a pH of

(i) 2, **(ii)** 3.7, **(iii)** 5,
(iv) 6, **(v)** 8 **(vi)** 11.

Table 16.4 Acid–base indicators

Indicator	Extreme colours		pH range
	Lower pH	Higher pH	
Alizarin yellow	colourless	yellow	10.1–12.0
Bromocresol green	yellow	blue	4.0–5.6
Litmus	red	blue	4.7–8.2
Methyl orange	red	yellow	3.2–4.4
Methyl red	yellow	red	4.8–6.0
Phenolphthalein	colourless	pink	8.2–10.0

How acid–base indicators work

Acid–base indicators consist of molecules that have different molecular structures at lower pH and at higher pH. These higher-pH and lower-pH forms have different colours.

Consider methyl orange. Above pH 4.4, methyl orange exists entirely as the ion

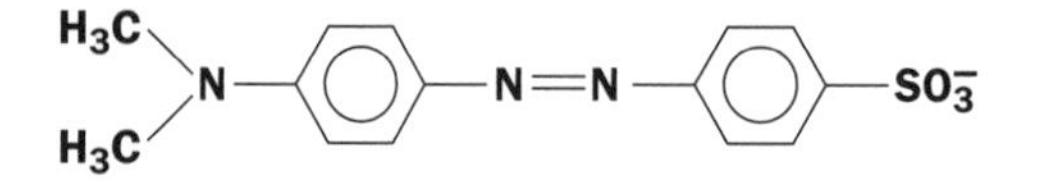

This ion is the higher-pH form. It is coloured yellow. It is represented as $R{-}SO_3^-$.

Below pH 3.2, methyl orange exists entirely as

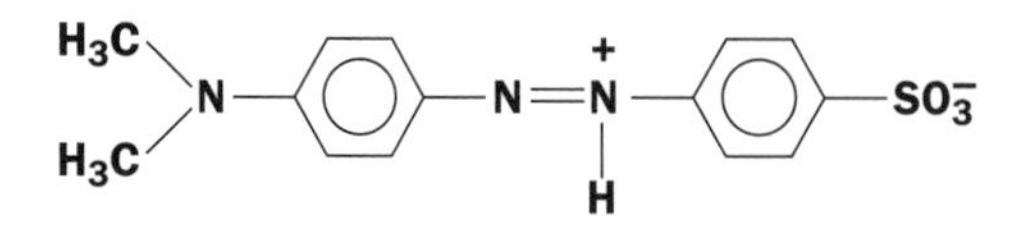

in which one of the nitrogen atoms has been bonded to a proton (i.e. protonated) as a result of reaction with $H_3O^+(aq)$. This ion is the lower-pH form. It is coloured red and is represented as $HR^+{-}SO_3^-$. These changes are summarized by the equilibrium equation

$$\underset{\text{yellow}}{R{-}SO_3^-(aq)} + H_3O^+(aq) \rightleftharpoons \underset{\text{red}}{HR^+{-}SO_3^-(aq)} + H_2O$$

All acid–base indicators undergo similar molecular changes with changing pH. The general equation representing these changes is

$$\text{higher-pH form} + H_3O^+(aq) \rightleftharpoons \underset{\text{('acid form')}}{\text{lower-pH form}} + H_2O$$

16.6 Variation of pH during an acid–base titration

In an acid–base titration, base solution is added to the acid solution until the acid has been exactly neutralized. This is known as the **equivalence point** (or **stoichiometric point**) of the reaction.

If we are using indicators in an acid–base titration, we are making the assumption that the **end point** of the indicator (the point at which it appears to suddenly change colour) is the same as the equivalence point. In order to look at this assumption further, we look at the way that the pH of a reaction mixture changes during an acid–base titration.

Titration of a strong acid and a strong base (SA–SB)

Suppose that we slowly add 0.100 mol dm^{-3} NaOH to 25.00 cm^3 of 0.100 mol dm^{-3} HCl, and use a pH meter to record the pH of the mixture after each addition of base. The reaction is

$$HCl(aq) + NaOH(aq) \rightarrow NaCl(aq) + H_2O(l)$$

Table 16.5 pH changes during the titration of 0.100 mol dm^{-3} NaOH with 25.00 cm^3 0.100 mol dm^{-3} HCl

Vol of added NaOH (cm^3)	pH
0.00	1.00
5.00	1.18
10.00	1.37
15.00	1.60
20.00	1.95
24.00	2.69
24.90	3.70
24.95	4.00
25.00	7.00
25.05	10.00
25.10	10.30
25.50	11.00
26.00	11.29
30.00	11.96
35.00	12.22
40.00	12.36
45.00	12.46
50.00	12.52

Table 16.5 shows the changes in pH during the titration. The initial pH of the acid is 1.00, and the pH of the mixture at the equivalence point (after addition of 25.00 cm^3 of NaOH, when only pure water and sodium chloride are present) is 7.00. When this data is plotted, Fig. 16.3(a) is obtained – virtually identical curves are obtained using any pair of strong acids and strong bases. Upon addition of NaOH, the pH rises slowly and, suddenly, shoots up near the equivalence point. After this, the pH slowly rises as the amount of excess NaOH increases.

Points to note:

- The very rapid change in pH near the equivalence point has a simple explanation. We have already noted (page 298) that the addition of even one drop of acid to a neutral solution causes a massive change in pH, and the same applies to the addition of base to a neutral solution. This is exactly what happens during the titration *just before* and *just after* the equivalence point. This is seen from the highlighted data in Table 16.5, where the addition of only 0.10 cm^3 of base (about 2 drops) causes the pH to change from 4 to 10.

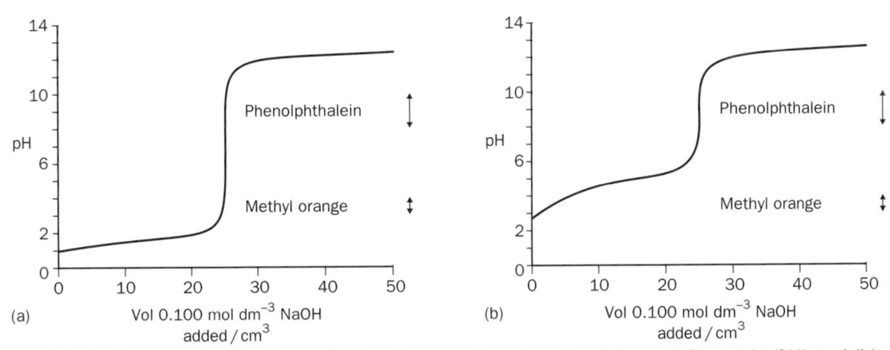

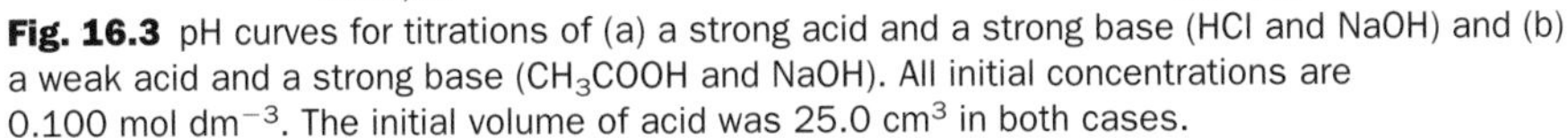

Fig. 16.3 pH curves for titrations of (a) a strong acid and a strong base (HCl and NaOH) and (b) a weak acid and a strong base (CH_3COOH and NaOH). All initial concentrations are 0.100 mol dm^{-3}. The initial volume of acid was 25.0 cm^3 in both cases.

- The approximate pH ranges of phenolphthalein (8–10) and methyl orange (3–4) are also shown on Fig. 16.3(a). Methyl orange begins to change colour fractionally before, and the phenolphthalein very fractionally after, the equivalence point. In practice, because the pH changes so rapidly near the equivalence point, the colour changes for phenolphthalein will start and finish over the addition of less than one drop ($<0.05\,cm^3$) of added base. The error involved in assuming that the indicated end point of phenolphthalein is the same as the equivalence point is negligible. The use of methyl orange will produce a slightly greater titration error with an endpoint at about $24.95\,cm^3$.

Titration curve for a weak acid and a strong base (WA–SB)

Figure 16.3(b) shows the pH curve for the titration of a strong base (NaOH) against a weak acid (ethanoic acid):

$$CH_3COOH(aq) + NaOH(aq) \rightarrow CH_3COONa(aq) + H_2O(l)$$

Similar pH curves are obtained with all titrations involving weak acids and strong bases. Beyond the equivalence point ($25.00\,cm^3$ of NaOH), the CH_3COOH–NaOH pH curve is identical to the HCl–NaOH curve, but before the equivalence point the curve differs from Fig. 16.3(a) in two important respects:

- The pH at the equivalence point is 8.7 (not 7) because the salt, sodium ethanoate, hydrolyses in solution.
- The pH of the acid at the beginning of the titration is higher than for the HCl–NaOH case because $CH_3COOH(aq)$ is a weak acid and is only partially ionized. A mixture of ethanoic acid and sodium ethanoate is also a buffer solution. This has the effect of shortening the vertical part of the graph in the region of the equivalence point. This, in turn, means that *methyl orange is unsuitable for use in this titration, since it changes colour well before the equivalence point.* However, phenolphthalein may still be used.

Exercise 16N

Strong acid–strong base titration

(i) Which of the indicators listed in Table 16.4 are unsuitable for use in a strong acid–strong base titration?

(ii) Which of the following pairs will give a pH curve virtually identical to Fig. 16.3(a):

(a) HNO_3–KOH, **(b)** HBr–NaOH **(c)** HCl–NH_3?

Exercise 16O

Weak acid – strong base titration

(i) Citric acid, a weak acid, is a constituent of fizzy drinks. Would you use bromocresol green indicator in titrations involving citric acid and potassium hydroxide?

(ii) Refer to the NaOH–CH_3COOH titration shown in Fig. 16.3(b). What simplified form does the Henderson–Hasselbalch equation take when $12.50\,cm^3$ of NaOH has been added? Use this relationship to estimate $K_a(CH_3COOH)$ from Fig. 16.3(b).

16.7 Buffering action of carbon dioxide in water

Carbon dioxide in water

Carbon dioxide is fairly soluble in water, and a small amount of the dissolved CO_2 reacts with water as follows:

$$CO_2(aq) + 2H_2O(l) \rightleftharpoons \underset{\text{hydrogencarbonate ion}}{HCO_3^-(aq)} + H_3O^+(aq) \quad \textbf{(16.12)}$$

where $K_{c(25\,°C)} = 4.3 \times 10^{-7}\,\text{mol dm}^{-3}$. The production of hydronium ions causes water in contact with air (for example rain water, tap water and laboratory deionized water) to be *slightly acidic*, and experiments show that such water possesses (at 25 °C) a pH of about 5.6.

However, many natural waters (lakes, ponds and reservoirs) are basic due to the presence of dissolved solids. For example, water saturated with CO_2 and limestone (calcium carbonate) at 25 °C contains a higher concentration of hydrogencarbonate ion than water saturated with CO_2 alone, and is *slightly basic* with a pH of 8.3.

If acid is added to a pond, the added H_3O^+ ions causes the equilibrium concentrations of water and carbon dioxide to increase. The CO_2 then bubbles off into the air. Without such buffering, natural waters would be easily acidified by acid rain. However, if too much acid is present, the buffer is overwhelmed and the pH of the water does change.

Buffers in the body

The CO_2–water equilibrium is also the main buffer present in blood. Excess $H_3O^+(aq)$ in blood, due to the presence of lactic acid produced by the metabolism of food, is removed by the reaction of the back reaction of equation (16.12). In this way, the pH of blood is maintained at 7.4. Variations in arterial blood pH of more than about 0.4 can be fatal.

The hydrogencarbonate buffer in blood would soon become exhausted if there was not a way for the body to get rid of waste. Excess carbon dioxide and excess acid are removed by the lungs and kidneys, respectively. In this way, the hydrogencarbonate ions in the blood are released to act as a buffer once again.

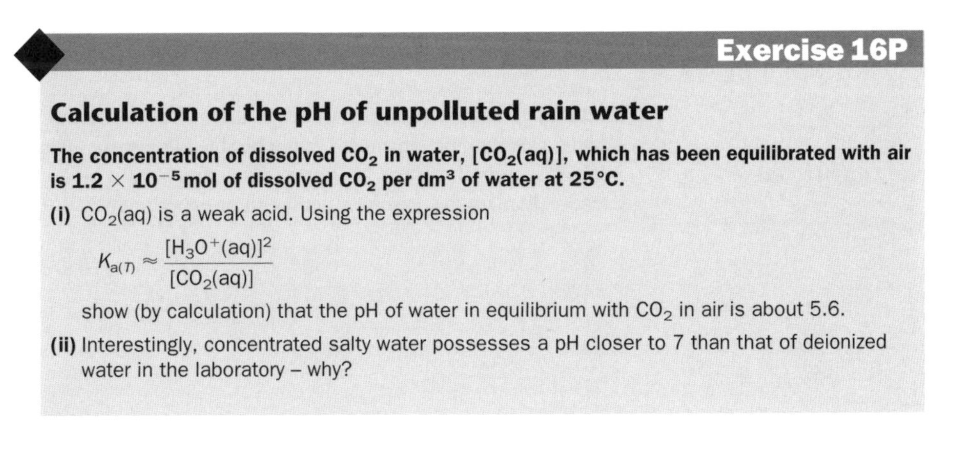

Exercise 16P

Calculation of the pH of unpolluted rain water

The concentration of dissolved CO_2 in water, $[CO_2(aq)]$, which has been equilibrated with air is 1.2×10^{-5} mol of dissolved CO_2 per dm^3 of water at 25 °C.

(i) $CO_2(aq)$ is a weak acid. Using the expression

$$K_{a(T)} \approx \frac{[H_3O^+(aq)]^2}{[CO_2(aq)]}$$

show (by calculation) that the pH of water in equilibrium with CO_2 in air is about 5.6.

(ii) Interestingly, concentrated salty water possesses a pH closer to 7 than that of deionized water in the laboratory – why?

Revision questions

16.1. Label the following as true (T) or false (F):

(i) The pH range of an indicator is the band of pH in which the indicator is coloured.

(ii) If methyl orange shows a yellow colour, this indicates that the solution is basic.

(iii) The pH of a solution of hydrochloric acid depends upon its concentration.

(iv) The acidity constant of an acid is the equilibrium constant for its ionization in solution.

(v) 25 cm^3 of 0.02 mol dm^{-3} H_2SO_4(aq) requires 50 cm^3 of 0.02 mol dm^{-3} NaOH for neutralization.

16.2. Calculate the pH at 25 °C of **(i)** 0.080 mol dm^{-3} HNO_3, **(ii)** 0.080 mol dm^{-3} KOH

16.3. **(i)** Calculate the pH of water at 100 °C.

(ii) 11.0 cm^3 of 0.040 mol dm^{-3} sodium hydroxide and 9.0 cm^3 of 0.040 mol dm^{-3} HCl are mixed in a beaker. Calculate the pH of the solution at 25 °C.

16.4. **(i)** Write down an equation showing the ionization of the (weak base) morphine ($C_{17}H_{19}O_3N$) in water. Calculate the pOH and pH of an aqueous solution containing 1.0 mg of morphine per dm^3 of water at 25 °C (K_b(morphine) = 1.6×10^{-6} mol dm^{-3}; ignore dissolved CO_2).

(ii) What volume of 0.0010 mol dm^{-3} HCl would be required to exactly neutralize 100 cm^3 of the morphine solution? Write an equation for the neutralization reaction in-volved.

16.5. The ionization of phenol (a weak acid) in water is represented by the equation:

$$C_6H_5OH(aq) + H_2O(l) \rightleftharpoons C_6H_5O^-(aq) + H_3O^+(aq)$$

Estimate the pH at 25 °C of a solution of phenol of concentration 5.00×10^{-3} mol dm^{-3} (ignore dissolved CO_2).

16.6. The reaction of ammonia with water is represented by the equation

$$NH_3(aq) + H_2O(l) \rightleftharpoons NH_4^+(aq) + OH^-(aq)$$

Show, by substitution, that the **acidity constant** of the conjugate acid NH_4^+(aq), symbolized $K_a(NH_4^+(aq))$, is related to $K_b(NH_3(aq))$ by the expression

$$K_a(NH_4{}^+(aq)) \times K_b(NH_3(aq)) = K_w$$

Calculate $K_a(NH_4^+(aq))$ at 25 °C ($K_b(NH_3(aq)$ = 1.8×10^{-5} mol dm^{-3}).

16.7. **(i)** Solutions of carbon dioxide are weakly acidic:

$$CO_2(aq) + 2H_2O(l) \rightleftharpoons HCO_3^-(aq) + H_3O^+(aq)$$

Derive the following expression for a solution containing dissolved CO_2(g):

$$K_{a(T)} \approx \frac{[H_3O^+(aq)]^2}{[CO_2(aq)]}$$

(ii) Carbonated water (provided by home fizzy drink dispensers) consists of water saturated with CO_2. Water *saturated* with CO_2 at 20 °C contains 4.0×10^{-2} mol of dissolved gas per dm^3 of solution. Calculate the pH of carbonated water at this temperature.

16.8. Predict whether solutions of the following salts will be basic, acidic or neutral: **(i)** phenylamine hydrochloride ($C_6H_5NH_4^+$,Cl^-), **(ii)** ammonium ethanoate, **(iii)** ammonium nitrate, **(iv)** potassium bromide, **(v)** sodium carbonate.

16.9. 0.100 g of benzoic acid, C_6H_5COOH, was added to 0.080 g of sodium benzoate salt (C_6H_5COONa), the mixture was dissolved in water and made up to 100 cm^3. Estimate the pH of the resulting buffer, and use equations to explain how the buffer responds upon **(i)** the addition of acid and **(ii)** the addition of alkali.

16.10. The following figure shows the way the pH changes when 0.100 mol dm^{-3} NH_3(aq), a weak base, is titrated with 25.00 cm^3 of 0.100 mol dm^{-3} HCl(aq) at room temperature. Briefly discuss the suitability of the indicators listed in Table 16.4 for this titration.

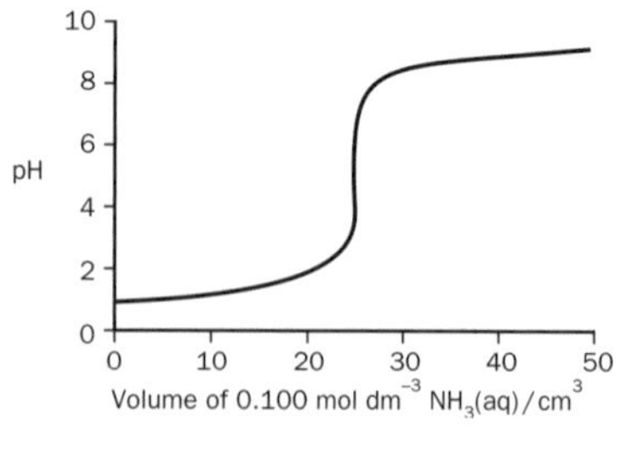

16.11. Scientists have carried out experiments to see whether athletes (for example, sprinters) might be able to improve their performance by ingesting sodium hydrogencarbonate before exercise. Explain the reasoning behind these experiments.

Organic Chemistry: Hydrocarbons

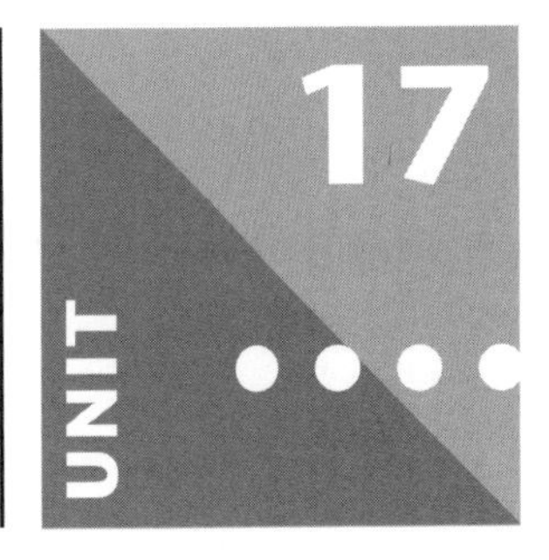

Objectives

- ▶ Explains what is meant by 'organic chemistry'
- ▶ Introduces you to some families of hydrocarbons
- ▶ Describes the sources of hydrocarbons and that hydrocarbons form the starting materials for many synthetic organic compounds
- ▶ Distinguishes between 'aliphatic' and 'aromatic' compounds

Contents

One element in the periodic table, carbon, forms so many compounds that it has an entire branch of chemistry devoted to it. Most of the chemistry of this element comes under the heading **organic** chemistry. The reason that carbon forms such a huge number of compounds is that it has the ability to form chains of carbon atoms (or the ability to **catenate**) – the carbon atoms are covalently bonded to one another and to other elements, such as hydrogen, oxygen and nitrogen.

The name 'organic' originates from the days when chemical compounds were divided into two classes, depending on their origin: 'inorganic' and 'organic'. Inorganic compounds were obtained from mineral sources, whereas organic chemicals were derived from living things. It was believed that organic compounds contained a 'vital force' and could not be made from inorganic compounds, until F. Wöhler (1800–1882) made the organic compound urea, $CO(NH_2)_2$, from the inorganic salt ammonium cyanate, NH_4CNO. Urea is formed as a waste product when proteins are metabolized. Today, organic chemistry refers to most of the chemistry of carbon compounds. Some exceptions include the chemistry of metal carbonates, carbon dioxide and carbon monoxide, which are included within inorganic chemistry.

Organic compounds can be organized into families of compounds with similar structural formulae and similar properties. This makes the study of organic chemistry easier, since there is such a large number of individual organic compounds.

17.1 Alkanes

The first family we shall look at is called the **alkanes**. They were originally called the paraffins (which comes from the Latin 'little affinity') because, as you will see, apart

Table 17.1 The first four alkanes

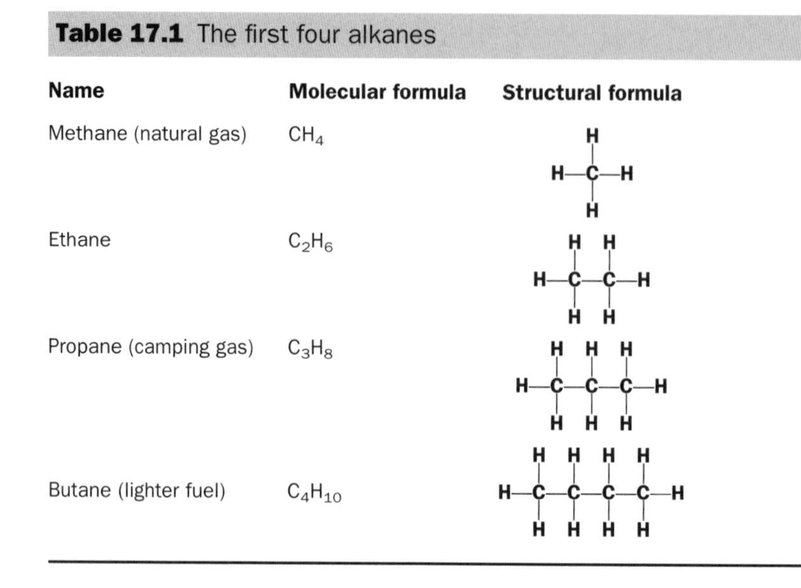

Name	Molecular formula	Structural formula
Methane (natural gas)	CH_4	
Ethane	C_2H_6	
Propane (camping gas)	C_3H_8	
Butane (lighter fuel)	C_4H_{10}	

from combustion reactions they are not particularly chemically reactive. They are found in crude oil and natural gas and are used extensively as fuels, or as starting materials for a wide variety of organic compounds. The first four members of the alkane family, which are all gases at 25 °C, are shown in Table 17.1. The alkanes belong to a larger classification of compounds, the **hydrocarbons**. Hydrocarbons are organic compounds that contain the elements carbon and hydrogen *only*. The alkanes have the general formula C_nH_{2n+2} where n, the number of carbon atoms, is 1, 2, 3, 4, 5 and so on. A family of compounds whose formulae differ by $-CH_2$ is called a **homologous series.**

After butane, the names of higher alkanes are taken from the Latin or Greek prefix for the number of carbon atoms in one molecule of the alkane:

Name	Number of carbon atoms
Pentane	5
Hexane	6
Heptane	7
Octane	8
Nonane	9
Decane	10
Undecane	11
Dodecane	12

Exercise 17A

Alkanes

(i) The alkanes are covalent compounds. Write Lewis structures for ethane and propane.

(ii) Write the molecular and structural formulae for octane.

(iii) What is the molecular formula for dodecane, which has 12 carbon atoms?

Writing the structural formulae of the higher alkanes can take up a great deal of space, so we often write **condensed structural formulae**. For example, the structural formula of pentane can be written in four different ways. These are shown below:

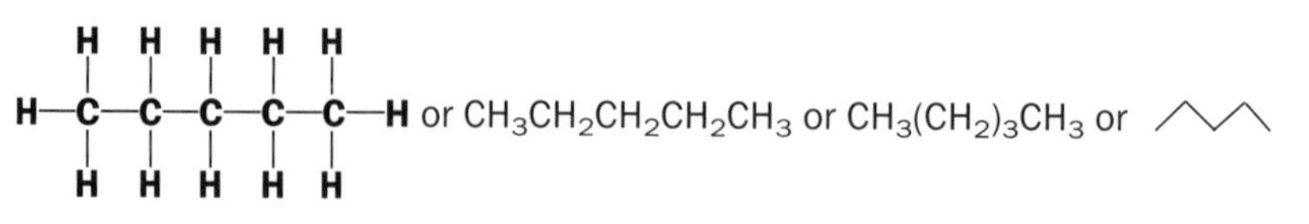

The last is a stick formula, where each end represents a carbon joined to three hydrogens and each point a carbon joined to two hydrogens.

Shapes of alkane molecules

Although we often draw the methane molecule as being flat, the shape of the molecule is actually *tetrahedral*, as drawn in Fig. 17.1.

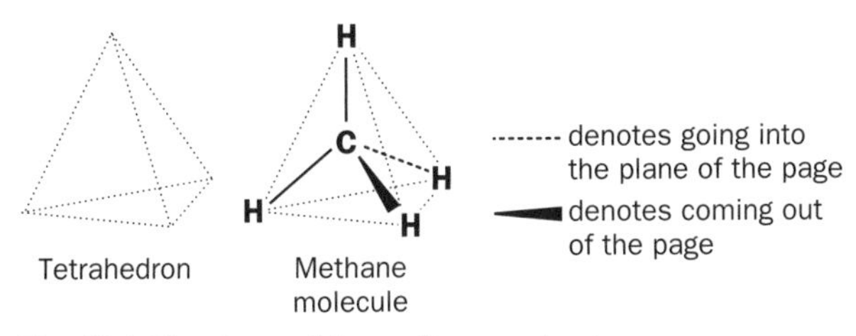

Fig. 17.1 The shape of the methane molecule.

Space-filling model of methane.

Exercise 17B

Shapes of the alkanes

Draw the three-dimensional structure of ethane (C_2H_6).

Isomers

There are two possible structures for C_4H_{10}:

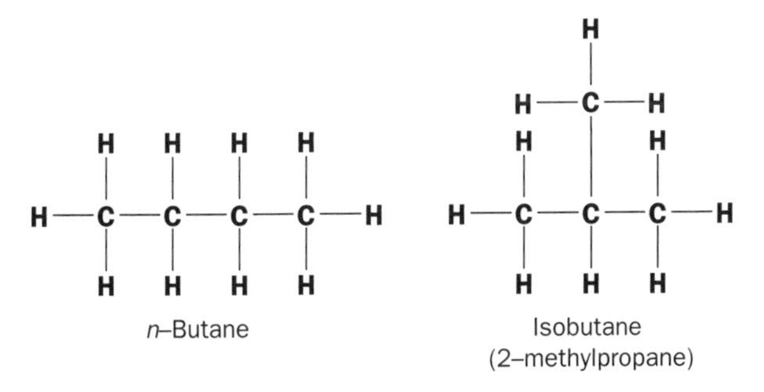

They have a different structure, although they both have the same molecular formula. One is called 'normal' butane, abbreviated to *n*-butane; the other may be called isobutane or 2-methylpropane – the latter name describes the structure of the compound, as you will see later. These compounds are **isomers** of one another. They are *not the same compounds* and have different melting points, boiling points and solubilities. **Isomers are compounds which have the same molecular formula, but different molecular structures.** After butane, the longer the carbon chain of an alkane, the more **structural isomers** are possible for a particular molecular formula. For example, there are 75 decanes ($C_{10}H_{22}$) and over three-hundred-thousand eicosanes ($C_{20}H_{42}$)!

Exercise 17C

Isomers of pentane

Draw the structures of the three isomers of pentane. It may help to make models of the isomers. If you do not have a molecular modelling kit, use straws to represent bonds and plasticine balls, of different colours, to represent atoms of hydrogen and carbon.

Naming alkanes

The rules for naming alkanes are as follows:

1. Identify the longest carbon chain.
2. Identify the 'branches' on the longest carbon chain and name them according to the number of carbon atoms they contain, together with the ending 'yl': methyl (1), ethyl (2), propyl (3), butyl (4) and so on. Note that these groups are called **alkyl** groups and have the general symbol R–. For example, R–H could mean CH_3–H, CH_3CH_2–H or $CH_3CH_2CH_2$–H, etc.
3. Number the carbon atoms on the longest carbon chain to describe the positions of the branches – *use the lowest numbers possible.*
4. Write the branches in alphabetical order.
5. If there are more than one branch with the same name, use the prefixes di-, tri-, tetra- etc.

Example 17.1

Name the compound

$$\begin{array}{l} CH_3CH_2CH_2\underset{\displaystyle CH_3}{\underset{|}{C}}HCH_2CH_3 \end{array}$$

► Answer

1. Identify the longest carbon chain:

$$CH_3CH_2CH_2\underset{\displaystyle CH_3}{\underset{|}{C}}HCH_2CH_3 \qquad \text{hexane}$$

So, the compound is derived from hexane, the alkane with six carbon atoms.

2. Identify the branch:

$$CH_3CH_2CH_2\underset{\displaystyle CH_3 \leftarrow}{\underset{|}{C}}HCH_2CH_3 \qquad \text{methylhexane}$$

The chain has a branch with one carbon atom – a methyl group.

3. Number the position of the branch:
This can be done as

$$\begin{array}{l} 1\ \ 2\ \ 3\ \ 4\ \ 5\ \ 6 \\ CH_3CH_2CH_2\underset{\displaystyle CH_3}{\underset{|}{C}}HCH_2CH_3 \end{array}$$

or as

$$\begin{array}{l} 6\ \ 5\ \ 4\ \ 3\ \ 2\ \ 1 \\ CH_3CH_2CH_2\underset{\displaystyle CH_3}{\underset{|}{C}}HCH_2CH_3 \qquad \text{3-methylhexane} \end{array}$$

In the last structure, the methyl group is on the carbon atom with the lowest number. The name of the compound is therefore **3-methylhexane** *and not* **4-methylhexane**.

Example 17.2

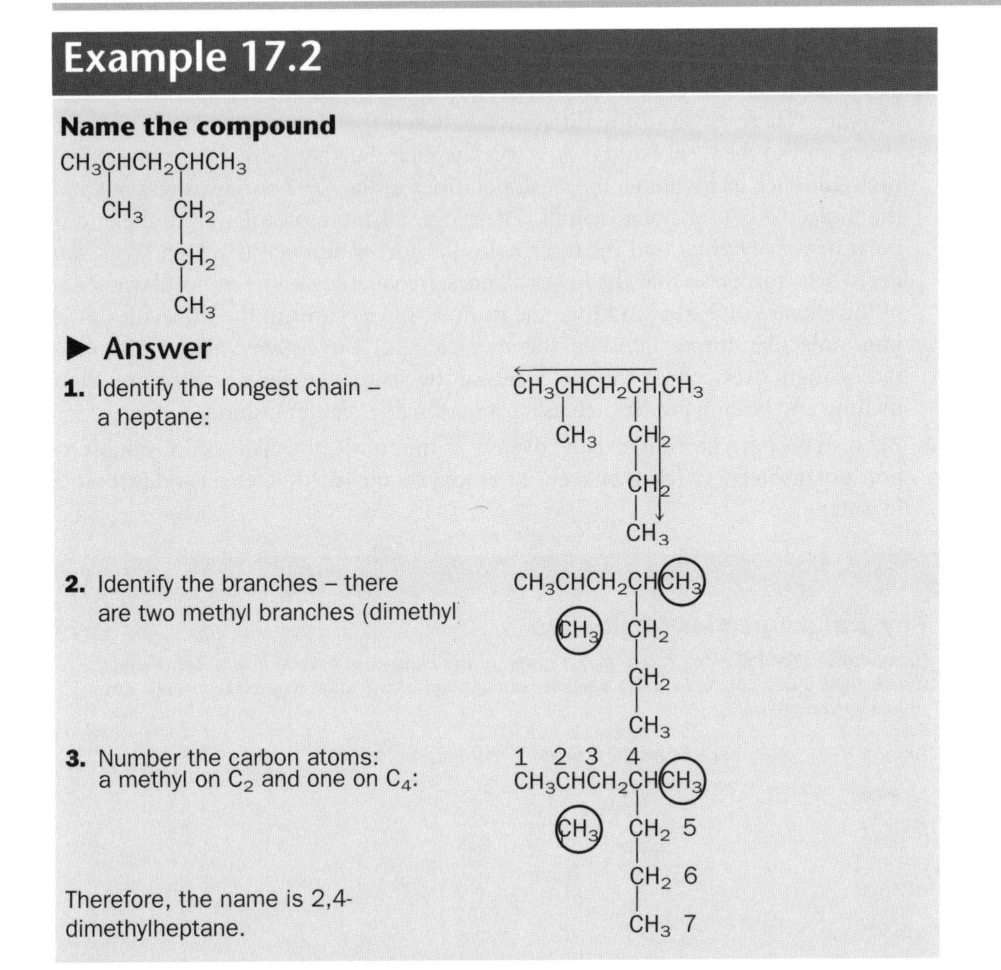

Therefore, the name is 2,4-dimethylheptane.

Exercise 17D

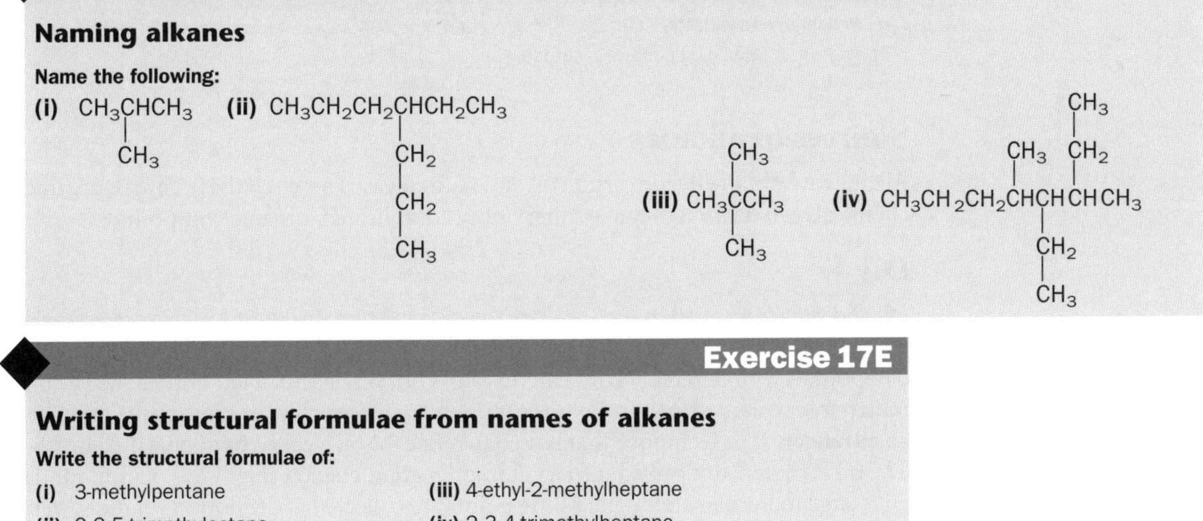

Exercise 17E

Writing structural formulae from names of alkanes

Write the structural formulae of:

(i) 3-methylpentane

(ii) 2,2,5-trimethyloctane

(iii) 4-ethyl-2-methylheptane

(iv) 2,3,4-trimethylheptane.

Physical properties of the alkanes

1. The alkane molecules contain carbon and hydrogen atoms covalently bonded together. There is very little difference between the electronegativities of these atoms and so the C–H bond is only weakly polar. Furthermore, in a symmetrical molecule such as methane, the weak polarities of the C–H bonds cancel out and the molecule is non-polar overall. Therefore, alkane molecules are either non-polar or weakly polar and the intermolecular forces between them are weak van der Waals' forces, so that the lower alkanes are volatile. As the molecular masses of the alkanes increase (and thus the number of electrons in the molecules), the intermolecular forces between the molecules of the heavier alkanes become increasingly stronger. As the molecules of the alkanes get larger, therefore, their melting and boiling points increase in a reasonably regular fashion.
2. Because they are non-polar and covalent compounds, the alkanes are soluble in non-polar solvents such as benzene or ethoxyethane (diethyl ether) and insoluble in water.

Exercise 17F

Physical properties of alkanes

Using data in the following table, plot a graph of the number of carbon atoms contained in the straight chain alkane (*x* axis) against boiling point of the alkane (*y* axis). Draw a curve to join up the points.

Alkane	Number of carbon atoms	Boiling point/K
Methane	1	109
Ethane	2	185
Propane	3	231
n-Butane	4	273
n-Pentane	5	309
n-Hexane	6	342
n-Heptane	7	372
n-Octane	8	***
n-Nonane	9	424
n-Decane	10	447

(i) Which alkane with the most number of carbon atoms is a gas at room temperature?
(ii) Estimate the boiling point of *n*-octane.

Sources of alkanes

Alkanes are obtained from crude oil and natural gas. These relatively simple organic chemicals can be used to make other, more complicated, organic compounds.

Oil

Oil and gas are trapped in rock and are the decayed remains of tiny marine plants and animals. Natural gas contains 60–90% methane. Crude oil is a mixture of many carbon compounds with different chain lengths and is not very useful if left untreated. The oil is therefore separated into groups of compounds with roughly the same chain length in an **oil refinery**. The technique that is used to refine the oil is called **fractional distillation**. The oil is heated and passed into a tall **fractionating column** (Fig. 17.2). Compounds with large molecules and high boiling points are collected at the bottom of the tower, whereas the more volatile components, with smaller molecules, are collected near

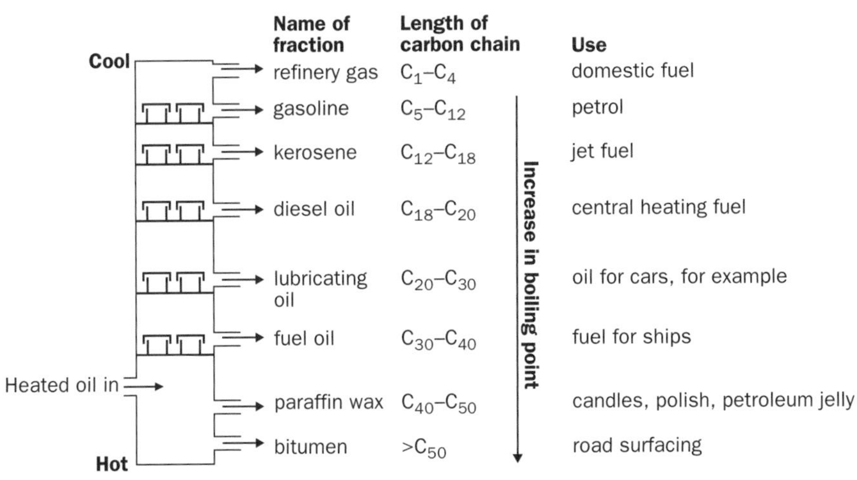

Fig. 17.2 A fractionating column.

the top. Each group of compounds collected is called a **fraction**. Each fraction is still a complicated mixture, since it contains alkanes of a range of carbon numbers and isomers of each carbon number. In addition to alkanes, crude oil contains some **cycloalkanes** and **aromatic compounds** – the proportion of each present depends upon the source of the oil. Aromatic compounds will be discussed later. Most synthetic chemicals produced by industry originate from the alkanes derived from crude oil: plastics, pharmaceuticals, adhesives and synthetic fibres are just a few examples.

Natural gas is used to produce hydrogen and carbon monoxide by reacting it with steam:

$$CH_4(g) + H_2O(g) \rightleftharpoons \underbrace{CO(g) + 3H_2(g)}_{\text{synthesis gas}}$$

The resulting gas mixture is called **synthesis gas**. The hydrogen from the mixture is used to manufacture ammonia in the Haber–Bosch process.

Coal

Coal is formed by the decomposition of plants under pressure. Some coals contain 2–6% sulfur. Burning these coals has led to pollution of the atmosphere and acid rain.

When coal is heated in the absence of air, a process called **destructive distillation**, it yields three products:

1. **coal gas** – mainly CH_4 and H_2 with some toxic carbon monoxide;
2. **coal tar** – this contains a wide range of organic materials which can be separated by distillation;
3. **coke** – a useful fuel and used in the making of steel.

At present, most organic chemicals are derived from petroleum – petroleum refining is cheaper and cleaner than refining coal. As the cost of petroleum rises and reserves fall, however, conversion of coal into useful organic chemicals becomes a more attractive prospect.

Cracking

The very long-chain alkanes are less useful than the smaller alkanes. For example, there is much more of a demand for petrol than for lubricating oil. Long-chain alkanes

BOX 17.1

Cycloalkanes

Molecules such as $CH_3CH_2CH_2CH_3$ have their carbon atoms connected in a chain. Carbon atoms can also be joined together in rings, in which case a **cyclic** molecule is formed. Cyclic molecules are often represented by polygons – the corners of the polygon represent a carbon atom together with the hydrogens joined to it. Some examples are:

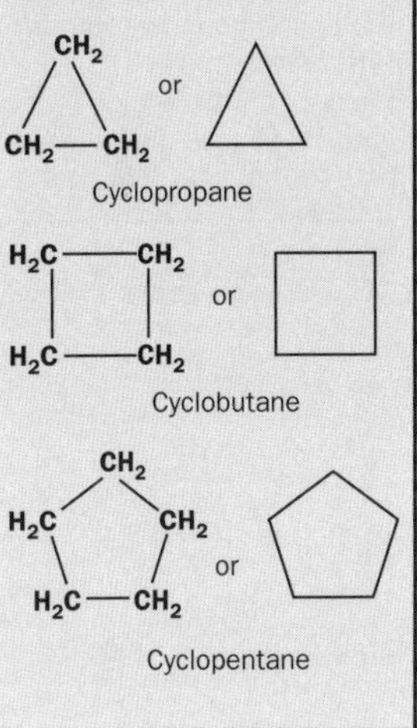

Exercise 17G

Cycloalkanes

(i) Can you write a general molecular formula for the cycloalkanes?

(ii) Cyclopentane is a much more stable and less reactive compound than cyclopropane. Can you give a reason why? (**Hint** Make models of both molecules.)

(iii) What are the molecular and structural formulae of *n*-hexane and cyclohexane?

BOX 17.2

Quality of petrol

Long-chain hydrocarbons tend to burn unevenly in car engines and so cause a rattling noise (knocking). Branched alkanes, produced from straight-chain alkanes by catalytic cracking, are added to the fuel to prevent knocking because they burn more evenly.

The alkane 2,2,4-trimethylpentane has good antiknock properties, whereas the antiknock properties of *n*-heptane are poor.

The **octane number** of petrol is a measure of its quality – an octane number of 0, for a particular petrol, means that it burns as well as pure heptane. If the octane number is 100, then the fuel is equivalent to pure 2,2,4-trimethylpentane. Many petrols have an octane number of 70; they burn as would a mixture of 70% 2,2,4-trimethylpentane and 30% heptane.

The burning of fuel in a car engine produces a variety of pollutants, such as carbon monoxide, unburned hydrocarbons and nitrogen oxides. Catalytic converters change these compounds into more environmentally friendly compounds. Although tetraethyllead ($Pb(C_2H_5)_4$) has been used as an additive which can be used to improve the antiknock properties of fuel (leaded petrol), it stops the platinum catalyst in a catalytic converter from functioning. Leaded petrol should therefore not be used in a car that contains a catalytic converter.

BOX 17.3

Putting a smell in dangerous gases

Crude propane is also known as LP (liquid petroleum) gas. It is generally used in rural areas that are not served by gas pipelines. LP contains about 80% propane. It can be stored as a liquid under normal temperatures, provided it is stored under pressure. The liquid propane occupies a much smaller volume than the gaseous propane. When pressurized tanks containing LP are opened, the gaseous fuel is released.

Propane is extremely flammable and has no taste, colour or smell. To warn users of an accidental leak, it is mixed with a strong-smelling substance such as **ethyl mercaptan**, C_2H_5SH (found in the aroma of skunks and rotting meat).

can be broken into smaller molecules, using a process called **cracking**. In **thermal cracking**, alkanes are heated to a high temperature. Smaller alkanes, hydrogen and small **alkenes**, such as **ethene**, are produced. Alkenes are members of another family of organic compounds, which are discussed later on in this unit; they are useful starting materials for the synthesis of many organic chemicals. **Catalytic cracking** involves heating higher boiling point petroleum fractions in the presence of a silica–alumina catalyst. Shorter alkanes which are suitable for use in petrol are made.

Chemical properties of alkanes

Alkanes are termed **saturated** because their molecules contain the maximum number of hydrogen atoms and other atoms or groups cannot 'add on' to alkane molecules.

1. Combustion

Alkanes are not particularly reactive; they do not readily react with dilute acids, alkalis, or oxidizing agents such as potassium permanganate and sodium dichromate. They do, however, burn in a plentiful supply of oxygen to form carbon dioxide and water:

$$CH_4(g) + 2O_2(g) \rightarrow CO_2(g) + 2H_2O(l)$$

The reaction is very exothermic, which is why alkanes are used as fuels. Another advantage of using alkanes as fuels is that 'clean' products are formed when they burn. The heavier alkanes, however, tend to burn in air with a sooty flame because there is insufficient oxygen in the air for all the carbon in the alkane to be converted to carbon dioxide (**complete combustion**), so some carbon and carbon monoxide are formed (**incomplete combustion**).

2. Halogenation

In ultraviolet light, chlorine or bromine react with alkanes to form chloroalkanes or bromoalkanes (the general term is **halogenoalkanes**). The reaction with halogens is called **halogenation** and, since it takes place in the presence of light, it is termed a **photochemical** reaction.

When methane, for example, reacts with chlorine gas in sunlight, chlorine atoms substitute for hydrogen atoms and **chloroalkanes** are formed:

$$CH_4(g) + Cl_2(g) \xrightarrow{h\nu} CH_3Cl(g) + HCl(g)$$

Depending on the amounts of methane and chlorine present, further substitution can occur to form CH_2Cl_2, $CHCl_3$ and eventually CCl_4. If propane is chlorinated, then the first substitution may be on an end carbon atom or a middle carbon atom so that the first product is a mixture of two possible **isomers**, namely

$CH_3CH_2CH_2Cl$ (1-chloropropane) or $CH_3CHClCH_3$ (2-chloropropane)

Because a mixture of products may be produced by the halogenation of alkanes, comprising different isomers and/or molecules with different degrees of substitution, the reactions are not really suitable for the laboratory preparation of halogenoalkanes. The process is more useful on an industrial scale where separation of individual compounds can be economically worthwhile.

BOX 17.4

Naming halogenoalkanes

The rules for naming halogenoalkanes are very similar to those for naming alkanes; the name of halogenoalkane is derived from the parent alkane. Remember, however, *to choose the longest carbon chain that also contains the halogen(s)* as a starting point.

The halogenoalkane is prefixed fluoro-, chloro-, bromo-, iodo-, dichloro-, tribromo- etc. Substituents are again arranged in alphabetical order:

$CHCl_3$
trichloromethane

$CH_3CH_2CHCH_2CHCH_3$ (with CH_3 on C4 and Br on C2)
2-bromo-4-methylhexane

Exercise 17H

Combustion of alkanes

The enthalpies of combustion for the first six straight-chain alkanes are given below:

Alkane	Formula	Enthalpy of combustion $\Delta H_c^\ominus$/kJ mol^{-1}
Methane	CH_4	−890
Ethane	C_2H_6	−1560
Propane	C_3H_8	−2220
Butane	C_4H_{10}	−2880
Pentane	C_5H_{12}	−3500
Hexane	C_6H_{14}	−4160

(i) Work out the difference in $\Delta H_c^\ominus$ between each successive alkane. Does $\Delta H_c^\ominus$ increase by a regular amount?

(ii) What is the average increase in $\Delta H_c^\ominus$ between successive alkanes?

(iii) The answer to **(ii)** represents the average enthalpy of combustion for what additional group of atoms?

(iv) Estimate the enthalpy of combustion for heptane, C_7H_{16}.

17.2 Alkenes

Another family of hydrocarbons is the **alkenes**. Alkene molecules contain a **double bond** between carbon atoms. They have the general formula C_nH_{2n} and their names end in 'ene' (Table 17.2). The first three are gases at 25 °C.

Shape of the ethene molecule

In ethene, each carbon atom lies at the centre of a triangle. The corners of each triangle are the two attached hydrogen atoms and the other carbon atom. The bond angles are 120°:

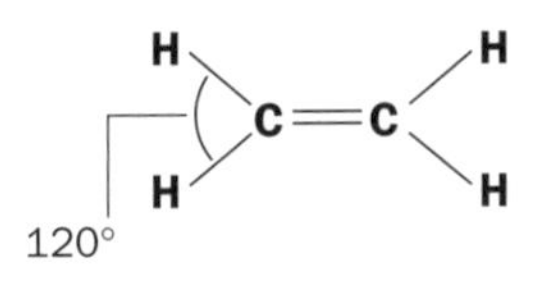

Exercise 17I

Ethene molecule

(i) Draw a Lewis structure to show the arrangement of the outer electrons of the atoms in an ethene molecule.

(ii) Make a model of the ethene molecule. Is there free rotation about the carbon–carbon double bond? How does this compare with ethane?

Table 17.2 The first three alkenes

Name	Molecular formula	Structural formula
Ethene	C_2H_4	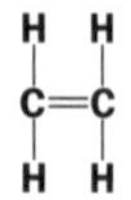
Propene	C_3H_6	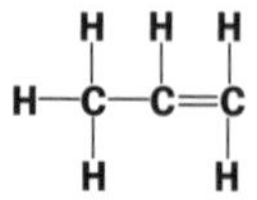
But-1-ene	C_4H_8	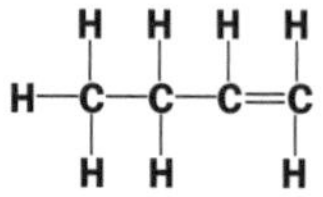

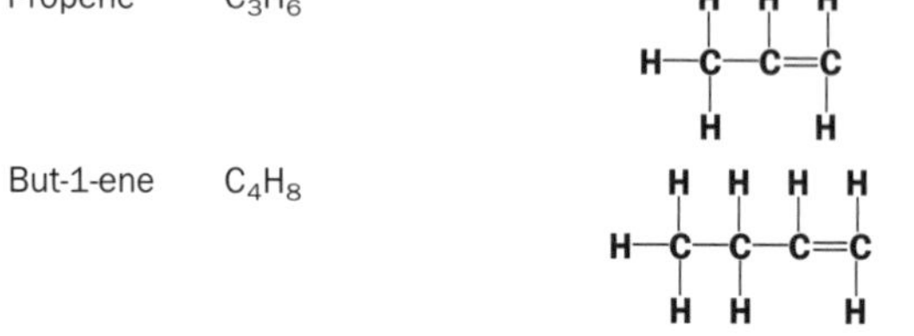

Naming alkenes

The following additional rules apply when naming alkenes:

1. Choose the longest chain of carbon atoms *that contains the double bond.*
2. Number the carbon atoms in the chain from the end nearest the double bond.
3. Pick the carbon atom with the lowest number to describe the position of the double bond.

Example 17.3

4 3 2 1

$CH_3CH_2CH{=}CH_2$ is called but-1-ene *not* but-2-ene

Example 17.4

CH_3

$CH_3CCH{=}CH_2$ is called 3,3-dimethylbut-1-ene

CH_3

Exercise 17J

Naming alkenes 1

Name the following:

(i) $CH_3CH_2CH_2CH_2CH{=}CH_2$

(ii) $CH_3CH{=}CHCH_3$

(iii) CH_3 / $CH_3CCH{=}CHCH_3$ / CH_3

(iv) $CH_2{=}CHCH_2Br$

Exercise 17K

Naming alkenes 2

Write structural formulae for:

(i) hex-3-ene

(ii) 2-methylpropene

(iii) 2,3-dimethylhex-2-ene

(iv) 3-chloro-2-methylpropene

(v) 3-bromohex-1-ene.

Isomerism in alkenes

There are a number of possible structural arrangements for a compound of formula C_4H_8:

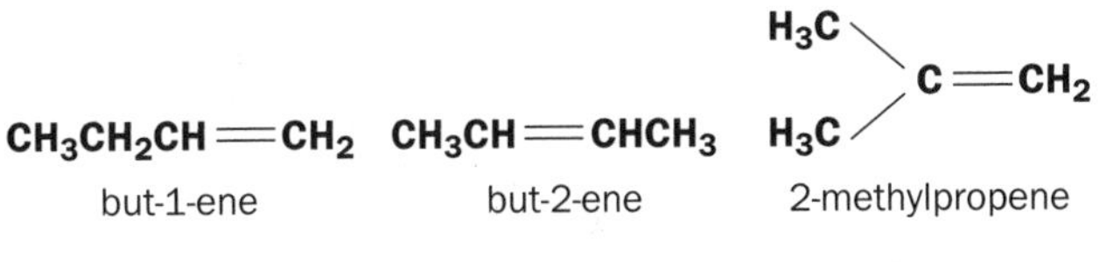

Experiments show that there are *four* isomers of C_4H_8. How can we account for the fourth isomer? In fact, there are two types of but-2-ene – they have different boiling points and melting points. In each butene, the atoms are arranged as follows:

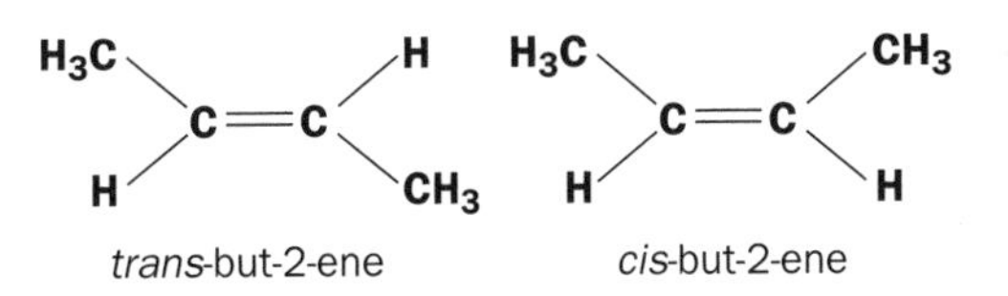

Both molecules are flat (planar). In *trans*-but-2-ene the methyl groups are on opposite sides of the double bond (*trans* means 'across'), whereas in *cis*-but-2-ene, the methyl groups are on the same side of the double bond (*cis* means 'on the side'). The two molecules are not the same because *no free rotation can occur around the carbon–carbon double bond.* These two butenes only differ from one another in the way that the atoms are orientated in space. They are **geometric** isomers.

BOX 17.5

More about isomerism

There are two major types of isomerism, *structural isomerism* and *stereoisomerism*.

In **structural isomerism**, first met in the discussion of alkanes and which exists between but-1-ene, but-2-ene and 2-methylpropene, the compounds have their atoms bonded to different groups of atoms. For example, consider the underlined carbon atoms in the structural isomers of butene (C_4H_8):

$CH_3CH_2\underline{C}H{=}CH_2$

$\underline{C}$ attached to: —H, —CH_2CH_3, =CH_2

$CH_3CH{=}\underline{C}HCH_3$

$\underline{C}$ attached to: —H, —CH_3, =$CHCH_3$

$CH_3\underline{C}(CH_3){=}CH_2$

$\underline{C}$ attached to: —CH_3, =CH_2

In **stereoisomerism** all atoms in the compounds are attached to the same partners, but the partners are arranged differently in space. Geometric isomerism, the isomerism shown by some alkenes because there is no free rotation about the double bond, is one type of isomerism in this class. The geometric isomers of 1,2-dichloroethene are shown below:

Cl(H)C=C(H)Cl — Cl and H on first carbon; H above, Cl below on second carbon

trans–1,2–dichloroethene

Cl(H)C=C(Cl)H — both Cl on the same side

cis–1,2–dichloroethene

Here, equivalent atoms in the two molecules are attached to the same atoms, but these atoms are arranged differently in space.

Another type of stereoisomerism is **optical isomerism**, which we will meet later.

Exercise 17L

Structural and geometric isomerism

Below are two isomers with the formula $C_2H_2Br_2$:

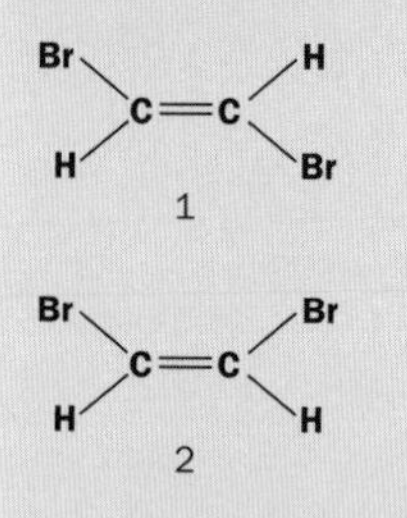

(i) A third isomer exists. Draw its structural formula and label it '3'.

(ii) What type of isomerism exists between the isomers 1 and 2?

(iii) What type of isomerism exists between isomers 2 and 3?

(iv) Name compounds 1, 2 and 3.

Chemical properties of alkenes

The physical properties of the alkenes are very similar to those of the alkanes. Chemically, however, they are much more reactive than the alkanes because they contain double bonds. Alkenes are said to be **unsaturated** hydrocarbons, because of the double bond, their molecules can add on more atoms of hydrogen. The unsaturation of alkenes gives rise to reactions with substances other than hydrogen. The chemical properties of alkenes are as follows:

1. Combustion

Like the alkanes, they burn in a plentiful supply of oxygen to form carbon dioxide and water:

$$C_2H_4(g) + 3O_2(g) \rightarrow 2CO_2(g) + 2H_2O(l)$$

2. Addition

Because alkenes are unsaturated, they will undergo reactions in which the carbon–carbon double bond becomes a single bond and other atoms bond to the carbons of the original double bond. A reaction in which two molecules combine to yield a single molecule of product is called an **addition reaction**. For ethene the general reaction may be written as

$$CH_2{=}CH_2 + X{-}Y \rightarrow \underset{X}{CH_2}{-}\underset{Y}{CH_2}$$

Some examples of the reagents 'X–Y' are given below:

Bromine (Br_2) Br–Br

Alkenes react rapidly with bromine at room temperature to form colourless dibromoalkanes. For example, if ethene gas is bubbled into **bromine water** (an aqueous solution of bromine) the colour of the solution turns from reddish-brown to colourless. This reaction is often used to test for unsaturation in a hydrocarbon:

$$CH_2{=}CH_2 + Br{-}Br \rightarrow \underset{Br}{CH_2}{-}\underset{Br}{CH_2}$$

Chlorine (Cl_2) Cl–Cl

A similar reaction occurs to that with bromine but without the distinctive colour change:

$$CH_2{=}CH_2 + Cl{-}Cl \rightarrow \underset{Cl}{CH_2}{-}\underset{Cl}{CH_2}$$

Hydrogen (H_2) H–H

The addition of hydrogen to an alkene is called hydrogenation and the product is an alkane. Hydrogenation is carried out in the presence of a catalyst (nickel, palladium or other platinum metals) and at a high pressure of hydrogen:

$$RCH{=}CH_2 + H{-}H \xrightarrow[500\,°C]{Ni} RCH(H){-}CH_2(H)$$

Hydrogen halides (HCl) H–Cl, (HBr) H–Br and (HI) H–I

The gaseous hydrogen halides react directly with the alkene to produce halogenoalkanes:

$$RCH{=}CH_2 + H{-}Cl \rightarrow RCH(Cl){-}CH_2(H)$$

Sulfuric acid (H_2SO_4) H–OSO_3H

Alkenes react with cold, concentrated sulfuric acid to form alkyl hydrogensulfates:

$$RCH{=}CH_2 + H{-}OSO_3H \rightarrow RCH(OSO_3H){-}CH_2(H)$$

3. Oxidation

Alkenes can be oxidized by cold, slightly alkaline potassium permanganate (manganate(VII)) to make substances called 1,2 diols, containing two OH groups on adjacent carbon atoms. The potassium permanganate solution turns from purple to brown. This is another test (the **Baeyer test**) that is often used to detect unsaturation in a hydrocarbon:

$$CH_2{=}CH_2 \xrightarrow{KMnO_4} CH_2(OH){-}CH_2(OH)$$

ethane-1,2-diol

4. Polymerization

Under the right conditions, alkenes will undergo addition reactions with each other. The reaction is called **polymerization**. For example, many ethene molecules will join together to form polythene – the double bonds break and the molecules link up:

$$n(CH_2{=}CH_2) \xrightarrow[\text{or catalyst}]{O_2,\ \text{heat, pressure}} \cdots CH_2{-}CH_2{-}CH_2{-}CH_2{-}CH_2{-}CH_2\cdots$$

where n = a large number (typically 20 000)

The product can also be written as

$$\cdots({-}CH_2{-}CH_2{-})_n\cdots$$

Long chain molecules called **polymers** are formed from the small ethene molecules, the **monomers**. The long, thin molecules of a polymer may also be called **linear macromolecules**. Polythene is an unreactive solid; it can be easily moulded and is used for plastic bags, bottles, washing-up bowls and plastic piping. The word 'plastic' means 'easily moulded'. Substituted ethene molecules can undergo the same addition reaction to give other polymers:

Styrene

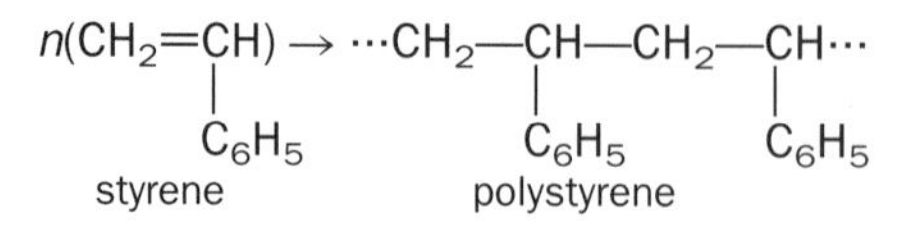

Vinyl chloride

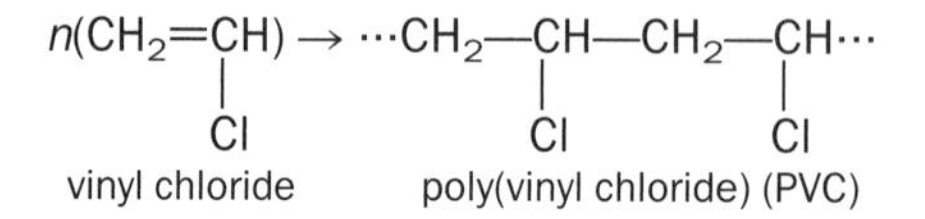

Note that the starting molecules (monomers) are named by their common, older names so that you can see how the names of the polymers arose.

BOX 17.6

Better model for bonding in alkenes

Although Lewis structures can be used to describe bonding in alkenes, they do not really explain why alkenes are so much more chemically reactive than alkanes. In order to explain the reactivity of alkenes, a more sophisticated model of bonding must be used.

In the Lewis structure model for ethene, two electron pairs are shared by the carbon atoms:

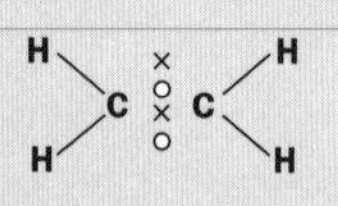

The Lewis structure suggests that both bonds in this double bond are of the same type. However, in our 'improved' model for the bonding in ethene, you will see that they are not the same.

First, we need to consider that the bonding electrons in the atoms concerned are in orbitals. Covalent bonds are formed by overlap of these orbitals. Each carbon in ethene forms three 'normal' covalent bonds by overlap with the orbitals containing electrons on the other carbon and two hydrogen atoms. The electron cloud between the nuclei of the atoms in each bond 'glues' the atoms together. Normal covalent bonds of this type are called σ (sigma) bonds (Fig. 17.3).

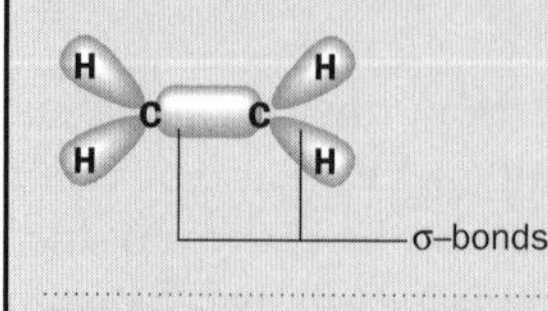

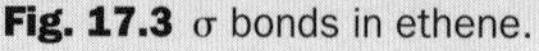

Fig. 17.3 σ bonds in ethene.

One electron is left over on each carbon atom. The 'left over' electrons are in *p* orbitals, as shown in Fig. 17.4.

Fig. 17.4 Formation of a π bond in ethene.

A π (pi) bond is formed by sideways overlap of these *p* orbitals on the carbon atoms. In this type of bond, the electron cloud is concentrated above and below the horizontal plane joining the two carbon atoms. The double bond between the carbon atoms in ethene therefore consists of a σ bond and a π bond. The exposed negatively charged electron cloud in the π bond is open to attack by positively charged species and this is why alkenes are so reactive.

Exercise 17M

Reactions of alkenes

(i) Write an equation for the reaction of ethene with hydrogen bromide. Name the product.

(ii) Write an equation for the reaction of ethene with cold, concentrated sulfuric acid. Name the product.

(iii) Write an equation for the reaction of propene with bromine. Name the product.

(iv) Write the structural formula of the product of the reaction of propene with cold, alkaline potassium permanganate.

(v) Write an equation for the formation of the polymer Orlon, formed by polymerization of the monomer $CH_2{=}CHCN$ (acrylonitrile).

17.3 Alkynes

The alkynes are a family of hydrocarbons that contain a carbon–carbon triple bond, with the general formula C_nH_{2n-2}. The first member of the family is ethyne (old name acetylene). Ethyne is a linear molecule:

$$H{-}C{\equiv}C{-}H$$

When it burns in oxygen, a great deal of heat is generated – **oxy-acetylene torches** are used for cutting and welding metals:

$$2C_2H_2(g) + 5O_2(g) \rightarrow 4CO_2(g) + 2H_2O(l)$$

The next member of the family, propyne, is obtained by replacing one of the hydrogen atoms in ethyne by a $-CH_3$ group:

$$CH_3{-}C{\equiv}C{-}H$$

Alkynes are unsaturated and undergo addition reactions in a similar fashion to alkenes. This time it is possible to obtain two addition products, depending on the number of moles of reagent reacting with the alkyne. If one mole of hydrogen, for example, is added to one mole of ethyne, the product is ethene:

$$H{-}C{\equiv}C{-}H(g) + H_2(g) \xrightarrow[150\,°C]{\text{Ni catalyst}} H_2C{=}CH_2(g)$$

Two moles of hydrogen produce ethane:

$$H{-}C{\equiv}C{-}H(g) + 2H_2(g) \xrightarrow[150\,°C]{\text{Ni catalyst}} H_3C{-}CH_3(g)$$

Exercise 17N

Alkynes

(i) Give the structural formula of but-1-yne, a third member of the family of alkynes.

(ii) What are the names and formulae of the two possible products formed when propyne is reacted with hydrogen in the presence of a nickel catalyst?

(iii) Write an equation for the reaction that occurs when ethyne is reacted with *excess* bromine. Name the product.

Exercise 170

Quantitative analysis of compounds

(i) A 7.02 mg sample of a hydrocarbon X gave 21.99 mg of carbon dioxide and 8.95 mg of water on combustion. Calculate the percentage composition of the elements in the hydrocarbon and determine its empirical formula.

(ii) X has molecular mass 42. X reacts with excess hydrogen in the presence of a nickel catalyst to form Y. X decolorises bromine water to form a compound Z. Write structural formulae for X, Y and Z.

BOX 17.7

Analysis of organic compounds

Often, one of the the first steps in the identification of an unknown organic compound is to submit the compound for **quantitative elemental analysis**. This type of analysis will determine the percentage by mass of the elements present in the compound. From the results of the analysis, the empirical formula of the compound can be calculated. To find out the relative amounts of carbon and hydrogen in a hydrocarbon, a weighed sample of the hydrocarbon is passed through a tube packed with copper(II) oxide at a temperature of about 700 °C. The copper(II) oxide oxidizes the carbon in the hydrocarbon to carbon dioxide and the hydrogen to steam:

$$\text{'CH}_2\text{' from organic compound} + 3\text{CuO} \rightarrow 3\text{Cu} + \text{CO}_2 + \text{H}_2\text{O}$$

The gases pass out of the combustion tube and through two weighed tubes containing first a drying agent, to absorb the water, and then a strong base (such as sodium hydroxide) to absorb the carbon dioxide formed. The increase in the mass of each tube gives the mass of each product from the combustion of the hydrocarbon.

For example, a sample of methane with a mass of 7.25 mg produced 19.90 mg carbon dioxide and 16.17 mg of water.

$$\text{Mass of carbon present} = \text{mass of carbon dioxide} \times \frac{M(\text{C})}{M(\text{CO}_2)}$$

$$= 19.90 \times \frac{12}{44} = 5.4 \text{ mg}$$

$$\text{Mass of hydrogen present} = \text{mass of water} \times \frac{M(2\text{H})}{M(\text{H}_2\text{O})}$$

$$= 16.17 \times \frac{2}{18} = 1.8 \text{ mg}$$

The percentage composition of the methane is therefore

$$\text{Carbon: } \frac{5.4}{7.25} \times 100 = 75\%; \quad \text{Hydrogen: } \frac{1.8}{7.25} \times 100 = 25\%$$

The empirical formula of an unknown compound can be worked out from its percentage composition using the calculations in Unit 8.

Note that it is possible, using various methods, to analyse for other elements in an organic compound, such as nitrogen, sulfur and halogens.

17.4 Aromatic hydrocarbons

The families of hydrocarbons studied so far, those with linear chains of carbon as the 'backbone' of the compounds, are classed as **aliphatic** hydrocarbons. There is another class of hydrocarbons – **aromatic** hydrocarbons. These are compounds that contain a benzene ring, or behave chemically like benzene. Benzene has the molecular formula C_6H_6, and its structural formula was first suggested by Kekulé in 1865:

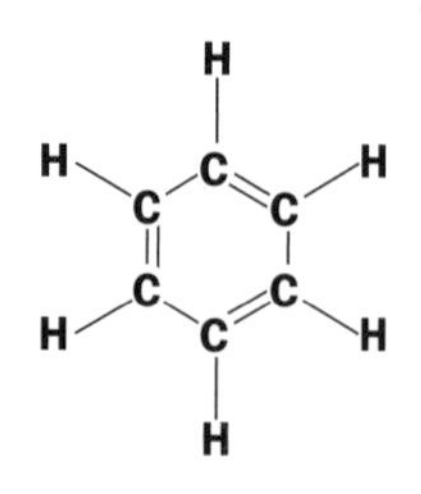

This was usually abbreviated to:

or

In fact, none of these structures correctly describes benzene. All the carbon-carbon bond lengths in benzene have been found to be the same, but the structures above do not predict this since double bonds are shorter than single bonds between the same atoms. Also, benzene is relatively unreactive towards addition, which we would not expect in a compound that contains three double bonds. These days the benzene ring is written as:

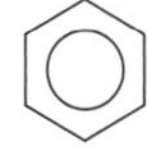

This structure represents the fact that electrons from the carbon–carbon double bonds are 'spread out' or **delocalized** over the whole molecule, which makes the molecule more stable. A Lewis structure is not sophisticated enough to describe the bonding in benzene. In order to obtain a more satisfactory description of the bonding, the electrons must be treated as 'clouds' (rather than 'crosses' or 'dots'). For a brief description of the bonding model, see Box 17.8.

BOX 17.8

Bonding in benzene

The observed properties of benzene can be explained using a more sophisticated model, than that proposed by Kekulé, for the bonding in benzene. Each carbon atom in the ring is bonded to a hydrogen and to one carbon on either side of it by σ bonds. Every carbon has one unused *p* orbital, containing a single electron, perpendicular to the plane. This gives a *planar* 'skeleton' as shown in Fig. 17.5.

The *p* orbitals overlap sideways with their neighbours to form π clouds or 'doughnuts' of electron density, above and below the molecule, as shown in Fig. 17.6. The electrons no longer 'belong' to the carbon atom from which they originate; they are free to 'wander' around the π clouds and are **delocalized**.

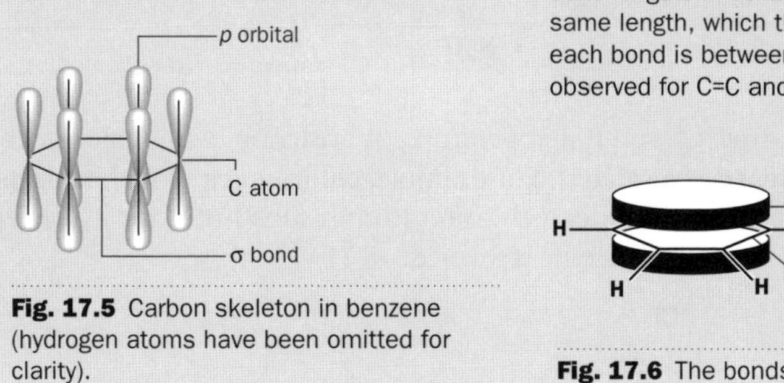

Fig. 17.5 Carbon skeleton in benzene (hydrogen atoms have been omitted for clarity).

Delocalization of the 'left over' *p* electrons is believed to make benzene more stable than would be expected if the Kekulé structures were correct, and explains why benzene does not react as if it had three double bonds. The model predicts that all the carbon–carbon bond lengths in benzene would be of the same length, which they are. The length of each bond is between those normally observed for C=C and C–C.

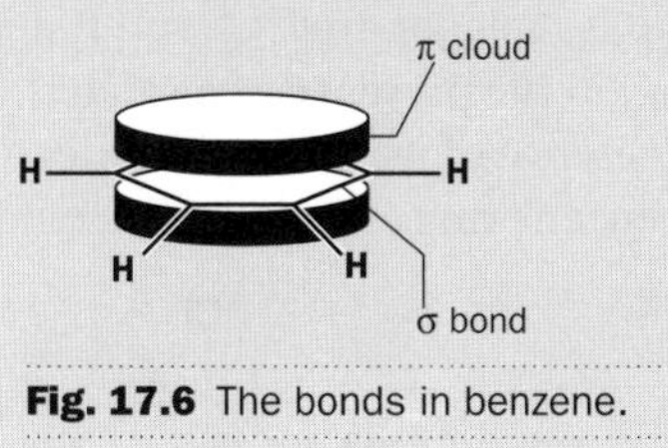

Fig. 17.6 The bonds in benzene.

Note that 'Ar–' stands for **aryl** and is used to represent an aromatic group, including a benzene ring. For example, Ar–Cl could mean:

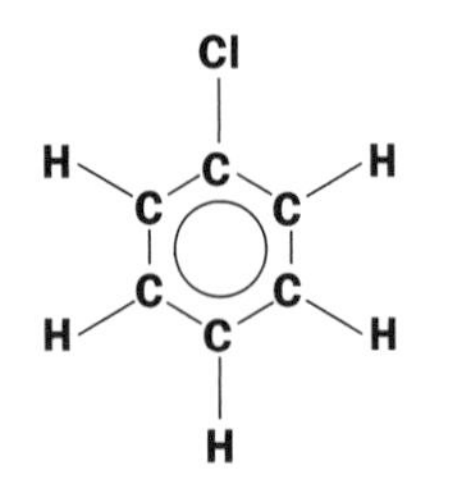

The symbol 'Ph–' stands specifically for the **phenyl** group (C_6H_5–), so the compound above might also be written C_6H_5–Cl or Ph–Cl.

Physical properties of benzene

Benzene is a colourless liquid which is insoluble in water. It was widely used as a solvent, but now this use is limited since benzene has been found to be toxic – inhalation of its vapour can cause leukaemia. Toluene (see next section) is used as a safer alternative.

Chemical properties of benzene

1. Combustion

Benzene burns in air with a smoky, luminous flame. In addition to carbon dioxide and water, particles of carbon are formed because of incomplete combustion.

2. Substitution

Benzene undergoes **substitution reactions** rather than addition reactions (it does not behave chemically like the alkenes). In these reactions, hydrogens attached to the ring carbons are substituted by other atoms or groups. Some examples are shown below:

- Benzene can be **nitrated** by reaction with a mixture of concentrated sulfuric and nitric acids at 60 °C. Nitrobenzene, a pale yellow liquid is formed:

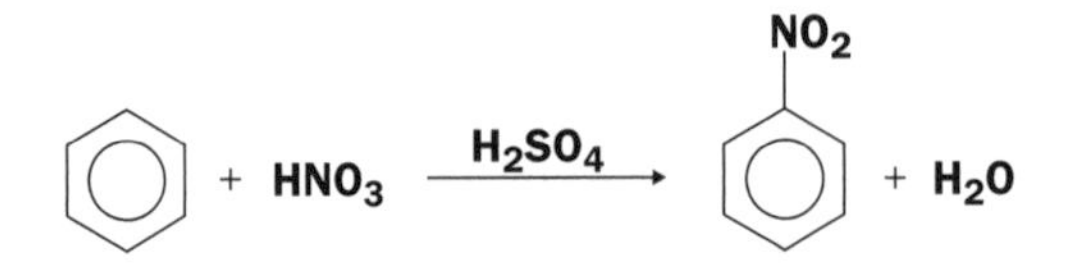

- Benzene is **halogenated** at room temperature by reaction with chlorine or bromine, in the presence of a catalyst. For example, chlorine reacts with benzene in the presence of iron(III) chloride. A chlorine atom is substituted for hydrogen in the benzene ring and chlorobenzene is formed:

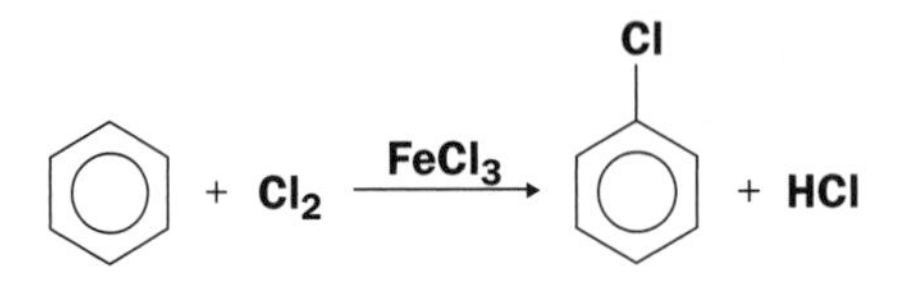

- Benzene is **alkylated** by reaction with an halogenoalkane, in the presence of a catalyst. For example, chloroethane reacts with benzene in the presence of the catalyst aluminium chloride to form ethylbenzene:

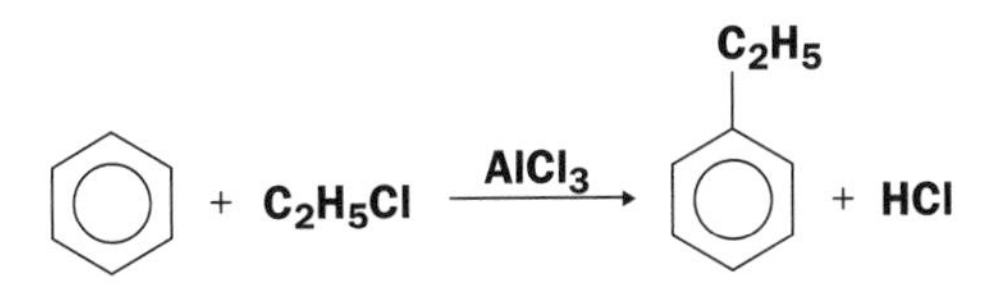

This type of reaction is often referred to a **Friedel–Crafts alkylation.**

- Benzene is **acylated** by reaction with an acid halide in the presence of a catalyst. An example of an acyl group is

RCO
|

where R is an alkyl group such as methyl, ethyl etc. The reaction of ethanoyl chloride with benzene in the presence of aluminium chloride gives phenylethanone:

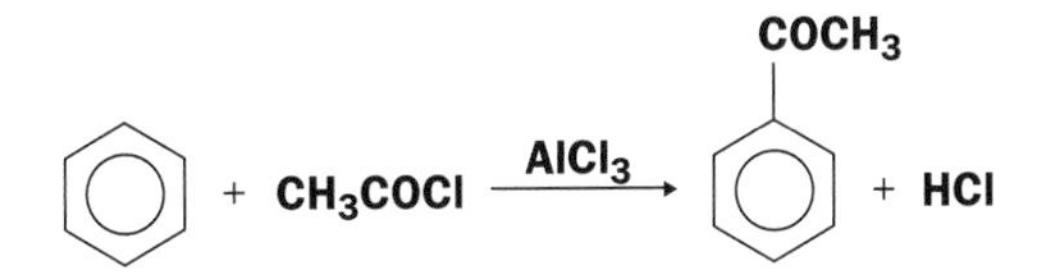

This type of reaction is called a **Friedel–Crafts acylation.**

3. Addition

Benzene does undergo some addition reactions:

- It can be hydrogenated to cyclohexane by reaction with hydrogen in the presence of a nickel catalyst, but at a higher temperature than that needed to hydrogenate simple alkenes.

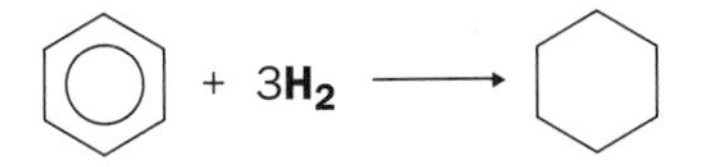

- In ultraviolet light, chlorine adds to benzene and eventually 1,2,3,4,5,6-hexachlorocyclohexane is formed. The numbers show that one chlorine atom is attached to each carbon atom of the ring.

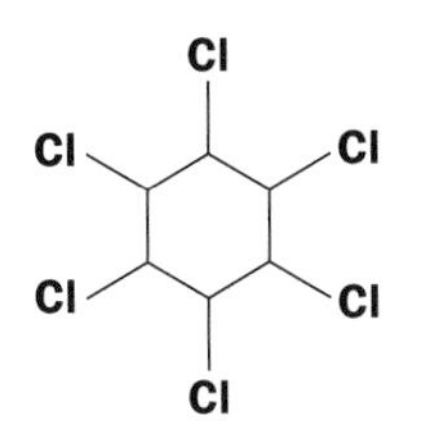

Sources of benzene

The main source of benzene is petroleum, although it can be obtained from coal tar.

Exercise 17P

Stability of benzene

The catalytic hydrogenation of cyclohexene, in the presence of a nickel catalyst, proceeds according to the equation

$$\text{cyclohexene} + H_2 \longrightarrow \text{cyclohexane}$$

$\Delta H = -119 \text{ kJ mol}^{-1}$

(i) If benzene had the Kekulé structure, write an equation that would represent the complete hydrogenation of benzene.

(ii) According to your equation, what would be the expected value of its standard enthalpy change of hydrogenation?

(iii) The experimental value of the enthalpy of hydrogenation of benzene to cyclohexane is -208 kJ mol^{-1}. Why is this value different from the one you have calculated?

Arenes

Hydrocarbons that contain both aromatic and aliphatic groups are called arenes. The structures of some arenes are

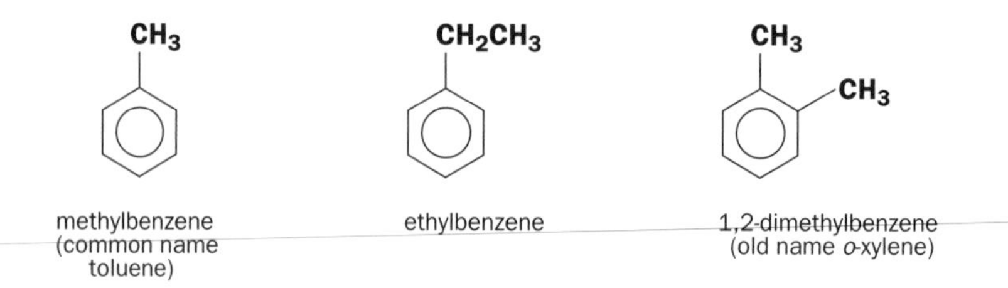

methylbenzene (common name toluene) | ethylbenzene | 1,2-dimethylbenzene (old name *o*-xylene)

Notice that there can be different structural isomers where more than one carbon atom in the ring is attached to separate groups. There are three possible xylenes, for example, the other two being

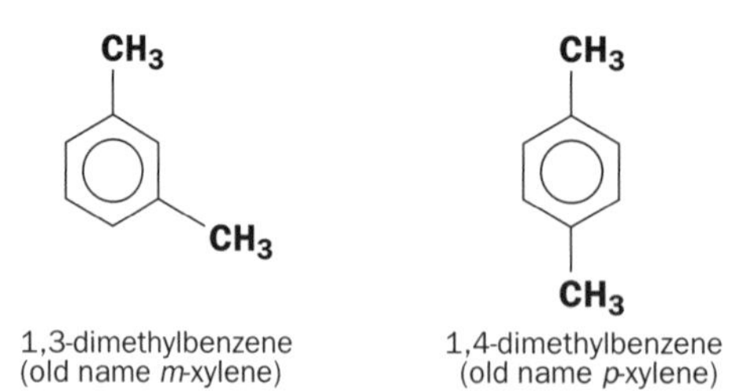

1,3-dimethylbenzene (old name *m*-xylene) | 1,4-dimethylbenzene (old name *p*-xylene)

Exercise 17Q

Arenes

Write down the structural formulae of all arenes with the molecular formula C_9H_{12} (there are eight).

Notice that they are named by numbering the different carbon atoms in the ring. The first carbon atom to be attached to an atom or group is given the number 1:

1 2 3 4 5 6

Fused ring aromatic compounds

Aromatic rings that share a pair of carbon atoms are said to be fused. Some examples are

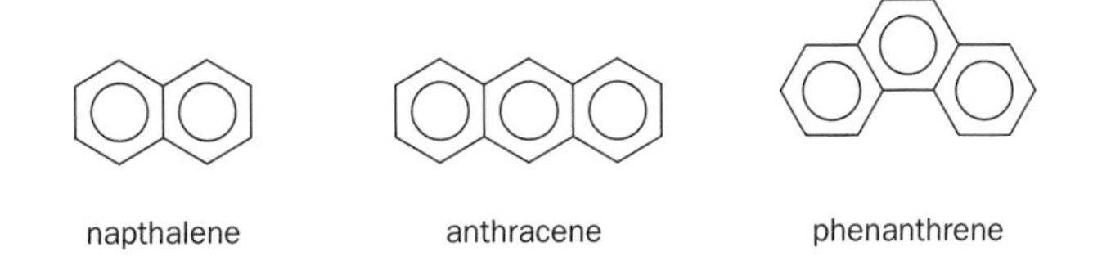

Napthalene has been used in mothballs. All these compounds are obtained from coal tar. They are also called **polyaromatic hydrocarbons** (PAHs). Polyaromatic hydrocarbons are found abundantly in the environment. They originate from many sources including cigarette smoke and car exhaust gases. There is cause for concern because some PAHs are considered to be toxic.

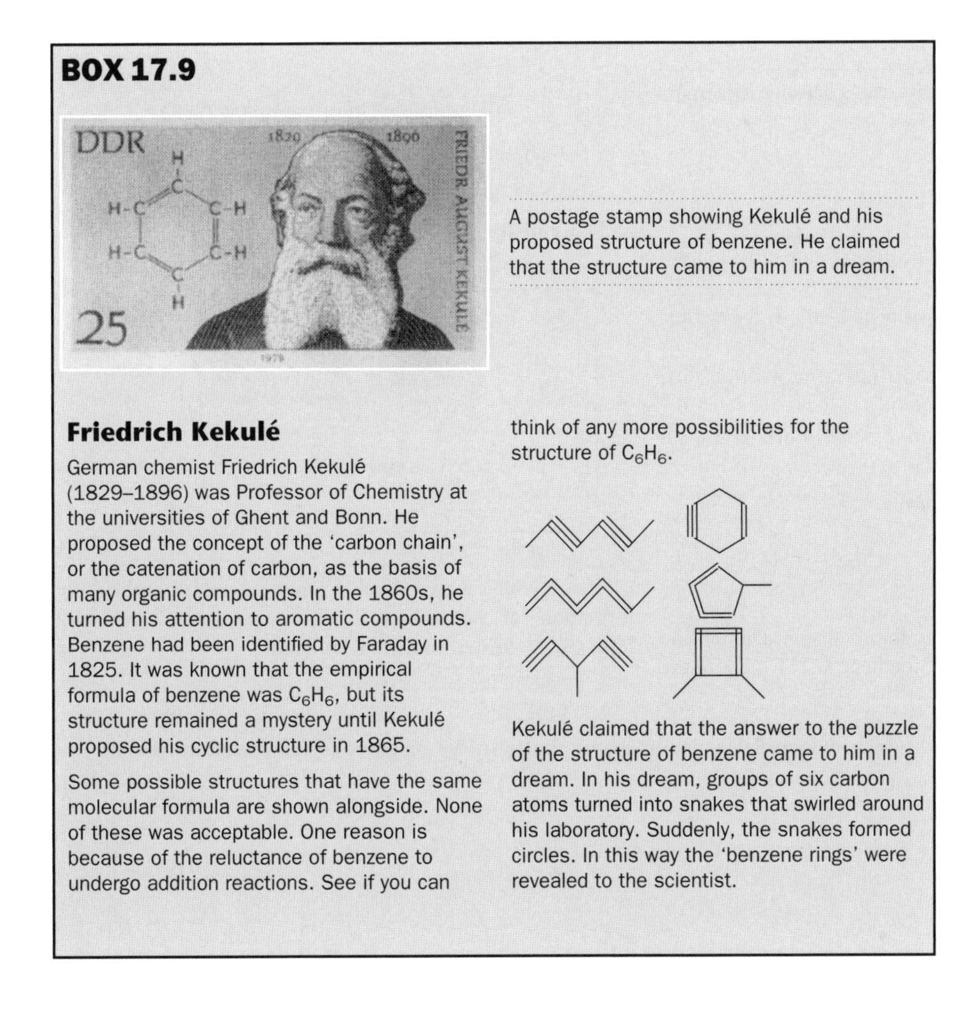

BOX 17.9

A postage stamp showing Kekulé and his proposed structure of benzene. He claimed that the structure came to him in a dream.

Friedrich Kekulé

German chemist Friedrich Kekulé (1829–1896) was Professor of Chemistry at the universities of Ghent and Bonn. He proposed the concept of the 'carbon chain', or the catenation of carbon, as the basis of many organic compounds. In the 1860s, he turned his attention to aromatic compounds. Benzene had been identified by Faraday in 1825. It was known that the empirical formula of benzene was C_6H_6, but its structure remained a mystery until Kekulé proposed his cyclic structure in 1865.

Some possible structures that have the same molecular formula are shown alongside. None of these was acceptable. One reason is because of the reluctance of benzene to undergo addition reactions. See if you can think of any more possibilities for the structure of C_6H_6.

Kekulé claimed that the answer to the puzzle of the structure of benzene came to him in a dream. In his dream, groups of six carbon atoms turned into snakes that swirled around his laboratory. Suddenly, the snakes formed circles. In this way the 'benzene rings' were revealed to the scientist.

Revision questions

17.1. Write the formula of *n*-hexane in three different ways.

17.2. Write the structural formulae of the five isomers of hexane and name them.

17.3. Name the following compounds:

(i) $CH_3CH{=}CHCH_2CH_3$

(ii) $(H_3C)_2C{=}CHCH_2CH(CH_3)_2$

17.4. Write structural formulae for:

(i) pent-1-ene
(ii) 5-chlorohex-2-ene
(iii) 4,4-dimethylpent-2-ene
(iv) buta-1,3-diene
(v) methylcyclopentane
(vi) 5-methyloct-3-yne.

17.5. Which of the following can exist as geometric isomers (*cis* and *trans* forms)?

(i) $(CH_3)_2C{=}CH_2$
(ii) $CH_3CH{=}CHCH_3$
(iii) $CH_3CH{=}CHCH_2CH_3$.

17.6. Write the structural formulae of, and name, the following compounds:

(i) The product of the reaction of pent-1-ene with bromine water.
(ii) The product of the reaction of but-1-ene with alkaline potassium permanganate solution.
(iii) The product of the reaction of *trans*-but-2-ene with hydrogen.

17.7. Give the structural formulae of:

(i) *trans*-1,2-dibromoethene
(ii) *trans*-1-chloroprop-1-ene
(iii) *cis*-hex-3-ene.

17.8. Write the structure of tetrafluoroethene. This alkene polymerizes to form polytetrafluoroethene (PTFE). Write the equation for the reaction. See if you can find what PTFE is used for in the home.

17.9. A compound is composed of 88.9% C and 11.1% H by mass.

(i) What is the empirical formula of the compound?
(ii) If the molecular mass of the compound is 54 u, what is the molecular formula of the compound?
(iii) Draw the structure of, and name, all possible isomers of the compound.

17.10. Benzopyrene, $C_{20}H_{12}$, is a fused ring aromatic compound. Write its structural formula.

17.11. Use the average bond enthalpies listed in Table 13.4 to calculate the standard enthalpy of formation of benzene – assume that the atoms are arranged according to the Kekulé structure

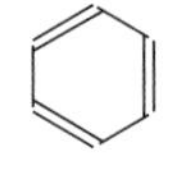

You are given the additional information (at 298K)

$$C_6H_6(l) \rightarrow C_6H_6(g) \quad \Delta H^{\ominus} = 31\ \text{kJ mol}^{-1}$$

$$C(s) \rightarrow C(g) \quad \Delta H^{\ominus} = 717\ \text{kJ mol}^{-1}$$

and the enthalpy diagram

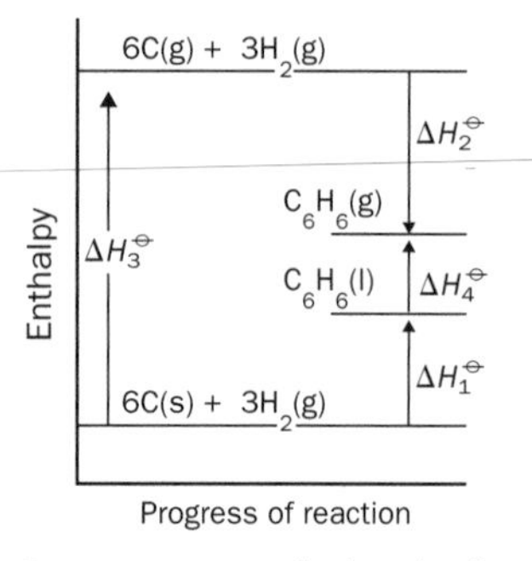

Compare your calculated value for the standard enthalpy of formation of benzene with the experimental value of 50 kJ mol^{-1}. Why is your value different?

Extension material to support this unit is available on our website at **http://www.palgrave.com/foundations/lewis**.

Common Classes of Organic Compounds

Objectives

- Explains what is meant by functional group
- Shows you how organic compounds can be grouped into families, depending upon the functional groups they contain
- Describes the main characteristics of common organic families
- Introduces optical isomerism

Contents

A great many organic chemicals can be derived from hydrocarbons, by replacing one or more or hydrogens by other atoms or groups (called **functional groups**). To try and bring some order into this vast number of compounds, we can again divide the organic compounds into 'families'. Members of each family have the same functional group(s) and the reactions that occur because of the presence of that group are characteristic of the family. It is a good idea to *learn* these functional groups before you go any further. See Table 18.1.

Table 18.1 Common functional groups

General formula of compounds	Name of functional group	Example
R–**X** (X = a halogen)	halogenoalkane	CH_3Cl chloromethane
R–**OH**	alcohol	CH_3CH_2OH ethanol
R–**NH₂**	amine	CH_3NH_2 methylamine
R–**COOH**	carboxylic acid	CH_3COOH ethanoic acid
R–**COCl**	acid chloride	CH_3COCl ethanoyl chloride
R–**CONH₂**	amide	CH_3CONH_2 ethanamide
R–**COOR′**	ester	CH_3COOCH_3 methyl ethanoate
R–**CHO**	aldehyde	CH_3CHO ethanal
R**CO**R′	ketone	CH_3COCH_3 propanone
R**O**R′	ether*	$CH_3CH_2OCH_2CH_3$ diethyl ether
R–**CN**	nitrile*	CH_3CN ethanenitrile

* The chemistry of these functional groups is not discussed in this book.

18.1 Halogenoalkanes (or alkyl halides)

We have already met these compounds in Chapter 17. They are formed by replacing the hydrogen atom of a hydrocarbon by a halogen (F, Cl, Br or I) and have the general formula R–X where X = a halogen. Thus:

CH_3F fluoromethane

CH_3Cl chloromethane

CH_3Br bromomethane

CH_3I iodomethane

Halogenoalkanes are insoluble in water. A mixture of halogenoalkane (if liquid) and water forms two layers and the halogenoalkanes tend to form the bottom layer, because they are quite dense.

Exercise 18A

Halogenoalkanes

(i) Write the structural formulae of the following alkyl halides:

(a) chloroethane

(b) 2-bromopropane

(c) 1,2-dichloro-3-fluorobutane

(d) 1,3,5-tribromobenzene.

(ii) Name the following:

(a) $CH_3CH(Br)Cl$

(b) $CH_3CH_2CHICH_2CH_3$

(c) CCl_3F

(d)

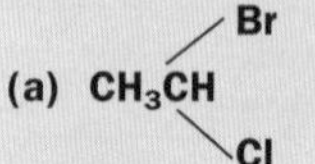

Important halogenoalkanes

Trichloromethane (or chloroform, $CHCl_3$) is non-flammable and was used as an anaesthetic, before it was found to cause liver damage. Tetrachloromethane (or carbon tetrachloride, CCl_4) also has anaesthetic properties, but it is even more toxic. Bromochlorotrifluoroethane (common name halothane) is now widely used. It also has anaesthetic properties and is much safer to use.

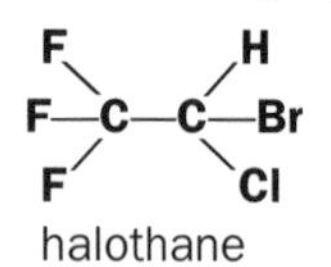

halothane

Substitution of the halogen in halogenoalkanes

A carbon–halogen bond is polar ($C^{\delta+}–X^{\delta-}$) and the carbon can be attacked by groups that carry an unshared pair of electrons. The result is a substitution reaction, in which one atom, ion or group is substituted for another. If some halogenoalkanes are heated with aqueous sodium hydroxide, the halogen is substituted by an –OH group, producing an alcohol:

$$CH_3CH_2CH_2Br + OH^- \rightarrow CH_3CH_2CH_2OH + Br^-$$

1-bromopropane + hydroxide ion → propan-1-ol + bromide ion

Box 14.1 gives more detail on how this reaction occurs.

18.2 Alcohols

The formulae of these compounds can be represented by R–OH. Their names end in 'ol'. The ending is sometimes prefixed by a number to indicate the carbon atom to which the –OH group is attached. Examples of alcohols are shown in Table 18.2.

Alcohols with more than one –OH group are named using diol, triol etc., depending on how many such groups are present. A 1,2 diol is often also referred to as a **glycol**.

Table 18.2 Some members of the alcohol family

Name	Formula	Boiling point/°C	State at room temparature
Methanol	CH_3OH	65	l
Ethanol	CH_3CH_2OH	79	l
Higher alcohol members can exist in isomeric forms – notice how they are named			
Propan-1-ol	$CH_3CH_2CH_2OH$	98	l
Propan-2-ol	$CH_3CHOHCH_3$	83	l
Butan-1-ol	$CH_3CH_2CH_2CH_2OH$	117	l
2-Methylpropan-2-ol	$CH_3C(CH_3)OHCH_3$	82	s/l

Example 18.1

Name the compound: $CH_2(OH)CH_2(OH)$

▶ Answer

ethane-1,2-diol (old name ethylene glycol)

Exercise 18B

Nomenclature of alcohols

(i) Name and write the structural formulae for the different isomers of pentanol, $C_5H_{11}OH$. There are eight.

(ii) Glycerol, or propane-1,2,3-triol, is used in many cosmetic preparations to make them feel soft. Write its structural formula.

Physical properties of alcohols

1. Ability to hydrogen bond

The –OH group in an alcohol is polar:

$$R—O^{\delta-}—H^{\delta+}$$

Alcohols are therefore able to **hydrogen bond** with themselves or with water:

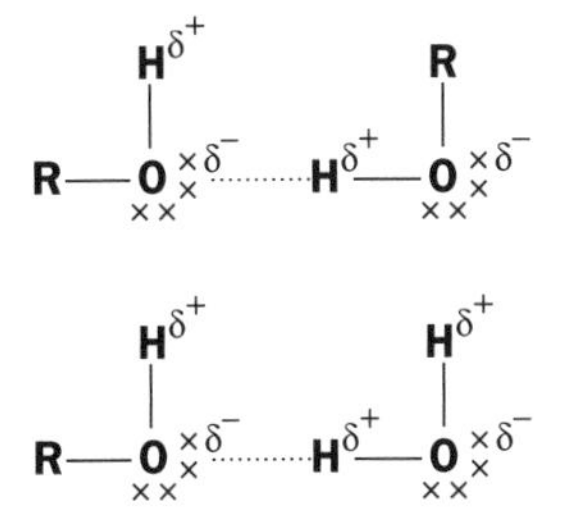

Because alcohols form hydrogen bonds with other alcohol molecules, the boiling points of the alcohols tend to be higher than other organic compounds of similar molecular mass. Their ability to hydrogen bond with water allows the lower molecular mass alcohols to dissolve in, or be miscible with, water. Ethane-1,2-diol, or ethylene glycol, is used as a component in antifreeze because it has a high solubility in water, it contains two hydroxy groups which can hydrogen bond with the water, and it has a high boiling point and low freezing point.

2. Alcohols are good solvents

The presence of an –OH group means that they have some of the solvent properties of water and can dissolve ionic compounds. The R– group allows them to behave as solvents to organic, covalent substances.

Exercise 18C

Properties of alcohols

Chloroethane, butane and propan-1-ol have boiling points of 286 K, 272 K and 371 K, respectively.

(i) Calculate the molecular mass of each compound.

(ii) Assuming all their melting points are lower than room temperature, what is the state of each substance under conditions of 20 °C and 1 atm pressure?

(iii) Explain why the boiling point of propan-1-ol is much higher than those of the other two compounds.

Ethanol

Although there are many alcohols, when the non-chemist refers to 'alcohol' they generally mean ethanol. Ethanol can be made by the decomposition of **carbohydrates**, a process known as **fermentation**.

Fruit juices, such as grape juice, contain the sugar glucose (molecular formula $C_6H_{12}O_6$). Glucose can be fermented into alcohol by the addition of yeast. Yeast is a fungus that contains the enzyme **zymase**, which catalyses the decomposition of glucose into ethanol. **Enzymes** are biological catalysts. The fermentation reaction is carried out at about 20 °C – high temperatures kill the yeast and at lower temperatures the process is too slow. The equation for the reaction may be written as

$$\underset{\text{glucose}}{C_6H_{12}O_6(aq)} \rightarrow \underset{\text{ethanol}}{2C_2H_5OH(aq)} + 2CO_2(g)$$

Fermentation ceases when the ethanol content of the mixture reaches about 14% (v/v); higher concentrations of ethanol kill the yeast. If a higher concentration of alcohol is required, the mixture is **fractionally distilled**. 'Spirits' such as whisky or brandy may contain about 40% ethanol. Although fermentation is used to produce alcoholic beverages, it is rarely used to synthesize the chemical ethanol on a large scale. Alcohols, such as ethanol, can be produced commercially by the addition of water across an alkene double bond (a **hydration reaction**). The reaction is catalysed by phosphoric acid:

$$\underset{\text{ethene}}{CH_2{=}CH_2} + \underset{\text{steam}}{H_2O} \xrightarrow[\text{300 °C, 70 atm}]{H_3PO_4} \underset{\text{ethanol}}{CH_3CH_2OH}$$

Dilute aqueous solutions of ethanol, when drunk as beverages, cause a pleasant relaxed feeling; however, too much 'alcohol' leads to headaches and vomiting. Long-term abuse of ethanol can lead to liver damage.

Methanol

Originally, methanol was produced when wood was distilled and, for this reason, was known as wood alcohol. Today, methanol can be produced from the catalytic hydrogenation of carbon monoxide:

$$CO(g) + 2H_2(g) \rightarrow CH_3OH(l)$$

Methanol is used as a solvent and as an antifreeze for car radiators. It is much more toxic than ethanol – drinking methanol can lead to blindness and then to death. Methanol can be used as fuel for cars.

Methylated spirits

Alcoholic beverages are heavily taxed in many countries, but besides being used to make drinks, ethanol has many useful industrial applications. It is used to make other organic chemicals and is an extremely useful industrial solvent. In order to avoid heavy taxation of the ethanol used for industrial purposes it is rendered unfit to drink (**denatured**). Small amounts of methanol are added to ethanol to produce 'industrial methylated spirit'. A purple dye is often added to methylated spirit sold for domestic use. The dye serves as a warning that the mixture is undrinkable.

Chemical properties of alcohols

1. They burn in air, giving out heat, and can be used as fuels:

$$C_2H_5OH(l) + 3O_2(g) \rightarrow 2CO_2(g) + 3H_2O(l)$$

2. Alcohols can be oxidized to carboxylic acids. For example, ethanol is slowly oxidized to ethanoic acid (*acetic acid* found in *vinegar*) by bacteria present in the air. This is why an open bottle of wine tastes sour when left in the air for a few days.

$$C_2H_5OH(l) + O_2(g) \xrightarrow{\text{bacteria}} CH_3COOH(l) + H_2O(l)$$

Strong oxidizing agents, such as acid potassium permanganate (manganate(VII)) solution or chromic acid (acidified sodium dichromate(VI) solution) will also oxidize alcohols.

3. Alcohols react with carboxylic acids to make **esters** (see some reactions of carboxylic acids page 339).

Primary, secondary and tertiary alcohols

Apart from methanol, alcohols may be classified as primary, secondary or tertiary:

primary alcohols contain a —CH_2OH group

secondary alcohols contain a >CHOH group

tertiary alcohols contain a ⋝COH group

Exercise 18D

Primary, secondary and tertiary alcohols

Classify the isomers of pentanol in Exercise 18B(i) as primary, secondary or tertiary.

18.3 Carbonyl compounds

Carbonyl compounds contain the carbonyl group, >C=O. The carbonyl group is part of the structural formulae of a number of organic families. The name of the family is determined by the other atoms bonded to the carbonyl carbon. If one of the atoms bonded is hydrogen, the compound belongs to the class of organic compounds known as **aldehydes**. However, if two carbon atoms are bonded to the carbonyl carbon, the compound is a **ketone**. To summarize,

aldehydes contain the R—C(H)=O group (sometimes written R—CHO)

ketones contain the R—C(R′)=O group (where R and R′ are alkyl groups)

The name of an aldehyde is derived from the number of carbon atoms present and ends in 'al'. For example:

HCHO	methanal (old name formaldehyde)
CH_3CHO	ethanal (old name acetaldehyde)
CH_3CH_2CHO	propanal

Ketones are named by changing the end of the name of the parent alkane (the one with the same number of carbon atoms) to 'one'. Sometimes it is necessary to number the position of the carbonyl group to avoid confusion:

CH_3COCH_3	propanone (commonly called acetone)
$CH_3COCH_2CH_3$	butanone
$CH_3COCH_2CH_2CH_3$	pentan-2-one

Exercise 18E

Aldehydes and ketones 1

Classify each of the following compounds as an aldehyde or a ketone:

(i) C_6H_5CHO

(ii) $CH_3CH_2COCH_2CH_3$

(iii) $CH_3CH_2CH_2CHO$

(iv) $CH_3CH_2C(=O)CH_3$

(v) $C_6H_5C(=O)CH_3$

Aldehydes and ketones can be isomeric with one another. For example, CH_3CH_2CHO and CH_3COCH_3 both have the molecular formula C_3H_6O. These **structural isomers** have some different chemical properties as well as different physical properties.

Physical properties of aldehydes and ketones

The carbonyl group is polar and has unshared electron pairs on the oxygen atom:

$$>C^{\delta+}=O^{\delta-}$$

The group can therefore take part in hydrogen bonding with water molecules:

$$R'—C^{\delta+}(=O^{\delta-})—R'' \cdots H^{\delta+}—O^{\delta-}—H^{\delta+}$$

Aldehydes and ketones with a low molecular mass are therefore soluble in water. Propanone (acetone) is widely used as a solvent. It is soluble in water and dissolves many polar and non-polar molecules. Examples include paint, nail varnish and plastics.

A 37% solution of methanal (common name formaldehyde) dissolved in water, with some methanol added, is called **formalin** and is used for preserving biological specimens.

Ethanal is formed when ethanol is oxidized in the liver. A buildup of ethanal in the bloodstream causes a hangover.

Ketones can have distinctive sweet smells. Spearmint-flavoured chewing gum gets its flavour from the ketone carvone:

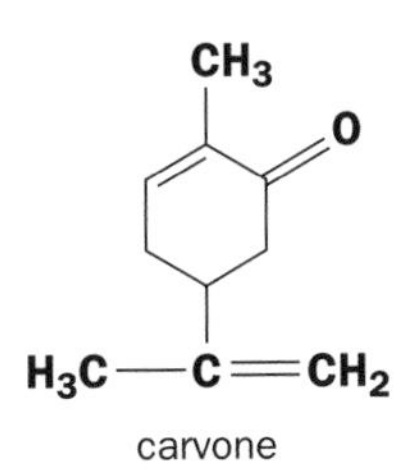

carvone

Aldehydes and ketones 2

(i) Why is there no such compound as ethanone?

(ii) Write down the structural formula of the aldehydes that are isomeric with $CH_3CH_2CH_2COCH_3$. What are their names?

(iii) Write down the names and structures of two aldehydes and one ketone with the molecular formula C_4H_8O.

Preparation of aldehydes and ketones by oxidation of alcohols

Primary or secondary alcohols can be oxidized to aldehydes or ketones. Aldehydes are easily further oxidized to the corresponding carboxylic acid and oxidation of a **primary alcohol** usually continues through the aldehyde to the acid. The oxidation of a **secondary alcohol** usually yields a ketone. Ketones are more resistant to further oxidation because C–C bonds would have to be broken and can only be further oxidized using very strong oxidising conditions. **Tertiary alcohols** are resistant to any oxidation, again because C–C bonds would have to be broken. These changes can be summarized as follows:

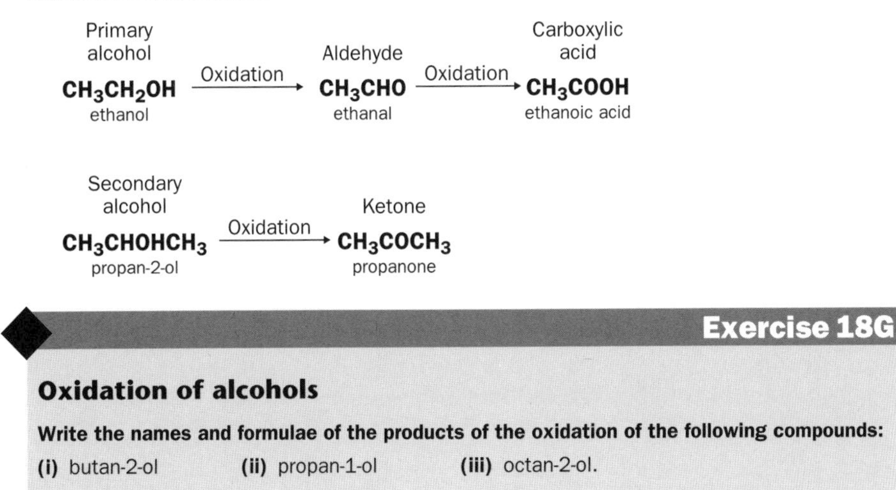

Oxidation of alcohols

Write the names and formulae of the products of the oxidation of the following compounds:

(i) butan-2-ol **(ii)** propan-1-ol **(iii)** octan-2-ol.

Acidified sodium dichromate is an example of an oxidizing agent which will bring about the oxidation of primary alcohols through to carboxylic acids and secondary alcohols to ketones. If an aldehyde is required as the product of oxidation of a primary alcohol, the apparatus must be designed so that the aldehyde will be distilled off as it is formed.

Carbonyl compounds can be reduced back to alcohols by using the reducing agents lithium aluminium hydride ($LiAlH_4$) or sodium borohydride ($NaBH_4$).

Tests to distinguish between aldehydes and ketones

The fact that aldehydes are easily oxidized to carboxylic acids, whereas ketones are not, can be used to distinguish between them:

1. Fehling's solution may be regarded as containing Cu^{2+} ions in basic solution. When this solution is boiled with an aldehyde, a brick-red precipitate of copper(I) oxide is formed – the aldehyde is oxidized to a carboxylic acid by the copper(II) ions. Ketones are ***not*** oxidized by this reagent.

   $$RCHO + 6H_2O + 2Cu^{2+} \rightarrow RCOOH + 4H_3O^+ + Cu_2O$$

2. Tollens' reagent contains Ag^+ ions dissolved in aqueous ammonia. If the reagent is warmed with an aldehyde, the silver ions are reduced to silver metal and a distinctive 'silver mirror' is deposited on the reaction container. Again, ketones do not react with this reagent.

   $$RCHO + 3H_2O + 2Ag^+ \xrightarrow{\text{aqueous ammonia}} RCOOH + 2H_3O^+ + 2Ag$$

Identification of aldehydes and ketones

The carbonyl group in aldehydes and ketones will react with **hydrazine** derivatives. Hydrazine is related to ammonia:

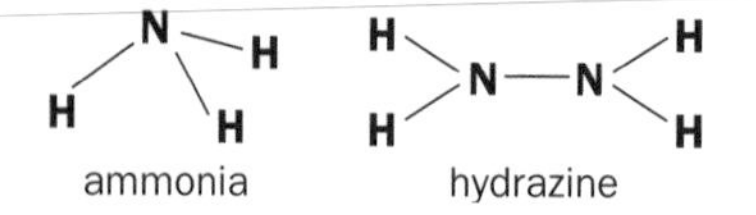

The products of the reaction of hydrazine with an aldehyde or ketone are known as a **hydrazones**:

Because water is 'eliminated' in the reaction, the reaction is known as a **condensation** reaction. A slight variation of this reaction can be used to identify aldehydes or ketones using 2,4-dintrophenylhydrazine because the hydrazones formed are orange solids that can be recrystallized easily.

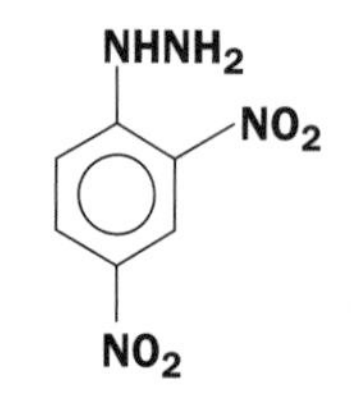

2,4-dinitrophenylhydrazine

An aldehyde or ketone is converted to the corresponding dinitrophenylhydrazone (DNP). The DNP is purified by recrystallization and its melting point compared with those reported in data books. In this way the original carbonyl compound can be identified.

Exercise 18H

Reactions of aldehydes and ketones

Draw the structural formulae of the organic products of the following reactions:

(i) The reaction of butanal on heating with Fehling's solution.

(ii) The reaction of propanone with 2,4-dinitrophenylhydrazine.

(iii) The reaction of propanal with hydrazine.

(iv) The reaction of propanal on warming with Tollens' reagent.

BOX 18.1

Carbohydrates

Carbohydrates are naturally occurring compounds of carbon, hydrogen and oxygen, with hydrogen and oxygen atoms in the ratio 2:1 as in water. They are **polyhydroxy** (they contain many –OH groups) aldehydes and ketones, or compounds that can be hydrolysed to make polyhydroxy aldehydes or ketones. The chemical names of many carbohydrates end in 'ose'. Starch, table sugar (sucrose) and cotton and paper (cellulose) are all composed of carbohydrates.

The simplest carbohydrates are the **monosaccharides** (or simple sugars). They cannot be split up into simpler carbohydrate units. Glucose ($C_6H_{12}O_6$) and fructose ($C_6H_{12}O_6$) are monosaccharides:

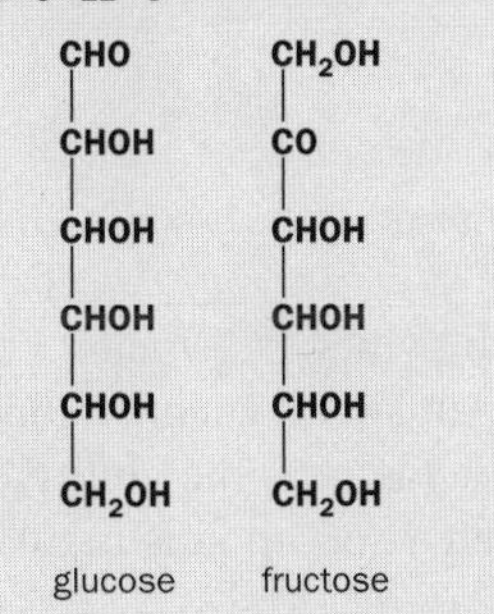

glucose fructose

Glucose is a sugar found in grapes, whereas fructose is the very sweet-tasting sugar found in honey and fruit. Because simple carbohydrates have a large number of –OH groups they are able to hydrogen bond and are therefore very soluble in water. Sugars also undergo some reactions that are typical of carbonyl compounds; for example, glucose reacts with DNP to give a crystalline derivative.

In addition to the 'open chain' form described above, glucose can exist in a ring form:

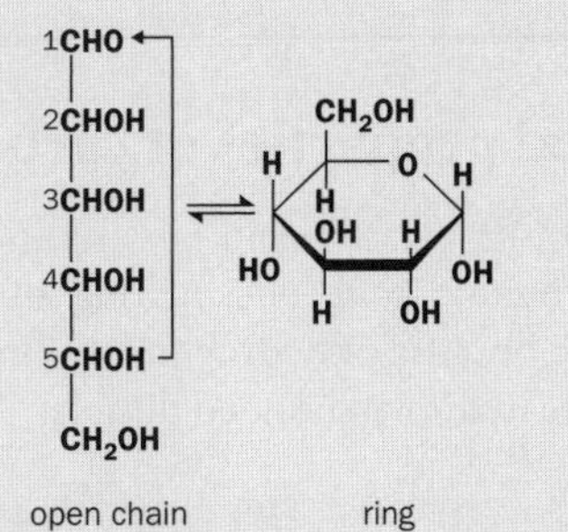

open chain ring

The –OH group on carbon atom 5, reacts with the carbonyl group 1, and a ring is formed. In aqueous solution an equilibrium exists between the two forms.

Monosaccharide units can join together by eliminating water: in twos (**disaccharides**), between two and eight units (**oligosaccharides**) and more than eight units (**polysaccharides**). Sucrose is a disaccharide that can be broken up into glucose and fructose units, whereas starch (from flour) and cellulose (from plants) are both polysaccharides. Our bodies convert sucrose and starch to glucose, which we either use for energy or store as **glycogen** (another polysaccharide). Glycogen is reconverted to glucose when energy is needed.

18.4 Carboxylic acids

Exercise 18I

Naming carboxylic acids

Name the following acids:

(i) $CH_3CH_2CH_2COOH$

(ii) $CH_3(CH_2)_5COOH$

(iii) COOH–CH_2–CH_2–COOH

These compounds contain a **carboxyl** group. The carboxyl group is written as –COOH or $-CO_2H$ and has the structure

—C=O
 |
OH

The systematic names of the acids end in 'oic acid'. Many of the acids, however, are still referred to by their old names. Some simple carboxylic acids are shown in Table 18.3.

The prefixes 'di', 'tri' etc. are used to show more than one carboxylic acid group. Ethanedioic acid (common name **oxalic acid**) is found in rhubarb leaves:

COOH
 |
COOH
oxalic acid

Table 18.3 Some carboxylic acids and their common names

Formula	Systematic name	Common name	Where found
HCOOH	methanoic acid	formic acid	ant or nettle stings
CH_3COOH	ethanoic acid	acetic acid	vinegar
CH_3CH_2COOH	propanoic acid	propionic acid	milk, butter and cheese
C_6H_5COOH	benzoic acid	benzoic acid	preservatives

Physical properties of carboxylic acids

Pure ethanoic (acetic) acid is a liquid, at room temperature, and is referred to as **glacial ethanoic acid**. The melting and boiling points of carboxylic acids are relatively high because of hydrogen bonding. Pairs of carboxylic acids molecules link up to produce hydrogen bonded dimers:

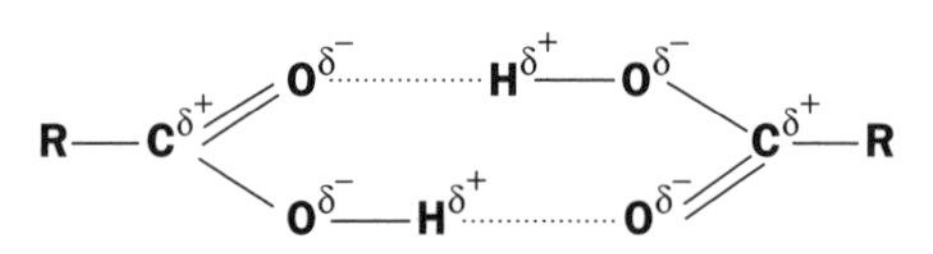

Because of their ability to hydrogen bond, the lower molecular mass carboxylic acids are soluble with water.

Methanoic, ethanoic and propanoic acids all have sharp, 'vinegary' smells. The smell of butanoic acid is detected in rancid butter and human sweat. Each person's sweat glands produce a characteristic blend of carboxylic acids, and the ability of dogs to track humans stems from the fact that the animals can detect and distinguish these different combinations of carboxylic acids.

Acid strengths of carboxylic acids

Carboxylic acids are weak acids compared with dilute sulfuric, nitric or hydrochloric acids:

$$CH_3COOH(l) + H_2O(l) \rightleftharpoons CH_3COO^-(aq) + H_3O^+(aq)$$

If an electronegative species, such as chlorine, is present in the acid molecule, close to the carboxyl group, chlorine pulls electrons towards it. This stabilizes the anion produced when the acid ionizes. For example, consider the reaction of chloroethanoic acid, $CH_2ClCOOH$, with water:

$$Cl \leftarrow CH_2COOH + H_2O \rightleftharpoons Cl \leftarrow CH_2COO^- + H_3O+$$

The $Cl \leftarrow CH_2COO^-$ ion is more stable than CH_3COO^-, so K_a for chloroethanoic acid is much greater than that for ethanoic acid. Additional electron-withdrawing groups increase this effect, but the influence of the electron-withdrawing group on acid strength falls off with distance.

Some reactions of carboxylic acids

1. Carboxylate salts are formed when a carboxylic acid reacts with base:

$$CH_3COOH(aq) + NaOH(aq) \rightarrow \underset{\text{sodium ethanoate}}{CH_3COO^-Na^+(aq)} + H_2O(l)$$

2. Carboxylic acids react with alcohols in an acid catalysed, reversible reaction. The compounds formed are called **esters**. An ester contains the $-COOR$ (or $-CO_2R$) group:

```
—C=O
 |
 OR
```

In a typical example of an esterification reaction, glacial ethanoic acid is heated with ethanol in the presence of concentrated sulfuric acid, which acts as the catalyst. The sweet-smelling ester ethyl ethanoate is formed:

$$CH_3COOH(l) + CH_3CH_2OH(l) \rightleftharpoons \underset{\text{ethyl ethanoate}}{CH_3COOCH_2CH_3(l)} + H_2O(l)$$

The general reaction is

$$RCOOH + R'OH \rightleftharpoons RCOOR' + H_2O$$

Note that these reactions are reversible. In order to maximize the yield of ester, excess of one of the reactants is used (usually the cheaper one!). Concentrated sulfuric acid, besides acting as the catalyst, removes the water produced, so shifting the equilibrium composition so that the concentration of ester increases. Because the reaction is reversible an ester can be **hydrolysed**, or reacted with water, in the presence of an acid catalyst, to make the parent alcohol and acid.

3. Carboxylic acids are reduced to primary alcohols using lithium aluminium hydride:

$$CH_3COOH \xrightarrow{LiAlH_4} CH_3CH_2OH$$

BOX 18.2

Fatty acids

Carboxylic acids with long hydrocarbon chains can be obtained from hydrolysis of fats or oils. They are called fatty acids. **Stearic acid** is a fatty acid and its sodium salt (sodium stearate) is used as soap. The hydrocarbon chain in a fatty acid may be **saturated**, as in stearic acid, or it may be **unsaturated** (contain a double bond). **Oleic acid**, obtained from corn oil, contains one double bond, whereas fatty acids with more than one double bond can be prepared from vegetable oils.

Common name	Formula	Systematic name
Stearic acid	$CH_3(CH_2)_{16}COOH$	octadecanoic acid
Oleic acid	$CH_3(CH_2)_7CH{=}CH(CH_2)_7COOH$	octadec-9-enoic acid

Exercise 18J

Strength of carboxylic acids

Arrange the following in order of *increasing* acid strength:

$CHCl_2COOH$, CF_3COOH, CH_3COOH, CCl_3COOH, $CH_2BrCOOH$, $CH_2ClCOOH$.

Exercise 18K

Esterification

(i) Name and write the structural formulae of the esters produced from the reactions of the following acids and alcohols:

(a) ethanoic acid and methanol

(b) octanoic acid and ethanol

(c) methanoic acid and methanol

(d) oxalic acid and ethanol.

(ii) Which alcohols and acids would be obtained on hydrolysis of the following esters?

(a) methyl propanoate

(b) ethyl methanoate

(c) $CH_3COOCH_2CH_2CH_3$

(d) $CH_2ClCOOCH_3$.

Derivatives of carboxylic acids

Esters are one family of compounds that can be derived from carboxylic acids. These and some other derivatives are shown below:

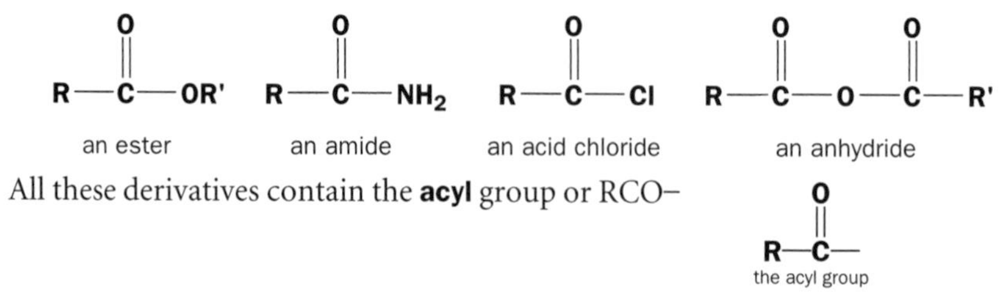

All these derivatives contain the **acyl** group or RCO–

BOX 18.3

Some examples of carboxylic acid derivatives

Nylon

The man-made fibre nylon 6,6 is an example of a polymer that consists of long-chain molecules, containing repeating –CONH– groups. Its formula is

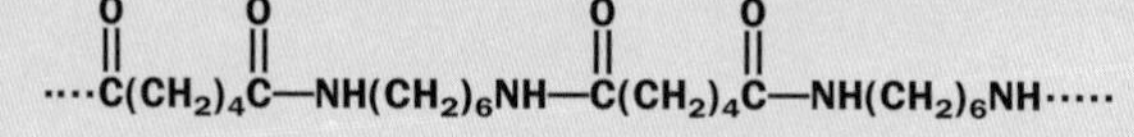

The –CONH– group is also present in amides:

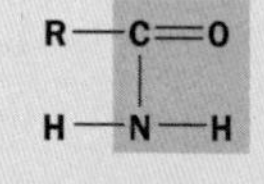

Nylon is therefore a **polyamide**. Proteins are also examples of polyamides.

Nylon 6,6 chains are formed by the reaction of many hexanedioic acid molecules with hexan-1,6-diamine molecules (a member of the amine family). Molecules of water are produced so that the chains can form. The formation of long chain molecules from two different monomers, with elimination of water, is called a **condensation polymerization** reaction:

$$n(HO_2C(CH_2)_4CO_2H) + n(H_2N(CH_2)_6NH_2) \longrightarrow$$

$$\cdots\left[C(=O)(CH_2)_4C(=O)—NH(CH_2)_6NH\right]_n\cdots + nH_2O$$

where n = a large number.

Fats and oils

These are naturally occurring esters of the triol glycerol (propan-1,2,3-triol). A fat is a substance that is solid at room temperature, whereas an oil is liquid. In vegetable oils, the hydrocarbon chains have many double bonds (they are **polyunsaturated**). Fats, however, tend to have very few, or no double bonds. Removal of most of the double bonds in a vegetable oil, by reaction with hydrogen (**hardening**), will convert it to a solid fat. Hardened corn oil, for example, is used to make margarine.

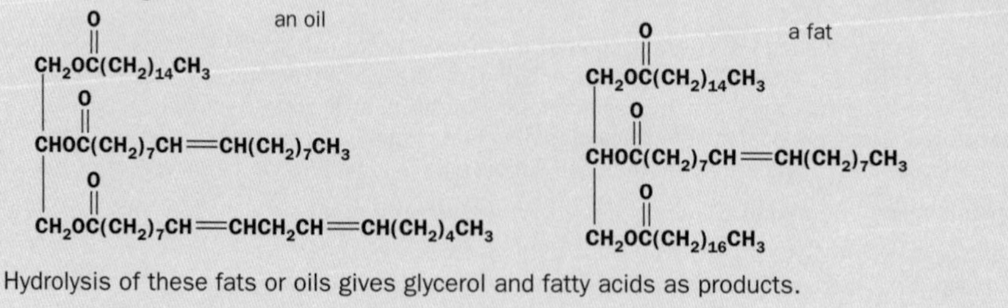

Hydrolysis of these fats or oils gives glycerol and fatty acids as products.

18.5 Amines

The formulae for primary amines are of the form R–NH_2, where –NH_2 is the amino group. Examples of primary amines are

CH_3NH_2	methylamine
$CH_3CH_2NH_2$	ethylamine
$CH_3CH_2CH_2NH_2$	propylamine
$H_2NCH_2CH_2CH_2NH_2$	propan-1,3-diamine

Secondary amines have the general formula

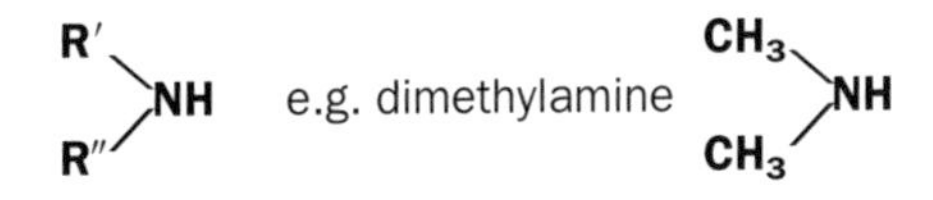

Tertiary amines have the structure

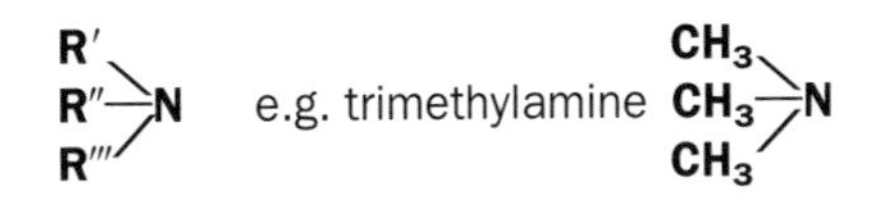

Physical properties of amines

1. Amines have very offensive fishy smells (trimethylamine smells like rotting salmon!). Amines occur in decomposing protein material. The amines $NH_2(CH_2)_4NH_2$ and $NH_2(CH_2)_5NH_2$, putrescine and cadaverine, are found in decaying animal flesh.
2. Primary and secondary amines have N–H bonds and therefore undergo hydrogen bonding with each other and with water. Tertiary amines cannot hydrogen bond with each other, but the lone pair of electrons present on the nitrogen atom enables them to hydrogen bond with water. Amines with low molecular mass are therefore miscible with water:

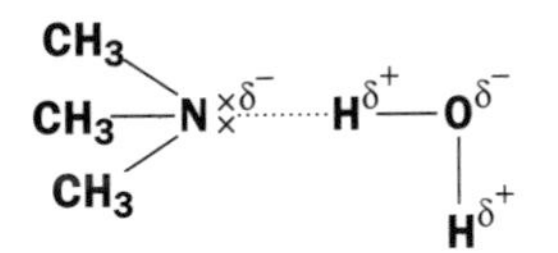

Basic properties of amines

An unshared pair of electrons on the nitrogen of an amine can accept a proton from an acid. The product is an amine salt:

$$CH_3NH_2 + HCl \rightleftharpoons \underset{\text{methyl ammonium chloride}}{CH_3NH_3^+, Cl^-}$$

In this reversible reaction, the amine is acting as a weak base. This reaction can be likened to the reaction of ammonia with hydrogen chloride. However, amines tend

Exercise 18L

The basicity of amines

Arrange the following compounds in order of *increasing* basicity:

CH_3NH_2, CH_3NHCH_3, NH_3.

to be more basic than ammonia because alkyl groups push electrons on to the nitrogen atom; this makes them better proton acceptors.

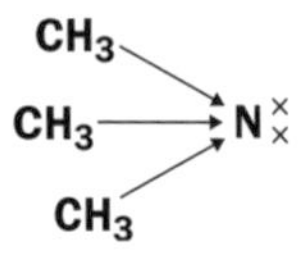

The 'electron push' of alkyl groups is known as an **inductive effect**.

18.6 Optical isomerism

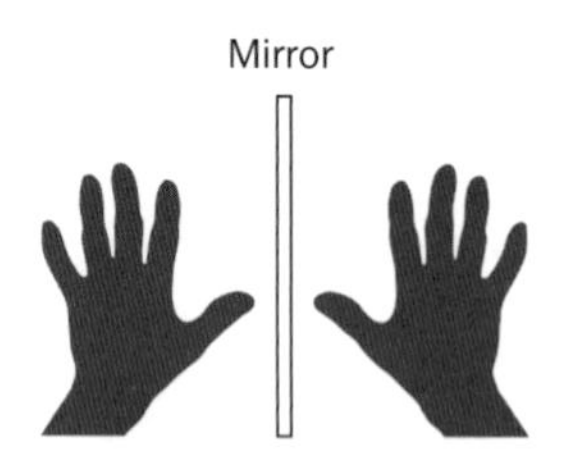

Fig. 18.1 A glove and its mirror image.

Some organic molecules can exist in two forms that are mirror images of one another. Think of a pair of gloves – the left and right hand gloves cannot 'fit over' each other or be **superimposed**. The right glove has the shape of the image of the left glove as seen in a mirror (Fig. 18.1).

Similarly, where a carbon atom is bonded to four *different* atoms or groups, then two such non-superimposable isomers are possible, as shown in Fig. 18.2 (make two models and try to superimpose one upon the other). The carbon atom is known as a **chiral centre**.

The mirror image isomers are called **enantiomers**. They generally have very similar chemical and physical properties, but differ in the way that they rotate **plane polarized light**. The isomers are **optically active** – they will rotate the light by equal amounts, but in opposite directions.

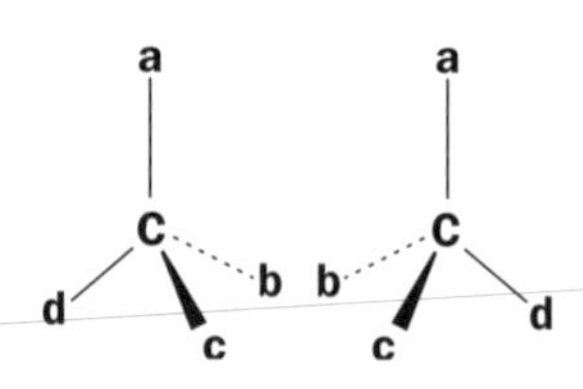

Fig. 18.2 Two non-superimposable isomers (a, b, c and d are different atoms or groups).

BOX 18.4

Plane polarized light

Ordinary light travels in waves and the wave vibrations are at right angles to the direction of travel. Plane polarized light is light that vibrates in only one plane and is obtained by passing ordinary light through a 'Polaroid' lens or polarizer. If the plane polarised light is passed through a solution of one enantiomer of a compound, the light will be rotated, say to the right, by a certain angle. This is shown in Fig. 18.3. The light is rotated by an equal angle, but to the left, by a solution of the other enantiomer of the compound with the same concentration.

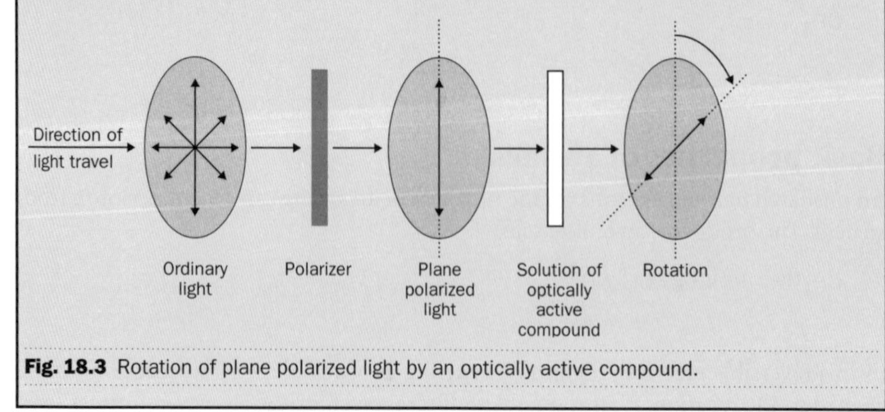

Fig. 18.3 Rotation of plane polarized light by an optically active compound.

Exercise 18M

Optical isomers

(i) Draw, or make models of the two isomeric forms of lactic acid, the substance that gives milk its sour taste. Mark the chiral centre with an asterisk.

$$\begin{array}{c} COOH \\ | \\ H-C-OH \\ | \\ CH_3 \end{array}$$

(ii) Which of the following compounds can exist in two enantiomeric forms?
2-methylbutan-1-ol, 1-chloropentane, 2-bromo-1-chlorobutane.

18.7 Amino acids and proteins

Amino acids

Amino acids contain an amino group ($-NH_2$) and a carboxylic acid group ($-COOH$). Since their formulae can be quite complicated they usually are referred to by their common names rather than their systematic names. For example, H_2NCH_2COOH is called **glycine** rather than aminoethanoic acid. Some common amino acids are shown in Table 18.4.

Amino acids are non-volatile crystalline solids, which are very soluble in water. Many of them are optically active. They also have large dipole moments. Because the molecules contain both an acidic and a basic group, the following equilibrium exists in solution:

$$H_2NCHRCOOH \rightleftharpoons {}^{+}H_3NCHRCOO^{-}$$

The 'acidic end' of the molecule provides a proton to protonate the 'basic' end. The ion that is formed is known as a **dipolar ion** or **zwitterion** and, because of the ionic charges at both ends, the amino acid has many properties of a salt.

Table 18.4 Amino acids

Name	Formula	Abbreviation
Alanine	$CH_3CH(NH_2)COOH$	Ala
Phenylalanine	$C_6H_5-CH_2CH(NH_2)COOH$	Phe
Cysteine	$HSCH_2CH(NH_2)COOH$	Cys
Glutamine	$H_2NC(=O)CH_2CH_2CH(NH_2)COOH$	Gln
Lysine	$H_2NCH_2CH_2CH_2CH_2CH(NH_2)COOH$	Lys

Exercise 18N

Structure of amino acids

(i) Most naturally occurring amino acids can exist in two optically active forms. Draw the two possible structures of the amino acid alanine, CH_3CHNH_2COOH.

(ii) What is the formula of the dipolar ion formed by alanine?

Exercise 180

Peptides

(i) Write the structural formula of the dipeptide glycylalanine (Gly-Ala).

(ii) How many different tripeptides can be made from different combinations of the three amino acids Gly, Ala and Cys?

Peptides

Peptides are formed when the amino group of one amino acid interacts with the carboxyl group of another. Water is eliminated in the reaction:

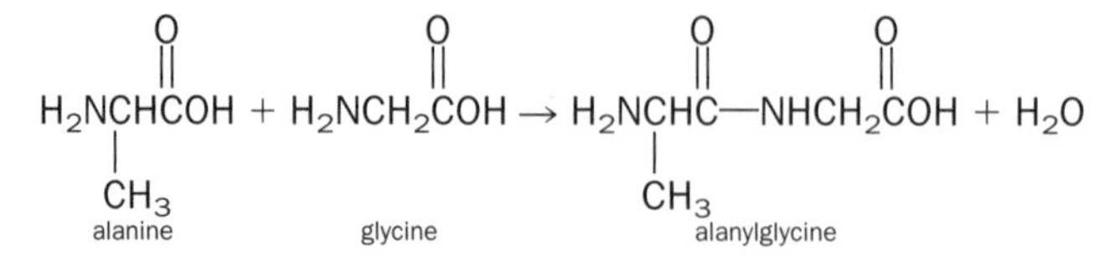

The group –CONH– that is formed in such compounds is a **peptide linkage** or **peptide bond**. **Dipeptides** and **tripeptides** are derived from the combinations of two and three amino acids respectively. **Polypeptides** is a term that describes peptides of molecular mass up to 10 000 u. Molecules larger than this are known as **proteins**. Polypeptides and proteins can also be thought of as polyamides made from amino acids. The name of a peptide is constructed from the names of the amino acids units as they appear left-to-right, starting with the amino acid fragment that has a free $-NH_2$ group:

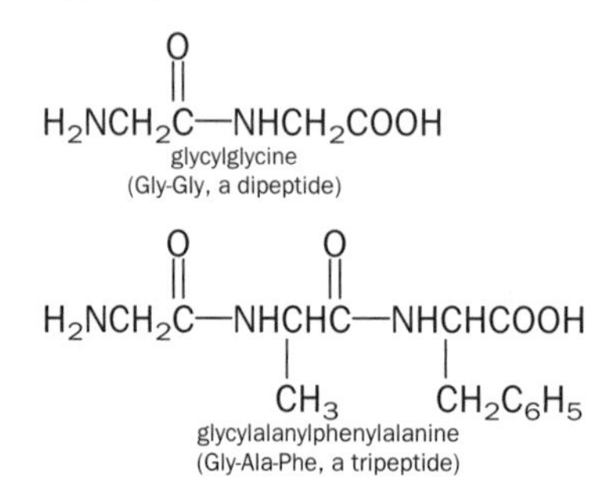

Proteins are found in all living cells. They make up a large part of an animal's body and are found, for example, in skin, blood, enzymes and hormones. A single protein molecule contains many, many amino acid units joined by peptide linkages. Proteins are divided into two groups:

1. Fibrous proteins, in which the molecules are long, tough and thread-like. They make up keratin in nails, wool, hair and feathers. These proteins are insoluble in water.
2. Globular proteins consist of molecules that are folded into compact shapes. They are found, for example, in enzymes and hormones, and in haemoglobin which transports oxygen in the body.

Structure of proteins

The amino acids found in proteins are α-aminocarboxylic acids. The **prefix α** refers to the fact that the amino group is attached to the same carbon atom bonded to the carboxylic acid group:

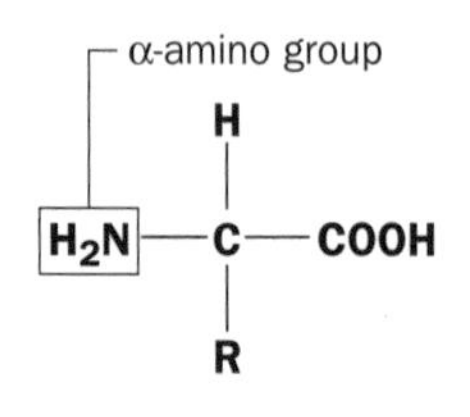

The structures of proteins can be studied on a number of levels:

1. The **primary structure** of a protein is the order in which the amino acid residues are linked together to form long chains.
2. The **secondary structure** of a protein is the way in which the chains are arranged in space to form coils, sheets or almost spherical shapes. The chains are held together by hydrogen bonds (see page 77). X-ray diffraction has proved to be a valuable technique in determining the often complicated secondary structure of proteins.

   An example of a protein which has a well studied secondary structure is keratin, found in hair. Each protein molecule in keratin is arranged in the shape of a spiral, called a **helix**. Hydrogen bonding holds the helix together, by linking to different sections of the same chain. The arrangement of the spiral in keratin is called an **alpha helix**. This means that it is a helix with a *right-handed* turn, as in Fig. 18.4.
3. The **tertiary structure** of a protein is a description of further interactions that occur, such as further folding of the secondary structure.

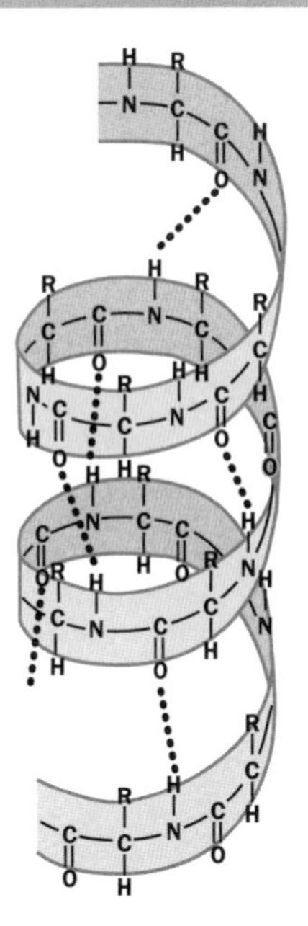

Fig. 18.4 The arrangements of the protein chains in keratin.

18.8 Substituted benzene derivatives

So far, we have discussed families of aliphatic compounds containing different functional groups. Benzene derivatives containing the same functional groups often show very different properties when the functional group is attached directly to the benzene ring. They are also sometimes named differently from their aliphatic counterparts.

Phenols

Phenols are compounds with hydroxy groups bonded to an aromatic ring (Ar–OH). The simplest member of this family is phenol, C_6H_5OH.

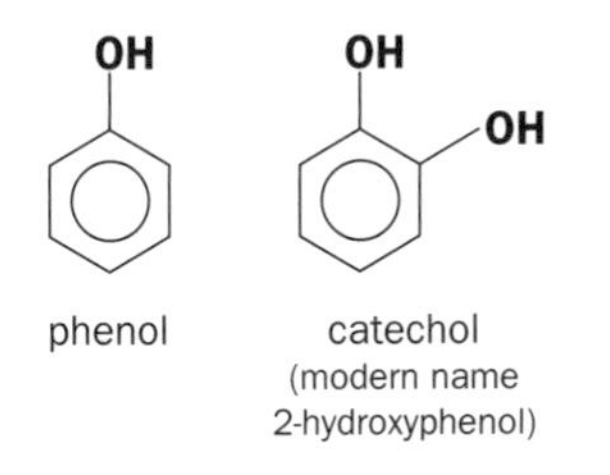

phenol

catechol (modern name 2-hydroxyphenol)

Some phenols are obtained from coal tar, but nowadays most phenols are made by oxidation of cumene C_9H_{12}. Cumene is produced from benzene and propene.

Phenols have acidic properties. For example, they react with aqueous sodium hydroxide to produce sodium phenoxides:

$$C_6H_5OH(s) + NaOH(aq) \rightarrow C_6H_5O^-,Na^+(aq) + H_2O(l)$$

Phenol is therefore much more soluble in sodium hydroxide solution than it is in water. Phenol is only weakly acidic, however, and it *does not react with carbonates to form carbon dioxide.*

BOX 18.5

Joseph Lister proposed the use of phenol as an antiseptic. He was a Scottish surgeon who rightly suspected that germs might be responsible for infection in wounds.

Phenol

Phenol was originally called carbolic acid. In the nineteenth century, Joseph Lister (1827–1912) used it to prevent wounds going septic and recommended that the compound should be used as an antiseptic in hospitals. Before this, amputees often had their fresh wounds dipped in coal tar (which contains phenol) and about half of patients operated on died from infections gained during the operations. The compound is a severe irritant to the skin, however, so it had to be replaced by alternatives. Many modern antiseptic molecules, however, still contain phenolic groups:

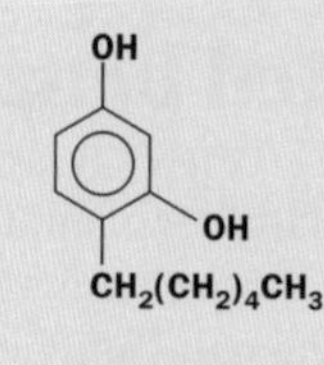

n-hexylresorcinol
used in antiseptics

Benzoic acid

In benzoic acid, a carboxyl group is attached to the benzene ring:

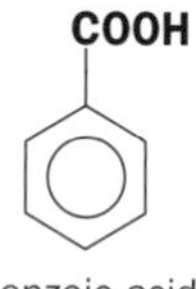

benzoic acid

The solid acid is slightly soluble in water and undergoes the typical reactions of carboxylic acids described previously. It is used as a preservative in food and drinks including iced lollies.

Aniline

Aniline is an aromatic amine with the structure

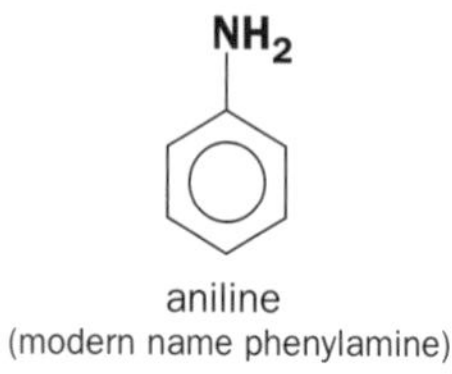

aniline
(modern name phenylamine)

The presence of the aromatic ring makes aniline a much weaker base than ammonia. Aniline is an *extremely toxic* liquid that readily enters the body through the skin or by inhalation and should be handled with extreme care.

Aniline undergoes an interesting reaction with sodium nitrite and ice-cold aqueous acid. The compound formed is called a **diazonium salt**:

$$C_6H_5NH_2(l) + NaNO_2(aq) + 2HCl(aq) \rightarrow \underbrace{C_6H_5{-}N{\equiv}N^+, Cl^-(aq)}_{\text{diazonium salt}} + NaCl(aq) + 2H_2O(aq)$$

Diazonium salts can be reacted to form many classes of compounds but decompose, even at cold temperatures, so they must be used immediately. They also react with certain aromatic compounds to form **azo compounds**. The reaction is called a coupling reaction, for which the general equation is

$$Ar{-}N{\equiv}N^+ + Ar'H \rightarrow \underbrace{Ar{-}N{=}N{-}Ar'}_{\text{azo compound}} + H^+$$

The double bond between the nitrogen atoms makes azo compounds strongly coloured. Because of their intense colours they are widely used as dyes. Many acid–base indicators are azo dyes.

Some examples of azo dyes are

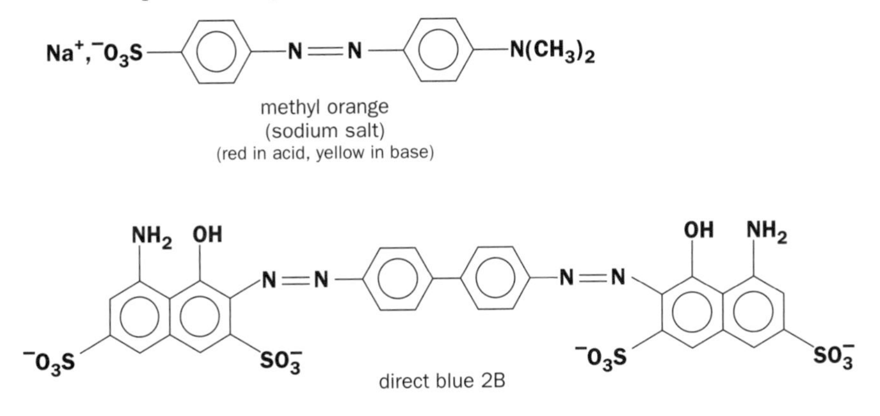

methyl orange
(sodium salt)
(red in acid, yellow in base)

direct blue 2B

Benzyl compounds

The benzyl group, $C_6H_5CH_2-$, is an **aralkyl group**. It contains an aromatic group (the benzene ring) and an alkyl group. If functional groups are attached to the alkyl part of the group and *not* to the ring, the behaviour of the new compound tends to resemble similar alkyl compounds. For example, benzyl alcohol is classed as an **alcohol** and does not behave chemically as a phenol.

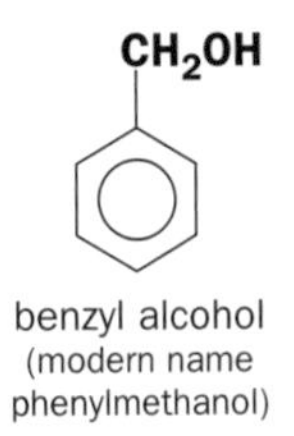

benzyl alcohol
(modern name
phenylmethanol)

BOX 18.6

Examples of reactions of ethanol, phenol and benzyl alcohol

Phenol	Ethanol	Benzyl alcohol
No ester formed when reacted with an organic acid	ester formed	ester formed
No simple product formed on oxidation	aldehyde/acid formed on oxidation	aldehyde/acid formed on oxidation
Reacts with neutral $FeCl_3$ to give a violet complex	no colour produced with the same reagent	no colour produced with the same reagent

Revision questions

18.1. Write the structural formulae of the following:
(i) hexan-2-ol
(ii) 2-methylpropan-2-ol
(iii) butane-1,3-diol
(iv) 3-chlorophenol.

18.2. Arrange in order of *increasing* acidity:
ethanoic acid, ethanol, phenol, water.

18.3. Write the structural formulae for
(i) 3-methylhexanal
(ii) pentane-2,4-dione
(iii) phenylethanone
(iv) 2-methylpropanoic acid
(v) butane-1,4-dioic acid.

18.4. What would be the likely product of the oxidation of ethane-1,2-diol with excess acidified dichromate?

18.5. Write the structures of the organic products of the following reactions:
(i) The reaction when propanoic acid is heated with ethanol in the presence of concentrated sulfuric acid.
(ii) The reduction of propanoic acid with lithium aluminium hydride.
(iii) The oxidation of pentanal with sodium dichromate.
(iv) The oxidation of ethanal with Fehling's solution.

18.6. Write the structural formula of the following derivatives of methanoic acid:
(i) the propyl ester
(ii) the amide
(iii) the acid chloride.

18.7. The compound 2-amino-3-hydroxypropanoic acid is more commonly known as serine and is an amino acid:
(i) Write the structural formula of serine.
(ii) Does serine exist as two different optical isomers?
(iii) What would be the structure of the product formed when serine is treated with acidified sodium dichromate solution?
(iv) Write the structural formula of the dipeptide Ser-Ser.

18.8. Write an equation for the acid-catalysed reaction of methanol with benzoic acid.

18.9. Write the structural formula of a fat derived from the esterification reaction of one molecule of glycerol and three molecules of octadecanoic acid ('octadec' = 18).

18.10. Identify the functional groups in the following well-known compounds:
(i) the synthetic fibre Dacron

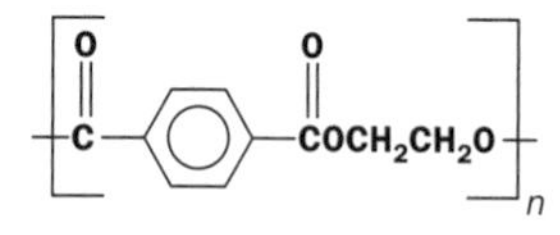

(ii) amphetamine

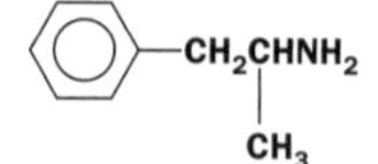

(iii) aspirin

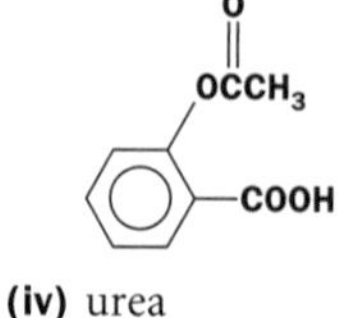

(iv) urea

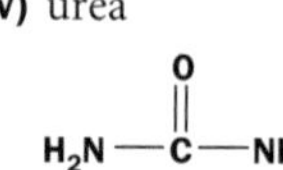

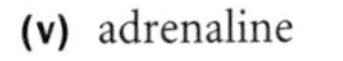

(v) adrenaline

OH
HO $CHCH_2NHCH_3$
HO

(vi) the polymer Saran (used for kitchen wrap).

$$\left[CH_2CHCl - CH_2CCl_2 \right]_n$$

18.11. A carbonyl compound X has been found by analysis to have the molecular formula C_3H_6O:

(i) Describe a chemical test to prove that X contains a carbonyl group.

(ii) X is an aldehyde. Describe how you would experimentally confirm this fact.

(iii) Write down the structural formulae of the compound(s) that X could be. How could you confirm the identity of X?

Extension material to support this unit is available on our website at **http://www.palgrave.com/foundations/lewis.**

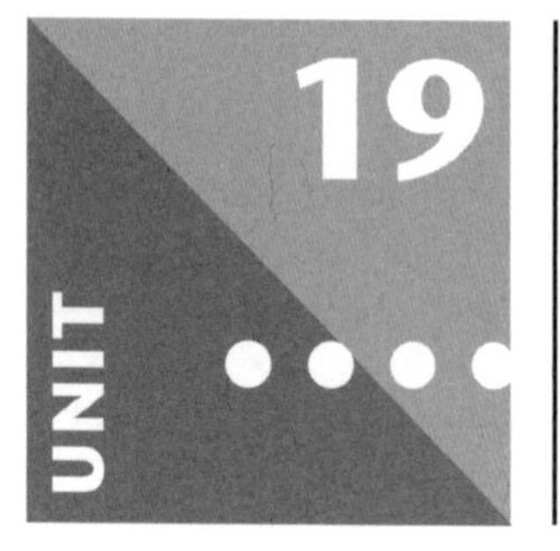

Separating Mixtures

Contents

Objectives

- Describes the techniques used by chemists to separate mixtures
- Demonstrates how to separate solids from solutions or suspensions, liquids from mixtures of liquids and solids from solid mixtures
- Explains what is meant by the terms ion exchange and solvent extraction
- Discusses the various types of chromatography

Often in practical chemistry, only one substance is needed from a mixture of substances. The substance has to be **separated** from the mixture.

19.1 Separating a solid from a liquid

A solid that is mixed with a liquid may be **insoluble** in the liquid, or it may have **dissolved.** When an insoluble solid floats in a liquid, the mixture is called a **suspension.**

Insoluble solids

Insoluble solids can be separated by **filtration**. The solid is trapped by the filter paper, while the liquid passes through it. A simple example of this technique is the separation of a mixture of fine sand in water. A residue of sand collects in the filter paper and the liquid that drips through (in this case water) is called the **filtrate**. This type of filtration is called **gravity filtration** (Fig. 19.1).

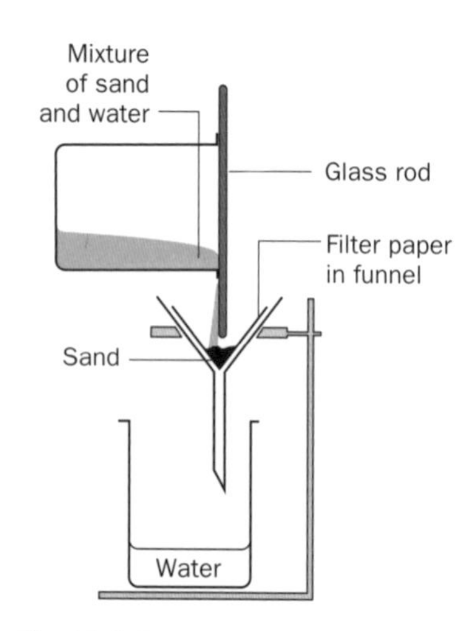

Fig. 19.1 Gravity filtration.

Small amounts of a fine suspension can be separated in a **centrifuge**. Tubes of the suspension are spun around very quickly and the solid collects at the bottom of the tube, leaving clear liquid at the top. The clear liquid can then be removed with a pipette, leaving the solid in the tube (Fig. 19.4).

BOX 19.1

Folding a filter paper

The diagrams in Fig. 19.2 show how a filter paper is folded for gravity filtration:

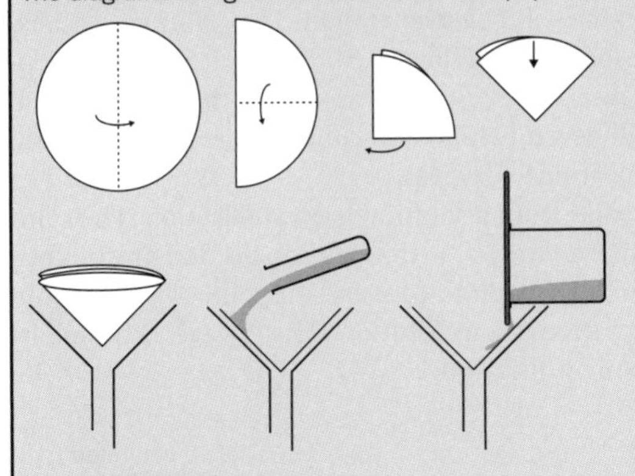

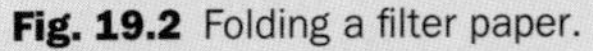

Fig. 19.2 Folding a filter paper.

Gravity filtration has the disadvantage of being rather slow. The process of filtration is speeded up using **vacuum filtration**, but more sophisticated equipment is required. A vacuum is used to 'pull' the liquid through the filter paper (Fig. 19.3).

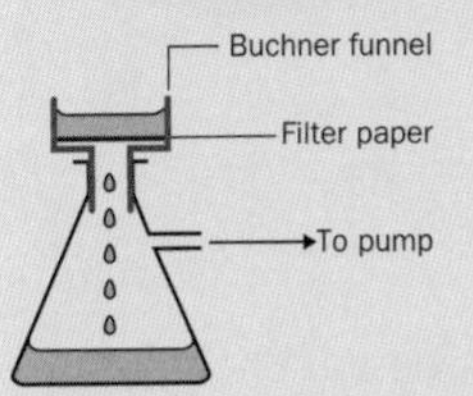

Fig. 19.3 Vacuum filtration.

A special funnel (called a **Buchner funnel**) is used which is connected, via a rubber seal, to a thick-walled flask with a side arm. A pump is attached to the side arm of the flask, and when the vacuum is applied the solution in the funnel is filtered rapidly. The filter paper lies flat inside the bottom of the funnel. The liquid is sucked through the filter paper, leaving insoluble solid behind. If the liquid is volatile, for example ethoxyethane (diethyl ether), sucking air through the filter paper for a few minutes after all the liquid has been pulled through is sufficient to dry the solid.

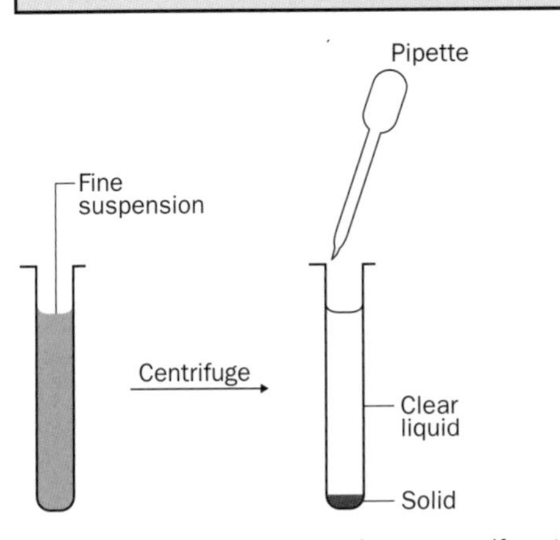

Fig. 19.4 Removing the liquid from a centrifuge tube.

Solid dissolved in a liquid (in solution)

The solid cannot be separated from the solvent by filtering or centrifuging, because now the solid particles are too small.

The solvent can be removed from the solid by **evaporation**. The solution is heated until the solvent boils off, leaving the solid behind (Fig. 19.5). This is often used for separating salt from salt solution. However, there are disadvantages with this method. If impurities are in the solvent they will be left behind and contaminate the solid. Also, some solids may decompose during heating to dryness.

A better method of recovering a solid from a solution is **crystallization**. The solution is heated and some of the solvent evaporates, so the solution becomes more concentrated. The concentrated solution is then left to cool and the solid crystallizes out (Fig. 19.6), with impurities tending to remain in solution. The crystals can then be separated from the remaining solution by filtration.

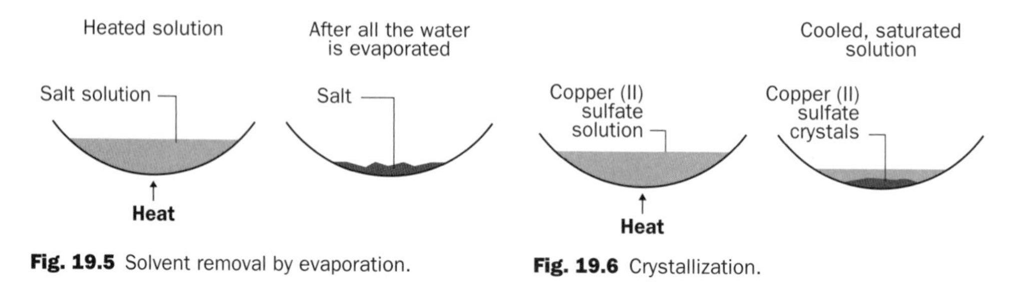

Fig. 19.5 Solvent removal by evaporation.

Fig. 19.6 Crystallization.

In order to recover the solvent from a solution, **distillation** is used. The solution is heated in a flask, until it boils, then the solvent vapour is passed through a **condenser** where it is cooled and condensed. The apparatus is shown in Fig. 19.7. The condenser consists of a glass tube surrounded by a glass jacket through which cold water is circulated. In this example, steam from the boiling salt solution passes through the condenser and condenses into water which drips into the beaker. Salt is left behind in the flask.

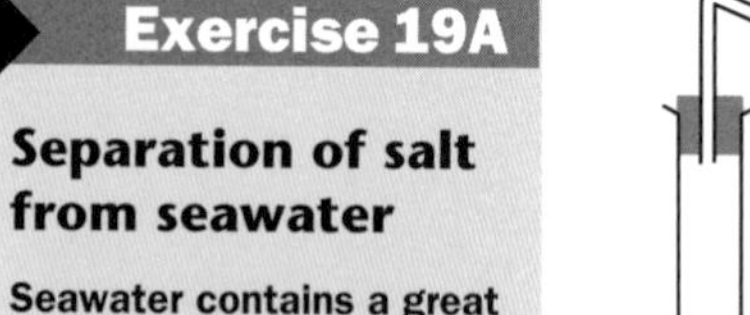

Exercise 19A

Separation of salt from seawater

Seawater contains a great deal of salt. Describe how you would separate a sample of salt from seawater. Do you think the sample would be pure NaCl?

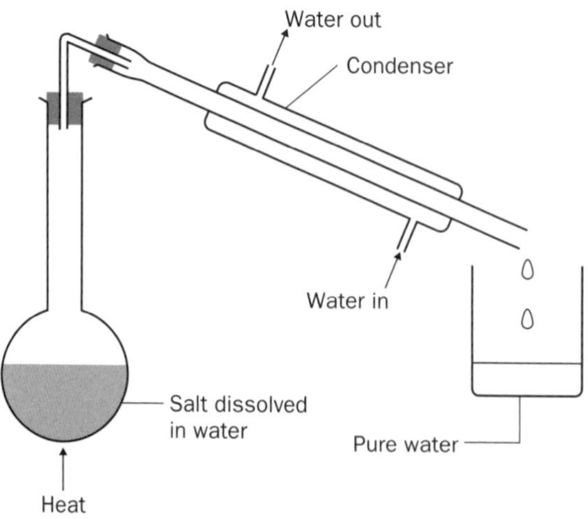

Fig. 19.7 Simple distillation.

BOX 19.2

Recrystallization

Recrystallization is used to purify a product. Many chemical preparations give 'dirty' products in which the product is contaminated with, for example, excess reactants or products from side reactions. A solvent is found that dissolves the desired product when the solvent is hot, but the product is not soluble when the solvent is cold. This choice of solvent is the difficult part and involves some 'trial and error' experiments. The impure product is dissolved in the **minimum** of the hot solvent and the hot solution is filtered very quickly to remove insoluble impurities before the mixture cools down – Buchner filtration (*using warm apparatus*) is useful here. The filtered solution is allowed to cool and the product crystallizes out. Any soluble impurities (since they are present at low concentration) tend to remain in solution and the desired product crystallizes out in a more pure form. If the product is required to be very pure, several recrystallizations may be necessary.

Marie Sklodowska Curie, discoverer of radium. She eventually died of radiation poisoning.

Marie Curie, Nobel prize discoverer of the elements radium and polonium, carried out laborious recrystallization experiments in order to eventually isolate very small amounts of a radium compound from the uranium ore pitchblende, obtained from Bohemian mines. The work was carried out in an old shed and took years of effort – the ore had to be crushed and stirred with various solvents to remove the many impurities present. Eventually, in 1902, about 0.1 g of radium chloride ($RaCl_2$) was obtained from several tonnes of ore. The radioactive compound ionized the surrounding air and glowed in the dark!

19.2 Separating two liquids

Two liquids may be **immiscible** (they do not mix and two layers are formed) or **miscible** (they mix completely).

Immiscible liquids

The liquids can be separated using a **separating funnel**. For example, consider the separation of oil and water. This mixture will separate out so that the oil floats on top of the water. If the mixture is poured into a separating funnel (Fig. 19.8) and allowed to settle, two layers are formed. The tap is opened to let the water run out and closed before the oil layer runs throught the tap. In this way, the two liquids are separated.

Fig. 19.8 A separating funnel.

Miscible liquids

Two miscible liquids can be separated by simple distillation (Fig. 19.7) if one liquid boils at least 25 °C higher than the first. If the difference between their boiling points is less than 25°C, then miscible liquids can be separated by **fractional distillation**. The apparatus is similar to that for ordinary distillation, except that a **fractionating column** is inserted between the flask and the condenser (Fig. 19.9). A fractionating column is a glass column packed with glass beads, or some other unreactive material that has a high surface area, upon which vapour can condense.

The less volatile a liquid is, the higher is its boiling point. When the mixture of liquids is heated, the vapour of both liquids passes up the column but the vapour of the liquid with the *higher boiling point* (water in this example) will tend to condense in the column and run back down into the flask. The vapour of the *lower boiling point liquid* tends to enter the condenser, condense and pass into the receiving beaker. If the boiling points of the two liquids are not too close, good separations can be achieved. In the example chosen (ethanol and water), a liquid much richer in ethanol (the **distillate**) is collected in the receiving flask, because ethanol (boiling point 78 °C) is the more volatile component.

In distillation, the separation depends on equilibria being established between descending liquid and rising vapour and is best achieved by heating the mixture as slowly as possible. Fractional distillation can be used to separate more than two components. It is used in industry to separate crude oil into several fractions. Each fraction contains components with boiling points that are close together.

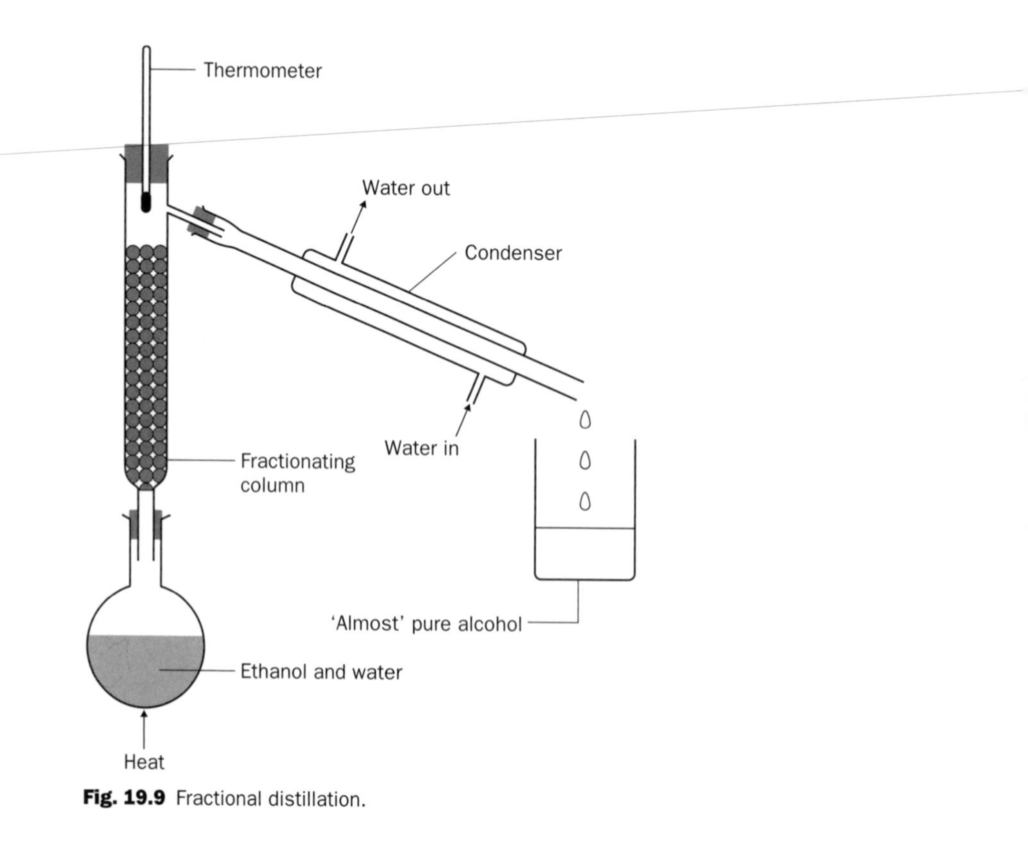

Fig. 19.9 Fractional distillation.

Exercise 19B

Separations

(i) Water and ethoxyethane (diethylether) are immiscible liquids. What apparatus would you use to separate them?

(ii) A mixture of sugar and sand needs to be separated. What techniques would you use?

(iii) The solubilities of potassium chloride and potassium chlorate in water are as follows:

Substance	Solubility in water/grams per 100 cm^3 water	
	at 20 °C	at 100 °C
Potassium chloride	35	57
Potassium chlorate	8	54

Given a mixture of 8 g potassium chloride and 25 g of potassium chlorate, how would you obtain a pure sample of potassium chlorate?

(iv) How would you separate two miscible liquids, with boiling points of 80 °C and 60 °C, respectively?

(v) What technique would you use to obtain pure water from seawater?

19.3 Separating solids

If a solid **sublimes** (passes from the solid state directly to the vapour state), this fact can be used to separate the solid from a mixture in which there are other solids. The mixture is heated and the sublimed material condensed on a cool surface.

A simple form of the apparatus, used to separate a mixture of salt and iodine, is shown in Fig. 19.10.

Often the process is carried out under reduced pressure, so that the substance can be sublimed at a lower temperature (Fig. 19.11).

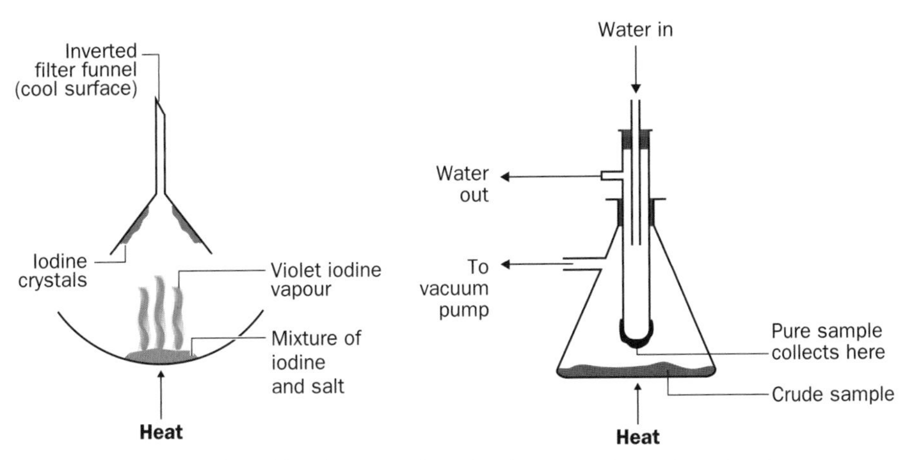

Fig. 19.10 Separation by sublimation.

Fig. 19.11 Sublimation under reduced pressure.

Ammonium chloride is another example of a solid that is *said* to sublime, but this is untrue; it actually decomposes when heated and reforms on a cold surface:

$$NH_4Cl(s) \rightleftharpoons NH_3(g) + HCl(g)$$

19.4 Steam distillation

Steam distillation is used to separate volatile organic substances that are immiscible with water from a mixture of involatile substances. It makes use of the fact that the total vapour pressure of two **immiscible** liquids is the sum of the vapour pressures of the pure liquid components, so that the boiling point of the mixture is lower than that of either of the components. This is useful for separating organic compounds that decompose at temperatures near their boiling point and therefore cannot be separated from a mixture by ordinary distillation. Steam from a steam generator is passed into the impure mixture. Vapours of the desired organic component and water condense in the condenser and are collected in the receiving flask (Fig. 19.12). The organic component can be separated from the water layer using a separating funnel.

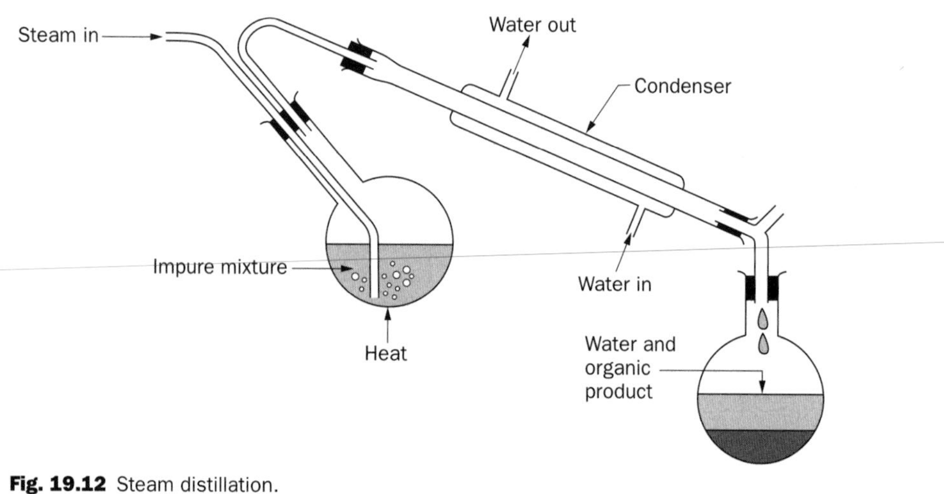

Fig. 19.12 Steam distillation.

19.5 Ion exchange

Ion exchangers remove ions from a solution and replace them with other ions from a solid phase. An ion exchanger is a column filled with small insoluble beads made of ion exchange resin (Fig. 19.13). The beads have ions loosely bound to them which are exchangeable with ions in a solution when it is passed through the column. The ions in the solution are replaced with the ions which were originally bound to the resin. After a time, the resin becomes exhausted as all its exchangeable ions have been replaced by ions from the solution. At this point the resin has to be **regenerated** and this is usually done by passing a concentrated solution of the ions which were originally bound to the resin through the column. Ions in the concentrated solution replace the ions now bound to the exhausted resin and the column is ready for use again.

For example, *hard water* is caused by the presence of calcium and magnesium ions in solution. Sodium aluminosilicate is a water softener which replaces the calcium and magnesium ions in hard water with sodium ions:

$$2NaAlSi_2O_6(s) + Ca^{2+}(aq) \rightleftharpoons Ca(AlSi_2O_6)_2(s) + 2Na^+(aq)$$

To regenerate the aluminosilicate a concentrated solution of sodium chloride is poured through the resin. Sodium ions replace the calcium and magnesium ions. The reaction is now the reverse of the one shown in the above equation.

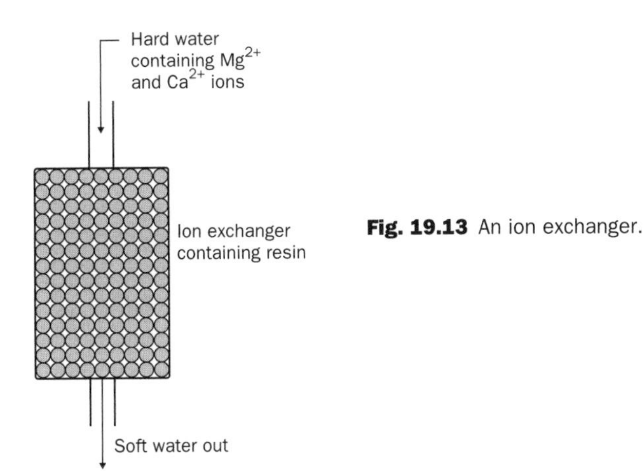

Fig. 19.13 An ion exchanger.

19.6 Solvent extraction

Frequently in organic preparations a mixture of compounds is produced. Inorganic compounds may be present and, to dissolve these, it is necessary to wash the reaction mixture with water. Now the required organic product may be in solution or suspension in water. If this mixture is shaken with an organic solvent, which is immiscible with water but in which the organic product is soluble, the organic product can be extracted from the water. Ethoxyethane (diethyl ether) is a good organic solvent for extracting organic products from aqueous solutions for the following reasons:

1. It is immiscible with water;
2. It is a good solvent;
3. It has a low boiling point and can therefore be removed easily by evaporation.

Ethoxyethane is shaken with the aqueous mixture in a separating funnel. The desired organic product moves from the aqueous layer into the ethoxyethane layer. The ethoxyethane layer can then be separated off (Fig. 19.15(a)). Note that after one separation, the water layer often still contains some of the desired organic product and, for this reason, the water layer is usually extracted with ethoxyethane a further one or two times with fresh solvent in order to collect as much organic product as possible (Fig. 19.15(b)). The separated layers are combined, then dried by shaking with, for example, a little anhydrous magnesium sulfate and then filtered (Fig. 19.15(c)). The pure, dry organic product is recovered by evaporating off the ethoxyethane (boiling point 35 °C).

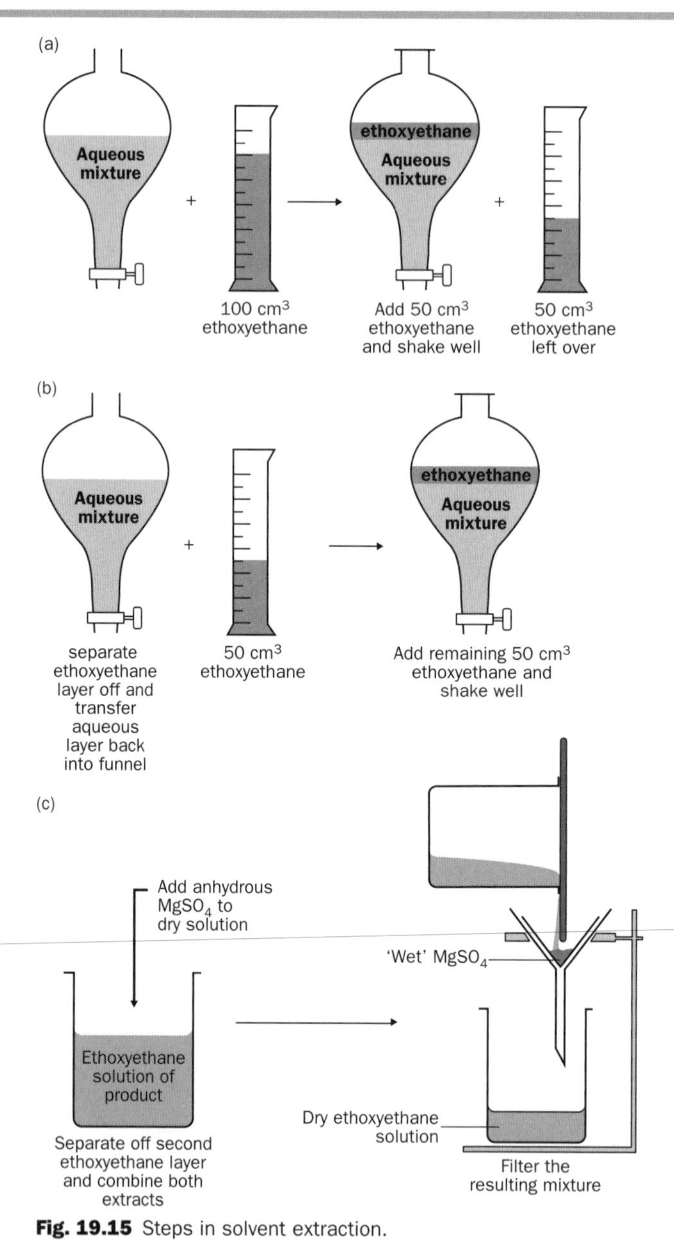

Fig. 19.15 Steps in solvent extraction.

BOX 19.3

Zeolites

Some aluminosilicates are known as **zeolites**. Zeolites are part of a class of compounds known as **molecular sieves** and consist of an aluminosilicate frame with group 1 or 2 cations trapped inside tunnels or cages. In addition to their ability to act as ion exchangers, they can selectively remove small molecules, such as water, from mixtures – hence the term 'molecular sieves' (Fig. 19.14).

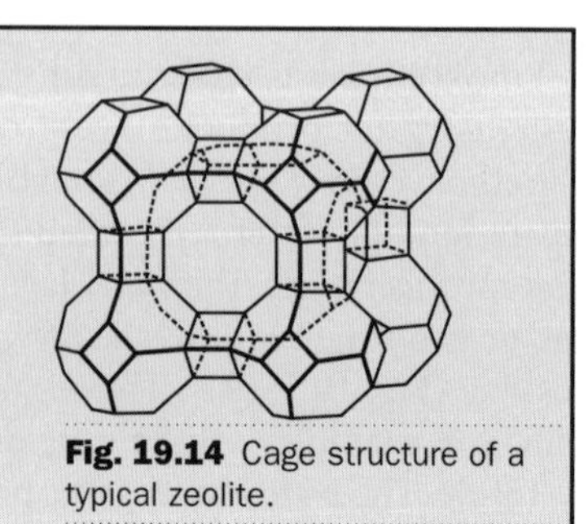

Fig. 19.14 Cage structure of a typical zeolite.

BOX 19.4

Reasoning behind multiple extractions

If a compound is shaken in a separating funnel with two immiscible solvents, such as ethoxyethane and water, some of the compound will dissolve in the ethoxyethane and some in the water. The amount of compound that dissolves in each depends upon how soluble it is in the solvents. K_d is defined as the distribution ratio

$$K_d = \frac{\text{concentration of compound in ethoxyethane}}{\text{concentration of compound in water}}$$

Note that K_d is a constant at a particular temperature. for the ethoxyethane–water system, $K_d = 4$ at 20 °C.

Supposing 10.0 g of organic product are in 100 cm^3 of water at 20 °C and 100 cm^3 of ethoxyethane are available for extraction of the compound. Is it better to use the 100 cm^3 of ethoxyethane in one go, or to perform two extractions using 50 cm^3 of ethoxyethane each time?

1. All 100 cm^3 of ethoxyethane used in one extraction

Let x g of the organic product be extracted into the ethoxyethane layer. The amount of organic product remaining in the aqueous layer is now $(10 - x)$ g.

Using the equation

$$K_d = \frac{\text{concentration of compound in ethoxyethane}}{\text{concentration of compound in water}}$$

$$4 = \frac{x/100}{(10 - x)/100}$$

Solving this equation gives $x = 8$ g. So, 8 g of product can be recovered from the ethoxyethane layer.

2. Two extractions are performed, each using 50 cm^3 ethoxyethane

For the first extraction let y g of the organic product be extracted into the ethoxyethane layer. The amount of organic product left in the water layer will be $(10 - y)$.

As before,

$$4 = \frac{y/50}{(10 - y)/100}$$

Therefore, $y = 6.7$ g.

Before the second extraction, $10 - 6.7 = 3.3$ g of product remains in the aqueous layer. Let z g of this be extracted into the ether layer. Then, the amount of organic product left in the water layer will be $(3.3 - z)$.

As before,

$$4 = \frac{z/50}{(3.3 - z)/100}$$

Therefore, $z = 2.2$ g.

The total amount of organic product obtained from the two extractions is

$(y + z)$ OR $6.7 + 2.2 = 8.9$ g.

This is nearly one more gram of product than that obtained from one extraction. More product would be obtained with more extractions, i.e. with smaller amounts of ethoxyethane, but the increase in amount of product obtained becomes smaller with each successive extraction.

Exercise 19C

Multiple extractions

(i) The distribution ratio of an organic compound X between toluene (methylbenzene) and water is 7.0. An aqueous mixture (100 cm^3) contains 1.0 g of X. The mixture is extracted once with 200 cm^3 of toluene. What mass of X is extracted into the organic layer?

(ii) Repeat the calculation but for a two-stage extraction, using 100 cm^3 of toluene each time. Do you extract more X using 200 cm^3 of toluene in one extraction, or 2 × 100 cm^3 of toluene in two successive extractions?

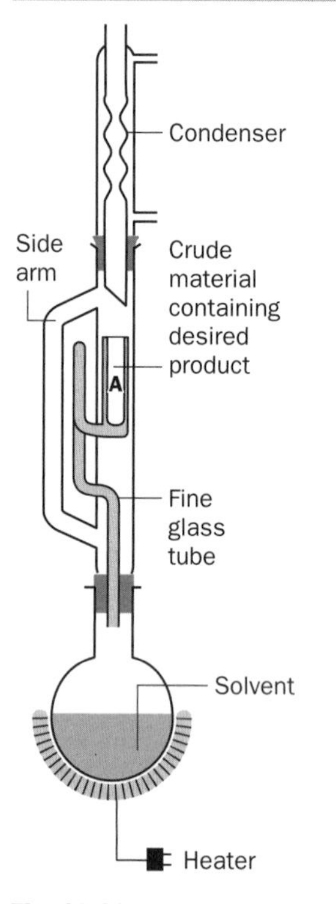

Fig. 19.16 Soxhlet extraction apparatus.

Soxhlet extraction

A compound can be isolated from a crude, solid product by repeated extraction using a hot solvent. The apparatus usually used for this purpose is a Soxhlet extraction apparatus (Fig. 19.16).

The crude material is put into a porous thimble at A, and the extracting solvent is placed in a flask at the bottom of the apparatus. The solvent in the flask is gently boiled and its vapour passes into the condenser via the side arm of the apparatus. Solvent vapour condenses and drips into the thimble, where the hot solvent dissolves some of the desired product present. When the level of solvent reaches the level of the top of the fine glass tube, the whole volume of solvent, containing dissolved product, siphons into the solvent flask at the bottom of the apparatus. The process continues automatically, until all the desired product has been extracted and is present, in solution, in the solvent flask. If the substances to be extracted are coloured, the process is stopped when the liquid being siphoned into the flask becomes pale or colourless.

BOX 19.5

Extraction of caffeine using supercritical fluid extraction

Caffeine has the molecular formula $C_8H_{10}N_4O_2$ and is present in tea, coffee and cola. It can be extracted from tea or coffee by Soxhlet extraction, using trichloromethane (chloroform) as a solvent. A disadvantage of Soxhlet extraction is that it takes *time*, often many hours. **Supercritical fluid extraction** (SFE) is a recently developed technique that can shorten extraction times considerably. The solvent used is a **supercritical fluid,** a substance that is above its critical temperature and pressure (see page 168). Carbon dioxide, for example, becomes supercritical at a temperature of 31 °C and a pressure of about 73 atm. Supercritical CO_2 is so dense that it has excellent solvating powers. It is also inert and will not react with the substances it is used to extract. For example, using this solvent, caffeine can be extracted from coffee beans in minutes. Decaffeinated coffee beans are preferentially prepared by removing the caffeine using non-toxic CO_2 as a solvent, rather than potentially toxic organic solvents such as hexane, dichloromethane or chloroform.

19.7 Chromatography

Chromatography has become one of the most important methods of separating substances from a solution. In chromatography, the solution is passed over (or through) a carefully chosen solid, such as paper or aluminium oxide. Chromatography depends upon the ability of the solid to slow down the movements of the substances in the solution. Different substances are separated because they are 'held back' by differing amounts. The 'holding back' is caused by the differing degrees of adsorption between the surface of the solid and the different substances in the mixture, or by the differing solubilities of the substances in a liquid around the surface of the solid, or a mixture of both effects. Chromatography is like a race, with the runners being the substances in the mixture. At the beginning of the race, or the start of the separation, the competitors are together, whereas by the end of the race the field is split and the individual runners are well separated.

The word chromatography literally means 'colour writing'. Chemists and biologists both use chromatography. For example, chromatographic separations are used extensively in forensic science to detect and identify trace amounts of substances. There are many types of chromatography, probably the simplest is **paper chromatography**.

Paper chromatography

Black ink is a mixture of several coloured substances. It can be separated into its components by **paper chromatography**. A small drop of black ink is spotted onto a rectangular sheet of filter paper. When the ink has dried the paper is folded into a cylinder shape and secured with a clip. The paper is then placed into a beaker containing a solvent, in this case propanone (acetone), making sure that the level of propanone is below the ink spot (Fig. 19.17). The solvent gradually rises up the filter paper by capillary action and separates the ink into a number of coloured spots. This separation happens because the different dyes present in the ink have different solubilities in the water surrounding the cellulose particles of the paper (the **stationary** phase) and the solvent (the **moving** phase). The dyes most soluble in the moving phase are carried furthest up the filter paper in a given period of time, whereas those most soluble in the water on the cellulose tend to lag behind. The different dyes distribute (**partition**) themselves between the water on the cellulose and the solvent moving over the cellulose. When the solvent reaches the top of the filter paper the different dyes in the ink will have separated out into different coloured spots. The paper can be taken out and dried – the resulting pattern is called a **chromatogram** (see Fig. 19.18).

This technique is generally used to identify the components in a mixture. For example, the dyes that it is thought an ink *might* contain can be spotted on the filter paper along with the ink.

The chromatogram obtained in Fig. 19.18 shows that the ink contains components A, B and C but not D because the ink has separated into three spots, which are at the same height as the spots obtained for A, B and C.

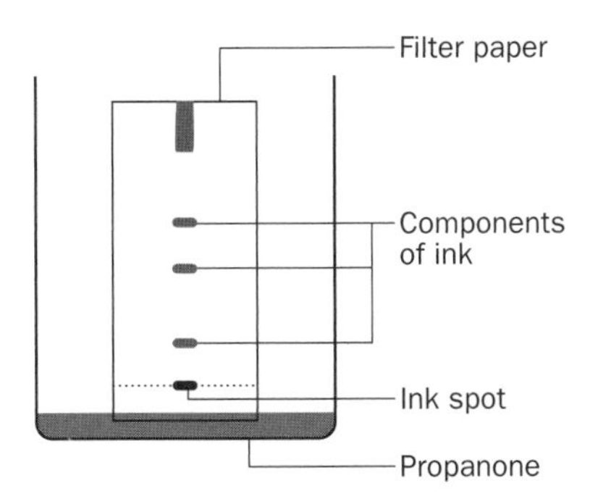

Fig. 19.17 Paper chromatography.

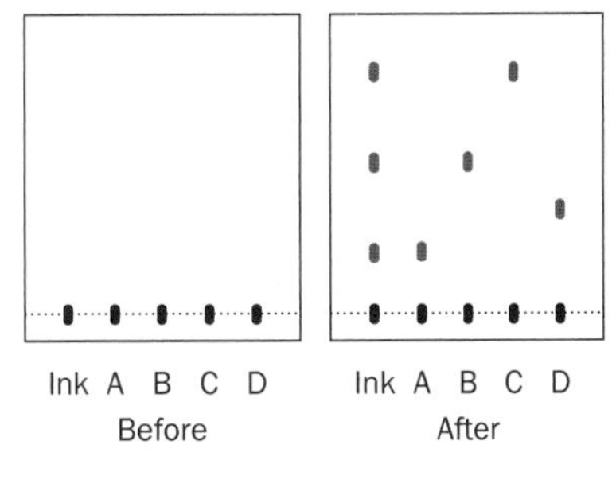

Fig. 19.18 Running a chromatogram.

Column chromatography

This is used to separate mixtures of compounds, so that each compound can be collected. A glass column is packed with an **adsorbent**, through which a solution of the mixture is passed. **Alumina** (**aluminium oxide**) or **silica gel** (prepared by adding concentrated hydrochloric acid with sodium silicate) are commonly used as adsorbents. The column must be uniformly packed so that solvent flows freely through it. The mixture is added at the top of the column and solvent is passed through the column to wash (or **elute**) the mixture down through the column. As the mixture moves through the

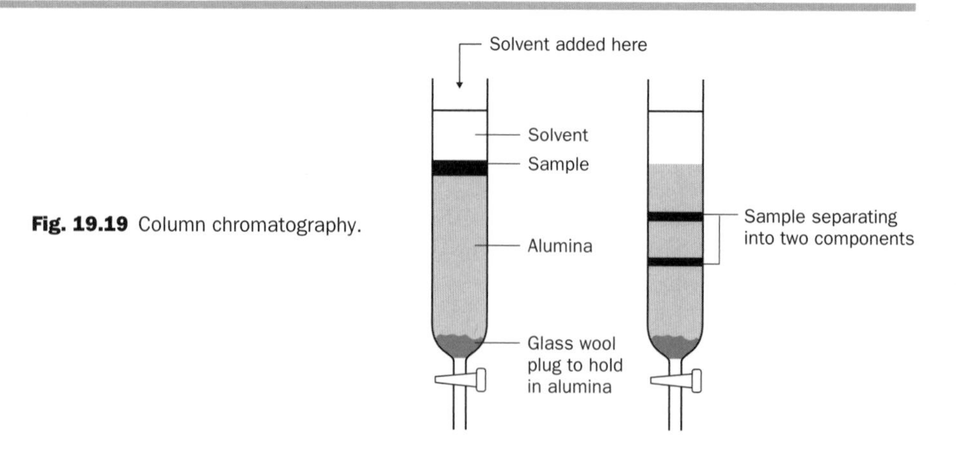

Fig. 19.19 Column chromatography.

column it separates into coloured bands of its components because the components move through the column at different rates. For example, a compound that is not strongly adsorbed but is very soluble in the eluting solvent, will move through the column rapidly. The bands are finally washed out at the bottom of the column and each one is collected in a different flask (Fig. 19.19).

Column chromatography becomes a little more difficult if the separation involves colourless compounds because the different bands cannot be detected by the eye. A series of fractions of the same volume can be collected (say 25 cm^3) and the purity of the fractions monitored by other chromatographic methods, such as TLC or GC.

Thin-layer chromatography (TLC)

Thin-layer chromatography (TLC) is a variation of column chromatography. A strip of glass is coated on one side with a thin layer of adsorbent (e.g. alumina). The substance to be tested is spotted near one end of the plate and the plate is put into a **developing jar** with a small amount of the solvent (to a level below the spot) and a cap put on the jar (Fig. 19.20). The solvent rises up the plate, separating the sample into its components. At the end of the experiment, a spot for each component is seen on the plate. When the solvent is near the top of the plate, *the position the solvent has reached is marked with a pencil* and the plate removed and dried. If the components are colourless, a number of techniques can be used to show them up:

1. Some compounds show up under ultraviolet light.
2. The dry plate can be placed in a stoppered jar with a few crystals of iodine – the crystals sublime and most organic compounds show up as coloured spots when exposed to iodine vapour.
3. The plate can be sprayed with a reagent that will react with a compound on the plate to form a coloured product.

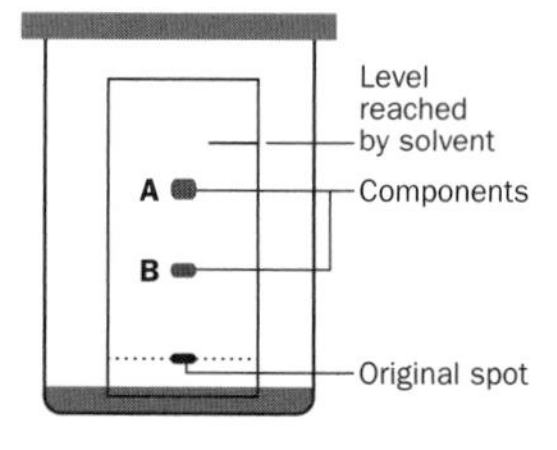

Fig. 19.20 Running a TLC plate.

TLC needs only small amounts of sample and is generally used to check the purity of the product of a chemical reaction. TLC can also be used to confirm the identity of an unknown sample. The **retention factor** (R_f value) of a compound is

$$R_f = \frac{\text{the distance travelled by the compound}}{\text{the distance travelled by the solvent}}$$

The R_f value is constant for a compound, on the same plate, if the temperature and solvent are kept the same. If a spot from an unknown substance is developed on a TLC plate together with a spot from a substance that is suspected to be the unknown, and the two substances are found to have the same R_f value, they are probably the same substance. A substance that just produces one spot when subjected to TLC is pure.

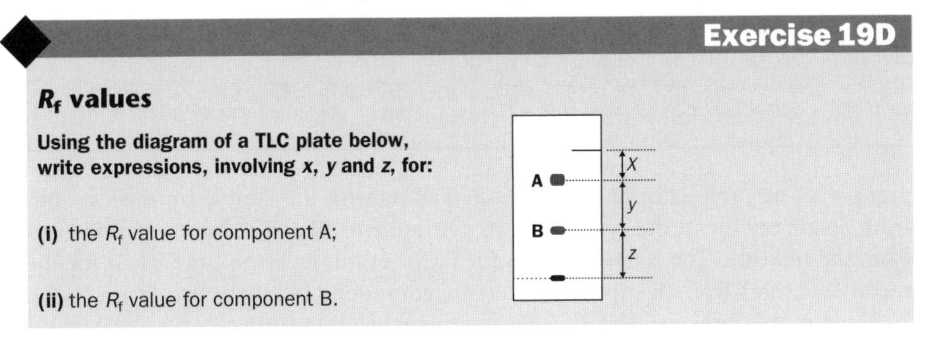

Exercise 19D

R_f values

Using the diagram of a TLC plate below, write expressions, involving *x*, *y* and *z*, for:

(i) the R_f value for component A;

(ii) the R_f value for component B.

Gas chromatography (GC)

Gas chromatography (GC) is used for checking the purity or identity of volatile liquid samples. In most experiments small quantities of sample are used and mixtures are separated, and the components identified, rather than collected. The sample is injected into the instrument, vaporized and carried in a stream of carrier gas (helium or nitrogen) on to a chromatography column. The components of the sample move through the column at different rates and reach a detector at different times. When a component hits the detector an electric signal is sent to a chart recorder, or VDU, which produces a graph, or chromatogram, of the different components passing through the detector. The time taken for a compound to pass through the apparatus and reach the detector is called the compound's **retention time**. The retention times of the components of a mixture can be compared with the retention times of known compounds obtained under the same conditions. In this way, the components of the mixture may be identified. The trace produced for each component also gives a rough idea of the relative quantity of the component. The *area* under each peak (*not* the height of the peak) is proportional to the percentage of that component present in the sample.

The chromatograph in the diagram (Fig. 19.21) shows three peaks. One of these

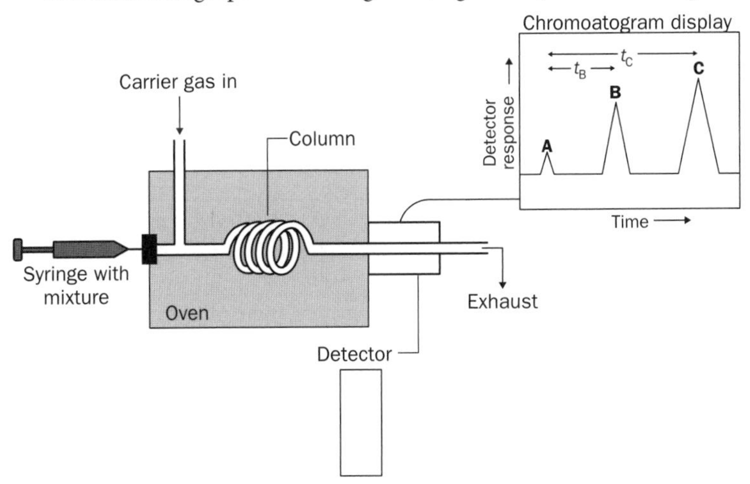

Fig. 19.21 Gas chromatogram.

BOX 19.6

Gas chromatography–mass spectrometry (GC–MS)

It is possible to 'add on' another technique to analyse the components once they are separated out from a mixture using GC. GC–MS combines gas chromatography with mass spectrometry (MS). The previously unknown organic components of a mixture are separated by GC, then passed into a MS unit where they are ionized and produce a characteristic fragmentation pattern. The fragmentation pattern of a substance can be used as a 'fingerprint' for that substance. The fingerprint of each component can be compared with a computer library of fingerprints of different substances, hopefully leading to the identification of the components of the original mixture. For example, this technique has been successfully employed in identifying trace amounts of drugs in the blood of athletes.

peaks – A – is a reference material, injected along with the sample (in this case air, from an air bubble in the syringe). Two components, B and C have separated out from the mixture. The retention time for B, for example, is the time taken for the recorder to mark from the tip of the reference compound (A) to the tip of peak B (t_B).

A summary of separation techniques is shown in Table 19.1.

Table 19.1 Separation techniques

Mixture	Separation technique
Insoluble solid in solvent	filtration (or, if a fine precipitate, centrifugation)
Solution of solid in liquid	evaporation or (better) crystallization to recover solid; distillation to recover liquid
Miscible mixture of liquids	fractional distillation
Immiscible mixture of liquids	separating funnel
Mixture of two solids, one of which sublimes	sublimation
An organic substance that is immiscible with water from a mixture of involatile substances	steam distillation
Mixture of substances in solution	paper chromatography or TLC for identification; column chromatography to collect the substances
Removal of ions (e.g. calcium or magnesium) from aqueous solution	ion exchange
Separation and identification of a mixture of volatile liquids	gas chromatography
Separation of organic components from an aqueous mixture	solvent extraction

Revision questions

19.1. What technique would you use to separate the following?

(i) Solid potassium chloride from an aqueous solution of potassium chloride.

(ii) Pure water from lemonade.

(iii) Hexane from crude oil.

(iv) Oxygen from liquid air.

(v) The colouring substance in a coloured drink.

(vi) Petrol from petrol and water.

(vii) Salt from a mixture of salt and chalk.

19.2. A green substance in grass dissolves in propanone (acetone). Describe how you would show that this green pigment contains more than one substance.

19.3. How would you separate a fine mixture of ammonium chloride and potassium chloride solids?

19.4. Screened methyl orange is a mixture of dyes which are soluble in ethanol:

(i) What techniques could you use to show that the substance is a mixture?

(ii) What technique could you use to separate and collect the dyes?

19.5. Amino acids, which are colourless and water soluble, react with a substance called ninhydrin according to the reaction:

amino acid + ninhydrin → blue/lilac product
(colourless) (colourless)

What technique could you employ to show that urine contains a mixture of amino acids?

19.6. Whisky contains a number of chemicals that belong to the alcohol family of organic compounds. If you were given a bottle of a particular whisky, what technique(s) could you use to separate and, possibly identify, the constituent alcohols?

19.7. You are a forensic scientist and have been asked to find out whether the ink from a pen of a suspect could have been used to write the signature on a forged cheque. How would you proceed?

19.8. At room temperature, the distribution ratio of butanoic acid between ethoxyethene (diethyl ether) and water is 3.5. In an experiment, 150 cm^3 of water containing 0.50 g of butanoic acid was shaken with 50 cm^3 of ethoxyethane. What mass of acid did the ethoxyethane extract?

19.9. The following diagram shows the gas chromatogram obtained when a mixture of heptane and octane was analysed:

(i) Assume the peaks are approximately triangular in shape, calculate the areas ABC and PQR. *The area of a triangle* = ½ × *base of triangle* × *height of triangle*.

(ii) What can you say about the relative amounts of heptane and octane in the original mixture?

(iii) Can you think of a method of comparing the area under the peaks that does not involve calculation and could be used even if they were not shaped like triangles?

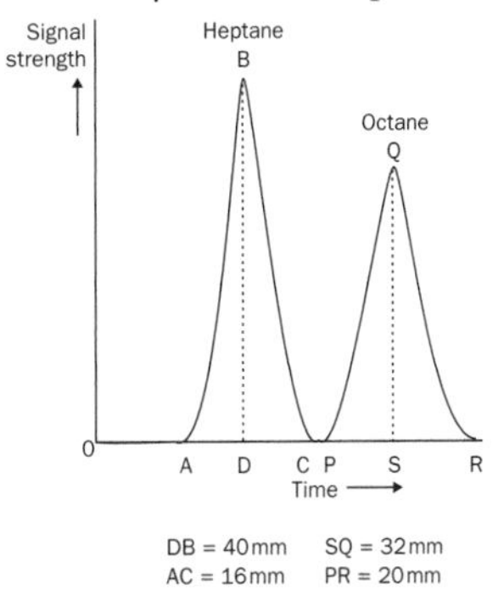

19.10. How would you show that the colours of Smarties are one single dye, or a mixture of dyes?

Extension material to support this unit is available on our website at **http://www.palgrave.com/foundations/lewis.**

Light and Spectroscopy

Contents

Objectives

- ▶ Looks at the absorption and emission of light
- ▶ Explains how spectrometers work
- ▶ Discusses the emission spectrum of the hydrogen atom
- ▶ Outlines the uses of IR and NMR spectroscopy in analysis
- ▶ Introduces the Beer–Lambert law

The interaction of light with matter may result in physical changes (such as reflection, absorption, diffraction or emission) or in chemical changes (such as dissociation or isomerization). Absorption and emission form the basis of the **spectroscopic analytical techniques** and the subject specialism is called **spectroscopy**. The chemical changes are no less important – the formation of smog, the depletion and formation of ozone, and photosynthesis itself, are all initiated by light.

20.1 Electromagnetic spectrum

We use the word 'light' to mean any radiation which is part of the electromagnetic spectrum (Fig. 20.1), although only a small wavelength region of the spectrum (roughly between 400 and 700 nm) is visible to the human eye.

All light travels at 3.00×10^8 metres per second (186 000 miles per second) in a vacuum. This velocity is given the symbol c.

Frequency and wavelength

Light may be thought of as being made up of waves or, alternatively, as being composed of particles of pure energy called **photons**.

The **wavelength** of a wave is literally the 'length of one wave' – see Fig. 20.2. It is symbolized λ and has the units of metres.

The **frequency** of a wave is the number of waves passing a point every second. Frequency is symbolized ν (nu) and has units of Hertz (Hz). (1 Hz is one wavelength

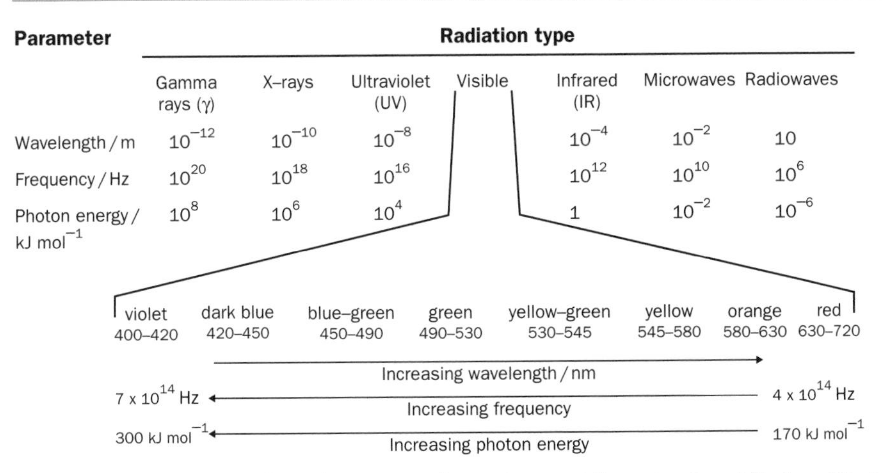

Fig. 20.1 The main parts of the electromagnetic spectrum and the order of magnitudes of their wavelengths, frequencies and photon energies. The wavelengths in the visible spectrum are listed in nanometres ($1\,\text{nm} = 10^{-9}\,\text{m}$).

per second.) The relationship between the wavelength, frequency and the velocity of light in a vacuum is

$$c = \nu\lambda$$

or

$$\nu = \frac{c}{\lambda}$$

This equation shows that

a wave with a high frequency possesses a short wavelength (and vice versa)

The **quantum theory** regards light as consisting of particles of energy, called **photons**. The energy E of one photon is given by the equation

$$E = h\nu$$

where h is a universal constant called the **Planck constant** which has the value

$$h = 6.626 \times 10^{-37}\,\text{kJ s}$$

It is sometimes more convenient to consider the energy of one mole of photons, and to use the Planck constant per mol of photons:

$$h \text{ per mol} = h \times N_A = 6.626 \times 10^{-37} \times 6.022 \times 10^{23} \approx 3.99 \times 10^{-13}\,\text{kJ s mol}^{-1}$$

Substituting

$$\nu = \frac{c}{\lambda}$$

for ν in the equation $E = h\nu$ gives

$$E = \frac{hc}{\lambda}$$

It follows from these equations that

High-energy photons correspond to high frequency (short wavelength) light.

Low-energy photons correspond to low frequency (long wavelength) light.

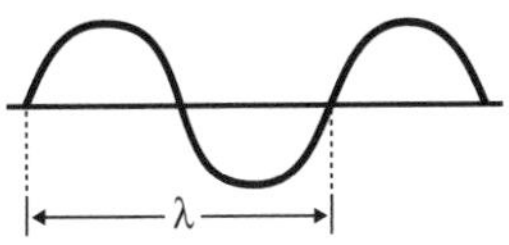

λ

Fig. 20.2 Definition of wavelength.

Exercise 20A

Frequency and wavelength

To receive BBC Radio 1 you require a radio tuned to about 98 megahertz (MHz). What wavelength is this?

Exercise 20B

Photon energies

Calculate the energy of a mole of photons of **(i)** red light of wavelength 700 nm and **(ii)** yellow light of wavelength 560 nm.

Example 20.1

Violet light has a wavelength of about 400 nm. What is the energy of one mole of photons of violet light?

▶ Answer

If we substitute hN_A for h in the equation $E = h\nu$, then E will have the units of kJ per mole of photons.

$$E = \frac{h\,N_A\,c}{\lambda} = \frac{3.99 \times 10^{-13}\,\text{kJ s mol}^{-1} \times 3.00 \times 10^{8}\,\text{m s}^{-1}}{400 \times 10^{-9}\,\text{m}} = 299\,\text{kJ mol}^{-1}$$

Therefore, the energy of the photons is about 300 kJ mol^{-1}

▶ Comment

The energy of chemical bonds are typically 200–400 kJ mol^{-1}. So we should not be surprised if the taking in of violet (and the even more energetic UV light) by molecules breaks chemical bonds and so brings about *chemical* change.

Energy transitions

The quantum theory restricts the energy of a molecule or atom to certain definite energy levels (Fig. 20.3). Movements between levels are called **transitions.**

Movement up the 'ladder' of energy levels (an upward transition) is only possible if the molecule or atom **absorbs** a photon whose energy *exactly* equals the energy gap (ΔE) involved in the jump. (This requirement is called the **Bohr condition.**) This process (termed **absorption**) is shown in Fig. 20.3. During absorption, the photon vanishes. The relationship between the energy jump and the frequency ν of the absorbed light is

$$\Delta E = h\nu$$

Emission, also shown in Fig. 20.3, is the loss of energy by the emission of photons. The equation $\Delta E = h\nu$ also applies to emission, with ΔE being the energy lost in the emission and ν being the frequency of the emitted light.

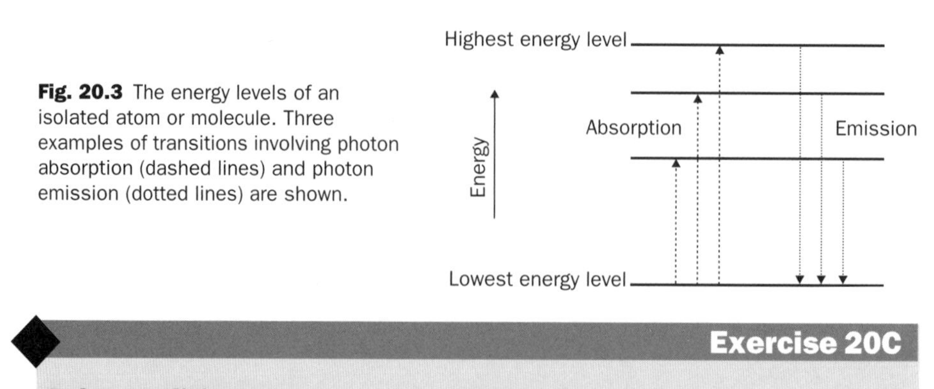

Fig. 20.3 The energy levels of an isolated atom or molecule. Three examples of transitions involving photon absorption (dashed lines) and photon emission (dotted lines) are shown.

Exercise 20C

Bohr condition

Two energy levels in an atom are 100 kJ mol^{-1} apart. What frequency of light needs to be absorbed in order for 1 mol of atoms to move from the lower to the upper level?

20.2 Energy levels of atoms and molecules

We now look at the types of energy possessed by molecules in more detail (Fig. 20.4). Atoms and molecules possess **electronic energy** – energy due to the electrons in the atomic or molecular orbitals. In addition to electronic energy, molecules (but not isolated atoms) also possess **vibrational energy** – energy which causes the atoms in molecules to vibrate.

Most of the energy of an atom or molecule is in the form of electronic energy. Making an analogy with money, electronic energy is 'big money' (pounds), whereas vibrational energy is 'small money' (pence).

The quantum theory only allows the electronic and vibrational energy of molecules to hold certain values. The energy of a molecule may be looked upon as a 'ladder' of allowed electronic energy levels or 'states' (labelled E_0, E_1 . . .) with sub-levels according to the vibrational energy (labelled v_0, v_1, v_2 . . .) possessed by the molecule. For example, a molecule in its second electronic energy level and its third vibrational level would be labelled E_1,v_2.

The lowest energy level of a molecule or atom is called its **ground state**, whereas higher levels are referred to as **excited states**. The making of excited states (by photon absorption or by raising the temperature of the sample) is called **excitation**. At room temperature nearly all atoms and molecules lie in their ground electronic and (for molecules) in their ground vibrational energy levels, i.e. they lie at E_0 and v_0. This means that, except at very high temperatures, molecular or atomic transitions involving absorption always start from the lowest energy level.

Transitions involving changes in electronic energy result in electrons moving from one orbital to another, the electrons being promoted to a higher energy orbital in absorption and demoted to a lower energy orbital in emission. The patterns observed in a **spectrometer** and which are caused by these transitions are referred to as **electronic spectra**. The transitions involve so much energy (typically $>$ 100 kJ per mole of absorbing atoms or molecules) that photons of ultraviolet or visible light are involved, and so the spectra are also called **UV–visible spectra**. In Fig. 20.4, transition 'a' involves the absorption of UV–visible light – this is what happens when a molecule of dye absorbs visible light. Transition 'c' involves the emission of UV–visible light.

Transitions in which the vibrational energy (only) of molecules is changed involve photons of infrared light, and the resulting spectroscopic patterns are referred to as **infrared spectra**. The energy jumps (ΔE) involved in infrared spectra are smaller than those involved in UV–visible spectra, and are typically 10 kJ per

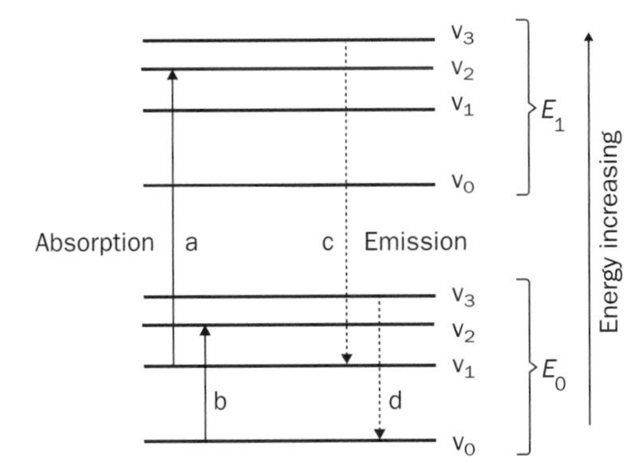

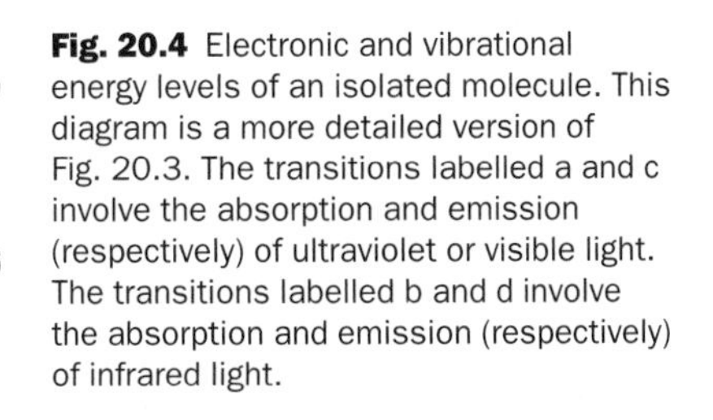

Fig. 20.4 Electronic and vibrational energy levels of an isolated molecule. This diagram is a more detailed version of Fig. 20.3. The transitions labelled a and c involve the absorption and emission (respectively) of ultraviolet or visible light. The transitions labelled b and d involve the absorption and emission (respectively) of infrared light.

mole of absorbing molecules. As a consequence of infrared absorption, the atoms in the molecule move farther in each vibration (the *amplitude* of the vibration increases). Emission of infrared light causes the atoms in a molecule to vibrate with a reduced amplitude. In Fig. 20.4, transition 'b' involves the absorption of infrared light, and transition 'd' involves the emission of infrared light. All hot objects emit infrared light. The warming of a beaker of water placed close to a heated electric filament occurs partly by absorption of infrared light.

20.3 Spectrometers

Instruments which measure and record the wavelengths at which samples emit or absorb light are called **spectrometers**. The basic components of a UV–visible or infrared spectrometer are shown in Fig. 20.5. The components include

- A **source** which continuously emits a broad range of wavelengths of light.
- A **diffraction grating** or prism which separates the wavelengths of light by spreading them at different angles and which (ideally) allows only one wavelength of light to strike the cell.
- A **cell** which is transparent to the incident light (quartz for UV, glass or perspex for visible light and solid sodium chloride for infrared). The cell contains the sample whose spectrum is required.
- An electronic **detector** (or photographic emulsion) which measures the intensity of the light beam emerging from the cell. Most commercial spectrometers are 'computer driven', and spectra are displayed on a monitor.

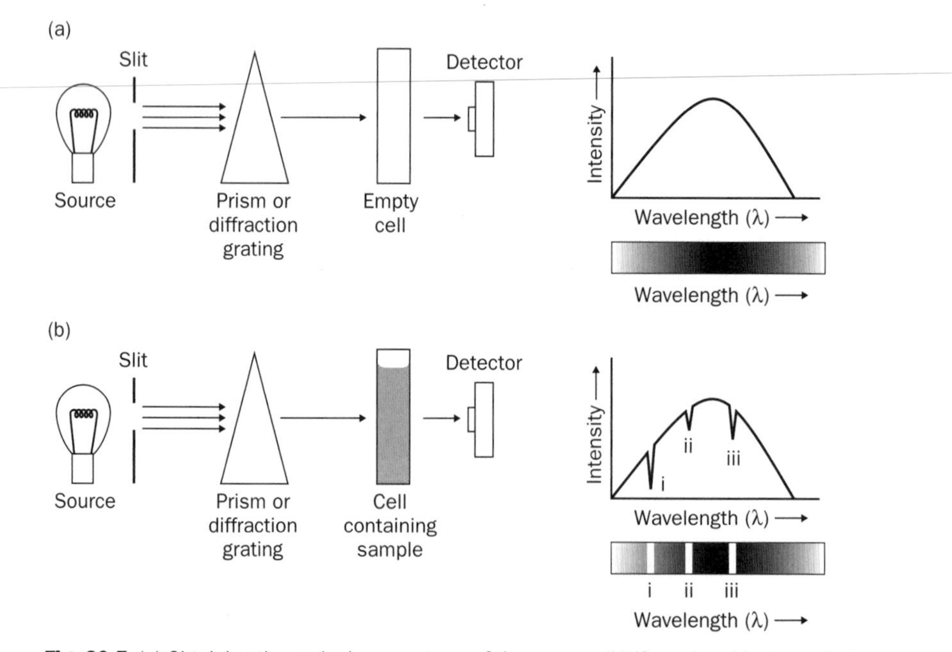

Fig. 20.5 (a) Obtaining the emission spectrum of the source. (b) If an absorbing sample is placed in the light beam, intensity measurements over the wavelength range of the source show that only certain wavelengths are absorbed. These appear as troughs in the graphical profile or as lines in the photographic emulsion.

The range and intensity of wavelengths emitted by the light source (its **emission spectrum**) is found using an uninterrupted beam (i.e. an empty cell, Fig. 20.5(a)) and may be displayed either as a graph of light intensity versus light wavelength, or in the form of a photographic emulsion in which *dark* areas register the emitted light to which it has been exposed.

If a sample is exposed to the light source, and some of the photons emitted by the light source have energies equal to the gaps between the energy levels found in the atoms or molecules in the sample, absorption may take place. The spectrum in Fig. 20.5(b) shows troughs at three wavelengths – (i), (ii) and (iii) – indicating that three different energy transitions have taken place within the atoms or molecules of the sample. The light-sensitive photographic emulsion registers the absorptions as *light* areas, i.e. wavelength regions in which the intensity of the transmitted light is small. (The light areas appear black when the emulsion is developed and printed as a photograph.)

20.4 Absorbance and transmittance of a sample

The light source in the spectrometer continuously bombards the sample cell with photons. If the sample cell is empty of absorbing material, the maximum intensity of light (at that wavelength) reaches the detector. We symbolize the maximum intensity as $I_{0(\lambda)}$ where λ is the wavelength of light selected by the grating or prism. If an absorbing sample is introduced into the beam and the sample absorbs light at that wavelength, excited states may be formed. The excited states lose their energy by collisions or by light emission, and the intensity of light reaching the detector along the cell axis is *reduced*. We symbolize the new value of intensity as $I_{(\lambda)}$. The **percent transmittance** of the sample, symbolized $\%T$, at wavelength λ is defined as

$$\%T = \frac{100 \times I_{(\lambda)}}{I_{0(\lambda)}}$$

Alternatively, the reduction in beam intensity may be expressed as an **absorbance** A_λ. Absorbance is a unitless quantity defined as

$$A_\lambda = \log\left(\frac{I_{0(\lambda)}}{I_{(\lambda)}}\right)$$

In commercial spectrometers, the instrument adjusts the diffraction grating so that successively lower (or higher) wavelengths are isolated, and then allowed to irradiate the sample. This is called a **scan** over the desired wavelength range and gives values of I at each λ. Spectrometers also obtain I_0 at each λ, either (as in **double-beam instruments**) by periodically diverting the incident beam through an empty cell during the scan, or (as in **single-beam instruments**) by computer storage of the variation of I with λ from a previous scan using an empty cell. Either way, the spectrometer has now sufficient information to produce a plot of either $\%T$ or A against wavelength. Such a plot is called an **absorption spectrum**. The absorption spectrum of the empty sample cell will be a straight line since, by definition, I equals I_0 at each λ(Fig. 20.6(a)). Scanning an absorbing sample will result in the spectrum showing peaks (Fig. 20.6(b)) which indicate the wavelengths of light absorbed.

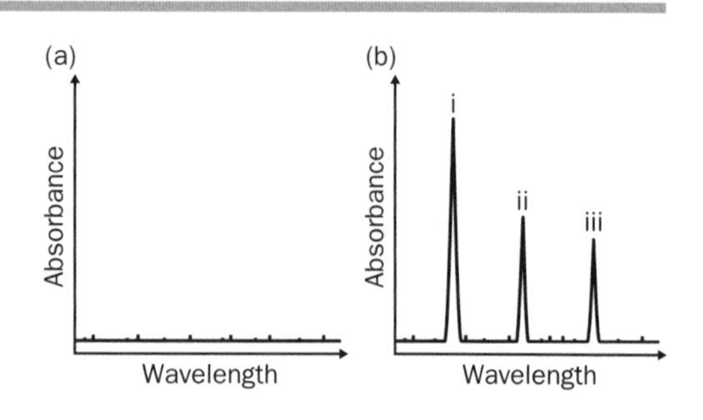

Fig. 20.6 Absorbance spectra for the case shown in Fig. 20.5. (a) The absorption spectrum of the spectrometer with an empty sample chamber (or with a non-absorbing sample). The small peaks are due to electronic 'noise' from the detector. (b) The absorption spectrum of a sample which absorbs at three wavelengths.

Use of spectra

Spectra are useful for two reasons:

1. They may be used as a 'fingerprint' of a compound or element. This is useful in chemical identification (**qualitative analysis**). Infrared and NMR absorption spectra are widely used in the laboratory for this purpose.
2. The absorbance of a substance at a particular wavelength is proportional to the concentration of that substance – a relationship called the Beer–Lambert law (see p. 392). Similarly, the intensity of light emission by a substance increases with the concentration of emitter. These relationships are useful in finding out the concentration of compounds or elements in mixtures (**quantitative analysis**).

For example, in the steel industry steel is analysed by subjecting it to a powerful electric spark. This causes the atoms in the steel to emit certain wavelengths of light ('spectral lines') which are used by the operators to deduce the elements (e.g. Mn, Cr), in addition to iron, which are present in the steel. The intensity of the lines depends upon the percentage of the element present in the sample of steel.

The remainder of this chapter discusses types of spectra (and their applications) in greater detail.

Example 20.2

40% of the light emitted by a spectrometer source at a wavelength of 460 nm is absorbed by a sample. What is the absorbance of the sample at 460 nm?

▶ Answer

If 40% of the light intensity is absorbed, 60% is transmitted. Since

$$\%T = \frac{100 \times I_{(\lambda)}}{I_{0(\lambda)}}$$

then

$$60\% = \frac{100 \times I_{(\lambda)}}{I_{0(\lambda)}}$$

or,

$$\frac{I_{0(\lambda)}}{I_{(\lambda)}} = \frac{100}{60} = 1.7$$

Therefore, the absorbance is log(1.7) = 0.23.

Example 20.2 *(continued)*

▶ Comment

It is usually unwise to deal with samples whose absorbances are greater than 2.0. An absorbance of 2.0 corresponds to a percent transmittance of 1% so that 99% of the incident light has been absorbed at that wavelength. At higher absorbances, the percentage of transmitted light is so small that it becomes comparable in size with light losses in the spectrometer optics, and experimental measurements are therefore subject to considerable error. (Now try Exercise 20D.)

Exercise 20D

Absorbance

(i) Calculate the absorbance of a sample where 10% of the light is absorbed at the monitoring wavelength.

(ii) The absorbance of a solution is 3.00. What percentage of the incident light is transmitted by the sample at this wavelength?

20.5 More about ultraviolet and visible spectra

Lyman emission series of the hydrogen atom

The simplest type of electronic spectra result from transitions within the simplest atom, hydrogen. The emission spectrum of the hydrogen atom at ultraviolet wavelengths consists of a series of emission peaks (or in a photographic emulsion, *dark* lines) called the **Lyman series**.

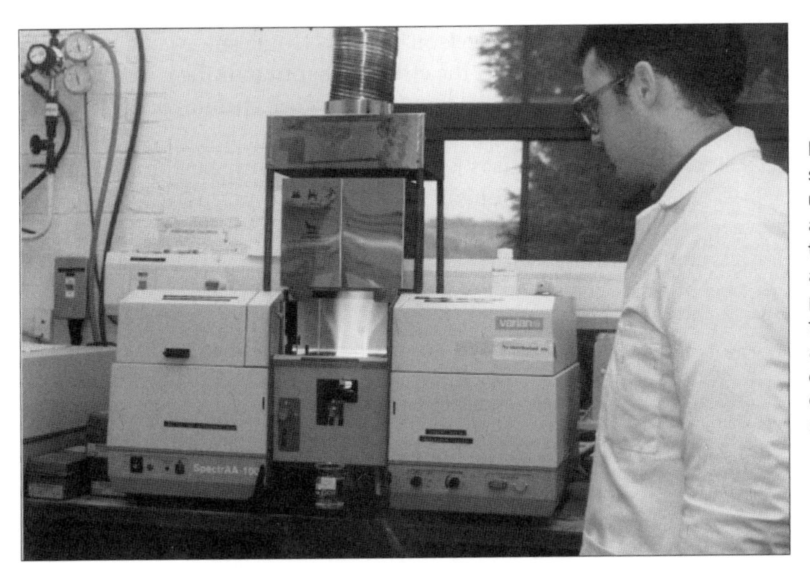

In an atomic absorption spectrometer, a sample is turned into atoms (atomized) using a hot flame. UV or visible radiation at a suitable wavelength is passed through the atomized sample. The concentration of absorbing atoms in the sample is proportional to the observed absorbance at that wavelength. The technique allows accurate measurements to be made of the concentrations of trace metals (e.g. Cu, Pb, Cr or Mn) even where the concentrations are below 1 mg per dm^3.

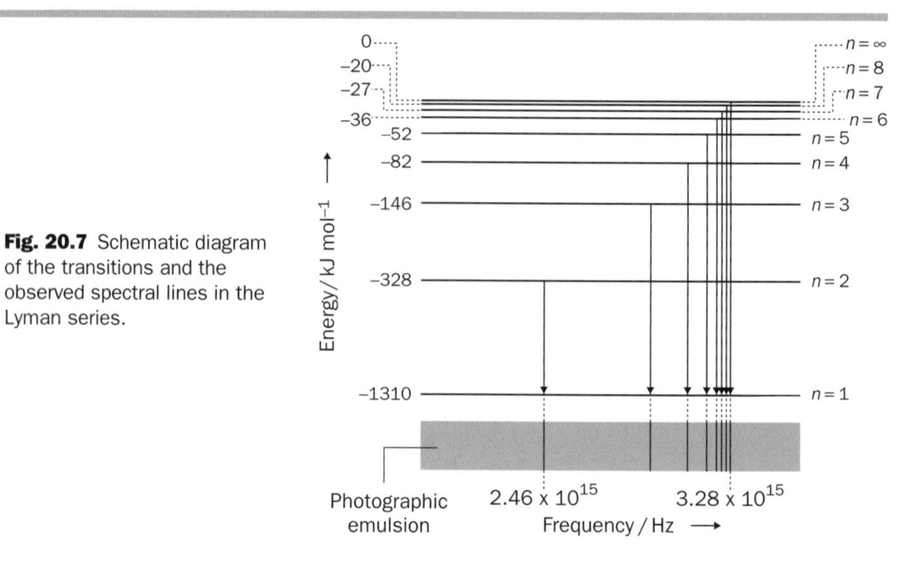

Fig. 20.7 Schematic diagram of the transitions and the observed spectral lines in the Lyman series.

Experimentally, the emission spectrum of the hydrogen atom is obtained by passing an electric spark through a quartz tube containing hydrogen gas. The energy from the spark dissociates some of the hydrogen molecules into atoms. Some of the dissociated hydrogen atoms have so much energy that they are in an excited electronic state. Although many of these excited atoms lose their excess energy by collisions with other atoms or molecules, some lose the radiation as light. The emission spectrum is obtained by passing the emitted light through a spectrometer. Measurements of the Lyman series must be made in equipment from which air has been evacuated. This is because the emission lines appear in the range 90–120 nm and both nitrogen and oxygen gases absorb UV radiation at these wavelengths.

The spectrum of the Lyman series, obtained using a photosensitive emulsion as detector, is shown at the bottom of Fig. 20.7. For clarity, all emission lines are shown as being of equal intensity.

Transitions that cause the Lyman series

The hydrogen atom contains only one electron. The electronic energy of a hydrogen atom depends upon the orbital in which the electron is found. If the electron lies in the lowest energy orbital ($n = 1$), the atom is in its ground electronic energy level (its ground state). The energy of 1 mol of hydrogen atoms in this level is -1310 kJ. (A negative energy seems strange, but this is just a convention. It means that a stationary electron just outside the attraction of the hydrogen nucleus will have zero energy.) If the electron lies in higher orbitals ($n > 1$), the atom is in an excited electronic state and its energy becomes bigger (i.e. less negative). At very high values of n the energy of 1 mol of hydrogen atoms approaches zero (see Fig. 20.7).

The transitions involved in the Lyman series involve hydrogen atoms in higher electronic energy levels losing energy by emitting light so that the hydrogen atom ends up in its ground state. We can represent these transitions as follows:

$$\underset{\text{electronically excited state}}{H_{(n>1)}} \rightarrow \underset{\text{ground state}}{H_{(n=1)}} + \underset{\text{emitted light}}{h\nu}$$

where n in the excited state takes the values 2, 3, 4, 5 etc.

These transitions are shown by downward arrows in Fig. 20.7.

Key facts about the Lyman series

- As with all spectra, the greater the energy gap involved in a transition, the higher is the frequency (and the lower is the wavelength) of the emitted light.
- Successive energy levels in the H atom get closer together. This is shown in Fig. 20.7. In consequence, the frequencies at which the lines appear in the spectrum get closer together. Eventually, the emission lines converge at 3.283×10^{15} Hz. This is the frequency of the light that would be emitted from the transition

  $$H_{(n=\infty)} \rightarrow H_{(n=1)} + h\nu$$

  The energy gap for this transition is equal in size to the (**first) standard ionization energy** of the H atom and is easily calculated:

  $$\Delta E = h\, N_A \nu$$
  $$= 3.99 \times 10^{-13} \times 3.283 \times 10^{15} = 1310\,\text{kJ mol}^{-1}$$

  If a mole of hydrogen atoms in their ground states are supplied with exactly 1310 kJ of energy, the electron in each atom will (just) be pushed out of range of the attraction of the nucleus, and all the atoms will be ionized:

  $$H(g) \rightarrow H^+(g) + e^- \qquad \Delta H^\ominus = +1310\,\text{kJ/mol}^{-1}$$

  If more than 1310 kJ of energy is supplied, the extra energy is channelled into the kinetic energy of the escaping electrons.
- There are other emission series for the H atom, one of which (the **Balmer series**) involves sufficiently small energy jumps that it produces a spectrum in the visible region. In the Balmer series, all the transitions result in the hydrogen atoms ending up in the $n = 2$ level:

  $$H_{(n>2)} \rightarrow H_{(n=2)} + h\nu$$

  The transition from $n = 3$ to $n = 2$ gives rise to a spectral line at 656.3 nm called the hydrogen atom 'alpha line' (Hα). The Hα line is easily observed in the **emission** from a hydrogen discharge lamp as a red line, but it is *not* easily observed in the laboratory in **absorption** ($n = 2$ to $n = 3$) because so few hydrogen atoms populate the $n = 2$ level at room temperature that the line is too weak to detect (Exercise 20E).

Exercise 20E

Absorption spectra

The Balmer series (including the Hα line) is readily observed in *absorption* in the spectra of stars (including our sun) – why?

Exercise 20F

Lyman series

Use the energies given in Fig. 20.7 to calculate the frequency of the light emitted in the Lyman transition $H_{(n=4)} \rightarrow H_{(n=1)} + h\nu$.

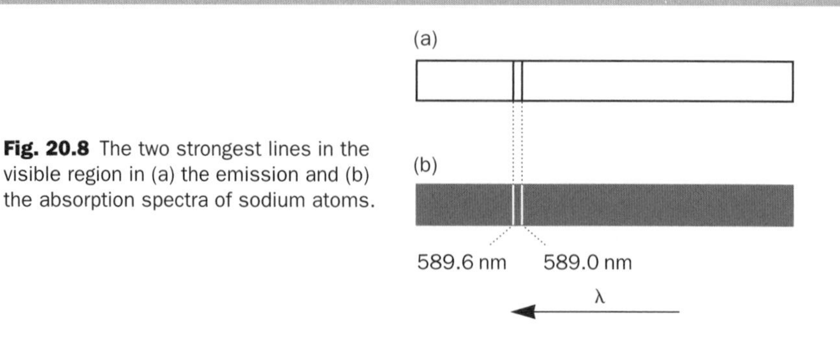

Fig. 20.8 The two strongest lines in the visible region in (a) the emission and (b) the absorption spectra of sodium atoms.

Electronic spectra of other atoms

The emission spectrum of sodium atoms is produced by passing an electric discharge through sodium vapour (sodium vapour is easily produced by heating sodium metal). Figure 20.8(a) shows how the two main emission lines in the sodium atom spectrum would appear when recorded on a photographic emulsion. When the human eye is used as a detector (as in a hand-held **spectroscope**), both lines are seen to be an intense yellow colour. The lines are responsible for the yellow colour of sodium street lights and the yellow colour observed when sodium compounds (such as salt) are sprinkled into flames. Although the lines are separated only by 0.6 nm, they can be distinguished (**resolved**) even with cheap spectroscopes.

The absorption spectrum (Fig. 20.8(b)) of sodium atoms is produced by measuring the light wavelengths absorbed when white light is passed through sodium vapour. The absorption spectrum as recorded on a photographic emulsion contains lines at the same frequency as those found in the emission spectrum, but they are light (not dark) because the lines represent a *reduction* in the intensity of the white light source at those wavelengths.

The absorption and emission spectra of metals such as copper, lithium, caesium, calcium and potassium involve lines at different wavelengths to those observed in the sodium spectrum. As with sodium, a few lines in each spectrum dominate the colours of compounds of these metals in the gas flame. This is the basis of the *flame tests* which are used to indicate the presence of compounds of these metals (see page 195).

Electronic spectra of molecules

Changes of electronic energy in *simple molecules*, such as H_2O, HCl or H_2, generally give rise to absorption and emission spectra in the UV region. Large organic molecules (including dyes) and some inorganic ions display part of their electronic spectra in the visible region. Figure 20.9 shows the electronic spectrum of potassium manganate(VII) (potassium permanganate, $KMnO_4$), in aqueous solution between 400 and 600 nm. Water and potassium ions do not absorb in this wavelength range – the absorption is solely due to the MnO_4^- ion.

Fluorescence

Some molecules absorb ultraviolet light and immediately emit light at a slightly longer wavelength. This is called **fluorescence**. For example, quinine (which gives tonic water its bitter taste) is colourless, but upon irradiation with UV light it emits a

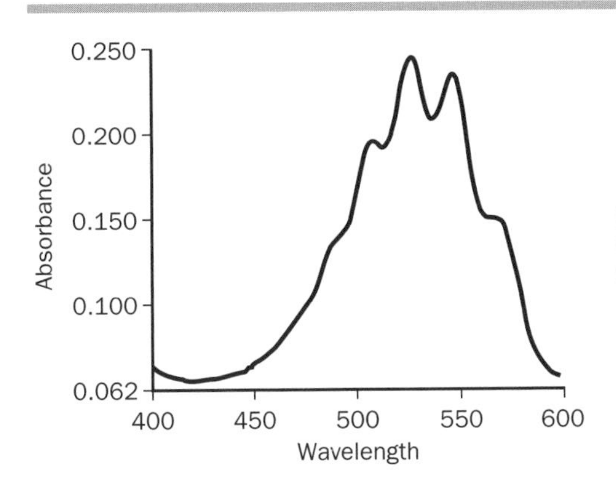

Fig. 20.9 Electronic spectrum of the manganate(VII) ion between 400 and 600 nm (perspex 1.0 cm cell).

purple glow. The emission stops as soon as the light is turned off. Other common fluorescent molecules include caffeine, benzene and 'fluorescein'. Fluorescent materials are used to label valuables ('security tagging') – the written information can only be seen in UV light.

Chemiluminescence

When fuels are burned in a flame, both heat *and* light are produced. There are some chemical reactions which produce light *but no heat.* These reactions are said to be **chemiluminescent**, and the light is emitted by electronically excited molecules or atoms produced during the reaction mechanism. Examples of chemiluminescence in nature are to be found in the fire-fly, glow-worms and in some types of fish found in the (otherwise dark) ocean depths. The familiar glow of air when subjected to high voltage discharges (including lightning) is also due to chemiluminescence.

Chemiluminescent 'fun sticks' contain reactants in separate tubes. On bending the stick, the tubes are broken and the reactants mix, producing the chemiluminescent emission. The observed colour of the emission depends upon the type of dye in the reaction mixture. The same principle is used in 'light sticks' designed to be used in emergencies by cavers and walkers, some of which continue to emit light for several hours.

> **Exercise 20G**
>
> **Chemiluminescence**
>
> The rate of chemiluminescent emission is controlled by the speed at which the excited electronic states are produced by the chemical reaction. With this in mind, what would you expect to happen to the intensity of emission from a funstick if the stick was placed in **(i)** iced water, **(ii)** warm water?

20.6 Absorption spectra and colour

Colour of compounds

The colour of a solution is controlled by the light that is *transmitted* through the solution.

Think of the manganate(VII) ion. Its solution, viewed under white light, is violet. From its spectrum, Fig. 20.9, we see that the most intense absorptions in the visible wavelengths lie between 520 and 550 nm. This is yellow-green light. The remaining colours of white light pass through the solution, and it happens that the human eye interprets this combination of transmitted wavelengths (and their relative intensities) as violet in colour (Fig. 20.10).

Similarly, a coloured solid absorbs only some wavelengths of light. The observed colour of the solid is due to the wavelengths of light that are unabsorbed, and which are *reflected* into the eyes of the observer.

It follows that the colour of a compound can be *predicted* from its absorption spectrum in the visible region.

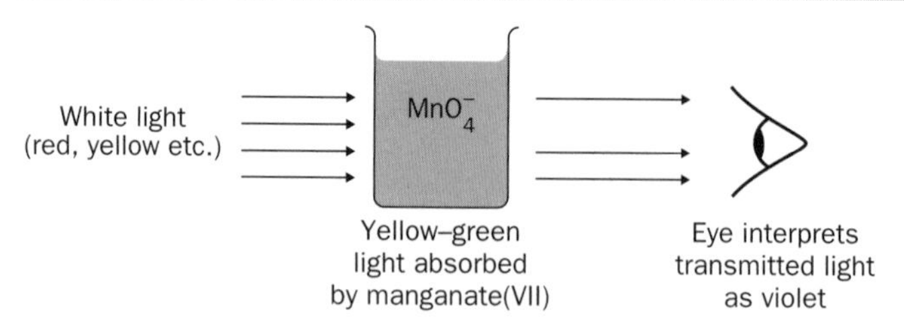

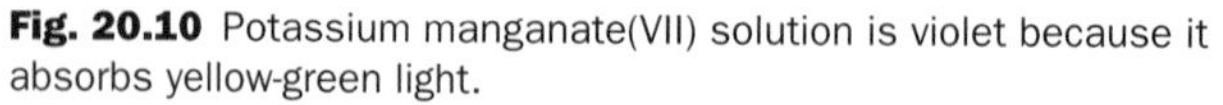

Fig. 20.10 Potassium manganate(VII) solution is violet because it absorbs yellow-green light.

Colour cheese

The 'colour cheese' summarizes the relationship between the wavelength of light most strongly absorbed by a substance or solution, and its colour in white light. To use the 'colour cheese' (Fig. 20.11):

1. Locate the approximate wavelength of light that is absorbed. The colour of the compound is shown in the *opposite* segment. Applying this to manganate(VII) ions, the absorbed light is yellow-green and so the solution appears violet. Reversing the procedure, the coloured component of carrots (β-carotene) must absorb blue-green light in order to appear orange.
2. Remember that absorptions outside the visible range (i.e. not listed in the cheese) are irrelevant to the colour of the compound. It follows that colourless substances (such as water) do not absorb in the visible wavelength range (400–720 nm).
3. Note that the colour cheese only works if *one* colour is absorbed more strongly than the others. If two or more colours are absorbed equally strongly, the observed colour depends upon the sensitivity of our eyes to the colours involved.

Exercise 20H

Predicting colours using the colour cheese

(i) Chlorophyll compounds are green coloured. What colour of light do they strongly absorb?

(ii) Copper(II) sulfate solution absorbs orange light strongly. What colour would you expect its solution to be?

(iii) Benzene shows a strong absorption between 200 and 250 nm, but does not absorb between 400 and 720 nm. Predict its colour.

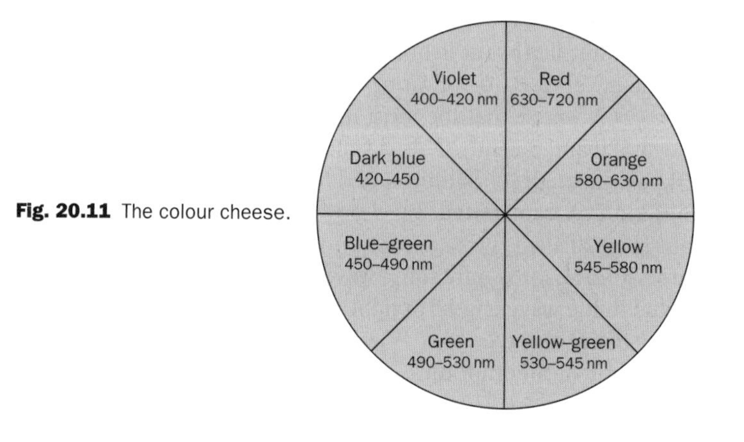

Fig. 20.11 The colour cheese.

Sunscreens are creams containing compounds which absorb the ultraviolet radiation emitted by the Sun. Skin protection is also provided by UV reflective materials (such as zinc oxide paste).

Exercise 20I

Predicting colour from spectra

Predict the colour of the dichromate(VI) ion from its spectrum (Fig. 20.12).

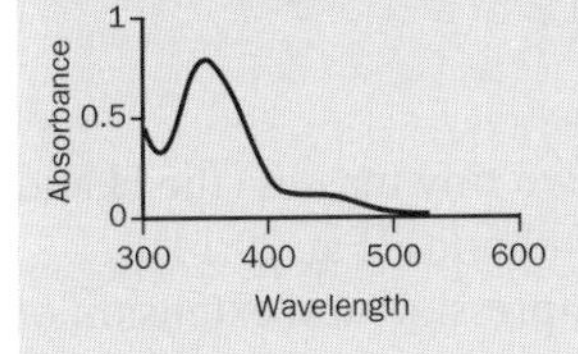

Fig. 20.12 Ultraviolet–visible spectrum of 2.5×10^{-4} mol dm^{-3} potassium dichromate solution (1.0 cm quartz cell).

Optical brighteners

Fluorescent molecules known as **optical brighteners** are used in washing powders to make clothes 'whiter than white'. Ageing fibres from white garments acquire a yellowish tinge which cannot be removed by cleaning. The fluorescent molecules in the washing powder attach themselves to the clothes during the washing cycle. The fluorescent molecules absorb UV light from the sun and emit blue light, and such emissions (when added to the reflected light of the fabric) cause the fabric to appear whiter than it would otherwise be. This is sometimes called 'blue whiteness'!

20.7 Infrared spectroscopy

Absorption of infrared radiation by compounds

To illustrate infrared absorption, we require a simple IR detector connected to a voltmeter (a suitable detector is the 'infrared sensor', sold by the Philip Harris Company). There is usually sufficient IR radiation reflected by benches and walls to provide a substantial background laboratory source of infrared light (Fig. 20.13(a)).

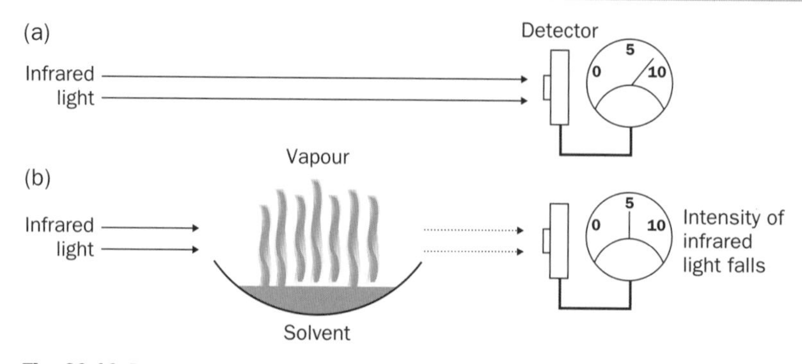

Fig. 20.13 Demonstration of infrared absorption. Vapour from the solvent absorbs some of the infrared light so that the intensity of light recorded by the detector falls.

If a watch glass of volatile solvent (e.g. ethoxyethane or methanol) is placed below the path of the infrared light, the intensity of detected infrared light falls (Fig. 20.13(b)). This reduction is caused by invisible solvent molecules in the vapour absorbing some of the infrared light. It is customary to use *wavenumber* instead of wavelength in infrared spectroscopy (Box 20.1).

'Spring model' of molecules

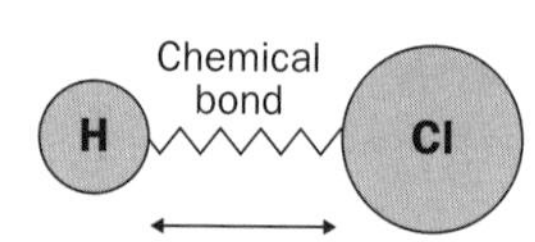

Fig. 20.14 The vibration of a hydrogen chloride molecule is caused by the stretching of its chemical bond. The molecule vibrates about 10^{14} times a second. Absorption of a photon increases the amplitude of the vibration, but does not alter its frequency.

A diatomic molecule, such as HCl, can be thought of as two tiny masses (the H and Cl atoms) connected by a microscopic spring (the covalent bond) (Fig. 20.14).

The vibration of a diatomic molecule involves the compression and extension of its covalent bond. The number of 'to and fro' movements per second is called the **vibrational frequency** of the molecule. The greater is the vibrational energy of a molecule, the higher is the amplitude of its vibration i.e. the greater is the distance travelled by the atoms during the vibration.

All molecules consist of atoms in continuous vibration – even at 0 K. Infrared absorption spectra are produced when molecules *increase* their vibrational energy by absorbing infrared light. The *electronic* energy of molecules does not change in these transitions (see Fig. 20.4). When a hydrogen chloride molecule absorbs infrared

BOX 20.1

Wavenumbers

In the infrared region, chemists depart from the practice of plotting absorbance against frequency or wavelength. Instead, spectra usually consist of absorbance values plotted against *wavenumber*. This gives more conveniently sized values. The wavenumber, symbolized $\bar{\nu}$, is the number of complete wavelengths per centimetre. This means that it is connected to wavelength by the relationship

$$\bar{\nu} = \frac{1}{\lambda}$$

When λ is expressed in centimetres, $\bar{\nu}$ is in cm^{-1} – its usual units.

Wavenumbers are *directly proportional* to the frequency ν (in Hz):

$$\bar{\nu}\ (cm^{-1}) = 3.333 \times 10^{-11} \times \nu$$

Rearrangement of this equation shows that $1\ cm^{-1}$ is equivalent to light of frequency 3.000×10^{10} Hz. A high frequency means a high wavenumber. In fact, some chemists in conversation speak of frequency when they mean wavenumber! Although incorrect, this serves to remind us that only a simple conversion factor separates the two quantities.

light, it increases its vibrational energy from the ground vibrational energy level to the second vibrational energy level (v_1). The transition involved is

$$HCl(E_0,v_0) + \text{infrared photon} \rightarrow HCl(E_0,v_1)$$

The $v_0 \rightarrow v_1$ transition is the only important transition in the infrared spectra of most diatomic molecules at room temperature. This, and the fact that diatomic molecules have only one way of vibrating – a simple stretching of the bond between the two atoms – explains why diatomics have only one major absorption peak in their infrared spectrum (Fig. 20.15).

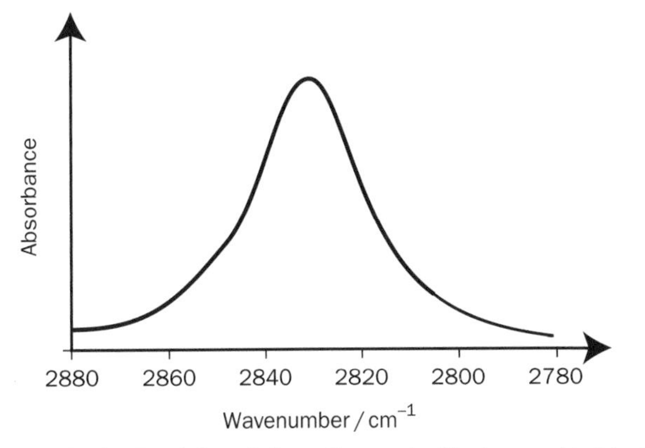

Fig. 20.15 The infrared absorption peak of hydrogen chloride gas at low resolution. (The gas is dissolved in CCl_4, but the solvent does not absorb in this wavenumber range).

Polyatomic molecules

Molecules containing more than two atoms (polyatomic molecules) may also be thought of as consisting of atoms connected by springs. For example, the propanone molecule

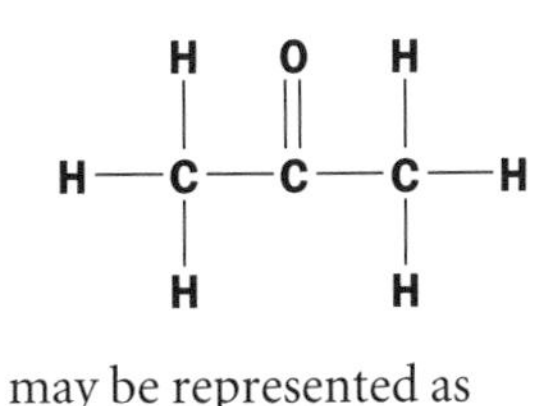

may be represented as

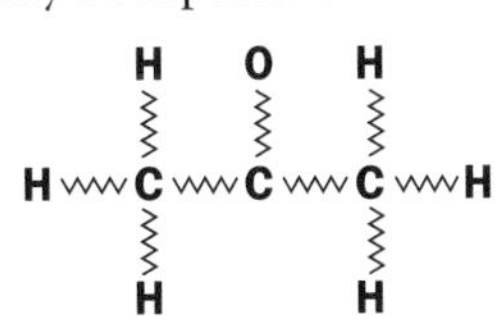

where ww are the 'springs'. The atoms on either side of bonds in polyatomic molecules may undergo **stretching vibrations** (Fig. 20.16(a)), but they may also bend against the skeleton of the molecule. These are **bending vibrations** (Fig. 20.16(b)).

If a polyatomic molecule absorbs an infrared photon, the whole molecule (like some microscopic jelly) vibrates with a higher amplitude. However, it is often found

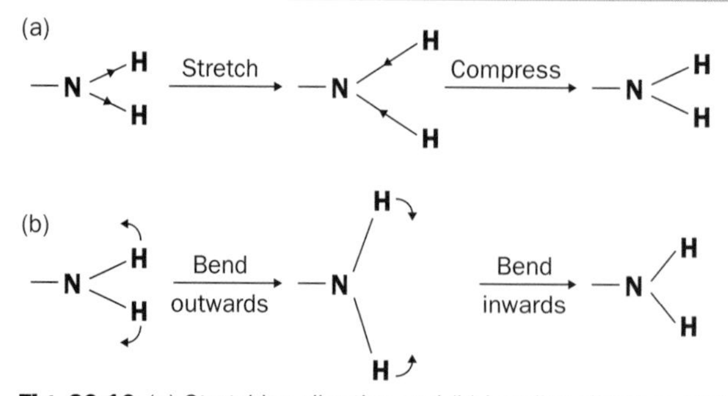

Fig. 20.16 (a) Stretching vibration and (b) bending vibration of the $-NH_2$ group.

that only a few atoms in the molecule vibrate appreciably (i.e. suffer *large* compressions, extensions or bends). (Returning to the jelly analogy, although the whole jelly is vibrating at a particular frequency, only parts of it move substantial distances during a single vibration.)

For example, propanone molecules absorb infrared light at precisely 1715 cm^{-1}. As a result of this absorption, the C and O atoms of the C=O group undergo a stretching motion at a higher amplitude. The motion of the other atoms in the molecule is largely unaffected by photon absorption, i.e. the vibration of the molecule is more or less localized around the carbonyl group. However, if propanone molecules absorb infrared light at exactly 2950 cm^{-1}, the C–H stretching motion becomes more pronounced, and the rest of the molecule – including the C=O group – moves very little.

An infrared spectrometer exposes the molecule under investigation to a range of infrared light of differing wavenumbers during a 'scan'. Peaks in the infrared spectrum occur at frequencies at which the different groups of atoms within the molecule become 'activated' at their higher amplitudes. This is evident from the spectra in Figs 20.17–20.25, where the vibrations associated with key peaks are labelled.

A simple pattern is often observed in infrared spectra in that *molecules containing the same groups of atoms give rise to peaks which are roughly at the same wavenumber irre-*

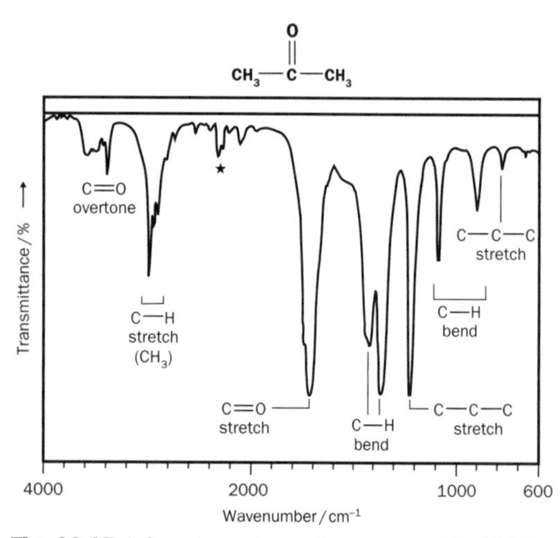

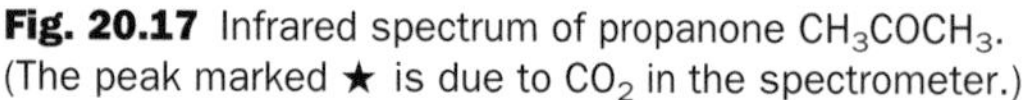
Fig. 20.17 Infrared spectrum of propanone CH_3COCH_3. (The peak marked ★ is due to CO_2 in the spectrometer.)

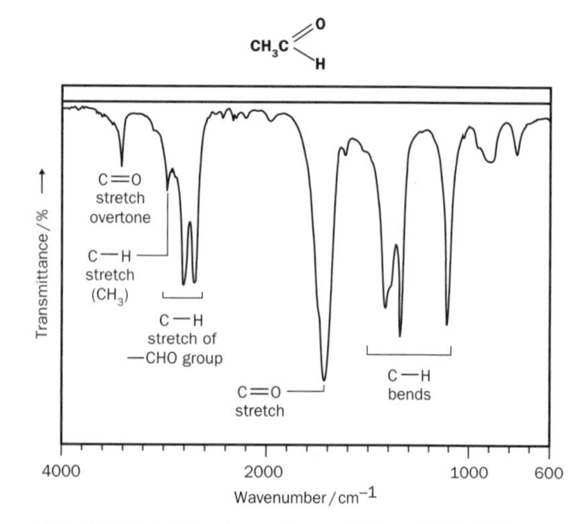

Fig. 20.18 Infrared spectrum of ethanal CH_3CHO.

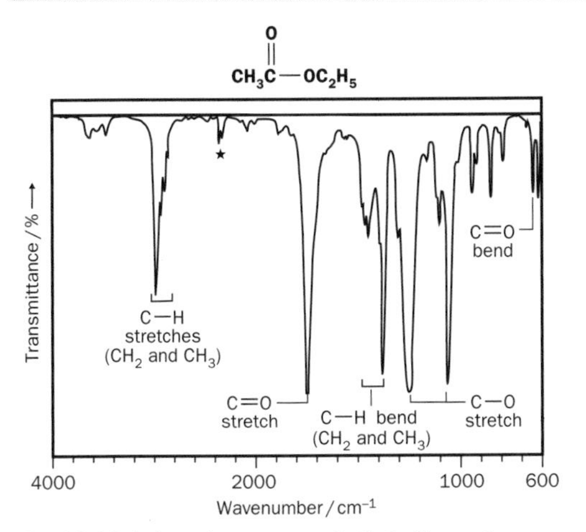

Fig. 20.19 Infrared spectrum of ethyl ethanoate $CH_3CO_2C_2H_5$.

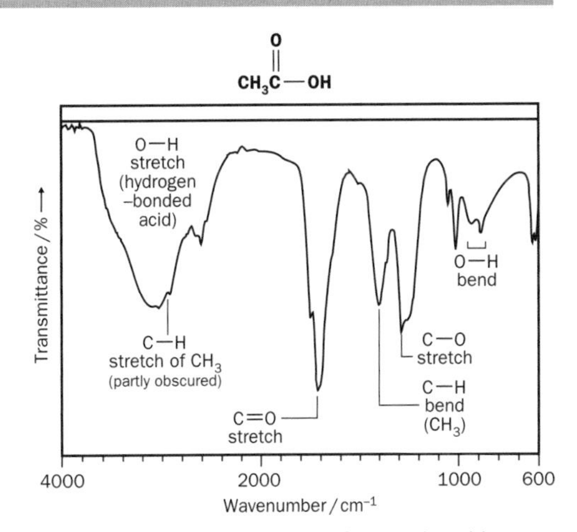

Fig. 20.20 Infrared spectrum of ethanoic acid CH_3COOH.

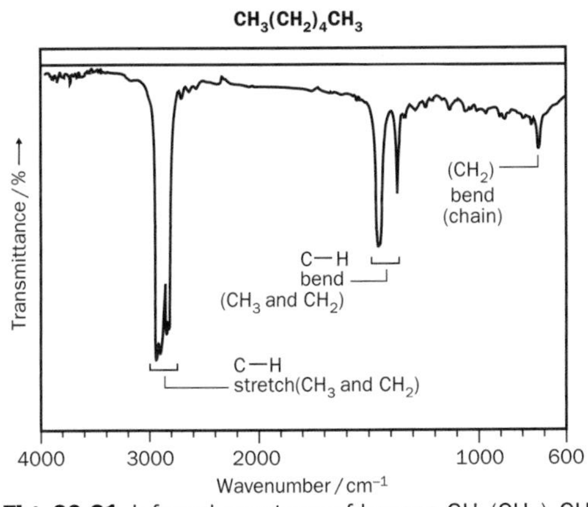

Fig. 20.21 Infrared spectrum of hexane $CH_3(CH_2)_4CH_3$.

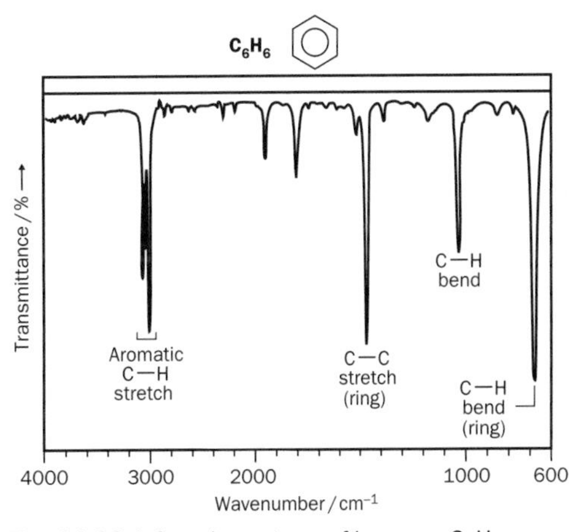

Fig. 20.22 Infrared spectrum of benzene C_6H_6.

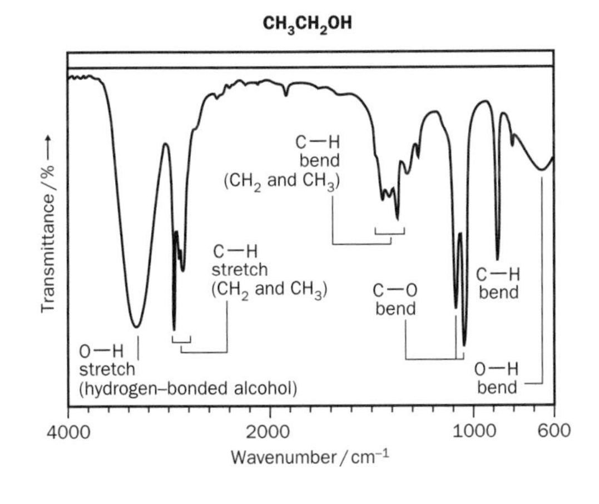

Fig. 20.23 Infrared spectrum of ethanol CH_3CH_2OH.

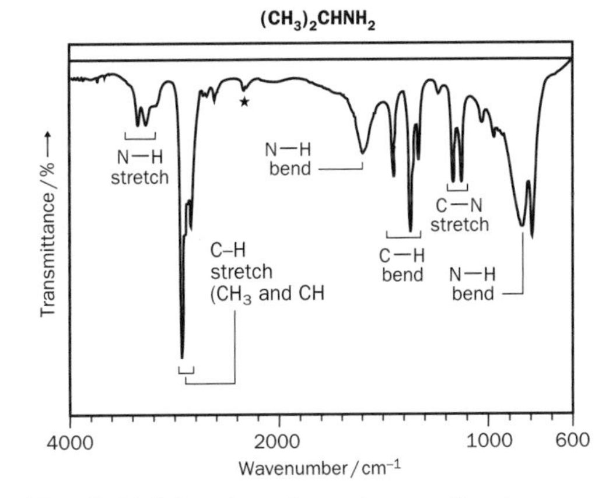

Fig. 20.24 Infrared spectrum of propan-2-amine $(CH_3)_2CHNH_2$.

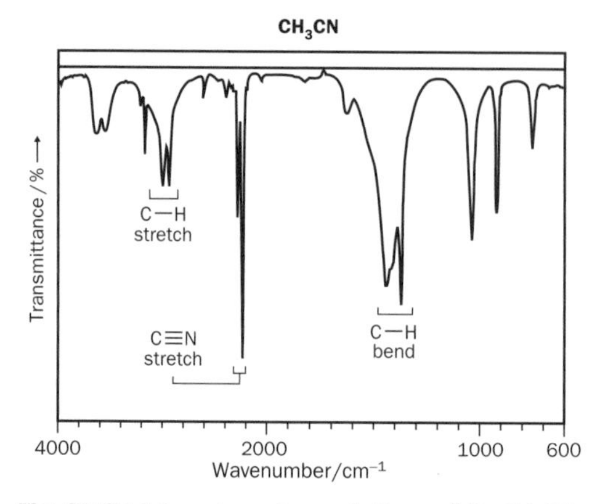

Fig. 20.25 Infrared spectrum of ethanenitrile CH_3CN.

spective of the molecule in which they are found. Typical of such groups are C=O, C–H, $-NH_2$ and –OH. For example, the higher amplitude vibration of the carbonyl group (C=O), due to the transition v_0 to v_1, is activated by infrared radiation between 1700 and 1800 cm^{-1}, the exact wavenumber varying with the molecule in question. This transition gives rise to a 'carbonyl peak' at 1715 cm^{-1} in propanone (CH_3COCH_3, Fig. 20.17), 1725 cm^{-1} in ethanal (CH_3CHO, Fig. 20.18), 1742 cm^{-1} in the ester ethyl ethanoate ($CH_3COOC_2H_5$, Fig. 20.19) and 1720 cm^{-1} in ethanoic acid (CH_3COOH, Fig. 20.20). The high intensity (peak height) and relatively narrow wavenumber range over which the C=O peak is observed in compounds makes it easy to recognise in the infrared spectrum of an unknown compound. This gives us a very useful way of identifying compounds that contain carbonyl groups. For example, photographic film consists of the ester cellulose triacetate and a spectrum of film shows an intense absorption at about 1745 cm^{-1}.

Other groups of atoms also give identifiable absorptions:

- C–H vibrations in molecules where the carbon atom of the C–H group is linked to other carbon atoms using single C–C bonds, as in

```
  H
  |   |
—C—C—
  |   |
```

  are observed as sharp peaks in a relatively narrow wavenumber band 3000–2850 cm^{-1} (see hexane, Fig. 20.21). The exceptions are C–H vibrations within the CHO group of aldehydes, which usually give two peaks between 2850 and 2750 cm^{-1} (Fig. 20.18). These are useful to distinguish aldehydes from ketones.
- C–H vibrations involving carbon atoms bonded to other carbon atoms by double or triple bonds, such as

```
H      /
 \C=C<
 /      \
```
and $H{-}C{\equiv}C{-}$

  give rise to peaks *above* 3000 cm^{-1}. C–H peaks above 3000 cm^{-1} are also observed where the vibrating C–H group is part of a benzene ring (see benzene, Fig. 20.22).
- The hydrogen-bonded –OH group of alcohols and phenols gives rise to a characteristic and very broad absorption in the range 3500–3200 cm^{-1} (see ethanol, Fig.

20.23). The –OH bond in carboxylic acids, also very broad, is found between 3500 and 2400 cm^{-1} (see ethanoic acid Fig. 20.20). The spectra of carboxylic acids is easily distinguished from that of alcohols and phenols, since acid spectra also contain the characteristic C=O peak.

- The NH_2 group gives rise to two weak absorptions (due to the N–H stretch) in the range 3500–3100 cm^{-1} (see propan-2-amine, Fig. 20.24).
- Nitriles (also known as cyanides), show a C≡N stretch near 2250 cm^{-1} (see ethanenitrile, Fig. 20.25).

The spectra in Figs 20.17–20.25 show the potential of infrared spectroscopy for distinguishing between common types of organic molecule, particularly if combined with information from other techniques such as mass spectrometry, nuclear magnetic resonance (NMR) and with the results of chemical tests.

Example 20.3

A compound P contains the following elements (percentage by mass):

C (30.6%), H (3.80%), O (20.3%), Cl (45.2%)

Its mass spectrum contains a parent ion at $m/e = 78$. Its infrared spectrum contains a very strong peak at 1800 cm^{-1}. P does not reduce Fehling's solution. Suggest a molecular formula for P.

▶ Answer

The simplest formula of the compound is first calculated from the percentage of elements by mass:

Element	Carbon	Hydrogen	Oxygen	Chlorine
Percentage by mass	30.6%	3.80%	20.3%	45.2%
Molar ratio	$\frac{30.6}{12.0} = 2.55$	$\frac{3.80}{1.0} = 3.80$	$\frac{20.3}{16.0} = 1.27$	$\frac{45.2}{35.5} = 1.27$
Simplest molar ratio	$\frac{2.55}{1.27} = 2.00$	$\frac{3.80}{1.27} = 3.00$	$\frac{1.27}{1.27} = 1.00$	$\frac{1.27}{1.27} = 1.00$
Simplest whole number ratio	2	3	1	1

Therefore, the simplest formula of the substance is C_2H_3OCl. The mass spectrum gives the mass of the parent ion as 78 u (^{35}Cl isotope), and we conclude that the molecule has the formula C_2H_3OCl.

The peak at 1800 cm^{-1} suggests the presence of a carbonyl group. Taking 'C=O' out of the formula C_2H_3OCl gives CH_3Cl. P does not reduce Fehling's solution, which excludes the aldehyde CH_2ClCHO. The structure for P is then most likely to be CH_3COCl:

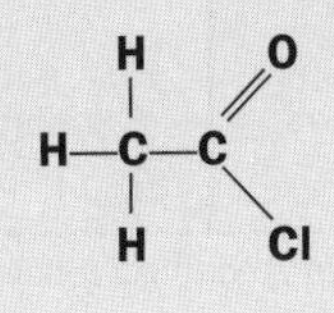

Exercise 20J

Using IR spectra in organic analysis 1

Which of the following compounds would be expected to show **(i)** a peak due to C=O, **(ii)** a broad band due to –OH, **(iii)** a peak due to C–H?

(a) butanol

(b) propanoic acid

(c) ethanal.

Exercise 20K

Using IR spectra in organic analysis 2

A compound Y contains (by mass) 19.2% nitrogen, 15.1% hydrogen and 65.8% carbon. Its mass spectrum contained an intense peak, due to the parent ion, at $m/e = 73$. Its infrared spectrum contained two peaks of medium intensity at 3300 cm^{-1} and 3370 cm^{-1}. Suggest a formula for Y.

20.8 Nuclear magnetic resonance spectroscopy

In **Nuclear magnetic resonance** (NMR) the absorption of radio waves by atomic nuclei surrounded by a powerful magnet produces spectra which are used to help decide the structural formulae of compounds.

Nuclei which have an odd number of protons or an odd number of neutrons (or both of these) behave like tiny bar magnets; they have a **magnetic moment** (they are also said to have **spin**). Just as when a small bar magnet is placed in the field of a second magnet it will be attracted or repelled, in an applied magnetic field such nuclei will have their magnetic moments aligned with the field or opposed to it. These arrangements will be of different energies as shown in Fig. 20.26, the difference in energy between the two will depend upon the strength of the external field.

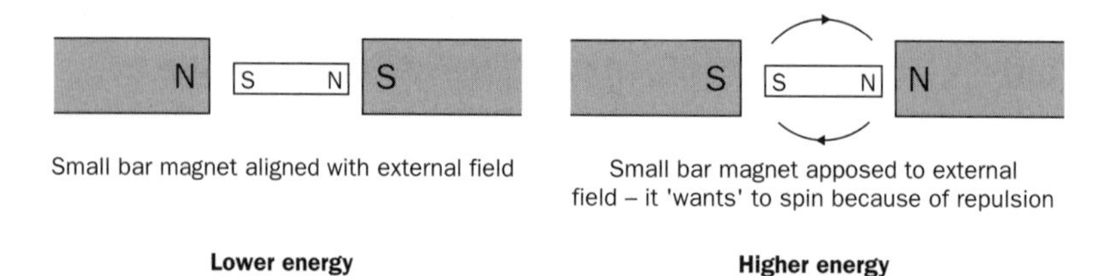

Fig. 20.26 Different alignments of bar magnets lead to different energy states.

NMR depends upon the fact that energy in the radiofrequency region of the spectrum can be absorbed by the sample and cause the energy of the nuclei in the lower energy state to jump to a higher energy; the frequencies at which the nuclei absorb radiation are referred to as **resonance frequencies** or (simply) as **resonances or signals**. The earliest (and still most important) nuclei to be studied by NMR were hydrogen nuclei (i.e. protons) and it is proton NMR (symbolized as ^{1}H-NMR) that we will deal with here.

Chemical shifts

Protons are sensitive to their chemical environment – electrons moving near them produce their own magnetic field, that changes the external field experienced by the proton. Protons in different chemical environments therefore experience slightly different magnetic fields and absorb at different frequencies.

Figure 20.27 shows the low-resolution NMR spectrum of ethanol, CH_3CH_2OH.

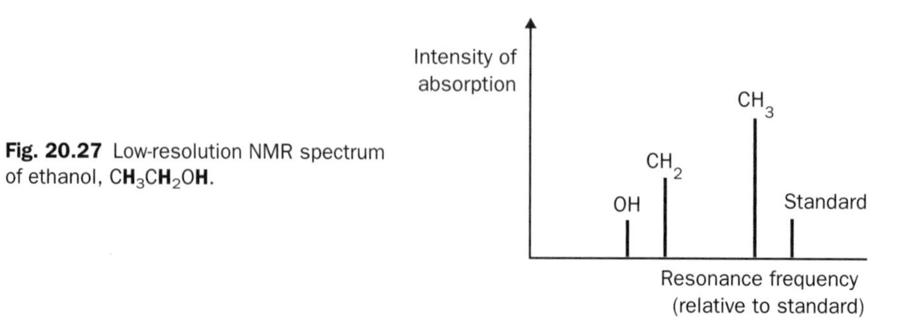

Fig. 20.27 Low-resolution NMR spectrum of ethanol, $\mathbf{CH_3CH_2OH}$.

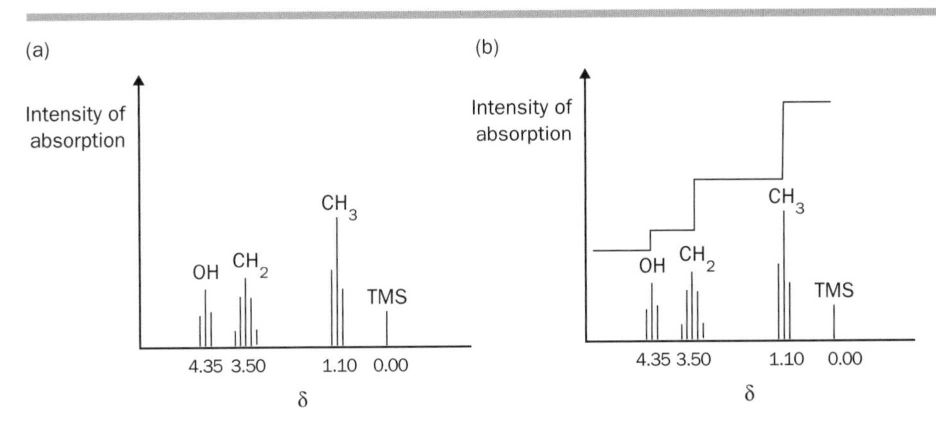

Fig. 20.28 (a) High-resolution NMR spectrum of pure ethanol, $\mathbf{CH_3CH_2OH}$. (b) The same spectrum with integration.

There are three signals because there are protons in three different environments, those present in $\mathbf{CH_3}$, in $\mathbf{CH_2}$ and the single **H** attached to the oxygen.

The resonance frequencies of the different protons are expressed as **chemical shifts** (δ) relative to a standard. **Tetramethylsilane**, (TMS), or $(CH_3)_4Si$, is widely used as a standard because it is inert and has a spectrum with a single absorption (there is only one type of hydrogen in the molecule). Different NMR spectrometers have magnets that generate different fields so, in order that scientists can compare data from different instruments, the chemical shift of a proton (δ) is calculated as

$$\delta = \frac{\text{shift from TMS for a particular proton in Hz}}{\text{spectrometer frequency in MHz}}$$

On this scale the resonance of TMS is exactly 0.00. As a result of using this system values of δ for a given proton will always be the same, no matter what instrument is used, and data is easily compared.

If the spectrum of ethanol is run under conditions of high resolution, the original peaks are seen to be split into multiplets as shown in Fig. 20.28(a). This is called **spin–spin splitting.** *Each type of proton experiences the protons on* neighbouring atoms and its resonance peak is split into $n+1$ components, where n is the number of protons on the nearest neighbouring atom(s).

Look again at Fig. 20.28(a). The resonance at δ 0.00 is due to the protons from TMS and the resonances centered at δ 4.35, δ 3.50 and δ 1.10 are due to the protons of ethanol. The ethanol resonances are split; at δ 1.10 the resonance is split into three (**a triplet**); this is because the CH_3 protons experience two protons on the next carbon atom (the CH_2 group and $n = 2$) and so their resonance splits into $n + 1$ or 2 + 1 = 3 peaks. At δ 3.50 the resonance is split into five (**a quintet**): the CH_2 protons experience four neighbouring protons (three from CH_3 and one from OH) so the resonance is split into $n + 1$, or five, peaks.

Exercise 20L

NMR signals

How many different signals will there be in the low resolution ^{1}H-NMR spectrum of:

(i) CH_3CH_2CHO

(ii) $(CH_3)_2CH{-}Br$ (two CH_3 groups attached to CH—Br)

You should now be able to work out why the resonance for the OH proton is a triplet. Notice that the intensity of the peaks of the split resonances are not all the same – these approximately follow a set of numbers known as Pascal's triangle in which each number is the sum of the left and right numbers above it (Table 20.1 gives the first five lines of the triangle).

Table 20.1 Pascal's triangle

Number of peaks	Intensities as shown by Pascal's triangle
1 (singlet)	1
2 (doublet)	1 1
3 (triplet)	1 2 1
4 (quartet)	1 3 3 1
5 (quintet)	1 4 6 4 1

The area under each resonance (whether split or not) is proportional to the number of protons responsible for that resonance. The NMR instrument will **integrate** each resonance and produce a vertical line over it, the height of which is proportional to its area. We are now in a position to consider the complete NMR spectrum of ethanol, with integration, in Fig. 20.28(b). Notice that the heights of the vertical lines over the OH, CH_2 and CH_3 resonances are in a ratio of 1:2:3, i.e. in accordance with the number of protons on each group.

Table 20.2 shows typical values of chemical shifts for protons in organic compounds. Notice that protons attached to a carbon atom also bonded to a halogen tend to have relatively high chemical shift (δ) values; the electronegative atoms reduce the electron density around the proton so that it experiences more of the external magnetic field (the proton is **deshielded**) and absorbs radiation at a higher frequency. Similarly, if a group or atom pushes electron density on to a proton (**shields** it), the proton absorbs radiation at a lower frequency (its chemical shift is lower).

Figures 20.29 to 20.33 show some typical NMR spectra.

Table 20.2 Typical values of chemical shifts for protons in organic compounds

Group	Type of compound	Approximate chemical shift /δ
CH_3–C	alkane	0.9
C–**CH_2**–C	alkane	1.3
CH_2=C	alkene	2.6
C_6**H_5**–	arene	7.3
CH_3–N	amine	2.3
CH_3COO–	ester, acid	2.0
CH_3CO–	ketone	2.1
CH_3Br	bromoalkane	2.6
CH_3Cl	chloroalkane	3.1
CH_3F	fluoroalkane	4.3

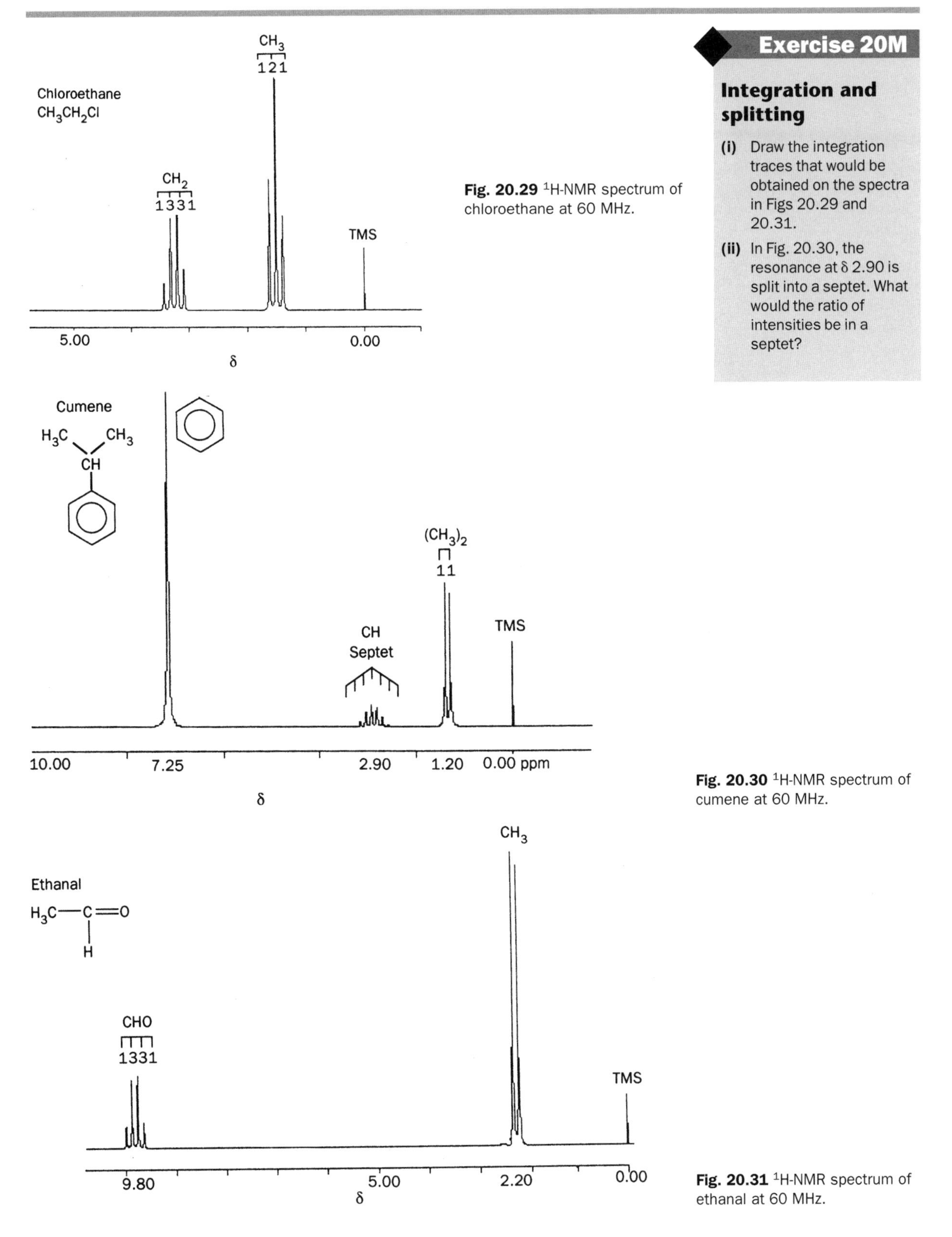

Fig. 20.29 ^{1}H-NMR spectrum of chloroethane at 60 MHz.

Fig. 20.30 ^{1}H-NMR spectrum of cumene at 60 MHz.

Fig. 20.31 ^{1}H-NMR spectrum of ethanal at 60 MHz.

Exercise 20M

Integration and splitting

(i) Draw the integration traces that would be obtained on the spectra in Figs 20.29 and 20.31.

(ii) In Fig. 20.30, the resonance at δ 2.90 is split into a septet. What would the ratio of intensities be in a septet?

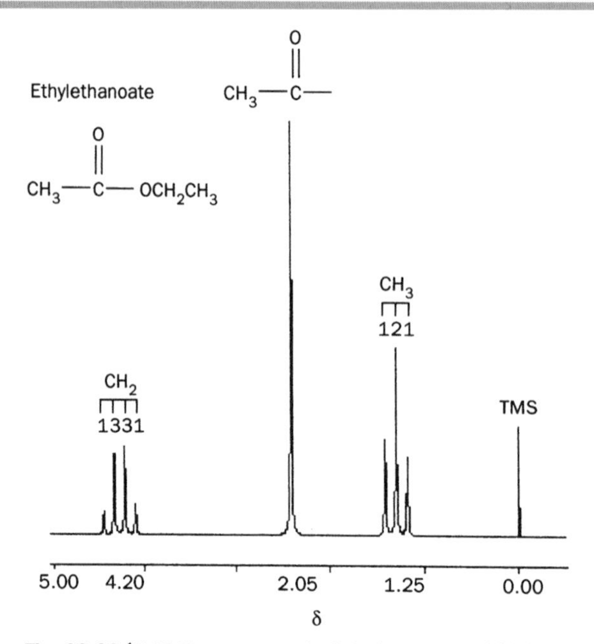

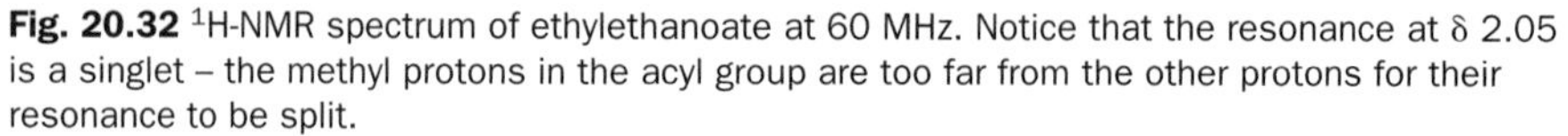

Fig. 20.32 ^{1}H-NMR spectrum of ethylethanoate at 60 MHz. Notice that the resonance at δ 2.05 is a singlet – the methyl protons in the acyl group are too far from the other protons for their resonance to be split.

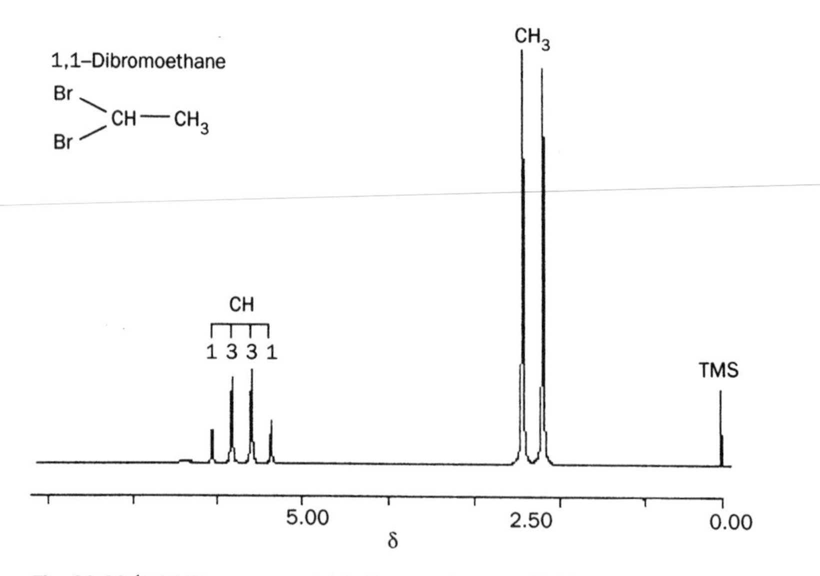

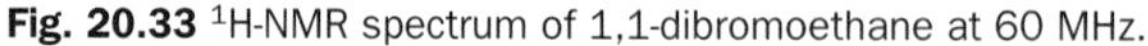

Fig. 20.33 ^{1}H-NMR spectrum of 1,1-dibromoethane at 60 MHz.

Human beings are composed of organic material, and therefore contain a large amount of combined hydrogen in many different compounds. This makes the human body a possible 'sample' for an NMR instrument. In hospitals, the phrase 'nuclear magnetic resonance' has been dropped in favour of **magnetic resonance** (because patients might wrongly believe the technique involves ionizing radiation) and the instruments used are commonly simply called **scanners**, but the principle of the instruments is the same as their laboratory-based cousins.

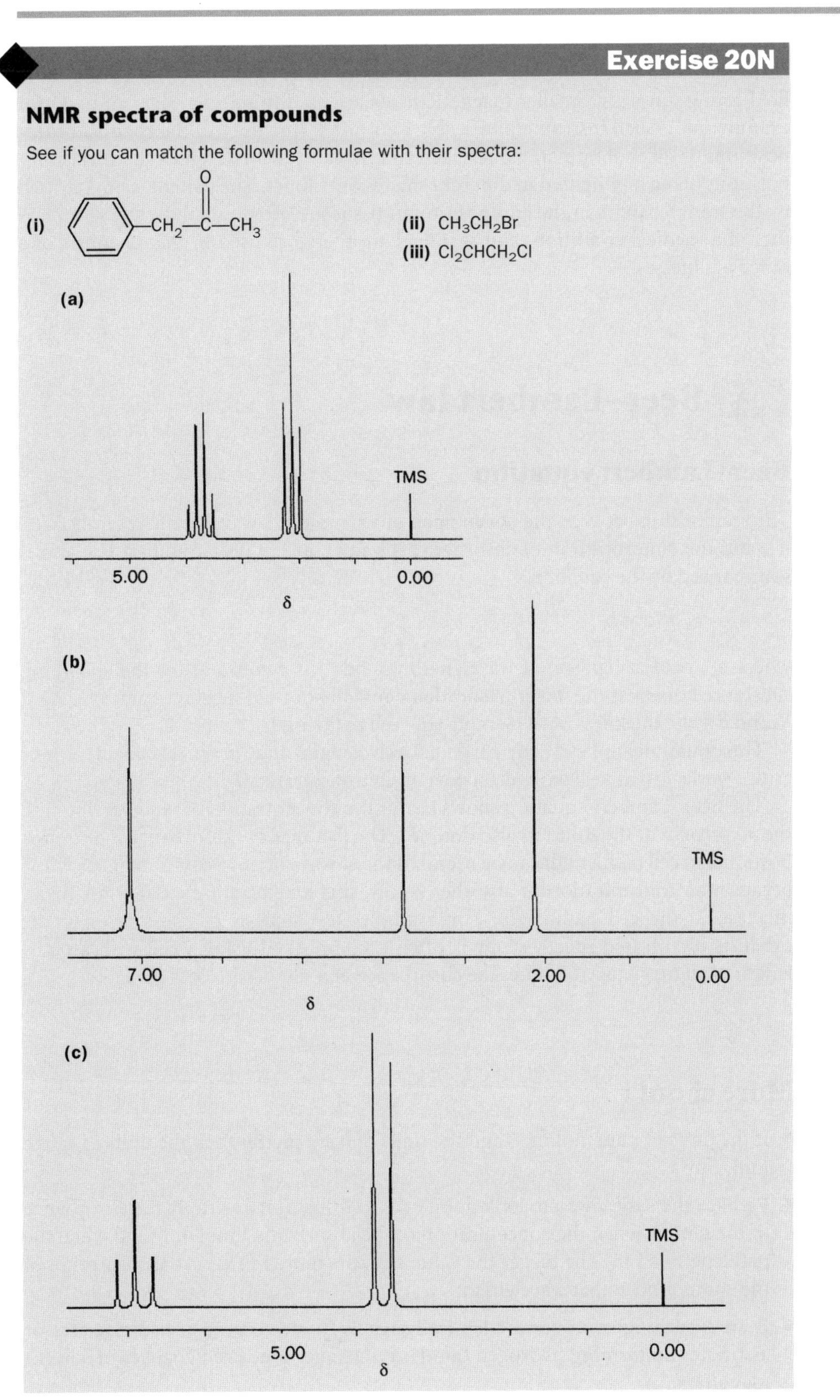
Exercise 20N
NMR spectra of compounds
See if you can match the following formulae with their spectra:
(i) C_6H_5–CH_2–C(=O)–CH_3
(ii) CH_3CH_2Br
(iii) Cl_2CHCH_2Cl
(a)
TMS
5.00
0.00
δ
(b)
TMS
7.00
2.00
0.00
δ
(c)
TMS
5.00
0.00
δ

In 'whole body NMR imaging systems', the patient lies inside a tube with the main magnetic field lying along the longitudinal axis of the body, with the radio field acting upwards. Smaller magnetic fields are used to select a 'slice' of body for examination. Such instruments are able to produce 'pictures' of the chemical compositions of slices of human tissue, (including organs), with different chemical compositions being highlighted in different shades or colours. The patient is not harmed by the scan. Changes in the composition of tissue are often valuable clinical indicators of a medical condition, and scanners are therefore used in the diagnosis of a range of illnesses.

20.9 Beer–Lambert law

Beer–Lambert equation

The relationship between the **absorbance** of a sample at a wavelength λ, symbolized A_λ, and the concentration of absorber c, is known as the Beer–Lambert law and is summarized by the equation

$$A_\lambda = \epsilon_\lambda \times c \times b$$

where ϵ_λ (read as 'epilson at wavelength lambda') is a constant for the absorbing substance known as the **molar absorption coefficient** of the substance at wavelength λ, and b is the thickness of the sample (the cell pathlength).

This equation applies at any particular wavelength, and (practical considerations aside) works just as well with ultraviolet, visible or infrared light.

The Beer–Lambert equation shows that if the concentration of *absorber* doubles, the *absorbance* of the solution also doubles. The fact that the absorbance is also proportional to cell pathlength can be useful. If the absorbance of a solution or gas is low because the concentration of absorber is low, this may be compensated for by an increase in the cell pathlength. For example, the analysis of trace gases in car exhausts by infrared spectroscopy is often accomplished using a cell with a pathlength of 10 m, giving 100 times the absorbance of a standard 10 cm gas cell.

More about ϵ

- If the units of c are mol m^{-3} and the units of b are metres then the units of ϵ_λ are mol^{-1} m^2.
- Perhaps the easiest way to look upon ϵ_λ is to think of it as simply the absorbance of the sample when the concentration of the absorber is 1 mol m^{-3} and when the pathlength is 1 m. The bigger the value of ϵ, the more *intense* is the absorption of the compound at that wavelength.
- If an absorbing compound chemically reacts in any way (e.g. it dissociates or associates by forming hydrogen bonds) ϵ_λ changes too. Even changes of solvent may alter ϵ_λ.

- Alternative units of ϵ_λ in common use are $mol^{-1}\,dm^3\,cm^{-1}$. These units are convenient because the concentration term in the Beer–Lambert equation is now expressed in $mol\,dm^{-3}$ and the cell pathlength is in centimetres. The conversion is

$$1\,mol^{-1}\,m^2 = 10\,mol^{-1}\,dm^3\,cm^{-1}$$

For example, $90\,mol^{-1}\,m^2 = 900\,mol^{-1}\,dm^3\,cm^{-1}$.

Beer–Lambert plots

The Beer–Lambert equation is of the form $y = mx$, and predicts that a plot of the absorbance of a compound at a particular wavelength against the concentration of compound should be a straight line, provided we use the same cell (i.e. same b) throughout the experiment. The slope of the graph is equal to $\epsilon_\lambda \times b$.

Example 20.4

Use the spectrum in Exercise 20I to estimate the molar absorption coefficient of potassium dichromate(VI) at 350 nm.

▶ Answer

Figure 20.12 shows that the absorbance of the sample at 350 nm is approximately 0.78:

$$c = 2.5 \times 10^{-4}\,mol\,dm^{-3} = 0.25\,mol\,m^{-3}; \qquad b = 1.0\,cm = 0.010\,m$$

Rearranging the Beer–Lambert equation,

$$\epsilon_{350} = A_{350}/cb = 0.78/(0.25 \times 0.010) = 310\,mol^{-1}\,m^2 \text{ (to two significant figs)}$$

Therefore, ϵ_{350} for potassium dichromate(VI) solution is $310\,mol^{-1}\,m^2$.

Exercise 200

Beer–Lambert law calculations

(i) Chlorophyll 'a' shows an absorption maximum at 680 nm with $\epsilon \approx 1.1 \times 10^4\,mol^{-1}\,m^2$. Estimate the concentration of chlorophyll 'a' in a leaf assuming (a) that chlorophyll 'a' is evenly distributed throughout a leaf of thickness 0.010 cm, (b) that the leaf possesses an absorbance of 1.1, and (c) that all the absorption at this wavelength is entirely due to chlorophyll 'a'.

(ii) The molar absorption coefficient of DNA (deoxyribonucleic acid) is about $1000\,mol^{-1}\,m^2$ (expressed per mole of base) at 260 nm. What concentration of the compound would be required to produce a solution of absorbance 0.010 in (a) a 1.0 cm cell and (b) a 4.0 cm cell?

(iii) A solution of a drug (concentration $0.0020\,mol\,dm^{-3}$) was placed in a 1.0 cm cell. The sample absorbance at 250 nm was 0.40. Calculate the molar absorption coefficient of the drug at 250 nm.

Example 20.5

An example of a Beer–Lambert plot is shown in Fig. 20.34. This shows the aborbance of 4-nitrophenol ($C_6H_4OHNO_2$) in ethanol solution at a wavelength of 312 nm. The cell pathlength was 1.0 cm. Calculate the molar absorption coefficient of 4-nitrophenol at 312 nm.

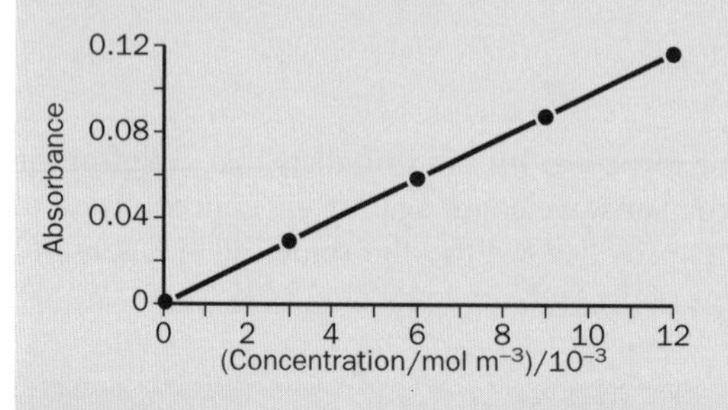

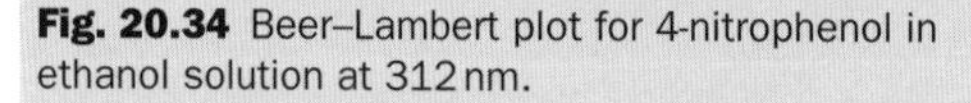

Fig. 20.34 Beer–Lambert plot for 4-nitrophenol in ethanol solution at 312 nm.

▶ Answer

The slope of the graph is

$$\frac{0.115 - 0.000}{12.0 \times 10^{-3} - 0.00} = 9.60\,\text{mol}^{-1}\,\text{m}^{-3} = \epsilon \times b$$

Since $b = 1.0\,\text{cm} = 0.010\,\text{m}$,

$$\epsilon = \frac{9.6\,\text{mol}^{-1}\,\text{m}^{3}}{0.010\,\text{m}} = 9.6 \times 10^{2}\,\text{mol}^{-1}\,\text{m}^{2}$$

Therefore, ϵ_{312} for 4-nitrophenol is $9.6 \times 10^{2}\,\text{mol}^{-1}\,\text{m}^{2}$.

Exercise 20P

Beer–Lambert law with two components

The molar absorption coefficient of 2-nitrophenol at 312 nm is $1.2 \times 10^{2}\,\text{mol}^{-1}\,\text{m}^{2}$. A solution contains 2-nitrophenol and 4-nitrophenol, both at a concentration of $1.0 \times 10^{-3}\,\text{mol m}^{-3}$. What is the absorbance of the mixture at 312 nm in a 1.0 cm cell? Assume that the two phenols do not react with each other. (Hint The mixture absorbance is the sum of the individual absorbances.)

Exercise 20Q

Beer–Lambert plot

The absorbance of different concentrations of manganate(VII) ions at 526 nm in a 1.0 cm glass cell was found by experiment:

$[MnO_4^-(aq)]/mol\ dm^{-3}$	Absorbance at 526 nm
0.0	0.00
8.7×10^{-5}	0.25
12.2×10^{-5}	0.35
14.0×10^{-5}	0.40
17.5×10^{-5}	0.50

(i) Convert the concentrations to $mol\ m^{-3}$.

(ii) Plot absorbance against concentration and hence calculate the molar absorption coefficient of the $MnO_4^-(aq)$ ion at 526 nm.

20.10 Photosynthesis

What is photosynthesis?

Photosynthesis is a complicated sequence of chemical reactions, driven by absorbed sunlight, in which carbohydrates are made. Photosynthesis is one of the most important of all reactions which involve light, because plants (either directly or indirectly) are the main source of our food.

The overall reaction involved in photosynthesis may be represented as

$$6CO_2(g) + 6H_2O(l) + h\nu \rightarrow \underset{\text{glucose}}{C_6H_{12}O_6(s)} + 6O_2(g)$$

Key points about photosynthesis

- This reaction is highly endothermic with $\Delta H^{\ominus} \approx +2800\ \text{kJ mol}^{-1}$. The reaction does not occur directly but in many steps, and in effect the energy for the conversion is obtained from sunlight ($h\nu$). This energy is initially trapped (i.e. absorbed) by **chlorophyll**, forming excited chlorophyll molecules. The excited chlorophyll molecules then pass on this energy to enable the sequence of reactions to occur.
- Cellulose (which provides plants with their bulk and mass) is manufactured within the plant by bonding together a large number (typically 300–2500) of glucose molecules in a straight chain, with the elimination of one water molecule from each glucose unit. (Taking H_2O from the formula for glucose, $C_6H_{12}O_6$, gives $C_6H_{10}O_5$, so that cellulose can be represented by the formula $(C_6H_{10}O_5)n$.)
- Starch (which is used by plants as a 'fuel' to supply energy for respiration and growth) is also made from glucose molecules linked together although, unlike cellulose, both straight and branched chains are present.
- In photosynthesis, the carbon (cellulose-based) skeleton of plant tissue and its stored carbohydrates are 'chemically assembled' from carbon dioxide gas. The

production of carbon-based solids in plants from a carbon-containing gas was something of a mystery to early scientists, who referred to photosynthesis as 'carbon fixation' – a phrase that is still in use today. Because of the widespread occurrence of plants using photosynthesis on land and sea, the scale of carbon fixation is enormous, with about 10^{11} tonnes of carbon being taken out of the atmosphere per year.

- Photosynthesis consumes CO_2 and also returns O_2 to the atmosphere. Photosynthesis is the most important way in which CO_2 levels (increased by the burning of fuels and by respiration) are reduced, and in which oxygen levels in the air are replenished.

Revision questions

20.1. Convert the following to frequencies (Hz): **(i)** 2×10^{-7} m, **(ii)** 500 nm, **(iii)** 1000 cm^{-1}.

20.2. Complete the table below (the first entry has been completed).

Definition	Word
Gain of energy of an atom or molecule by collision with a photon	absorption
Loss of energy by an atom or molecule by the giving out of light	
The logarithm of the ratio of the incident to transmitted intensities	
The production of light by chemical reaction	
The absorbance of a sample at a wavelength at unit pathlength and unit concentration	

20.3. Calculate the energy of 1 mol of photons of light with an equivalent wavelength of 1000 nm.

20.4. The energy gap between two energy levels in a molecule is 30 kJ mol^{-1}. What is the frequency of the light required to be absorbed or emitted in a transition? To which part of the electromagnetic spectrum (UV, visible, infrared or microwave) does this radiation belong?

20.5. **(i)** In a spectrometer, 84% of the incident light at 240 nm is absorbed by a solution. What is the absorbance of the solution?

(ii) The molar absorption coefficient of a dye at 450 nm is 1000 mol^{-1} m^2. Calculate the percentage of light transmitted through exactly 1 cm of dye solution of concentration 1.00×10^{-5} mol dm^{-3} at the same wavelength. By how much will the dye solution need to be diluted to give a solution with an absorbance of 0.010?

20.6. The (first) standard ionization energy of Na(g) is 494 kJ mol^{-1}. What wavelength of light would need to be absorbed in order to ionize sodium atoms?

20.7. The energy E (in kJ mol^{-1}) of the electronic levels in the hydrogen atom are easily calculated using the equation

$$E = -\frac{1310}{n^2}$$

where n is the principal quantum number. (Test this formula by applying it to $n = 2$. As Fig. 20.7 shows, $E = -328$ kJ mol^{-1}.)

(i) Calculate the energy of the $n = 50$ level.

(ii) What frequency of light will be emitted in the transition $n = 50 \rightarrow n = 3$?

20.8. **(i)** The sun consists of an incredibly hot centre (above 10^6 K) surrounded by a relatively cool layer at about 5800 K. If the sun is examined by a hand spectroscope, the solar spectrum is seen to contain some dark lines, Fraunhofer lines, superimposed upon a continuous blend of 'rainbow' colours. What causes the dark lines?

(ii) The following diagram shows the simplified electronic energy levels of an atom or molecule:

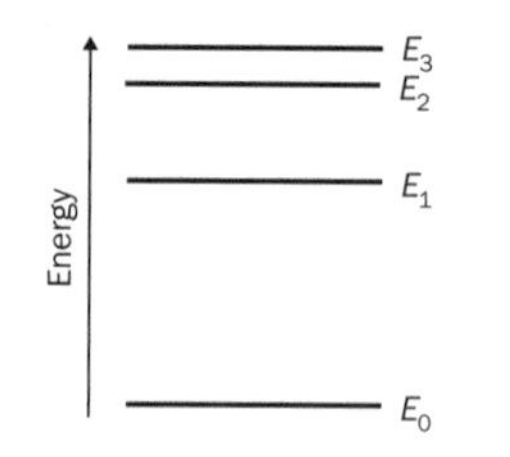

Draw in all the possible transitions corresponding to absorption and emission at room temperature. Use this diagram to explain why the emission spectra of atoms and molecules generally contain more lines than their absorption spectra.

20.9. **(i)** Sketch the expected absorption spectrum for a red filter.

(ii) Study the following UV–visible spectrum of a compound and predict the colour of its solution.

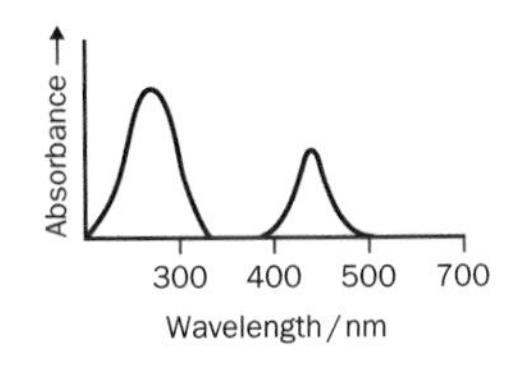

(iii) The colour of paints may also be explained using the colour cheese to predict which frequencies of light are absorbed and which are reflected. Explain why when lots of different paints are mixed, a black colour results.

20.10. The following table lists the molar absorption coefficient of ozone, $O_3(g)$ at 298 K in part of the ultraviolet region.

Wavelength/nm	Molar absorption coefficient/$mol^{-1}\,m^2$
210	11
220	35
230	101
240	204
250	298
254	369
260	298
270	204
280	101
290	35
300	11

Plot ϵ against wavelength. Calculate the absorbance at 254 nm of a 10.0 cm gas cell containing pure ozone at a pressure of 100.0 Pa at 298 K. (**Hint** Use the ideal gas equation to convert gas pressure to gas concentration.)

20.11. The simplified (a) mass spectrum and (b) infrared spectrum of an unknown compound are shown below.

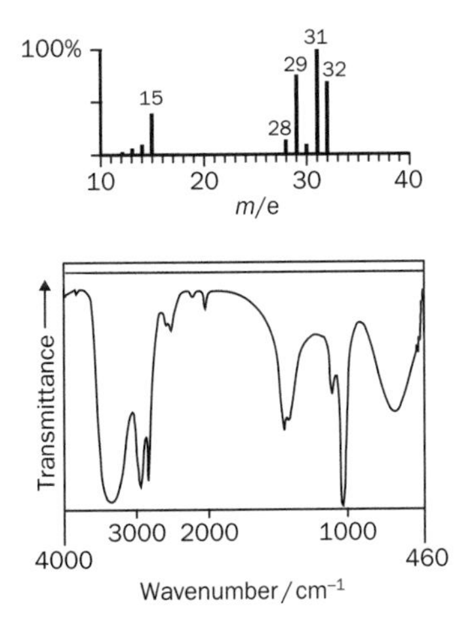

Additional information:

(a) The low resolution proton NMR spectrum (not shown) contains two signals.

(b) The compound contains 37.5% C, and 12.5% H; testing for other elements was not carried out.

Suggest a structure for the unknown compound.

20.12. Hydrocarbon concentrations in water can be measured by extracting the hydrocarbons from the water with a suitable solvent (such as trifluorotrichloroethane (CF_3CCl_3), and subsequently measuring the absorbance of the hydrocarbon solution in the C–H stretch region at 2860 cm^{-1}.

(i) Why should chloroform $CHCl_3$ not be used as the solvent in such measurements?

(ii) In an experiment, octane from a water sample was completely extracted using an equal volume of CF_3CCl_3 solvent. The absorbance of octane in solution in a 20 mm cell was found to be 0.050 at 2860 cm^{-1}. Given that $\epsilon(C_8H_{18})$ in CF_3CCl_3 at 2860 cm^{-1} is 5.0 $mol^{-1}\,m^2$, calculate the concentration of octane in the original water sample in (a) $mol\,m^{-3}$ and (b) $mg\,dm^{-3}$.

20.13. Chromate(VI) ions ($CrO_4^{2-}(aq)$) may be completely reduced to chromium(III) ions ($Cr^{3+}(aq)$) in the presence of thiourea catalyst and a sodium ethanoate–ethanoic acid buffer. The reaction may be simply expressed as

$$CrO_4^{2-}(aq) + 8H^+(aq) + 3e^- \rightarrow Cr^{3+}(aq) + 4H_2O(l)$$

(i) By working out the oxidation numbers of chromium in the two species, show that reduction has taken place.

(ii) At 440 nm, $\epsilon(CrO_4^{2-}(aq)) = 200.0\,mol^{-1}\,m^2$ and $\epsilon(Cr^{3+}(aq)) = 1.0\,mol^{-1}\,m^2$. In an experiment, a solution initially containing $5.0 \times 10^{-4}\,mol\,dm^{-3}$ chromate(VI) ion was completely reduced to chromium(III) ions. Assuming a cell pathlength of 1.0 cm, what will be the initial and final absorbance of the reaction mixture at 440 nm? (Assume that the thiourea and buffer do not absorb at 440 nm.)

20.14. Flame atomic absorption spectroscopy (AAS) is a specialized spectroscopic technique widely used in analytical laboratories to find the levels of trace metals in solution. By consulting a textbook on analytical chemistry, find out **(i)** what light sources are used in AAS and **(ii)** the purpose of the flame.

20.15. Identify the compound, of molecular formula $C_2H_4Br_2$, whose ^{1}H-NMR spectrum is shown below:

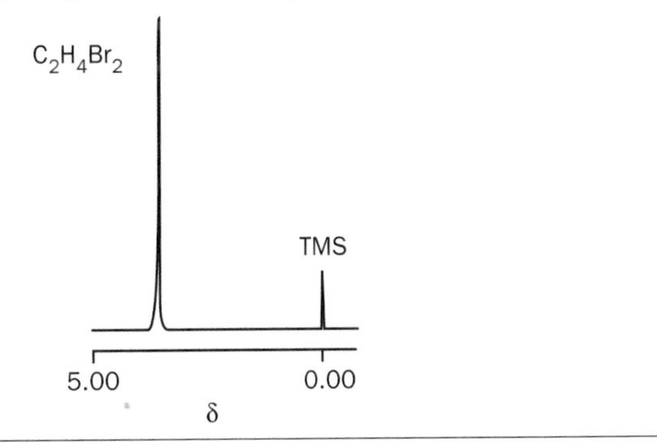

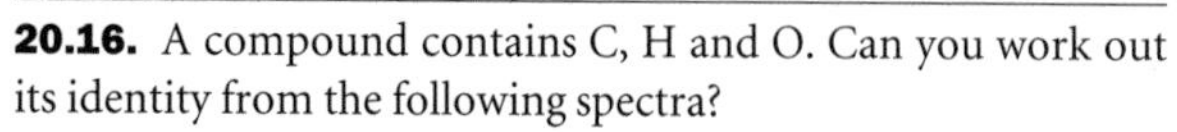

20.16. A compound contains C, H and O. Can you work out its identity from the following spectra?

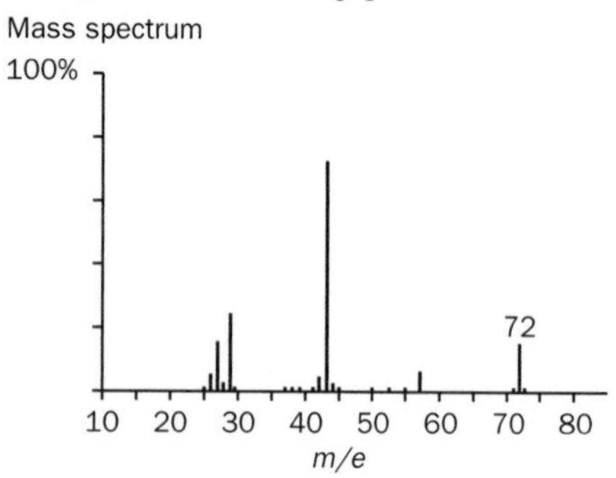

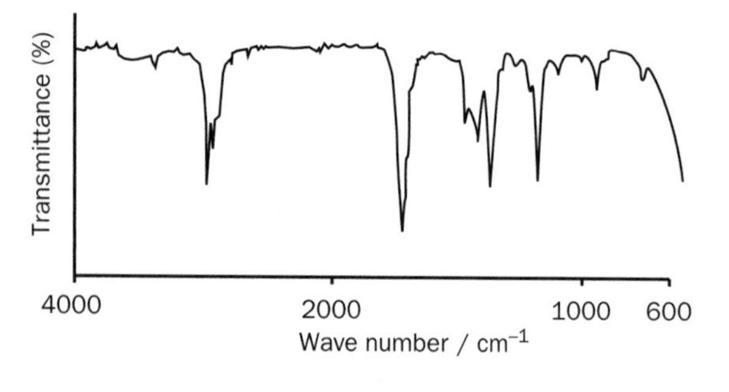

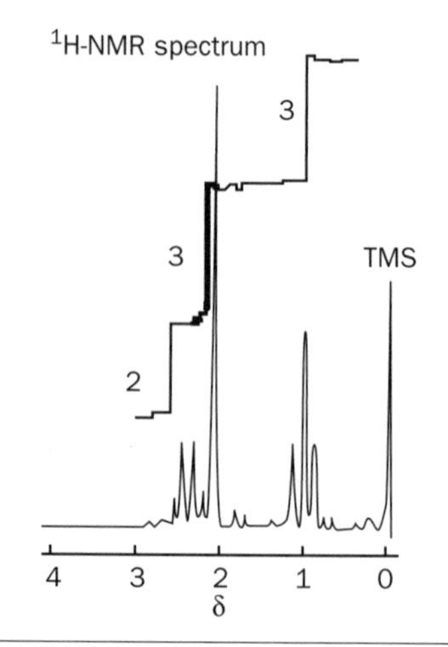

Nuclear and Radiochemistry

Objectives

- ▶ Defines radioactivity
- ▶ Looks at the properties of nuclear radiation
- ▶ Gives examples of calculations involving radionuclides and isotopic half-lifes
- ▶ Discusses the uses of radioisotopes

Contents

The radioactivity of naturally occurring and artificially produced isotopes, nuclear fission (the reactions involved in commercial nuclear reactors) and nuclear fusion (the source of the sun's power) are all examples of nuclear reactions (Fig. 21.1). Radioactivity started as a scientific curiosity. From these humble beginnings developed nuclear power (the main provider of power for the generation of electricity in some countries) and the awesome power of nuclear warheads.

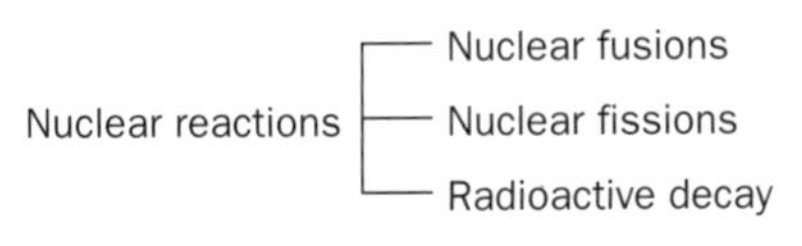

Fig. 21.1 The main types of nuclear reaction.

21.1 Radioactivity

Radioactivity was discovered accidentally in 1896, when Antoine Henri Becquerel (1852–1908) noticed that an uranium ore caused a photographic plate to blacken. The uranium gave off invisible 'rays' which were similar to the X-rays discovered the previous year. Forty-nine years later, doctors in Hiroshima reported that previously unexposed X-ray plates in hospital vaults had been fogged by the radiation emitted by the first atomic bomb.

Pioneering work in radioactivity was carried out by the husband and wife team of Pierre and Marie Curie, who (in 1898) extracted the elements polonium and radium from an ore called pitchblende. The harmful effects of nuclear radiation were then unknown, and even today their laboratory notebooks remain dangerously radioactive.

Definition of radioactivity

Atomic nuclei are said to be **unstable** when they *spontaneously disintegrate* and simultaneously give off **nuclear radiation**. This phenomenon is called **radioactivity**. The key ideas in this definition are:

1. Spontaneous

Radioactivity is spontaneous. This means that it does not require any help to start or to continue. The rate of disintegration does *not* depend upon temperature. This is in contrast to *chemical* reactions, whose rates are often drastically affected by changes in temperature. The radioactive decay of the nucleus of an atom is unaffected by the presence of other atoms and cannot be catalysed. For example, uranium-235 decays at the same speed (and into the same products) whether it is pure uranium, combined as uranium oxide (UO_2) or as uranium fluoride (UF_6).

2. Disintegration

In radioactive decay, one **nuclide** changes into another. We can represent this as

parent nuclide → daughter nuclide + nuclear radiation

where the term 'nuclide' means an **atom** of a particular isotope. As a result of this change, the number of protons or neutrons (or both) in the parent nuclide changes. Where the number of protons changes, a new element is formed. Radioactive changes are therefore very different from chemical reactions, in which atoms are neither destroyed nor created and where reaction simply involves the rearrangement of atoms to form new molecules.

3. Nuclear radiation

The term 'nuclear radiation' includes particles (such as alpha particles, high-energy electrons and nuclei), and also high-frequency light known as gamma rays. See Table 21.1 and Fig. 21.2.

Table 21.1 The main types of nuclear radiation

Name	Symbol	Nature	Penetration
Alpha particle	${}^{4}_{2}He^{2+}$ or α	helium nuclei (i.e. helium ions)	relatively massive but travel only a few cm in air and are stopped by thin card
Beta particle	${}^{0}_{-1}e^{-}$ or β	highly energetic (and fast moving) electrons	travel up to 5 m in air; stopped by thin metal plate
Positron	${}^{0}_{-1}e^{+}$ or β^{+}	highly energetic positively charged electrons	as beta particle
Gamma rays	γ	very short wavelength electromagnetic radiation; travels at the speed of light	very penetrating; requires thick lead plate and/or thick concrete shielding
Neutrons*	${}^{1}_{0}n$	relatively massive particles	as gamma rays
Fission fragments*	–	highly ionized nuclei of medium and large atoms, e.g. U^{20+}	relatively massive and possess enormous energies; stopped by thick lead and concrete shielding

* Produced during nuclear fission.

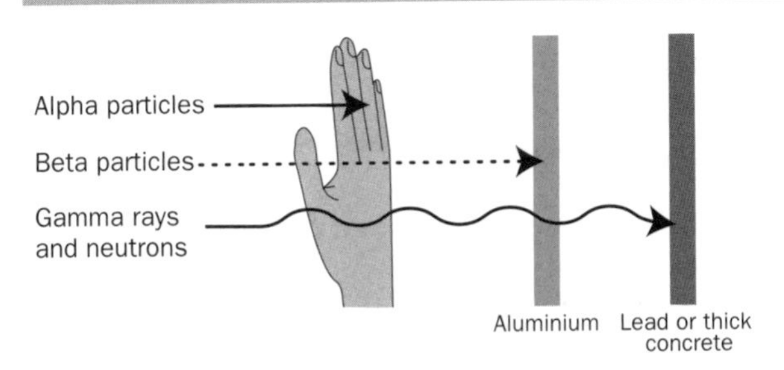

Fig. 21.2 Penetrating powers of nuclear radiation.

21.2 Radionuclides and radioisotopes

The term **radionuclide** means an atom which undergoes radioactive decay; such an atom is said to be (radioactively) **unstable**. Isotopes which are radioactive are called **radioisotopes**.

There are 365 different nuclides in nature. Of these, 55 are radionuclides. In addition, there are hundreds of radionuclides which have been artificially manufactured, mainly by smashing nuclei together at very high speed in giant devices called **particle accelerators**. For example, in 1937, technetium (Tc) was made by firing deuterium atoms at molybdenum (Mo) atoms:

$$^{96}_{42}\text{Mo} + ^{2}_{1}\text{H} \rightarrow ^{97}_{43}\text{Tc} + ^{1}_{0}\text{n}$$

Elements with atomic numbers greater than that of uranium (92) are known as **transuranium elements**. They do not occur naturally and have all been made in particle accelerators or in nuclear reactors.

Table 21.2 shows the half-lives and the mode of decay of selected radionuclides.

Table 21.2 Data on selected radionuclides

Nuclide	Main type of radiation emitted	Half-life
$^{137}_{55}\text{Cs}$	β, γ	30.17 years
$^{14}_{6}\text{C}$	β	5730 years
$^{131}_{53}\text{I}$	β, γ	8.04 days
$^{226}_{88}\text{Ra}$	α, γ	1600 years
$^{60}_{27}\text{Co}$	β, γ	5.27 years
$^{241}_{95}\text{Am}$	α, γ	432 years
$^{90}_{38}\text{Sr}$	β	28 years
$^{32}_{15}\text{P}$	β	14.3 days
$^{40}_{19}\text{K}$	β, γ	1.25×10^9 years
$^{238}_{92}\text{U}$	α	4.46×10^9 years
$^{235}_{92}\text{U}$	α	7.04×10^8 years
$^{3}_{1}\text{H}$ (tritium, T)	β	12.26 years
$^{222}_{86}\text{Rn}$	α, γ	3.82 days
$^{35}_{16}\text{S}$	β	87.2 days
$^{239}_{94}\text{Pu}$	α, γ	2.41×10^4 years

21.3 More about nuclear radiation

Alpha particles

Alpha particles are doubly ionized helium atoms, i.e. helium nuclei. Since a helium nucleus consists of two protons and two neutrons, the loss of an alpha particle from a nucleus reduces the number of protons by two and the number of neutrons by two. Therefore the mass number *A* falls by four and the atomic number *Z* falls by two. The emission of alpha particles is represented by the general equation

$$^{A}_{Z}X \rightarrow \, ^{A-4}_{Z-2}Y^{2-} + \, ^{4}_{2}He^{2+}$$

where X is the parent nuclide and Y is the daughter nuclide. Both Y^{2-} and $^{4}_{2}He^{2+}$ become electrically neutral by losing or gaining electrons from the container or the surrounding air.

An example of alpha emission is provided by the decay of radium-226:

$$^{226}_{88}Ra \longrightarrow \, ^{222}_{86}Rn^{2-} + \, ^{4}_{2}He^{2+}$$

88 protons	86 protons	2 protons
138 neutrons	136 neutrons	2 neutrons
88 electrons	88 electrons	0 electrons

Exercise 21A

Alpha decay

Complete the equation

$$^{218}_{84}Po \rightarrow ?? + \, ^{4}_{2}He^{2+}$$

Notice that the total charge, the sum of the mass numbers, and the sum of the atomic numbers, are equal on both sides. This is true of all nuclear equations.

Alpha particles are so bulky that they do not usually travel more than 10 cm in air. They are easily stopped by card, brick or thin metal sheet. However, their mass and speed (typically about $10^{7}\,ms^{-1}$) means that they possess considerable kinetic energy and so cause substantial ionisation, and if brought into close contact with human tissue (for example, by ingestion) they usually cause much more damage than beta or gamma radiation.

Beta particles

Beta particles are high-energy electrons which are ejected from the **nucleus**. Since there are not normally any electrons in the nucleus, the beta particles must have been produced in the nucleus during radioactive decay. It is now known that during this process, a neutron changes into a proton and an electron. From this it follows that the atomic number (the number of protons) of the nuclide increases by one, while the mass number (the number of protons and neutrons) remains the same. The electron escapes from the nucleus and is now called a beta particle. The general equation is

$$^{A}_{Z}X \rightarrow \, _{Z+1}^{\;\;A}Y^{+} + \, _{-1}^{\;\;0}e^{-}$$

where the atomic number of the electron is (for the purposes of a balanced nuclear equation) regarded as −1. An example of beta emission is the radioactive decay of strontium-90:

$$^{90}_{38}Sr \longrightarrow \, ^{90}_{39}Y^{+} + \, _{-1}^{\;\;0}e^{-}$$

38 protons	39 protons	0 protons
52 neutrons	51 neutrons	0 neutrons
38 electrons	38 electrons	1 electron

Exercise 21B

Beta decay

Complete the equation

$$^{215}_{83}Bi \rightarrow ?? + \, _{-1}^{\;\;0}e^{-}$$

The positively charged yttrium ion will gain an extra electron from the surroundings.

Beta particles vary greatly in their energies and speeds, and (therefore) in their penetration of matter. Some beta particles travel at speeds at 99% the speed of light, and as a consequence penetrate several metres of air.

Exercise 21C

Alpha and beta decay

An element P, radioactively decays to an element Q with the emission of an alpha particle. Q decays to element R by beta emission and R decays to element S by beta emission. What can you state about P and S?

Gamma radiation

Gamma radiation is very short wavelength electromagnetic radiation, very similar to X-rays. Their wavelengths are usually 10^{-10}–10^{-13} m, corresponding to about 10^6–10^9 kJ per mol of photons.

Gamma rays are usually emitted by radionuclides following the emission of alpha or beta particles. Just as the brightness of a torch falls as the observer moves away from the torch, an observer armed with a Geiger counter will detect a lower intensity of gamma rays as he or she moves away from a radioactive source. In principle, the intensity of gamma rays never falls to zero, although in practice the observed intensity does fall to an undetectable level. This 'zero level' may also be reached by placing concrete or lead between the source and the detector.

Exercise 21D

Energy of gamma rays

Confirm that a gamma ray wavelength of 10^{-10} m corresponds to an energy of about 10^6 kJ per mole of photons

21.4 Mathematics of radioactive decay

The number of atoms of a radionuclide N_t remaining after an interval of t seconds may be calculated from the formula

$$N_t = N_0 \times e^{-kt}$$

where N_0 is the number of atoms at $t = 0$ and k is the first-order rate constant (often called the **decay constant**) with the units of s^{-1}.

Since the number of atoms of a radionuclide is proportional to its mass, N_t *and* N_0 may be replaced by the mass of radionuclide at time t (symbolized m_t) and $t = 0$ (symbolized m_0) respectively:

$$m_t = m_0 \times e^{-kt}$$

These equations show that the fall in the number of atoms of radionuclide with time is exponential (Fig. 21.3(a)). In fact, any quantity which is proportional to the number of atoms (such as the number of counts per minute registered on a geiger counter) may also be substituted in these equations.

The half-life ($t_{1/2}$) of a radionuclide is the time which elapses before half the nuclei

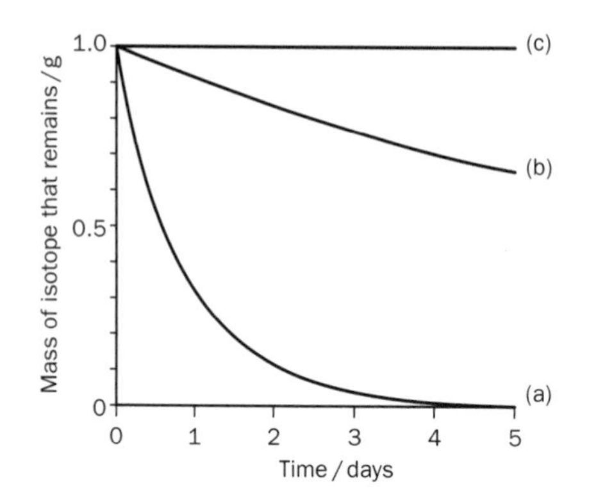

Fig. 21.3 The decay of 1 g of each of three radionuclides: curve (a) $^{24}_{11}Na$($t_{1/2}$ = 15.0 h); curve (b) $^{131}_{53}I$ ($t_{1/2}$ = 8.04 days); curve (c) $^{137}_{55}Cs$ ($t_{1/2}$ = 30.17 yr).

Exercise 21E

Half-life 1

Sketch a graph showing the decay of 1 g of uranium-238 on the same time scale as shown in Fig. 21.3.

of the radionuclide have disintegrated. The factor k is related to the half-life of the decay by the equation

$$t_{1/2} = \frac{0.693}{k}$$

(see page 258). If k is required in the units of s^{-1}, $t_{1/2}$ must be in the units of seconds.

Key points about half-life

- A long half-live indicates that it will be a long time before the radioactivity of a sample containing atoms of that nuclide fades to a near-zero level. Such a radionuclide is said to be **long-lived**. Similarly, a radionuclide with a short half-life is said to be **short-lived**. Most of the world is slowly becoming less radioactive, as the radioactive elements brought together when the earth was formed (about 4.5 billion years ago) decay away.
- Short-lived radionuclides emit nuclear radiation more rapidly than long-lived radionuclides, and a sample of a short-lived radionuclide will (statistically) undergo more disintegrations per second than the same number of nuclei of a stable radionuclide. It follows that short-lived radionuclides give out *more intense* nuclear radiation than long-lived radionuclides, but for a shorter time.
- The disintegration of radionuclides is *random*. Think of 100 radium-226 atoms decaying with a half-life of 1600 years. Although it is statistically likely that 50 Ra atoms will have disintegrated after 1600 years, we cannot predict *which* individual Ra atoms *will* disintegrate during this time.

Plot of radionuclide mass against time

Figure 21.3 shows the fall in mass of samples of three radionuclides (with an initial mass of 1 g each) over 5 days, calculated using the equation $m_t = m_0 \times e^{-kt}$.

- All three curves are drawn on the same scale. The longest lived radionuclide ($^{137}_{55}Cs$) has undergone a negligible amount of decay over 5 days.

- In general, the early part of an exponential plot is straight (linear). The decay of the radionuclide of intermediate half-life ($^{131}_{53}I$) is small during the 5 days, and its plot remains in the straight part of the exponential curve.

Exercise 21F

Half-life 2

(i) In nature, uranium atoms consist of approximately 0.7% uranium-235 and the rest uranium-238. From Table 21.2, explain why the percentage of uranium-235 was much higher than this millions of years ago.

(ii) $^{201}_{87}Fr$ decays by alpha emission, with a half-life of 0.0048 s. $^{231}_{90}Th$ decays by beta emission, with a half-life of 25 h. **(a)** Write equations showing the decay of Fr and Th. What elements are produced in the decay of francium and thorium, respectively? **(b)** Suppose that a mixture contains 0.0100 g of $^{201}_{87}Fr$ and 0.0100 g of $^{231}_{90}Th$. How much of each radionuclide will be left after 100 h?

21.5 Uses of radionuclides

1. Nuclear fission of uranium-235

Natural uranium contains about 99.3% $^{238}_{92}U$ and 0.7% $^{235}_{92}U$, usually in the form of uranium oxide (UO_2). Purified uranium oxide is used as the 'fuel' for most types of nuclear reactor, with the percentage of the $^{235}_{92}U$ artificially raised (**enriched**) to 2–3% in order to achieve sufficient fission.

Nuclear fission involves the splitting of the nucleus into two nuclei of roughly equal mass. Uranium-235 and uranium-238 naturally undergo a type of fission (known as **spontaneous fission**) in which the uranium nucleus, without assistance, breaks up into two nuclei and produces a neutron, but this process is incredibly slow – even slower than the radioactive decay of these radioisotopes by alpha emission.

Neutrons may be absorbed by uranium-235 nuclei in a process called **induced nuclear fission**, commonly simply referred to as **nuclear fission**. (Initially, the

Nuclear fission provides about 20% of the electricity generated within the UK. Economic worries about the decommissioning of old nuclear stations, and major accidents in Windscale UK (1957), Three Mile Island USA (1979) and Chernobyl Ukraine (1986) have caused many people to question whether or not more nuclear plants should be built. This is a photograph of the fourth reactor at the Chernobyl nuclear power plant where an explosion resulted in the world's worst nuclear accident.

neutrons originate from the spontaneous fission of uranium or from cosmic rays which come from outer space). The equations

$$^{235}_{92}U + ^{1}_{0}n(slow) \rightarrow ^{139}_{54}Xe + ^{95}_{38}Sr + 2^{1}_{0}n$$

$$^{235}_{92}U + ^{1}_{0}n(slow) \rightarrow ^{142}_{56}Ba + ^{92}_{36}Kr + 2^{1}_{0}n$$

$$^{235}_{92}U + ^{1}_{0}n(slow) \rightarrow ^{139}_{56}Ba + ^{94}_{36}Kr + 3^{1}_{0}n$$

are a few of the many observed induced fission reactions involving ^{235}U. Note that neutrons are made in the fission reactions. Gamma rays are also produced in each case, together with about 10^{10} kJ of energy per mole of ^{235}U consumed. (This energy is distributed in the kinetic energy of the new nuclei and in the energy of the accompanying gamma rays.) It is this type of fission reaction which is exploited in commercial nuclear power stations.

Not all the induced fission reactions involving ^{235}U are equally likely, and experiments have shown that, on average, there are 2.5 neutrons produced for every ^{235}U atom that is broken up. This means that *more neutrons are generated in nuclear fission than are used up* (Fig. 21.4). It is this fact that allows, in principle, the fission of ^{235}U to be self-sustaining, the fission occurring at a faster and faster rate as the neutrons produced induce fission in the remaining U-235 nuclei, causing a **chain reaction**.

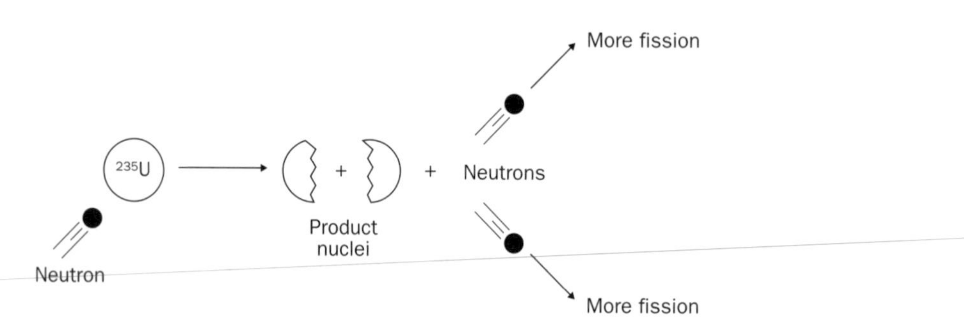

Fig. 21.4 Nuclear fission can be thought of as the breaking of a big ball (the ^{235}U nucleus) into two roughly equal halves (the product nuclei) using a small ball (a neutron) as ammunition. 2–3 neutrons are also produced, and these may cause fission in more ^{235}U nuclei, so setting up a chain reaction. The chain reaction is controlled in a nuclear power station, and uncontrolled in a nuclear bomb.

Key points about nuclear fission reactors

- In practice, a chain reaction will only occur if the absorption of neutrons by ^{235}U nuclei is efficient. If neutrons are travelling too fast, they are not absorbed, they simply pass through the ^{235}U nuclei and no fission occurs. For this reason, the neutrons must be slowed down using a **moderator** (commonly graphite or water) which absorbs the excess energy of the neutrons (Fig. 21.5). Because ^{235}U undergoes fission using 'slow' neutrons, it is said to be **fissile**.
- Slow neutrons will not cause fission in uranium-238 nuclei, which only undergo fission if highly energetic neutrons are used, i.e. ^{238}U is **non-fissile**. Fission using high-energy neutrons is less efficient than fission using slow neutrons. This means that the overall gain in energy from the use of ^{238}U as fuel is much less than that for ^{235}U, and so ^{238}U fission is not commercially viable.

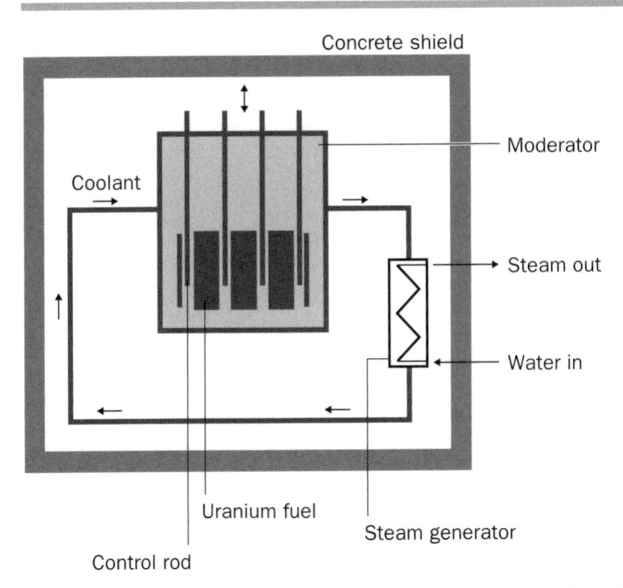

Fig. 21.5 The main parts of a nuclear reactor. Heat is transferred outside the nuclear reactor using a liquid or gas as coolant, such as water or carbon dioxide. The coolant then transfers the heat to a heat exchanger which converts water to steam to drive turbines which produce electricity.

- The uranium is stored inside the reactor core in alloy tubes called **fuel rods**. A minimum mass of ^{235}U is needed in the reactor in order to ensure that there are enough ^{235}U nuclei to absorb the neutrons emitted during the induced fission, so that further fission may continue. This **critical mass** of ^{235}U depends upon the shape of the uranium fuel. It is usually several kilograms.
- Many of the fission products (such as $^{142}_{56}Ba$ and $^{92}_{36}Kr$) are intensely radioactive in their own right. This makes the intensity of nuclear radiation in the nuclear core thousands of millions of times more intense than that generated by the natural radioactivity of the uranium radioisotopes alone. (A typical fission reactor may contain about 200 tonnes of uranium. Once fission has started, the radiation intensity of the reactor core increases by a factor of about 100 000 000 000.) Elaborate shielding (including thick concrete walls) is therefore needed to protect workers and members of the public.
- A chain reaction cannot be allowed to proceed unchecked in a nuclear reactor because, although there is no danger of a nuclear explosion, the tremendous heat generated by uncontrolled fission might cause structural damage to the reactor itself. This would make it very likely that some of the intensely radioactive material in the core of the reactor would escape into the surrounding environment – this happened at Chernobyl. For this reason, **control rods** made of neutron-absorbing material (such as boron or cadmium) are lowered into the nuclear reactor. To speed up fission, some of the control rods are withdrawn. When demand for electricity is low (as at night), extra control rods are lowered into the reactor.
- The nuclei and neutrons produced as fission products possess considerable kinetic energy and, due to collisions with the reactor coolant, this energy is utimately

converted to heat. This heat is utilized to produce electricity using a conventional steam turbine.

- In any nuclear reactor only about 0.1% of the initial mass of fuel is converted into energy, and it becomes necessary to replace the fuel rods 2–3 times a year. The main reason for changing the fuel rods is to remove the fission products which would otherwise absorb too many neutrons and slow down nuclear fission.

2. Nuclear fusion

Nuclear fusion reactions are the source of the sun's energy, on which life on earth ultimately depends. The possibility of obtaining energy from controlled fusion reactions on earth is now a major research interest in several countries, with most projects using deuterium and tritium as the fuels. The fusion reaction is

$$^{2}_{1}H + ^{3}_{1}H \rightarrow ^{4}_{2}He + ^{1}_{0}n$$

About 5×10^9 kJ of energy is released per mole of tritium consumed.

Positive nuclei repel each other, and it has been found that deuterium and tritium nuclei must possess huge amounts of kinetic energy before they can 'fuse' together. Such kinetic energies require a temperature of about 100 million °C. At such temperatures all atoms are ionized, and the mixture of positively charged nuclei and free electrons is called a **plasma**. If the particles in the plasma were to strike the walls of the reactor vessel, other nuclei would mix with the plasma and any fusion would stop. The remarkable solution to this problem is to confine the plasma in the form of a ring using a powerful magnetic field. To date, fusion has been achieved only for a fraction of a second, and the commercial production of electricity from nuclear fusion has yet to be realized.

'Cold fusion', i.e. nuclear fusion at ambient temperatures, has been claimed but not proved (see page 426).

3. Radioactive dating

The fact that the half-life of a radionuclide is a constant (and independent of the molecule containing the nuclide or the temperature of the radioactive source) is used to determine the age of rocks or archaeological relics. An example is provided by the use of 'carbon-14 dating' to find the age of dead plants. Although most carbon atoms in nature are stable $^{12}_{6}C$, a tiny amount of radioactive $^{14}_{6}C$ is also present and the $^{14}_{6}C$:$^{12}_{6}C$ ratio remains approximately constant during the life of the plant. The level of $^{14}_{6}C$ in the plant begins to fall immediately after death. $^{14}_{6}C$ has a half-life of 5700 yr. Mass spectrometers are used to measure the $^{14}_{6}C$:$^{12}_{6}C$ ratio in a sample of the dead plant. Knowing the ratio for a living plant of a similar type, the time that has elapsed since the death of the organism may be calculated.

4. Treatment of cancer

Although high levels of X-ray and nuclear radiation cause cancer, such radiation may also be used to kill cancer cells, which are particularly sensitive to radiation because of their high rate of growth. The use of nuclear radiation in this way is called **radiotherapy**. The aim of radiotherapy is to deliver as high a dose as possible to the malignant tissue without causing severe injury to the surrounding healthy tissue.

5. Radioisotopes as tracers

Radioisotopes have the same *chemical* reactions as non-radioactive isotopes of the same element, but they have the advantage that their position may be located using suitable detection equipment, i.e. they can be *traced*. In medicine, tracers consisting of compounds containing technetium $^{99}_{43}Tc$, are used to locate brain tumours. The use of salt which contains radioactive $^{24}_{11}Na$ allows doctors to follow movements of sodium ions in the kidney. In **positron emission tomography (PET)** positron-emitting nuclides, such as $^{15}_{8}O$, are used to provide an image of the flow of blood in the brain.

6. Miscellaneous uses

These include the use of the beta emitter $^{241}_{95}Am$ ($t_{1/2} = 432$ yr) in smoke detectors, the sterilization of food using nuclear radiation, the use of gamma sources to estimate the thickness of metal pipes, and the use of radioisotopes in an analytical technique called **neutron activation analysis** (see Box 21.1).

BOX 21.1

Was Napoleon poisoned?

Emperor Napoleon was confined to the island of St Helena in 1816. He died in 1821, convinced that he was being poisoned by his captives, the British authorities. Although a post-mortem examination suggested that Napoleon suffered from an advanced cancer and from hepatitis, many people have since speculated that he was slowly poisoned with arsenic in order to prevent him from re-emerging as a political force in Europe.

In more recent times it has been realized that hair contains many chemical deposits which reflect the health and diet of the individual. Fortunately, several people obtained hair cuttings from Napoleon's body shortly after his death, and these were carefully preserved from generation to generation. In the 1960s these samples were analysed by **neutron activation analysis**. In this technique, the sample is bombarded with neutrons, in this case causing any arsenic-75 atoms to be converted to radioactive arsenic-80 atoms:

$$^{75}_{33}As + 5^{1}_{0}n \rightarrow {}^{80}_{33}As$$

Because arsenic-80 is radioactive, it may be detected in tiny concentrations. The concentration of arsenic-80 present is equal to the initial concentration of arsenic-75 present in the hair sample.

Analysis of the arsenic content in the Emperor's hair samples did indeed show much higher arsenic levels (about 10 mg kg^{-1} of hair) than would be regarded as healthy today. However, the highest As levels (about 40 mg kg^{-1}) were found in hair cuttings that were almost certainly taken *before* his imprisonment, and were probably caused by the ingestion of arsenic-containing medicines (which were popular in those days) before his capture. In addition, hair cuttings from unrelated individuals of the same period suggest that the arsenic levels in many people were very much higher than is the case today. It was therefore concluded that it was unlikely that the Emperor was poisoned with arsenic. It is possible that some other poison was used, but it seems most likely that Napoleon died of natural causes.

In 1991 it was suggested that Zachary Taylor, the 12th president of the United States, had been murdered using arsenic because of his antislavery views. However, neutron activation analysis of Taylor's hair revealed insignificant levels of arsenic.

Revision questions

21.1. In 1919, Lord Rutherford made the first artificial radioisotope by bombarding nitrogen atoms with alpha particles:

$$^{14}_{7}N + ^{4}_{2}He^{2+} \rightarrow ?? + ^{1}_{1}H$$

Complete the equation and identify the radioisotope he made.

21.2. Tritium decays by emitting beta particles with $t_{1/2}$ = 12.26 yr.

(i) Write an equation for the radioactive decay of tritium.

(ii) Tritium is prepared commercially by reacting lithium-6 with neutrons from a nuclear reactor. Write an equation for the reaction.

21.3. The time for 99% of a sample of a radioisotope to decay is given by the equation

$$t_{0.99} = \frac{4.605}{k} \text{ seconds}$$

Similarly, the time for 99.9% of a sample of a radioisotope to decay is given by

$$t_{0.999} = \frac{6.908}{k} \text{ seconds}$$

A sample of pure polonium-210 is used in a laboratory experiment. Calculate the time it takes for **(a)** 99%, **(b)** 99.9% of the mass of the polonium to decay.

21.4. France generates over 70% of its electricity from nuclear power. (The figure for the USA and for the UK is just over 20%.) Is there any 'turning back' for such countries that depend so heavily upon nuclear power? Find out what would be the advantages of adopting nuclear fusion if (and when) it becomes available.

21.5. **(i)** From goverment publications, obtain a list showing the various contributions to the background radiation to which the population is exposed. What is the biggest source of background radiation?

(ii) List four factors which control the degree of harm caused by nuclear radiation to an individual.

Extension material to support this unit is available on our website at **http://www.palgrave.com/foundations/lewis.**

Environmental Chemistry

Objectives

- Introduces the topic of environmental chemistry
- Defines the meaning of the word pollutant
- Discusses selected topics under the headings air, water and land pollution
- Includes an introduction to waste management

Contents

22.1 Introduction

Environmental chemistry is the study of the behaviour of chemical species within the earth's environment. This Unit contains a brief introduction to some of the topics of concern to the environmental chemist: **air, water and land pollution**; **energy generation**; and **waste management**.

Pollutants

A substance becomes a **pollutant** when it is present in a concentration that is high enough for it to have a harmful effect on the natural environment.

Often, only highly toxic substances are thought of as pollutants, but even substances that are normally considered harmless may pollute if they are present in high enough concentrations and in the *wrong place* at the *wrong time*. For example, nitrate is added to soil in order to increase plant growth, but an excessive concentration of nitrate present in drinking water can be toxic, especially to young children.

Pollution originates from a **source**. The pollutant is then **transported** by air, water or dumped on land by man. Some of the pollutant may be absorbed (**assimilated**) or **chemically changed** by the environment; the rest builds up to a concentration that enables it to damage organisms or buildings, or to upset the balance of environmental processes (Fig. 22.1).

Technologically advanced cultures tend to produce more pollution than more primitive ones. Their high rates of emission of sewage, waste products from energy generation and exhaust fumes from transport, for example, may exceed the capacity of the environment to prevent the buildup of pollutants arising from such emissions.

Exercise 22A

Nitrate in drinking water

The maximum safe limit of nitrate in drinking water is a concentration of 45 ppm (mg dm^{-3}). A sample of water (volume 250 cm^3) was analysed and found to contain 0.012 g of NO_3^-.

(i) What is the concentration of NO_3^- in the sample in (a) mg dm^{-3} (b) mol dm^{-3}?

(ii) Would the water be considered safe to drink?

(iii) Ammonium nitrate (NH_4NO_3) is a common, solid fertilizer. It supplies nitrogen to the soil which is an essential plant nutrient. What percentage of nitrogen by mass does it contain?

Exercise 22B

Carbon dioxide pollution

Using the environmental pollution model in Fig. 22.1, give as many examples as you can think of, under each heading, for CO_2.

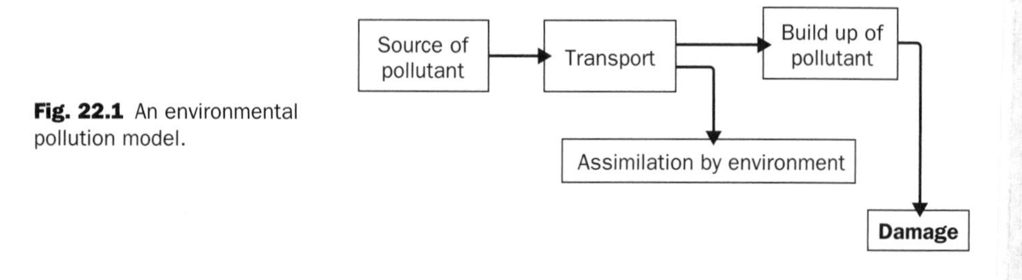

Fig. 22.1 An environmental pollution model.

22.2 Atmospheric pollution

Atmosphere

The lower part of the atmosphere in which we live is called the **troposphere**. Above the troposphere, between 20 and 50 km above sea level lies the **stratosphere.**

The energy that causes the mixing of gases and the pollutants that we produce, in the air, comes from sunlight. This energy also causes evaporation and cloud formation which are part of the **water cycle.**

Mechanism of ozone layer depletion

Ozone (O_3) exists in the stratosphere. Ozone absorbs harmful UV radiation which would otherwise harm living things. The effects of ozone loss could include increased human cataracts and skin cancer, reduction of plankton in ocean waters and destruction of plants, including crops. Ozone layer destruction has been reported in recent years and a major cause of this is believed to be the release of chlorofluorocarbon compounds (CFCs), sometimes known as 'freons'. These compounds are chemically unreactive, non-toxic and odourless, properties which have caused them to be used as solvents, aerosol propellants, refrigerant fluids and blowing agents for expanded plastic foams. They are so stable, however, that they persist

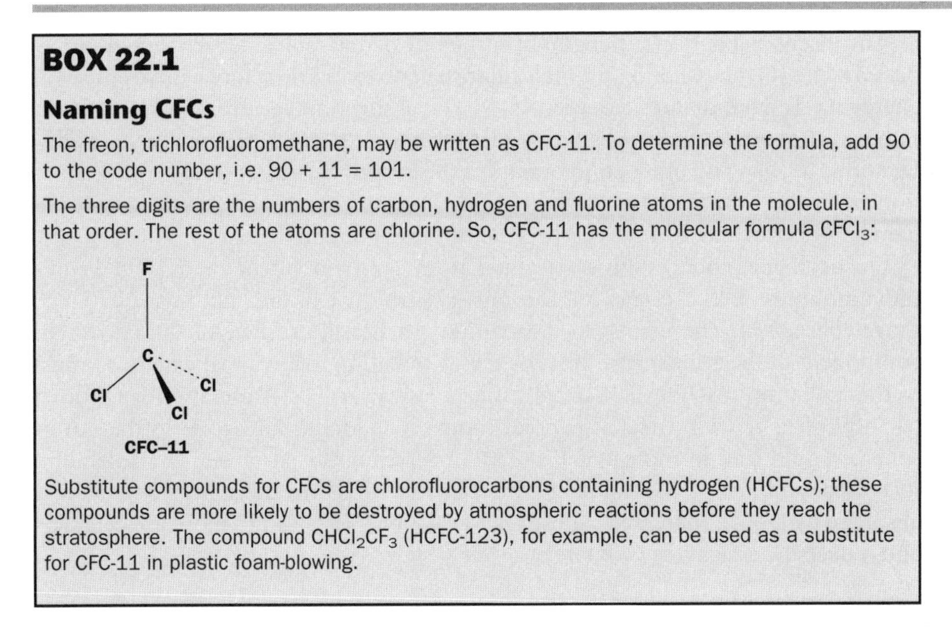

BOX 22.1

Naming CFCs

The freon, trichlorofluoromethane, may be written as CFC-11. To determine the formula, add 90 to the code number, i.e. 90 + 11 = 101.

The three digits are the numbers of carbon, hydrogen and fluorine atoms in the molecule, in that order. The rest of the atoms are chlorine. So, CFC-11 has the molecular formula $CFCl_3$:

Substitute compounds for CFCs are chlorofluorocarbons containing hydrogen (HCFCs); these compounds are more likely to be destroyed by atmospheric reactions before they reach the stratosphere. The compound $CHCl_2CF_3$ (HCFC-123), for example, can be used as a substitute for CFC-11 in plastic foam-blowing.

in the atmosphere for years and eventually enter its upper layers, where they are broken down by the powerful UV radiation emitted by the sun. Their decomposition products can then destroy ozone. The equations for these reactions (starting with a typical freon, CF_2Cl_2) are as follows:

1. UV radiation causes the chlorofluorocarbon to dissociate:

$$CF_2Cl_2 + h\nu \rightarrow Cl\bullet + \bullet CClF_2$$

2. A highly reactive chlorine atom (**free radical**) is produced. This reacts with ozone:

$$Cl\bullet + O_3 \rightarrow ClO\bullet + O_2$$

3. Oxygen molecules split up into atomic oxygen in the upper atmosphere:

$$O_2 + h\nu\ (<240\ \text{nm}) \rightarrow 2O$$

4. Reaction with atomic oxygen produces more chlorine:

$$ClO\bullet + O \rightarrow Cl\bullet + O_2$$

Notice that Cl• radicals are regenerated in this reaction, and are free to react with more ozone in step 2. Many O_3 molecules can be destroyed for each Cl• produced. It has been shown that over one thousand ozone molecules can be destroyed by one Cl•. The greatest losses of ozone have occurred above the South Pole, producing a massive ozone 'hole'.

The filtering of UV light by ozone

Ultraviolet radiation (roughly any radiation in the wavelength range 50–400 nm) can be classified into one of three regions, depending upon its wavelength:

Wavelength/nm	Sub-classification
<290	UV-C
290–320	UV-B
320–400	UV-A

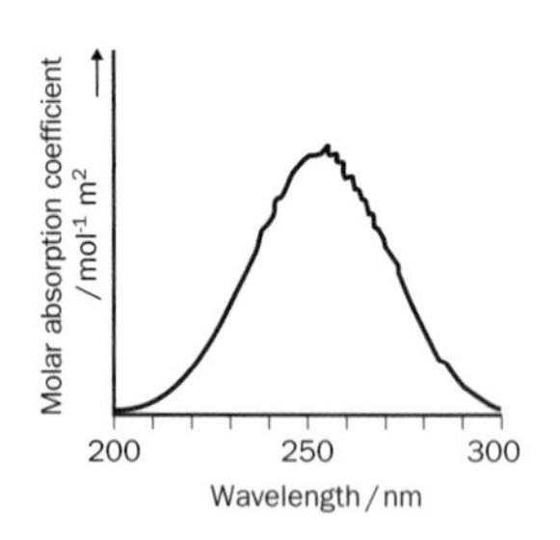

Fig. 22.2 The absorption spectrum of ozone.

The ultraviolet light wavelengths absorbed by ozone molecules are shown in Fig. 22.2. Ozone (O_3) has a different molecular structure, and therefore a different set of energy levels from diatomic oxygen (O_2). The absorption spectrum of O_2 is therefore not the same. O_2 absorbs ultraviolet radiation of much shorter wavelengths. Diatomic oxygen and other components of the atmosphere filter out harmful radiation from the sun, in the ultraviolet region, up to 220 nm. This radiation would damage our eyes and skin if it reached the surface of the Earth.

Ozone, helped to a certain extent by diatomic oxygen, filters out all of the sun's radiation in the UV-C range, but can only absorb *some* of the light in the 290–320 range (Fig. 22.2). The unabsorbed radiation reaches the surface of the Earth. No component of the atmosphere absorbs UV-A radiation to any great extent, so most of this radiation reaches the Earth's surface. However, UV-A radiation is the least harmful of light within the ultraviolet range. A reduction in ozone in the atmosphere, would lead to more UV-B radiation reaching the surface of the Earth and increased sunburn of human skin (this can lead to skin cancer because UV-B is absorbed by DNA). Increased exposure to UV-B may also lead to more eye cataracts and reduce the effectiveness of our immune systems.

Exercise 22C

CFCs and the ozone layer

(i) Write a Lewis structure to show the outer electron arrangement and bonding in CFC-11

(ii) What are the molecular formulae and names of HCFC-22 and HCFC-141?

(iii) Comment on the following cartoon:

Humans with light skin are more susceptible to the adverse effects of UV-B exposure, because their skins do not contain a great deal of the protective pigment melanin which absorbs UV-B. Sunscreens are designed to block UV-B radiation, but not UV-A, so that it is possible to expose the skin for longer periods without it burning.

A reduction in ozone in the atmosphere, would lead to more UV-B radiation reaching the surface of the Earth and increased sunburn of human skin. Over-exposure of the sun's radiation can also increase the risk of skin cancer.

Acid rain

This is a term used to describe all precipitation (rain, snow or fog) which is made acidic by acids stronger than aqueous CO_2. The gases SO_2 and NO_2, mainly emitted from fossil fuel combustion, are major contributors towards acid rain since these gases are oxidized to acids by the oxygen in the atmosphere:

$$SO_2(g) + \tfrac{1}{2}O_2(g) + H_2O(l) \rightarrow H_2SO_4(aq)$$

and

$$2NO_2(g) + \tfrac{1}{2}O_2(g) + H_2O(l) \rightarrow 2HNO_3(aq)$$

Acid rain is toxic to vegetation and aquatic life, damages buildings and statues, and dissolves **heavy metals** from soils, rocks and sediments. The concentration of SO_2 has fallen continuously since 1970 as the amount of coal burned as domestic fuel has fallen. However, the concentration of NO_2 has increased slightly during the same period due to increased traffic.

BOX 22.2

Heavy metals and aluminium concentrations in river water

The term 'heavy metal', to a chemist, means one of a group of the metals in the lower right hand corner of the Periodic Table (metals in Groups 3–16 that are in periods 4 and greater). Examples include Fe, Cu, Pb, Cd and Hg. One reason that these metals are considered toxic is that they have a great affinity for sulfur and attack sulfur bonds in enzymes, making them unable to function properly.

Aluminium is considered to be a light metal, but it can still be toxic at high concentrations. Concentrations of the aqueous aluminium ion have been found to be higher in 'acid' (pH 4–5) river and lake waters and this has been found to adversely affect the trout population in those waters, particularly if the calcium concentration is low.

Global warming and the greenhouse effect

The glass of a greenhouse 'traps' infrared radiation. Similarly, greenhouse gases in the atmosphere are so-called because they can absorb infrared radiation and reduce heat loss from the Earth, as shown in Fig. 22.3. Such gases include water vapour, CO_2, CH_4 and CFCs. Although the concentration of water vapour in the atmosphere has not changed appreciably in recent years, there has been a marked increase in the levels of the other three. Deforestation and the large-scale burning of fossil fuels, for example, have contributed to the steady rise in the atmospheric concentration of CO_2 this century. If these trends continue, in the future there could be an increase in the temperature of the atmosphere resulting in climatic change. For example, there may be less rainfall in temperate zones and more rainfall in the drier areas of the world. However, increased CO_2 levels in the atmosphere should cause plants undergoing photosynthesis to take up the gas at a greater rate, so that plants living in a warmer climate with adequate rainfall should grow faster.

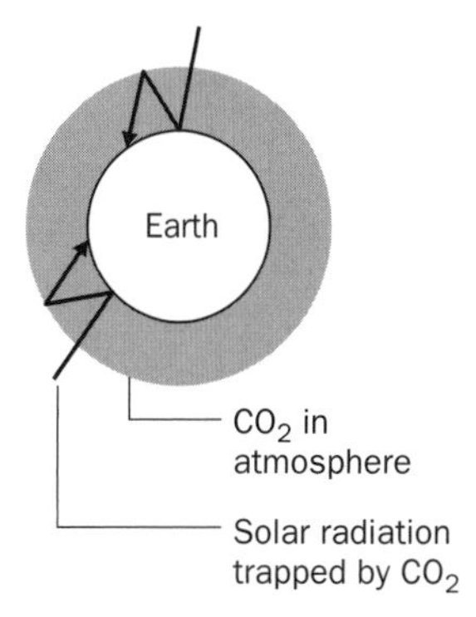

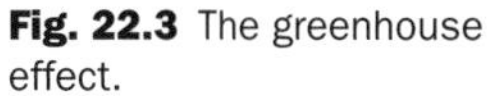

Fig. 22.3 The greenhouse effect.

Predictions of climatic change have been made using complex computer models, which incorporate many assumptions. The future situation is still by no means clear.

Radon

Pollution caused by radionuclides can be divided into natural pollution (caused chiefly by radon gas) and manmade pollution (mainly caused by the nuclear power industry).

Exercise 22D

Greenhouse effect

(i) List possible ways in which an increase in world population could contribute to the greenhouse effect.

(ii) Suggest some alternatives to fossil fuels that would **not** increase the greenhouse effect.

Radon gas (mainly $^{222}_{86}Rn$) is the biggest single source of radiation for most of the population. Its half-life is 3.82 days. Radon is continually being produced as a daughter nuclide from the radioactive decay of uranium in rocks. The gas passes through rocks and enters buildings. Well insulated houses retain more radon gas than draughty ones. The gas itself does not cause damage, but its radioactive disintegration products cling to dust in the air; the contaminated dust can then stick to the lungs and damage them. People living in Cornwall, Somerset, Northamptonshire and Derbyshire are particularly affected because the local rocks contain more uranium.

Smog and soot from cars

The infamous 'smog' of Los Angeles and Tokyo (and of many large cities) is caused mainly by the existence of a vast number of motor cars whose emissions are trapped in a relatively small area. The cars emit NO_2 which (in the presence of sunlight and unburned petrol) produces ozone and other chemical irritants. The NO_2 often gives the air a hazy brown coloration. Some cities give the predicted daily ozone levels with the daily weather forecast! The ozone is purely destructive – it cannot reach the stratosphere sufficiently fast enough to reduce ozone depletion.

Recent medical studies suggest that fine particles of soot emitted by vehicles (especially diesel lorries) may be the cause of many illnesses and premature deaths.

Infrared spectroscopy (page 379) can be used to study the composition of exhaust gases. See Fig. 22.4.

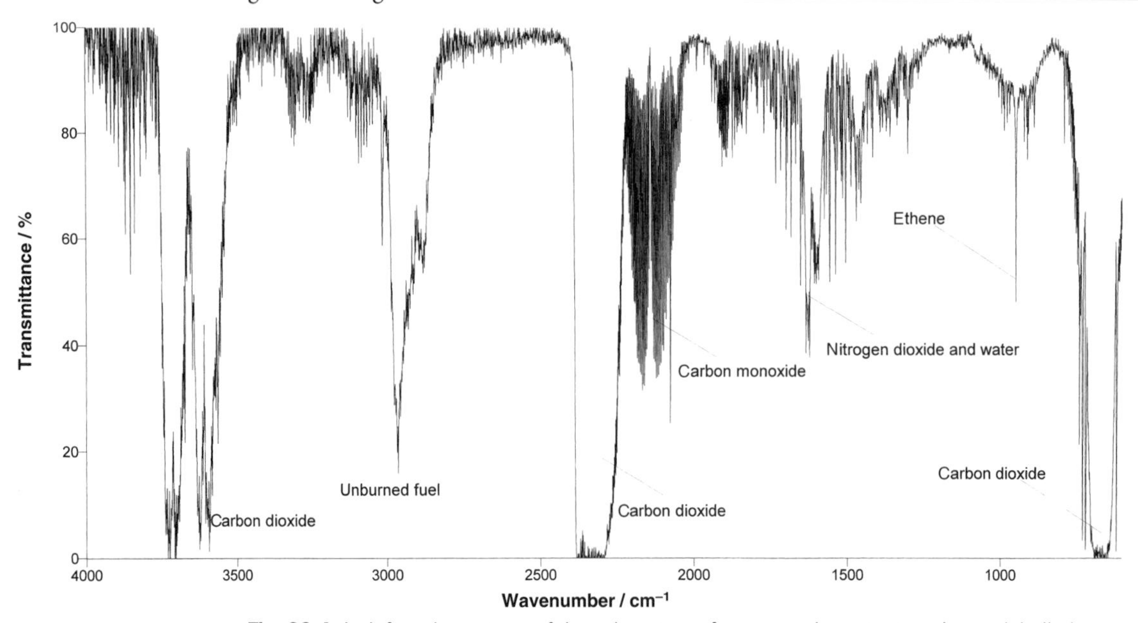

Fig. 22.4 An infrared spectrum of the exhaust gas from a car. Important peaks are labelled.

22.3 Water pollution

The quality of drinking water is very important to human welfare. The pollution of water by sewage has been linked to the spread of diseases such as cholera and typhoid fever. Elimination of these diseases in developed countries has been a direct result of purification of such water, principally by disinfection using chlorine.

However, waste from industry in developed countries can contaminate water supplies with toxic chemicals:

1. **Heavy metals**

   Metals such as Cd, Pb and Hg may be present in industrial or mining waste. These metals can prove poisonous to humans – cadmium and mercury can cause kidney damage, and lead poisoning can cause damage to the kidneys, liver, brain and central nervous system. All of these metals are **cumulative poisons** – the body does not excrete them and their concentration builds up.

2. **Detergents and fertilizers**

   These may contain **phosphates** as additives. The addition of phosphorus to water, in the form of the phosphate anion PO_4^{3-}, encourages the formation of algae which reduces the dissolved oxygen concentration of the water. This process is known as **eutrophication** and threatens the development of higher life forms, such as fish.

3. **Acid-polluted water (pH < 3)**

   This is deadly to most forms of aquatic life. Water downstream from a mine may be contaminated by **acid mine drainage**, the result of microbial oxidation of discarded waste material at the mine site. Acid mine water principally contains sulfuric acid produced by the oxidation of iron pyrites (FeS_2). Industrial wastes and acid rain may also contribute to the acidity of natural waters.

4. **Polychlorinated biphenyls (PCBs)**

   These chemicals are relatively recent contaminants of water, with peak production occurring in 1970. PCBs have high stabilities and this led to their being used in many applications, for example as fluids in transformers and capacitors. PCBs are resistant to oxidation when released into the environment and can cause skin disorders in humans. They may be carcinogenic.

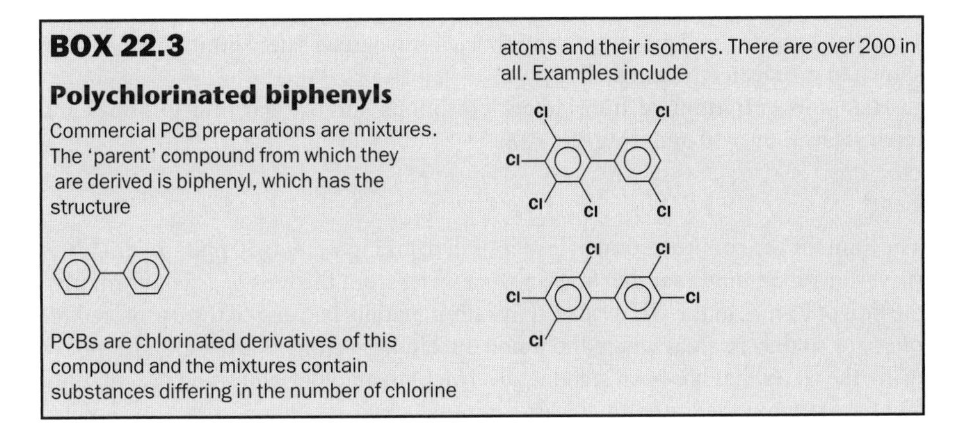

BOX 22.3

Polychlorinated biphenyls

Commercial PCB preparations are mixtures. The 'parent' compound from which they are derived is biphenyl, which has the structure

PCBs are chlorinated derivatives of this compound and the mixtures contain substances differing in the number of chlorine atoms and their isomers. There are over 200 in all. Examples include

Importance of dissolved oxygen in water

The dissolved oxygen concentration in water is vital for the support of fish. The lower the level, the more polluted is the sample. The dissolved gas is used by microorganisms to oxidize organic matter in sewage. Unless the water can restore its dissolved oxygen concentration – for example by the turbulent flow of shallow waters – the water will no longer support many organisms. Typically, fish growth is inhibited below a dissolved oxygen concentration of 6 mg dm^{-3}.

It is not generally realized how little oxygen is present dissolved in a sample of 'pure' water. A litre of water at 5 °C, in free contact with the atmosphere, contains only about 9 cm^3 of oxygen, weighing 13 mg. As the temperature rises, the oxygen concentration falls so that, at 20 °C, it is only about two-thirds of the level it was at 5 °C (see page 185).

Oxygen reaches the water in two main ways. First, it dissolves at the surface from the atmosphere. Still water takes up oxygen slowly, whereas turbulent water takes it up more rapidly since bubbles are often submerged. The second source of oxygen in water is from photosynthesis. Where there are many aquatic green plants present, the water often becomes **supersaturated** with oxygen during the hours of daylight. However, after dark, photosynthesis stops but the plants continue to respire and actually reduce the amount of dissolved oxygen. Therefore, during a 24 hour period, some waters have a considerable range of dissolved oxygen levels. A single determination is therefore of little value in assessing the general condition of a body of water. Several determinations, at various depths, locations and times need to be made.

The amount of oxygen used up in microbial oxidation is – the **biochemical oxygen demand** (BOD) – another important water-quality indicator. The BOD is taken as a realistic measure of water quality – a 'clean' river would have a BOD value of less than 5 ppm, whereas a very polluted river could have a BOD value of 17 ppm or more. A BOD determination takes a few days, so another parameter called the **chemical oxygen demand** (COD) is sometimes measured. In a COD determination, acid dichromate is used to oxidize the organic matter in a sample of water. The measurement takes only a couple of hours.

Drinking water

Fluoride

Soluble fluoride is often added to drinking water to bring it up to a concentration of 1 ppm or 1 mg dm^{-3}. This concentration is within agreed safety limits and has been shown to protect teeth against decay.

High concentrations of fluoride are poisonous and are harmful to bones and teeth at levels over 10 ppm (mg dm^{-3}).

Lead

The limit for the concentration of lead ions in drinking water is 50 ppb ($\mu g\ dm^{-3}$). A survey in 1990 found that the 'first draw' of water from the taps of a significant proportion of homes in the UK exceeded this limit, mainly because of the use of lead for pipes, or solder, in areas where the water is relatively acidic. It is a good idea not to drink the water that has been standing overnight in an older plumbing system, until you have run out the water for at least a minute.

BOX 22.4

Winkler method for determining dissolved oxygen levels in water

The Winkler method is a chemical way of 'fixing' the oxygen gas in a water sample so that the dissolved oxygen level may be found by titration later.

The chemistry may be summarized as:

1. The chemical trapping of dissolved oxygen as insoluble manganese(IV) dioxide:

$$\underset{\text{manganese(II) ion}}{2Mn^{2+}(aq)} + \underset{\text{alkali}}{4OH^-(aq)} + \underset{\text{dissolved oxygen}}{O_2(g)} \rightarrow \underset{\text{brown precipitate}}{2MnO_2(s)} + 2H_2O(l)$$

2. Manganese dioxide reacts with iodide ions, oxidizing them to solid iodine:

$$MnO_2(s) + 4H^+(aq) + 2I^-(aq) \rightarrow \underset{\text{yellow brown}}{I_2(s)} + Mn^{2+}(aq) + 2H_2O(l)$$

3. The iodine is titrated with thiosulfate solution, using starch as indicator:

$$2S_2O_3^{2-}(aq) + I_2(s) \rightarrow S_4O_6^{2-}(aq) + 2I^-(aq)$$

From the above equations,

$$4S_2O_3^{2-}(aq) \equiv O_2(g)$$

$$O_2(g) \equiv 2MnO_2(s) \equiv 2I_2(s) \equiv 4S_2O_3^{2-}(aq)$$

That is, for every four moles of thiosulfate used up in the iodine titration, there was originally one mole of dissolved oxygen in the water sample.

As an example calculation, assume the following measurements:

volume of water sample = $272\,cm^3$

concentration of sodium thiosulfate solution = $0.0200\,mol\,dm^{-3}$

volume of thiosulfate solution used up in titration = $15.10\,cm^3$

Therefore, the number of moles of thiosulfate ($S_2O_3^{2-}(aq)$) used is

$$(15.10/1000) \times 0.0200 = 3.02 \times 10^{-4}\,mol$$

The moles of O_2 present in the original water sample is

$$(1/4) \times 3.02 \times 10^{-4} = 7.55 \times 10^{-5}\,mol$$

Therefore the concentration of dissolved oxygen is

$$7.55 \times 10^{-5} \times (1000/272) = 2.78 \times 10^{-4}\,mol\,dm^{-3}$$

or

$$2.78 \times 10^{-4} \times 32 = 8.88 \times 10^{-3}\,g\,dm^{-3}$$

(this is the same as $8.88\,mg\,dm^{-3}$ or 8.88 ppm).

Exercise 22E

Food waste

Food wastes, although they are considered 'natural' can have a very damaging effect on the environment if they are simply discharged into rivers and lakes. This is because they have a high biochemical oxygen demand.

Suppose that the effluent from a food manufacturing industry has a glucose ($C_6H_{12}O_6$) content of 600 mg dm^{-3}.

(i) Write the chemical equation for the complete oxidation of glucose.

(ii) What is the oxygen requirement (in mg dm^{-3}) for the complete oxidation of $1.0 \times 10^5\,dm^3$ of this waste?

pH

The pH of drinking water should be between 5.5 and 9.5. A decrease in the pH of the water increases the solubility of metal ions.

Other metals

The maximum recommended levels of impurities for drinking water, set by the EU, are as follows

Impurity	Concentration/ppm or mg dm^{-3}
Zn	5
Fe	0.2
Mn	0.05
Cu	3
Cd	0.005
Al	0.2

Sulfate

Sulfate is harmless at moderate levels, but excessive sulfate (> 500 ppm) is thought to have a laxative effect.

Nitrate

Excess nitrate in drinking water can lead to methemoglobinemia ('blue-baby' syndrome). It also may be linked to stomach cancer, although this link has not been proved. The EU have set a maximum limit of 50 ppm in drinking water.

Quality of river water

The **Environmental Agency** – formerly the National River Authority (NRA) is responsible for the conservation of water resources in England and Wales. Their duties include monitoring and improving river quality. Rivers are classified according to their quality as shown in Table 22.1.

Table 22.1 River classification (Source: *Understanding our Environment*, Edited by R. M. Harrison, Royal Society of Chemistry)

River quality class	Dissolved O_2 saturation/%	BOD /ppm	NH_4 as N /ppm	Comment
Class 1A	>80 (non-toxic to trout and coarse fish)	3	0.4	supports high-class fisheries high quality
Class 1B	>60 (non-toxic to trout and coarse fish)	5	0.9	lower quality than 1A but still high quality
Class 2	>40	9	–	moderate quality
Class 3	>10	17	–	low grade; fish rarely present
Class 4	very low or anaerobic	–	–	very polluted; no fish

22.4 Land pollution

Landfill

Landfill involves disposal of hazardous waste on, or within a hole in, the ground. When full, the site is then covered with soil and planted with green cover. Problems have been associated with landfill sites in the past – volatile waste may lead to unpleasant odours, and the decomposition of covered garbage may release methane and other gases for many years. Also, rain may seep into the site and dissolve (**leach**) buried toxic material – the resulting solution can contaminate drinking water supplies via underground water courses. Modern sites are constructed so that the leachate is collected and contained and evolved gases are treated. The latter is particularly important because generation of methane could lead to explosions. Although landfill is a relatively inexpensive way of containing waste, fewer suitable sites are becoming available.

Pesticides

Pesticides are substances that are used to kill, or block the reproductive processes, of unwanted organisms. Synthetic pesticides are of concern to us, because of the possible effect upon human health of eating food, or drinking water, contaminated with these chemicals. Most pesticides can be put into one of three categories:

1. Insecticides

Control of insects by insecticides helps to curb disease (for example malaria and yellow fever) and protect crops. **Organochlorines** are a group of compounds which have been developed and used as insecticides since the 1950s. Organochlorines are stable in the environment, toxic to insects in small amounts, but much less so to humans; and, because they are organic compounds, not very soluble in water. Probably the best-known organochlorine compound is DDT (1,4-dichlorodiphenyltrichloroethane):

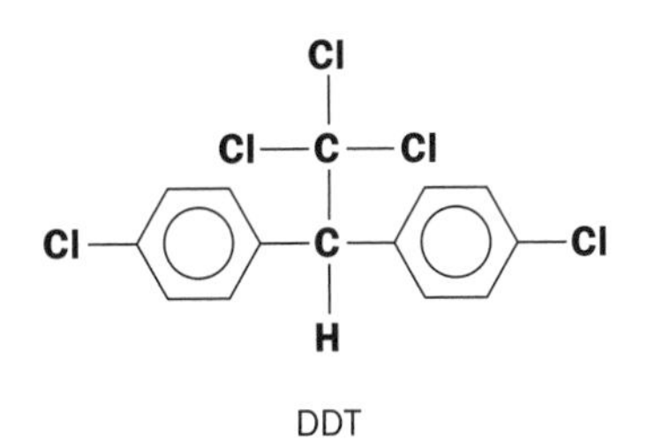

DDT

DDT was discovered in 1939 and hailed as a miraculous insecticide. It was widely used during the Second World War, and afterwards, to combat body lice that carry typhoid, malaria-carrying mosquitoes and pests attacking cotton and food crops. DDT was so effective that it began to be overused and, because it does not degrade easily, its concentration in the environment rose quickly. Bird populations in the United States began to decline because it had affected their ability to reproduce efficiently and legal measures were introduced to curb the use of the insecticide. Small doses of DDT are not immediately harmful to humans, but recent research suggests

BOX 22.5

Problem with hazardous waste disposal – the Love Canal

The Love Canal waste tip in Niagara Falls (USA) was originally a site that had been excavated to build a canal. During the 1930s and 1940s, a chemical works had used it as a site for the disposal of many different chemicals, including toxic chlorinated organic compounds.

In the 1950s, a housing estate and school were built on the land. During a particularly wet winter in the mid-1970s, leachate containing many toxic species flooded the basements of homes and drums containing chemical waste surfaced through the soil. Many people were evacuated and over $100 million was spent in an attempt to rectify the situation.

that DDT may increase the incidence of breast cancer. DDT is still in use in developing countries.

2. Herbicides

Herbicides are used to kill plants. Sodium chlorate, $NaClO_3$ and sodium arsenite, Na_3AsO_3 were commonly used as weed killers in the first half of this century, but inorganic arsenic compounds, in particular, are toxic to mammals. Organic herbicides are now used. They are much more toxic to certain types of plants than to others, so they can be used as *selective* weedkillers. **Atrazine**, which is a member of a class of herbicides called the **triazines**, is widely used to kill weeds in cornfields. The triazines contain six-membered rings with alternating carbon and nitrogen atoms. As yet, atrazine has not been proved harmful to human health.

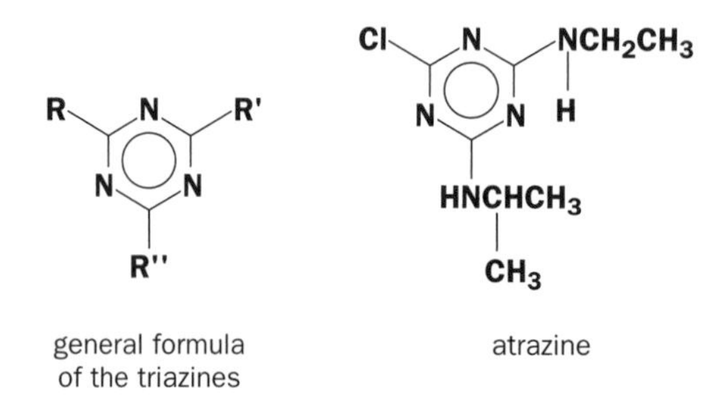

general formula of the triazines

atrazine

Paraquat, another herbicide, is quite water soluble and has been used to destroy marijuana crops. Paraquat poisoning can occur by skin contact, inhalation or ingestion and large doses of the poison can damage human vital organs with fatal consequences.

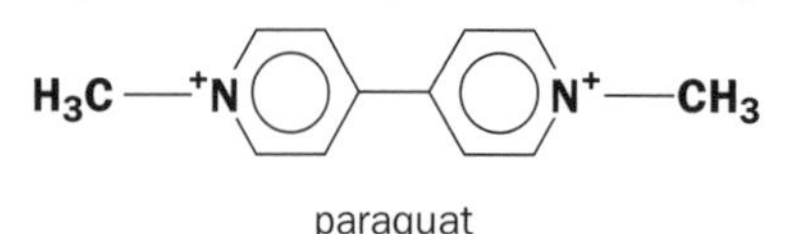

paraquat

3. Fungicides

Organic compounds of mercury have been used as fungicides, which are used to check the growth of fungi. The compounds break down in soil and this has had disastrous

BOX 22.6

Toxicity of mercury

The uses of the element mercury are many, mainly because it is liquid at room temperature and can be used as an electrical conductor in a variety of electrical switches. When given energy, mercury atoms emit light in the visible region and can be used for lighting. The vapour of mercury, however, is highly toxic and when breathed it damages the central nervous system. Since the metal is very volatile, it should be handled only in a well ventilated area. Liquid mercury is not particularly toxic and can be ingested and excreted (for example if you accidently break the bulb of a thermometer in your mouth and swallow the mercury) without too much danger.

The Romans mined the mercury ore mercury sulfide, HgS or cinnibar, and roasted the ore in air to obtain the metal, by condensing the mercury vapour given off. Presumably because of the high toxicity of the vapour, the life expectancy of the slaves that worked on the process was about six months!

Mercury forms solutions or alloys with many different metals, called **amalgams**. The dental amalgam used to fill cavities in teeth contains mercury, silver and tin. During chewing, it has been suggested that a tiny amount of mercury is vaporized and may cause a long-term health hazard. The matter has not been definitively resolved, but alternative mercury-free fillings are being developed.

Methylmercury derivatives are the most toxic form of the element. Because they are organometallics they will dissolve in fatty tissue in animals and accumulate. Most of the mercury present in humans is in the form of methylmercury, formed in rivers by microorganisms which convert Hg^{2+} into CH_3Hg^+ and $(CH_3)_2Hg$. Fish absorb methylmercury through their gills and humans eat the fish.

consequences – many human deaths in Iraq (1971–1972) resulted from the population eating bread made from grain that had been treated with methylmercury to prevent it being destroyed by fungus. The methylmercury cation is an example of an **organometallic** species; it contains an organic group (methyl) directly bonded to a metal:

CH_3Hg^+
methylmercury cation

22.5 Bioconcentration and biomagnification of pesticides

Bioconcentration of pesticides in fish

Pesticides are often washed from land into river or lake water. The concentrations of pesticide in the water are usually tiny, but pesticides are found at much higher concentrations in the oily tissue of the fish present. The concentration of chemicals in living things in this way is called **bioconcentration.**

Bioconcentration occurs when the polluted water passes through the gills of fish. The gills allow oxygen from the water to enter the bloodstream of the fish, but they also allow organochlorine compounds to enter the fatty tissue. The organochlorine compounds are distributed between the fatty (oily) tissue and water. As we would expect, organochlorine compounds are more soluble in oily tissue than in water (like dissolves like).

One way of assessing whether or not an organic compound is likely to bioconcentrate in fish is to determine its distribution ratio (see page 182) between a suitable oil and water. A convenient oil is octan-1-ol (usually known simply as octanol), $CH_3(CH_2)_6CH_2OH$. The distribution ratio is symbolized K_{ow} (o for octanol and w for water):

$$K_{ow} = \frac{\text{concentration of compound in octanol}}{\text{concentration of compound in water}}$$

The higher the K_{ow} value, the more likely it is that the compound will build up in the fish. For example, K_{ow} of DDT at 20 °C is about 100 000. This is an enormous ratio, and confirms that DDT has a tremendous potential for bioconcentration in fish.

Because the K_{ow} values for many compounds are so large, it is usually more convenient to express the ratio as a logarithm, log K_{ow}. For example,

$$\log K_{ow}(\text{DDT}) = \log (1 \times 10^5) = 5.0$$

Biomagnification

The concentration of pesticides at the top of the food chain is magnified because they consume large quantities of smaller animals. This effect is called **biomagnification**. Some examples are:

1. Birds of prey (such as falcons and eagles) have shown high concentrations of DDT in their bodies. High concentrations of DDT can cause bird eggs to have very thin shells and fewer eggs then survive to hatch. This has caused a dramatic fall in the population of some birds. Tighter controls on the use of DDT has resulted in a slow recovery in these populations.

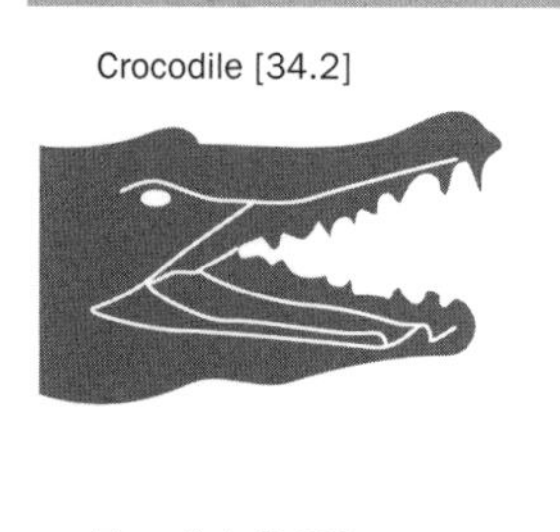

Fig. 22.5 Biomagnification in Lake Kariba. DDT levels are given in units of micrograms of DDT per gram of animal fat. (Adapted from H. Berg, M. Kiibue and N. Kautsky, *Ambio*, 1992, vol. 21, page 444.)

2. Studies in Lake Kariba in Africa, are illustrated in Fig. 22.5. By eating large numbers of smaller fish, tigerfish build up concentrations of DDT in their tissue. By eating tiger fish (and other prey) the crocodile accumulates even larger levels of DDT.

Exercise 22F

Use of K_{ow}

(i) The log K_{ow} (25 °C) values for lindane, dieldrin and vinyl chloride are approximately 4, 3.5 and 0.6, respectively. Which substance is least likely to be absorbed into fatty tissue?

(ii) The concentration of DDT in a badly polluted lake was estimated to be 3×10^{-6} mg dm^{-3} (ppm). Assuming that the fatty tissue of fish is in equilibrium with the DDT concentration in the lake, estimate the concentration of DDT in the fish. Compare your answer with the measured concentrations which were in the range 0.5–2 mg dm^{-3} (ppm).

22.6 Energy and the environment

The generation of most of the energy necessary to support our increasing energy needs, contributes to the reduction of natural resources and the pollution of the environment. Many attempts to solve one of these problems result in creating another. For example, catalytic converters were introduced to destroy pollutants (NO, hydrocarbons and CO) emitted from car exhausts. The catalysts present oxidize carbon-containing pollutants to CO_2 and reduce NO to harmless N_2. Continued use of these catalysts, however, uses up rare-metal resources (such as platinum) and increases petrol consumption.

Current energy options

At present, most of our energy comes from fossil fuels. It has been estimated that, in the future, more coal reserves will be available than those for petroleum and natural gas. The burning of coal with a high sulfur content, however, can contribute to environmental problems.

Nuclear emissions

The disposal of radioactive waste in most Western countries is strictly regulated. Compared with the volume of waste produced by conventional industry and by oil, gas and coal power stations, the volume of highly radioactive waste generated by the nuclear power industry is tiny.

Of all the radioactive waste, the disposal of highly radioactive waste from used nuclear reactor fuel (**high level waste**) is the most controversial. The waste is first allowed to cool, and during this period some of the short-lived isotopes completely decay away. Most of the waste contains uranium which is eventually extracted to make new fuel, and about 1% of it is plutonium. The remainder is stored in special containers. It is planned to turn this radioactive waste into a tough glass-like material (**vitrification**) for long-term storage. The long half-lives of some of the radionuclides in the waste mean that the storage areas will always need to be monitored and protected.

Exercise 22G

Sulfur emission from coal burning

Coal typically contains 2.5% sulfur by mass. What volume of sulfur dioxide is released into the atmosphere, measured at 20 °C and 1 atm, when 1 tonne of coal is burned? (1 tonne = 1000 kg)

Affluent countries (such as the USA, Canada and the European states) have, in recent times, produced a nuclear industry which (on average) produces less pollution and harm than many conventional industries (such as the coal, oil and chemical industries). However, poorer countries have a less satisfactory record. In addition to the Chernobyl accident (see Information Box 22.7), there have been a series of disturbing incidents in the former USSR in which relatively large amounts of radionuclides have been disposed of recklessly. The dumping of old submarine nuclear reactors at sea is one example.

BOX 22.7

Chernobyl

At 1.23 am on 26 April 1986, one of the nuclear reactors at Chernobyl, Ukraine, went into an uncontrolled chain reaction which produced a massive pressure of steam. Two seconds later, the enormous steam pressure created an explosion which blew the thousand tonne safety cover off the top of the reactor. Huge amounts of radioactive material were blown into the air and the graphite moderator caught fire. Local firemen showed immense bravery in tackling the fire and equally brave helicopter pilots dropped dolomite (a form of calcium carbonate containing magnesium) on to the flames. Under the intense heat, the calcium carbonate decomposed into carbon dioxide (and calcium oxide), with the CO_2 produced helping to extinguish the flames. Many of the firemen and pilots died within a few weeks of being exposed to such high levels of radiation.

Away from the immediate area surrounding Chernobyl, the most damaging radionuclides were strontium-90 ($t_{1/2}$ = 28 yr), caesium-137 ($t_{1/2}$ = 30 yr), and iodine-131 ($t_{1/2}$ = 8 days). These radionuclides produced a 'fallout' (Fig. 22.6) which caused extensive contamination over much of Europe, particularly on high grounds where the rainfall is greater such as Scotland, North Wales and Cumbria.

Strontium is in the same group as calcium in the Periodic Table, and both metals have similar chemical reactions. It is not surprising therefore, to find radioactive strontium taking the place of calcium during pregnancy and infant growth, and groups of individuals in these categories are at the greatest risk.

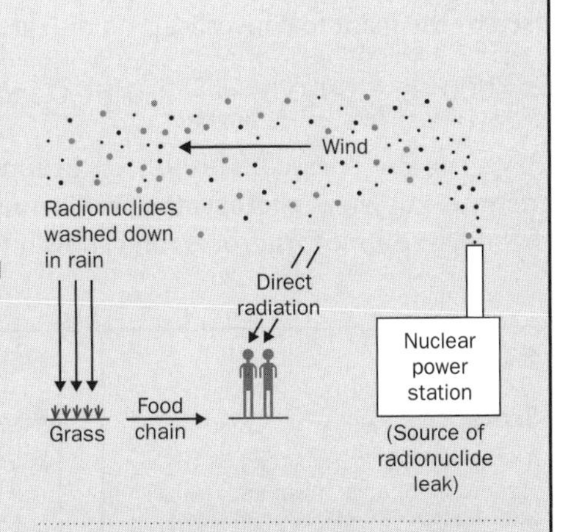

Fig. 22.6 Radioactive materials emitted from an accident at a nuclear power station can directly irradiate individuals, plants and animals, or it can be washed down by rain.

Fallout iodine is deposited on plants. Cows and sheep consume contaminated grass and some ends up in the animals' milk. This explains why milk from contaminated areas was banned during the Chernobyl crisis.

Caesium-137 has the longest half-life of the three contaminants and therefore remains present in the environment longest. It also ends up in animal milk. It was caesium-137 which caused milk and lifestock restrictions to be kept in force long after the hazard from iodine-131 had receded.

Possible future sources of energy

Wind and tidal energy can be made use of by direct conversion of these energies to electricity. Hydroelectric and geothermal power, which use waterpower to generate electricity and underground heat to produce steam, respectively, are possible future energy sources. Although these are 'clean' energy options, even large-scale usage of all of them could only supply a limited amount of our total energy requirement.

An ideal energy source is one that is widely available, cheap and does not add to the Earth's pollutants. Two possibilities for the future are **solar energy** and **nuclear fusion**.

Solar energy

Energy from the sun can be captured by using solar power cells that directly convert sunlight to electricity. These cells have already been used in space vehicles. On Earth, solar energy is intermittent, but the energy could be stored. For example, hydrogen gas could be produced by the electrolysis of suitable salt solutions. Hydrogen can be stored, piped and burned without pollution or used in a **fuel cell**.

Green plants already utilize solar energy. They contain chlorophyll molecules that absorb some of the electromagnetic radiation emitted by the sun, and this energy is used by the plant to convert carbon dioxide and water to carbohydrates:

$$6CO_2(g) + 6H_2O(l) \xrightarrow[\text{chlorophyll}]{\text{solar energy}} C_6H_{12}O_6(aq) + 6O_2(g)$$

Oxygen is also released. This process of converting solar energy into stored chemical energy is known as **photosynthesis**. Chemists are interested in producing chemicals that might mimic this process and allow it to be controlled.

BOX 22.8

Solar cells

A solar battery, or solar cell, is a 'sandwich' of silicon wafers. The silicon wafers have chemicals such as boron or arsenic added to them which make them either have a deficiency of electrons or an excess of electrons (Fig. 22.7).

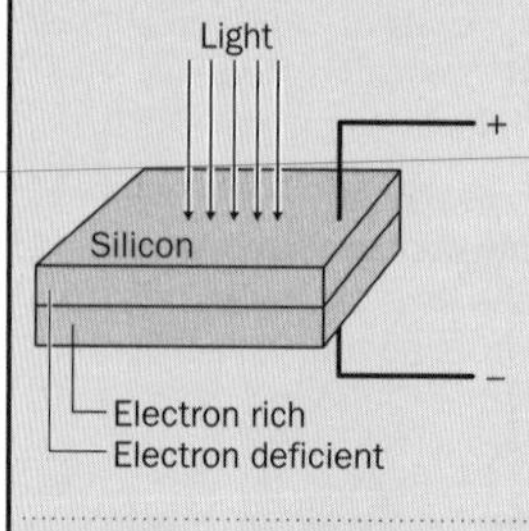

Light
+
Silicon
−
Electron rich
Electron deficient

Fig. 22.7 A solar cell.

When sunlight falls on the cell, energy is absorbed and electrons can move from one silicon layer to another. If the cell is part of a circuit, an electric current flows.

BOX 22.9

Fuel cells

Many fuel cells use the reaction between hydrogen and oxygen to produce electricity. The gases are fed into the cell, where they react at electrodes (Fig. 22.8). The best electrodes are made of platinum. The product, water, proves useful in a space shuttle.

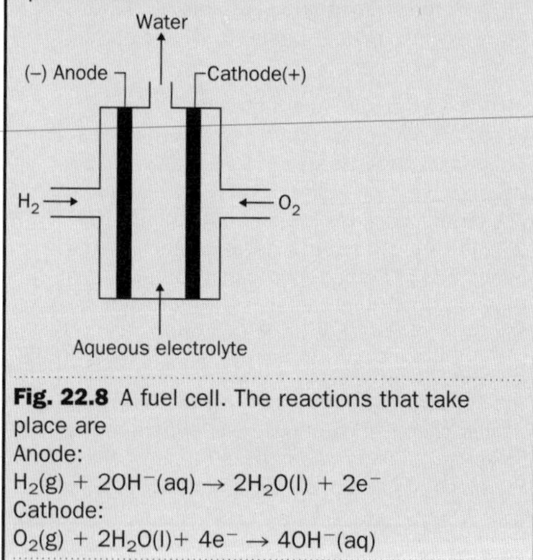

Fig. 22.8 A fuel cell. The reactions that take place are
Anode:
$H_2(g) + 2OH^-(aq) \rightarrow 2H_2O(l) + 2e^-$
Cathode:
$O_2(g) + 2H_2O(l) + 4e^- \rightarrow 4OH^-(aq)$

Nuclear fusion

The fusion of two light nuclei into a heavier nucleus, would release a great deal of energy (similar reactions produce the sun's energy release). Fusion reactors are generally recognized as posing fewer safety hazards than nuclear fission reactors.

In 1989, 'cold fusion' was announced by two scientists in the USA. They claimed that during the electrolysis of deuterium oxide, using palladium electrodes, the deuterium became sufficiently concentrated in the palladium electrode to fuse and unexpectedly large amounts of energy were generated. This report caused a great deal of excitement amongst many researchers but, to date, the claim has not been substantiated and many scientists remain sceptical of the idea.

Energy conservation

Although it is desirable to exploit new, non-polluting sources of energy, energy conservation can significantly reduce both the rate of depletion of current resources and the level of environmental pollution. Examples of how such energy savings might be made include the use of trains or buses instead of cars for transportation (thus saving fuel), the use of fossil fuels for heating rather than electricity, and the improved insulation of buildings.

22.7 Managing waste

The production and disposal of waste is the cause of a great deal of environmental pollution. As well as household waste, which includes sewage and municipal garbage, many industrial wastes from manufacturing processes (some of which may be highly toxic) require treatment and/or safe disposal.

Recycling

When materials are recycled, there is often a twofold benefit – as well as saving on the cost of raw material, waste disposal costs may be reduced. Examples of recycling by industry are:

1. the collection and recycling of glass (in bottle banks);
2. the use of scrap metal in the manufacture of steel;
3. recovery of energy from burning combustible waste.

Sewage treatment

The main stages in the treatment of sewage are as follows:

1. The removal of large solids that get into the system by filtering the wastewater through screens. The solids that are removed are disposed of in landfill sites.
2. Settlement in tanks to allow the removal of solids that settle out (called **sludge**). This process also allows removal of grease, which floats to the surface and can be skimmed off.
3. The organic content of wastewater will have a high biochemical oxygen demand. This is substantially reduced by degradation of the organic matter using microbial oxidation.
4. Various physical and chemical processes are applied, in order to improve the quality of the wastewater. These include chemical removal of phosphate, coagulation, further filtration and disinfection. Disinfection may be carried out using chlorine; however, ozone is increasingly used because of concern over the possible production of toxic organic chlorine compounds during water chlorination.

The safe disposal of the sludge produced during water treatment is a problem. The sludge is dried and then may be **incinerated**, **digested** or **dumped**.

Exercise 22H

Phosphate

Phosphate (PO_4^{3-}) may be precipitated from wastewater as its calcium salt by treatment with limewater ($Ca(OH)_2(aq)$).

(i) Write a balanced ionic equation for this reaction.

(ii) If 1 dm^3 of a solution of wastewater contains a concentration of 0.0003 mol dm^{-3} of phosphate, what mass of Ca^{2+} must be added before calcium phosphate is precipitated? Assume there is no increase in volume upon adding the solid calcium salt and that it is completely dissociated in water. Calcium phosphate has $K_s = 1 \times 10^{-24}$ (mol $dm^{-3})^5$ at room temperature.

Exercise 22I

Chlorination

These reactions occur when chlorine dissolves in water:

$$\underline{Cl_2} + H_2O \rightarrow H^+ + \underline{Cl^-} + \underline{HOCl}$$

$$HOCl \rightleftharpoons H^+ + \underline{OCl^-}$$

(i) Work out the oxidation number for chlorine in the underlined species.

(ii) 'Active chlorine' kills bacteria. It is present in chlorine-containing species that can act as oxidizing agents. Which species in **(i)** are oxidizing agents? Which one is not and why?

BOX 22.10

Coagulation

Coagulation is used to aggregate fine suspended material, so that it may have sufficient size and density to settle out or **flocculate**. For example, $Al_2(SO_4)_3$ or $Fe_2(SO_4)_3$ are added to the water, whereupon the insoluble hydroxides are formed at pH values $\geqslant 7$:

$$Al^{3+}(aq) + 3OH^-(aq) \rightarrow Al(OH)_3(s)$$

The **gelatinous** (jelly-like) precipitate of hydroxide formed carries suspended material with it as it settles. The suspended material consists of colloidal particles, that would not otherwise settle out. The precipitate that forms can be filtered off and consumes hydroxide ions, neutralizing alkaline waters and resulting in a decrease of pH.

Exercise 22J

Sewage treatment

A sewage treatment plant processes 100 000 dm^3 of wastewater per day. The waste contains 400 ppm (mg dm^{-3}) of biodegradable $\{CH_2O\}$.

(i) If all of the $\{CH_2O\}$ is converted to methane by anaerobic digestion, how many dm^3 of methane (at 0 °C and 1 atm) could be produced on a daily basis? (Assume the molar volume of any gas at 0 °C and 1 atm to be 22.4 dm^3.)

(ii) Write an equation for the burning of methane in oxygen.

(iii) The standard enthalpy of combustion of methane, $\Delta H_c^{\ominus}$ (298 K) = −890 kJ mol^{-1}. How much energy (at 298 K and 1 atm) could be generated by the burning of one day's production of methane?

Incineration

Incineration converts organic materials to CO_2 and H_2O. It may serve to destroy household waste, chemical waste and biological waste (e.g. from hospitals). A high temperature is required, usually in excess of 1000 °C, and a plentiful supply of oxygen. Exhaust gases must be filtered. The process greatly reduces the volume of waste – an inorganic ash is left behind, which is disposed of by landfill. Incineration provides a means to dispose of the relatively inert PCBs, and the high temperatures generated allow endothermic reactions, such as the destruction of C–Cl bonds in organochlorine compounds, to take place.

The chief disadvantage of incineration is that it might lead to air pollution. The ash from municipal incinerators is very finely divided and can be ingested into the lungs.

Incomplete combustion of PCBs can cause formation of highly toxic chloro compounds such as polychlorodibenzodioxins (PCDDs) and polychlorodibenzofurans (PCDFs). This is most likely to happen in older municipal waste incinerators, when a combination of insufficient oxygen and too low a temperature for safe incineration might occur.

Digestion

Anaerobic digestion occurs when microorganisms degrade wastes in the absence of oxygen. It may be used to treat sewage sludge, but the process can also be used to degrade a variety of toxic organic wastes. Carbon dioxide and methane, which may be used as a fuel, are the products. The overall process is described by the equation

$$2\{CH_2O\} \rightarrow CO_2 + CH_4$$

where $\{CH_2O\}$ is a general formula for organic waste.

Dumping

Ocean dumping of sewage sludge has been widely practised in the seas around the UK. However, the practice of application of sludge to the land is increasing. The sludge contains nitrogen and phosphorus which make it useful as a fertilizer. Urban areas produce sludge with high toxic metal content, so the amount of such sludge dumped in this way must be carefully controlled.

Revision questions

22.1. A factory discharges waste containing lead ions [Pb^{2+}(aq)] into a river at the rate of 10 ppm (mg dm^{-3}) per minute. Convert this rate to mol dm^{-3} min^{-1}.

22.2. (i) Catalytic converters in car exhausts allow complete oxidation of octane to occur at a low temperature. Write an equation for this reaction.

(ii) A converter operates at 95% efficiency. How much CO_2 is formed from the combustion of 200 g of octane?

22.3. The release of aluminium ions as a result of acid rain is less significant in areas containing limestone rock – why?

22.4. Carbon monoxide is a very poisonous product of the incomplete combustion of hydrocarbons and is present in car exhausts. A flask, of volume 1 dm^3 contains pure CO at a pressure of 1.5 atm at 298 K. The flask is dropped and broken by a careless laboratory worker and the gas is released into a small sealed room of dimensions 5 m × 5 m × 4 m which contains air at 1 atm pressure and 298 K.

(i) Estimate the concentration of CO in the room in mg m^{-3}.

(ii) Assuming that the concentration of CO is hazardous above 55 mg m^{-3}, will the laboratory worker be exposed to a dangerous level of the gas?

22.5. The vapour of tetrahydrofuran (C_4H_8O), a solvent for PVC, is detectable by odour alone at concentrations of 10 mg m^{-3} or above. What mass of the solvent would need to be allowed to evaporate in a room of volume 100 m^3 in order to achieve this concentration? (Assume no loss of vapour.)

22.6. Chemical waste is often destroyed by incineration. Why is this not possible with radioactive waste?

22.7. Background radioactivity in Europe showed a considerable rise in the early 1960s and in 1986 before falling back to its current level. Suggest a reason for these peaks.

22.8. Estimate the time it takes for 99% of samples of $^{137}_{55}Cs$ and $^{131}_{53}I$ to decay away.

22.9. A bottled water had the following analysis on its label:

Ca^{2+}	80 ppm
Mg^{2+}	15 ppm
K^+	4 ppm
Na^+	6 ppm
SO_4^{2-}	85 ppm

(i) If the volume of the bottle was 1.25 L, how many grams of Mg^{2+} did it contain?

(ii) What is the concentration of sulfate in the bottle, in units of $mol\,dm^{-3}$? (Remember that $1\,L = 1\,dm^3$.)

22.10. A sample of water had a dissolved oxygen value of $6\,mg\,dm^{-3}$ O_2. What volume of $0.010\,mol\,dm^{-3}$ sodium thiosulfate solution would be needed to react with $200\,cm^3$ of sample in the Winkler method?

22.11. Study the following data for two toxic insecticides:

Insecticide	log K_{ow}	$t_{1/2}$ (soil)
Chlordane	5.5	3 yr
Parathion	3.8	<3 weeks

Comment upon the importance of these figures.

22.12. Fish do not always grow as well in warm water as cold water – why? (**Hint** dissolved oxygen.)

22.13. It is often asked whether or not the depleted ozone layer could be replenished artificially, by dumping ozone from high-flying aircraft into the stratosphere. Assess the viability of such an operation by estimating the mass of ozone (O_3) that would be needed to restore the ozone concentration in the stratosphere from $1 \times 10^{-11}\,mol\,dm^{-3}$ to $2 \times 10^{-11}\,mol\,dm^{-3}$. Take the volume of the stratosphere as $1 \times 10^{22}\,dm^3$. Assuming that it costs £200 to make and transport 1 kg of O_3 into the stratosphere, what would be the cost of the operation?

Valencies of Common Ions

Name*	Valency	Symbol	Name*	Valency	Symbol
Ammonium	1	NH_4^+	Ethanoate or *acetate*	1	CH_3COO^-
Copper(I) or *cuprous*	1	Cu^+	Hydrogencarbonate or *bicarbonate*	1	HCO_3^-
Caesium	1	Cs^+	Hydrogensulfate (*hydrogensulphate*)	1	HSO_4^-
Potassium	1	K^+	Hydrogensulfite (*hydrogensulphite* or *bisulphite*)	1	HSO_3^-
Silver	1	Ag^+	Bromide	1	Br^-
Sodium	1	Na^+	Bromate or *bromate(V)*	1	BrO_3^-
Lithium	1	Li^+	Chloride	1	Cl^-
Hydrogen	1	H^+	Chlorate or *chlorate(V)*	1	ClO_3^-
Barium	2	Ba^{2+}	Cyanide	1	CN^-
Calcium	2	Ca^{2+}	Fluoride	1	F^-
Cobalt(II)	2	Co^{2+}	Hydride	1	H^-
Copper(II) or *cupric*	2	Cu^{2+}	Hydroxide	1	OH^-
Iron(II) or *ferrous*	2	Fe^{2+}	Iodide	1	I^-
Lead(II)	2	Pb^{2+}	Nitrate or *nitrate(V)*	1	NO_3^-
Magnesium	2	Mg^{2+}	Nitrite or *nitrate(III)*	1	NO_2^-
Mercury(II) or *mercuric*	2	Hg^{2+}	Manganate(VII) or *permanganate*	1	MnO_4^-
Nickel	2	Ni^{2+}	Carbonate	2	CO_3^{2-}
Strontium	2	Sr^{2+}	Chromate or *chromate(VI)*	2	CrO_4^{2-}
Tin(II) or *stannous*	2	Sn^{2+}	Dichromate or *dichromate(VI)*	2	$Cr_2O_7^{2-}$
Zinc	2	Zn^{2+}	Oxide	2	O^{2-}
Aluminium	3	Al^{3+}	Sulfide or *sulphide*	2	S^{2-}
Chromium (III)	3	Cr^{3+}	Sulfite, *sulfate(IV)* or *sulphite*	2	SO_3^{2-}
Iron(III) or *ferric*	3	Fe^{3+}	Sulfate, *sulfate(VI)* or *sulphate*	2	SO_4^{2-}
Tin(IV) or *stannic*†	4	Sn^{4+}	Nitride	3	N^{3-}
			Phosphate	3	PO_4^{3-}

*Common or alternative names are shown in italics.

†Does not exist as an ion, but exhibits the valency tabulated.

Electronic Structures

This table gives the electronic configurations of the elements in their ground state

Shell		1	2		3			4				5					6			7
Subshell		1s	2s	2p	3s	3p	3d	4s	4p	4d	4f	5s	5p	5d	5f	5g	6s	6p	6(f, g, h)	7
1	H	1																		
2	He	2																		
3	Li	2	1																	
4	Be	2	2																	
5	B	2	2	1																
6	C	2	2	2																
7	N	2	2	3																
8	O	2	2	4																
9	F	2	2	5																
10	Ne	2	2	6																
11	Na	2	2	6	1															
12	Mg	2	2	6	2															
13	Al	2	2	6	2	1														
14	Si	2	2	6	2	2														
15	P	2	2	6	2	3														
16	S	2	2	6	2	4														
17	Cl	2	2	6	2	5														
18	Ar	2	2	6	2	6														
19	K	2	2	6	2	6		1												
20	Ca	2	2	6	2	6		2												
	d-block elements																			
21	Sc	2	2	6	2	6	1	2												
22	Ti	2	2	6	2	6	2	2												
23	V	2	2	6	2	6	3	2												
24	Cr	2	2	6	2	6	5	1												
25	Mn	2	2	6	2	6	5	2												
26	Fe	2	2	6	2	6	6	2												
27	Co	2	2	6	2	6	7	2												
28	Ni	2	2	6	2	6	8	2												
29	Cu	2	2	6	2	6	10	1												
30	Zn	2	2	6	2	6	10	2												
31	Ga	2	2	6	2	6	10	2	1											
32	Ge	2	2	6	2	6	10	2	2											
33	As	2	2	6	2	6	10	2	3											
34	Se	2	2	6	2	6	10	2	4											
35	Br	2	2	6	2	6	10	2	5											
36	Kr	2	2	6	2	6	10	2	6											
37	Rb	2	2	6	2	6	10	2	6			1								
38	Sr	2	2	6	2	6	10	2	6			2								
	d-block elements																			
39	Y	2	2	6	2	6	10	2	6	1		2								
40	Zr	2	2	6	2	6	10	2	6	2		2								
41	Nb	2	2	6	2	6	10	2	6	4		1								
42	Mo	2	2	6	2	6	10	2	6	5		1								
43	Tc	2	2	6	2	6	10	2	6	6		1								
44	Ru	2	2	6	2	6	10	2	6	7		1								
45	Rh	2	2	6	2	6	10	2	6	8		1								
46	Pd	2	2	6	2	6	10	2	6	10										
47	Ag	2	2	6	2	6	10	2	6	10		1								
48	Cd	2	2	6	2	6	10	2	6	10		2								
49	In	2	2	6	2	6	10	2	6	10		2	1							
50	Sn	2	2	6	2	6	10	2	6	10		2	2							
51	Sb	2	2	6	2	6	10	2	6	10		2	3							
52	Te	2	2	6	2	6	10	2	6	10		2	4							
53	I	2	2	6	2	6	10	2	6	10		2	5							
54	Xe	2	2	6	2	6	10	2	6	10		2	6							

Shell		1	2	3	4				5					6			7
Subshell					4s	4p	4d	4f	5s	5p	5d	5f	5g	6s	6p	6(f, g, h)	7
55	Cs	2	8	18	2	6	10		2	6				1			
56	Ba	2	8	18	2	6	10		2	6				2			
Lanthanoids																	
57	La	2	8	18	2	6	10		2	6	1			2			
58	Ce	2	8	18	2	6	10	2	2	6				2			
59	Pr	2	8	18	2	6	10	3	2	6				2			
60	Nd	2	8	18	2	6	10	4	2	6				2			
61	Pm	2	8	18	2	6	10	5	2	6				2			
62	Sm	2	8	18	2	6	10	6	2	6				2			
63	Eu	2	8	18	2	6	10	7	2	6				2			
64	Gd	2	8	18	2	6	10	7	2	6	1			2			
65	Tb	2	8	18	2	6	10	9	2	6				2			
66	Dy	2	8	18	2	6	10	10	2	6				2			
67	Ho	2	8	18	2	6	10	11	2	6				2			
68	Er	2	8	18	2	6	10	12	2	6				2			
69	Tm	2	8	18	2	6	10	13	2	6				2			
70	Yb	2	8	18	2	6	10	14	2	6				2			
d-block elements																	
71	Lu	2	8	18	2	6	10	14	2	6	1			2			
72	Hf	2	8	18	2	6	10	14	2	6	2			2			
73	Ta	2	8	18	2	6	10	14	2	6	3			2			
74	W	2	8	18	2	6	10	14	2	6	4			2			
75	Re	2	8	18	2	6	10	14	2	6	5			2			
76	Os	2	8	18	2	6	10	14	2	6	6			2			
77	Ir	2	8	18	2	6	10	14	2	6	7			2			
78	Pt	2	8	18	2	6	10	14	2	6	9			1			
79	Au	2	8	18	2	6	10	14	2	6	10			1			
80	Hg	2	8	18	2	6	10	14	2	6	10			2			
81	Tl	2	8	18	2	6	10	14	2	6	10			2	1		
82	Pb	2	8	18	2	6	10	14	2	6	10			2	2		
83	Bi	2	8	18	2	6	10	14	2	6	10			2	3		
84	Po	2	8	18	2	6	10	14	2	6	10			2	4		
85	At	2	8	18	2	6	10	14	2	6	10			2	5		
86	Rn	2	8	18	2	6	10	14	2	6	10			2	6		
87	Fr	2	8	18	2	6	10	14	2	6	10			2	6		1
88	Ra	2	8	18	2	6	10	14	2	6	10			2	6		2
Actinoids																	
89	Ac	2	8	18	2	6	10	14	2	6	10			2	6	1	2
90	Th	2	8	18	2	6	10	14	2	6	10			2	6	2	2
91	Pa	2	8	18	2	6	10	14	2	6	10	2		2	6	1	2
92	U	2	8	18	2	6	10	14	2	6	10	3		2	6	1	2
93	Np	2	8	18	2	6	10	14	2	6	10	4		2	6	1	2
94	Pu	2	8	18	2	6	10	14	2	6	10	6		2	6		2
95	Am	2	8	18	2	6	10	14	2	6	10	7		2	6		2
96	Cm	2	8	18	2	6	10	14	2	6	10	7		2	6	1	2
97	Bk	2	8	18	2	6	10	14	2	6	10	9		2	6		2
98	Cf	2	8	18	2	6	10	14	2	6	10	10		2	6		2
99	Es	2	8	18	2	6	10	14	2	6	10	11		2	6		2
100	Fm	2	8	18	2	6	10	14	2	6	10	12		2	6		2
101	Md	2	8	18	2	6	10	14	2	6	10	13		2	6		2
102	No	2	8	18	2	6	10	14	2	6	10	14		2	6		2
103	Lr	2	8	18	2	6	10	14	2	6	10	14		2	6	1	2
104	Rf	2	8	18	2	6	10	14	2	6	10	14		2	6	2	2
105	Db	2	8	18	2	6	10	14	2	6	10	14		2	6	3	2
106	Sg	2	8	18	2	6	10	14	2	6	10	14		2	6	4	2

Beyond $_{94}$Pu the assignments are conjectural

Answers to Exercises and Revision Questions

Unit 1

Exercises

1A

(i) 3.45×10^{-5} **(ii)** 3×10^{8}
(iii) $8.205\ 75 \times 10^{-2}$ **(iv)** 3.5×10^{0}
(v) 6.022×10^{23} **(vi)** 1.7×10^{1}

1B

(i) 379.75 (or 380) **(ii)** 17.34 **(iii)** −1.921 (or −1.92)
(iv) 2.75×10^{-5} **(v)** 0.409
(vi) 5.517×10^{-3} (or 5.5×10^{-3})
(vii) 5.52×10^{-6}

1C

(i) 0.040 nm **(ii)** 10 g **(iii)** 6×10^{-7} m.

1D

(i) Rearranging, $C = \frac{q}{m \times \Delta T}$. The units of C are $\frac{J}{g \times K} = J\,g^{-1}\,K^{-1}$.

(ii) $M = \frac{m}{n}$ and the units are $\frac{g}{mol} = g\,mol^{-1}$.

1E

Some of the alcohol in the wine has been oxidized.

1F

(i) (a) 1, (b) 4, (c) 2 or 3 (compare the example of 150 in Box 1.3), (d) 3.
(ii) No, to four significant figures the atomic mass is 15.99 atomic mass units.
(iii) (a) 0.0347, (b) 0.035, (c) 0.03.

1G

(i) $[H^+(aq)] = 1.484 \times 10^{-5}\ mol\,dm^{-3}$, which becomes $1.5 \times 10^{-5}\ mol\,dm^{-3}$ (two significant figures).
(ii) 10.432 mg ≈ 10.4 mg to one decimal place.
(iii) $pK_a = -\log K_a = -\log(8.4 \times 10^{-4}\ mol) = 3.08$

Revision questions

1.1 Cr^{3+} 6.9×10^{-11} m, F^- 1.36×10^{-8} m, O 1.40×10^{-8} m.
1.2 Uncertainty is at least ± 0.1 °C.
1.3 (i) 0.1235 V, **(ii)** 12.45 m, **(iii)** 0.003 558, **(iv)** 1201 K.
1.4 (i) 3.29×10^{-4}, **(ii)** 0.449.
1.5 Total mass is 0.49 g (to two decimal places).
1.6 Units are $mol^2\,dm^{-6}$. $K_w = 1 \times 10^{-14}$ ($[H_3O^+]$ is supplied to only one significant figure).
1.7 ε has units of $mol^{-1}\,m^2$.
1.8 (i) random
(ii) systematic (too big to be explained by draughts)
(iii) systematic (pressure is falling).
1.9 (i) pH = −log(**8.987** × 10^{-6}) = −(−5.046 38) = 5.**0464** (four decimal places and rounded up).
(ii) $[H^+(aq)] = 10^{-11.344} = 4.528\,976 \times 10^{-12} = \mathbf{4.53} \times 10^{-12}\ mol\,dm^{-3}$ (three significant figures and rounded up).

Unit 2

Exercises

2A

(i) chemical **(ii)** physical **(iii)** physical
(iv) physical **(v)** physical.

2B

(i) physical **(ii)** physical **(iii)** chemical
(iv) chemical **(v)** chemical.

2C

(i) extensive **(ii)** intensive **(iii)** extensive
(iv) intensive.

2D

1. (i) barium **(ii)** silicon **(iii)** argon
(iv) fluorine **(v)** lithium **(vi)** mercury
(vii) tin **(viii)** silver **(ix)** boron **(x)** titanium.
2. (i) caesium (1) bromine (1)
(ii) copper (1) chlorine (2)
(iii) iron (3) oxygen (4)
(iv) carbon (1) hydrogen (4)
(v) lead (1) sulfur (1) oxygen (4)
(vi) potassium (2) chromium (2) oxygen (7)
(vii) sodium (3) phosphorus (1) oxygen (4)
(viii) manganese (1) oxygen (2)

2E

carbon, hydrogen, oxygen, phosphorus, potassium, iodine, nitrogen, sulfur, calcium, iron, sodium, chlorine.

2F

1. (i) Natrium (*suda* is Arabic for 'headache'; sodium carbonate was considered a remedy)
(ii) *ferrum* (Latin) **(iii)** *plumbum* (Latin)
(iv) *stannum* (Latin) **(v)** *kalium* (Latin)
(vi) *argentum* (Latin)
2. (i) Lise Meitner worked on uranium fission and realized that atomic fission could produce vast quantities of energy. **(ii)** Glenn T Seaborg, an American nuclear chemist and Nobel prize winner (1951) for investigating the chemistry of heavy elements.
(iii) Niels Bohr and **(iv)** Ernest Rutherford made major contributions to our understanding of atomic structure.

2G

(i) CaO **(ii)** H_2S **(iii)** MgF_2 **(iv)** $MgCl_2$
(v) AlF_3 **(vi)** Al_2O_3 **(vii)** HCl **(viii)** LiBr
(ix) MgS **(x)** Mg_3N_2.

2H

(i) Cu_2O **(ii)** CuO **(iii)** PbO_2 **(iv)** Cr_2S_3
(v) MnO_2 **(vi)** CrF_6 **(vii)** Co_2S_3 **(viii)** Cr_2O_3
(ix) V_2O_5 **(x)** TiF_3 **(xi)** NO **(xii)** $AsCl_3$
(xiii) SF_6 **(xiv)** N_2 **(xv)** S_2Cl_2 **(xvi)** N_2O_5
(xvii) O_2F_2 **(xviii)** TeO_3 **(xix)** Cl_2O_6
(xx) $SiCl_4$.

2I

(i) $CuSO_4$ **(ii)** $Cu(NO_3)_2$ **(iii)** NH_4Cl
(iv) Na_3PO_4 **(v)** $Ca_3(PO_4)_2$ **(vi)** H_2SO_4
(vii) HNO_3 **(viii)** $Al(NO_3)_3$ **(ix)** Li_2CO_3
(x) $(NH_4)_2CO_3$ **(xi)** $Ca(OH)_2$ **(xii)** $KHCO_3$
(xiii) $Ca(HCO_3)_2$ **(xiv)** $NaHSO_4$ **(xv)** $Fe(OH)_3$.

2J

(i) $N_2 + 3H_2 \rightarrow 2NH_3$
(ii) $2Na + Cl_2 \rightarrow 2NaCl$
(iii) $4Al + 3O_2 \rightarrow 2Al_2O_3$
(iv) $2Al + 2H_3PO_4 \rightarrow 2AlPO_4 + 3H_2$
(v) $4Na + O_2 \rightarrow 2Na_2O$
(vi) $Mg_3N_2 + 6H_2O \rightarrow 3Mg(OH)_2 + 2NH_3$
(vii) $2Cr_2O_3 + 3Si \rightarrow 4Cr + 3SiO_2$
(viii) $CS_2 + 3O_2 \rightarrow CO_2 + 2SO_2$
(ix) $Fe_2O_3 + 3CO \rightarrow 2Fe + 3CO_2$
(x) $4FeS_2 + 11O_2 \rightarrow 2Fe_2O_3 + 8SO_2$.

2K

(i) $4K + O_2 \rightarrow 2K_2O$
(ii) $2Fe + 3Cl_2 \rightarrow 2FeCl_3$
(iii) $2Li + 2H_2O \rightarrow 2LiOH + H_2$
(iv) $3Mg + N_2 \rightarrow Mg_3N_2$
(v) $Ca(OH)_2 + CO_2 \rightarrow CaCO_3 + H_2O$
(vi) $2NaOH + H_2SO_4 \rightarrow Na_2SO_4 + 2H_2O$
(vii) $2Au_2O_3 \rightarrow 4Au + 3O_2$
(viii) $Cu + 4HNO_3 \rightarrow Cu(NO_3)_2 + 2NO_2 + 2H_2O$

2L

(i) $CaCO_3(s) \rightarrow CaO(s) + CO_2(g)$
(ii) $Pb(NO_3)_2(aq) + 2NaI(aq) \rightarrow PbI_2(s) + 2NaNO_3(aq)$
(iii) $2Al(s) + 3Cl_2(g) \rightarrow 2AlCl_3(s)$
(iv) $CH_4(g) + 2O_2(g) \rightarrow CO_2(g) + 2H_2O(g)$.

Revision questions

2.1 (i) mixture **(ii)** compound **(iii)** element
(iv) mixture **(v)** compound **(vi)** mixture
(vii) element **(viii)** mixture **(ix)** element
(x) compound.
2.2 m.p. physical,
b.p. physical,
soft physical,
corrodes chemical,
conducts electricity physical,
reacts with water chemical.
2.3 liquid.
2.4 gas.
2.5 (i) *cuprum* (Latin for Cyprus, famous for its copper mines)
(ii) *stibium* (Latin for 'mark'; once used for eyebrow pencils)
(iii) *hydrargyrum* (Latin).
2.6 (i) O **(ii)** 2O **(iii)** $2O_2$ **(iv)** $3SO_3$
(v) 4Ne.
2.7 (i) lithium (2), carbon (1), oxygen (3).
(ii) calcium (1), nitrogen (2), oxygen (6).
(iii) nitrogen (3), hydrogen (12), phosphorus (1), oxygen (4).
2.8 (i) BaF_2 **(ii)** $SnBr_2$ **(iii)** $SnCl_4$
(iv) $NaCH_3CO_2$ **(v)** Al_2S_3 **(vi)** KOH
(vii) $NaBrO_3$ **(viii)** $K_2Cr_2O_7$ **(ix)** $(NH_4)_3PO_4$
(x) $Pb(CN)_2$.
2.9 As a bleach (hair, textiles, paper, straw, leather etc.) or as a mild antiseptic.
2.10 $2H_2O_2(l) \rightarrow 2H_2O(l) + O_2(g)$.
2.11 (i) $KClO_4 \rightarrow KCl + 2O_2$
(ii) $2S_8 + 3AsF_5 \rightarrow S_{16}(AsF_6)_2 + AsF_3$
(iii) $Hg + 2NH_4I \rightarrow HgI_2 + H_2 + 2NH_3$

(iv) $Al_4C_3 + 12H_2O \rightarrow 4Al(OH)_3 + 3CH_4$
(v) $3Zn + 2H_3PO_4 \rightarrow Zn_3(PO_4)_2 + 3H_2$
(vi) $Fe_2O_3 + Na_2CO_3 \rightarrow 2NaFeO_2 + CO_2$.
2.12 (i) $2Zn(s) + O_2(g) \rightarrow 2ZnO(s)$
(ii) $2K(s) + 2H_2O(l) \rightarrow 2KOH(aq) + H_2(g)$
(iii) $CO_2(g) + 2Mg(s) \rightarrow 2MgO(s) + C(s)$.

Unit 3

Exercises

3A

Six neutrons and six protons in the nucleus, with six electrons outside.

3B

(i) 1836 times (using exact masses)
(ii) 3.97 (about 4).

3C

Isotope	Protons	Neutrons	Electrons
U-235	92	143	92
U-238	92	146	92
H-1	1	0	1
H-2	1	1	1
H-3	1	2	1

3D

(i) (a) uranium-238 (b) fluorine (c) hydrogen-3 (tritium) (d) tritium.
(ii) 49.31%.

3E

$m(O) = [(15.9949 \times 99.759) + (16.9991 \times 0.0374) + (17.9992 \times 0.2039)]/100 = 15.9994\,u = 16.00\,u$ (to four significant figures).

3F

The intensities add up to 100, and so correspond to percentage abundances:

$$m(Ne) = [(91 \times 20) + (0.3 \times 21) + (8.7 \times 22)]/100 = 20.18\,u \approx 20\,u$$

3G

(i) m/e values are 44, 15, 14 (i.e. 28/2).
(ii) Fragmentation follows ionization:

$HF(g) + e^- \rightarrow HF^+(g) + e^- + e^-$
$HF^+(g) \rightarrow F^+(g) + H(g)$

m/e for the F^+ ion is 19, and for HF^+ is 20.
(iii) The parent ions $C_6H_6^+$, $C_6H_5NO_2^+$ and $C_6H_5OH^+$ are produced in the ionization stage. They then fragment producing a $C_6H_5^+$ ion e.g:

$C_6H_5NO_2^+(g) \rightarrow C_6H_5^+(g) + NO_2(g)$
$m/e = 77$

3H

$Be^{3+}(g) \rightarrow Be^{4+}(g) + e^-$
electrons: 1 0

There are only four electrons in the beryllium atom and so there is no fifth ionization energy.

3I

Element and symbol	Atomic number	Bohr structure	s,p,d,f structure
Hydrogen H	1	1.	$1s^1$
Helium He	2	2.	$1s^2$
Lithium Li	3	2.1	$1s^2 2s^1$
Beryllium Be	4	2.2	$1s^2 2s^2$
Boron B	5	2.3	$1s^2 2s^2 2p^1$
Carbon C	6	2.4	$1s^2 2s^2 2p^2$
Nitrogen N	7	2.5	$1s^2 2s^2 2p^3$
Oxygen O	8	2.6	$1s^2 2s^2 2p^4$
Fluorine F	9	2.7	$1s^2 2s^2 2p^5$
Neon Ne	10	2.8	$1s^2 2s^2 2p^6$
Sodium Na	11	2.8.1	$1s^2 2s^2 2p^6 3s^1$
Magnesium Mg	12	2.8.2	$1s^2 2s^2 2p^6 3s^2$
Aluminium Al	13	2.8.3	$1s^2 2s^2 2p^6 3s^2 3p^1$
Silicon Si	14	2.8.4	$1s^2 2s^2 2p^6 3s^2 3p^2$
Phosphorus P	15	2.8.5	$1s^2 2s^2 2p^6 3s^2 3p^3$
Sulfur S	16	2.8.6	$1s^2 2s^2 2p^6 3s^2 3p^4$
Chlorine Cl	17	2.8.7	$1s^2 2s^2 2p^6 3s^2 3p^5$
Argon Ar	18	2.8.8	$1s^2 2s^2 2p^6 3s^2 3p^6$
Potassium K	19	2.8.8.1	$1s^2 2s^2 2p^6 3s^2 3p^6 4s^1$
Calcium Ca	20	2.8.8.2	$1s^2 2s^2 2p^6 3s^2 3p^6 4s^2$

3J

(i)

Element and symbol	1s	2s	$2p_x$	$2p_y$	$2p_z$	s,p,d,f structure
Hydrogen H	↑					$1s^1$
Helium He	↑↓					$1s^2$
Lithium Li	↑↓	↑				$1s^2 2s^1$
Beryllium Be	↑↓	↑↓				$1s^2 2s^2$
Boron B	↑↓	↑↓	↑			$1s^2 2s^2 2p^1$
Carbon C	↑↓	↑↓	↑	↑		$1s^2 2s^2 2p^2$
Nitrogen N	↑↓	↑↓	↑	↑	↑	$1s^2 2s^2 2p^3$
Oxygen O	↑↓	↑↓	↑↓	↑	↑	$1s^2 2s^2 2p^4$
Fluorine F	↑↓	↑↓	↑↓	↑↓	↑	$1s^2 2s^2 2p^5$
Neon Ne	↑↓	↑↓	↑↓	↑↓	↑↓	$1s^2 2s^2 2p^6$

(ii)

Element and symbol	3d	3d	3d	3d	3d	4s	s,p,d,f structure
Manganese (Mn)	↑	↑	↑	↑	↑	↑↓	$[Ar]\,3d^5 4s^2$
Iron (Fe)	↑↓	↑	↑	↑	↑	↑↓	$[Ar]\,3d^6 4s^2$
Copper (Cu)	↑↓	↑↓	↑↓	↑↓	↑↓	↑	$[Ar]\,3d^{10} 4s^1$

Revision questions

3.1 Following are false:
(ii) should be $^{12}_{6}C$,
(iv) atoms of isotopes are not identical,
(v) a mass spectrometer does not detect neutral atoms.

3.2

Isotope	Number of neutrons	Number of electrons	Number of protons	Abundance/%
$^{32}S_{16}$	16	16	16	95.0
$^{33}S_{16}$	17	16	16	0.76
$^{34}S_{16}$	18	16	16	4.2
$^{36}S_{16}$	20	16	16	0.021

$m(S)$ is given by the equation:

$$\frac{(95 \times 32) + (0.76 \times 33) + (4.2 \times 34) + (0.021 \times 36)}{100}$$

$$= \frac{3208.6}{100} = 32\,u \text{ (to two significant figures)}$$

Sources of approximation are **(i)** mass numbers have been used instead of isotopic masses and **(ii)** some of the abundances are only given to two significant figures. (This explains why the total of the percentage abundances is 99.981%, not 100.000%.)

3.3

$Mg(g) + e^- \rightarrow Mg^+(g) + e^- + e^-$
$m/e \approx 24$
(2.8.1)

$Mg(g) + e^- \rightarrow Mg^{2+}(g) + e^- + e^- + e^-$
$m/e \approx 12$
(2.8)

3.4

(i)
$F_2(g) + e^- \rightarrow F_2^+(g) + e^- + e^-$
$m/e = 38$

$F_2^+(g) \rightarrow F^+(g) + F(g)$ (fragmentation)
$m/e = 19$

(ii)
$HCN(g) + e^- \rightarrow HCN^+(g) + e^- + e^-$
$m/e = 27$

$HCN(g) + e^- \rightarrow HCN^{2+}(g) + e^- + e^- + e^-$
$m/e = 13.5$

and

$HCN^+(g) \rightarrow CN^+(g) + H(g)$ (fragmentation)
$m/e = 26$

(iii)
$C_6H_5Cl(g) + e^- \rightarrow C_6H_5Cl(g)^+ + e^- + e^-$
$m/e = 114$ (due to $C_6H_5{}^{37}Cl^+$)
and 112 (due to $C_6H_5{}^{35}Cl^+$)

$C_6H_5Cl(g)^+ \rightarrow C_6H_5^+(g) + Cl(g)$ (fragmentation)
$m/e = 77$

(iv) $N_2^+(g)$ from air gives a line at $m/e = 28$.
(v)
$Cl_2(g) + e^- \rightarrow Cl_2^+(g) + e^- + e^-$
$m/e = 70$ (due to $^{35}Cl^{35}Cl^+$)
$m/e = 72$ ($^{35}Cl^{37}Cl^+$)
$m/e = 74$ ($^{37}Cl^{37}Cl^+$)

$CH_4(g) + e^- \rightarrow CH_4^+(g) + e^- + e^-$
$m/e = 16$

If some chlorine and methane react:

$CH_4(g) + Cl_2(g) \rightarrow CH_3Cl(g) + HCl$

$CH_3Cl(g) + e^- \rightarrow CH_3Cl^+(g) + e^- + e^-$
$m/e = 50$ (for $CH_3{}^{35}Cl^+$)
$m/e = 52$ (for $CH_3{}^{37}Cl^+$)

3.5
$2\,^1_1H_2(g) + O_2(g) \rightarrow 2\,^1_1H_2O(l)$
$2\,^2_1H_2(g) + O_2(g) \rightarrow 2\,^2_1H_2O(l)$

Heavy water (deuterium oxide) will be more dense than ordinary water.
3.6 $m(^2_1H_2{}^{16}_{8}O) = (2 \times 2.0140) + (15.9949) = 20.023\,u$ (to five significant figures).
Since $1\,u = 1.660\,54 \times 10^{-24}\,g$,

$m(^2_1H_2{}^{16}_{8}O) = 1.660\,54 \times 10^{-24} \times 20.0229 = 3.3249 \times 10^{-23}\,g$

or $3.325 \times 10^{-23}\,g$ to four significant figures .

$m(H_2O) = (2 \times 1.008) + (16.00) = 18.02\,u$.
$m(H_2O) = 18.02 \times 1.660\,54 \times 10^{-24}\,g = 2.992 \times 10^{-23}\,g$ to four significant figures.
3.7 See text.
3.8 A plot is seen to contain three groups containing eight electrons (the outer electrons are removed first), eight more electrons and two more electrons, respectively.
Sum of ionization energies = 1 389 373 kJ mol^{-1}. This is the energy required to strip all the electrons from the atom.
3.9 For a definition of orbital see the text. The size of a 100% orbital is the size of the universe itself.
3.10 (i) 2.8.7 **(ii)** 2.8.8 **(iii)** 2.8.6 **(iv)** 2
(v) 2.8 **(vi)** [Ar]9.
(ii), **(iv)**, and **(v)** are electronic structures in which all the shells are filled.
3.11 (i) $1s^2 2s^2 2p^6 3s^2 3p^5$
(ii) $1s^2 2s^2 2p^6 3s^2 3p^6$
(iii) $1s^2 2s^2 2p^6 3s^2 3p^4$ **(iv)** $1s^2$
(v) $1s^2 2s^2 2p^6$
(vi) [Ar] $3d^9$. Using the box notation for **(iii)**:

$1s^2$	$2s^2$	$2p_x^2$	$2p_y^2$	$2p_z^2$	$3s^2$	$3p_x^2$	$3p_y^1$	$3p_z^1$
↑↓	↑↓	↑↓	↑↓	↑↓	↑↓	↑↓	↑	↑

Unit 4

Exercises

4A

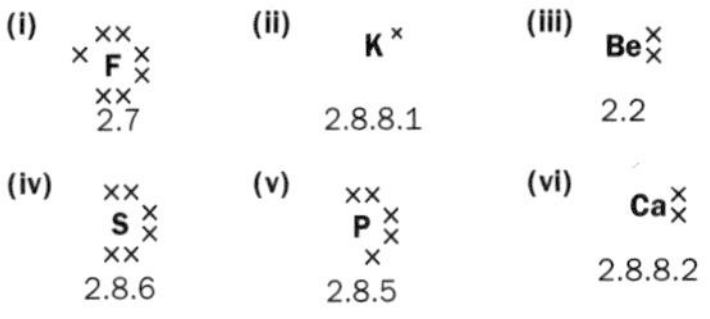

(i) F (×× dot-and-cross) 2.7 **(ii)** K× 2.8.8.1 **(iii)** Be (××) 2.2
(iv) S (××) 2.8.6 **(v)** P (××) 2.8.5 **(vi)** Ca (××) 2.8.2

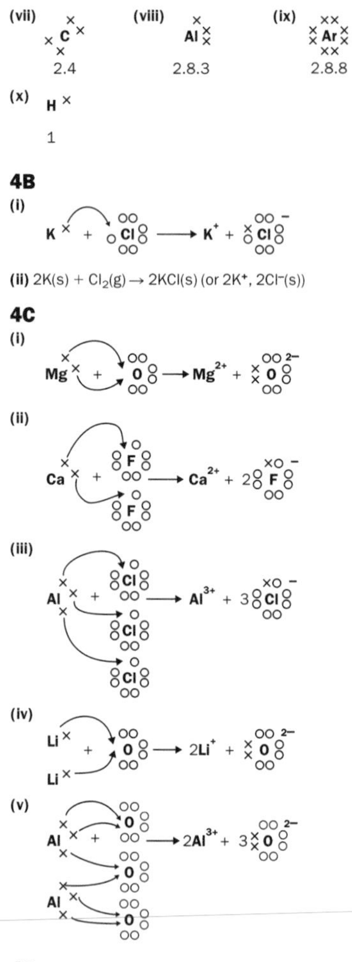

(ii) $2K(s) + Cl_2(g) \rightarrow 2KCl(s)$ (or $2K^+, 2Cl^-(s)$)

4D

Lewis structures:

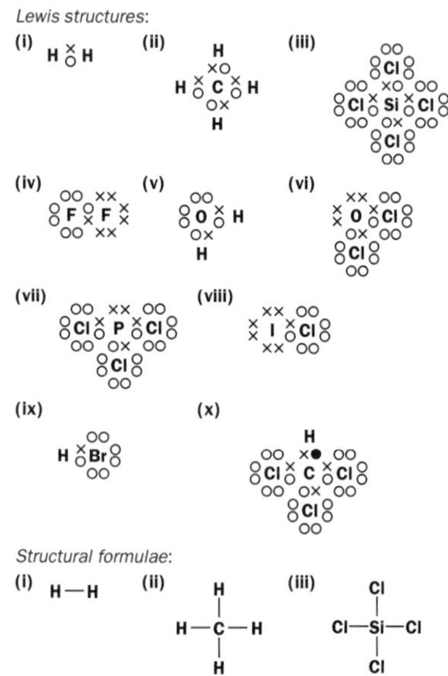

Structural formulae:

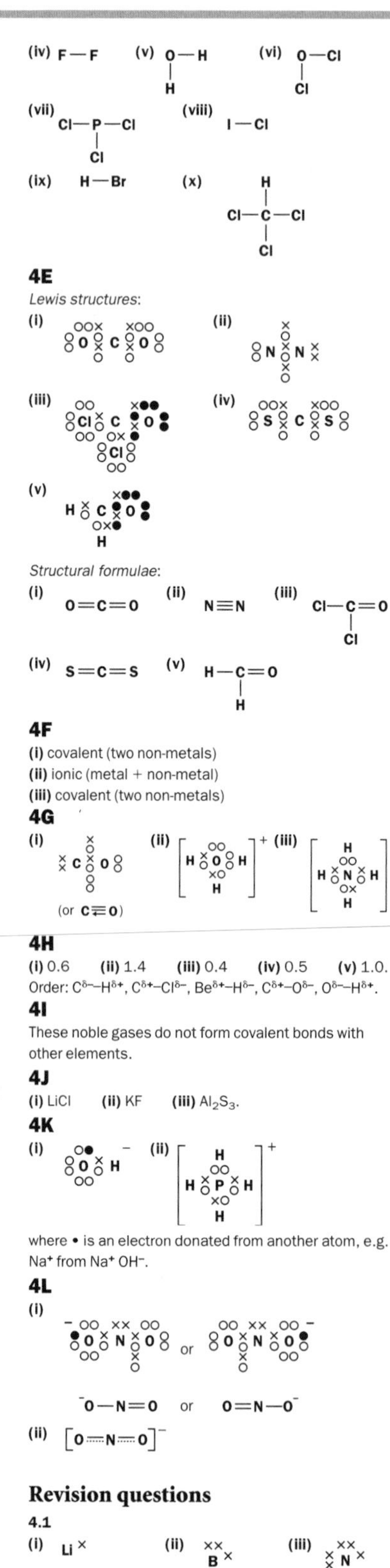

4E

Lewis structures:

Structural formulae:

4F

(i) covalent (two non-metals)
(ii) ionic (metal + non-metal)
(iii) covalent (two non-metals)

4G

(or C≡O)

4H

(i) 0.6 **(ii)** 1.4 **(iii)** 0.4 **(iv)** 0.5 **(v)** 1.0.
Order: $C^{\delta-}$–$H^{\delta+}$, $C^{\delta+}$–$Cl^{\delta-}$, $Be^{\delta+}$–$H^{\delta-}$, $C^{\delta+}$–$O^{\delta-}$, $O^{\delta-}$–$H^{\delta+}$.

4I

These noble gases do not form covalent bonds with other elements.

4J

(i) LiCl **(ii)** KF **(iii)** Al_2S_3.

4K

where • is an electron donated from another atom, e.g. Na^+ from $Na^+ OH^-$.

4L

Revision questions

4.1

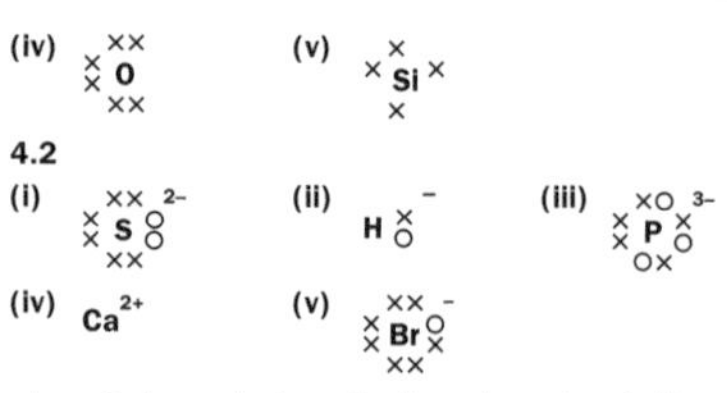

where ○ shows electrons that have been donated by other atoms.

4.3

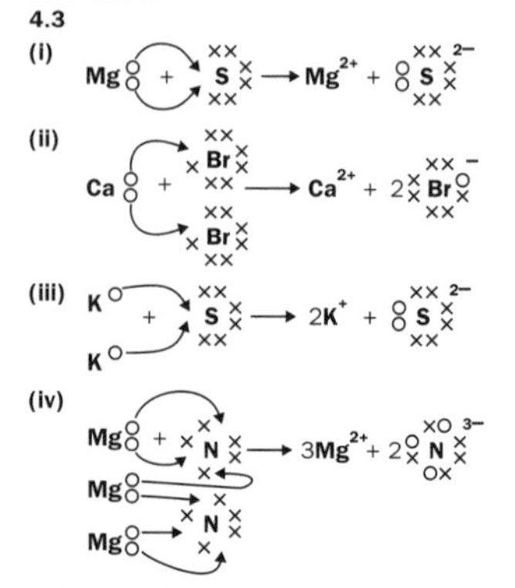

4.4 Because the sodium atom has the electronic structure 2.8.1, it loses only one electron to achieve the stable electronic structure of neon.

4.5

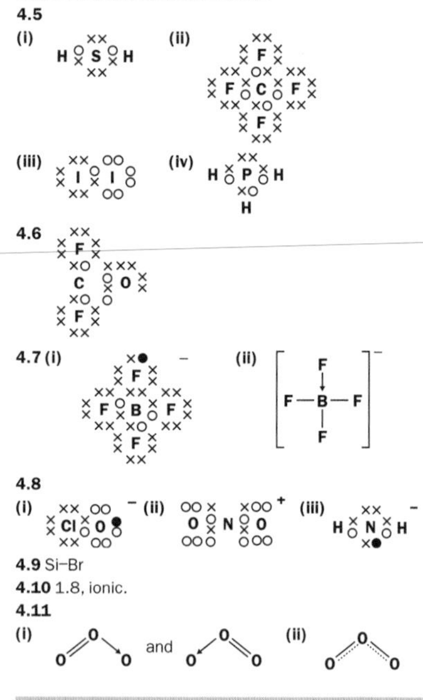

4.9 Si–Br

4.10 1.8, ionic.

4.11

Unit 5

Exercises

5A

Li has to lose one electron only in order to achieve a stable electronic structure. Be and B have to lose two and three electrons, respectively. This takes a great deal more energy than for Li.

5B

(i) Cl–Be–Cl **(ii)** 4

(iii)

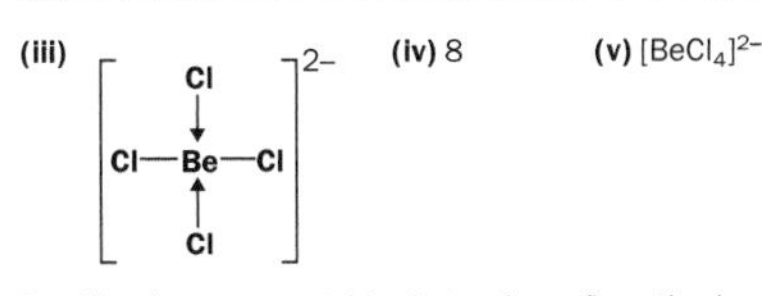

(iv) 8 (v) $[BeCl_4]^{2-}$

Beryllium has a more stable electronic configuration in $[BeCl_4]^{2-}$, because the atom is surrounded by eight electrons.

5C

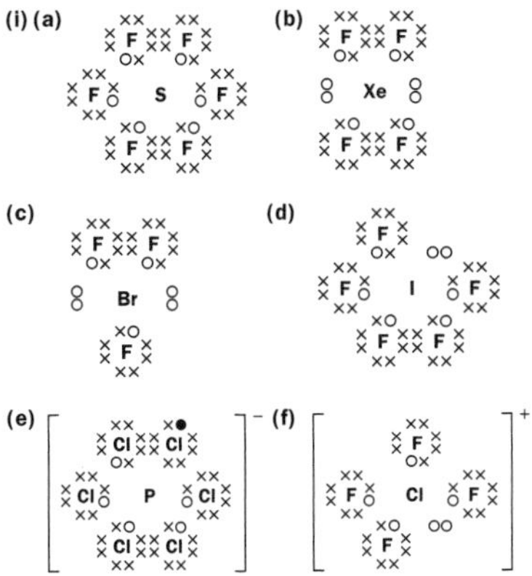

(ii) Nitrogen ($1s^2\ 2s^2\ 2p^3$) cannot expand its octet as there are no available d orbitals to do so.
(iii) Chlorine can expand its octet, but it is too small an atom to accommodate seven atoms of fluorine around it. Iodine is a much bigger atom.

5D

(i) trigonal planar **(ii)** tetrahedral
(iii) roughly tetrahedral (not all the bond lengths are the same)
(iv) tetrahedral **(v)** octahedral.

5E

(i) Distorted tetrahedral with respect to the electron pairs; 'bent' with respect to H–S bonds (similar to water).
(ii) Distorted tetrahedral with respect to the electron pairs; trigonal pyramidal with respect to its bonds (similar to ammonia).
(iii) Distorted octahedral with respect to the electron pairs.
(iv) Distorted tetrahedral with respect to the electron pairs; 'bent' with respect to its bonds.
(v) As for **(iv)**.

5F

(i) trigonal planar (remember that all the bond lengths are the same because of resonance)
(ii) trigonal planar **(iii)** linear **(iv)** tetrahedral
(v) trigonal planar.

5G

1. (i) no – linear **(ii)** yes **(iii)** no – tetrahedral
(iv) yes **(v)** yes.
2. (i) yes **(ii)** $H^{\delta+}–F^{\delta-}$.

5H

(i) two
(ii) As Fig. 5.6, with '2+' ions and two free electrons for each ion in the structure.
(iii) three
(iv) Na, Mg, Al (the more valence electrons that are free, the stronger the bonding).
(v) Al (it has the strongest metallic bonding, therefore it requires most energy to break down the structure).

5I

····C≡C–C≡C–C≡C–C····

5J

(i) Kr, Ar, Ne, He (the smaller the atom, the fewer the electrons and the weaker the forces that attract the atoms together).
(ii) Again, I_2 is the largest molecule and has the greatest number of electrons, so the forces that attract I_2 molecules together are the strongest and it has the highest melting and boiling points. The other halogens follow this trend.

Revision questions

5.1

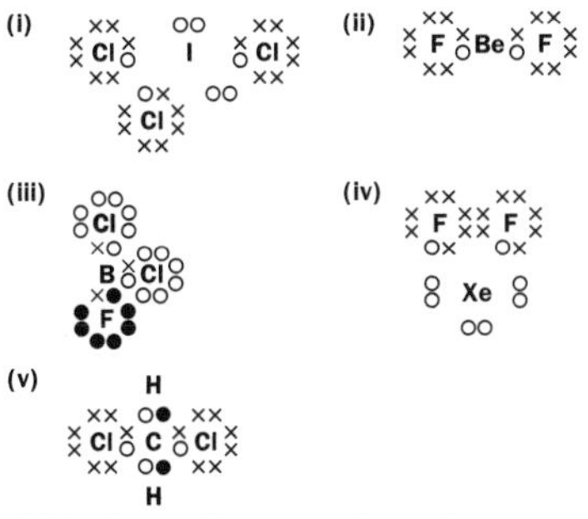

5.2 The shapes with respect to the electron pairs around the central atom are roughly the following:
(i) trigonal bipyramidal **(ii)** linear
(iii) trigonal planar
(iv) trigonal bipyramidal **(v)** tetrahedral.

5.3

linear

5.4 (i) no **(ii)** yes **(iii)** no **(iv)** yes **(v)** yes.
5.5 Substances which exist as atoms or non-polar molecules (such as the noble gases or iodine), liquefy or solidify at appropriate temperatures.
5.6 Ca because it has two valence electrons and stronger metallic bonding (more electron density to 'glue' the positive ions together).
5.7 Each carbon atom in graphite has one free, mobile electron. There are no free electrons in diamond, they are all involved in covalent bonding.
5.8 A CO_2 molecule is linear and non-polar. Only London forces attract the molecules together in the solid state.
5.9 The oxygen atom of water has two lone pairs of electrons and can therefore form two hydrogen bonds per molecule. The nitrogen atom in ammonia has only one lone pair and ammonia molecules can only form one hydrogen bond per molecule. In addition, oxygen is more electronegative than nitrogen.
5.10 No, there is not enough electronegativity difference between carbon and hydrogen in the C–H bond to produce a hydrogen that is sufficiently positively charged to hydrogen bond to the oxygen that is bonded to carbon in a neighbouring molecule.
5.11 If you make a model you will find there are 8 hexagons and 12 pentagons.

Unit 6

Exercises

6A

(i)

$Ca^{2+},2NO_3^-(s) \xrightarrow{H_2O} Ca^{2+}(aq) + 2NO_3^-(aq)$

(ii)

$2K^+,SO_4^{2-}(s) \xrightarrow{H_2O} 2K^+(aq) + SO_4^{2-}(aq)$

(iii)

$2Na^+,CO_3^{2-}\cdot10H_2O(s) \xrightarrow{H_2O} 2Na^+(aq) + CO_3^{2-}(aq) + 10H_2O(l)$

6B

The nitrate ion is the spectator ion. Overall ionic reaction is:

$Cu^{2+}(aq) + \cancel{2NO_3^-}(aq) + Zn(s) \rightarrow Zn^{2+}(aq) + \cancel{2NO_3^-}(aq) + Cu(s)$

or

$Cu^{2+}(aq) + Zn(s) \rightarrow Zn^{2+}(aq) + Cu(s)$

This is the same ionic equation as for the reaction between copper sulfate solution and zinc solid.

6C

(i) carbonate, sulfide and hydroxide.
(ii) We need to mix a solution of a soluble barium salt (such as barium ethanoate, $(CH_3COO^-)_2\ Ba^{2+}$) with a solution of a soluble sulfate (such as magnesium sulfate, Mg^{2+},SO_4^{2-}).
(iii) We need to mix a solution of a soluble iron salt (such as iron(II) sulfate, Fe^{2+},SO_4^{2-}) with a solution of a soluble hydroxide (such as sodium hydroxide, Na^+,OH^-).

6D

Coordinate bond.

6E

$HBr(g) + H_2O(l) \rightarrow H_3O^+(aq) + Br^-(aq)$

or

$HBr(g) \xrightarrow{H_2O} H^+(aq) + Br^-(aq)$

6F

(i) Calcium chloride will form a white precipitate, whereas magnesium nitrate will not. Potassium sulfate may also give a cloudiness of (slightly soluble) silver sulfate depending upon the concentrations of reactants used.
(ii) The precipitates are silver bromide (AgBr) and silver iodide (AgI):

$Ag^+(aq) + Br^-(aq) \rightarrow AgBr(s)$

$Ag^+(aq) + I^-(aq) \rightarrow AgI(s)$

These precipitates are useful in confirming the presence of bromides and iodides in solution.

6G

(i) The rest do not contain a sulfate ion.

6H

(ii) Ammonium carbonate:

$NH_4^+(aq) + OH^-(aq) \rightarrow NH_3(g) + H_2O(l)$

6I

1. (i) two **(ii)** one.
2. (i) ethanoates (acetates) **(ii)** nitrates.

6J

$K^+,OH^-(s) \xrightarrow{H_2O} K^+(aq) + OH^-(aq)$

$Ca^{2+},2OH^-(s) \xrightarrow{H_2O} Ca^{2+}(aq) + 2OH^-(aq)$

6K

(i) $Fe(s) + H_2SO_4(aq) \rightarrow FeSO_4(aq) + H_2(g)$
(ii) $Fe(s) + 2H^+(aq) \rightarrow Fe^{2+}(aq) + H_2(g)$
($SO_4^{2-}(aq)$ is the spectator ion).

6L

(i) D is Zn^{2+} (solid ZnO turns yellow upon heating). The white precipitate suggests that C is Cl^-.
(ii) Dilute sulfuric acid. Confirm this by adding $BaCl_2$/HCl (white precipitate observed).
(iii) Mix the solutions together one by one. The copper solution is coloured blue. Both sodium carbonate and sodium hydroxide will form a blue precipitate with copper ions since both copper(II) hydroxide and copper(II) carbonate are insoluble (see Table 6.2). However, only sodium carbonate will 'fizz' with the remaining solution, dilute hydrochloric acid.

Revision questions

6.1 (i) Pure nitric 'acid' is a covalent liquid and does not contain any $H^+(aq)$ ions.
(ii) It produces $OH^-(aq)$ ions when it dissolves in water.
(iii) The water present in the ammonia would react with the concentrated acid producing dilute sulfuric acid which would neutralize the ammonia.
6.2 (i) $CuSO_4 \cdot 5H_2O(s) \rightarrow Cu^{2+}(aq) + SO_4^{2-}(aq) + 5H_2O(l)$
(ii) $Li^+,OH^-(s) \rightarrow Li^+(aq) + OH^-(aq)$
6.3 (i) $(COOH)_2(aq) + 2KOH(aq) \rightarrow (COOK)_2(aq) + 2H_2O(l)$,
or $H^+(aq) + OH^-(aq) \rightarrow H_2O(l)$
(ii) $(COOH)_2(aq) + Zn(s) \rightarrow (COO)_2Zn(aq) + H_2(l)$
or $2H^+(aq) + Zn(s) \rightarrow H_2(g) + Zn^{2+}(aq)$
6.4 (i) $MgCO_3(aq) + H_2SO_4(aq) \rightarrow MgSO_4(aq) + H_2O(l) + CO_2(g)$,
or $2H^+(aq) + CO_3^{2-}(aq) \rightarrow CO_2(g) + H_2O(l)$
(ii) $(NH_4)_2SO_4(aq) + 2NaOH(aq) \rightarrow 2NH_3(g) + 2H_2O(l) + Na_2SO_4(aq)$,
or $NH_4^+(aq) + OH^-(aq) \rightarrow H_2O(l) + NH_3(g)$
(iii) $Na_2S(aq) + 2HNO_3(aq) \rightarrow 2NaNO_3(aq) + H_2S(g)$,
or $S^{2-}(s) + 2H^+(aq) \rightarrow H_2S(g)$
(iv) $Na_2SO_3(aq) + 2CH_3CO_2H(aq) \rightarrow 2CH_3CO_2Na(aq) + SO_2(g) + H_2O(l)$
or $SO_3^{2-}(aq) + 2H^+(aq) \rightarrow SO_2(g) + H_2O(l)$
6.5 (i) Both gases react with water, producing acidic solutions. SO_2 forms a more strongly acidic solution than CO_2, i.e. it produces more $H^+(aq)$ ions per mole of dissolved gas in aqueous solution. (SO_2 is also more soluble than CO_2.) The presence of greater numbers of $H^+(aq)$ ions means that more limestone can be destroyed.
(ii) The $OH^-(aq)$ ions force any dissolved CO_2 to produce carbonate ions, $CO_3^{2-}(aq)$, which react with Ca^{2+} producing insoluble calcium carbonate:

$Ca^{2+}(aq) + CO_3^{2-}(aq) \rightarrow CaCO_3(s)$

6.6 The blue colour suggests that the substance contains copper(II) ions ($K = Cu^{2+}$). The fizzing with acid suggests the presence of a carbonate ion ($J = CO_3^{2-}$). In the fizzing reaction, the carbonate is converted to a salt and to CO_2, while the copper(II) ion is unaffected. The copper(II) ion forms a blue precipitate of copper(II) hydroxide with $OH^-(aq)$.
6.7 $G = Fe^{3+}(aq)$ (confirmed by precipitate of $Fe(OH)_3$);
$H = NO_3^-$ (confirmed by brown ring test).
6.8 $S = S^{2-}$ and $T = Pb^{2+}$:

$S^{2-}(s) + 2H^+(aq) \rightarrow H_2S(g)$

This reaction causes Pb^{2+} ions in the solid PbS to enter aqueous solution where they may now be identified by the addition of sulfate ions which precipitate white lead(II) sulfate:

$Pb^{2+}(aq) + SO_4^{2-}(aq) \rightarrow PbSO_4(s)$

and by the addition of hydroxide ions:

$Pb^{2+}(aq) + 2OH^-(aq) \rightarrow Pb(OH)_2(s)$

$Pb(OH)_2(s) + 2OH^-(aq) \rightarrow [Pb(OH)_4]^{2-}(aq)$

6.9 Under alkaline conditions, any free aluminium ions are precipitated as $Al(OH)_3$ or as higher hydroxides:

$Al^{3+}(aq) + 3OH^-(aq) \rightarrow Al(OH)_3(s)$.

The precipitate 'locks up' $Al^{3+}(aq)$. In acidic conditions, free aluminium ions are produced:

$Al(OH)_3(s) + 3H^+(aq) \rightarrow Al^{3+}(aq) + 3H_2O(l)$
toxic

Unit 7

Exercises

7A

(i) $Cu^{2+}(aq) + 2e^- \rightarrow Cu(s)$
(ii) $Zn(s) \rightarrow Zn^{2+}(aq) + 2e^-$
(iii) $Cu^{2+}(aq) + Zn(s) \rightarrow Cu(s) + Zn^{2+}(aq)$
(iv) The sulfate ions remain the same throughout the reaction – they do not take part in the redox reaction. The sulfate ions are **spectator** ions.

7B

(i) Zn oxidized (gains O), O_2 reduced (loses O!)
(ii) Mg oxidized (loses electrons), Cu^{2+} reduced (gains electrons)
(iii) Cl_2 reduced (gains electrons), I^- oxidized (loses electrons)
(iv) Since sodium chloride is Na^+Cl^-, Na oxidized (loses electron), Cl_2 reduced (gains electrons)
(v) Cl_2 reduced (gains H), H_2 oxidised (loses H!).

7C

(i) +4 **(ii)** −3 **(iii)** 0 **(iv)** +5 **(v)** +5
(vi) +7 **(vii)** +2 **(viii)** +5 (contains VO_3^-)
(ix) +5 **(x)** +6.

7D

(i)

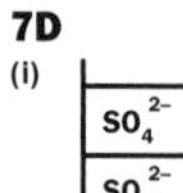
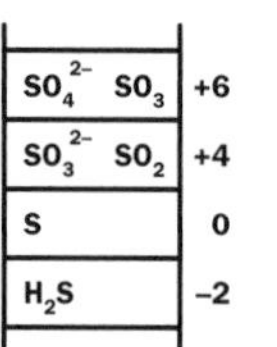

(ii) No, the oxidation number of sulfur does not change.
(iii) SO_2 is reduced to S because the oxidation number of sulfur decreases +4 to 0. H_2S is oxidised to S because the oxidation number of sulfur increases from −2 to 0.

7E

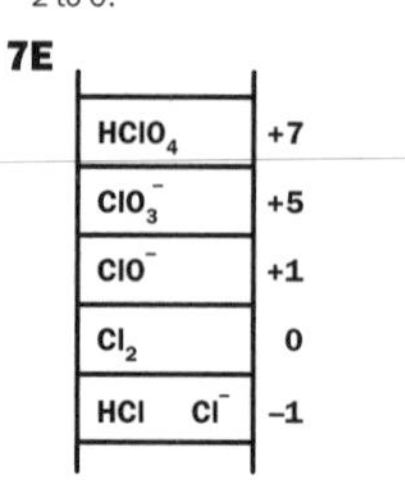

7F

(i) Those with oxidation numbers +4 and 0, because they can be oxidized (increase their oxidation number) and reduced (decrease their oxidation number).
(ii) Those with oxidation number +6, because they can be reduced only.
(iii) H_2S, because it can be oxidized only.

7G

1. (i) $I^- + 3H_2O \rightarrow IO_3^- + 6H^+ + 6e^-$
(ii) $NO_3^- + 4H^+ + 3e^- \rightarrow NO + 2H_2O$
(iii) $Cr_2O_7^{2-} + 14H^+ + 6e^- \rightarrow 2Cr^{3+} + 7H_2O$
(iv) $H_2O_2 + 2H^+ + 2e^- \rightarrow 2H_2O$
(v) To balance $S^{2-} \rightarrow SO_4^{2-}$
Balance the Os with H_2O:
$S^{2-} + 4H_2O \rightarrow SO_4^{2-}$
Balance the hydrogens by adding $8H_2O$ to the opposite side and $8OH^-$ to the same side:
$S^{2-} + 4H_2O + 8OH^- \rightarrow SO_4^{2-} + 8H_2O$
Balance the charges:
$S^{2-} + 4H_2O + 8OH^- \rightarrow SO_4^{2-} + 8H_2O + 8e^-$
Simplify the equation:
$S^{2-} + 8OH^- \rightarrow SO_4^{2-} + 4H_2O + 8e^-$

2. (i) $Cu + 2Ag^+ \rightarrow Cu^{2+} + 2Ag$
(ii) $Cu + 2NO_3^- + 4H^+ \rightarrow Cu^{2+} + 2NO_2 + 2H_2O$
(iii) $I_2 + 2S_2O_3^{2-} \rightarrow S_4O_6^{2-} + 2I^-$
(iv) $2I^- + H_2O_2 + 2H^+ \rightarrow I_2 + 2H_2O$
(v) $2H_2O_2 + ClO_2^- \rightarrow 2O_2 + Cl^- + 2H_2O$

7H

(i)

Pt | $H_2(g)$ | $H^+(aq)$ || $Cu^{2+}(aq)$ | Cu(s)

(ii) $+0.34\,V = E^\ominus(Cu^{2+}(aq)/Cu(s)) - 0$, therefore $E^\ominus(Cu^{2+}(aq)/Cu(s)) = +0.34\,V$
(iii) $H_2(g) + Cu^{2+}(aq) \rightarrow 2H^+(aq) + Cu(s)$

7I

(i) yes

$I_2(s) + 2e^- \rightleftharpoons 2I^-(aq)$ +0.54
$Br_2(l) + 2e^- \rightleftharpoons 2Br^-(aq)$ +1.09

(ii) no

$2H^+(aq) + 2e^- \rightleftharpoons H_2(g)$ 0.00
$Cu^{2+}(aq) + 2e^- \rightleftharpoons Cu(s)$ +0.34

(iii) yes

$Zn^{2+}(aq) + 2e^- \rightleftharpoons Zn(s$ −0.76
$Fe^{3+}(aq) + e^- \rightleftharpoons Fe^{2+}(aq)$ +0.77

(iv) yes

$I_2(s) + 2e^- \rightleftharpoons 2I^-(aq)$ +0.54
$Fe^{3+}(aq) + e^- \rightleftharpoons Fe^{2+}(aq)$ +0.77

7J

Work out the reactions by using anticlockwise arrows:
(i) $Mg(s) + 2Ag^+(aq) \rightarrow Mg^{2+}(aq) + 2Ag(s)$
(ii) no reaction
(iii) $Ni(s) + Cu^{2+}(aq) \rightarrow Ni^{2+}(aq) + Cu(s)$
(iv) $Zn(s) + 2H^+(aq) \rightarrow Zn^{2+}(aq) + H_2(g)$
(v) no reaction.

7K

(i) There is no dissolved oxygen in boiled water.
(ii) Water is more highly conducting if it carries dissolved salt and rusting occurs more rapidly. A great deal of salt is around at coastal locations.
(iii) Acid rain provides H^+ ions, making rusting occur more readily (see the equations for rusting).

7L

The standard electrode potential of iron is more negative than that for zinc, showing that iron loses electrons to form $Fe^{2+}(aq)$ more readily than tin forms $Sn^{2+}(aq)$.
Tin plating is not as effective as galvanizing. Provided that the coating is undamaged, the tin protects the iron underneath from water and oxygen. Once the surface is scratched, however, iron loses electrons more readily than tin to form Fe^{2+} ions, which then are oxidized to rust.

7M

Species	Oxidation number
NO_3^- HNO_3	+5
NO_2	+4
NO_2^- HNO_2	+3
NO	+2
N_2O	+1
N_2	0
NH_3 NH_4^+	−3

Revision questions

7.1 (i) C +4
(ii) Mn +4
(iii) S +6
(iv) Sn +4
(v) N −3

7.2
(ii) yes

$Cu^{2+}(aq) + 2e^- \rightleftharpoons Cu(s)$ +0.34
$Fe^{3+}(aq) + e^- \rightleftharpoons Fe^{2+}(aq)$ +0.77

7.3 (i) $Cu^{2+}(aq) + Fe(s) \rightarrow Fe^{2+}(aq) + Cu(s)$
(ii) no reaction

7.4 (i) Au (oxidation number 0 to +1)
(ii) manganese in MnO_4^- (+7 to +2)
(iii) MnO_4^- it is reduced to Mn^{2+}.

7.5 $SO_2(g) + 2H_2O(l) + 2Fe^{3+}(aq) \rightarrow SO_4^{2-}(aq) + 4H^+(aq) + 2Fe^{2+}(aq)$

7.6 Find out the reaction:

$Al^{3+}(aq) + 3e^- \rightleftharpoons Al(s)$ −1.67
$Fe^{2+}(aq) + 2e^- \rightleftharpoons Fe(s)$ −0.44

(i) From the Al electrode to the Fe electrode.
(ii) The reactions occurring are:
$Al(s) \rightarrow Al^{3+}(aq) + 3e^-$
$Fe^{2+}(aq) + 2e^- \rightarrow Fe(s)$
(iii) $Al^{3+}(aq) + 3e^- \rightarrow Al(s)$ $E^\ominus = -1.67$
$Fe^{2+}(aq) + 2e^- \rightarrow Fe(s)$ $E^\ominus = -0.44$
$E_{cell} = E^\ominus_R - E^\ominus_L$
$= -0.44 - (-1.67)$
$= 1.23V$
(iv) Al(s) I Al^{3+}(aq) II Fe^{2+}(aq) I Fe(s).

7.7 (i) Ag^+, S^{2-}
(ii) Yes, silver ions are reduced to silver by aluminium, which is oxidized to aluminium ions.
(iii) Polish rubs off the sulfide layer, this method 'regenerates' the silver metal and it is not lost.

7.8 $MnO_2(s) + 4H^+(aq) + 2Cl^-(aq) \rightarrow Mn^{2+}(aq) + 2H_2O(l) + Cl_2(g)$

7.9 $Zn + 4OH^- \rightarrow [Zn(OH)_4]^{2-} + 2e^-$ (oxidation)
$NO_3^- + 6H_2O + 8e^- \rightarrow NH_3 + 9OH^-$ (reduction).
Multiply the oxidation reaction by 4 and add both half equations then simplify.
$4Zn(s) + NO_3^-(aq) + 7OH^-(aq) + 6H_2O(l) \rightarrow 4[Zn(OH)_4]^{2-}(aq) + NH_3(g)$

7.10 (i) Ni^{2+} (gains electrons and is reduced)
(ii) Zn (loses electrons and is oxidized).

7.11 Chromium plate provides physical protection for the iron against water and oxygen. Once it is scratched, the chromium will lose electrons in preference to the iron (check using anticlockwise arrows) forming a protective layer of chromium oxide that covers the chromium metal.

Unit 8

Exercises

8A
(i) 63 u **(ii)** 120 u **(iii)** 26 u **(iv)** 256 u
(v) 88 u **(vi)** 404 u **(vii)** 20 u.

8B
(i) 20 g (1 mol) **(ii)** 12×10^{23} **(iii)** 2.4 g (0.1 mol)
(iv) 159.5 g.

8C
1. (i) 14 g **(ii)** 224 g **(iii)** 53.3 g **(iv)** 2520 g.
2. (i) 1 mol **(ii)** 0.80 mol **(iii)** 5.1×10^{-3} mol
(iv) 0.50 mol.
3. Na_2CO_3 (0.01 mol).
4. (i) 3.8×10^{22} (0.063 mol)
(ii) 6.6×10^{20} (5.5×10^{-4} mol Al_2O_3).

8D
Volume of oil in drop = $\pi \times (d/2)^2 \times$ thickness of layer (1 molecule):

$2.54 \times 10^{-5} = \pi \times 10^2 \times h$
$h = 8.1 \times 10^{-8}$ cm

Volume of 1 molecule = $h^3 = (8.1 \times 10^{-8})^3$
$= 5.3 \times 10^{-22}$ cm^3.
$N_A = 282/(0.891 \times 5.3 \times 10^{-22}) = 6 \times 10^{23}$ mol^{-1}.

8E
(i) 23% **(ii)** 16% **(iii)** 68% **(iv)** 80%
(v) 73%.

8F
(i) CH_4 **(ii)** Al_2O_3 **(iii)** Fe_3O_4 **(iv)** Na_2SO_4.

8G
(i) C_2H_6 **(ii)** C_5H_7N (empirical formula)
$C_{10}H_{14}N_2$ (molecular formula)

8H
$CuSO_4 \cdot 5H_2O \rightarrow CuSO_4 \cdot 3H_2O + 2H_2O$
$CuSO_4 \cdot 3H_2O \rightarrow CuSO_4 \cdot H_2O + 2H_2O$
$CuSO_4 \cdot H_2O \rightarrow CuSO_4 + H_2O$
$CuSO_4 \rightarrow CuO + SO_3$.

8I
(i) 2 g **(ii)** 0.32 g **(iii)** 38 g
(iv) $C(s) + O_2(g) \rightarrow CO_2(g)$, 0.60 g
(v) $CuCO_3(s) \rightarrow CuO(s) + CO_2(g)$, 0.82 g.

8J
(i) 6 dm^3 **(ii)** 1.6 g **(iii)** 4.8 dm^3 **(iv)** 10 dm^3
(v) 80 cm^3 **(vi)** 7.5 g.

8K
56%.

8L
(i) oxygen, 0.1 mol
(ii) magnesium, 2.1 g
(iii) copper(II) oxide, 2.4 g.

Revision questions

8.1 (i) 28 u **(ii)** 134.5 u **(iii)** 98 u **(iv)** 164 u
(v) 322 u.
8.2 (i) 35.5 g **(ii)** 71 g **(iii)** 124 g **(iv)** 127 g
(v) 720 g.
8.3 (i) 0.94 mol **(ii)** 0.25 mol **(iii)** 0.25 mol
(iv) 0.051 mol **(v)** 10 mol.
8.4 (i) 29.3 g **(ii)** 11 g **(iii)** 4 g **(iv)** 624 g
(v) 6300 g.
8.5 7.5 g
There are 2.00/17 mol NH_3. 1 mol NH_3 has the same number of molecules as there are in 1 mol SO_2.
2.00/17 mol NH_3 has the same number of molecules as there are in 2.00/17 mol SO_2 or 2.00/17 × 64 g SO_2 = 7.5 g SO_2.
8.6 (i) 6 g
1 mol Mg reacts with 1 mol S; 24 g Mg reacts with 32 g S; 6 g Mg reacts with 8 g S.
8.7 (i) 0.8 g:
2 mol $NaNO_3$ produces 1 mol O_2; 170 g $NaNO_3$ produces 32 g O_2; 4.25 g $NaNO_3$ produces 32/170 × 4.25 g O_2.
(ii) 0.6 dm^3:
2 mol $NaNO_3$ produces 24 dm^3 O_2; 170 g $NaNO_3$ produces 24 dm^3 O_2; 4.25 g $NaNO_3$ produces 24/170 × 4.25 dm^3 O_2.
8.8 (i) $2NaHCO_3(s) \rightarrow Na_2CO_3(s) + H_2O(l) + CO_2(g)$
(ii) 168 g **(iii)** 44 g, 24 dm^3
(iv) 0.60 dm^3 (600 cm^3).
8.9 51%:

$$\frac{7 \times \text{molecular mass of water}}{\text{molecular mass of } MgSO_4 \cdot 7H_2O} \times 100 = \frac{7 \times 18}{246} \times 100$$

8.10 C_2H_4 (the simplest, or empirical, formula is CH_2).
8.11 Amount of Al in moles = 13.50/27 = 0.50 mol.
Mass of Cl combined with Al = 66.75 − 13.50 = 53.25 g.
Amount of Cl in moles = 53.25/35.5 = 1.50 mol.
Ratio of the number of moles Al atoms to number of moles Cl atoms is 0.50:1.50 or 1:3. Formula is $AlCl_3$.

Unit 9

Exercises

9A
(i) 10.0 mol dm^{-3} **(ii)** 0.008 mol dm^{-3}
(iii) 50 dm^3 **(iv)** 28.6 g.

9B
(i) K^+ and Cl^- both 0.1 mol dm^{-3}
(ii) Na^+ and CO_3^{2-} are 0.250 mol dm^{-3} and 0.125 mol dm^{-3}, respectively
(iii) NH_4^+ and SO_4^{2-} are 0.50 mol dm^{-3} and 0.25 mol dm^{-3}, respectively
(iv) Cu^{2+} and NO_3^- are 0.15 mol dm^{-3} and 0.30 mol dm^{-3}, respectively.

9C
(i) 0.05 mol dm^{-3} **(ii)** 0.10 mol dm^{-3}
(iii) 0.20 mol dm^{-3} **(iv)** 0.40 mol dm^{-3}
(v) 1.00 mol dm^{-3}.

9D
(i) 0.53 g
Amount in mol = vol × molar conc = 500/1000 × 0.01
$= 5.0 \times 10^{-3}$ mol
$= 5.0 \times 10^{-3} \times M(Na_2CO_3)$ g = 0.53 g.
(ii) it would be more concentrated than the value calculated
(iii) 0.02 mol dm^{-3}
amount in mol present = vol × molar conc = 50/1000 × 0.1 = 5×10^{-3} mol
if this amount is made up to 250 cm^3, then

molar conc = amount in mol/vol
$= 5 \times 10^{-3}/0.25 = 0.02$ mol dm^{-3}

(iv) 125 cm^3
Number of moles needed = vol × conc = 1 × 2 = 2 mol
vol = number mol/conc = 2/16 = 0.125 dm^3
(v) 0.05 dm^3.

9E
(i) 50 cm^3
1 mol $H_2SO_4 \equiv$ 2 mol KOH

$$\frac{\text{vol} \times \text{molar concentration } H_2SO_4}{\text{vol} \times \text{molar concentration KOH}} = \frac{1}{2}$$

$$\frac{V \times 1.0}{25/1000 \times 4.0} = \frac{1}{2}$$

$V = 0.050$ dm^3 (50 cm^3)

(ii) 0.071 mol dm^{-3} **(iii)** 0.263 mol dm^{-3}
(iv) 20 cm^3.

9F
(i) $6Fe^{2+}(aq) + Cr_2O_7^{2-}(aq) + 14H^+(aq) \rightarrow 6Fe^{3+}(aq) + 2Cr^{3+}(aq) + 7H_2O(l)$

6 mol $Fe^{2+} \equiv$ 1 mol $Cr_2O_7^{2-}$, so that

$$\frac{\text{vol} \times \text{molar concentration } Fe^{2+}}{\text{vol} \times \text{molar concentration } Cr_2O_7^{2-}} = \frac{6}{1}$$

$$\frac{25/1000 \times C}{13.5/1000 \times 0.02} = \frac{6}{1}$$

C = 0.065 mol dm^{-3}
(ii) 50 cm^3
(iii) 210 cm^3
The equation is:

$3MnO_4^- + 5Fe^{2+} + 5C_2O_4^{2-} + 24H^+ \rightarrow 3Mn^{2+} + 5Fe^{3+} + 10CO_2 + 12H_2O$

$$\frac{\text{vol} \times \text{molar concentration } MnO_4^-}{\text{amount in moles of } FeC_2O_4} = \frac{3}{5} = \frac{V \times 0.020}{0.007}$$

$V = 210\ cm^3$

9G

(i) 67 cm^3

$$\frac{\text{vol} \times \text{molar concentration } H_2SO_4}{\text{amount in moles of CuO}} = \frac{1}{1} = \frac{V \times 1.5}{7.95/(63.5 + 16)}$$

$V = 0.067\ dm^3$

(ii) 0.29 g **(iii)** 417 cm^3 **(iv)** 3.1 g.

9H

(i) 0.1128 g **(ii)** to ensure they were completely dry
(iii) 3.9595×10^{-3} mol dm^{-3} (molar mass of Ni $(DMG)_2$ = 284.882 and amount of $Ni(DMG)_2$ in 100 cm^3 = 0.1128/284.882 mol. Multiply this amount of $Ni(DMG)_2$ by 10 to get the concentration)
(iv) 0.4648 g (mass of Ni^{2+} in original solution of volume 2000 cm^3 = $3.9595 \times 10^{-3} \times 58.69 \times 2$).

9I

(i) 1.2 mg **(ii)** 400 ppm.

9J

(i) 0.014 mol dm^{-3}
500 mg in 1 dm^{-3} is $500/1000 \times 1/35.5$ mol dm^{-3}
(ii) 200 ppm
(the solution has been diluted 250/50 = 5 times)
(iii) 0.5 ppb.

9K

(i) 13.4% **(ii)** 42 g **(iii)** 180 cm^3
(iv) 425 cm^3.

9L

(i)

$$X_R = \frac{n_R}{n_P + n_Q + n_R + n_S}$$

(ii)

$$\frac{n_P}{n_P + n_Q + n_R + n_S} + \frac{n_Q}{n_P + n_Q + n_R + n_S} + \frac{n_R}{n_P + n_Q + n_R + n_S} + \frac{n_S}{n_P + n_Q + n_R + n_S}$$

$$= \frac{n_P + n_Q + n_R + n_S}{n_P + n_Q + n_R + n_S} = 1$$

(iii) pure P.

9M

(i) 3/10 or 0.3
(ii) 0.0089, from

$$\frac{37.25/74.5}{37.25/74.5 + 1000/18}.$$

9N

(i) weakly acidic **(ii)** weakly acidic
(iii) strongly basic **(iv)** strongly basic
(v) weakly basic **(vi)** weakly acidic
(vii) strongly basic.

Revision questions

9.1 (i) 4.20 g **(ii)** 126 g **(iii)** 4.7 g
9.2 It absorbs water and carbon dioxide from the air and therefore would increase in mass when handled in air.
9.3 (i) 50 cm^3 **(ii)** 100 cm^3.
9.4 (i) 100 cm^3 **(ii)** 23.1 g.
9.5 (i) 0.35 g

$KOH + HNO_3 \rightarrow KNO_3 + H_2O$

Amount in mol HNO_3 = vol × molar conc, i.e.

$25/1000 \times 0.250 = 6.25 \times 10^{-3}$ mol

The same amount in mol of KOH is needed, i.e.

$6.25 \times 10^{-3} \times (39 + 16 + 1)$ g

(ii) 0.39 g, from

$CuCO_3 + 2HNO_3 \rightarrow Cu(NO_3)_2 + H_2O + CO_2$

9.6 (i) 0.100 mol dm^{-3} **(ii)** approx. 6 g dm^{-3}.
9.7 100 cm^3.
9.8 20 g.
9.9 KBr: 2000 ppm of K^+

0.050 mol dm^{-3} of K^+ is 0.050×39
= approx. 2 g dm^{-3} or 2000 mg dm^{-3}

Na_2SO_4: 460 ppm of Na^+

0.010×2 mol dm^{-3} Na^+ is $0.020 \times 23 = 0.46$ g dm^{-3}

$Pb(NO_3)_2$: approx. 1200 ppm of Pb^{2+}
9.10 0.020 g
9.11 0.12 g
20 mg dm^{-3} of Na^+ is $20/1000 \times 1/23$ mol dm^{-3} Na^+.
This corresponds to
2 × the number of mol of $Na_2CO_3{\cdot}10H_2O$
or $20/1000 \times 1/23 \times 1/2$ mol dm^{-3} $Na_2CO_3{\cdot}10H_2O$
or $20/1000 \times 1/23 \times 1/2 \times 286$ g dm^{-3} $Na_2CO_3{\cdot}10H_2O$
9.12 (i) pH = $-\log[H^+(aq)] = -\log[4.48 \times 10^{-9}] = 8.348$
(ii) pH = $-\log[H^+(aq)]$; $5.93 = -\log[H^+(aq)]$
$[H^+(aq)] = 10^{-5.93}$
$[H^+(aq)] = 1.2 \times 10^{-6}$ mol dm^{-3}

Unit 10

Exercises

10A

Changes in phase are reversible. The key to reversing such changes is temperature.

10B

(i) Temperature may be thought of as the average kinetic energy of a large group of molecules. If the same amount of energy is absorbed by fewer water molecules, the resulting temperature change is higher.
(ii) Hot days mean a higher temperature, so that more molecules possess enough energy to break away from the liquid. Windy days push the air (containing water vapour) away so that fewer water molecules can condense back on to the clothes.
(iii) Molecules of a gas are always moving and they spread out to fill any container (diffusion).

10C

No. Cooking is a chemical reaction and the rate of reaction depends upon the temperature. The temperature of boiling water is the same (normally 100°C) whether the water boils slowly or rapidly.

10D

(i) $(20/760) \times 101\,325 = 2666$ Pa ≈ 2700 Pa
(ii) $10 \times 101\,325 = 1\,013\,250$ Pa **(iii)** 107 000 Pa,
(iv) 1000 Pa.

10E

The temperature has increased by a factor of (2500/500) = 5.00. This means that the volume increases by fivefold, from 73.0 cm^3 to 365 cm^3 (370 cm^3 to two sig figs).

10F

Pressure has increased by a factor of 90.0/30.0 = 3.00, causing the volume to reduce by a factor of one-third:

$$\frac{1}{3.00} \times 200.0 = 66.7\ cm^3$$

10G

The number of moles of Zn = (mass/molar mass) = 0.50/65 = 0.0077 mol. The equation is:

$Zn(s) + H_2SO_4(aq) \rightarrow ZnSO_4(aq) + H_2(g)$

Therefore, 1 mol of Zn produces 1 mol of H_2, so that 0.0077 mol of Zn produces 0.0077 mol of H_2. 1 mol of H_2 occupies 24 dm^3 at room temperature and pressure, so that 0.0077 mol of H_2 occupies 0.0077×24 = 0.185 dm^3 (≈190 cm^3).

10H

Volume of argon in 1.0 dm^3 = $(0.94/100) \times 1.00$ = 0.0094 dm^3 (9.4 cm^3). Mass of argon in 1 kg of air is $(1.28/100) \times 1.0 = 0.0128$ kg (≈13 g)

10I

$T = 20.0 + 273.15 = 293.2$ K
$V = 0.0020\ m^3$
$P = nRT/V = (1.0 \times 8.3145 \times 293.2)/0.0020$
$= 1\,218\,698$ Pa $= 1.2 \times 10^6$ Pa (to two sig figs).

10J

The greater the intermolecular forces, the greater the deviations from ideal behaviour. CO_2 is a larger molecule than H_2 or N_2, and so the London dispersion forces between its molecules will be greater. (The vapour of polar molecules, such as ethanol or water, show considerable deviations from ideal behaviour because even stronger intermolecular forces are present.)

10K

(i) A liquid boils when its vapour pressure equals the external (atmospheric) pressure. Atmospheric pressure decreases with increasing altitude, so that the vapour pressure required in order for boiling to occur is lower than is needed at sea level. The (reduced) vapour pressure is therefore achieved at a lower liquid temperature, i.e. the boiling point is lower.
(ii) Salty water contains ions from the salt at the surface of the solution. The ions occupy sites that would be taken by water molecules in pure water, and so the number of water molecules in the surface that can escape is reduced in comparison with pure water at the same temperature. A higher temperature than 100 °C is needed in order for the water molecules in the surface of the solution to provide a vapour pressure of 1 atm (boiling point).

10L

Total gas pressure = air pressure + vapour pressure of ethanol, so

101 = ? + 7.9

and air pressure = 101 − 7.9 ≈ 93 kPa.

10M

(i) Treat the gases separately. −119 °C is at the critical temperature of oxygen so that liquid oxygen will condense if subjected to its critical pressure. However, −119 °C is above the critical temperature of hydrogen; hydrogen will not condense out, whatever the applied pressure.
(ii) The problem with liquefying hydrogen gas is getting its temperature at (or below) the critical temperature of −240 °C. This is well below the normal boiling point of liquid oxygen or liquid nitrogen. Liquid helium (bp −269 °C) could be used as coolant.

Revision questions

10.1 (i) $(0.124/760) = 1.63 \times 10^{-4}$ atm
(ii) $1.5 \times 101\,325 = 151\,988$ Pa
(iii) $89 \times 1000 = 89\,000$ N m^{-2}
(iv) 100.0 kPa is 1 bar so that 101.325 kPa is 1.013 25 bar
(v) −100 + 273.15 = 173.15 K.
10.2 Avogadro's law states that equal volumes of gases (at the same pressure and temperature) contain the same number of molecules. Therefore, if (within experimental error) equal volumes of H_2 and Cl_2 react together, producing twice that volume of HCl, the ratio in which the molecules react must be 1:1 and the ratio of HCl product molecules to H_2 or Cl_2 molecules must be 2:1. This confirms the ratio contained in the equation.
10.3 $M(KClO_3) = 122.5$ g mol^{-1}, so that the number of moles of $KClO_3$ = 10.0/122.5 = 0.0816.
According to the chemical equation, 2 mol of $KClO_3$ gives 3 mol of O_2. Therefore, 0.0816 mol of $KClO_3$ gives $(3/2) \times 0.0816 = 0.122$ mol of O_2.
1 mol of O_2 occupies 24 dm^3 at room temperature and

pressure. Therefore, 0.122 mol of O_2 occupies $0.122 \times 24 = 2.9$ dm^3. Summarizing, 10.0 g of potassium chlorate decomposes to make about 2.9 dm^3 of oxygen gas.

10.4 $P_T = 25 + 50 = 75$ kPa $= 75\,000$ Pa.
Rearrangement of the equation $PV = nRT$ gives

$$n = \frac{PV}{RT}$$

To work out the number of moles of chlorine gas and oxygen gas, substitute into this equation using the partial pressure of each gas for P:

number of moles of Cl_2 =
$(25\,000 \times 0.0100)/8.3145 \times 293 = 0.103$

number of moles of O_2 =
$(50\,000 \times 0.0100)/8.3145 \times 293 = 0.205$

Total number of moles of gas $= 0.103 + 0.205 = 0.308$, so

$$\text{mole fraction of } O_2 = \frac{0.205}{0.308} = 0.666$$

(Quicker way! The mole fraction equals the ratio $p_{O_2}/P_T = 50.0$ kPa/75.0 kPa = 0.666.)

10.5 Number of moles of $N_2 = 2.0/28 = 0.071$
number of moles of $H_2 = 6.0/2.0 = 3.0$
total number of moles of gas = 3.071

$$P_T = \frac{nRT}{V} = \frac{3.071 \times 8.3145 \times 473}{0.0020}$$
$$= 6.0 \times 10^6 \text{ Pa} = 60 \text{ atm}$$

10.6 Start by calculating the number of moles of He in the balloon:

$$n = \frac{PV}{RT} = \frac{1.00 \times 10^5 \times 1.00}{8.3145 \times 293} = 41.05$$

Then apply the equation

$$T = \frac{PV}{nR}$$

At the new altitude

$$T = \frac{5.00 \times 10^4 \times 1.66}{41.05 \times 8.3145} = 243 \text{ K}$$

We conclude that the temperature at the new altitude was 243 K.

10.7 First, calculate the molar volume of an ideal gas at 273 K and 1 atm pressure:

$$V = \frac{nRT}{P}$$
$$= \frac{1.000 \times 8.3145 \times 273}{101\,325} \approx 0.0224 \text{ m}^3 \text{ or } 22.4 \text{ dm}^3$$

The closer the experimentally determined molar volume of a real gas is to 22.4 dm^3, the more ideally behaved is the gas. The experimental molar volumes suggest that H_2 behaves most ideally, followed by CO_2 and (least ideally) NH_3. This reflects the greater degree of intermolecular force between NH_3 molecules.

10.8 $P = e^{-(34 \times 80/173)} = 1.5 \times 10^{-7}$ atm.

10.9 At 40 °C, the vapour pressure of water = 7370 Pa. The vapour occupies half the flask so that $V = 1.0$ dm^3 = 0.0010 m^3.

$$n = \frac{PV}{RT} = \frac{7370 \times 0.0010}{8.3145 \times 313} = 2.8 \times 10^{-3} \text{ mol}$$

To convert the number of moles of water to molecules of water, we multiply by N_A:

number of molecules =
$2.8 \times 10^{-3} \times 6.022 \times 10^{23} \approx 1.7 \times 10^{21}$

10.10 If a solvent has a high vapour pressure at room temperature, it is likely it has a sufficiently high vapour pressure at lower temperatures to be ignited by a naked flame.

Unit 11

Exercises

11A

Oil floats on water – this was concealed by the opaque earthenware jug. The captain poured the oil layer into his opponent's tankard.

11B

Cyclohexane is less dense than water (see Table 11.1) and floats to the top. Table 11.1 shows the density of cyclohexane as 0.77 g cm^{-3}, so that 130 cm^3 of cyclohexane has a mass of 100 g. Table 11.3 shows that 100 g of cyclohexane saturated with water contains 0.01 g of water. Therefore, 130 cm^3 of cyclohexane contains 0.01 g of water.

11C

All the molecules are very similar and differ only in their chain length. Butan-1-ol, possessing the shortest chain, is predicted to be the most soluble in water.

11D

1 dm^3 of saturated solution contains 1.6×10^{-2} mol, so that 100 cm^3 of solution contains 1.6×10^{-3} mol. $M(PbCl_2) = 278$ g mol^{-1}. The mass of 1.6×10^{-3} mol is $(1.6 \times 10^{-3} \times 278) = 0.45$ g. The equation for the dissolving of solid lead chloride is

$$PbCl_2(s) \rightleftharpoons Pb^{2+}(aq) + 2Cl^-(aq)$$

Any lead chloride that dissolves subsequently dissociates into ions. The concentration of dissolved lead chloride is 1.6×10^{-2} mol dm^{-3}, so that the concentration of lead ions is also 1.6×10^{-2} mol dm^{-3}. The concentration of chloride ions is double this, i.e. 3.2×10^{-2} mol dm^{-3}.

11E

For copper(I) iodide,

$$CuI(s) \rightleftharpoons Cu^+(aq) + I^-(aq)$$

$K_s = [Cu^+(aq)] \times [I^-(aq)]$ mol^2 dm^{-6}

For aluminium hydroxide,

$$Al(OH)_3(s) \rightleftharpoons Al^{3+}(aq) + 3OH^-(aq)$$

$K_s = [Al^{3+}(aq)] \times [OH^-(aq)]^3$ mol^4 dm^{-12}

11F

$$BaSO_4(s) \rightleftharpoons Ba^{2+}(aq) + SO_4^{2-}(aq)$$

$$K_s = [Ba^{2+}(aq)]\,[SO_4^{2-}(aq)] = s \times s = s^2$$

or,

$$s = \sqrt{K_s} = \sqrt{(1.1 \times 10^{-10})} = 1.0 \times 10^{-5} \text{ mol dm}^{-3}$$

$$MgF_2(s) \rightleftharpoons Mg^{2+}(aq) + 2F^-(aq)$$

The concentration of fluoride ions is twice the molar solubility of MgF_2, i.e.

$$[F^-(aq)] = 2s$$

and

$$[F^-(aq)]^2 = [2s]^2 = 4s^2$$

Apart from this complication we follow the same pattern as for $BaSO_4$:

$$K_s = [Mg^{2+}(aq)]\,[F^-(aq)]^2 = s \times 4s^2 = 4s^3$$

or,

$$s = \sqrt[3]{\left(\frac{K_s}{4}\right)} = \sqrt[3]{\left(\frac{6.4 \times 10^{-9}}{4}\right)}$$
$$= \sqrt[3]{(1.6 \times 10^{-9})} = 1.2 \times 10^{-3} \text{ mol dm}^{-3}$$

11G

(i) The product of ion concentrations is

$$[Mg^{2+}(aq)][F^-(aq)]^2 = [0.0010][0.0030]^2 = 9.0 \times 10^{-9}$$

Table 11.4 shows that $K_s = 6.4 \times 10^{-9}$ mol^3 dm^{-9}. Therefore, the solubility product is exceeded and magnesium fluoride precipitates.

Addition of 90 cm^3 of water to 10 cm^3 of mixture dilutes the ion concentrations by a factor of ten. Then, the product of ion concentrations is

$$[Mg^{2+}(aq)][F^-(aq)]^2 = [0.00010][0.00030]^2$$
$$= 9.0 \times 10^{-12}$$

This is well below the solubility product and so the magnesium fluoride will not precipitate out.

(ii) $[Ba^{2+}(aq)]$ from $BaCl_2$ is 2.0×10^{-4} mol dm^{-3} before mixing and 1.0×10^{-4} mol dm^{-3} after mixing. $[SO_4^{2-}(aq)]$ from Na_2SO_4 is 4.0×10^{-4} mol dm^{-3} before mixing and 2.0×10^{-4} mol dm^{-3} after dilution.
The product of ion concentrations is

$$[Ba^{2+}(aq)]\,[SO_4^{2-}(aq)] = [1.0 \times 10^{-4}][2.0 \times 10^{-4}]$$
$$= 2.0 \times 10^{-8}$$

This is greater than K_s (1.1×10^{-10} mol^2 dm^{-6}), and so the barium sulfate will precipitate out.

11H

The concentration of SO_2 in the water layer is $0.90 \times 0.10 = 0.090$ mol dm^{-3}.

11I

$\pi = \Delta cRT$, so that

$$\Delta c = \pi/RT = 91/8.3145 \times 300$$
$$= 0.036 \text{ mol m}^{-3} = 3.6 \times 10^{-5} \text{ mol dm}^{-3}$$

Therefore, 1 dm^3 of solution contains 3.6×10^{-5} moles. Using

$$\text{moles} = \frac{\text{mass}}{\text{molar mass}}$$

the molar mass is

$$\frac{\text{mass}}{\text{moles}} = \frac{0.25}{3.6 \times 10^{-5}} = 6900 \text{ g mol}^{-1}$$

The average molecular mass is therefore 6900 u.

Revision questions

11.1 The true answers are **(i)** and **(iii)**.

11.2 (i) Miscible, presumably because strong hydrogen bonding occurs between the water and propan-2-ol molecules.

(ii) Miscible, presumably because the intermolecular forces between propanone and methyl benzene (a big molecule) are at least as strong as those between pairs of molecules of methyl benzene and between pairs of propanone molecules.

11.3 C is the least polar solvent. Ionic solids are at best only slightly soluble in non-polar solvents.

11.4 (i)
(a) $K_s = [Ca^{2+}(aq)][CO_3^{2-}(aq)] = s^2$ where s is the molar solubility.

$$s = \sqrt{K_s} = 9.3 \times 10^{-5} \text{ mol dm}^{-3}$$

(b) The solubility in g dm^{-3} is obtained by multiplying the molar solubility by the molar mass of calcium carbonate (100 g mol^{-1}), giving the solubility as 9.3×10^{-3} g dm^{-3} at that temperature.

(ii) The mass of $CaCl_2$ in 20.0 cm^3 of solution $= 2.0 \times 10^{-5}$ g. The molar mass of $CaCl_2$ is 111 g mol^{-1}, so the number of moles of $CaCl_2$ is

$$2.0 \times 10^{-5}/111 = 1.8 \times 10^{-7} \text{ mol}$$

$[CaCl_2(aq)] = [Ca^{2+}(aq)] =$
$1.8 \times 10^{-7} \times 1000/20 = 9.0 \times 10^{-6}$ mol dm^{-3}.

The sodium carbonate solution has been diluted to 1.0×10^{-5} mol dm^{-3}.

$[Na_2CO_3(aq)] = [CO_3^{2-}(aq)] = 1.0 \times 10^{-5}$ mol dm^{-3}.

The ionic product is

$[Ca^{2+}(aq)]\,[CO_3^{2-}(aq)]$
$= (9.00 \times 10^{-6})\,(1.0 \times 10^{-5}) = 9.0 \times 10^{-11}$

This is less than K_s, so that calcium carbonate will not precipitate out.

11.5 One mole of calcium oxalate dissolves in water producing an equal concentration of calcium and oxalate ions:

$$K_s(Ca(C_2O_4)) = [Ca^{2+}(aq)][C_2O_4^{2-}(aq)] = [Ca^{2+}(aq)]^2$$
$$= (4.47 \times 10^{-5})^2 = 2.00 \times 10^{-9}\ mol^2\ dm^{-6}$$

11.6

$$[drug\ in\ octanol] = [drug\ in\ water] \times 1000$$
$$= 0.01 \times 1000 = 10\ mg\ dm^{-3}\ of\ wall$$

(We are assuming that octanol behaves similarly to the lipids in the stomach wall.)

11.7 $4.6\ m^3 = 4600\ dm^3$. This volume of gas contains $4600/22.4 = 205$ mol of Cl_2. Since $1\ m^3 = 1000\ dm^3$, the concentration of gas is

moles of gas/volume of solution
$= 205/1000 \approx 0.21\ mol\ dm^{-3}$.

11.8 The molar concentration of O_2 at 0 °C is

$$\frac{14.6}{1000 \times 32} = 4.6 \times 10^{-4}\ mol\ dm^{-3}\ O_2$$

The volume of O_2 at 0 °C is $22.4 \times 4.6 \times 10^{-4}$
$= 0.0100\ dm^3$ or $10.0\ cm^3\ O_2$.

Similarly at 30 °C, the molar concentration of O_2 is

$$\frac{7.6}{1000 \times 32} = 2.4 \times 10^{-4}\ mol\ dm^{-3}$$

Therefore, the volume of O_2 at 30 °C is

$24.9 \times 2.4 \times 10^{-4} = 0.0060\ dm^3$ or $6.0\ cm^3\ O_2$

11.9 (i) $5.0 \times 10^{-3}\ mol\ dm^{-3} = 5.0\ mol\ m^{-3}$

$$\pi = \Delta cRT$$
$$= 5.0 \times 8.3145 \times 298 = 12\,400\ Pa\ (\approx 0.12\ atm)$$

(ii)

$$P = \frac{nRT}{V} = cRT = 5.0 \times 8.3145 \times 298 = 12\,400\ Pa$$

In other words, the osmotic pressure of an ideal solution (ideal means that it obeys $\pi = \Delta cRT$) is the same as the pressure of an ideal gas at the same pressure and temperature.

11.10 (i) $0.0175\ atm = 1773\ Pa$

$$\Delta c = \pi/RT = 1773\,/\,(8.3145 \times 298)$$
$$= 0.716\ mol\ m^{-3} = 7.16 \times 10^{-4}\ mol\ dm^{-3}$$

$$c_2 - c_1 = \Delta c$$

Since the osmotic pressure was measured against pure solvent, $c_1 = 0$. Therefore,

c_2 = concentration of polystyrene
$= 7.16 \times 10^{-4}\ mol\ dm^{-3}$

moles of polystyrene = volume × molar concentration
$= 0.20 \times 7.16 \times 10^{-4}$
$= 1.43 \times 10^{-4}$

Molar mass = mass / moles = $5.0 / 1.43 \times 10^{-4}$
$= 35\,000\ g\ mol^{-1}$.

Molecular mass of polystyrene = 35 000 u.

(ii)

moles = mass/molar mass
$= 0.100/100\,000 = 1.00 \times 10^{-6}$

molar concentration $= 1.00 \times 10^{-6}/1.00$
$= 1.00 \times 10^{-6}\ mol\ dm^{-3}$
$= 1.00 \times 10^{-3}\ mol\ m^{-3}$

$$\pi = cRT = 1.00 \times 10^{-3} \times 8.3145 \times 293$$
$$= 2.44\ Pa\ (2.4 \times 10^{-5}\ atm)$$

11.11 Tyndall effect (see text). In soap, colloidal particles are ionic micelles. Such particles cannot be separated using filter paper.

11.12 The vinegar produces an egg without its shell. The outerpart of the remaining egg is a semipermeable membrane. The inner part of the egg contains protein, and water travels across the membrane (driven by osmosis) in an attempt to equalize concentrations. This causes the egg to swell. In syrup, the reverse occurs – water passes out of the egg and the egg shrinks.

Unit 12

Exercises

12A

(i) Na 2.8.1 or $1s^22s^22p^63s^1$
K 2.8.8.1 or $1s^22s^22p^63s^23p^64s^1$
Rb 2.8.18.8.1 or $1s^22s^22p^63s^23p^63d^{10}4s^24p^65s^1$
[inert gas]ns^1 where *n* is a whole number greater than 1

(ii) F 2.7 or $1s^22s^22p^5$
Cl 2.8.7 or $1s^22s^22p^63s^23p^5$
Br 2.8.18.7 or $1s^22s^22p^63s^23p^63d^{10}4s^24p^5$
I 2. 8.18.18.7 or
$1s^22s^22p^63s^23p^63d^{10}4s^24p^64d^{10}5s^25p^5$
outer electronic structure ns^2np^5 where *n* is a whole number greater than 1

12B

(i) No – the second ionization energy is too high.
(ii) K^+ has the electron arrangement of an inert gas, argon.
(iii) The second electron is nearer the nucleus and less shielded from the nucleus by electrons in shells between it and the nucleus. To take away a second electron would involve 'disruption' of a stable octet of electrons.

12C

(i) The boiling points decrease as the group is descended. The metallic bonds become longer and weaker as the atoms get larger.
(ii) About 960 K.
(iii) The alkali metals have relatively large atomic radii and their outer electrons are weakly held – the outer electron is well shielded from the nucleus by the full inner shells. The large atomic radii of the atoms results in longer, weaker metallic bonding as compared with the bonding in other metals, making them relatively 'soft'.
(iv) The large atomic radii of the elements results in them taking up more 'room' (volume), per unit mass, than other metals. They therefore have lower densities.

12D

(i) Extremely reactive:

$Fr(s) + O_2(g) \rightarrow FrO_2(s)$
$2Fr(s) + 2H_2O(l) \rightarrow 2FrOH(aq) + H_2(g)$

(ii) Fr_2CO_3, very soluble, very thermally stable.
(iii) Thermally the most stable. Likely to decompose to the nitrite with strong heating:

$2FrNO_3(s) \rightarrow 2FrNO_2(s) + O_2(g)$

12E

The metallic bonds are shorter and stronger because the atomic radii of the Group 2 metals are smaller than those of the corresponding alkali metals. Also two valence electrons are released per ion, to 'glue' the cations together in the structure.

12F

(i) The atom has lost its outer shell electrons and the overall positive charge 'pulls' the remaining electrons closer to the nucleus.
(ii) Although there is an increase in the positive charge on the nucleus, the outer electrons are farther away from the nucleus and shielded by more full shells, as the group is descended.
(iii) The atomic and ionic radii are similar.

12G

(i) $SiH_3–SiH_2–SiH_2–SiH_2–SiH_2–SiH_3$
(ii) Covalent, similar shape to ethane; each Ge atom has four bonds arranged tetrahedrally around it.

12H

(i)

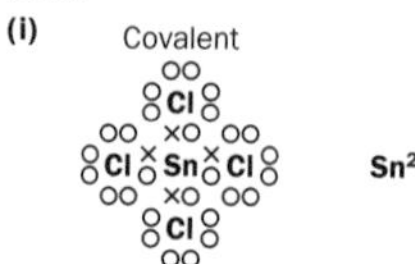

(ii) Pb^{II} is the more stable oxidation state of lead.
(iii) Pb–H bonds are longer and weaker than C–H bonds, because Pb is a much larger atom than C.
(iv) CO.
(v) GeO and GeO_2. You would expect GeO_2 to be the more stable, because Ge^{IV} is the more stable oxidation state (GeO is readily oxidized to GeO_2 in air).

12I

(i) $SiCl_4(l) + 2H_2O(l) \rightarrow SiO_2(s) + 4HCl(aq)$
(ii) It is longer and weaker. Si is a larger atom than C.

12J

(i) Si–Si bonds are longer and weaker than C–C bonds. The shared electrons can 'jump out' of the Si–Si bonds more easily.
(ii) Only a small proportion of the bonding electrons are set free; each silicon atom still has three covalent bonds to keep it in place if one bonding electron is lost.

12K

(i) Black/dark.
(ii) Solid.
(iii) The least stable hydride with respect to decomposition into its elements.
(iv) Would be the halogen with the least oxidizing power.
(v) At^- as the largest halide ion would be the most highly polarised – you would expect the salt to have a great deal of covalent character.
(vi) Dark yellow or a darker colour, like orange.

12L

(i) $Xe(g) + 2F_2(g) \rightarrow XeF_4(s)$
(ii)

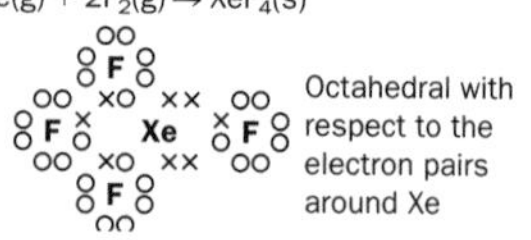

(iii) In XeF_4, Xe has oxidation state +4 it would 'prefer' to be reduced to oxidation state 0, i.e. atoms of the element Xe, which have a stable octet.

12M

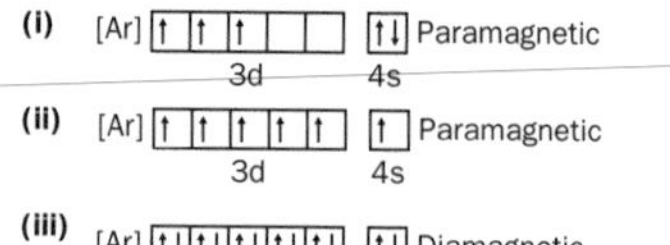

Metal **(ii)** has the most unpaired electrons.

12N

(i) $1s^22s^22p^63s^23p^63d^5$ **(ii)** $1s^22s^22p^63s^23p^63d^4$
(iii) +7, MnO_4^-.

12O

(i) Sc^{3+} has no d electrons.
(ii) Zn^{2+} has a full subshell of d electrons – there are no 'gaps' for d electrons to be promoted.

12P

(i) 2–
(ii) $[CuCl_4]^{2-}$
(iii) 4
(iv) tetrachlorocuprate(II).

12Q

1. **(i)** hexacyanoferrate(III)
   (ii) tetraamminezinc(II)
   (iii) hexaaquatitanium(III)
   (iv) tetraamminediaquacopper(II)
   (v) pentaamminebromocobalt(III)
   (vi) diamminetetrachloroplatinum(IV)
2. **(i)** $[PtCl_6]^{2-}$
   (ii) $[Cr(H_2O)_4Cl_2]^+$
   (iii) $[Pd(CN)_6]^{2-}$
   (iv) $[CoBr_4]^{2-}$
   (v) $[Fe(H_2O)_6]^{3+}$

12R
(i) 6 **(ii)** 2−.

12S
(i) A Sodium atom loses its outer shell electron when it becomes an ion. The positive charge on the ion draws the remaining electrons closer into the nucleus. A chlorine atom gains an electron to its outer shell when it becomes an ion. The negative charge on the ion makes the electrons 'spread out'.
(ii) The 2+ charge on the magnesium ion draws the electrons closer to the nucleus.
(iii) The greater the negative charge on the ion, the more the electrons repel each other and 'spread out'.

Revision questions

12.1 (i) H **(ii)** X, L or Z **(iii)** Y or G **(iv)** X or D **(v)** Z, M or K **(vi)** I **(vii)** J **(viii)** H **(ix)** D **(x)** L.
12.2 (i) violently $Ra(s) + 2H_2O(l) \rightarrow Ra(OH)_2(aq) + H_2(g)$
(ii) the most thermally stable and insoluble carbonate of Group 2. The carbonate should decompose to the oxide and carbon dioxide, on strong heating.
12.3 Tin reverts to the crumbly α allotrope at low temperatures.
12.4 CS.
12.5 (i) Li_2O

$Li_2O(s) + H_2O(l) \rightarrow 2LiOH(aq)$

(ii) NO_2 **(iii)** BeO.
12.6 (i) ionic **(ii)** covalent.
12.7 (ii) C (network solid) and Al (strong metallic bonding).
(iii) Any of H, He, N, O, F, Ne, Ar; London dispersion forces between simple molecules or atoms.
(iv) Li, Na, Mg or Al. See Unit 5.
(v) Yes. There is a pattern of low boiling points then a gradual rise to very high boiling point(s), then a drop.
12.8 (i) 6 **(ii)** +2 **(iii)** tetraaquadichloronickel(II).
12.9 (i) $[Cr(H_2O)_4Cl_2]^+$ **(ii)** $[PtBr_6]^{2-}$.
12.10 (i) $[H_3N\text{–}Cu\text{–}NH_3]^+$ linear
(ii)

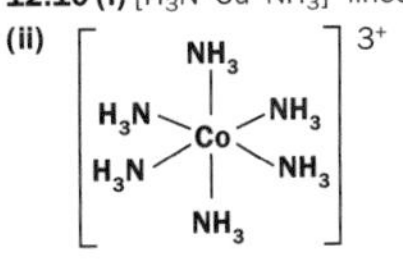

octahedral.

Unit 13

Exercises

13A

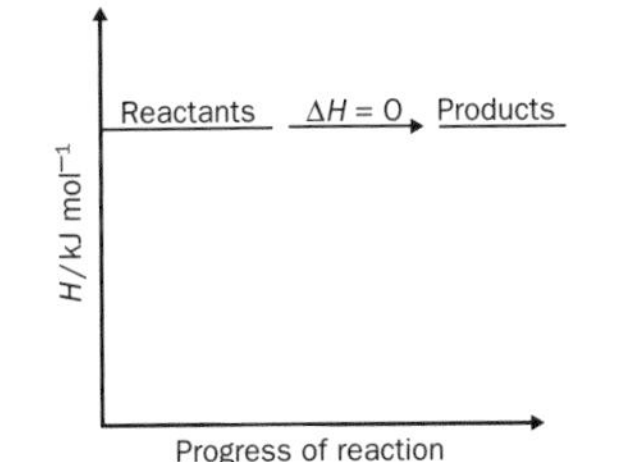

13B
(i) A balanced equation should be also written. Only then do we know how much sodium and chlorine have reacted.
(ii) Number of moles of carbon
$= 120\ g / 12\ g\ mol^{-1} = 10$.
The equation shows that complete combustion of 1 mol of carbon produces 393.15 kJ of heat. Therefore, the complete burning of 10 mol of carbon produces 3931.5 kJ of heat energy.

13C
601.7 kJ is required to decompose 2 mol of MgO, so 300.9 kJ is required to decompose 1 mol of MgO.

13D
Eqn (13.14) is obtained by adding eqns (13.12) and (13.13). By Hess's law,

$$\Delta H^{\ominus}_{(13.14)} = \Delta H^{\ominus}_{(13.12)} + \Delta H^{\ominus}_{(13.13)} = -296.8 + (-98.9) = -395.7\ kJ\ mol^{-1}$$

13E
(i) To speed up the reaction so that the reaction was complete in a short time.
(ii) The number of moles of lead(II) nitrate is

concentration × volume
$0.10\ mol\ dm^{-3} \times (100/1000) = 0.010$

The number of moles of Zn used = 0.007 7 moles. The equation

$Zn(s) + Pb^{2+}(aq) \rightarrow Zn^{2+}(aq) + Pb(s)$

shows that 1 mol of solid zinc reacts with 1 mol of $Pb^{2+}(aq)$, so that lead(II) nitrate is in excess.

13F
The solutions are dilute and may be assumed to be water ($C = 4.18\ J\ g^{-1}\,°C^{-1}$). Therefore the total mass of water ($1\ cm^3 \equiv 1\ g$) is 50.0 g + 50.0 g = 100.0 g.

$\Delta T = 23.60 - 17.40 = +6.20\,°C$

Enthalpy change of reactants
$= -mC\Delta T = -100.0 \times 4.18 \times +6.2 = -2600\ J$.
Numbers of moles of HCl initially present is

$$\frac{50.0}{1000} \times 1.00 = 0.0500$$

The number of mol of NaOH initially present is double this. The reaction

$H^+(aq) + OH^-(aq) \rightarrow H_2O(l)$

($H^+(aq)$ from hydrochloric acid, $OH^-(aq)$ from sodium hydroxide) shows that 1 mol of water is produced when 1 mol of acid and 1 mol of alkali react. Therefore, HCl is the limiting reagent, and the production of 0.0500 mol of water is accompanied by the production of 2591 J of heat. When **one** mole of water is made, the enthalpy change is

$$\Delta H = \frac{1}{0.0500} \times -2591 = -51820\ J, \text{ or } -51.8\ kJ\ mol^{-1}$$

The accepted value is $\Delta H^{\ominus} = -57.3$ kJ per mole of water produced. The most important source of error is caused by the loss of heat from the calorimeter, which results in an underestimation of the temperature rise.

13G
To complete the equations, add the elements in their standard states. The physical states of the elements and compounds at 298 K should also be included because we are assuming that $\Delta H^{\ominus}_f$ refers to this temperature.

$H_2(g) + S(s) + 2O_2(g) \rightarrow H_2SO_4(l)$

$6C(s) + 6H_2(g) + 3O_2(g) \rightarrow C_6H_{12}O_6(s)$

13H
(i) $2CO(g) + O_2(g) \rightarrow 2CO_2(g)$
Using the standard enthalpy data in Table 13.2,
$\Delta H^{\ominus}_f(CO_2(g)) = -393.51\ kJ\ mol^{-1}$,
$\Delta H^{\ominus}_f(CO(g)) = -110.53\ kJ\ mol^{-1}$ and
$\Delta H^{\ominus}_f(O_2(g)) = 0\ kJ\ mol^{-1}$. Therefore,

$$\Delta H^{\ominus} = [2\Delta H^{\ominus}_f(CO_2(g))] - [2\Delta H^{\ominus}_f(CO(g)) + \Delta H^{\ominus}_f(O_2(g))]$$
$$= [\,2(-393.51)] - [\,2(-110.53) + 0]$$
$$= -565.96\ kJ\ mol^{-1}$$

(ii) $Fe_2O_3(s) + 3CO(g) \rightarrow 2Fe(s) + 3CO_2(g)$
$\Delta H^{\ominus}_f(CO_2(g)) = -393.51\ kJ\ mol^{-1}$;
$\Delta H^{\ominus}_f(CO(g) = -110.53\ kJ\ mol^{-1}$;
$\Delta H^{\ominus}_f(Fe_2O_3)(s) = -824.2\ kJ\ mol^{-1}$;
remember that
$\Delta H^{\ominus}_f(Fe(s)) = 0\ kJ\ mol^{-1}$

$$\Delta H^{\ominus} = [3\Delta H^{\ominus}_f(CO_2(g)) + 2\Delta H^{\ominus}_f(Fe(s))] - [3\Delta H^{\ominus}_f(CO(g) + \Delta H^{\ominus}_f(Fe_2O_3)]$$
$$= -1180.5 - (-1155.8) = -24.74\ kJ\ mol^{-1}$$

(iii) $NH_3(g) \rightarrow \frac{1}{2}N_2(g) + \frac{3}{2}H_2(g)$
$\Delta H^{\ominus}_f(NH_3(g)) = -46.11\ kJ\ mol^{-1}$, and
$\Delta H^{\ominus}_f(H_2(g)) = 0\ kJ\ mol^{-1}$
$\Delta H^{\ominus}_f(N_2(g)) = 0\ kJ\ mol^{-1}$

$$\Delta H^{\ominus} = [\tfrac{1}{2}\Delta H^{\ominus}_f(N_2(g)) + \tfrac{3}{2}\Delta H^{\ominus}_f(H_2(g))] - [\Delta H^{\ominus}_f(NH_3(g))]$$
$$= -(-46.11) = +46.11\ kJ\ mol^{-1}$$

The positive sign shows that energy is absorbed in order to decompose the molecule.

13I
(i) $M(C_{12}H_{22}O_{11}) = 342\ g\ mol^{-1}$. 5644 kJ of heat is given out when 342 g of sucrose are completely burned in oxygen.
Therefore, 1 g of sucrose gives out 5644/342 = 16.5 kJ; energy value = $16.5\ kJ\ g^{-1}$.
(ii) 1 g of H_2 produces 142 kJ of heat. Therefore, the mass of H_2 that produces 100 kJ of heat is

$$\frac{100}{142} \times 1 = 0.704\ g$$

0.704 g of H_2 contains (0.704/2.00) = 0.352 mol of H_2; 1 mol of H_2 occupies 24 dm^3 at room temperature and pressure. Therefore, 0.352 mol of H_2 occupies

$0.352 \times 24 = 8.5\ dm^3$

i.e. 8.5 dm^3 of gas burns to give 100 kJ of heat.

13J
20 g produces 120 Cal (= 120 kcal) when burned. The energy value of the chocolate bar is
$4.18 \times 120 \approx 500$ kJ. The energy value of 1 g of chocolate is
$(1/20) \times 500 \approx 25\ kJ\ g^{-1}$ (compare with the materials in Table 13.3).

13K
Consumption of $O_2 = 0.27 \times 60 \times 24 = 390\ dm^3\ day^{-1}$.
Energy requirement per day = 390 × 20 = 7800 kJ (1870 Cal).

13L
2237 × 0.1 = 223.7 kJ (endothermic)

13M
(i) The process $Na(s) \rightarrow Na^+(g) + e^-$ is the sum of the atomisation and ionisation steps in the Born–Haber cycle. Therefore, the energy requirement is

$\Delta H^{\ominus}_1 + \Delta H^{\ominus}_2 = 109 + 494 = +603\ kJ\ mol^{-1}$

(ii) The process $Na^+,Cl^-(s) \rightarrow Na(g) + \frac{1}{2}Cl_2(g)$ involves the sum of $(-\Delta H^{\ominus}_5) + (-\Delta H^{\ominus}_4) + (-\Delta H^{\ominus}_3) + (-\Delta H^{\ominus}_2)$
$= 771 + 364 - 121 - 494 = 520\ kJ\ mol^{-1}$

13N
The formation of magnesium fluoride is

$Mg(s) + F_2(g) \rightarrow Mg^{2+},F^-_2(s) \quad \Delta H^{\ominus}_f = -1121\ kJ\ mol^{-1}$

This may also be split up into steps, as with NaCl in the text:

$$\Delta H^{\ominus}_5 = \Delta H^{\ominus}_f - \Delta H^{\ominus}_1 - \Delta H^{\ominus}_2 - \Delta H^{\ominus}_3 - \Delta H^{\ominus}_4 \quad (A)$$

Note that:
(a) $\Delta H^{\ominus}_2$ involves two ionization steps:

$Mg(g) \rightarrow Mg^+(g) + e^- \quad \Delta H^{\ominus} = 744\ kJ\ mol^{-1}$

$Mg^+(g) \rightarrow Mg^{2+}(g) + e^- \quad \Delta H^{\ominus} = 1456\ kJ\ mol^{-1}$

so that $\Delta H^{\ominus}_2 = 744 + 1456 = 2200\ kJ\ mol^{-1}$.
(b) Step 3 involves the production of two fluorine atoms:

$F_2(g) \rightarrow 2F(g) \quad \Delta H^{\ominus}_3 = 158\ kJ\ mol^{-1}$

(c) Two fluorine atoms accept electrons:

$2F(g)+2e^- \rightarrow 2F^-(g) \quad \Delta H_4^\ominus=(2\times-339)=-678 \text{ kJ mol}^{-1}$

Substituting into equation (A),

$\Delta H_5^\ominus = -1121 - 148 - 2200 - 158 + 678$
$= -2949 \text{ kJ mol}^{-1}$

or, $\Delta H_L^\ominus = -\Delta H_5^\ominus = +2949 \text{ kJ mol}^{-1}$.

13O

Average = (531 + 1075)/2 = 803 kJ mol^{-1}.

13P

(i) The standard enthalpy change for the process $CH_4(g) \rightarrow CH_3(g) + H(g)$ will approximately equal $\Delta H_{C-H}^\ominus$. Therefore the enthalpy change for the reverse process:

$CH_3(g) + H(g) \rightarrow CH_4(g)$

will be $-\Delta H_{C-H}^\ominus$, i.e. -412 kJ mol^{-1}.

(ii) Comparing the bond lengths with the bond enthalpy data in Table 13.4, the trend supports the proposition that long bonds are weak bonds (or, conversely, that short bonds are strong bonds).

(iii) Double and triple bonds are stronger than single bonds. The general order of bond strength is triple bonds > double bonds > single bonds.

13Q

(i) $CH_2{=}CH_2(g) + Cl_2(g) \rightarrow CH_2Cl{-}CH_2Cl(g)$

To completely break up the molecules on the left-hand side of this equation we need to dissociate: one C=C bond, four C–H bonds and one Cl–Cl bond. From Table 13.4,

energy absorbed = 612 + 4(412) + 242 = 2502 kJ mol^{-1}

i.e. the energy change due to bond destruction is + 2502 kJ.

To form the reactant molecules, we need to make four C–H bonds, two C–Cl bonds and one C–C bond. From Table 13.4,

energy given out = 4(412) + 2(338) + 348 = 2672 kJ mol^{-1}

i.e. the energy change due to bond formation = −2672 kJ. Therefore, the net change in energy is

$\Delta H^\ominus(298 \text{ K}) = 2502 - 2672 = -170 \text{ kJ mol}^{-1}$

(ii) Bond energies can be used directly only if all the reactants and products are gases. If some reactants or products are not gases then the enthalpy change for the phase change (solid → gas or liquid → gas) must be allowed for.

We start by working out the enthalpy change for the reaction in which all the reactants and products are gaseous:

$2H_2O(g) \rightarrow 2H_2(g) + O_2(g)$ (A)

We now apply the average bond energies in Table 13.4 for the heat absorbed in breaking bonds in reactant molecules, this involves breaking four O–H bonds:

$4 \times 463 = 1852 \text{ kJ mol}^{-1}$

From Table 13.4 the heat released when bonds in product molecules are assembled, which involves making two H–H bonds and one O=O bond is

$(2 \times 436) + 497 = 1369 \text{ kJ mol}^{-1}$

The overall change in enthalpy is therefore

$\Delta H_A^\ominus = 1852 - 1369 = +483 \text{ kJ mol}^{-1}$

However, we want the enthalpy change for the reaction

$2H_2O(l) \rightarrow 2H_2(g) + O_2(g)$ (B)

and we are given

$H_2O(l) \rightarrow H_2O(g) \quad \Delta H^\ominus = +44 \text{ kJ mol}^{-1}$

or

$2H_2O(l) \rightarrow 2H_2O(g) \quad \Delta H^\ominus = +88 \text{ kJ mol}^{-1}$

$H_2O(g)$ possesses a higher enthalpy than $H_2O(l)$: this means that the enthalpy change for reaction (B) will be larger than that of reaction (A) by 88 kJ. (If you can't see this, draw an energy diagram for the processes involved):

$\Delta H_B^\ominus(298 \text{ K}) = 483 + 88 = +571 \text{ kJ mol}^{-1}$

Revision questions

13.1 Following are false: **(i)**, **(ii)**, **(iv)** (2 mol $SO_3(g)$ made) and **(vi)**.

13.2 (i) See text.

(ii) Combustion of methyl benzene:

$C_7H_8(l) + 9O_2(g) \rightarrow 7CO_2(g) + 4H_2O(l)$

$\Delta H_c^\ominus = [4\Delta H_f^\ominus(H_2O(l)) + 7\Delta H_f^\ominus(CO_2(g))] - [\Delta H_f^\ominus(C_7H_8(l)) + 9\Delta H_f^\ominus(O_2(g))]$

From Table 13.2:

$= [(-285.83 \times 4) + (-393.51 \times 7)] - [12.0 + 0]$
$= (-3897.89) - (12.0) = -3909.9$ kJ per mole of methyl benzene

(iii) 1 mol of methyl benzene has a molar mass of approximately 92 g mol^{-1}.

Heat given out in the complete combustion of 1 g of methyl benzene is $3909.9/92 \approx 42 \text{ kJ g}^{-1}$.

1 cm^3 of methyl benzene has a mass of 0.86 g.

The heat given out in the complete combustion of 1 cm^3 of methyl benzene is approximately $42 \times 0.86 = 36 \text{ kJ cm}^{-3}$.

13.3 (i) $\Delta H^\ominus = [\Delta H_f^\ominus(P(s)\,(red))] - [\Delta H_f^\ominus(P(s)\,(white))]$
$= -18 - 0 = -18 \text{ kJ mol}^{-1}$

i.e. red phosphorus is the more energetically stable form of phosphorus.

(ii)

$\Delta H^\ominus = [\Delta H_f^\ominus(H_2O(l)) + \Delta H_f^\ominus(CH_3CO_2C_2H_5(l))] - [\Delta H_f^\ominus(C_2H_5OH(l)) + \Delta H_f^\ominus(CH_3CO_2H)(l)]$
$= (-285.83 - 481) - (-277.69 - 485)$
$= (-766.83) - (-762.69) \approx -4 \text{ kJ mol}^{-1}$

(iii)

$\Delta H^\ominus = [3\Delta H_f^\ominus(N_2(g)) + 9\Delta H_f^\ominus(H_2O(g)) + 5\Delta H_f^\ominus(Al_2O_3(s)) + 6\Delta H_f^\ominus(HCl(g))] - [6\Delta H_f^\ominus(NH_4ClO_4(s)) + 10\Delta H_f^\ominus(Al(s))]$

The $\Delta H_f^\ominus$ of elements in their reference states is zero. $\Delta H_f^\ominus(NH_4ClO_4(s))$ is obtained from Table 13.2.

$\Delta H^\ominus = [(-241.82 \times 9) + (-1675.7 \times 5) + (-92.31 \times 6)] - [-295.31 \times 6]$
$= [-2176.38 - 8378.5 - 553.86] + [1771.86]$
$\approx -9337 \text{ kJ mol}^{-1}$

(Notice that the effect of producing a highly exothermic compound ($Al_2O_3(s)$), is to make the overall reaction highly exothermic.)

13.4 Ignoring the heat capacity of the calorimeter and of the reactants and products,

$q = -m_{mixture}c_{mixture}\Delta T$
$= -100 \times 4.18 \times +0.41 = -1.7 \times 10^2 \text{ J}$
$\therefore \Delta H = -1.7 \times 10^2 \text{ J}$

Neither reactant is in excess. The number of moles of $Ag^+(aq)$ (which equals the number of moles of $Cl^-(aq)$) is

$\frac{50}{1000} \times 0.050 = 2.5 \times 10^{-3}$

The enthalpy change when 1 mol of either reactant is consumed is

$\Delta H = \frac{1}{2.5 \times 10^{-3}} \times -1.7 \times 10^2$
$= -6.8 \times 10^4 = -68 \text{ kJ mol}^{-1}$

13.5 $HCOOH(l) + \frac{1}{2}O_2(g) \rightarrow CO_2(g) + H_2O(l)$

$\Delta H_c^\ominus = -277.2 \text{ kJ mol}^{-1}$

$\Delta H_c^\ominus = [\Delta H_f^\ominus(COO(g)) + \Delta H_f^\ominus(H_2O(l))] - [\Delta H_f^\ominus(HCOOH(l)) + \frac{1}{2}\Delta H_f^\ominus(O_2(g))]$

where $\Delta H_f^\ominus(HCOOH(l))$ is the only unknown, $\Delta H_f^\ominus(CO_2(g)) = -393.51 \text{ kJ mol}^{-1}$ and $\Delta H_f^\ominus(H_2O(l)) = -285.83 \text{ kJ mol}^{-1}$ (from Table 13.2) and $\Delta H_f^\ominus(O_2(g)) = 0$. Substitution and rearrangement gives:

$\Delta H_f^\ominus(HCOOH(l) = (-393.51 - 285.83) - (-277.2)$
$= -402.1 \text{ kJ mol}^{-1}$

13.6 (i) $C_6H_{12}O_6(s) + 6O_2(g) \rightarrow 6CO_2(g) + 6H_2O(l)$

(ii) $M(C_6H_{12}O_6) = 180 \text{ g mol}^{-1}$. The complete combustion of 180 g of glucose releases 2816 kJ of energy. The complete combustion of 10 g of glucose would release $(10/180) \times 2816 = 160$ kJ (2 significant figures). If the combustion of glucose within the body were only 40% efficient, only 64 kJ of energy would be available for work for every 10 g of glucose consumed.

13.7 (i) The third standard ionization energy of aluminium is the standard enthalpy change for the process

$Al^{2+}(g) \rightarrow Al^{3+}(g) + e^-$

$\Delta H_{ion(3)}^\ominus = 2751 \text{ kJ mol}^{-1}$

(ii) The equation

$Al(s) \rightarrow Al^{3+}(g) + 3e^-$ (A)

is the sum of the equations:

$Al(s) \rightarrow Al(g)$ (B)

$Al(g) \rightarrow Al^+(g) + e^-$ (C)

$Al^+(g) \rightarrow Al^{2+}(g) + e^-$ (D)

$Al^{2+}(g) \rightarrow Al^{3+}(g) + 3e^-$ (E)

The enthalpy change for reaction (A) is the sum of the enthalpy changes for reactions (B)–(E) (If you are unsure about this, sketch an energy diagram):

$\Delta H_A^\ominus = \Delta H_B^\ominus + \Delta H_C^\ominus + \Delta H_D^\ominus + \Delta H_E^\ominus$
$= 326 + 584 + 1823 + 2751 = 5484 \text{ kJ mol}^{-1}$

13.8

(a) $Ca(s) \rightarrow Ca(g)$	$\Delta H_1^\ominus$
(b) $Ca(g) \rightarrow Ca^+(g) + e^-$	$\Delta H_{2A}^\ominus$
(c) $Ca^+(g) \rightarrow Ca^{2+}(g) + e^-$	$\Delta H_{2B}^\ominus$
(d) $O_2(g) \rightarrow O(g) + O(g)$	$\Delta H_3^\ominus$
(e) $O(g) + 2e^- \rightarrow O^{2-}(g)$	$\Delta H_4^\ominus$
(f) $Ca^{2+},O^-(s) \rightarrow Ca^{2+}(g) + O^{2-}(g)$	$\Delta H_L^\ominus$

The formation of calcium oxide,

$Ca(s) + \frac{1}{2}O_2(g) \rightarrow Ca^{2+},O^-(s) \quad \Delta H_f^\ominus$

may also be split up into steps, as for NaCl. We follow the pattern for NaCl in the text, with $\Delta H_2^\ominus = \Delta H_{2A}^\ominus + \Delta H_{2B}^\ominus$. Therefore,

$\Delta H_f^\ominus = \Delta H_1^\ominus + \Delta H_2^\ominus + \frac{1}{2}\Delta H_3^\ominus + \Delta H_4^\ominus + (-\Delta H_L^\ominus)$

Notice that only half of $\Delta H_3^\ominus$ is required because only one atom of O is required to combine with one atom of Ca.

$\Delta H_f^\ominus = 193 + 1740 + 248.5 + 703 - 3513$
$= -629 \text{ kJ mol}^{-1}$

13.9

$2F_2(g) + S(g) \rightarrow SF_4(g) \quad \Delta H^\ominus = -994 \text{ kJ mol}^{-1}$

In breaking bonds in the reactant molecules, we break two F–F bonds:

heat absorbed = $2 \times 158 = 316$ kJ

The heat released when bonds in product molecules are assembled involves making four S–F bonds, each of bond enthalpy x. The total enthalpy change is $-4x$ (which is negative because heat is given out when bonds are formed). Therefore, the overall change in enthalpy is

$-994 = -4x + 316$

which gives

$x = \Delta H_{S-F}^\ominus = (994 + 316)/4$
$= 328$ kJ per mole of S–F bonds.

Unit 14

Exercises

14A

(i) NO disappears at twice the rate that H_2 disappears $= 0.02$ mol dm^{-3} s^{-1}.
(ii) Nitrous oxide appears at the same rate at which H_2 disappears $= 0.01$ mol dm^{-3} s^{-1}.
(iii) Water appears at the same rate at which H_2 disappears $= 0.01$ mol dm^{-3} s^{-1}.

14B

$$\text{initial rate} = \frac{0.60 - 0.10}{10} = 0.050 \text{ mol dm}^{-3}\text{ s}^{-1}$$

$$\text{rate at 25 s} = \frac{0.27 - 0.15}{30} = 4.0 \times 10^{-3} \text{ mol dm}^{-3}\text{ s}^{-1}$$

(Your answers might be slightly different due to the difficulty of drawing tangents.)

14C

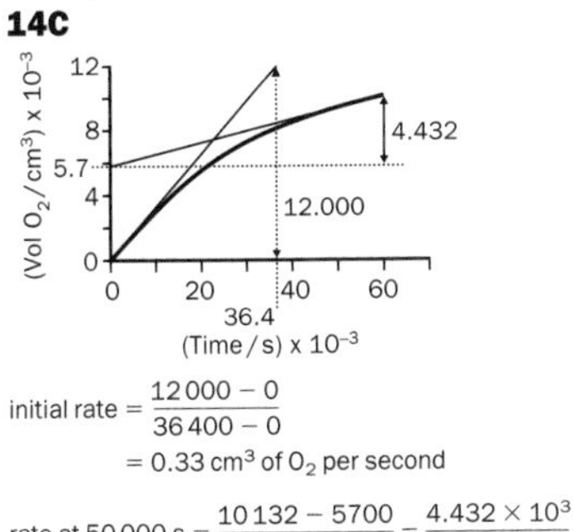

$$\text{initial rate} = \frac{12\,000 - 0}{36\,400 - 0}$$
$= 0.33$ cm^3 of O_2 per second

$$\text{rate at 50 000 s} = \frac{10\,132 - 5700}{60\,000 - 0} = \frac{4.432 \times 10^3}{60\,000}$$
$= 0.074$ cm^3 of O_2 per second

14D

The cut divides the marble into two rectangular blocks of dimensions $0.5 \times 1 \times 1$ cm. The surface area of each block is $(0.5 + 1 + 1 + 0.5 + 0.5 + 0.5) = 4$ cm^3. The total surface area is the sum of the surface area for both blocks $= 8$ cm^3. This is 33% greater than the surface area of the *undivided* block (6 cm^3). Dividing the block into even smaller pieces greatly increases the total surface area available for attack by the acid.

14E

(i) $100 - 20\,°C = 80\,°C$ so that the increase in rate is $2 \times 2 \times 2 \times 2 \times 2 \times 2 \times 2 \times 2$ (i.e. 2^8) $= 256$ times.
(ii) The heat generated by the exothermic reaction raises the temperature of the reaction mixture, so increasing its rate.

14F

(i) Nails have a lower surface area than powder.
(ii) (a) Water dilutes the concentration of acid, so reducing the number of acid–zinc collisions per second.
(b) Ice cools the mixture, thereby reducing reaction rate.
(c) Sodium carbonate reacts with the acid, so drastic ally reducing the concentration of hydrogen ions in solution.

14G

(i) 1 **(ii)** 2 **(iii)** 1 **(iv)** 3.

14H

(i) Third order. **(ii)** Rate $= 7000 \times (0.060)^3 = 1.5$ mol dm^{-3} s^{-1}.

14I

(i) Second order.
(ii)

$$k = \frac{\text{initial rate}}{[C_{12}H_{22}O_{11}][H^+]} = \frac{4.0 \times 10^{-6}}{(0.20)(0.10)}$$
$= 2.0 \times 10^{-4}$ mol^{-1} dm^3 s^{-1}

14J

Rate of reaction $= k[\text{M}][\text{Y}]^0 = k[\text{M}]$

14K

rate of reaction $= k[Br^-(aq)]^x[BrO_3^-(aq)]^y[H^+(aq)]^z$

Comparison of experiments 1 and 2 shows that doubling $[Br^-(aq)]$ doubles the initial rate; therefore, $x = 1$. Comparing experiments 1 and 3 shows that increasing $[BrO_3^-(aq)]$ by a factor of four increases the initial rate by four; therefore, $y = 1$. Comparing experiments 1 and 4 shows that doubling $H^+(aq)$ increases the rate by four; therefore, $z = 2$ and the rate expression is

rate $= k[Br^-(aq)]\,[BrO_3^-(aq)]\,[H^+(aq)]^2$

14L

10 h $= 10 \times 60 \times 60 = 36\,000$ s.

$[A]_t = [A]_0\, e^{-kt} = 0.01 \times e^{-(6.5 \times 10^{-6} \times 36\,000)}$
$= 0.01 \times 0.791 = 0.0079$ mol dm^{-3}.

14M

$$t_{1/2} = \frac{0.693}{k} = \frac{0.693}{9.6 \times 10^{-3}} = 72 \text{ s}$$

14N

$$t_{1/2} = \frac{0.693}{5.73 \times 10^{-7}} = 1\,209\,424 \text{ s (14.0 days)}.$$

42 days is three half-lives. During this period the concentration of pesticide will have fallen to one-eighth of its initial concentration, i.e. from 0.10 mg dm^{-3} to 0.0125 mg dm^{-3}.

Revision questions

14.1 See text.
14.2 (i).
14.3 (i) II **(ii)** I.
14.4 (i) 1
(ii) initial rate
$= 2.00 \times 10^{-5} \times 0.100 = 2.00 \times 10^{-6}$ mol dm^{-3} s^{-1}
(iii) rate $= 2.00 \times 10^{-5} \times 0.0523 = 1.05 \times 10^{-6}$ mol dm^{-3} s^{-1}
(iv) half life $= 0.693/2.00 \times 10^{-5}$
$= 3465$ s ≈ 9.6 h

14.5 (i) 1
(ii) plot as follows:

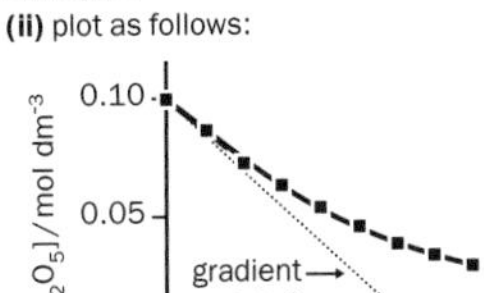

(iii) $$\text{initial rate} = \frac{0.10 - 0.0}{3.3 \times 60 \times 60}$$
$= 8.4 \times 10^{-6}$ mol dm^{-3} s^{-1}
(iv) initial rate $= k[N_2O_5(g)]_0$
$k(65\,°C) = 8.4 \times 10^{-6}/0.10 = 8.4 \times 10^{-5}$ s^{-1}
(v)

$$t_{1/2} = \frac{0.693}{8.4 \times 10^{-5}} = 8300 \text{ s (2.3 h)}$$

(Inspection of the graph and table confirms that the reaction half-life is roughly 2 h.)

14.6 (i) Consumption of hydroxide ion is negligible and rate of reaction depends upon the concentration of the bromomethane only:

rate of reaction $= k'[CH_3Br(l)]$

with $k' = k[OH^-(aq)]$. If the bromomethane were in excess, the reaction would also be pseudo first order and the rate expression would be

rate of reaction $= k''[OH^-(aq)]$

with $k'' = k[CH_3Br(l)]$
(ii) (a) Initial rate of reaction $= k[CH_3Br(l)][OH^-(aq)]$
$= 3.0 \times 10^{-4} \times 1.0 \times 10^{-2} \times 1.0 \times 10^{-4}$
$= 3.0 \times 10^{-10}$ mol dm^{-3} s^{-1}
(b) pseudo-rate constant $= k' = k[OH^-(aq)]$
$= 3.0 \times 10^{-4} \times 1.0 \times 10^{-2} = 3.0 \times 10^{-6}$ s^{-1}
(c) pseudo half-life of reaction $= 0.693 / 3.0 \times 10^{-6}$
$= 2.3 \times 10^5$ s.

14.7 (i)

rate $= k[ClO]^x[NO_2]^y[N_2]^z$

Comparison of experiments 1 and 2 shows that halving the concentration of ClO halves the initial rate; therefore, $x = 1$. Experiments 1 and 3 show that doubling $[NO_2]$ doubles the rate; therefore, $y = 1$. Similarly, comparison of experiments 2 and 4 shows that doubling $[N_2]$ doubles the initial rate; therefore, $z = 1$. The rate expression becomes

rate $= k[ClO][NO_2][N_2]$

(ii) From the data of experiment 1,

$$k = \frac{\text{initial rate}}{[ClO][NO_2][N_2]}$$

$$= \frac{3.5 \times 10^{-4}}{[1.0 \times 10^{-5}][2.0 \times 10^{-5}][3.0 \times 10^{-5}]}$$

$= 5.8 \times 10^{10}$ mol^{-2} dm^6 s^{-1}

From the data of the remaining experiments:

experiment 2: $k = 6.0 \times 10^{10}$
experiment 3: $k = 5.9 \times 10^{10}$
experiment 4: $k = 6.0 \times 10^{10}$

Average k(298 K) $= 5.9 \times 10^{10}$ mol^{-2} dm^6 s^{-1}.
(iii) If the reaction between ClO, NO_2 and N_2 were elementary, we would expect the rate expression to be $K[ClO][NO_2][N_2]$, as above. The rate expression is therefore *consistent* with a one-step mechanism.

14.8

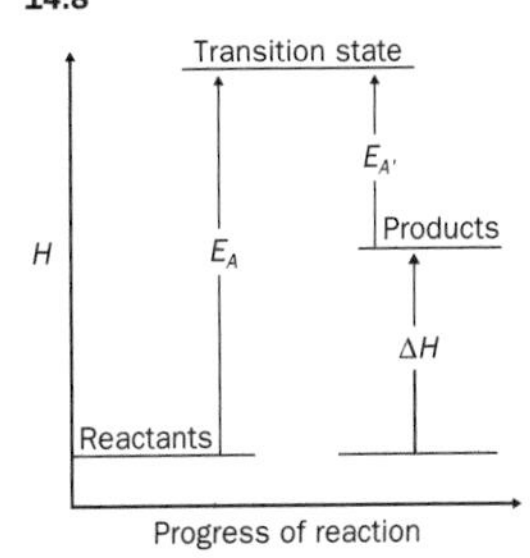

14.9 First derive the expression

$$\ln\left(\frac{[A]_0}{[A]_t}\right) = kt$$

as in the text. Symbolize the time at which 99% of the concentration of A has disappeared as $t_{0.99}$. After this time, the concentration of A will have fallen to one hundredth of its initial value:

$$[A]_{0.99} = \frac{[A]_0}{100}$$

Substitution gives

$$\ln\left(\frac{[A]_0}{[A]_0/100}\right) = \ln(100) = k \times t_{0.99}$$

Since $\ln(100) = 4.605$,
$4.605 = k \times t_{0.99}$

or $$t_{0.99} = \frac{4.605}{k}$$

14.10

(a) Starting materials

O=C O=O C=O

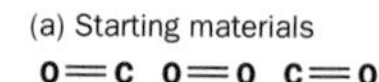

Platinum

(b) Adsorption

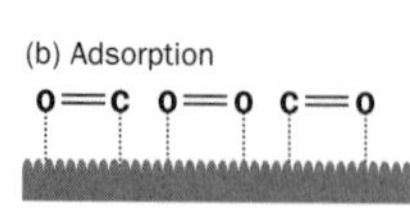

(c) Rearrangement

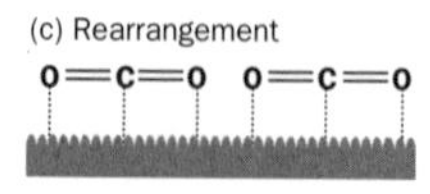

(d) Desorption

O=C=O O=C=O

Unit 15

Exercises

15A

Statement (b) is correct. Statement (a) is incorrect because the double arrows show that the reaction is an equilibrium reaction and complete conversion of reactants to products does not occur under the conditions of the experiment.

15B

(i)

$$K_c = \frac{[Fe^{3+}(aq)]^2[Cl^-(aq)]^2}{[Fe^{2+}(aq)]^2[Cl_2(g)]}$$ the units are mol dm^{-3}

(ii)

$$K_c = \frac{[ClONO_2(g)]}{[NO_2(g)][ClO(g)]}$$ the units are mol^{-1} dm^3

(iii)

$$K_c = \frac{[NO_2(g)]^2}{[N_2O_4(g)]}$$ the units are mol dm^{-3}

15C

(i) Reactions in Table 15.1 where $K_c > 10^3$ are 2, 7, 8 (at 298 K) and 10.

(ii) The reaction between sodium ions and EDTA

$Na^+(aq) + EDTA^{4-}(aq) \rightleftharpoons NaEDTA^{3-}(aq)$

does not go to completion. This means that the $EDTA^{4-}$ will not complex all the sodium ions in a mixture and the $Na^+(aq)$–$EDTA^{4-}$ reaction is therefore worthless in quantitative analysis. (See also Exercise 12R)

15D

Yes. The equilibrium constant for the reverse reaction at 1100 K is $1/(7 \times 10^{-9}) = 1.4 \times 10^8$. We conclude that decomposition goes to completion.

15E

(i) K_c (average) = 45 at 490 °C.

$$K_{c(T)} = \frac{[HI(g)]^2}{[H_2(g)][I_2(g)]}$$

or,

$[HI(g)] = \sqrt{(K_{c(T)}[H_2(g)][I_2(g)])}$
$= \sqrt{(45 \times 3.00 \times 3.00)} = 20$ mol dm^{-3}

In the reaction between hydrogen and iodine, $1H_2(g) \equiv 1I_2(g) \equiv 2HI(g)$. Therefore, if 20 mol dm^{-3} of HI is produced, (20/2) = 10 mol dm^{-3} of each reactant is used up. This means that initially there must have been 10 + 3.00 = 13 mol dm^{-3} of $H_2(g)$ and $I_2(g)$.

(ii)

$$K_c\,(800\text{ K}) = \frac{[HI(g)]^2}{[H_2(g)][I_2(g)]} = \frac{[19.1]^2}{[3.50][2.50]} = 41.7$$

This value is lower than the value at 490°C (=45), suggesting that the equilibrium constant decreases with increasing temperature.

15F

$$\frac{[N_2(g)][H_2(g)]^3}{[NH_3]^2} = \frac{0.04 \times (0.03)^3}{(0.02)^2} \approx 3 \times 10^{-3}\text{ mol}^2\text{ dm}^6$$

This is not equal to the equilibrium constant at this temperature. Therefore, the reaction is not at equilibrium.

15G

One of the products, water vapour, is driven away (by heating and by the flow of unused hydrogen gas). This prevents the reaction reaching equilibrium. In an attempt to restore the concentration of water, more iron(III) oxide reacts with hydrogen. Eventually, all the Fe_2O_3 is consumed.

15H

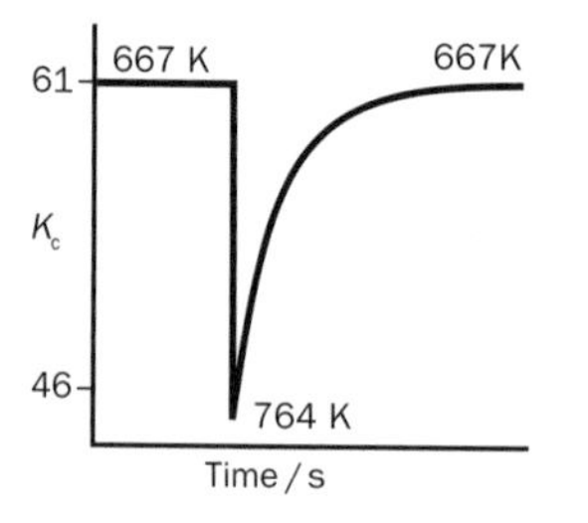

15I

Raising the temperature increases the equilibrium concentration of NO_2. This suggests that the formation of NO_2 is endothermic since, for endothermic reactions, K_c increases with increasing temperature. (It is useful to note that chemical decompositions are generally endothermic.)

15J

(i) Equilibrium concentration of ethane ($C_2H_6(g)$) rises; K_c is unaffected.

(ii) This increases the concentration of hydrogen reactant. Therefore the equilibrium concentration of ethane increases; K_c is unaffected.

(iii) No effect on composition or upon K_c.

(iv) Exothermic reaction, so K_c increases. Thus the concentration of ethane increases.

15K

Applying the equilibrium law,

$$K_{c(T)} = \frac{[H_2O(g)]}{[H_2O(l)]}$$

However, the concentration of a pure liquid is fixed at that temperature. We can now write

$K_{c(T)}[H_2O(l)] = [H_2O(g)]$

By the ideal gas equation,

$P = cRT$

where c is the concentration of gas (= n/V) and P is its partial pressure. Therefore,

$P_{(H_2O)} = K_{c(T)}[H_2O(l)] \times RT$

where R is the universal gas constant. All the quantities on the right-hand side of this equation are fixed at a particular temperature. We conclude that the partial pressure of water vapour in equilibrium with its liquid is also fixed at a particular temperature.

Revision questions

15.1 The chemical equation is missing.

15.2 (i) The completed table is

	(Concentration/mol dm^{-3})/10^{-3})			
Time(s)	[H_2(g)]	[I_2(g)]	[HI(g)]	Q
0	1.000	1.000	0.000	0.000
50	0.897	0.897	0.206	0.053
100	0.813	0.813	0.374	0.211
200	0.685	0.685	0.630	0.846
300	0.592	0.592	0.817	1.905
400	0.521	0.521	1.040	3.985
500	0.465	0.465	1.069	5.285
600	0.420	0.420	1.160	7.628
700	0.383	0.383	1.234	10.381
800	0.352	0.352	1.296	13.556
850	0.333	0.333	1.334	16.048
900	0.332	0.332	1.336	16.193
950	0.331	0.331	1.338	16.340

(ii) A plot of [H_2(g)] and [HI(g)] against time:

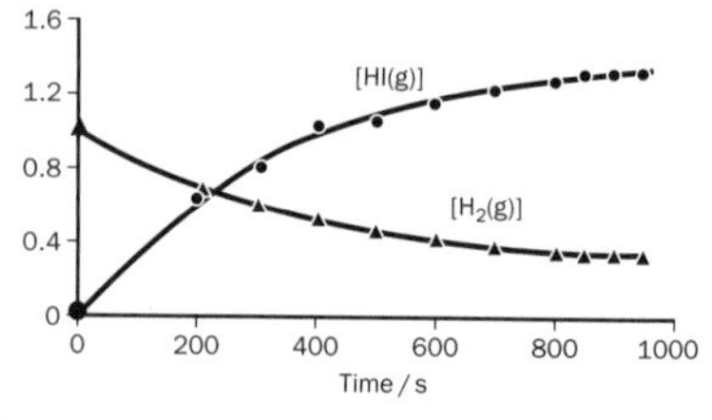

The initial rate of consumption of hydrogen is $(1.000 - 0.897) \times 10^{-3}/50 = 2.06 \times 10^{-6}$ mol dm^{-3} s^{-1}. The initial rate of HI production is $0.206 \times 10^{-3}/50 = 4.12 \times 10^{-6}$ mol dm^{-3} s^{-1}. This is double the rate at which H_2(g) is consumed, as expected by the reaction equation.

(iii) A plot of Q against time reveals that equilibrium has been reached at about 900 s. The equilibrium constant is very roughly 16.

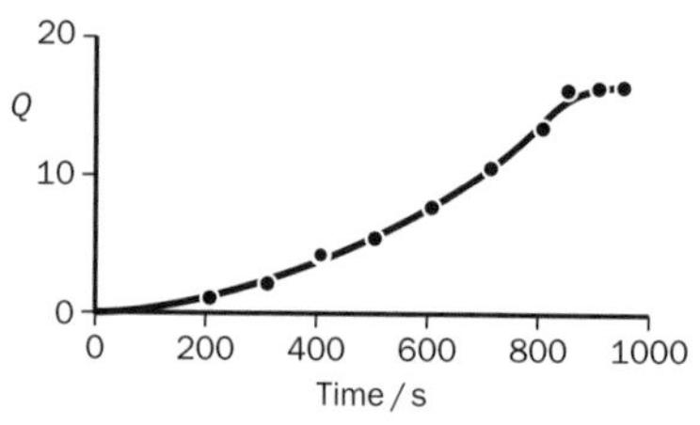

(iv) Figure 15.6 suggests a crude estimate of the temperature as being about 1000–1100 K.

15.3 (i) The presence of water in the initial mixture produces a lower yield of ester in the equilibrium mixture in order to maintain K_c at that temperature. (Le Chatelier's principle allows us to come to the same conclusion.)

(ii) (a)

$$K_{c(T)} = \frac{[CH_3COOC_2H_5(l)][H_2O(l)]}{[CH_3COOH(l)][C_2H_5OH(l)]} = \frac{(4.48)^2}{(2.67)(1.93)} = 3.89$$

(b) According to the chemical equation for this reaction, 1 mol of ester (and 1 mol of water) are produced using 1 mol of acid and 1 mol of ethanol. Therefore, the initial concentration of ethanol was $2.67 + 4.48 = 7.15$ mol dm^{-3}. Similarly, the initial concentration of ethanoic acid was $4.48 + 1.93 = 6.41$ mol dm^{-3}.

(iii)

$$K_{c(T)} = \frac{k_f}{k_b}$$

Here,

$$3.89 = \frac{2.4 \times 10^{-4}}{k_b}$$

giving,

$$k_b = \frac{2.4 \times 10^{-4}}{3.89} = 6.2 \times 10^{-5}\ \text{mol}^{-1}\ \text{dm}^{-3}\ \text{s}^{-1}$$

15.4

$$K_{p(T)} = \frac{(p_{NOCl})^2}{(p_{NO})^2 \times p_{Cl_2}} = \frac{(0.32)^2}{(0.22)^2(0.11)} = 19\ \text{atm}^{-1}$$

15.5

$$K_{c(T)} = \frac{[Sn^{4+}(aq)][Fe^{2+}(aq)]^2}{[Sn^{2+}(aq)][Fe^{3+}(aq)]^2}$$

The equilibrium concentration of iron(II) ions will be double that of tin(IV), i.e. $[Fe^{2+}(aq)] = 0.080$ mol dm^{-3}. The following equilibrium concentrations are also known: $[Sn^{4+}(aq)] = 0.040$ mol dm^{-3}; $[Sn^{2+}(aq)] = 0.050$ mol dm^{-3}. We also know that $K_c = 1.0 \times 10^{10}$. Substituting into the equilibrium constant expression,

$$K_c = \frac{0.040 \times (0.080)^2}{0.050 \times [Fe^{3+}(aq)]^2}$$

giving

$$[Fe^{3+}(aq)] = \sqrt{\left(\frac{0.040 \times (0.080)^2}{0.050 \times 1.0 \times 10^{10}}\right)} = 7.2 \times 10^{-7}\ \text{mol dm}^{-3}$$

15.6 No, K_c is so small that virtually no phosgene has decomposed at this temperature.

15.7 We make two observations before tackling this problem:

(a) We take 'yield' to mean the gain in concentration of methanol at equilibrium.

(b) This reaction involves a reduction in the number of molecules, going from left to right.

This allows us to make the following conclusions:

(i) Reduces gain in $[CH_3OH]$ at equilibrium.

(ii) The equilibrium concentration of CH_3OH increases.

(iii) This decreases the equilibrium concentration of CH_3OH.

(iv) No effect.

(v) Equilibrium composition contains more product (CH_3OH) at lower temperatures.

K_c is only affected in (v) – K_c increases with decreasing temperature for an exothermic reaction.

15.8 To see the relationship, substitute the following equilibrium expressions into the equation

$$K_c(3) = K_c(1) \times K_c(2):$$

$$K_c(1) = \frac{[NO(g)]^2}{[O_2(g)][N_2(g)]};\quad K_c(2) = \frac{[NO_2(g)]^2}{[O_2(g)][NO(g)]^2};$$

$$K_c(3) = \frac{[NO_2(g)]^2}{[N_2(g)][O_2(g)]^2}$$

15.9

(i) Applying the equilibrium law:

$$K_{c(T)} = \frac{[O_2(g)]_{water}}{[O_2(g)]_{air}}$$

or

$$[O_2(g)]_{air} = \frac{[O_2(g)]_{water}}{K_{c(T)}}$$

By the ideal gas equation, the partial pressure of oxygen, p_{O_2}, in the air is proportional to its concentration:

$$p_{O_2} = [O_2(g)]_{air}RT$$

so that

$$p_{O_2} = \frac{[O_2(g)]_{water} \times RT}{K_{c(T)}}$$

$$\text{or } [O_2(g)]_{water} = \frac{p_{O_2} \times K_{c(T)}}{RT}$$

which may be expressed in the form of Henry's law (see Chapter 11):

$$c = K_H \times p$$

where $p = p_{O_2}$, $c = [O_2(g)]_{water}$ and $K_H = \frac{K_{c(T)}}{RT}$

(ii) Applying the equilibrium law;

$$K_{c(T)} = \frac{[I_2]_{CCl_4}}{[I_2]_{H_2O}}$$

leads directly to the distribution law. $K_{c(T)}$ is the distribution ratio, $K_{d(T)}$.

Unit 16

Exercises

16A

K_w increases as the temperature is increased.

16B

(i) $[H_3O^+(aq)] = 0.10$ mol dm^{-3}

$$[OH^-(aq)] = 1.0 \times 10^{-14}/[H_3O^+(aq)] = 1.0 \times 10^{-14}/0.10 = 1.0 \times 10^{-13}\ \text{mol dm}^{-3}$$

$\text{pH} = -\log(0.10) = 1.0$

(ii) $[H_3O^+(aq)] = 10^{-2.20} = 6.3 \times 10^{-3}$ mol dm^{-3}

16C

(i) The number of moles of $Ca(OH)_2$ is

$$\frac{0.10\ \text{g}}{(40 + 17 + 17)\ \text{g mol}^{-1}} = 0.0014\ \text{mol}$$

The concentration of calcium hydroxide is

$$\frac{0.0014}{0.200} = 0.0070\ \text{mol dm}^{-3}$$

One $Ca(OH)_2$ provides two hydroxide ions in solution. Thus, $[OH^-(aq)] = 0.014$ mol dm^{-3}.
At room temperature,
$1.0 \times 10^{-14} = [0.014]\,[H_3O^+(aq)]$.
Rearranging;

$$[H_3O^+(aq)] = 1.0 \times 10^{-14}/0.014 = 7.1 \times 10^{-13}\ \text{mol dm}^{-3}$$

$\text{pH} = -\log[\,H_3O^+(aq)\,] = 12.15$

(ii) We answer (b) first:

(b)

$[H_3O^+(aq)] = 10^{-11.10} = 7.9 \times 10^{-12}$ mol dm^{-3}

(a)

$$[OH^-(aq)] = 1.0 \times 10^{-14}/[H_3O^+(aq)] = 1.0 \times 10^{-14}/\,7.94 \times 10^{-12} = 1.3 \times 10^{-3}\ \text{mol dm}^{-3}$$

16D

$\text{pOH} = 14 - \text{pH} = 14 - 5.80 = 8.20$;
$[OH^-(aq)] = 10^{-8.20} = 6.3 \times 10^{-9}$ mol dm^{-3}

16E

$[H_3O^+(aq)]$	$[OH^-(aq)]$	pH	pOH
10^{-14}	1	14	0
10^{-13}	10^{-1}	13	1
10^{-12}	10^{-2}	12	2
10^{-11}	10^{-3}	11	3
10^{-10}	10^{-4}	10	4
10^{-9}	10^{-5}	9	5
10^{-8}	10^{-6}	8	6
10^{-7}	10^{-7}	7	7
10^{-6}	10^{-8}	6	8
10^{-5}	10^{-9}	5	9
10^{-4}	10^{-10}	4	10
10^{-3}	10^{-11}	3	11
10^{-2}	10^{-12}	2	12
10^{-1}	10^{-13}	1	13

Note the pattern: the pH decreases steadily and the pOH increases steadily. A high pOH value indicates a strongly acidic solution. A high pH indicates a strongly basic solution.

16F

(i)

$[H_3O^+(aq)] \approx \sqrt{(K_a \times C_A)}$

$C_A = 0.020$ mol dm^{-3}

$K_a = 4.9 \times 10^{-10}$ mol dm^{-3} (from Table 16.2)

$$[H_3O^+(aq)] \approx \sqrt{(4.9 \times 10^{-10} \times 0.020)} = 3.1 \times 10^{-6}\ \text{mol dm}^{-3}$$

The percentage of HCN molecules that are ionised is calculated as follows:

$$\%\ \text{ionized} = \frac{[H_3O^+(aq)]}{C_A} \times 100 = \frac{3.1 \times 10^{-6} \times 100}{0.020} = 0.016\%$$

$\text{pH} = -\log[H_3O^+(aq)] = -\log(3.1 \times 10^{-6}) = 5.5$

(ii) Mass of benzoic acid in solution $= 4000 \times 0.050/100 = 2.0$ g.
The formula of benzoic acid is obtained from Table 16.2; $M(C_6H_5COOH) = 122$ g mol^{-1}.
Number of moles of benzoic acid $= 2.0/122 = 1.6 \times 10^{-2}$.
Concentration of benzoic acid $= 1.6 \times 10^{-2} / 1.0 = 1.6 \times 10^{-2}$ mol dm^{-3}.

$[H_3O^+(aq) \approx \sqrt{(K_a \times C_A)}$

$C_A = 1.6 \times 10^{-2}$ mol dm^{-3};
$K_a = 6.5 \times 10^{-5}$ mol dm^{-3} (from Table 16.2)

$$[H_3O^+(aq)] \approx \sqrt{(6.5 \times 10^{-5} \times 1.60 \times 10^{-2})} = 1.0 \times 10^{-3}\ \text{mol dm}^{-3}$$

$\text{pH} = -\log[H_3O^+(aq)] = -\log(1.0 \times 10^{-3}) = 3.00$

The percentage of benzoic acid molecules ionised is $[1.0 \times 10^{-3}/1.6 \times 10^{-2}] \times 100 \approx 6\,\%$. Solution of the equation

$$K_{a(T)} = \frac{[H_3O^+(aq)]^2}{C_A - [H_3O^+(aq)]}$$

gives an accurate value of $[H_3O^+(aq)]$ as 9.85×10^{-4} mol dm^{-3}, close to the approximate value obtained above.

16G

(i) $CH_3COOH(aq) + H_2O(l) \rightleftharpoons CH_3COO^-(aq) + H_3O^+(aq)$
As more hydronium ions are mopped up by $OH^-(aq)$ ions in the neutralisation reaction, more ethanoic acid molecules ionize. The titration volumes are then identical with those of a strong acid.

(ii) The members of the first two pairs are strong acids and strong bases. The CH_3COOH–NaOH pair contains a weak acid and most of its molecules must first ionize before neutralization can take place. The ionization process is endothermic and reduces the overall amount of heat given out in the reaction between ethanoic acid and sodium hydroxide.

16H

(i) $K_b = 10^{-5.98} = 1.1 \times 10^{-6}$ mol dm^{-3}

(ii)

$$K_{b(T)} \approx \frac{[OH^-(aq)]^2}{C_B}$$

Rearranging;

$$[OH^-(aq)] = \sqrt{(K_{b(T)}C_B)} = \sqrt{(1.8 \times 10^{-5} \times 0.200)} = 1.9 \times 10^{-3}\ \text{mol dm}^{-3}$$

$$[H_3O^+(aq)] = \frac{1.0 \times 10^{-14}}{1.90 \times 10^{-3}} = 5.3 \times 10^{-12}\ \text{mol dm}^{-3}$$

$\text{pH} = -\log(5.3 \times 10^{-12}) = 11.28$

(iii) $C_6H_5NH_2(aq) + H_2O(l) \rightleftharpoons \underset{\text{conjugate acid}}{C_6H_5NH_3^+(aq)} + OH^-(aq)$

$[H_3O^+(aq)] = 10^{-7.80} = 1.6 \times 10^{-8}$ mol dm^{-3}

$$[OH^-(aq)] \approx \frac{1.0 \times 10^{-14}}{1.6 \times 10^{-8}} = 6.3 \times 10^{-7} \text{ mol dm}^{-3}$$

$$C_B = \text{initial conc} \approx \frac{[OH^-(aq)]^2}{K_{b(T)}}$$

$$\approx \frac{[6.3 \times 10^{-7}]^2}{4.3 \times 10^{-10}} = 9.2 \times 10^{-4} \text{ mol dm}^{-3}$$

16I

Salt	Formula	Parent acid and base	Acidic	Basic	Neutral
Iron(III) nitrate	$Fe(NO_3)_3$	HNO_3 (SA) $Fe(OH)_3$ (WB)	✓		
Calcium chloride	$CaCl_2$	HCl (SA) $Ca(OH)_2$ (SB)			✓
Sodium sulfate	Na_2SO_4	H_2SO_4 (SA) NaOH(SB)			✓
Ammonium benzoate	$C_6H_5CO_2NH_4$	$C_6H_5CO_2H$ (WA) $NH_3(aq)$ (WB)	✓		

($K_{a\,(\text{benzoic acid})} > K_{b\,(\text{ammonia})}$).

16J

Measure pH at different acid and salt concentrations. Plot $[H_3O^+(aq)]$ against the ratio C_A:C_S. The slope of the graph equals $K_{a(T)}$.

16K

(i) Number of moles of ethanoic acid = 0.10 × 25/1000 = 2.5×10^{-3}.

Number of mol of sodium ethanoate = 0.10 × 50/1000 = 5.0×10^{-3}.

Concentration of ethanoic acid = $(2.5 \times 10^{-3})/75/1000 = 3.3 \times 10^{-2}$ mol dm^{-3} = C_A.

Concentration of sodium ethanoate = $(5.0 \times 10^{-3})/75/1000 = 6.6 \times 10^{-2}$ mol dm^{-3} = C_S.

$C_A/C_S = 0.50$

$$[H_3O^+(aq)] \approx \frac{C_A \times K_{a(T)}}{C_s}$$

$= 0.50 \times 1.8 \times 10^{-5} = 9.0 \times 10^{-6}$ mol dm^{-3}

pH = −log (9.0×10^{-6}) = 5.05

(ii) Similar calculations show that if C_A:C_s = 2.0,

$$[H_3O^+(aq)] \approx \frac{C_A \times K_{a(T)}}{C_s}$$

$= 2.0 \times 1.8 \times 10^{-5} = 3.6 \times 10^{-5}$ mol dm^{-3}

pH = −log (3.6×10^{-5}) = 4.44

16L

$C_s/C_A = 1$

$$pH = pK_a + \log\frac{C_s}{C_A} = 4.18 - 0, \text{ so } pK_a = 4.18$$

$K_a = 10^{-4.18} = 6.6 \times 10^{-5}$ mol dm^{-3}. This suggests that the acid is benzoic acid.

16M

pH	Phenolphthalein	Bromocresol green
2	colourless	yellow
3.7	colourless	yellow
5	colourless	blue/yellow
6	colourless	blue
8	colourless	blue
11	pink	blue

16N

(i) Alizarin yellow and (to a lesser extent) methyl orange.
(ii) Pairs (a) and (b); (c) involves a weak base.

16O

(i) Bromocresol green changes over the pH range 4.0–5.6. This is too soon, and its use in the titration of a weak acid and a strong base will lead to an underestimation of the equivalence point.

(ii) For the mixture in question, 12.50 cm^3 represents half-equivalence, i.e. the volume of NaOH required to neutralize half the ethanoic acid. At half-equivalence the concentration of ethanoic acid equals the concentration of sodium ethanoate, i.e. $C_s = C_A$ and log C_S/C_A = 0. The Henderson–Hasselbalch equation then simplifies to pH of solution = pK_a. This gives us an approximate method (in addition to the method outlined in Exercise 16J) of finding the acidity constant of a weak acid. For example, Fig. 16.3(b) shows that after the addition of 12.5 cm^3 of NaOH, pH ≈ 4.7 ≈ pK_a(CH_3CO_2H). Therefore,

$K_a \approx 2 \times 10^{-5}$ mol dm^{-3}.

16P

(i) From Table 16.2, $K_a = 4.3 \times 10^{-7}$ mol dm^{-3}. Applying the equation

$$K_{a(T)} \approx \frac{[H_3O^+(aq)]^2}{[CO_2(aq)]}$$

(you may recognize this as equation (16.5).

$[H_3O^+(aq)]^2 \approx 4.3 \times 10^{-7} \times 1.2 \times 10^{-5}$
$= 5.2 \times 10^{-12}$ mol^2 dm^{-6}

$[H_3O^+(aq)] \approx \sqrt{}(5.2 \times 10^{-12}) \approx 2.3 \times 10^{-6}$

pH = −log(2.3×10^{-6}) = 5.62

(ii) Salt occupies some of the space that would otherwise be taken up by dissolved carbon dioxide. Therefore, salty water is less acidic than deionized water.

Revision questions

16.1 (i) F
(ii) F (A pH of 6 would cause it to be yellow),
(iii) T, **(iv)** T, **(v)** T.

16.2 (i) HNO_3 provides 0.080 mol dm^{-3} of $H^+(aq)$.

pH = −log (0.080) = 1.10

(ii) KOH provides 0.080 mol dm^{-3} of $OH^-(aq)$ in solution.

$$[H^+(aq)] = \frac{1.0 \times 10^{-14}}{0.080} = 1.3 \times 10^{-13} \text{ mol dm}^{-3}$$

pH = −log$[1.3 \times 10^{-13}]$ = 12.89

16.3 (i) Table 16.1 shows that $K_w = 51.3 \times 10^{-14}$ mol^2 dm^{-6} at 100 °C. In pure water $[OH^-(aq)] = [H_3O^+(aq)]$; therefore $51.3 \times 10^{-14} = [H_3O^+(aq)]^2$ or $[H_3O^+(aq)] = 7.16 \times 10^{-7}$ mol dm^{-3}.

pH (100 °C) = −log$[7.16 \times 10^{-7}]$ = 6.145

(ii) Number of moles of NaOH = [11.0/1000) × 0.040 = 4.4×10^{-4}.

Number of moles of HCl = (9.0/1000) × 0.040 = 3.6×10^{-4}.

Since 1 HCl ≡ 1 NaOH, after neutralization we have 0.80×10^{-4} mol of $OH^-(aq)$ spread out over 20 cm^3 of solution. This is a concentration of

$(0.80 \times 10^{-4})/20/1000 = 4.0 \times 10^{-3}$ mol dm^{-3} of $[OH^-(aq)]$

At 25 °C,

$$[H^+(aq)] = \frac{1.0 \times 10^{-14}}{4.0 \times 10^{-3}} = 2.5 \times 10^{-12} \text{ mol dm}^{-3}$$

pH = −log$[2.5 \times 10^{-12}]$ = 11.60

16.4 (i)

$C_{17}H_{19}O_3N(aq) + H_2O(l) \rightleftharpoons C_{17}H_{19}O_3NH^+(aq) + OH^-(aq)$ (A)

Since:

$K_b = 1.6 \times 10^{-6}$ mol dm^{-3};
$M(C_{17}H_{19}O_3N)$ = 285 g mol^{-1};
1.0 mg = 0.0010 g

Number of mols of morphine = 0.0010 g / 285 g mol^{-1} = 3.5×10^{-6} mol.

Concentration of morphine = $C_B = 3.5 \times 10^{-6}$ mol per dm^3

We now use the equation

$$K_{b(T)} \approx \frac{[OH^-(aq)]^2}{C_B}$$

Rearranging and substituting values,

$[OH^-(aq)] \approx \sqrt{}(K_{b(T)} \times C_B)$
$= \sqrt{}(3.5 \times 10^{-6} \times 1.6 \times 10^{-6})$
$= 2.4 \times 10^{-6}$ mol dm^{-3}

pOH = −log (2.4×10^{-6}) = 5.62

Also, since pH + pOH = 14,
pH = 14 − 5.62 = 8.38

(ii) The $OH^-(aq)$ produced in reaction (A) reacts with the $H_3O^+(aq)$ from HCl(aq):

$H_3O^+(aq) + OH^-(aq) \rightarrow 2H_2O(l)$ (B)

Addition of equations (A) and (B) gives the overall reaction

$C_{17}H_{19}O_3N(aq) + H_3O^+(aq) \rightleftharpoons C_{17}H_{19}O_3NH^+(aq) + H_2O(l)$

The number of mols of morphine in 100 cm^3 of solution is 3.5×10^{-7}. This requires an equal number of moles of HCl(aq) for neutralization. We calculate the volume of 0.0010 mol dm^{-3} HCl(aq) required as follows:

$$\text{volume of HCl(aq)} = \frac{\text{moles}}{\text{concentration}} = \frac{3.5 \times 10^{-7}}{0.0010}$$

$= 3.5 \times 10^{-4}$ dm^3 or 0.35 cm^3

16.5 The ionization of phenol is represented as

$C_6H_5OH(aq) + H_2O(l) \rightleftharpoons C_6H_5O^-(aq) + H_3O^+(aq)$

using $[H_3O^+(aq)] \approx \sqrt{}(K_a \times C_A)$, $C_A = 5.00 \times 10^{-3}$ mol dm^{-3} and $K_a = 1.3 \times 10^{-10}$ mol dm^{-3} (from Table 16.2),

$[H_3O^+(aq)] \approx \sqrt{}(1.3 \times 10^{-10} \times 5.00 \times 10^{-3})$
$\approx 8.1 \times 10^{-7}$ mol dm^{-3}

pH = −log$[H_3O^+(aq)]$ = −log(8.06×10^{-7}) = 6.094

16.6 The acidity constant for $NH_4^+(aq)$ is the equilibrium constant that applies to the equation

$NH_4^+(aq) + H_2O(l) \rightleftharpoons NH_3(aq) + H_3O^+(aq)$

that is,

$$K_a(NH_4^+(aq)) = \frac{[NH_3(aq)][H_3O^+(aq)]}{[NH_4^+(aq)]}$$

For the equation

$NH_3(aq) + H_2O(l) \rightleftharpoons NH_4^+(aq) + OH^-(aq)$

the following is true:

$$K_b(NH_3(aq)) = \frac{[NH_4^+(aq)][OH^-(aq)]}{[NH_3(aq)]}$$

We also know that

$K_w = [H_3O^+(aq)][OH^-(aq)]$

The relationship

$K_a \times K_b = K_w$ (A)

is now proved by substitution and cancellation of terms:

$$\frac{[NH_3(aq)][H_3O^+(aq)]}{[NH_4^+(aq)]} \times \frac{[NH_4^+(aq)][OH^-(aq)]}{[NH_3(aq)]} = [H_3O^+(aq)][OH^-(aq)]$$

$K_a \quad \times \quad K_b \quad = \quad K_w$

Equation (A) applies for all pairs of conjugate acids and bases. Here,

$K_a(NH_4^+(aq)) = K_w/K_b(NH_3(aq))$
$= 1.0 \times 10^{-14}/(1.8 \times 10^{-5})$
$= 5.6 \times 10^{-10}$ mol dm^{-3}

16.7 (i) Assuming that the dissolved CO_2 concentration equals its initial concentration, and applying the equilibrium law,

$$K_a \approx \frac{[HCO_3^-(aq)][H_3O^+(aq)]}{[CO_2(aq)]}$$

If the only acidity is due to dissolved carbon dioxide, $[HCO_3^-(aq)] = [H_3O^+(aq)]$, giving

$$K_a \approx \frac{[H_3O^+(aq)]^2}{[CO_2(aq)]}$$

(ii) $[H_3O^+(aq)] \approx \sqrt{(4.3 \times 10^{-7} \times 4.0 \times 10^{-2})}$
$= 1.3 \times 10^{-4}$ mol dm^{-3}
(corresponding to 0.3% ionization).
pH $= -\log(1.3 \times 10^{-4}) = 3.89$

16.8 (i) WB–SA, therefore acidic **(ii)** WA–WB, $K_a = K_b$ so approximately neutral **(iii)** WB–SA, therefore acidic **(iv)** SA–SB, therefore neutral **(v)** SB–WA, so basic.

16.9 $M(C_6H_5COOH) = 122$ g mol^{-1},
$M(C_6H_5COONa) = 144$ g mol^{-1}
Number of moles of $C_6H_5COOH = 0.100/122$
$= 8.20 \times 10^{-4}$.
Number of moles of $C_6H_5COONa = 0.080/144$
$= 5.6 \times 10^{-4}$.
Concentration of $(C_6H_5COOH) = 8.2 \times 10^{-4} / 0.100$
$= 8.2 \times 10^{-3}$ mol dm^{-3}.
Concentration of $(C_6H_5COONa) = 5.6 \times 10^{-4} / 0.100$
$= 5.6 \times 10^{-3}$ mol dm^{-3}.

$$[H_3O^+(aq)] \approx \frac{C_A \times K_{a(T)}}{C_s} \approx \frac{8.20 \times 10^{-3} \times 6.5 \times 10^{-5}}{5.6 \times 10^{-3}}$$
$= 9.5 \times 10^{-5}$ mol dm^{-3}

pH $= -\log(9.5 \times 10^{-5}) = 4.02$

$C_6H_5COOH(aq) + H_2O(l) \rightleftharpoons C_6H_5COO^-(aq) + H_3O^+(aq)$

Explanation of buffer action is as follows. (a) Addition of $H_3O^+(aq)$ causes association of the $C_6H_5COO^-(aq)$ and $H_3O^+(aq)$, so restoring $[H_3O^+(aq)]$. (b) Addition of $OH^-(aq)$ neutralizes $H_3O^+(aq)$, and more benzoic acid ionizes to compensate.

16.10 Choice of indicators from Table 16.4 for a weak base–strong acid titration are bromocresol green and methyl red. The remaining indicators begin to change colour before or after the pH at the equivalence point (≈pH 4–8).

16.11 The ingested hydrogencarbonate increases the buffer capacity of the athlete.

Unit 17

Exercises

17A

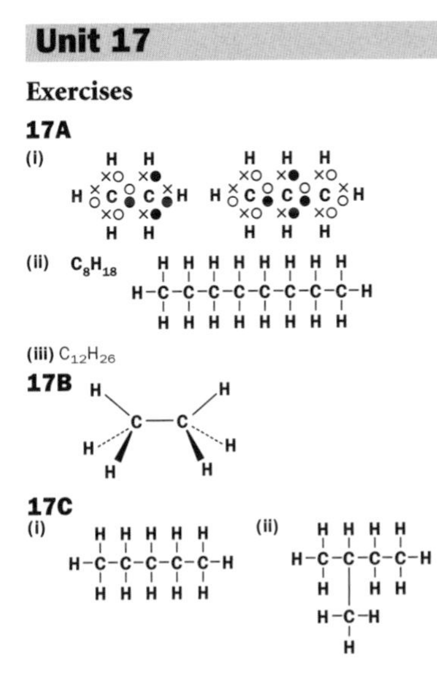

(i)

(ii) C_8H_{18}

(iii) $C_{12}H_{26}$

17B

17C

(i)

(ii)

(iii)

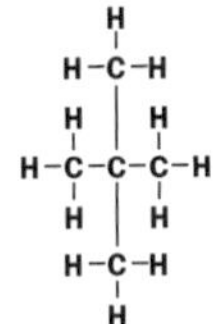

17D

(i) 2-methylpropane
(ii) 4-ethylheptane
(iii) 2,2-dimethylpropane
(iv) 4-ethyl-3,5-dimethyloctane

17E

(i)

(ii)

(iii)

(iv)

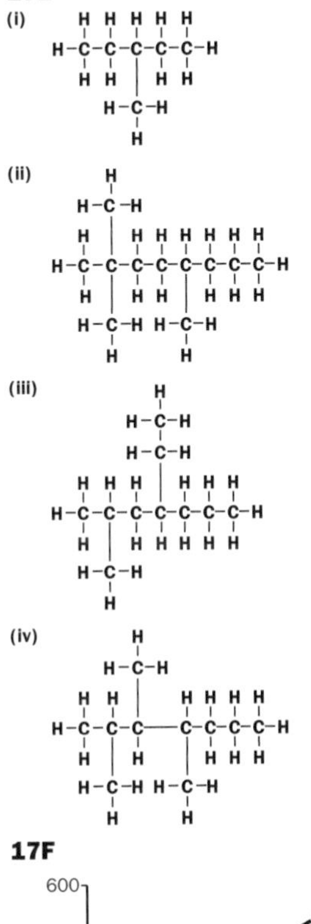

17F

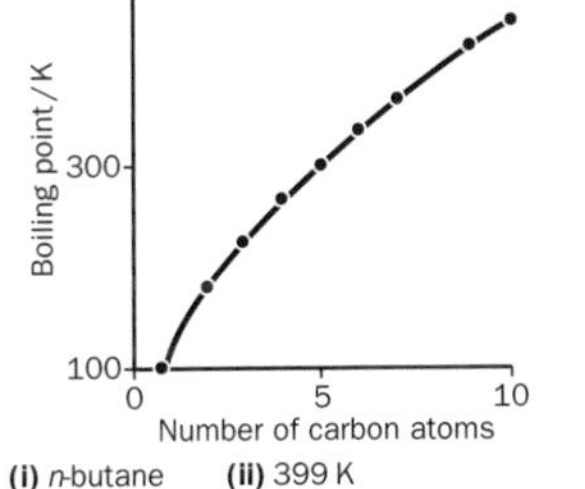

(i) *n*-butane **(ii)** 399 K

17G

(i) C_nH_{2n}
(ii) The ring is far less 'strained' – the bond angle is larger than in cyclopropane.
(iii) Hexane / Cyclohexane

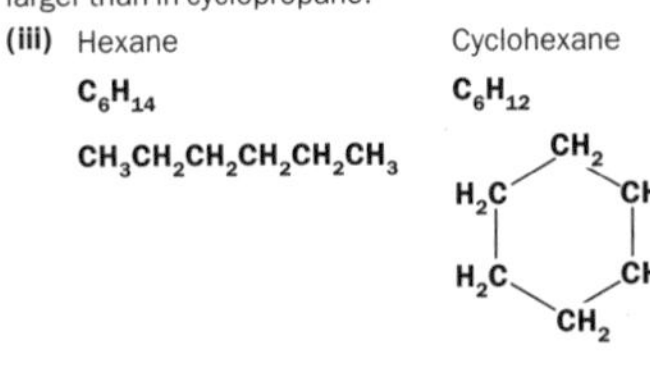

17H

(i) 670, 660, 660, 620, 660; yes, usually around 660 kJ mol^{-1}
(ii) 654 kJ mol^{-1}
(iii) $-CH_2$
(iv) -4814 kJ mol^{-1}.

17I

(i) See Box 17.6 **(ii)** No, unlike ethane.

17J

(i) hex-1-ene
(ii) but-2-ene
(iii) 4,4-dimethylpent-2-ene
(iv) 3-bromoprop-1-ene

17K

(i) $CH_3CH_2CH{=}CHCH_2CH_3$ (ii) $CH_3C(CH_3){=}CH_2$

(iii) $CH_3CH_2CH_2C(CH_3){=}C(CH_3)CH_3$ (iv) $CH_2ClC(CH_3){=}CH_2$

(v) $CH_3CH_2CH_2CHBrCH{=}CH_2$

17L

(i) 3 = $Br_2C{=}CH_2$ (Br, Br on one carbon; H, H on the other)

(ii) geometric **(iii)** structural
(iv) 1 is *trans*-1,2-dibromoethene, **2** is *cis*-1,2-dibromoethene, **3** is 1,1-dibromoethene.

17M

(i) $CH_2{=}CH_2 + HBr \rightarrow CH_3{-}CH_2Br$ (bromoethane)
(ii) $CH_2{=}CH_2 + H_2SO_4 \rightarrow CH_3{-}CH_2{-}OSO_3H$ (ethyl hydrogensulfate)
(iii) $CH_3{-}CH{=}CH_2 + Br_2 \rightarrow CH_3{-}CHBr{-}CH_2Br$ (1,2-dibromopropane)
(iv) $CH_3{-}CHOH{-}CH_2OH$
(v) $n(CH_2{=}CHCN) \rightarrow \cdots CH_2{-}CHCN{-}CH_2{-}CHCN{-}CH_2{-}CHCN{-}CH_2{-}CHCN{-}CH_2{-}CHCN\cdots$.

17N

(i) $CH_3{-}CH_2C{\equiv}CH$
(ii) propene, $CH_3{-}CH{=}CH_2$ and propane, $CH_3CH_2CH_3$
(iii) $CHBr_2{-}CHBr_2$ 1,1,2,2-tetrabromoethane.

17O

(i)

$$\text{mass of C present} = 21.99 \times \frac{12}{44} = 6.00 \text{ mg}$$

$$\text{percentage C} = \frac{6}{7.02} \times 100 = 85.5\%$$

$$\text{mass of H present} = 8.95 \times \frac{2}{18} = 1 \text{ mg}$$

$$\text{percentage H} = \frac{1}{7.02} \times 100 = 14.3\%$$

Empirical formula:

	C	H
no of moles	85.5/12	14.3/1
ratio	7.1 : 14.3 1 : 2	

Therefore, the empirical formula is CH_2.
(ii) molecular formula X = C_3H_6, structure $CH_3{-}CH{=}CH_2$
Y is $CH_3CH_2CH_3$
Z is $CH_3CHBrCH_2Br$

17P

(i) benzene + $3H_2$ ⟶ cyclohexane

(ii) $3 \times -119 = -357$ kJ mol^{-1}
(iii) Benzene has a lower enthalpy of formation than that suggested by the Kekulé structure.

17Q

(i)

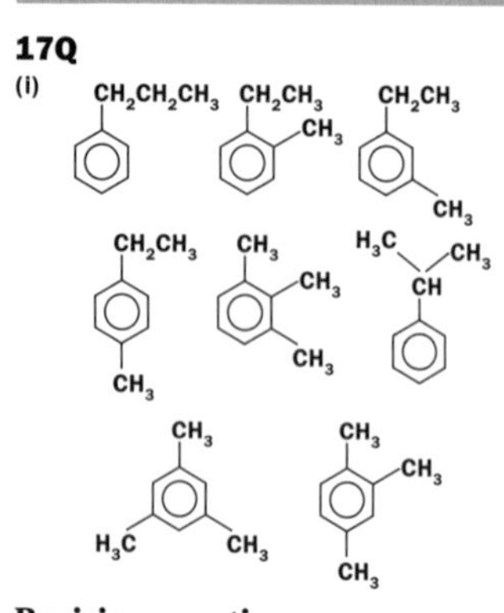

Revision questions

17.1 C_6H_{14}, $CH_3CH_2CH_2CH_2CH_2CH_3$, $CH_3(CH_2)_4CH_3$

17.2

$CH_3CH_2CH_2CH_2CH_2CH_3$ *n*-hexane

$CH_3CH_2CH_2CH(CH_3)CH_3$ 2-methylpentane

$CH_3CH_2CH(CH_3)CH_2CH_3$ 3-methylpentane

$CH_3CH(CH_3)CH(CH_3)CH_3$ 2,3-dimethylbutane

$CH_3C(CH_3)_2CH_2CH_3$ 2,2-dimethylbutane

17.3 (i) pent-2-ene **(ii)** 2,5-dimethylhex-2-ene

17.4 (i) $CH_2{=}CHCH_2CH_2CH_3$ **(ii)** $CH_3CH{=}CHCH_2CHClCH_3$

(iii) $CH_3CH{=}CHC(CH_3)_2CH_3$ **(iv)** $CH_2{=}CHCH{=}CH_2$

(v)

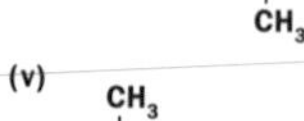

17.5 (ii) and **(iii)**

17.6 (i) $CH_2Br{-}CHBrCH_2CH_2CH_3$ (1,2-dibromopentane)

(ii) $CH_2OHCH(OH)CH_2CH_3$ (butan-1,2-diol)

(iii) $CH_3CH_2CH_2CH_3$ (butane)

17.7

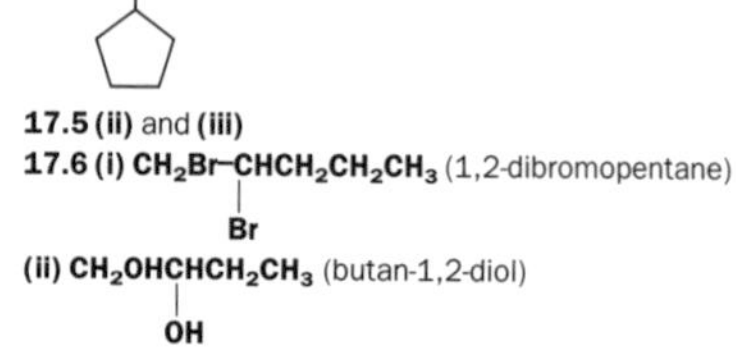

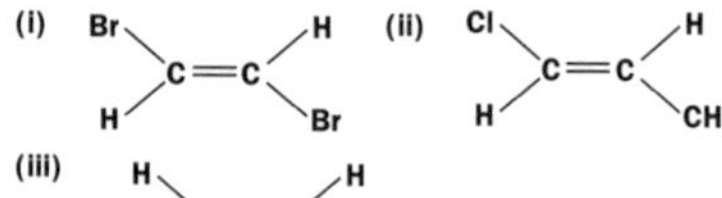

17.8 tetrafluoroethene: $CF_2{=}CF_2$

PTFE:

$\cdots CF_2{-}CF_2{-}CF_2{-}CF_2{-}CF_2{-}CF_2{-}CF_2{-}CF_2{-}CF_2{-}CF_2\cdots$

a non-stick coating for frying pans.

17.9 (i)

	C	H
no of moles	$\frac{88.9}{12}$	$\frac{11.1}{1}$
ratio	7.4 : 11.1 2 : 3	

The empirical formula is C_2H_3.

(ii) C_4H_6

(iii) $CH_3CH_2C{\equiv}CH$, but-1-yne

$CH_3C{\equiv}CCH_3$, but-2-yne

17.10

17.11 We used bond energies to find the standard enthalpy change for the reaction:

$6C(s) + 3H_2(g) \rightarrow C_6H_6(l) \quad \Delta H_1^{\ominus} = ?$ (equation 1)

We start by calculating the standard enthalpy change for the reaction in which gaseous carbon makes gaseous benzene:

$6C(g) + 3H_2(g) \rightarrow C_6H_6(g) \quad \Delta H_2^{\ominus} = ?$ (equation 2)

From Table 13.4,

$\Delta H^{\ominus}_{H-H} = 436$ kJ mol^{-1}
$\Delta H^{\ominus}_{C-H} = 412$ kJ mol^{-1}
$\Delta H^{\ominus}_{C=C} = 612$ kJ mol^{-1}
$\Delta H^{\ominus}_{C-C} = 348$ kJ mol^{-1}

Assuming that benzene has the Kelukè molecular structure, the heat absorbed in breaking bonds in reactant molecules, i.e. three H–H bonds is

heat absorbed $= 3 \times 436 = 1308$ kJ

The heat released when bonds in product molecules are assembled, which involves making three C=C bonds, three C–C bonds and six C–H bonds, is

heat given out $= (3 \times 612) + (6 \times 412) + (3 \times 348) = 5352$ kJ

The overall change in enthalpy is

$\Delta H_2^{\ominus} = -5352 + 1308 = -4044$ kJ mol^{-1}

The difference between equations (2) and (1) is that we have to allow for the production of gaseous carbon atoms and formation of gaseous benzene being endothermic:

$6C(s) \rightarrow 6C(g)\ \Delta H_3^{\ominus}(298\text{ K}) = (717 \times 6) = 4302$ kJ mol^{-1}

$C_6H_6(l) \rightarrow C_6H_6(g)\ \ \Delta H_4^{\ominus}(298\text{ K}) = 31$ kJ mol^{-1}

Use of the enthalpy diagram confirms that:

$\Delta H_1^{\ominus} + \Delta H_4^{\ominus} - \Delta H_2^{\ominus} = \Delta H_3^{\ominus}$ or

$\Delta H_1^{\ominus} = \Delta H_3^{\ominus} - \Delta H_4^{\ominus} + \Delta H_2^{\ominus}$

$\Delta H_1^{\ominus} = 4302 - 31 + (-4044) = +227$ kJ mol^{-1}

Therefore, the difference between the theoretical and experimental values is $(227 - 50) = 177$ kJ mol^{-1}. We explain this difference by assuming that the Kelukè structure is not the true structure of benzene – benzene has a lower enthalpy than this structure suggests.

Unit 18

Exercises

18A

(i) (a) CH_3CH_2Cl (b) $CH_3CHBrCH_3$

(c) $CH_2ClCHClCHFCH_3$

(d)

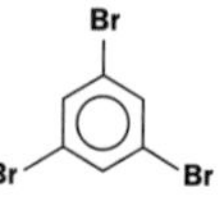

(ii) (a) 1-bromo-1-chloroethane (b) 3-iodopentane

(c) trichlorofluoromethane

(d) 3-chloromethylbenzene (or 3-chlorotoluene)

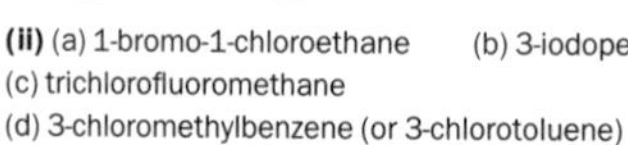

18B

(i) I: pentan-1-ol

$CH_3CH_2CH_2CH_2CH_2OH$

II: pentan-2-ol

$CH_3CH_2CH_2CHOHCH_3$

III: pentan-3-ol

$CH_3CH_2CHOHCH_2CH_3$

IV: 2,2-dimethylpropan-1-ol

$H_3C{-}C(CH_2OH)(CH_3){-}CH_3$

V: 2-methylbutan-2-ol

$CH_3CH_2C(CH_3)OHCH_3$

VI: 3-methylbutan-2-ol

$CH_3CHOHCH(CH_3)CH_3$

VII: 2-methylbutan-1-ol

$CH_3CH_2CH(CH_3)CH_2OH$

VIII: 3-methylbutan-1-ol

$CH_3CH(CH_3)CH_2CH_2OH$

(ii) $CH_2OHCHOHCH_2OH$

18C

(i) approx: 64.5, 58, 60

(ii) propan-1-ol is a liquid, the others are gases

(iii) hydrogen bonding between the alcohol molecules results in stronger intermolecular attractions than in the other compounds.

18D

I, IV, VII, VIII are primary; II, III, VI are secondary; V is tertiary

18E

(i) aldehyde **(ii)** ketone **(iii)** aldehyde
(iv) ketone **(v)** ketone.

18F

(i) 'eth' is a prefix that refers to two carbon atoms. The simplest ketone, propanone, contains three.

(ii)

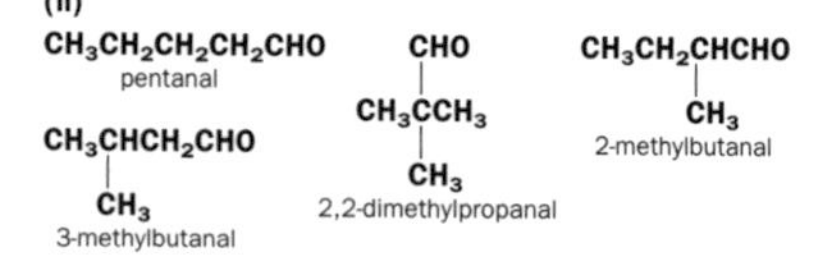

(iii)

$CH_3CH_2CH_2CHO$ butanal

$CH_3CH(CH_3)CHO$ 2-methylpropanal

$CH_3COCH_2CH_3$ butanone

18G

(i) butanone $CH_3COCH_2CH_3$

(ii) propanoic acid C_2H_5COOH

(iii) octan-2-one $C_6H_{13}COCH_3$.

18H

(i) $CH_3CH_3CH_2COOH$

(ii)

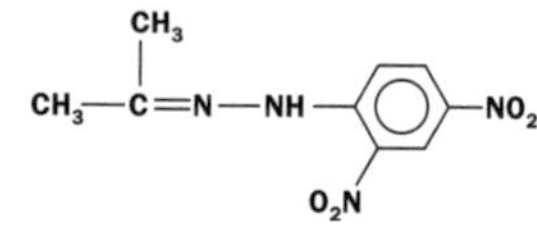

(iii) $CH_3—C(H)=N—NH_2$ **(iv)** CH_3CH_2COOH

18I

(i) butanoic acid **(ii)** heptanoic acid
(iii) butane-1,4-dioic acid.

18J

(i) CH_3COOH, $CH_2BrCOOH$, $CH_2ClCOOH$, $CHCl_2COOH$, CCl_3COOH, CF_3COOH
(chlorine is more electronegative than bromine).

18K

(i) (a) methyl ethanoate CH_3COOCH_3
(b) ethyl octanoate $C_7H_{15}COOCH_2CH_3$
(c) methyl methanoate $HCOOCH_3$
(d) diethyl oxalate (diethyl ethanedioate) $(COOCH_2CH_3)_2$.
(ii) (a) propanoic acid, methanol
(b) methanoic acid, ethanol
(c) ethanoic acid, propan-1-ol
(d) chloroethanoic acid, methanol.

18L

NH_3, CH_3NH_2, CH_3NHCH_3.

18M

(i) Middle C has an asterisk.
(ii) 2 methylbutan-1-ol, 2-bromo-1-chlorobutane.

18N

(i)

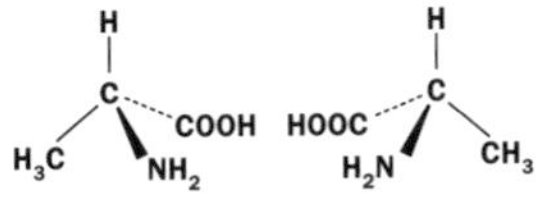

(ii) $CH_3CH(NH_3^+)COO^-$

18O

(i)

$H_2NCH_2C(=O)—NHCH(CH_3)COOH$

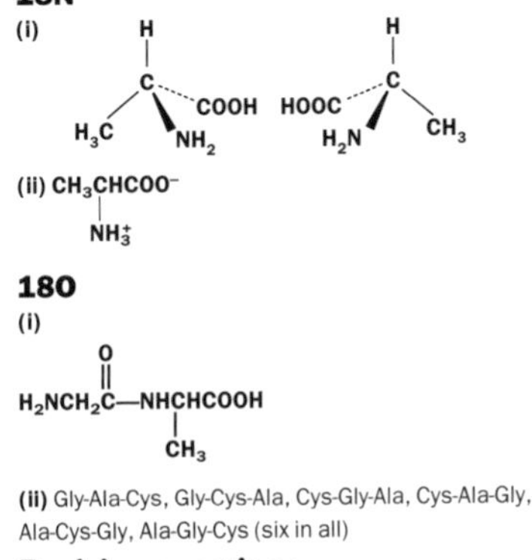

(ii) Gly-Ala-Cys, Gly-Cys-Ala, Cys-Gly-Ala, Cys-Ala-Gly, Ala-Cys-Gly, Ala-Gly-Cys (six in all)

Revision questions

18.1 (i) $CH_3CH_2CH_2CH_2CHOHCH_3$
(ii) $CH_3C(CH_3)OHCH_3$
(iii) $CH_2OHCH_2CHOHCH_3$
(iv)

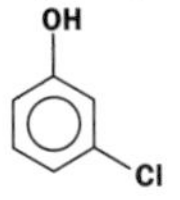

18.2 ethanol, water, phenol, ethanoic acid.
18.3
(i) $CH_3CH_2CH_2CH(CH_3)CH_2CHO$
Notice the aldehyde group contains the carbon atom numbered 1.
(ii) $CH_3COCH_2COCH_3$ **(iii)**

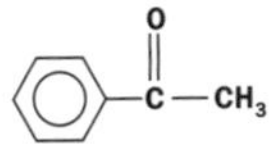

(iv) $CH_3CH(CH_3)COOH$ **(v)** $COOH—CH_2—CH_2—COOH$

18.4 oxalic acid
18.5 (i) $CH_3CH_2COOCH_2CH_3$ **(ii)** $CH_3CH_2CH_2OH$
(iii) $CH_3(CH_2)_3COOH$ **(iv)** CH_3COOH

18.6 (i) $HCOOCH_2CH_2CH_3$ **(ii)** $HCONH_2$
(iii) $HCOCl$.
18.7
(i) $CH_2OHC^*H(NH_2)COOH$
(ii) Yes, it has a carbon atom attached to four different groups, marked '*'.
(iii) $HO_2CCH(NH_2)COOH$
(iv) $CH_2OHCH(NH_2)C(=O)—NHCH(CH_2OH)COOH$
18.8

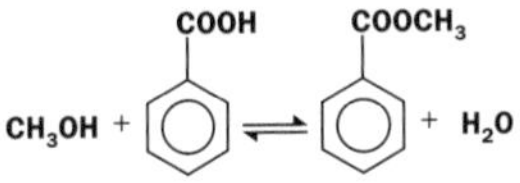

18.9

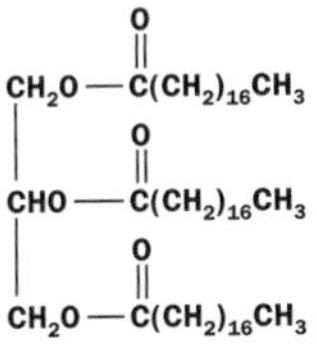

18.10 (i) ester **(ii)** amino group
(iii) ester and carboxylic acid **(iv)** amide
(v) phenol, alcohol, amino **(vi)** halogen.
18.11 (i) X would form an orange precipitate with 2,4-dinitrophenylhydrazine.
(ii) X would reduce Tollens' reagent or Fehling's solution.
(iii) CH_3CH_2CHO; to confirm this, check the melting point of the DNP derivative against that in a data book.

Unit 19

Exercises

19A

Heat the sample of sea water to dryness, so that all the water is evaporated off. The solid left would not be pure sodium chloride, because sea water contains other dissolved salts such as magnesium chloride and potassium chloride.

19B

(i) A separating funnel.
(ii) Add mixture to water and stir → separate sand off by gravity filtration → wash sand well and dry in an oven → filtrate contains dissolved sugar → evaporate water from filtrate to obtain concentrated solution (solid sugar decomposes when heated, so this is 'safer' than evaporating solution to dryness) → allow concentrated solution to cool and sugar crystals should separate out → filter solution to remove sugar crystals.
(iii) Dissolve the mixture in water at a temperature just less than 100 °C → cool solution to 20 °C → crystalline potassium chlorate separates out, which can be removed by filtering (potassium chloride remains in solution).
(iv) Fractional distillation will result in a mixture much more rich in the volatile component (the one that boils at 60 °C) collecting in the receiving container.
(v) Simple distillation.

19C

(i) Let the amount of X in the organic layer be x g, then

$$K_d = 7 = \frac{x/200}{(1 - x)/100}$$

or $x = 0.93$ g.
(ii) First extraction: let the amount of X in the organic layer be z g, then

$$K_d = 7 = \frac{z/100}{(1 - z)/100}$$

Therefore $z = 0.875$ g, i.e. 0.125 g left in the aqueous layer.
Second extraction: let amount of X in organic layer be y g, then

$$K_d = 7 = \frac{y/100}{(0.125 - y)/100}$$

Therefore, $y = 0.109$ g and the total amount extracted is

$0.875 + 0.109 = 0.98$ g

that is, 0.05 g more than for one extraction.

19D

(i)

$$R_f \text{ for A} = \frac{y + z}{x + y + z}$$

(ii)

$$R_f \text{ for B} = \frac{z}{x + y + z}$$

Revision questions

19.1 (i) evaporation
(ii) simple distillation
(iii) fractional distillation
(iv) fractional distillation
(v) column chromatography
(vi) separation using a separating funnel
(vii) stir mixture in water → filter to remove chalk → evaporate off water from filtrate to give salt.
19.2 Paper chromatography using propanone as solvent. If substances are not all coloured, use a suitable method to develop the spots.
19.3 Sublimation – ammonium chloride 'sublimes', potassium chloride does not.
19.4 (i) paper chromatography (or TLC)
(ii) column chromatography. Ethanol can be used as solvent in both techniques.
19.5 Paper chromatography → spot urine and spots of amino acids it is thought to contain on the paper → run chromatogram using a suitable solvent (water) → develop colourless spots by spraying with ninhydrin.
19.6 Gas chromatography → compare retention times of components of mixture with the retention times of known alcohols (or use GC–MS!).
19.7 Paper chromatography → compare pattern of spots of suspect ink with a sample of ink obtained from the forged cheque.
19.8 Let mass extracted be x g, then

$$3.5 = \frac{x/50}{(0.5 - x)/150}$$

i.e. $x = 0.27$ g.
19.9
(i) Area ABC = $1/2 \times 16 \times 40 = 320\ mm^2$
Area PQR = $1/2 \times 20 \times 32 = 320\ mm^2$
(ii) The same area under each of the peaks suggests there are equal amounts of each constituent in the mixture.
(iii) You could cut out each peak from the chart and weigh them, then compare masses.
19.10 Dissolve 'dyes' from smarties by stirring them in a *little* water. Use paper chromatography to see if more than one spot occurs.

Unit 20

Exercises

20A

Rearranging,

$$\nu = \frac{c}{\lambda} \text{ gives } \lambda = \frac{c}{\nu}$$

$\nu = 98$ MHz $= 98\,000\,000$ Hz, therefore

$$\lambda = \frac{3.00 \times 10^8}{9.8 \times 10^7} = 3.1 \text{ m}$$

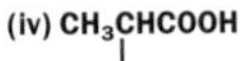

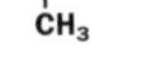

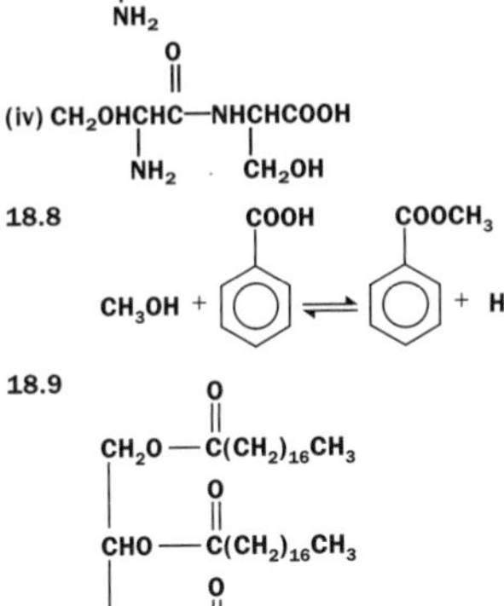

20B

(i)

$$E = \frac{3.00 \times 10^8 \times 3.99 \times 10^{-13}}{700 \times 10^{-9}} = 171 \text{ kJ mol}^{-1}$$

(ii)

$$E = \frac{3.00 \times 10^8 \times 3.99 \times 10^{-13}}{560 \times 10^{-9}} = 214 \text{ kJ mol}^{-1}$$

20C

$\Delta E = hN_A\nu$, therefore

$$\nu = \frac{\Delta E}{hN_A} = 100 \text{ kJ mol}^{-1}/3.99 \times 10^{-13} \text{ kJ s mol}^{-1}$$
$$= 2.51 \times 10^{14} \text{ Hz (i.e. s}^{-1})$$

20D

(i) If 10% of the light is absorbed, 90% of the light is transmitted, i.e.

$$\left(\frac{I_0}{I}\right) = \frac{100}{90} = 1.11$$

$$\text{absorbance} = A = \log\left(\frac{I_0}{I}\right) = \log(1.11) = 0.0453$$

(ii)

$$\left(\frac{I_0}{I}\right) = 10^{3.00} = 1.0 \times 10^3 \quad \therefore \left(\frac{I}{I_0}\right) = 0.0010$$

therefore, the percentage of light transmitted is

$$\frac{100 \times I}{I_0} = 100 \times 0.0010 = 0.10\%$$

20E

The higher temperatures mean that many hydrogen atoms are present in the $n = 2$ level.

20F

The energy gap between the $n = 4$ and $n = 1$ level is 1228 kJ mol^{-1}. Using the equation $\Delta E = hN_A\nu$

$$\nu = \frac{\Delta E}{hN_A} = 1228/3.99 \times 10^{-13} = 3.08 \times 10^{15} \text{ Hz}$$

20G

The rate at which the excited states is produced can be controlled like any other chemical reaction.

(i) Lowering the temperature of the reaction mixture will slow down the reaction and decrease the intensity of the emitted light.

(ii) Increasing temperature increases the emission intensity.

20H

(i) red

(ii) blue green

(iii) colourless (absorptions are outside the visible region).

20I

Only absorptions above 400 nm are relevant. The absorptions between 400 nm and 520 nm peak at about 430 nm, suggesting that the solution would be yellow (this is the observed colour).

20J

(i) propionic acid and ethanal

(ii) propionic acid and butanol

(iii) all three.

20K

Its simplest formula is $C_4H_{11}N$; infrared suggests $-NH_2$ present; $m(C_4H_{11}N) = 73$ u (as confirmed by mass spectrum). Possible structures include $(CH_3)_2CHCH_2NH_2$ and $CH_3(CH_2)_3NH_2$.

20L

(i) 3 (protons from CH_3, CH_2 and CHO)

(ii) 2 (protons from CH_3 and CH).

20M

(i)

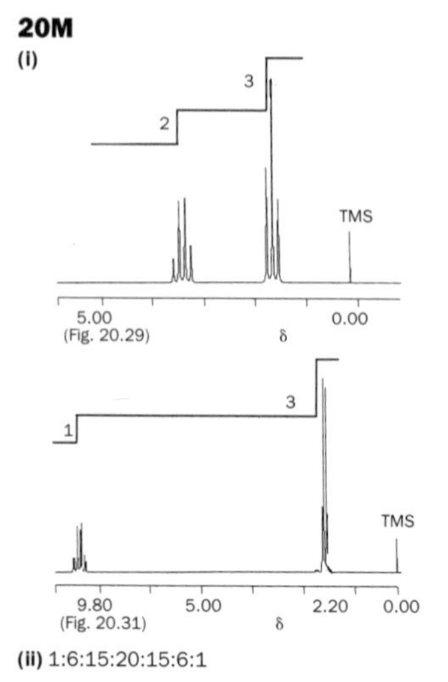

(ii) 1:6:15:20:15:6:1

20N

(i) (b)

(ii) (a)

(iii) (c)

20O

(i) $A_\lambda = \varepsilon_\lambda \times cb$, so that $c = A_\lambda/\varepsilon_\lambda b = 1.1/(1.1 \times 10^4 \times 0.00010) = 1.0$ mol m^{-3}, i.e. 0.0010 mol dm^{-3}.

(ii) (a) $c = A_\lambda/\varepsilon_\lambda b = 0.010/(1000 \times 0.010) = 0.0010$ mol m^{-3}

(b) 0.00025 mol m^{-3}.

(iii) $c = 2.0$ mol m^{-3}; $\varepsilon_\lambda = A_\lambda/cb = 0.40/(2 \times 1.0 \times 10^{-2}) = 20$ mol^{-1} m^2.

20P

For 2-nitrophenol, $A_2 = \varepsilon_2 c_2 b$ (where the subscript '2' stands for 2-nitrophenol).

For 4-nitrophenol, $A_4 = \varepsilon_4 c_4 b$ (where the subscript '4' stands for 4-nitrophenol).

ε_4 is given in Example 20.5 as 9.6×10^2 mol^{-1} m^2.

Therefore, the total absorbance at 313 nm is

$$\varepsilon_2 c_2 b + \varepsilon_4 c_4 b = (1.2 \times 10^2 \times 10^{-3} \times 10^{-2}) + (9.6 \times 10^2 \times 10^{-3} \times 10^{-2}) = 0.0012 + 0.0096 = 0.0108$$

20Q

(i) The concentrations are converted to mol m^{-3} by multiplying by 1000. The plot is:

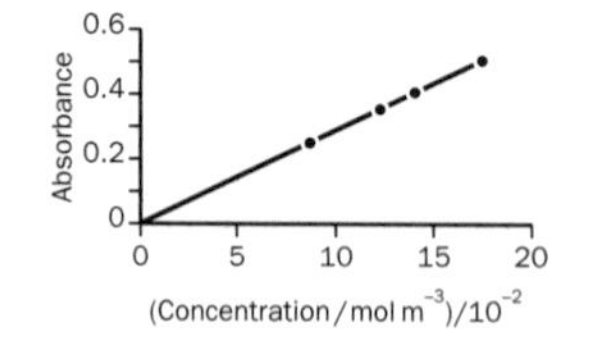

(ii) Slope $= \varepsilon \times b = (0.50 - 0.00)/(17.5 \times 10^{-2} - 0.0) = 2.9$ mol^{-1} m^3. Given that $b = 10^{-2}$ m,

$$\varepsilon_{526 \text{ nm}} = \text{slope}/b = 2.9 \times 10^2 \text{ mol}^{-1} \text{ m}^2.$$

Revision questions

20.1 (i) $\nu = c/\lambda = 3.00 \times 10^8/2 \times 10^{-7} = 1.5 \times 10^{15}$ Hz, **(ii)** 6.0×10^{14} Hz,

(iii) 3.00×10^{13} Hz.

20.2 The missing words are emission (or fluorescence), absorbance, chemiluminescence and molar absorption coefficient.

20.3 $\nu = c/\lambda = 3.00 \times 10^8/1000 \times 10^{-9} = 3.00 \times 10^{14}$ Hz

$E = hN_A\nu = 3.99 \times 10^{-13} \times 3.00 \times 10^{14} = 120$ kJ mol^{-1}.

20.4 $\Delta E = hN_A\nu$; $\nu = \Delta E/hN_A = 30/3.99 \times 10^{-13} = 7.5 \times 10^{13}$ Hz. Figure 20.1 shows that this corresponds to the infrared region of the electromagnetic spectrum.

20.5 (i) Percentage transmitted light = 100 − 84 = 16%, therefore

$$16\% = \frac{100 \times I}{I_0}$$

By rearrangement,

$$\frac{I_0}{I} = \frac{100}{16} = 6.25$$

$$A = \log(6.25) = 0.796$$

(ii) 1.00×10^{-5} mol dm^{-3} ≡ 1.00×10^{-2} mol M^{-3}

$A = \varepsilon cb = 1000 \times 1.00 \times 10^{-2} \times 10^{-2} = 0.100$

$$\frac{I_0}{I} = 10^{0.100} = 1.26$$

$$\frac{I}{I_0} = 0.79, \quad 100\frac{I}{I_0} = 79\%$$

To give A = 0.010, dilute by a factor of 10.

20.6 $\nu = \Delta E/hN_A = 494/3.99 \times 10^{-13} = 1.24 \times 10^{15}$ Hz.

$\lambda = c/\nu = 3.0 \times 10^8/1.24 \times 10^{15} = 2.4 \times 10^{-7}$ m (240 nm)

20.7 (i) $E = -1310/50^2 = -0.524$ kJ mol^{-1}.

(ii) E for $n = 3$ level is $-1310/3^2 = -145.5$ kJ mol^{-1}. Therefore,

$$\Delta E = -0.524 - (-)145.4 = 144.9 \text{ kJ mol}^{-1}$$

$$\nu = \Delta E/hN_A = 144.9/3.99 \times 10^{-13} = 3.63 \times 10^{14} \text{ Hz}$$

20.8 (i) Dark lines show wavelengths at which elements in the sun's cooler outer region absorb light. (Some lines are also produced by absorbing elements in the Earth's atmosphere.)

(ii)

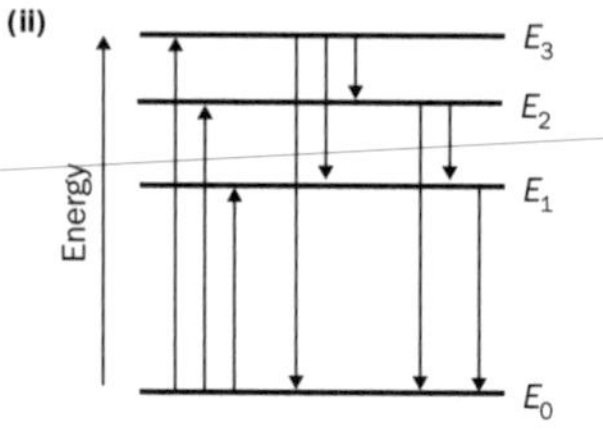

There are three possible transitions in absorption, and six in emission. (The smaller number in absorption follows from the fact that all molecules start at E_0.) Generalizing, an emission spectrum contains more lines than the corresponding absorption spectrum because the excited state does not have to fall to the ground state in 'one go'.

20.9 (i) It is likely that the absorption spectrum will contain a single strong peak between 490 and 530 nm.

(ii) The absorptions below 400 nm may be ignored. The colour cheese allows us to predict that the solution will absorb light at about 430 nm in wavelength and that it will therefore appear yellow.

(iii) The mixture will absorb virtually all the colours of white light. Very little light will be reflected and the paint will appear black.

20.10 The plot is very similar to Fig. 22.2 (see Chapter 22). Concentration of ozone c is

$$c = \frac{P}{RT} = \frac{100.0 \times 10^2}{8.3145 \times 298} = 4.04 \times 10^{-2} \text{ mol m}^{-3}$$

Therefore,

$$A(254\text{nm}) = \varepsilon cb = 369 \times 4.04 \times 10^{-2} \times 0.100 = 1.49$$

20.11

37.5% C and 12.5% H leads to the ratio of H to C as 4:1. Infrared spectrum contains a broad band at about 3400

cm^{-1} which is recognizable as characteristic of a hydrogen-bonded –OH group. There is no carbonyl absorption present, suggesting that the compound is an alcohol or phenol. The parent molecular ion possesses a molecular mass of 32 u (too low for a phenol). This suggests that the alcohol is CH_3OH. This contains only two types of protons (as indicated by the NMR spectrum) and the H:C ratio is 4:1 as required.

20.12 (i) Chloroform contains C–H bonds which would absorb in the same region as the hydrocarbons. CF_3CCl_3 contains no C–H bonds.

(ii) $b = 20\text{ mm} = 2.0 \times 10^{-2}$ m. $c = A/\varepsilon b = 0.050/(5.0 \times 2.0 \times 10^{-2}) = 0.50\text{ mol m}^{-3}$ or $5.0 \times 10^{-4}\text{ mol dm}^{-3}$. $M(C_8H_{18}) = 114\text{ g mol}^{-1}$. $c = 114 \times 5.0 \times 10^{-4} \times 10^3 = 57\text{ mg dm}^{-3}$.

20.13 (i)

$CrO_4^{2-}(aq) \rightarrow Cr^{3+}(aq)$
+6 +3 (reduced)

(ii) $5.0 \times 10^{-4}\text{ mol dm}^{-3} = 5.0 \times 10^{-1}\text{ mol m}^{-3}$. Initial absorbance $= \varepsilon cb = 200 \times 5.0 \times 10^{-1} \times 1.0 \times 10^{-2} = 1.0$. After reaction, the concentration of $Cr^{3+}(aq)$ is also $5.0 \times 10^{-1}\text{ mol m}^{-3}$; no $CrO_4^{2-}(aq)$ remains. Final absorbance $= \varepsilon c\, b = 1.0 \times 5.0 \times 10^{-1} \times 1.0 \times 10^{-2} = 5.0 \times 10^{-3}$.

20.14 The flame atomizes the sample so that atoms of the elements are produced. The flame does not produce the excited states of atoms, and this function is carried out by the light sources which are known as hollow cathode lamps. A different lamp is required for each element.

20.15 1,2-dibromoethane, CH_2BrCH_2Br (all protons in the same chemical environment, so only one resonance in the NMR spectrum).

20.16 Butanone, $CH_3COCH_2CH_3$:
The mass spectrum shows the molecular ion at 72, indicating a molecular formula of C_4H_8O.
The infrared spectrum shows a sharp peak at about 1700 cm^{-1}, indicating a carbonyl group.
The ^{1}H-NMR has a quartet at δ 2.5 (CH_2 protons split into a quartet by CH_3), a singlet at δ 2.0 (CH_3 attached to CO and too far away from the other protons for them to split the resonance) and a triplet at δ 1.0 (CH_3 split by CH_2 next to it).

Unit 21

Exercises

21A

$^{218}_{84}Po \rightarrow\ ^{214}_{82}X^{2-} +\ ^{4}_{2}He^{2+}$

Examination of the periodic table shows that X = Pb.

21B

$^{215}_{83}Bi \rightarrow\ ^{215}_{84}X^{+} +\ ^{0}_{-1}e^{-}$

Inspection of a periodic table shows that X = Po (polonium).

21C

P and S are isotopes of the same element.

21D

From Chapter 20,

$\nu = c/\lambda = 3.00 \times 10^8/10^{-10} = 3.0 \times 10^{18}$ Hz

$E = hN_A\nu = 3.99 \times 10^{-13} \times 3.0 \times 10^{18}$
$= 1.2 \times 10^6\text{ kJ mol}^{-1} \approx 10^6\text{ kJ mol}^{-1}$.

21E

^{238}U has an extremely long half-life (Table 21.2). Its decay would be represented by a horizontal straight line as for ^{137}Cs in Fig. 21.3.

21F

(i) ^{235}U decays faster than ^{238}U, so that more ^{235}U would have been present millions of years ago.

(ii) (a)

$^{201}_{87}Fr \rightarrow\ ^{197}_{85}At^{2-} +\ ^{4}_{2}He^{2+}$
astatine

$^{231}_{90}Th \rightarrow\ ^{231}_{91}Pa^{+} +\ ^{0}_{-1}e^{-}$
protactinium

(b)
For francium, it is easy to show that after 10 half-lives there is only a tiny fraction of radionuclide left. The reduction in mass after 10 half-lives is actually $1/(2 \times 2 \times 2 \times 2 \times 2 \times 2 \times 2 \times 2 \times 2 \times 2) = 1/2^{10} = 1/1024$. If 0.0100 g is reduced by 1024, only about 1×10^{-5} g of Fr would be left. In fact, 100 h represents 7.5×10^7 half-lives of Fr! We can safely assume that after 100 h, all the Fr would be gone.

For thorium, 100 h is four half-lives so that the original mass will fall by a factor of $2 \times 2 \times 2 \times 2 = 16$. One-sixteenth of 0.0100 g is 0.000 625 g.

These calculations may also be made using the formula $m_t = m_o \times e^{-kt}$, with k being calculated from the equation

$$t_{1/2} = \frac{0.693}{k}$$

For example, for thorium,

$$k = \frac{0.693}{t_{1/2}} = 0.693/(25 \times 60 \times 60) = 7.7 \times 10^{-6}\text{ s}^{-1}.$$

$m_t = m_o \times e^{-kt} = 0.0100 \times e^{-(7.7 \times 10^{-6} \times 3.6 \times 10^5)}$ (where 100 h $\equiv 3.6 \times 10^5$ s) $= 0.0100 \times 0.0625 = 0.000\,625$ g (as above).

Revision questions

21.1 $^{14}_{7}N +\ ^{4}_{2}He^{2+} \rightarrow\ ^{17}_{8}O^{2+} +\ ^{1}_{1}H$

21.2 (i) $^{3}_{1}H \rightarrow\ ^{3}_{2}He^{+} +\ ^{0}_{-1}e^{-}$

(ii) $^{6}_{3}Li +\ ^{1}_{0}n \rightarrow\ ^{3}_{1}H +\ ^{4}_{2}He$

21.3 From the half-life of $^{210}_{84}Po$ (140 days) we calculate that $k = 5.7 \times 10^{-8}\text{ s}^{-1}$.
(a) $t_{0.99} = 4.605/5.7 \times 10^{-8} \approx 8.0 \times 10^7$ s or ≈ 926 days
(b) $t_{0.999} = 6.908/5.7 \times 10^{-8} \approx 1.21 \times 10^8$ s or 1403 days

21.4 Lots to think about here. Alternative fuels and their availability, the pollution resulting from conventional fuels against the hazards of nuclear power. The safety aspects of nuclear fusion are discussed in *Environmental Physics*, by E. Boeker and R. van Grondelle, John Wiley, 1995.

21.5 (i) Radon gas.

(ii) The harm caused depends upon the activity of the source (i.e the half-life and mass of the radionuclide), the type of nuclear radiation, the distance of the individual from the source, the health and age of the individual, the area or part of body exposed, the duration of exposure, the degree of protection provided etc.

Unit 22

Exercises

22A

(i)
0.012 g in 250 cm^3 is 0.048 g in 1 dm^{-3} (or 48 ppm (mg dm^{-3})).
0.048 g in 1 dm^{-3} is

$$\frac{0.048}{62} = 7.7 \times 10^{-4}\text{ mol dm}^{-3}$$

(ii) No.

(iii)

$$\%N = \frac{2 \times 14 \times 100}{80} = 35\%$$

22B

Source: burning of fuels
Transport: diffusion through atmosphere
Assimilation: plant photosynthesis
Damage: greenhouse effect

22C

(i)

F
Cl C Cl
Cl

(ii) $22 + 90 = 112$ or CHF_2Cl
$141 + 90 = 231$ or $C_2H_3FCl_2$

(iii) Many aerosols used CFCs as propellants.The CFCs have been implicated in the destruction of the ozone layer. The modern trend is to use alternatives such as butane.

22D

(i) Increase in the burning of fossil fuels, more deforestation.

(ii) Nuclear power, wind energy, solar energy.

22E

(i) $C_6H_{12}O_6(aq) + 6O_2(g) \rightarrow 6CO_2(g) + 6H_2O(l)$

(ii) waste contains 600×10^5 mg glucose or $600 \times 10^2/180$ mol glucose. This needs

$$\frac{600 \times 10^2 \times 6}{180}\text{ mol } O_2$$

or

$$\frac{600 \times 10^2}{180} \times 6 \times 32\text{ g } O_2 \text{ in } 10^5\text{ dm}^3$$

which is 0.64 g in 1 dm^3, or 640 mg dm^{-3} (or ppm).

22F

(i) Vinyl chloride

(ii) $K_{oct} = 100\,000$ so that $[DDT]_{oct}$
$= 100\,000 \times 3 \times 10^{-6} = 0.3\text{ mg dm}^{-3}$ (ppm).
The observed concentration is higher, but the value of K_{oct} as a means of making a crude prediction remains.

22G

$S(s) + O_2(g) \rightarrow SO_2(g)$
1 tonne coal = 1000 kg which contains

$$\frac{2.5 \times 1000 \times 1000}{100}\text{ g S}$$

or

$$\frac{2.5 \times 1000 \times 1000}{100 \times 32}\text{ mol}$$

$$\equiv \frac{2.5 \times 1000 \times 1000 \times 24}{100 \times 32}\text{ dm}^3$$

i.e. 1.9×10^4 dm^3

22H

(i) $3Ca^{2+}(aq) + 2PO_4^{3-}(aq) \rightarrow Ca_3(PO_4)_2(s)$

(ii) $K_{sp} = [Ca^{2+}]^3[PO_4^{3-}]^2$

$$[Ca^{2+}]^3 = \frac{K_{sp}}{[PO_4^{3-}]^2} = \frac{1 \times 10^{-24}}{[3 \times 10^{-4}]^2}\text{ mol}^3\text{ dm}^{-9}$$

Therefore, $[Ca^{2+}] = 9 \times 10^{-5}$ g dm^{-3}.

22I

(i) Cl_2O
Cl^- −1
HOCl, OCl^- +1 (O is −2)

(i) All are oxidizing agents except Cl^-, it cannot be reduced.

22J

(i) $2\{CH_2O\} \rightarrow CO_2 + CH_4$
400 mg in 1 dm^3 is $400 \times 100\,000$ mg in the volume of waste, or

$$\frac{400 \times 100\,000}{1000 \times 30}\text{ mol of } \{CH_2O\}$$

From the equation, this is equivalent to

$$\frac{400 \times 100\,000}{100 \times 30 \times 2}\text{ mol } CH_4$$

or

$$\frac{400 \times 100\,000 \times 22.4}{100 \times 30 \times 2} \text{ dm}^3\ CH_4$$

$= 1.5 \times 10^4$ dm^3

(ii) $CH_4(g) + 2O_2(g) \rightarrow CO_2(g) + 2H_2O(l)$
(iii) 5.9×10^5 kJ

Revision questions

22.1

$[Pb^{2+}] = 10$ mg dm^{-3}

$$= \frac{10}{1000} \text{ g dm}^{-3} = \frac{10 \times 1}{1000 \times 207} \text{ mol dm}^{-3}$$

Rate $= 4.8 \times 10^{-5}$ mol dm^{-3} min^{-1}

22.2 (i) $C_8H_{18}(l) + 12\tfrac{1}{2}O_2(g) \rightarrow 8CO_2(g) + 9H_2O(l)$
(ii) 200 g octane is 200/114 mol.
This would produce 200/114 × 8 mol CO_2 at 100% efficiency. At 95% efficiency, 200/114 × 8 × 95/100 = 13.3 mol CO_2 (about 585 g).

22.3 Limestone is calcium carbonate, which reacts with the acid and neutralizes it.

22.4 Amount in mol of CO = PV/RT
$= 1.5 \times 10^5 \times 10^{-3}/8.3145 \times 298$
$= 0.0606$
The flask is released into a sealed room of volume 100 m^3. The concentration of gas (in mol m^{-3}) is
$n/v = 0.0606/100 = 0.000\,606$ mol m^{-3}
$= 6.06 \times 10^{-4}$ mol m^{-3}
$= 17$ mg m^{-3}
This is well below the dangerous level. However, higher concentration will be experienced near the bottle.

22.5 $\dfrac{100 \times 10}{1000} = 1$ g

22.6 Radioactive particles would be released into the atmosphere; radioactivity is *not* destroyed by chemical reactions.

22.7 (i) Nuclear testing during the 1950s.
(ii) Chernobyl.

22.8 1/100 of the sample would remain. In one half-life, 1/2 the sample would remain, in two half lives 1/2 × 1/2 would remain, and so on.
So, $(\tfrac{1}{2})^x$ is approx 1/100. Therefore, x is just under 7 half-lives, or 210 yr (Cs), and 56 days (I).

22.9 (i) $\dfrac{15 \times 1.25}{1000}$ g $= 1.88 \times 10^{-2}$ g
(ii) $\dfrac{85}{1000 \times 96} = 8.9 \times 10^{-4}$ mol dm^{-3}

22.10 Present in the sample is

$$\frac{6 \times 1 \times 200}{1000 \times 32 \times 1000} \text{ mol } O_2$$

This is equivalent to 4 × mol $S_2O_3^{2-}$, from the equations in Information Box 22.4 (i.e. $4S_2O_3^{2-}(aq) \equiv O_2(g)$). Therefore, moles of $S_2O_3^{2-}$ required is

$$\frac{6 \times 1 \times 200}{4 \times 1000 \times 32 \times 1000}$$

Substituting amount in moles = vol × molar conc
amount in mol = as above
vol = unknown
molar conc. = 0.010 mol dm^{-3}
Therefore, vol = 0.15 dm^3 = 15 cm^3

22.11 The high log K_{oct} values suggest that bioconcentration in fish is likely. The long residence time of chlordane indicates that it will persist for a much longer time in the environment with (other things being equal) a greater degree of environmental damage.

22.12 Oxygen is less soluble in hot water. If a river or lake is already polluted, the dissolved oxygen level in hot weather might go below 6 mg dm^{-3}.

22.13 We need to add ozone such that the concentration of the stratosphere increases by 1×10^{-11} mol dm^3.
Number of moles of O_3 = concentration × volume
$= 1 \times 10^{-11} \times 1 \times 10^{22} = 10^{11}$ mol. Given that $M(O_3)$ = 48 g mol^{-1},

mass of $O_3 = 48 \times 10^{11}$ g or 4.8×10^9 kg.
cost of $O_3 = 4.8 \times 10^9 \times 200$
= £960 000 000 000

Comment
The mass of ozone required is fantastically large. Ozone is made by passing electricity through oxygen. It would require about 2×10^{13} kJ of energy to make 4.8 million tonnes of ozone. To put this into perspective, we would require a 1000 MW nuclear reactor to operate for about 8 months in order to provide all the required energy! The rapid decomposition of ozone on surfaces makes the whole operation even more technically difficult.

Glossary

absolute temperature (0 K) The lowest possible temperature ($-273.15\,°C$).

absorbance A measure of the extent to which a sample in a cell absorbs light at a particular wavelength. Defined as $\log (I_o/I)$.

absorption (of light) The taking in of a photon of light by an atom, molecule or ion.

abundance The percentage (in terms of numbers of atoms) of an isotope in a sample of an element.

accuracy The closeness of measurement(s) to the true value.

acid A substance that produces hydronium ions (H_3O^+ or $H^+(aq)$) in solution.

acidic solution A solution with a pH below 7 at 25 °C.

acidity constant (K_a) The equilibrium constant for the ionisation of an acid in water.

activation energy The minimum energy required by colliding reactant molecules so that chemical change occurs.

activity The number of disintegrations of a radioactive sample per second (see the website).

addition reaction A reaction in which a carbon–carbon multiple bond is broken and parts of another molecule are added to each carbon atom of the multiple bond.

alcohol A series of organic compounds in which an alkyl group is bonded to –OH. e.g. CH_3OH

aldehyde One of a series of organic compounds which contain a –CHO group, e.g. CH_3CHO.

alkali A water-soluble base.

alkali metal A Group 1 element.

alkaline earth metal A Group 2 element.

alkaline solution A solution of a base – a solution with a pH above 7 at 25 °C.

alkane A hydrocarbon of general formula C_nH_{2n+2}, e.g. CH_4.

alkene A hydrocarbon of general formula C_nH_{2n}, e.g. C_2H_4.

alkyne A hydrocarbon of general formula C_nH_{2n-2}, e.g. C_2H_2.

allotropes Alternative forms of an element in which the atoms are joined together in different ways, e.g. diamond, graphite and buckminsterfullerene.

alpha particle A helium ion, produced in some types of radioactive decay.

amine Compounds in which organic groups replace the hydrogen atoms in ammonia, e.g. CH_3NH_2,$(CH_3)_2NH$ and $(CH_3)_3N$

amphoteric substance A substance is amphoteric when it shows reactions that are characteristic of both acids and bases.

anode In an electrochemical cell (or battery), the electrode at which oxidation takes place.

aromatic compound An organic compound that contains a benzene ring, e.g. C_6H_5Cl.

aryl group An aromatic group, e.g. $-C_6H_5$.

atmosphere (1) The gaseous envelope around a planet. (2) A unit of pressure, equal to 101 kPa.

atomic mass (of an element) The average mass of one atom of that element in atomic mass units; symbolized m(element).

atomic mass unit (u) A very small unit of mass ($1\,u \approx 1.66 \times 10^{-24}\,g$).

atomic number (Z) The number of protons in an atom.

Avogadro's constant (N_A) The number of objects per mol of objects ($N_A = 6.022 \times 10^{23}\,mol^{-1}$).

Avogadro's law (or principle) Equal volumes of all gases at the same pressure and temperature contain the same number of molecules (or moles).

base A substance that accepts a proton from an acid, forming a salt and water only.

basicity constant (K_b) The equilibrium constant for the ionisation of a base in water.

Beer–Lambert equation (law) An equation relating the absorbance A of a substance to the cell path length b and the molar absorption coefficient ε of the substance. Formally expressed as $A = \varepsilon cb$.

bond dissociation enthalpy ($\Delta H^{\ominus}_{A-B}$) A measure of the strength of a bond. The standard enthalpy change when 1 mol of bonds are broken, all species being in the gas phase.

buffer (solution) A solution which resists changes in pH upon the addition of acid, base or upon dilution.

carboxylic acid One of a series of organic compounds which contain a –COOH group, e.g. CH_3COOH.

catalyst A substance which, although it is not consumed, speeds up a reaction

cathode In a battery or electrochemical cell, the electrode at which reduction takes place.

chain reaction A sequence of self-sustaining reactions.

chemiluminescence The emission of light from a chemical reaction.

chromatography A technique used to separate mixtures. The separation takes place on a specially selected material, e.g. alumina.

colloid Particles of substances in solution which are smaller than those in suspensions but larger than individual molecules or ions.

common ion effect A reduction in the concentration of a species 'AB' in solution, due to the swamping of the mixture with A or B ions from another source.

complex (inorganic chemistry) Species consisting of ligands bonded to a central metal ion, e.g. $[Ag(NH_3)_2]^+$.

compound A substance consisting of atoms of different elements joined together in a definite ratio.

conjugate acid Species made when a base gains a proton.

conjugate base Species made when an acid loses a proton.

coordination number The number of nearest neighbours. These are ions of opposite charge in an ionic crystal, or the number of atoms that are attached to the central metal ion in a complex.

critical pressure The pressure required to liquefy a substance at its critical temperature.

critical temperature The temperature above which a substance cannot be liquefied, whatever the applied pressure.

decay constant (*k*) The rate constant (first order) for the decay of a radionuclide.

deionized water Pure water obtained by passing tap water through an ion-exchange resin.

delocalization The 'spreading' out of electron density over a molecule or ion.

density Mass per unit volume of substance; common unit $g\,cm^{-3}$.

diatomic molecule A molecule that consists of two atoms, e.g. HCl and O_2.

dipole A partial positive and equal partial negative charge separated by a distance.

dissociation Another word for 'break up'.

distillation The separation of a mixture, making use of the fact that its components have different boiling points.

electrochemical cell Two electrodes in electrical contact with each other and with a solution (the **electrolyte**).

electrode An experimental arrangement (normally involving a metal conductor) at which oxidation or reduction occurs, e.g. in an electrochemical cell.

electron A negatively charged sub-atomic particle.

electron-gain enthalpy ($\Delta H^{\ominus}_{eg}$) The standard enthalpy change when one mol of anions is formed, all species being in the gas phase.

electronegativity The measure of the ability of an atom to attract bonding electrons to itself when it is joined to another atom by a covalent bond.

element A substance in which all the atoms have the same atomic number. Matter that cannot be broken down into anything simpler by chemical reactions.

elementary reaction An individual reaction step; part of an overall reaction mechanism.

emission (of light) The giving out of a photon of light by an atom, molecule or ion.

empirical formula A chemical formula that uses the smallest whole number ratio of the atoms present in a compound.

emulsifying agent A substance that stabilizes an emulsion so allowing emulsification.

emulsion A colloidal solution consisting of one liquid dispersed in another.

end point In a titration, the point at which enough solution has been added so that the indicator begins to change colour. Ideally, the end point should equal the equivalence point.

endothermic reaction A reaction in which heat is taken in.

enthalpy The energy (or 'heat content') of a substance at constant pressure.

enthalpy of fusion ($\Delta H^{\ominus}_{fus}$) The standard enthalpy change when 1 g (or 1 mol) of a liquid is formed from 1 g (or 1 mol) of solid without a change in temperature; units are $J\,g^{-1}$ or $J\,mol^{-1}$. See the website.

enthalpy of vaporization ($\Delta H^{\ominus}_{vap}$) The standard enthalpy change when 1 g (or 1 mol) of a vapour is formed from 1 g (or 1 mol) of liquid without a change in temperature. Units are $J\,g^{-1}$ or $J\,mol^{-1}$. See the website.

equilibrium A dynamic equilibrium is a state of balance that is produced by two opposing processes (e.g. forward and back reactions) which are occurring at the same rate.

equilibrium composition A particular set of concentrations of reactants and products at equilibrium.

equilibrium constant (K_c) A ratio (defined by the reaction equation) involving the concentrations of reactants and products in an equilibrium reaction.

equivalence point In a titration, the point when exactly enough solution has been added from a burette to completely react with the solution in the flask.

error The difference between the measured value and the true value.

ester One of a series of organic compounds which contain the –COOR group, e.g. $CH_3COOC_2H_5$.

excited state An energy state other than the ground state.

exothermic reaction A reaction in which heat is given out.

explosion A violent increase in gas pressure in a confined space.

extensive property A property of a substance that depends on the amount of substance present, e.g. volume.

first-order reaction (decay) A reaction which follows first-order kinetics.

fissile Having the ability to undergo fission by slow neutrons; uranium-235 is fissile, uranium-238 is not.

fluorescence The emission of light from a substance.

frequency The number of waves of light passing a fixed point per second.

fusion (1) solidification; (2) in nuclear science, the formation of heavy nuclei from lighter nuclei (as in the sun).

ground state The lowest energy state of the atom, molecule or an ion.

half-life ($t_{1/2}$) The time taken for the concentration of a reactant to fall to half its original value.

hard water Water that contains dissolved magnesium or calcium salts, and which does not lather well with soap.

halogen A Group 17 element.

Henry's law This states that 'the solubility of a gas is proportional to the pressure of the gas above the liquid'.

hydration of ions The attachment of water molecules to a central ion.

hydrogen bond A hydrogen atom bridging two strongly electronegative atoms (N, F, Cl or O).

hydrolysis The reaction of a substance with water.

hydrophilic Water-attracting.

hydrophobic Water-repelling.

ideal gas A gas whose pressure, volume, and temperature may be predicted with perfect accuracy by the ideal gas equation.

immiscible liquids Liquids which do not dissolve in each other (i.e. they do not mix).

inert pair A pair of electrons that form part of the outer (or valence) shell of electrons in an atom, but do not take part in bond formation.

intensive property A property of a substance that does not depend on the amount of substance present, e.g. temperature.

intermolecular forces Forces between molecules.

ion An atom or group of atoms that possess a positive or negative charge.

ion exchange The exchange of one type of ion in solution with another.

ionic product constant (K_w) The product of the hydronium and hydroxide ion concentrations for any aqueous solution; the product varies with temperature.

ionization The formation of an ion

ionization energy (or enthalpy) The standard energy change when one electron is removed from a gaseous atom, molecule or ion.

isomers Compounds that have the same molecular formula, but different structures.

isotopes Atoms with the same number of protons but different numbers of neutrons.

isotopic mass The mass of one atom of an isotope in atomic mass units; symbolized m(isotope).

kelvin (K) The SI unit of temperature. 0 °C is equivalent to 273.15 K.

ketone Compounds that contain the carbonyl group (–C=O) between organic groups, e.g. CH_3COCH_3.

lattice enthalpy ($\Delta H_L^{\ominus}$) The standard enthalpy change when 1 mol of a crystal lattice is broken into isolated gaseous particles.

Le Chatelier's principle States that the concentrations of reactants and products in an equilibrium mixture will alter so as to counteract any changes in pressure, temperature or concentration.

ligand A molecule or anion that attaches itself to the central metal ion in a complex.

light A wave that travels at 3.00×10^8 metres per second.

London dispersion forces Weak attractive intermolecular or interatomic forces.

lone pair A pair of electrons in the outer (valence) shell of an atom not shared with another atom.

mass number (A) The number of protons added to the number of neutrons.

mass spectrum When a molecule or atom is ionised, a characteristic group of ions with different masses are made. A plot of the relative abundance of these ions against their mass by charge ratio (m/e) is called a mass spectrum.

mean bond enthalpy ΔH_{A-B} An average bond dissociation energy for a bond A–B, for example averaged over several different types of molecules.

melting point The temperature at which a solid melts at a particular pressure (see normal melting point).

metalloid Elements with properties that lie between those typical of metals and non-metals. Also known as semi-metals.

miscible Liquids which dissolve in each other (i.e. mix).

molar absorption coefficient Symbolized ε; A measure of the intensity of absorption of a substance at a particular wavelength.

molar solubility The concentration of a substance (in units of moles per cubic decimetre, i.e. $mol\,dm^{-3}$) in a saturated solution of that substance at that temperature.

mole (mol) The SI unit of amount of substance

mole fraction The number of moles of a substance in a mixture divided by the total number of moles of all the components of the mixture.

molecule A neutral particle consisting of two or more atoms combined in a definite ratio.

neutral solution A solution with equal numbers of hydronium and hydroxide ions, with a pH equal to 7 at 25 °C.

neutralization reaction The chemical reaction in which a base 'cancels out' an acid, producing a salt and water only.

neutron A neutral particle found in atoms of all elements (except for protium ^1_1H).

NMR Nuclear magnetic resonance – a type of spectroscopy.

noble gas A Group 18 element.

normal boiling point Also known as normal boiling temperature – the temperature at which the vapour pressure of a liquid equals 1 atm.

normal freezing point Also called **normal freezing temperature** – the temperature at which a liquid freezes under a pressure of 1 atm.

normal melting point Also called **melting temperature** – the temperature at which a solid melts under a pressure of 1 atm. Equals the normal freezing point.

nucleon A name for the particles in the nucleus (i.e. for neutrons and protons).

nuclide An *atom* of a particular isotope.

orbital The volume in which an electron is to be found with a certain probability (e.g. 90%). Each orbital holds up to two electrons, which may be thought of as spinning in opposite directions.

osmosis The flow of solvent (usually water) through a semipermeable membrane.

osmotic pressure The pressure which must be applied across a semipermeable membrane in order to stop osmosis.

oxidizing agent A substance which causes another substance to be oxidized in a chemical reaction and is itself reduced.

pascal (Pa) SI unit of pressure, 101 325 Pa = 1 atm.

percent transmittance The percentage of light at a particular wavelength which is allowed through a sample in a spectrometer; defined as $(I/I_o) \times 100$.

pH The negative logarithm of the hydronium ion concentration in an aqueous solution. At 25°C, pure water is neutral with a pH of 7.0.

phenol An organic compound in which an –OH group is directly attached to a benzene ring, e.g. C_6H_5OH.

photolysis The decomposition of a substance using light as the energy source.

photon A 'particle' of light; it has no mass and consists of pure energy.

pK_a The negative logarithm of an acidity constant.

pK_b The negative logarithm of a basicity constant.

polar molecule A molecule having one end that is positively charged while the other is equally negatively charged.

polyatomic molecule A molecule that contains more than two atoms, e.g. O_3, H_2O.

precipitation The formation of an insoluble substance in solution.

precision The closeness of repeat measurements.

proton A positively charged particle in the nucleus of all atoms; also an isolated hydrogen ion.

pyrolysis The decomposition of a substance using heat as the energy source.

radioactivity The spontaneous disintegration of nuclei.

radioisotope An isotope that is radioactive.

radionuclide An atom of a radioisotope.

random errors Errors which cause repeat measurements to be scattered (up and down) around the true value.

rate constant (k) The rate of a reaction (at a particular temperature) when the reacting substances are at a concentration of $1\,mol\,dm^{-3}$; k values measure the relative speeds of different reactions.

rate expression An expression, found by experiment, which relates the reaction rate to the rate constant and to the concentrations of the reactants.

rate of reaction The speed of a reaction at an instant; usual units are $mol\,dm^{-3}\,s^{-1}$.

reaction intermediate A species that is produced and consumed during an overall reaction.

reaction mechanism The collection of elementary reactions making up an overall reaction.

redox couple The oxidized and reduced species in a half-reaction.

reducing agent A substance which causes another substance to be reduced in a chemical reaction and it itself is oxidized.

resonance Mixing Lewis structures to obtain the actual structure of a species.

room temperature and pressure (RTP) In this book, taken as 20 °C and 1 atm pressure.

salt A substance made by the neutralisation of an acid with a base.

saturated hydrocarbon A hydrocarbon that contains no carbon–carbon multiple bonds and cannot react with hydrogen, e.g. an alkane.

semiconductor A substance that is not a good conductor of electricity at room temperature, but becomes conducting if certain other elements are added to it in small amounts.

specific heat capacity A quantity, symbolized *C*, which is equal to the amount of heat energy required to raise the temperature of 1 g of material by 1 °C; units are $J\,g^{-1}\,°C^{-1}$ or $J\,g^{-1}\,K^{-1}$. See the website.

spectrometer An instrument which measures the frequencies and intensities of emitted or absorbed light. (A simpler hand-held instrument is called a **spectroscope**.)

spectrum The pattern of frequencies (and the intensities) of light that is emitted (*emission spectrum*) or absorbed (*absorption spectrum*) by a species.

standard ambient temperature and pressure (SATP) 25 °C (298 K) and 100 kPa.

standard deviation A calculated quantity which measures the precision of a set of repeat results. See the website.

standard enthalpy of formation ($\Delta H_f^{\ominus}$) The standard enthalpy change that accompanies the formation of 1 mol of a substance from its constituent elements in their standard states.

standard state A pure form of a substance at 1 atm pressure.

standard temperature and pressure (STP) 0 °C and 1 atm pressure.

stoichiometric point Another name for *equivalence point.*

strong acid An acid that is completely ionized in solution.

strong base A base that is completely ionized in solution.

strongly acidic solution A solution with a pH below 3.

strongly basic solution A solution with a pH above 12.

sublimation Changing directly from a solid to a vapour.

substitution reaction A reaction in which an atom or group of atoms replaces another atom or group of atoms in a molecule.

systematic errors Errors which cause measurements either to be higher or lower than the true value, and which cannot be averaged out.

titration A technique in which a solution of known concentration (a standard solution) is reacted with a solution of unknown concentration, until equivalent quantities have reacted.

transition metal A group of elements found between groups 2 and 13 of the Periodic Table.

uncertainty (in a measurement) The part of a measurement given after the ± sign. For example, if a temperature is recorded as 25.5 ± 0.4°C, the uncertainty is ± 0.4°C.

unsaturated hydrocarbon A hydrocarbon that contains carbon–carbon multiple bonds and can react with hydrogen, e.g. an alkene or alkyne.

Van de Waal's forces A collective term for the weakest attractions between molecules or atoms. Two major Van de Waal's forces are London dispersion forces and dipole–dipole interactions.

vaporization The conversion of a liquid to a gas.

vapour Gas; commonly used to mean a gas above a liquid. Sometimes defined as a gas below its critical temperature.

vapour pressure The pressure of a vapour in equilibrium with a liquid.

wavelength The length of a single wave of light of a particular frequency.

wavenumber (cm^{-1}) The number of wavelengths of light per centimetre.

weak acids–weak bases Acids or bases that are incompletely ionized in solution.

Index

Page numbers followed by T or B denote tables or boxes, respectively. Entries followed by G are to be found in the Glossary